Geology of the country north of Derby

The Derby district lies at the south-eastern tip of the Pennines and includes areas of unspoilt beauty in the west, contrasting markedly with the industrial landscape of the East Midlands Coalfield to the east.

The importance and prosperity of the district is due largely to its mineral wealth. Lead has been exploited since Roman times; other economic products include coal, ironstone, limestone, moulding sands and refractory and building materials.

After an introductory chapter, the general stratigraphy of the Carboniferous, Permian and Triassic rocks is described and illustrated with numerous plates and figures. A brief account of the Neogene deposits is followed by chapters on the Quaternary era, the structure and the economic products of the area. A special chapter on geophysical investigations interprets basement structures of the district and places it in an overall regional setting.

Local details of the stratigraphy and palaeontology of the Carboniferous, Permo-Triassic, and Quaternary rocks, together with further information on their structure and geophysical aspects, are included in a miniprint section in an attempt to reduce the size and cost of this publication.

Appendices, also in miniprint, contain comprehensive summaries of boreholes, shafts and drifts, detailed recordings of sections of Westphalian strata, the locations of opencast sites and a complete list of Geological Survey photographs of the district.

Plate 1   Cromford Canal Opencast Site, near Eastwood (L 871)

The face shovel is excavating the Top Hard Coal which contains many old workings
(below and behind figure). The Comb Seam is largely concealed by debris on the bench
in the back wall. Sandstone overlying the Comb Coal is present in the upper part of the face.

GEOLOGICAL SURVEY OF GREAT BRITAIN
England and Wales

D. V. FROST and
J. G. O. SMART

Geology of the country north of Derby

Memoir for 1:50 000 geological sheet 125

CONTRIBUTORS

Stratigraphy
N. Aitkenhead

Geophysics
J. D. Cornwell

Palaeontology
M. Mitchell
W. H. C. Ramsbottom
M. A. Calver and
J. Pattison

Micropalaeontology
M. J. Reynolds and
G. Warrington

Petrography
R. K. Harrison
K. S. Siddiqui and
N. G. Berridge

INSTITUTE OF GEOLOGICAL SCIENCES
Natural Environment Research Council

LONDON   HER MAJESTY'S STATIONERY OFFICE   1979

© *Crown copyright 1979*

Bibliographical reference
FROST, D. V. and SMART, J. G. O. 1979. Geology of the country north of Derby. *Mem. Geol. Surv. G.B.*, sheet 125.

Authors
D. V. FROST, BSc, PhD and
J. G. O. SMART, BSc

Contributors
N. Aitkenhead, BSc, PhD, N. G. Berridge, BSc, PhD,
M. A. Calver, MA, PhD, J. D. Cornwell, MSc, PhD,
M. Mitchell, MA, J. Pattison, MSc,
W. H. C. Ramsbottom MA, PhD, M. J. Reynolds, BSc,
MPhil, K. S. Siddiqui, MSc and G. Warrington, BSc, PhD
Institute of Geological Sciences,
Ring Road Halton, Leeds LS15 8TQ

R. K. Harrison, MSc
Institute of Geological Sciences, London

Other publications of the Institute dealing with the geology of this and adjoining districts

BOOKS

British Regional Geology
The Pennines and adjacent areas, 3rd Edition 1954, reprinted 1978

Memoirs
Geology of the country around Buxton (in preparation)
Geology of the country around Chesterfield, Matlock and Mansfield, 1967
Geology of the country around Ollerton, 1967
Geology of the country around Ashbourne (in preparation)
Geology of the country around Burton upon Trent, Rugeley and Uttoxeter, 1955

MAPS

1 : 625 000
Sheet 2 Geological
Sheet 2 Quaternary
Sheet 2 Aeromagnetic

1 : 50 000 (and one inch to one mile)
Sheet 111 (Buxton) 1978
Sheet 112 (Chesterfield) 1963
Sheet 113 (Ollerton) 1966
Sheet 124 (Ashbourne) in preparation
Sheet 126 (Nottingham) 1974
Sheet 140 (Burton upon Trent) 1953
Sheet 141 (Loughborough) 1976
Sheet 142 (Melton Mowbray) 1977

1 : 25 000
Matlock

Geology of Derby

Typeset for the Institute of Geological Sciences by Raithby, Lawrence & Company Ltd. Leicester and London
Illustration films by U.D.O. Jenn Ltd, London
Printed in England for Her Majesty's Stationery Office by Ebenezer Baylis and Son, Limited, The Trinity Press, Worcester, and London   Dd 696803 K16

ISBN 0 11 884119 X

CONTENTS

viii

TABLES

PREFACE

This memoir describes the geology of the district covered by the Derby (125) Sheet 1:50 000 Geological Map of England and Wales. The district was originally surveyed on the scale of one-inch to one mile by W. W. Smyth, A. C. Ramsay, E. Hull, W. T. Aveline and T. R. Polwhele and the results published on Old Series sheets 71 NW and 71 NE between 1855 and 1858. Subsequent additions by A. H. Green, W. T. Aveline and J. R. Dakyns resulted in further publication in 1867 to 1878. A six-inch survey was carried out by W. Gibson, T. I. Pocock, C. B. Wedd, R. L. Sherlock, G. W. Lamplugh and C. Fox-Strangways between 1902 and 1907: the uncoloured one-inch sheet with drift lines was published in 1907 and completed in colour in 1908.

During the 1939–45 war Mr W. N. Edwards carried out a reconnaissance survey of the exposed coalfield areas of the district, based upon mining information compiled by himself and Mr A. J. Butler. Dr J. Shirley recorded information from opencast operations during this period. The northern margin of the Derby sheet was resurveyed on a six-inch scale by Messrs E. G. Smith, G. H. Rhys, P. McL. D. Duff and R. A. Eden during their work on the adjoining Chesterfield sheet from 1952 to 1960. Systematic fieldwork on the Derby sheet was started in 1960 by Dr D. V. Frost to be joined in 1961 by Mr J. G. O. Smart and later, by Dr N. Aitkenhead in 1965, with whose help the sheet was completed in 1966. The resurvey was under the supervision of Mr D. R. A. Ponsford as District Geologist. The six-inch maps covered by the various surveyors are listed on p. xi. The solid with drift edition of the 1:50 000 map was published in 1972. Part of the Derby district is covered by a special 1:25 000 sheet (Matlock) published in 1970.

The writing and compilation of the present memoir has been shared equally by Mr Smart and Dr Frost. Mr Smart was largely responsible for the Namurian, Westphalian (A and B), Quaternary and structural chapters. Dr Frost has written much of the Dinantian, Westphalian (C), Permo-Triassic and Economic products and water supply chapters. Dr Aitkenhead has contributed towards many of the chapters and in particular that on the Namurian. The memoir was edited by Mr D. R. A. Ponsford.

Mr M. Mitchell identified the fossils from the Dinantian and contributed an account of the stratigraphical palaeontology. Dr W. H. C. Ramsbottom and Dr M. A. Calver, assisted by Miss D. M. Gregory, identified fossils from the Namurian and Westphalian respectively and contributed to the relevant stratigraphical chapters. Mr M. J. Reynolds has provided micropalaeontological information from the Namurian. Mr J. Pattison has identified the Permian macrofossils and Dr G. Warrington has supplied palynological data. Mr R. K. Harrison, Dr N. Berridge and Mr K. S. Siddiqui have supplied mineralogical and petrological data for several chapters, and Dr J. D. Cornwell has written a geophysical account of the area. Mr M. Lock, Coal Scientist/Deputy Regional Geologist, of the National Coal Board has contributed an account on the properties of the coals and Mr J. E. Metcalfe, formerly Regional Opencast Manager (Development), has written a short history of opencast mining in the district.

Grateful acknowledgement is made to numerous organisations and individuals for generous help, advice and information, in particular to Mr R. E. Elliott and his staff of the Regional Geological Services of the National Coal Board, to Mr Metcalfe and his staff of the Opencast Executive and to the survey staff of the now superseded Nos. 4, 5 and 6 Areas of the National Coal Board, East Midlands Division. BP Limited has kindly sanctioned the inclusion of records of oil bores.

The stratigraphical details, shaft and borehole sections, etc., relative to the district are so numerous that it has been necessary to present them in 'miniprint' although they were originally prepared for full publication. This has resulted in some duplication between the two sections particularly in Chapters 4 and 10. The miniprint section is listed on p. vi and appears at the back of the memoir. Page-sized copies of this material can be provided by the Institute of Geological Sciences on application.

AUSTIN W. WOODLAND
Director

Institute of Geological Sciences
Exhibition Road, London SW7 2DE
26 January 1979

SIX-INCH MAPS

Published geological maps included wholly or in part in
1:50 000 Sheet 125 (Derby) are listed below with the
names of the surveyors and dates of survey. The surveyors
were N. Aitkenhead, P.McL. D. Duff, R.A. Eden,
D. V. Frost, G. H. Rhys, J. G. O. Smart and E. G. Smith.
All the maps are published.

SK 23 NE	Kirk Langley and Trusley	1966
	Smart	
SK 24 SE	Brailsford and Weston Underwood	1966
	Frost	
SK 24 NE	Turnditch and Idridgehay	1965
	Frost	
SK 25 SE	Wirksworth	1955–66
	Frost, Eden and Rhys	
SK 33 NW	Mackworth and Derby (NW)	1965–66
	Smart	
SK 33 NE	Breadsall and Derby (NE)	1964–65
	Smart	
SK 34 SW	Duffield	1966
	Aitkenhead	
SK 34 SE	Little Eaton and Morley	1965–66
	Smart	
SK 34 NW	Shottle and Belper	1965–66
	Aitkenhead	
SK 34 NE	Belper and Denby	1961
	Smart	
SK 35 SW	Alderwasley	1952–65
	Aitkenhead and Rhys	
SK 35 SE	Ambergate	1952–61
	Frost, Duff, Rhys and Smith	
SK 43 NW	Spondon and Dale	1964
	Frost	
SK 43 NE	Stanton and Stapleford	1964
	Frost	
SK 44 SW	Smalley and West Hallam	1964–65
	Smart	
SK 44 SE	Ilkeston	1963
	Smart	
SK 44 NW	Heanor	1961–62
	Smart	
SK 44 NE	Eastwood	1963
	Smart	
SK 45 SW	Ripley and Ironville	1952–61
	Frost, Duff and Smith	
SK 45 SE	Selston and Underwood	1953–63
	Frost, Duff and Smith	
SK 53 NW	Beeston and Wollaton	1963–64
	Frost	
SK 54 SE	Cinderhill	1963–64
	Frost	
SK 54 NW	Hucknall	1963
	Frost	
SK 55 SW	Newstead and Annesley	1953–63
	Frost, Duff and Smith	

GEOLOGICAL SEQUENCE

The rocks summarised below are present in the district

SUPERFICIAL DEPOSITS (DRIFT)

Quaternary
Landslips
Peat
Alluvium
River Terraces
Head
Glacial Sand and Gravel
Boulder Clay

SOLID

Tertiary
NEOGENE
Brassington Formation — sand in hollows in Dinantian Limestone

Thickness in feet

Triassic
CARNIAN
Edwalton Formation — red and green silty mudstones with Cotgrave Skerry at base — 20+

CARNIAN–LADINIAN
Harlequin Formation — alternating red silty mudstones and green siltstones up to — 180

LADINIAN
Carlton Formation — massively bedded red mudstones with Plains Skerry some 15 ft below top up to — 60

LADINIAN–ANISIAN
Radcliffe Formation — evenly bedded thinly laminated red and purple mudstones and pink siltstones up to — 45

ANISIAN
Waterstones Formation — alternating red-brown micaceous mudstone and siltstone with sandstone — 0 to 150

Woodthorpe Formation — up to six rhythmic alternations of sandstone and mudstone — 0 to 30

SCYTHIAN
Pebble Beds — yellow-brown pebbly sandstone with rare 'marl' bands — 0 to 170

Permian
Lower Mottled Sandstone (?Upper Permian in part) — red and yellow sandstone with mudstone partings — 0 to 100

Moira Breccia (?Lower Permian–Scythian) — red calcareous sand with rock fragments up to — 4

Middle Marl — red mudstone with sandstone, dolomitic in part — 0 to 25
Lower Magnesian Limestone — buff-red sandy dolomite — 0 to 40
Lower Marl — grey mudstone and dolostone — 0 to 80
Permian Basal Breccia — 0 to 7

Carboniferous
Westphalian (Coal Measures)
Westphalian C
(*phillipsii* and Upper *similis-pulchra* zones; Upper Coal Measures and upper part of Middle Coal Measures) — mudstones, siltstones, sandstones, seatearths and variable coals, mostly thin with the exception of the High Main. Mansfield Marine Band at base — 700+

Westphalian B
Lower *similis-pulchra* and upper part of *modiolaris* zones—lower part of Middle Coal Measures) — mudstones, siltstones, sandstones, seatearths and many economically important coals such as Mainbright, High Hazles and Second Waterloo. Clay Cross Marine Band at base — up to 1000

Westphalian A
(*modiolaris* (lower part), *communis* and *lenisulcata* zones—Lower Coal Measures) — mudstones, siltstones, sandstones, seatearths, ganisters and coals including Deep Soft, Deep Hard, Low Main, Blackshale and Kilburn. Pot Clay Marine Band at base — up to 1700

Namurian (Millstone Grit Series)
Yeadonian (G₁) — mudstones with Rough Rock at top and *Gastrioceras cancellatum* Marine Band at base — 120 to 190

Marsdenian (R₂) — mudstones and sandstones including Redmires Flags, Chatsworth Grit and Ashover Grit. Simmondley and Ringinglow coals near top and *Reticuloceras gracile* Marine Band at base — 450 to 1100

Kinderscoutian (R₁) — mudstones with *Homoceras magistrorum* Band at base — up to 160

Alportian (H₂) — mudstones, ?turbiditic in south-west, with *Hudsonoceras proteus* Band at base — up to c. 30

Chokierian (H₁) — mudstones, turbiditic in south-west, with *Homoceras subglobosum* Band at base — up to 120

Arnsbergian (E₂) — mudstones, turbiditic in south-west, with *Cravenoceras cowlingense* Band at base — up to 655

Pendleian (E₁) — mudstones, turbiditic in south-west, with *Cravenoceras leion* Band at base — up to 566

Dinantian (Carboniferous Limestone Series)
Shelf and marginal facies
Cawdor Formation (P₂) — dark grey thinly bedded limestone with shale partings — up to 60

Matlock Formation (D₂) — grey to dark grey, well-bedded limestone with Matlock Lower Lava at base — up to 150

Hoptonwood Formation (D₁) — pale grey massive limestone — 280+

Basin facies
Widmerpool Formation (B₂–P₂) — turbiditic mudstones, siltstones, sandstones and impure limestones with tuffs and sills — P₂ up to 366; P₁ up to 1104; B₂ 192+

NOTES

Throughout the memoir the word 'district' refers to the area covered by the Derby (125) Sheet.

National Grid references are given in square brackets. Unless otherwise stated all lie within the 100 km square SK.

Numbers preceded by L refer to photographs in the Institute's collections.

Letters preceding specimen numbers refer to Institute collections as follows:

E   English sliced rocks
GSM Leeds
Za   Leeds

CHAPTER 1

Introduction

PHYSICAL FEATURES AND DRAINAGE

This memoir describes the Derby (125) Sheet of the 1:50 000 (formerly One-inch) Geological Map of England and Wales. It completes the resurvey of the exposed part of the Yorkshire and East Midlands Coalfield, the southern limit of which is within this district (Figure 1).

The district lies chiefly in Derbyshire but includes part of Nottinghamshire to the east of the River Erewash, which forms the county boundary (Figure 2).

The highest ground, composed of the oldest rocks of the district (Dinantian limestone), lies to the north-west near Wirksworth at a little over 1000 ft. The lowest ground is in the south-east near Beeston where alluvium of the River Trent overlies Triassic rocks at an altitude of a little under 100 ft.

The two major rivers are the Derwent and Erewash, which drain southwards to join the Trent just beyond the southern margin of the district. The River Derwent and its tributary, the Ecclesbourne, drain the western hills, composed of Namurian rocks, including thick sandstones such as the Ashover Grit, which reaches a height of 1032 ft at Alport Hill. The River Erewash, in the east, traverses the gentler landscapes of the Westphalian rocks.

Along the southern and eastern margins of the district the Permo-Triassic rocks, cropping out in a belt some 3 miles wide, produce a varied topography of limestone dip-slopes, undulating heavy clay lands and well-drained, elevated sandy areas.

The south-eastern corner of the Derbyshire Dome is included in the district near Wirksworth in the north-west. Here the Dinantian outcrop is given over largely to upland pastures, and the once flourishing lead/zinc mining industry centred on Wirksworth has now been replaced by limestone quarrying. The village of Crich, 4 miles E of Wirksworth, stands on limestone which crops out on the southern flank of an anticlinal inlier.

Further beds of Dinantian age crop out in the Ecclesbourne Valley, west of Duffield, but they are composed of mudstones, siltstones and thin limestones so the topography is comparable with that of the Westphalian outcrop.

The Namurian rocks of the Derwent catchment are of little economic importance, their outcrop comprising an area of small farms, some of which are being replaced by low-density housing for the commuters of the large towns to the east and south. The Derwent Valley has for many years been a major road and rail route, with Ambergate an important junction. It is now better known for the numerous factories and mills strung along its length.

The exposed coalfield has been extensively exploited for its coal, ironstone, seatearth and brick clay since the beginning of the industrial revolution. Much of the ground is covered by buildings, tips and excavations, though pockets of mixed farming still remain.

The Permo-Triassic outcrops are extensively farmed, the Lower Magnesian Limestone producing the most fertile soil in the district. Some areas are covered by large tips and housing estates associated with deep mining in the concealed coalfield. Sand and clay have been quarried locally.

The large spreads of terrace gravels and alluvium in the Trent Valley have provided level and desirable sites for the spread of the large cities of Derby and Nottingham.

HISTORICAL

The importance and prosperity of the district is due largely to its mineral wealth. Pigs of smelted lead showing Latin inscriptions prove that lead ore was worked near Wirksworth during the Roman occupation. In 1086 Wirksworth possessed a church and a priest and was already a place of industrial prosperity. Its population of about 1000 was chiefly engaged in lead mining. The ores were placed in wood fires on the surrounding hills, such as Bole Hill, for smelting. They were then brought to the Moot Hall where the courts for the regulation of trade were, and still are held. The importance of lead has since declined, but the network of old workings and drainage levels, the hundreds of shafts and the lines of spoil heaps indicate the extent of this former mining industry (Rieuwerts, 1972).

In 1204 Derby (possibly derived from Derventio—the Roman Camp nearby) was granted privileges similar to those afforded to Nottingham but including the monopoly of dyeing cloth (Pendleton, 1886). The town was noted for its wool, malt marts and its Darby Ale. The brewing industry, now well established farther south at Burton, was probably attracted to the area by the excellent water supply from wells in the Permo-Triassic rocks.

The main industrial development of the region began with the discovery of coal and its allied deposits. The coal was first used by the smiths and lime-burners and later as a domestic fuel. Impetus was given to the coal production by the introduction of machines for shaft and underground haulage, the improvement of ventilation, and the extension of railways to convey the coal more rapidly and cheaply to the markets.

Cinderhill Colliery was one of the first Nottinghamshire collieries to adopt ventilation furnaces at the top of shafts in 1843. Despite such improvements, coal production was retarded by many explosions such as that at Annesley in 1877. By this date most large collieries had gone over to fan ventilation. Better ventilation removed the dangers of 'black damp', a concentration of carbon dioxide and nitrogen, and 'fire-damp', which was a particular problem at Pinxton Colliery (Griffin, 1971).

Willey Lane Pit, near Underwood, had the first winding engine in the district in 1838 and by 1841 engines had replaced 'whin gins' at most of the collieries, Swanwick being

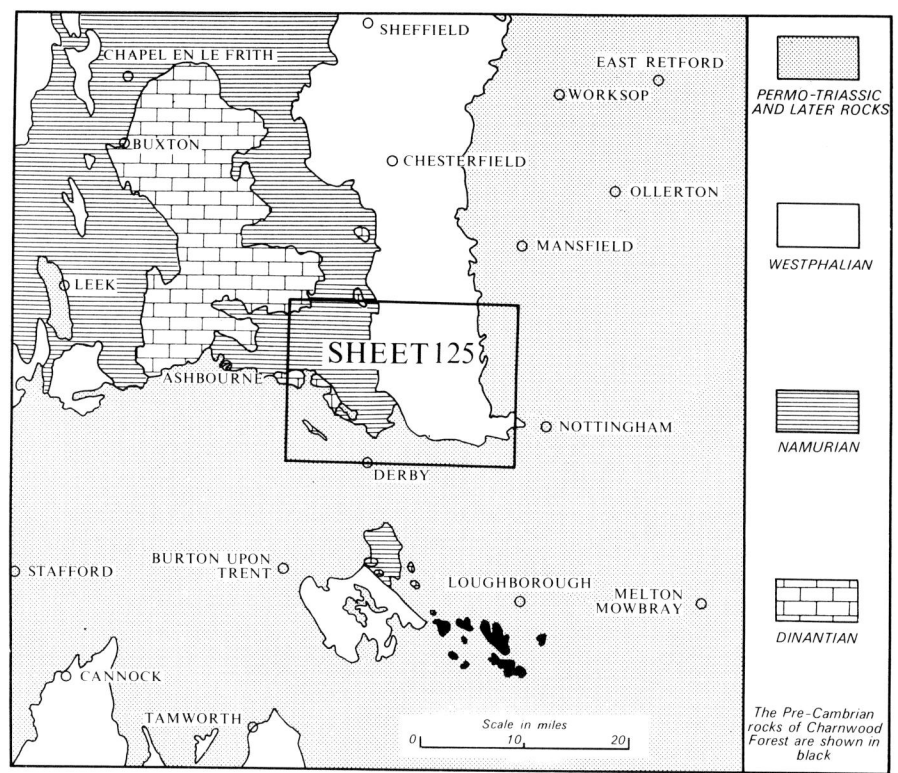

Figure 1 Sketch-map showing the general geological relations of the Derby district

Figure 2 Sketch-map showing the principal physical features and drainage of the district

a notable exception. Here, however, mechanical means of haulage were first used underground to extract coal from workings in a 2 ft 7 in seam, which were too low even for the employment of children.

The first definite references to rail transportation in connection with coal mining were in 1597. Rails were laid from the pits at Wollaton and Strelley to the River Trent to ease and speed the passage of coal from the mine to the barge, for the 'Strelley Cartway is so fowle as few cariadges can pass'. A 'plateway' built by the Derby Canal Company in 1797 from Little Eaton to Kilburn and Denby remained in use until 1908 and was known as the Little Eaton gangway.

Such technological advances resulted in greatly improved productivity in the twenty years after 1840 but as the workings extended the distances from the shafts and the depths increased. More pits were sunk and more men employed to work longer hours during day and night shifts to maintain output, particularly as the best and thickest seams such as the Top Hard were by then largely worked out.

In the 1939–45 war, production of coal from the collieries was supplemented by opencast mining, and millions of tons were extracted by this means.

Although 70 collieries were working within the Derby district in 1811, only two remain open in the exposed coalfield at the time of writing. Even in the more prosperous concealed coalfield the higher thicker seams, such as the High Main, are rapidly being worked out and further exploration of the deeper coals may become necessary. There are still many millions of tons of coal left in some abandoned areas of the Derby district, but modern sophisticated machinery is unable to operate in seams that are thin, variable or faulted.

Farey, in his General View of the Agriculture and Minerals of Derbyshire of 1811, lists 18 localities where ironstone 'rakes' had been worked. Remnants of this type of ironstone working, which paid little regard to replacing or levelling the ground afterwards are still visible at Morley Park and near Stanton. Later, the ore was worked at depth by bell pits. Most of the ironstones were smelted by charcoal furnaces until about 1770 when the country's timber was exhausted. The close proximity of coal to the ironstone deposits commonly resulted in a colliery being established near the iron furnaces and mines. By 1811 all the iron made in Derbyshire was smelted in coal or coke furnaces blown by steam-driven bellows. The first such furnace installed by Francis Hurt at Morley Park, made 700 pigs of iron annually. Gradually, however, the working of Westphalian ironstone became uneconomic and ceased, and Ironville was built on levelled tips dating from the days of local iron smelting.

The declining lead, iron and coal mining industries have been replaced by the extraction of sand and gravels and limestones for aggregates, and by pottery clays such as those at Denby. The industrial landscape is correspondingly changing. Old pit tips are being levelled and used as fill, and rows of terraced houses built for miners are being demolished. The railway lines at Ambergate and Annesley, now uneconomic, are being removed. The exposed coalfield is thus slowly returning to its former agricultural state except in those areas selected for housing employees of industries peripheral to coal mining. In the west of the district there remains much unspoilt open country—a valuable recreational outlet for the population.

BASEMENT (PRE-CARBONIFEROUS) ROCKS

No borings have penetrated the base of the Carboniferous rocks within the district, so the nature of the underlying rocks has to be inferred from outcrops and boreholes in surrounding areas. The information available at the time was summarised by Kent (1966, 1967, 1968), who envisaged that in Carboniferous times there was a shallow basement composed of crystalline rocks similar to those now exposed in Charnwood Forest; it formed part of the Midland Barrier, lying mainly south of the Derby district and extended across the East Midlands to north Norfolk (Kent, 1967, fig. 1, p. 131). The crystalline massif is flanked to the north by a belt of quartzitic conglomerates of possible Old Red Sandstone age, found underlying the Carboniferous rocks in oil boreholes at Eakring in Nottinghamshire and in Lincolnshire. At Eakring, Kent noted (1968, p. 142) 'heavily sheared mudstone which might be older Palaeozoic or (less probably) Pre-Cambrian' underlying the conglomerates.

A more recent borehole to the north-west of the district at Eyam, Derbyshire (Dunham, 1973), proved thick Lower Carboniferous rocks overlying steeply dipping Ordovician (?Llanvirn) mudstones. These sediments are more closely allied to the rocks at depth at Eakring than to the shallow basement of the English Midlands.

It is therefore inferred that the basement of the major part of the Derby district consists of folded Lower Palaeozoic rocks, possibly overlain in the east by a wedge of Old Red Sandstone which could account for the easterly decrease in Bouguer anomaly values discussed on p. 117. In the extreme south of the district Carboniferous rocks may rest directly on crystalline Pre-Cambrian rocks.

Dr Cornwell (Chapter 10) discusses geophysical evidence for the presence, within the Lower Palaeozoic and possible Precambrian basement sequences, of a belt of volcanic rocks extending south-eastwards from Wirksworth through Nottingham. The Aeromagnetic Map (Figure 65) shows a broad magnetic high in this area. More localised magnetic anomalies, east of Belper and at Hucknall, may indicate intrusions, possibly penetrating the Lower Carboniferous.

GEOLOGICAL HISTORY

Following the Caledonian earth movements, the Dinantian Sea transgressed over the eroded basement rocks into the Derby district. No Tournaisian rocks are known but their presence may be inferred at depth in the district. The oldest rocks seen at outcrop are Dinantian limestones which are exposed near Wirksworth; they were deposited slowly on a 'shelf', in relatively shallow seas rich in animal life. At the same time, greater thicknesses of mudstones, thinly bedded, dark grey limestones and thin sandstones were being laid down to the south-west around Duffield and Derby in deeper muddier waters of a basinal area known as the Widmerpool Gulf. Sedimentation here was at times so rapid that the deposits became unstable and, perhaps triggered off by earthquakes, slumped down the slopes of the Gulf to form turbidites. These are thought to have come mainly from the south.

Between basin and shelf deposition was a reef belt, in which sedimentation kept pace with subsidence. At times local uplift along this margin upset the delicate balance and erosion of the reefs gave rise to non-sequences and unconformities within the limestone formations.

Periodically during Dinantian times volcanic activity resulted in the extrusion of lavas from vents like that at Hopton and the formation of tuffs in both the Widmerpool Gulf and the Derbyshire shelf areas.

In late Dinantian times the rate of subsidence increased and basinal conditions extended northwards until, in the early Namurian, shales were deposited against the eroded limestone cliffs of the reef margin and eventually spread over them on to the shelf itself. However, deposition in the shelf area was still slow compared with the turbidite deposition which continued in the Widmerpool Gulf. Later in the Namurian, the pattern of sedimentation changed with the establishment of a large delta resulting in the deposition of the Ashover Grit. Carboniferous sedimentation for the first time outpaced subsidence and a coal swamp was established which is now represented by a coal seam overlying the Ashover Grit. Subsequent fluctuations in the rate of subsidence are reflected in distinct cycles of sedimentation, each beginning with the deposition of marine shales which pass up through mudstones into deltaic siltstones and sandstones.

In Westphalian times the Namurian pattern continued but with increasing dominance of shallow-water deposits of argillaceous composition. The cycles increased in number but became generally thinner. Emergence, with the establishment of land floras, became more frequent and in many cases prolonged, resulting in coal complexes. Volcanic activity, which persisted to the east, resulted in sporadic showers of fine ash falling in the Derby district.

The oncoming of the Hercynian Orogeny was heralded by the intrusion of dolerite sills into the Dinantian rocks. Regional uplift and increased erosion is reflected by fewer coals and more sandstones in the Westphalian C sequence. Eventually the whole of the Pennine area was uplifted, folded and faulted, much of the fracturing following old lines of weakness established in Caledonian times. A long period of denudation followed during the late Carboniferous and early Permian when many thousands of feet of Carboniferous rocks were removed. The resulting land surface was subjected to tropical weathering and the truncated Carboniferous rocks were superficially reddened to a depth of 30 ft or so.

In Upper Permian times the sea returned from the east and calcareous sediments were laid down on a peneplain in a south-western extension of the Zechstein Sea during a long period of desiccation and slow deposition under arid conditions. This ended with a period of fluvial activity and deposition of Triassic sands over a wider area than that covered by the Zechstein Sea.

There followed a period of minor uplift, when the Triassic deposits were locally removed and in some places the old Carboniferous surface was again laid bare. Further subsidence occurred and the later Triassic rocks were deposited in another period of increasing aridity.

During Permo-Triassic times, deep-seated migratory fluids invaded the rocks of the district, particularly the well-jointed and fractured Dinantian limestones and left ore deposits of lead, zinc and copper.

Most of the post-Triassic rocks of the district were removed by prolonged erosion during the Tertiary, but isolated remnants of reworked Triassic sands, dating from Neogene times, were preserved in hollows in the Dinantian limestone surface. The Alpine Orogeny produced only a slight easterly and southerly tilt, associated with minor faulting along pre-existing Hercynian fractures.

During the Pleistocene, a Wolstonian ice sheet overrode the district from the north-west; it met ice advancing from the east along the southern margin of the district and later melted, probably after a long period of stagnation. It left deposits of boulder clay and sand and gravel now found as isolated patches mainly on the higher ground.

Terraces of the River Trent were formed probably during the waning of the Wolstonian ice sheet and during the subsequent Ipswichian Interglacial. The succeeding Devensian ice sheet did not reach the district, which suffered periglacial conditions with the formation of frost wedges, valley bulges and cambering and the deposition of head, and further river gravels.

The Flandrian Stage saw the final evolution of the landscape into its present form with the deposition of the recent alluvium of rivers and streams. However, the earthquake activity which still continues on a small scale in the Derby district is a reminder of its instability in the past.

CHAPTER 2

Dinantian (Carboniferous Limestone Series)

INTRODUCTION

The Dinantian limestones of Derbyshire may be regarded as belonging to three main facies (Edwards and Trotter, 1954) all of which are represented within the district. They comprise a 'massif' or 'shelf' facies of beds deposited over an ancient relatively stable block, a 'basin' facies developed in more rapidly subsiding areas around the block, and a 'marginal' facies, usually associated with 'reef' beds, deposited between the block and the basin. This facies classification provides a convenient way of describing the palaeogeography and the related lithological types but is probably an oversimplification of the complex interrelationships of the processes of sedimentation. Availability and proximity to source of sediments, depth of water and strength of currents may be as important as the implied structural controls.

Limestones of the 'shelf' and 'marginal' facies crop out in the Wirksworth–Hopton area, where they form the south-eastern margin of the main limestone outcrop of the 'Derbyshire Dome'. Beds of 'shelf' facies also crop out in a small area around Crich and were proved at depth in Ironville oil bores Nos. 1 and 3 drilled through Westphalian and Namurian rocks farther east (see Figures 3 and 4).

The 'basin' facies is developed to the south of Wirksworth in the Widmerpool Gulf (Falcon and Kent, 1960, p. 18), where the Dinantian beds comprise a thick turbidite sequence of mudstones, siltstones, sandstones and dark impure limestones for which the name Widmerpool Formation is here introduced. South of Wirksworth (Figure 3) the Widmerpool Formation is concealed below Namurian beds, but farther south towards Derby it crops out over some six square miles in inliers surrounded by Triassic rocks. The information from these scattered surface exposures is supplemented by the Duffield Borehole [3428 4217], drilled in 1967 (Institute of Geological Sciences, 1968, p. 82; Aitkenhead, 1977), and it is now possible to trace the changes from 'shelf' to 'basin' facies over a distance of 8 miles from Wirksworth to Duffield and to interpret the complex geological history of the 'marginal' area between these two facies. Further information on the gulf was obtained from the Trusley Borehole [2548 3588] drilled in 1969 by the Institute of Geological Sciences. This borehole, sited in the south-western corner of the district some 11 miles S of Wirksworth, proved a reddened facies of the Widmerpool Formation.

GENERAL ACCOUNT

The Wirksworth–Derby area provides a cross-section through a range of typical 'block' and 'basin' sediments in the Carboniferous of northern England.

The Widmerpool Gulf is characterised by a thick monotonous sequence of successive graded units of calcareous mudstones, siltstones and sandstones contrasting faunally and lithologically with intercalated darker fissile mudstones. There is a complete absence of sedimentary structures and fossils positively indicative of a shallow-water environment, and the graded units are thought to have been deposited by turbidity currents in a marine environment deeper than that of the 'shelf' areas.

In the 'shelf' facies near Matlock, the Hoptonwood Limestone (D_1) and Matlock Limestone (D_2) comprise grey, massive, standard limestones, but southwards to Wirksworth the top of the sequence thins and passes into irregularly bedded dark-coloured limestones associated with non-sequences and breccias. A 'reef' facies, confined to the highest beds (Cawdor Limestone = P_2) at Matlock, is also present in the upper part of the Hoptonwood Limestone farther south in the Wirksworth–Hopton marginal area. Distinct knoll-reefs are uncommon, presumably due to the erosion to which limestones of D_1 and D_2 age were subjected. Unconformities are locally present between the various mapped subdivisions, i.e. between beds of D_1 and D_2, D_2 and P, and P_2 and E ages, as well as within P_2 beds.

The transition from 'shelf' to 'basin' facies through the 'marginal' reef belt probably occurs rapidly over a narrow zone which is commonly faulted and/or folded. The fourfold increase in thickness of Dinantian sediments into the Widmerpool Gulf is a measure of both the greater subsidence and the more rapid deposition in the gulf compared with the positive stable 'block' or 'shelf' area.

Using the criteria listed by Kent (1966, p. 325) it is possible to indicate the limits of the Widmerpool Gulf within the district. Assuming that the presence of thick Namurian sediments is a continuation of a thick Dinantian development beneath, then the north-eastern limits of the Widmerpool Gulf can be delineated as shown in Figure 3. The Hopton Borehole proved Namurian (E_2) shales at least three times thicker than the equivalent strata at Ashover some 6 miles to the north-east. Ramsbottom and others (1962, fig. 1) show the thinning of Namurian shales across the axes of the Ashover and Crich anticlines, suggesting that these areas remained positive throughout most of Namurian times. Conversely, in the intervening syncline there is a south-westwards thickening of some 200 ft towards Wirksworth and this may be accounted for by the small embayment on the northern edge of the 'gulf' (Figure 3).

The juxtaposition of Namurian thickening together with important fault lines, e.g. near Smalley (SK 44 SW), Ripley (SK 44 NW) and Wirksworth (SK 25 SE) facilitates further delineation.

It is assumed that Duffield Borehole occupies a central position in the Widmerpool Gulf, but little information is available regarding the gulf's south-western limits. Trusley Borehole, drilled for 177 ft below the base of the Trias, showed a Carboniferous sequence of reddened turbiditic sandstones and siltstones with interbedded pelagic mudstones. The sediments are of 'basin' facies but occupy a more distal position than those in Duffield Borehole and the Kirk Langley area. The sediments were unusual in containing thin horizontal beds and irregular discordant veinlets of gypsum.

The northernmost exposure, at Flower Lilies (see locality 4, Figure 3 and p. 130c), contains more limestone than any other recorded sequence in the Widmerpool Formation (Green and others, 1887, p. 88). This may indicate either a position nearer the margin of the gulf than Duffield, so that the gulf was less extensive in Dinantian than in Namurian times, or some local facies change within the gulf.

The north-westerly trend of the gulf in the present district compares well with that of the Edale Gulf and Gainsborough Trough. This and other evidence supports Kent's (1966) postulate that the gulfs originated as downwarped and/or faulted belts along Charnian lines of weakness.

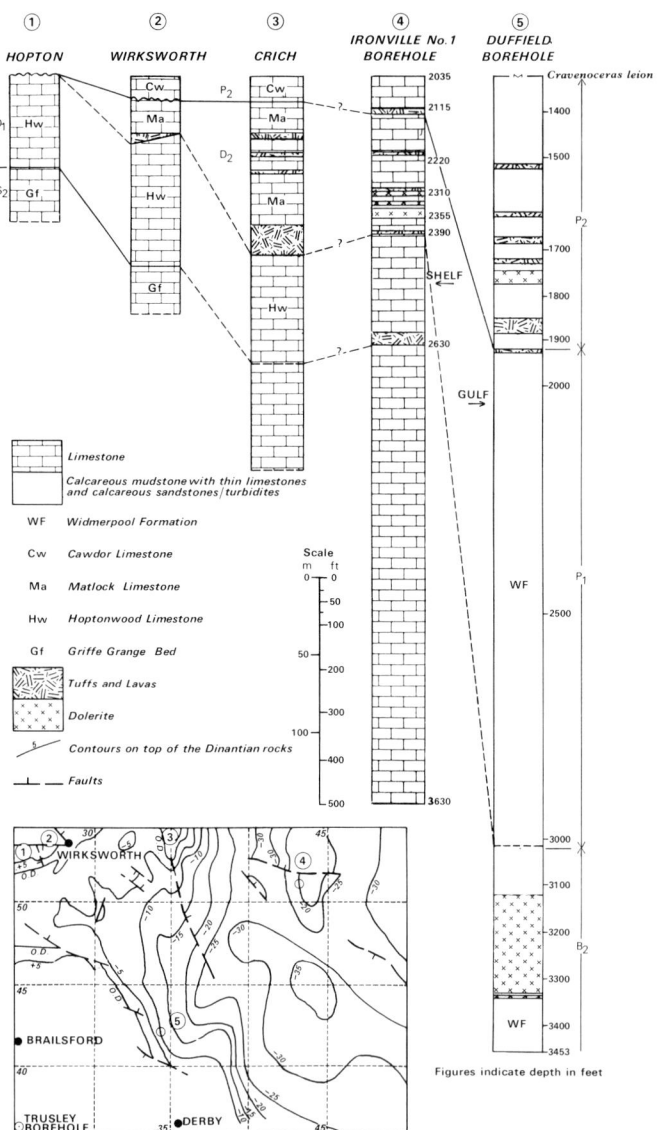

Figure 4 Section of the Dinantian rocks in the district with inset location map showing probable configuration of contours on the top of the Dinantian in hundreds of feet

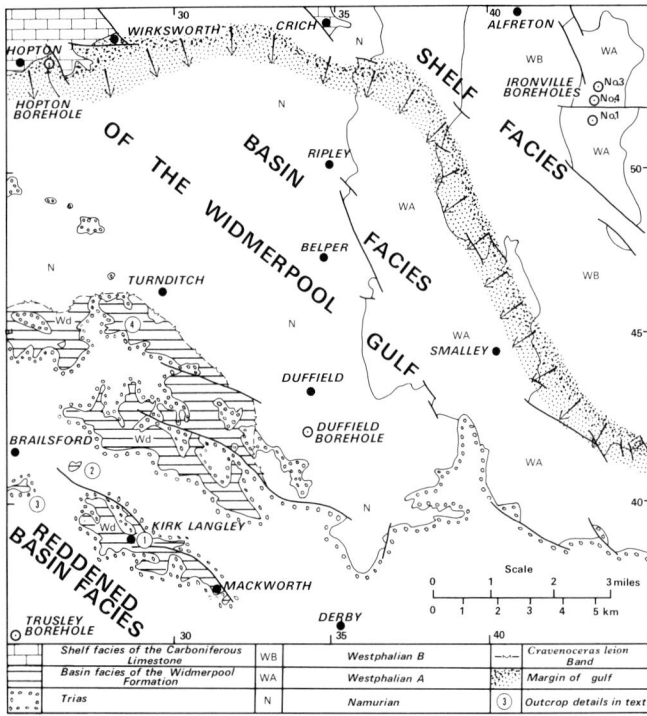

Figure 3 Outline geology of the Wirksworth–Derby area showing the north-eastern margin of the Widmerpool Gulf

SHELF AND MARGINAL FACIES

The main outcrop

Subdivisions erected by Eden and others (1959, p. 33) in the adjacent Matlock area are here modified in conformity with current stratigraphical practice so as to give the following classification:

Cawdor Group	Cawdor Shale Formation	} P_2
	Cawdor Limestone Formation	
Matlock Limestone Formation	Limestone with Matlock Lower Lava at base	} D_2
Hoptonwood Limestone Formation		D_1

Smith and others (1967, pp. 11–12) noted that in the Matlock area the Hoptonwood and Matlock limestones consist largely of grey massive beds typical of the 'shelf' facies, but that towards Wirksworth the Matlock Limestone thins and passes into irregularly bedded dark-coloured limestones with non-sequences and breccias. They infer that the latter form part of an apron-reef complex largely hidden beneath the Cawdor and Namurian shales to the south and east. Shirley (1959, p. 420) noted that south-west of Wirksworth, reef-like limestones occur in D_1 (Hoptonwood Limestone) and reef-like mounds, possibly of Cawdor Limestone, project through the overlying Namurian shales [2800 5300]. The present survey has confirmed these stratigraphical changes which occur as the limestone margin is approached. Boreholes beyond this margin, e.g. at Hopton, prove thick Namurian (E_2) shales which between Hopton and Carsington overstep westwards on to successively lower horizons of the Hoptonwood Limestone. There is therefore good evidence of a significant break or breaks within the sequence in P Zone or post-P Zone times with the removal or non-deposition of some 300 ft of sediments in this marginal area. The concealed apron-reef postulated to the south-east of Wirksworth (Smith and others, 1967, p. 13) may have been partly or wholly removed by late Dinantian-early Namurian erosion, following local uplift, as in the area to the south-west of Wirksworth.

Instability in the Wirksworth–Hopton area during much of early Carboniferous times is implied therefore by the transition from block to basinal deposition, by the upper Dinantian erosional breaks and associated lavas and tuffs (p. 96), and by the later ore mineralisation. The area shows many of the facies changes present in the Castleton area some 20 miles to the north-north-west (Eden and others, 1964).

HOPTONWOOD LIMESTONE

This formation, consists of up to 280 ft of limestones which are pale grey to off-white, fine-grained to porcellanous, partly oolitic and massively bedded. Macrofossils are not common but shelly concentrations occur along the southern margin of the outcrop and a small knoll-reef [2610 5332] near Hopton contains a D_1 assemblage. The Hoptonwood Limestone crops out over a large area to the north of Carsington and Hopton, where extensive dolomitisation often masks the original lithology and fauna. Its base is not exposed within the district although the limestones in the floor of Stone Dene [2580 5369] near Hopton must be close to the underlying Griffe Grange Bed (Smith and others, 1967, p. 8).

Beds at the top of the Hoptonwood Limestone crop out on a 'promontory' south of Godfreyhole and are also exposed as inliers in quarries on the west or upthrow side of the Gulf Fault north-west of Wirksworth. In these quarries the highest beds pass south-eastwards into disturbed and brecciated strata associated with the reef margin.

Interbedded with the limestones are lavas and at least seven clay bands or wayboards, some of which may be lateral equivalents of lavas. An agglomerate, which probably occupies a volcanic neck, is present at Hopton.

The Hoptonwood Limestone contains limited coral and brachiopod faunas including *Carcinophyllum vaughani*, *Dibunophyllum bourtonense*, *Palaeosmilia murchisoni* and *Davidsonina septosa*, an assemblage which is characteristic of the Lower *Dibunophyllum* (D_1) Zone. Outcrops of the reefs along the southern margin of the limestone yield rich brachiopod faunas typical of this zone including *Overtonia fimbriata*, *Proboscidella proboscidea*, *Pugnax pugnus* [small form] and *Pugnoides triplex*, forms which are restricted to beds of this age in the reef limestones at Treak Cliff, Castleton (Mitchell *in* Stevenson and Gaunt, 1971, p. 149). Information is insufficient however to indicate which part of the zone the fauna represents.

The Hoptonwood Limestone represents the Fifth Group of Minor Cycles of Ramsbottom (1973) and is of Asbian age (George and others, 1976, fig. 8). The general thickness and presence of unconformities suggest that parts of the sequence are probably missing.

MATLOCK LIMESTONE

This formation comprises limestones with the Matlock Lower Lava at the base.

The Lava is not exposed but it forms a well-marked feature 200 yd W of Hopton Tunnel and can be traced for one mile southwards to the Yokecliffe Rake where it apparently ends, but may possibly continue and be connected to a feeder down the Rake. Soft green clay some 7 in thick was proved in a borehole [2730 5434] near Four Lane Ends and is considered to be at the horizon of the Lava.

To the north, on the Chesterfield (112) Sheet, the Matlock Lower Lava, an amygdaloidal olivine-basalt, is an accumulation of flows up to 380 ft thick, but west of Wirksworth it has a maximum thickness of 15 ft and may be a single flow.

The overlying beds comprise variable dark grey and grey limestones up to 140 ft thick, with fossiliferous bands and locally much chert. The top shows sharp lithological changes. A small patch of fossiliferous reef limestone is preserved in Stonycroft Quarry [2862 5431] at the top of the formation. Traced south-eastwards over a distance of about 100 yd through the quarries north and west of Wirksworth, the whole of the Matlock Limestone (i.e. 140 ft) is progressively cut out so that in Baileycroft Quarry, Wirksworth (Plate 2; Figure 5) dark, thinly bedded Cawdor Limestone rests directly on the massive Hoptonwood Limestone. An unconformity is also present within the Matlock Limestone. In Dale Quarry, 300 yd W of Baileycroft Quarry, dark grey cherty Matlock Limestone is again present. Most of the quarry faces are inaccessible.

In Middlepeak Quarry, a 'Girvanella' band (possibly equivalent to the Upper 'Girvanella' Band of north Derbyshire) lies some 18 ft from the top of the Matlock Limestone, here totalling 80 ft. The band has not been found in Stonycroft Quarry 200 yd to the south, where 60 ft of D_2 beds were measured. 'Girvanella' nodules are present near the base of the Matlock Limestone in a small quarry [2729 5362] near Godfreyhole. This may be the correlative of the Lower 'Girvanella' Band of North Derbyshire (Eden and others, 1964, fig.1), and if so shows that the lower part of the sequence is complete.

The Matlock Limestone crops out north-west of Wirksworth and is well exposed in the approaches to Hopton Tunnel and in the Yokecliffe Rake fault scarp. It is dark grey and well-bedded but sporadic massive pale greyish brown beds provide valuable marker horizons. Bedded chert as well as irregular chert nodules are common, and there are several coral and shelly bands. Limestone breccias occur at the top of the sequence. Only a small part south of Hopton Tunnel has been dolomitised.

The eastern side of the Godfreyhole 'promontory' comprises some 50 ft of dark grey, well-bedded limestones which appear to be directly overlain by low Namurian (E) shales.

The fauna of the Matlock Limestone is more varied than that of the Hoptonwood Limestone and includes *Aulophyllum fungites pachyendothecum, Dibunophyllum bipartitum bipartitum, Diphyphyllum lateseptatum, Lonsdaleia floriformis, Palaeosmilia regia, Megachonetes siblyi, Pugilis pugilis* and *Striatifera striata*—all diagnostic of the Upper *Dibunophyllum* (D₂) Zone. The Matlock Limestone represents the Sixth Group of Minor Cycles of Ramsbottom (1973) and is of Brigantian age (George and others, 1976, fig. 8).

CAWDOR GROUP

This group consists of fossiliferous thinly bedded cherty limestones with shale partings (Cawdor Limestone) totalling 60 ft in thickness, overlain by Cawdor Shale some 20 ft thick. In the Derby district the basal 20 ft or so of Cawdor Limestone are well exposed, but the top is missing due to unconformable overlap by Namurian shales. The basal beds crop out north and west of Wirksworth and comprise grey and dark grey thinly bedded cherty limestones with shale partings. Most of these limestones have been extracted in the Wirksworth quarries (Middlepeak, Stonycroft and Baileycroft), but a few remaining isolated pillars and pockets provide excellent collecting grounds for the shelly reef-type faunas. In the accessible north-western corner of Dale Quarry the Matlock Limestone–Cawdor Limestone junction is well displayed.

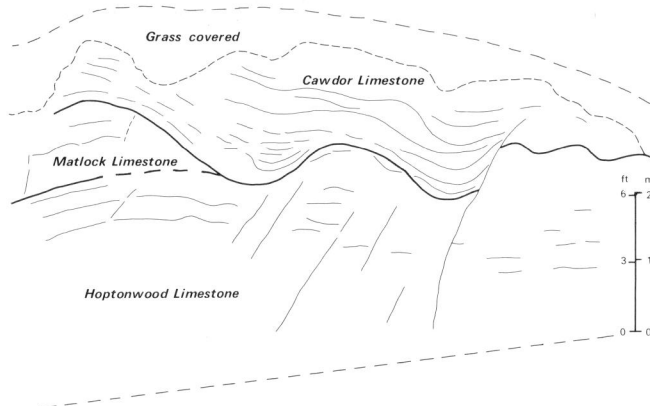

Figure 5 Interpretation of Plate 2.1 showing the unconformable relationship of the Cawdor Limestone to the Matlock and Hoptonwood limestones. Baileycroft Quarry, south-east face

South-west of Wirksworth and south of the Yokecliffe Rake the Cawdor Limestone is missing owing to Namurian overlap, but near Millers Green [2793 5305] a topographical

knoll, some 120 yd by 100 yd, of poorly bedded massive pale grey-cream limestone protrudes through the surrounding shales, and contains a reef fauna of the Cawdor Limestone (Shirley, 1959, p. 420). Traced north-eastwards, up dip, the limestone passes into pale grey brachiopod-rich and partly oolitic limestone. South-eastwards, a few feet of thinly bedded dark grey limestone are poorly exposed [2797 5300]. Apparently overlying the pale grey 'knoll' limestone, they are in turn overlain by laminated shales which have yielded E₂ₐ goniatites some 10 ft above the limestone.

There is a rich fauna of zaphrentoid corals and reef brachiopods in the Cawdor Limestone and Ramsbottom (*in* Smith and others, 1967, p. 52) has described the lithologies and faunas of the several distinctive facies that can be distinguished. No goniatites have been collected from the present district, but the age of the Cawdor Limestone is established by the records of *Goniatites (Mesoglyphioceras) granosus* and *Lyrogoniatites* aff. *georgiensis* from the Cawdor Limestone of Cawdor Quarry. These indicate an upper *Posidonia* (P₂) age—within the range P₂ₐ₋c (Ramsbottom op. cit.).

Volcanic rocks

The 'toadstones' in this part of Derbyshire were described by Arnold-Bemrose (1894, 1907), Geikie (1897) and Pocock (*in* Gibson and others, 1908) (Figure 6). More recently, Harrison (*in* Smith and others, 1967, p. 261) has discussed the petrography of the Matlock Lower Lava. Its apparent isotopic age of 185 ± 5 million years relates to intense hydrothermal alteration subsequent to its formation (Fitch and others, 1970). Its true age must exceed that of the Ible Sill (Smith and others, 1967, p. 26), emplaced in the Hoptonwood Limestone 4 miles N of the present district, and dated at 309 ± 21 million years (op. cit).

The Hopton Agglomerate, which is exposed along the east side of the Hopton–Middleton road, was originally mapped by the Geological Survey in the 1850s as interbedded with the limestones, and was estimated to be 1000 ft below the Lower Lava. Geikie (1897) considered it was a vent with transgressive relationships to the limestone. Pocock, during the resurvey of the area in 1903, pointed out that the limestone thicknesses were exaggerated and separated the agglomerate and the limestone from the 'Yoredales' (Namurian Shales) to the south by a fault. Arnold-Bemrose (1907) discovered agglomerate to the south of Pocock's fault and suggested that the vent cut across both the limestone and overlying shales so that a fault was unnecessary.

The present survey has found no evidence for a fault south of the vent and has confirmed the northern and eastern boundaries of the agglomerate as transgressive.

The agglomerate is cut off to the north by the Yokecliffe Fault. Between Newclose and Quickset Mine [2625 5379], however, 'toadstone' is commonly incorporated in the drystone walls of nearby fields. If this material was excavated from deep workings along the Rake, the latter may have acted as a feeder for the Matlock Lower Lava, the outcrop [2752 5384] of which appears to end at the Yokecliffe Rake near Godfreyhole. Moreover, the location of this possible

Plate 2

1 Baileycroft Quarry, Wirksworth

Irregular erosional base of Cawdor Limestone (P_2 age) cutting out the Matlock Limestone (D_2 age) and resting unconformably on the Hoptonwood Limestone (D_1 age). See Figure 5 (L 931)

2 Dale Quarry, Wirksworth

The Hoptonwood (D_1) and Matlock (D_2) limestones. The quarry remnant was left unworked because of past lead mining activity in Burton's Vein, which runs through its centre. This has rendered the rock unstable as well as dangerous for blasting. The background is composed of Ashover Grit (Main Bed)—the first sandstone overlying the limestones in this region (L 934)

feeder between the vent, now filled by the Hopton Agglomerate, on the west, and the lava on the east, suggests that the three features belong to the same volcanic episode.

The agglomerate is a breccia composed of small lapilli and subangular blocks of dolerite and basalt up to 2 ft across. The basalt contains augite and feldspar, often fresh; olivines are altered and set in a black glassy base.

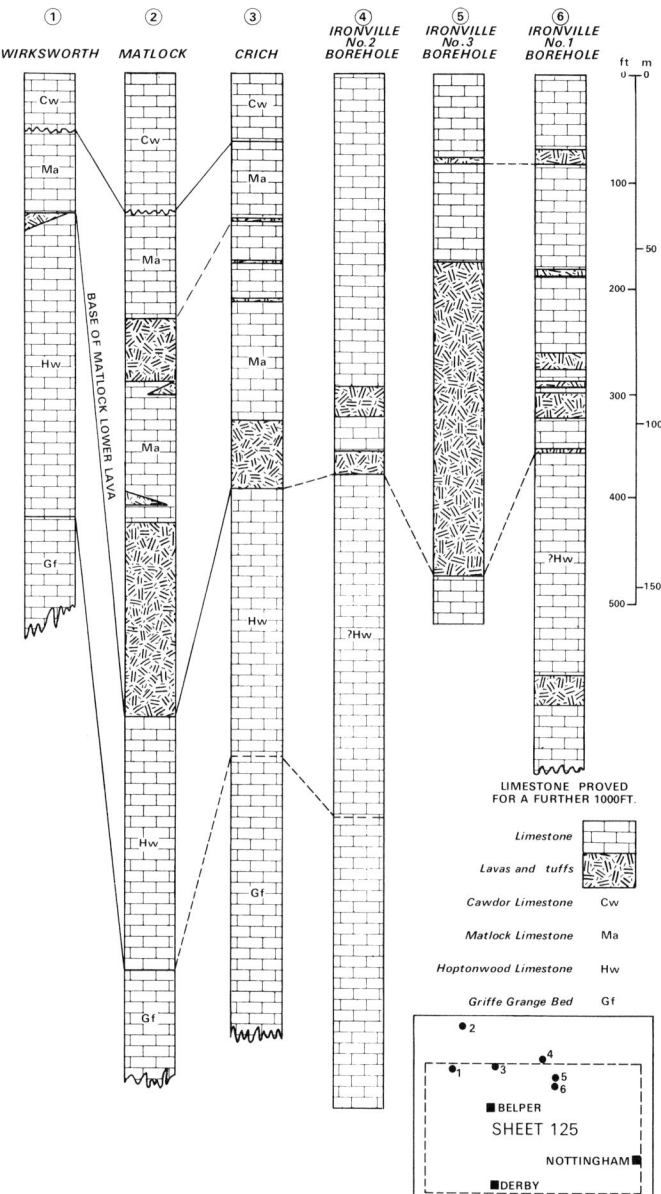

Figure 6 Suggested correlation of lavas and tuffs in the Dinantian

Crich

The northern half of this periclinal dome was described by Smith and others (1967, p. 35) who give the following section:

	Thickness ft
Cawdor Shale	
Dark shales with limestone bands	20 to 30
Cawdor Limestone	
Limestone	60
Clay	1
Matlock Limestone	
Grey and dark grey limestone with chert and interbedded clay bands	163
Limestone (not exposed)	100
Lava (not exposed) ?Matlock Lower Lava	average 60
Hoptonwood Limestone and ?Griffe Grange Bed	
Limestones (not exposed)	500+

The unexposed beds were proved in shafts now abandoned.

The village of Crich is built almost entirely on limestone, which is faulted against Ashover Grit along the Southern Crich Fault in the west. The limestone dips at angles of 4 to 20° under the Cawdor and Namurian shales to the east. To the south it plunges even more steeply, but the overlying beds are concealed by drift.

The 20 to 30 ft of Cawdor Shale overlying the limestones are not exposed, their presence being inferred from boreholes to the north.

Provings in deep boreholes

Dinantian limestone of the shelf facies has been reached in two deep boreholes within the district—Ironville Nos. 1 and 3—which were drilled into anticlines in the exposed Westphalian rocks, in the hope of finding oil.

Ironville No. 1 Oil Bore was sited on the southern end of the Ironville Anticline and the top of the Dinantian was reached at 2035 ft. Examination of the chipping samples by Mr K. S. Siddiqui showed concentrations of tuffaceous fragments at 2105 to 2220 ft, 2295 to 2300 ft and 2385 to 2390 ft. Olivine-dolerite was recorded between 2300 and 2355 ft.

The upper 70 ft of limestones are dark grey and contain a 5-ft shale band at 2075 ft—features comparable with the Cawdor Group. The beds associated with the tuffs and dolerites are probably the Matlock Limestone. Below 2390 ft the limestone is generally whitish in colour and suggestive of the Hoptonwood Limestone. Below 3000 ft the beds are more variable, dark grey alternating with pale grey-brown and white limestones which may be correlatives of the Griffe Grange Bed (p. 7).

Ironville No. 3 Oil Bore on the eastern flank of the anticline reached Dinantian limestone at 2200 ft. Spot cores of the top of the limestone showed dark grey limestones associated with black calcareous shales. *Zaphrentites sp.* and *Diphyphyllum lateseptatum*, genera recorded from the Cawdor Limestone at outcrop, occurred at 2250 ft. 'White trap' (bleached basic volcanic rock) was encountered at 2300 ft, and green and mottled tuff and agglomerate extended between 2400 and 2700 ft. The limestone beneath the volcanic rocks was pale brown to off-white and finely crystalline. The hole was completed at a depth of 2742 ft.

As at outcrop, darker limestones are common in the upper part of the succession, and paler grey or brown crystalline limestones occur towards the base. In Ironville No. 3 Oil Bore, the 300 ft of tuff and agglomerate separate the two types of limestone. Whilst accurate correlation is impossible, it seems reasonable to consider these extrusives as the approximate equivalents of the Matlock Lower Lava (see Figure 6) which separates the pale Hoptonwood Limestone from the Matlock and Cawdor limestones at outcrop.

BASIN FACIES

The Widmerpool Formation outcrop borders that of the Lower Namurian in the Turnditch, Quarndon and Weston Underwood districts and forms poorly exposed inliers in the valley of Mackworth Brook near Kirk Langley, at Wildpark Wood, one mile east of Brailsford, and in a valley immediately south of Brailsford (Figure 3).

Sporadic exposures in streams such as Blind Brook, Kedleston, and Cutler Brook, Weston Underwood, show that the formation consists of thin beds of dark grey argillaceous limestone and grey calcareous siltstone; sole markings are commonly present. The beds contain plant fragments, with occasional poorly preserved bivalves (see also Gibson and others, 1908, p. 24). A few of the coarser limestones contain conspicuous crinoid and brachiopod debris. An extensive microflora obtained from a sequence near Mackworth [3137 3768] included no diagnostic species. The Duffield Borehole [3428 4217] was therefore drilled to establish a detailed succession of the formation. This proved Dinantian rocks belonging to the Upper *Beyrichoceras* (B_2), Lower *Posidonia* (P_1) and Upper *Posidonia* (P_2) zones, totalling some 2100 ft of fine-grained sediments together with igneous rocks and tuffs corresponding to the top 500 ft or so of the 'shelf' limestones north of Wirksworth. The borehole demonstrated therefore not only the complete change in facies but also a four-fold increase in the thickness of the beds (Figure 4).

Apart from igneous rocks and two relatively thin calcareous mudstone sequences the succession consists largely of turbiditic graded silty mudstone/sandstone and muddy limestone units with dark grey mudstone intercalations. Further details are given by Aitkenhead (1977).

Provisional thicknesses of strata (excluding igneous rocks) attributed to the various zones are shown below:

Zone	Thickness			Depth	
	ft	in		ft	in
P_2	566	$6\frac{1}{2}$	to	1925	7
P_1	1104	$11\frac{1}{2}$	to	3030	$6\frac{1}{2}$
B_2 (pars)	192	$2\frac{1}{2}+$	to	3436	10

Six tuff bands and two dolerite sills were encountered. Details of the tuffs are as follows:

Zone	Thickness		Depth	
	ft	in	ft	in
P_2	6	4	1535	5
P_2	1	7	1629	6
P_2	10	3	1690	2
P_2	2	10	1735	0
P_2		1	1843	6
P_2		3	1864	10
P_2	24	5	1891	2
P_1	4	6	1930	1

The tuffs range from fine to coarse-grained and vary in colour from pale blue to greenish grey (p. 167d). They usually contain pumice fragments and in some beds lapilli are present. Some intermixing of tuff with calcareous mud has been noted, particularly in the thickest band.

At outcrop, white, tuffaceous, calcareous clay bands 3 ft and $2\frac{1}{2}$ ft occur in Blind Brook [3097 4233] and a temporary exposure [3037 4352] respectively. They are of volcanic origin but some reworking as well as mineral replacement seems to have occurred.

Two dolerite sills are 27 ft 9 in and 220 ft 3 in thick, their bases occurring at 1779 ft 7 in (P_1) and 3337 ft 3 in (B_2) respectively. Both have chilled upper and lower margins and have baked the adjacent sediments.

A second borehole [2548 3588] at Trusley in the southwest corner of the district passed through 331 ft of Triassic rocks into a sequence of purplish red sandstones, siltstones and mudstones of Dinantian age. This sequence, like Duffield, was composed largely of graded turbidite units and confirmed a southerly extension of the Widmerpool Formation. The turbidite sequence of red and purple sandstones, siltstones and silty mudstones showed grading, slumping and sole markings. The lower part of the sequence is calcareous. Interbedded with the turbidite units are red, brown and purple mudstones interlaminated with pale grey-green mudstones.

Detailed examination of representative lengths of Trusley core show that the sandstone-bearing turbidites have a proximality index (R. G. Walker, 1967) of 20 per cent compared with 96 per cent in the B_2 Zone sandstone-bearing turbidites of similar age in Duffield Borehole. Thus both sequences are of distal turbidites, but that in Duffield Borehole appears to be more distal than that in Trusley (p. 174c).

In thin sections made from the lowest and most calcareous parts of the sequence (504 ft) Dr W. H. C. Ramsbottom identified bryozoa and crinoid fragments and the following D_1 Zone algae and foraminifera—*Koninckopora inflata*, *Archaediscus krestovnikovi*, *A. moelleri*, *Endothyra spp.* and *Tetrataxis sp.*

Sub-Trias secondary reddening was detected throughout the 177 ft of strata proved. This is in marked contrast to surface exposures of the formation, in which reddening is restricted to patchy staining. At Trusley the Pebble Beds are absent and the Waterstones overlie the Widmerpool Formation; the thickness therefore of reddened beds may be partly a reflection of the large time interval between the Dinantian and the Anisian.

Both the Carboniferous and Triassic rocks in the Trusley Borehole are intruded by numerous gypsum veins and stringers, the thickest of which (up to 3½ in) were usually found along the bedding planes. Whilst the presence of penecontemporaneous gypsum and anhydrite is known within the Carboniferous of the Widmerpool Gulf (Llewellyn and others, 1969, p. 85), a Triassic origin for the sulphate minerals at Trusley is considered most likely. The numerous fault planes and fractures present in the Carboniferous rocks would have provided easy access for the migrating fluids.

MISCELLANEOUS ROCK-TYPES ASSOCIATED WITH DINANTIAN LIMESTONES

Dolomite

Dolomite is present mainly in parts of the Hoptonwood and Matlock limestones. The contact between dolomite and limestone is sometimes sharp, but the line on the published six-inch and 1:50 000 maps mostly represents a zone of gradual transition. The dolomitic limestone weathers into characteristic 'tors' the most prominent of which occur just outside the present district, north-west of Hopton, i.e. the Harborough Rocks.

Dunham (1952, p. 402) considered the dolomite to be a product of secondary alteration which preceded the ore-mineralisation. The field relationships were discussed in detail by Parsons (1922), who considered the dolomitisation to be pre-Triassic, a view reiterated by Kent (1957).

The identification of Triassic pebbly sandstones at an elevation of about 800 ft near Blackwall [2561 4959] adds weight to the conclusion that much of the Dinantian limestone in this district was once covered by Triassic sands—the deposits in the Brassington silica sand-pits being relics of this former cover (Kent, 1957). Purple-staining of the Namurian sandstones cropping out at high elevations, e.g. Black Rocks, Cromford, and the Kirk Ireton synclinal plateau together with remanié deposits of clay and Bunter-type pebbles over much of the limestone and shale outcrop near Hopton is additional evidence to support this view.

No Permian deposits are preserved, and it is unlikely that the Zechstein transgression reached this area. The Dinantian limestones may therefore have been altered by the action of the magnesium-rich Triassic waters percolating down from the sea floor. Evidence in favour of this theory is:

1 That the areas of maximum dolomitisation usually occur in regions of maximum elevation. The boundary between the dolomite and underlying limestone in Golconda Mine [2437 5505], just north-west of the present district, is in the form of an irregular dome, although the bedding in the limestones is roughly horizontal (Shirley, 1959). The 'dome' may therefore be the reflection of a local elevation in the Triassic sea floor.
2 The absence of dolomitisation below the Matlock lavas, e.g. at [2653 5405], suggests that they operated as a protective cap which prevented downward percolation. Similarly, notable areas of undolomitised limestone in other elevated areas may be due to the barrier afforded by local patches of Namurian shale, since removed.
3 The association of the silica-sand deposits with the dolomitisation in elevated areas.

4 Extrapolation from the area of outcrop near Turnditch suggests that the base of the Trias lay close above the dolomitised zone in the Brassington–Carsington–Yokecliffe Rake area (Figure 63).

Chert and silicified limestone

Chert is commonest in the Matlock and Cawdor limestones, where it occurs as bedded tabular sheets of limited lateral extent and as layers of irregular nodules. It decreases in importance towards the gulf margin. Pocock (in Gibson and others, 1908, p. 12) described black chert from the Cawdor Limestone in Dale Quarry and noted an abundance of foraminifera and sponge spicules. Crinoid debris is common.

The formation of chert predates the dolomitisation, and according to Sargent (1921) was an inorganic process which occurred penecontemporaneously with the formation of the surrounding limestones. The introduction of the silica was attributed to contemporary vulcanism.

Pettijohn (1957, pp. 439–440) preferred a post-depositional origin with an early segregation into nodules and bands of initially diffuse silica. Orme (1973) found that the majority of Derbyshire cherts were formed by replacement of carbonate prior to consolidation, and suggested that in many cases segregation of the inorganic silica may have been penecontemporaneous with sedimentation.

Silicified limestones or quartz rocks result from a later more intense replacement of the limestone, but are of restricted occurrence, usually associated with faults and/or mineral veins. They are considered to be products of low temperature metasomatism by hydrothermal fluids.

One large boulder of silicified limestone was noted near Sprink Wood [2732 5300] south-west of Wirksworth. It was partly hematitised and contained large quantities of fluorite. Smith and others (1967, p. 39) suggested that in mineralised rocks silicification probably post-dated fluoritisation. Orme (1973) considers the process to precede fluoritisation. Such conflicting views probably result from the silicification being associated with a complex and prolonged mineralising phase ranging from the Triassic to the Jurassic.

Bituminous limestone

Pocock (in Gibson and others, 1908, p. 13) noted the local occurrence of bitumen in limestone near Hopton and Wirksworth. It is common in the pale grey massive limestone which forms a small 'knoll' near Pittywood Farm [2790 5310] south-west of Wirksworth. In thin section the dull-black bituminous material is seen to form secondary impregnations filling the pore spaces, and dense localised concentrations completely obliterate and mask the host-rock. Tests with various solvents proved negative and no aromatic compounds were observed.

Fragments of black bitumen were common in the Hoptonwood Limestone proved in the site investigation borehole [2652 5340] near Hopton.

MINERALS ASSOCIATED WITH DINANTIAN STRATA

The Dinantian limestone exposed in the Wirksworth area is extensively mineralised, its anticlinal structure being favourable for ore deposition below the Namurian shale cover now partly removed by erosion. Orebodies may exist below the Upper Carboniferous cover, for fluorite veining was recorded at depths of 2300 to 2400 ft in Ironville No. 3 Oil Bore. Large areas of the outcrop are dolomitised and there are patches of silicification. The orebodies, principally of galena and fluorite with calcite and baryte as the main gangue minerals, were once of considerable economic importance (pp. 104–109).

Gypsum veining has been recorded at depth in the Widmerpool Formation at Trusley Borehole.

CHAPTER 3

Namurian (Millstone Grit Series)

INTRODUCTION

The outcrop of Namurian strata is between 5 and 7 miles wide and extends from Wirksworth in the north-west of the district south-eastwards to Breadsall. It forms ground rising to over 1000 ft at Alport Hill between the rivers Derwent and Ecclesbourne. The alternating sandstones and shales in the upper part of the sequence form an upland area to the west of the coalfield and have weathered into a series of impressive scarps or 'edges', particularly along the Derwent valley north of Duffield. The lowest part, consisting largely of shales, floors much of the Ecclesbourne valley in the west of the district.

The Namurian rocks of the Derby district, as elsewhere in Britain, reflect an upward transition from the wholly marine environment of the Dinantian to the paralic environment of the Westphalian. The 'block' or 'massif' and 'gulf' controls on Dinantian sedimentation continued to influence deposition in the Namurian though they are apparent only in the thickness variations shown by the upper part of the sequence, where the predominantly arenaceous rocks are of deltaic origin.

Previous research

Whitehurst (1778, p.147) introduced the term 'Millstone Grit' for a coarse sandstone separated by shales from the Dinantian limestone in the Derwent valley. Farey (1811) called this sandstone the 'First Grit Rock' in the Matlock area and calculated that it was separated from the limestone by 420 ft of 'Limestone Shale'.

The primary geological survey by Green and others (1869, 1887) grouped the strata now assigned to the Widmerpool Formation and the lower part of the Namurian into a lower division consisting of black shales with thin beds and nodules of earthy limestone, and the upper of shales with thin beds of hard close-grained sandstone (Yoredale Sandstones). Gibson and others (1908), unable to sustain this distinction or to determine a demarcation between Lower and Upper Carboniferous, named the beds 'Limestone Shales' up to the horizon of the lowest sandstone (now a lower leaf of the Ashover Grit) which they termed 'Shale Grit'. For the latter and the succeeding 'Kinderscout Grit' (also now Ashover Grit) they followed the terminology of Green and others (1869), but for the higher sandstones they used the local names Belper Grit (now Chatsworth Grit), Coxbench Grit (now Rough Rock) and Rough Rock (now Crawshaw Sandstone, of Westphalian age).

Classification

Based upon the work of Hind (1909) and Bisat (1924), the base of the Namurian is defined by the first entry of the goniatite *Cravenoceras leion* and the top is drawn at the base of the *Gastrioceras subcrenatum* Marine Band. It is therefore chronologically synonymous with the Millstone Grit Series, as used extensively in publications of the Geological Survey dealing with the southern Pennines. Because of poor exposures some of the detailed palaeontology, particularly in the upper parts of the sequence, remains unproved within the district. The zonation already established in the adjacent Chesterfield district (Smith and others, 1967) is applicable however, and is shown in Figure 7 relative to the generalised vertical section of the Namurian of the Derby district. Borehole sections in the district and adjacent areas are shown graphically on Plate 4, which includes the probable correlation of sections lacking palaeontological control.

Some modern sedimentological work in the district has been published in outline by Mayhew (1967).

Lithologies

Two general sedimentary environments are reflected in the lithologies of the succession, with the main change appearing near the base of the Marsdenian Stage. The underlying predominantly argillaceous sequence is widely, but poorly exposed and most of the information on it relates to the 'basin' sequence proved in the Duffield Borehole (Aitkenhead, 1977).

In the borehole, the pre-Marsdenian beds consist largely of mudstones with eleven separate sequences of predominantly thin quartzose sandstone–siltstone–silty mudstone units. Each unit displays irregular graded bedding, usually with a basal sandstone division which is massive except for faint parallel lamination at its base. Less commonly sandstone with strong parallel lamination, and even more rarely massive sandstone, underlies a ripple-laminated division. The bases of the unit are always sharp and often show sole-structures. These features are characteristic of the deposits of waning turbidity currents which are henceforward referred to as turbidites (Bouma, 1962, pp. 49–50). The turbidite units are normally less than 1 ft thick but may reach 4 ft. They usually contain poorly preserved plant debris, but are otherwise unfossiliferous. Dark grey fissile mudstone intercalations between individual turbidite units commonly contain fish-scales and represent slow background deposition in quiet bottom waters. Chemical and spectrographic analyses have been published by Read and others (1973).

The mudstones between the turbidite sequences in the E_{1a} to E_{2b} zones are generally calcareous, particularly in the goniatite bands, and are mainly unlaminated and silty with

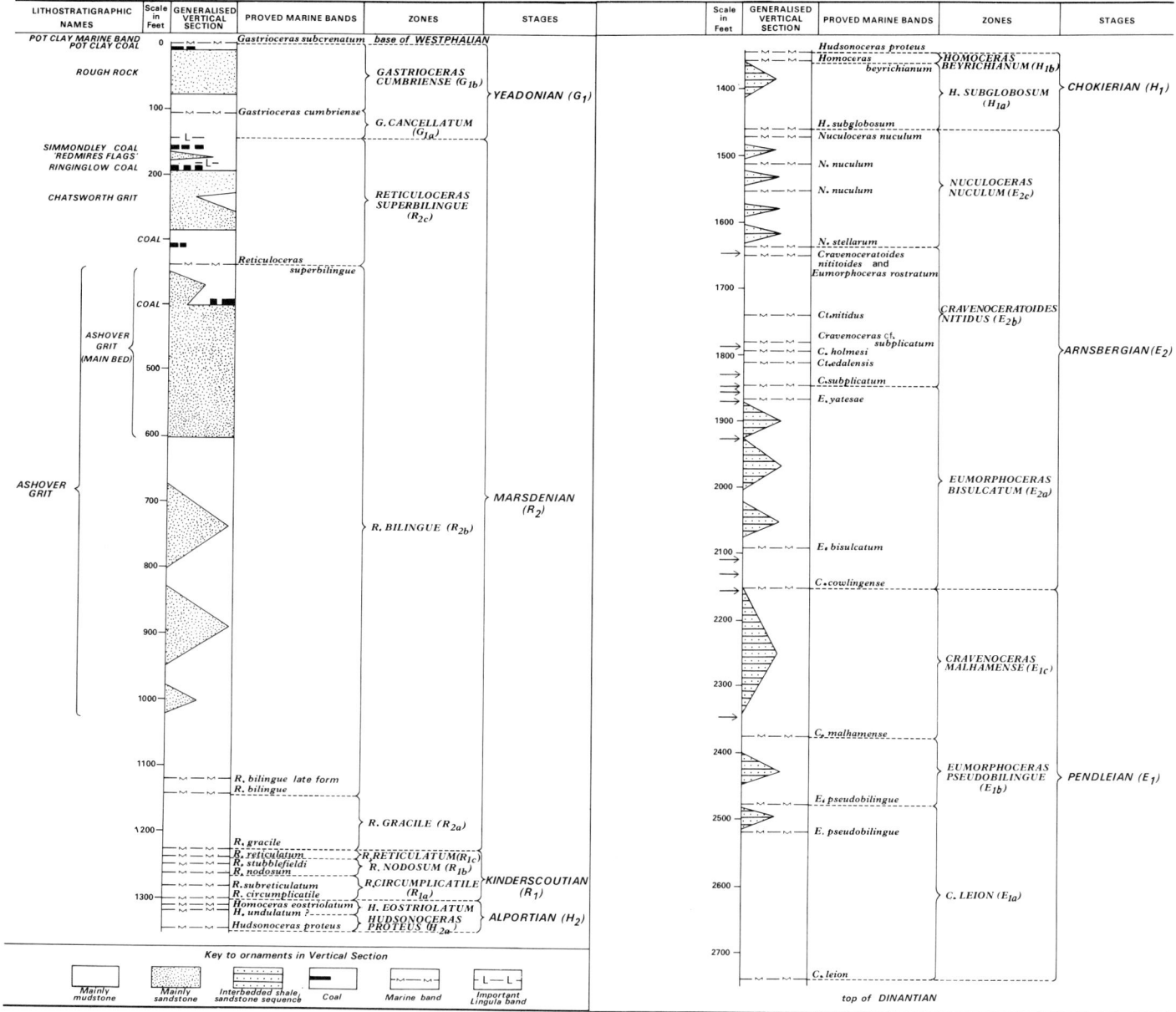

Figure 7   Generalised vertical section and classification of the Namurian in the Derby district

The arrows on the left-hand column margin indicate the horizons of K-bentonites formed as volcanic ash

fissile intercalations in which fossils, if any, are concentrated. The unlaminated beds may also be turbiditic in origin.

Above E_{2b} the mudstones are mainly shaly, non-calcareous and pyritic, but become increasingly silty above the *R. gracile* Marine Band, heralding the incoming of deltaic deposition that was to continue during the remainder of the Namurian and into the Westphalian.

There are no complete sections within the district of the mudstones deposited in the 'massif' or 'block' environment. The nearest are those described in the Tansley, Uppertown and Highoredish boreholes in the Chesterfield district (Ramsbottom and others, 1962), where the sequence is

largely composed of grey mudstones which are either calcareous or pyritous.

'Bullions' in the form of thin beds or concretions of hard calcareous siltstone with well-preserved goniatites are present in the fossiliferous mudstones of both 'massif' and 'gulf' environments. The concretions are normally discoid and may be several feet in diameter. Holdsworth (1966) described similar bullions from Staffordshire as calcite ferroan dolomite with an apparent complete absence of clay minerals. Besides goniatites, Holdsworth found they contained small sponges, abundant Radiolaria and spat of bivalves and gastropods.

Laminae of potassium bentonites rich in kaolinite are present in the mudstones of both environments. These have been identified by Trewin (1968) in Staffordshire and north Derbyshire as altered wind-blown volcanic ash; the laminae are brown or pale grey when unweathered or pale grey to white in surface exposure. Eleven such bands (Figure 7) proved by the Duffield Borehole can be correlated with those described by Trewin (Aitkenhead, 1977). None has been detected at outcrop within the district.

Ironstone nodules and bands are relatively rare in the 'gulf' facies sediments, although many of the turbiditic sandstones have a ferruginous cement.

The change to deltaic conditions during the Marsdenian and Yeadonian produced a rhythmic form of sedimentation. Cycles, which are not always complete, are typically as follows (Smith and others, 1967, p. 5):

Coal
Seatearth
Sandstone or siltstone
Grey mudstone becoming silty at top
Dark mudstone, commonly marine at or near base

The darker marine mudstones are very fossiliferous and pyritous and form discrete bands up to 1 or 2 ft thick. They also display a cyclic arrangement of their fauna (Ramsbottom and others, 1962, pp. 114–117). The grey mudstones contain rare bivalve horizons; plant remains are common as are ironstone nodules and bands. The mudstones normally grade upwards through silty strata to become interbanded with the basal parts of the overlying sandstones. Sharp contacts with the sandstones are indicative of erosion or 'washout' conditions.

There are three thick sandstone sequences in the district, i.e. the Ashover Grit, the Chatsworth Grit and, close to the top of the Namurian, the Rough Rock. Only the Ashover Grit is coarse enough to merit the term 'grit' but the name Chatsworth Grit, dating from Green and others (1869), is retained for historical reasons.

The Ashover Grit sequence comprises an uppermost persistent sandstone called the Main Bed because of its thickness, extensive outcrop and strong topographic expression, and at least three major lower sandstones of a less presistent nature. Lithologically the Ashover Grit is a coarse arkosic sandstone, pebbly in places, and of a pale grey to brown colour blotched with pink. The pebbles are mainly of vein quartz and less than 1 inch in size. Mayhew (1967, p. 94) has recognised two major facies—'a lower non-pebbly facies characterised by trough-shaped sets of cross-strata which accumulated under delta distributary channel-mouth bar conditions' and a 'transgressive, more coarse-grained pebbly facies dominated by planar sets of cross-strata which was deposited within fluvial channels'. Both facies are present in the Main Bed in this district, but the less persistent lower sandstones are characterised by the non-pebbly facies.

The Chatsworth Grit and Rough Rock are similarly arkosic, but less coarse than the Ashover Grit and with only rare pebbles; cross-bedding is common as are micaceous flaggy and shaly beds. Slumping, locally on a large scale, is present in the Rough Rock. There are commonly two leaves to the Chatsworth Grit.

Harrison (*in* Smith and others, 1967, pp. 267–276) has described and discussed petrographical and modal analyses of the Ashover and Chatsworth Grits from the Chesterfield area. He found broad similarities in the major and minor mineral suites with ubiquitous orthoclase and microcline and minor acid plagioclase. Altered muscovite and biotite are fairly constant minor constituents. Heavy minerals include zircon, rutile, apatite, tourmaline, ilmenite and monazite. The modal analyses show little difference between the ratios of quartz to feldspar in the two grits.

The deltaic deposits were laid down without any significant lithological differentiation in both 'gulf' and 'massif' areas, the locations of which are revealed only by thickness increases in the 'gulf' areas.

The coal seams which cap at least five of the Namurian cycles of the district are all impersistent and normally less than 1 ft thick. However near Alderwasley one unnamed coal above the *R. superbilingue* Marine Band is 2 ft thick (Wedd *in* Gibson and others, 1908, p. 42) and was formerly worked underground. Seatearths or ganisters are generally more persistent than the overlying coals.

Sedimentation

The Central Province area of Dinantian deposition which extended from the Craven Fault belt of Yorkshire to the Wales–Brabant Island (now the English Midlands) continued to receive sediment during Namurian times (Ramsbottom, 1970). A large river system draining from the north and north-east deposited deltaic sediments up to 6000 ft thick. Deposition of thick deltaic sandstones began in early Namurian times in the north, but did not move southwards to reach the Derby district until the Marsdenian Stage. Until then, the southern part of the province was receiving sediments derived from the Wales–Brabant Island and transported largely by turbidity currents (Holdsworth, 1963; Trewin and Holdsworth, 1973; Aitkenhead, 1977) into the Widmerpool Gulf (Falcon and Kent, 1960). Although its exact limits are difficult to define (pp. 5 and 25), the 'gulf' continued to have a marked influence on sedimentation in the Namurian.

The sediments in the 'gulf' are interpreted by Aitkenhead (1977) as mainly distal turbidites alternating with sequences of mudstones containing most of the goniatite marker bands which occur extensively throughout the Central Province. Aitkenhead (*ibid.*) has also tentatively suggested that the source of the turbidites was to the south-west of Duffield around which, and as far east as the Derwent valley and Breadsall, the turbidite sequences form mappable features, shown on the geological maps as 'sandstone/shale interbedded'. This distribution suggests proximity to a turbidite feeder channel.

The Duffield Borehole sequence shows that deposition from turbidity currents continued intermittently through much of the Pendleian, Arnsbergian and Chokierian. There is no evidence, however, that such deposition continued in the district during the Alportian and Kinderscoutian.

Plate 3
1 Slumping in the Rough Rock, Coxbench
Wood Quarry (L 909)

2 Alport Stone, a pillar of Ashover Grit
in Alport Hill Quarry (L 880)

There is less evidence of the 'block' or 'massif' sequences in the district. The Ironville oil bores provide little detail, but the sequence below the Ashover Grit appears to be similar to that of the Tansley Borehole (Ramsbottom and others, 1962). Ironville is considered therefore to lie on the 'block' or 'massif' area beyond the area dominantly influenced by turbidite flows from the Wales–Brabant Island. In Table 1 the Tansley–Ashover area is included to illustrate the probable thickness variations of the Kinderscoutian and lower stages for which only an overall figure is ascertainable for the Ironville oil bores. It shows that rocks of the early Namurian stages are thin and fairly constant over some 50 square miles of the 'block' between Ashover, Ironville and Ambergate. The increase in sediment thickness in the Widmerpool Gulf occurs largely in E_{1a} to H_{1a} strata, which are expanded by a factor of 5.

Table 1 Thickness variations of the Stages of the Namurian

	1	2	3	4
Yeadonian (G_1)	165	147*	182	190
Marsdenian (R_2)	800	458	c. 850	1100
Kinderscoutian (R_1)	119			100
Alportian (H_2)	14			?
Chokerian (H_1)	39			110+
Arnsbergian (E_2)	57			655
Pendleian (E_1)	32			566
Sub-total E_1 to R_1	261	326	c. 370	c.1561
Total	1226	931	1402	2551 to 2651

* Average figure.
1 Tansley Borehole and Ashover area. Ramsbottom and others (1962) and published six-inch maps SK 36 NW and SW, surveyed by G. H. Rhys, E. G. Smith and R. A. Eden.
2 Ironville Nos. 3 and 4 oil bores.
3 Nether Heage Borehole and Wirksworth area.
4 Duffield Borehole and adjacent outcrop.

Table 1 provides an indication of the relative amounts of subsidence between 'block' and 'gulf' in the Derby district. It also demonstrates the weakening influence of the Widmerpool Gulf during Namurian times.

During the Marsdenian and Yeadonian the district received sediments from the advancing delta, but the overall thickening into the 'gulf' was most evident in the former stage. This thickening is largely expressed in the sandstone beds of the Ashover Grit, the combined thickness of which expands from 100 to 600 ft and is illustrated by isopachs of the interval between the *R. bilingue* and *R. superbilingue* marine bands (Figure 8). The measured cross-bedding directions in the Ashover Grit (Figure 9) indicate that the depositing currents were in general flowing towards the north-west. In the Chatsworth Grit (p. 135b) cross-bedding directions so far measured indicate depositional currents flowing towards the south-west (Mayhew, 1967, p. 100) yet the sandstone is

thickest in the 'gulf' areas compared with the Ironville oil bores on the 'block'.

The overall thickness of the Yeadonian rocks (Tables 1 and 3) shows little variation between 'block and 'gulf' yet the Rough Rock itself appears to be thickest in the 'gulf'.

Since the completion of the Derby resurvey Chisholm (1977) has identified growth faulting in the Ashover Grit near Stanton to the north-west of this district. Similar growth faulting appears to be present in the Ashover Grit near Little Eaton and may also occur in the vicinity of Alport Hill.

GENERAL STRATIGRAPHY

Pendleian (E_1)

These beds are thickest in the Widmerpool Gulf sequence of Duffield Borehole where they consist of 566 ft of mudstones with turbidite mudstones, siltstones, sandstones and/ or limestones. There are three main suites of turbidites, of which the uppermost in the *C. malhamense* Zone (E_{1c}) spans about 175 ft of strata. A thin K-bentonite band was proved near the base of this Zone. In the 'gulf' marginal area to the north, near Wirksworth, the E_1 Stage is thin or absent due either to non-deposition or to subsequent erosion.

Arnsbergian (E_2)

Beds of this stage vary from 655 ft of mudstones with turbidites proved in Duffield Borehole to 200 ft of grey silty mudstones and calcareous siltstones bordering the Dinantian limestone in the north. As in other parts of Derbyshire, mudstones of the E_2 stage overlap the lowest Namurian sediments to rest directly on the limestone. In the present district E_2 shales are considered to rest on limestones ranging in age from D_1 (Hoptonwood Limestone) to P_2 (Cawdor Formation). K-bentonites were recorded in rocks of this stage in the 'gulf' facies (Figure 7).

Chokerian (H_1) and Alportian (H_2)

Few details are known of the H Zone sequence in the Derby district. A three-fold thickening of measures between the Ashover 'block' facies and the Duffield 'gulf' facies is indicated from borehole data. Much of this thickening would appear to be in the Chokerian which probably includes the uppermost turbidite suite proved in Duffield Borehole, where some 109 ft are assigned to the H_{1a} Subzone.

An estimate of the maximum thickness of the entire H Zone from surface observations is of the order of 130 ft.

Kinderscoutian (R_1)

Borehole evidence from Ashover (Ramsbottom and others, 1962) and thickness estimates from scattered exposures in the Franker Brook and near Breadsall show that the rate of deposition was fairly constant over a large area of Derbyshire. The measures of the Kinderscoutian Stage at outcrop are probably less than 100 ft thick in the Derby district and differ from the Chesterfield and other Pennine areas to the north in the absence of any sandstone representative of the Kinderscout Grit. Its recognition in Ironville No. 2 Oil

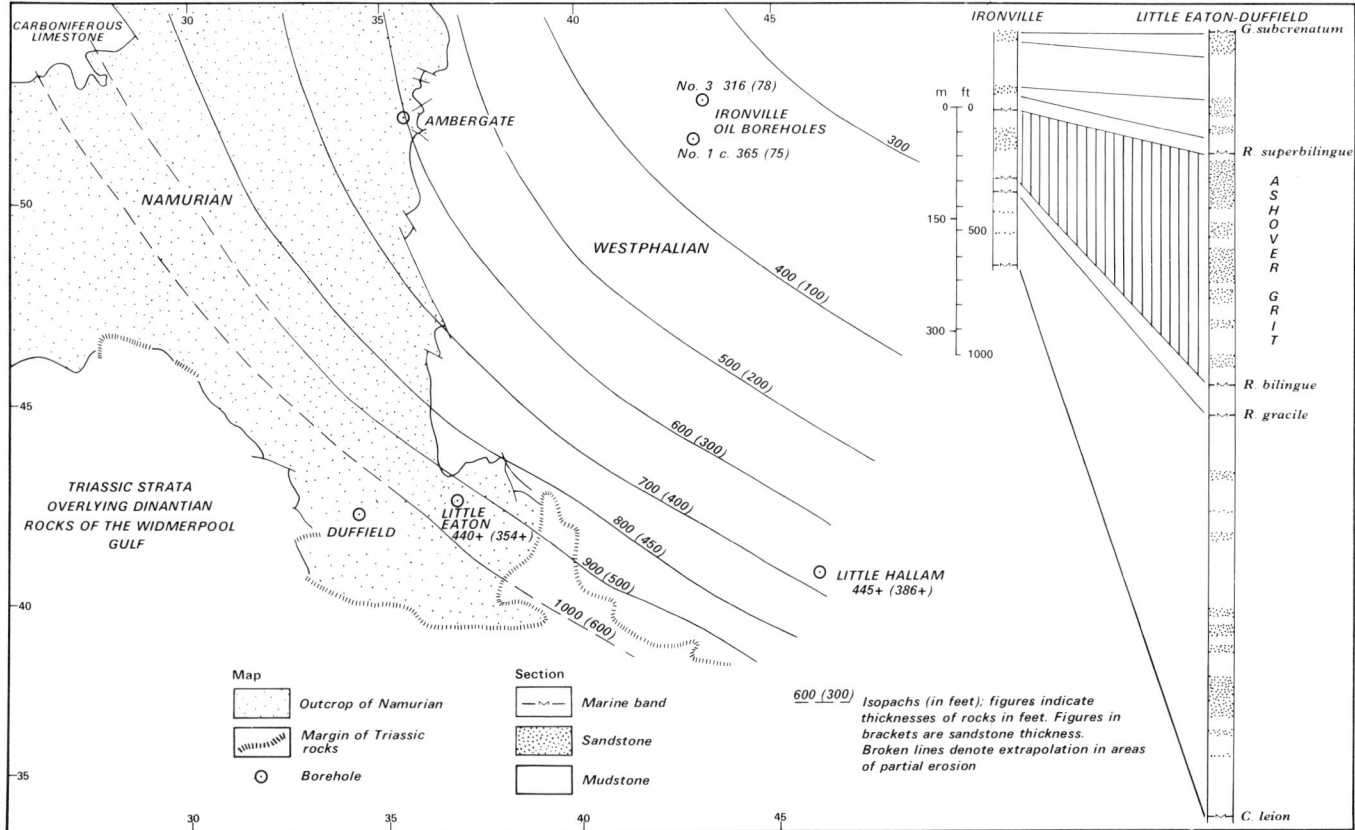

Figure 8 Generalised isopachs of the rocks between the *R. bilingue* and *R. superbilingue* marine bands

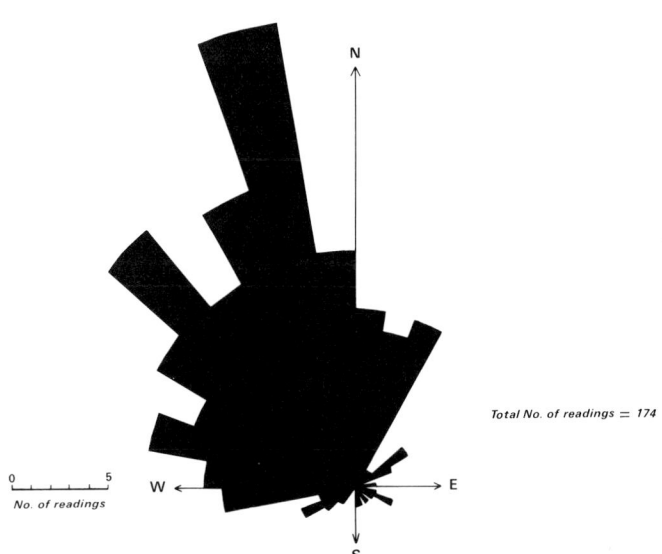

Figure 9 Rose diagram showing dip directions of cross-bedding in the Ashover Grit (Main Bed)

Bore (Smith and others, 1967, fig. 7) seems doubtful, for the lowest sandstone here is now considered to be the Ashover Grit.

Marsdenian (R₂)

The Marsdenian Stage consists of 500 to 1100 ft of strata lying between the base of the *Reticuloceras gracile* Marine Band and the base of the *Gastrioceras cancellatum* Marine Band (p. 134b). The following upward sequence of beds is known to be present in the district; the *Reticuloceras gracile* Marine Band, known only in the vicinity of Wirksworth and Shottlegate although its horizon can be inferred from the gamma logs of Ironville No. 3 Oil Bore (see also Downing and Howitt, 1969, fig. 3), is succeeded by argillaceous rocks, containing the *Reticuloceras bilingue* Marine Band. The interval between these marine bands however is not known. Early to late forms of *R. bilingue* have been found at outcrop below the Ashover Grit, but the information is too indefinite to enable separation into discrete bands. The Ashover Grit is divided into lower leaves of variable persistence and an upper persistent Main Bed. An unnamed and impersistent coal closely overlies the Main Bed and the succeeding argillaceous and arenaceous sequence contains *Lingula* beds (Plate 4) and is topped by the *Reticuloceras superbilingue* Marine Band. The overlying argillaceous rocks, which include another unnamed and impersistent coal, extend up to the Chatsworth Grit. This grit is in two leaves, with the upper capped by the Baslow or Ringinglow Coal (Smith and others, 1967; Stevenson and Gaunt, 1971). The overlying mudstones are in turn overlain by thin Redmires Flags, at least in boreholes remote from the outcrop. At the top of the stage a thin coal correlated with the Simmondley Coal of Stevenson and Gaunt (1971) is present.

The thicknesses of these individual beds are shown in Table 2.

Table 2
Thickness variations of the Marsdenian rocks

	1	2	3	4	5	6	7
R. gracile Marine Band to base of lowest Ashover Grit Sandstone	30	50+*	300–400*	—	—	156	—
Ashover Grit below Main Bed	700	340	375	490	320	0	—
Ashover Grit Main Bed	170	360	—	160	170	78	386
Ashover Grit (total)	870	700	—	650	490	78	—
Top of Ashover Grit to R. superbilingue Marine Band	—	65†	—	65	50–85	82	—
R. bilingue Marine Band to R. superbilingue Marine Band	—	815+	—	—	—	<316	—
Top of Ashover Grit to base of Chatsworth Grit	—	150	—	100	75	150	144
Chatsworth Grit	—	150	—	160	90	35	104
Marsdenian rocks overlying Chatsworth Grit	—	—	—	—	—	39	40
Total Marsdenian	—	Say 1100	—	—	—	458	—

* from R. bilingue Marine Band
† differs from vertical section on 6-in map

1 Wirksworth area.
2 Duffield–Little Eaton area.
3 Ferriby Brook section near Breadsall.
4 Belper area.
5 Ambergate–Crich area.
6 Ironville No. 3 Oil Bore.
7 Little Hallam and Trowell Moor Colliery boreholes.

Yeadonian (G₁)

The lower, argillaceous, part of the Yeadonian rocks include the inferred *Gastrioceras cancellatum* Marine Band, marking the base of the stage, and the *G. cumbriense* Marine Band. The upper sandy part contains the Rough Rock capped by a few feet of strata comprising the thin Pot Clay Coal and its seatearth or ganister. Table 3 illustrates the thickness variations of the Yeadonian strata in boreholes in this district; their locations are given in the 'Details'; thicknesses are to the nearest foot.

Within the present district the *G. cancellatum* and *G. cumbriense* marine bands have been proved in Ironville No. 4 Oil Bore and Nether Heage Borehole. The *G. cancellatum* Marine Band yields only *Lingula*. The Rough Rock is ubiquitous but varies markedly in thickness and in its sedimentation. The Pot Clay Coal is known only in two boreholes in the south-east of the district, but its seatearth or ganister is widespread.

The Rough Rock has been confused with the Crawshaw Sandstone of the Westphalian A by Stephens (1952) who described the 'peculiar lithological development of the Rough Rock' emphasising the upper coarse division and the finer lower division, separated at Ridgeway [3585 5154] by 20 ft of shale. It is in the basal part of this shale that the Pot Clay Marine Band has been discovered (p. 136b) thus identifying Stephens' upper division as the Crawshaw Sandstone.

Table 3 Thickness variations of the Yeadonian rocks

	1	2	3	4	5	6
Rough Rock and overlying strata	70	55	42	112	80	21
Strata below Rough Rock	58	112	140	46	52	20+
Total Yeadonian	128	167	182	158	132	—

1 Ironville No. 3 Oil Bore.
2 Ironville No. 4 Oil Bore.
3 Nether Heage Borehole.
4 Little Hallam Water Bore.
5 Trowell Moor Colliery Underground Borehole.
6 Beechdale Road Borehole.

WESTPHALIAN ROCK-TYPES

Coals

The coals form about 3.5 per cent[1] of the Westphalian A and B measures. In this account their classification follows that of the former Coal Survey Laboratory, which separated the following types of coal:

1 Brights (Softs)—layered, generally brittle coal formed mainly from the large Lycopod floras.
2 Dull (Hards)—less obviously banded, harder coal formed by decomposition of smaller herbaceous floras or a mixture of herbaceous and arborescent floras.
3 Banded dull and brights—a mixture of 1 and 2.
4 Cannel—finely comminuted plant debris with a variable content of mud, devoid of lamination.
5 Dirty coal—inferior admixture of bright and/or dull coals and dirt, with an ash content of 15 to 40 per cent.
6 Dirt—mudstone, commonly carbonaceous or with interbedded coal streaks and an ash content over 40 per cent.

Some of the diagrams that follow show the relative importance of the various coal types in the major seams and that the pattern of coal types within a seam is an important guide to its correlation. Typical 'bright' coals are the Blackshale, Deep Soft, High Hazles and Mainbright; 'hards' or dull coal is present as thick bands in the Low Main, Deep Hard and Top Hard seams.

The coals are considered to have been formed from vegetation which grew *in situ* in water which, though shallow, was deep enough and sufficiently de-oxygenated for the preservation of vegetable matter (Elliott, 1968, p. 359).

The phenomenon of split seams has been discussed by, among others, Raistrick and Marshall (1939, p. 63), and Elliott (1968, p. 366). In the present district there are two prevalent types. The first involves interdigitation on a regional scale with sediment derived mainly from the north and east and less commonly from the south and west. The second is local and linear in form due to local subsidence and sedimentation, possibly along a pre-existing water course.

Coal seams are subject to another facies-controlled variation known as swilleys. These take the form of linear hollows filled with coaly material; they occur at the base of seams and locally increase their thickness. Elliott (1965, p. 137) visualises levées of a river which ponded back a shallow lake in which cannel lenses were formed. Swilleys therefore record the course of rivers across the coal swamps. Good examples from the Derby district are known in the Deep Hard, Deep Soft and Top Hard coals of the Hucknall area and in the Lowbright Coal at Cinderhill Colliery.

Seatearths

The seatearths include all grades of sediment from sandstone to mudstone, but are distinguished by the general absence of bedding and the presence of rootlets. Once thought to be the soils in which the coal-forming forest grew, they are now believed to represent a sub-aqueous soil accumulation on which vegetation flourished without being preserved as coal. Coal formation normally took place later under more stagnant swamp conditions (Moore *in* Murchison and Westoll, 1968, pp. 113–122).

Elliott (1968, fig. 2 and p. 359) has shown that the seatearths between the Blackshale and Mainbright coals are of two main types—grey mudstones with nodular siderite, and brown mudstones and sphaerosiderite. The brown seatearths are usually overlain by thin coals. Elliott attributes the different coloration to varying groundwater conditions. The green colours noted in seatearths above the Mansfield Marine Band are a reflection of increased subaerial reaction in areas of better drainage.

Mudstones

Sediments of this grade form over 50 per cent of the Westphalian sequence and have been referred to as shales or bind, etc. by early workers. Mudstones devoid of any silt fraction are rare in this district and even the marine mudstones are slightly silty though darker grey than most of the non-marine beds.

Most mudstones contain evidence of bottom-living faunas, thus indicating a rate of sedimentation sufficiently gentle for living organisms to flourish, with bottom currents too slight to disturb the fine laminations now preserved. By an increase in silt content mudstone grades into siltstone or passes by intercalation and interlamination of sandy beds (striped beds) into sandstone.

Sandstones and siltstones

These sediments form up to 45 per cent of the total sequence and represent deposition in an increased energy environment. They range from rapidly deposited siltstones, locally containing upright tree trunks presumably swamped by sediment-laden floodwater, to washout sandstones, beneath which the underlying mudstones and coals have been removed, giving rise to breccio-conglomerates. Within this range there is a complex relationship between the sand and silt fractions, reflecting the varied conditions of deposition (Elliott, 1968, p. 355).

Miscellaneous rocks

Tonsteins are unlaminated layers or lenses of pale brown-grey to black kaolinitic mudstone usually associated with seatearths and coals. Strauss (1971), working mainly to the east and south of this district, has extended the earlier work of Eden and others (1963) and recognised tonsteins at 16 horizons within the Westphalian of the East Midlands. Of these only three have been located in the Derby district—within the First Piper and Deep Hard coals and below the High Main Coal. A fourth horizon—above the Brown Rake Coal—is now considered to be the related Fragmental Clayrock (see below). Tonsteins have diverse origins varying from diagenetic to pyroclastic (see e.g. Williamson, 1970).

---

[1] The percentages of rock-types in this account are calculated from the text-figures showing representative sections of the measures in the 'Details of Stratigraphy'.

Plate 4
Sections of the Namurian strata in
the Derby district compared with
the Tansley Borehole. The gamma
logs of the Tansley, Duffield,
Nether Heage, Ironville No. 3
and No. 4 boreholes are also shown
(gamma activity increases to the
right)

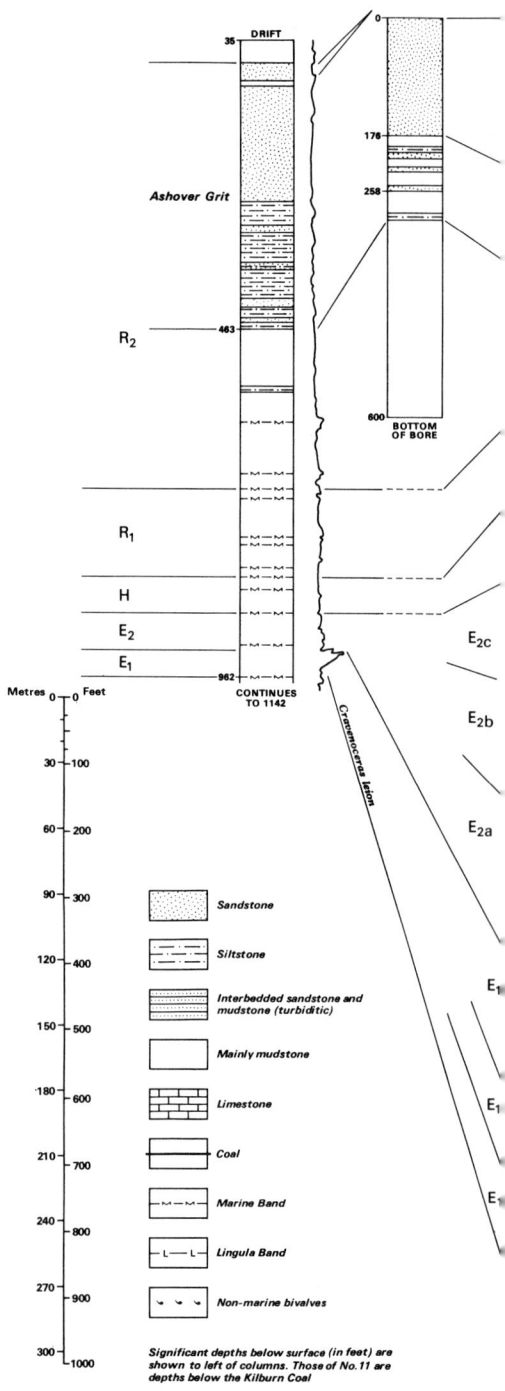

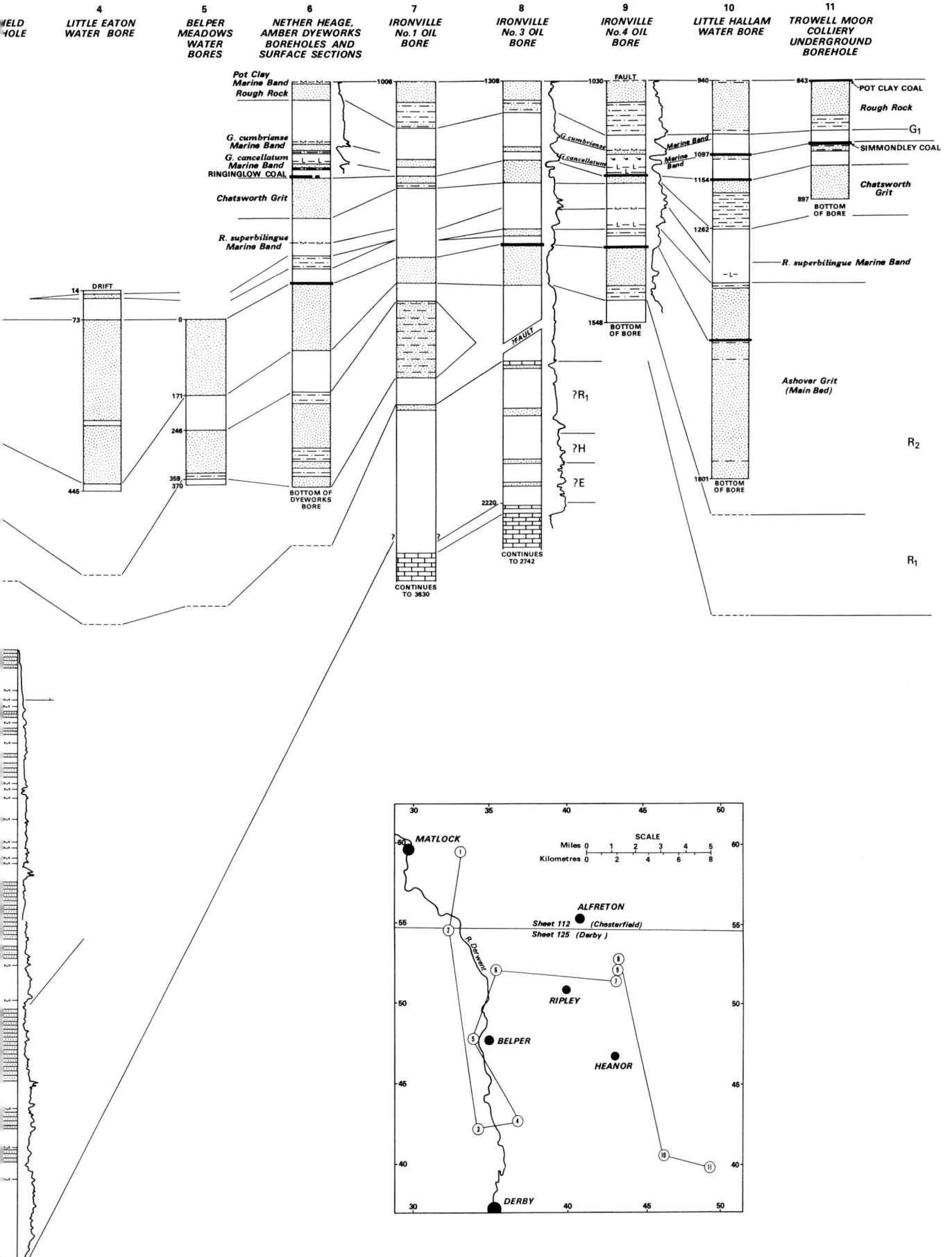

CHAPTER 4

Westphalian (Coal Measures)

INTRODUCTION AND CLASSIFICATION

Westphalian (Coal Measures) strata rest conformably upon the Namurian and are unconformably overlain by Permo-Triassic rocks in the east and south-east of the district. They crop out in a central belt some 8 miles wide and, with their extension below the Permo-Triassic rocks, comprise the south-western part of the Yorkshire and East Midlands Coalfield.

The maximum thickness of measures proved in this part of the Pennine Province is some 3700 ft. Westphalian A rocks (Lower Coal Measures), some 1600 ft thick, contain several important workable coals, mainly in the top 500 ft; Westphalian B rocks (in large part the Middle Coal Measures) are about 1400 ft thick and contain the majority of the economically important seams; Westphalian C (700 ft) is a sequence largely devoid of useful seams apart from the High Main Coal.

The measures consist chiefly of mudstones with subordinate siltstones, sandstones, seatearths and coals in repetitive sequences or cycles, commonly in that upward order and each usually between 30 and 50 ft thick. The thickest cycles are usually dominated by sandstone and are mainly of Westphalian C age.

The lithologies are closely comparable with those found elsewhere in the Pennine Province, comprehensive accounts of which have been given by Trueman (1954, pp. 1–29), Smith and others (1967, pp. 80–83) and Murchison and Westoll (1968, pp. 71–85).

The strata in Britain are subdivided into Lower, Middle and Upper Coal Measures with the boundaries at marine bands (Stubblefield and Trotter, 1957). Correspondence with the Westphalian classification is shown in Figure 10.

Further subdivisions are based on the non-marine bivalves (mussels) (Trueman and Weir, 1946–56; Weir, 1960–68) and these have been divided further into faunal belts (Calver, 1956; Eagar, 1947, 1952, 1954, 1956) which facilitate more detailed correlation, particularly in Westphalian A and B. Plants have long been employed as broad zonal indicators, and recently Smith and Butterworth (1967) have established seven miospore zones of value for coal seam identification and the wider aspects of inter-coalfield correlation. The main features of the palaeontology of the Westphalian have been described by Calver (*in* Smith and others, 1967, pp. 87–97; 1973, pp. 42–46).

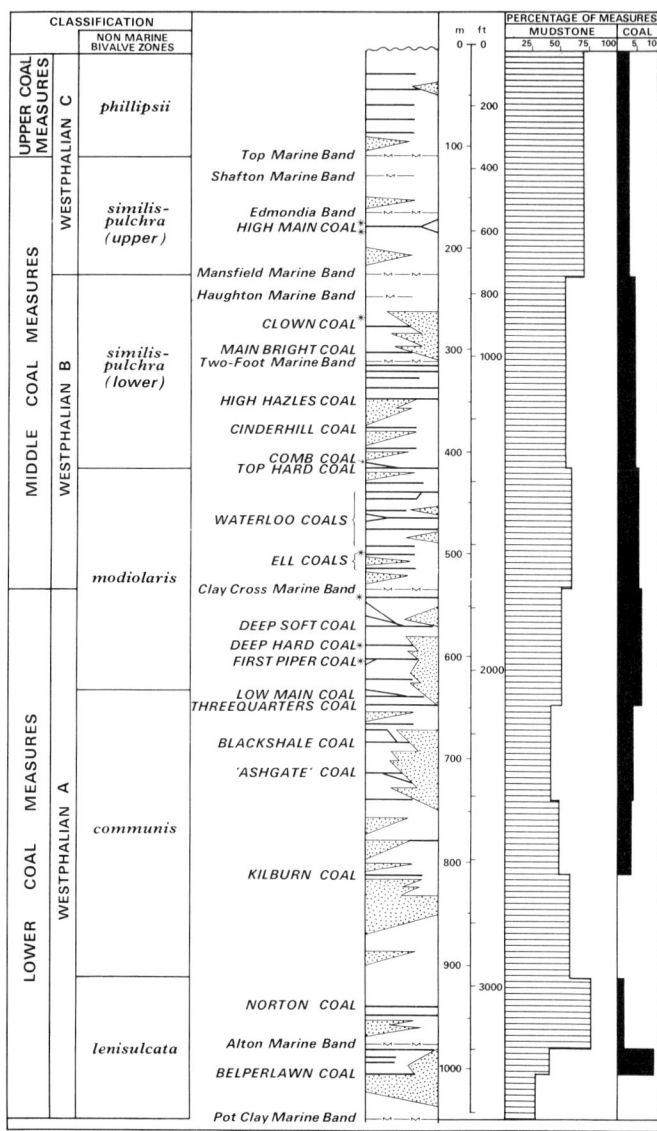

Figure 10 Generalised sequence and classification of the Westphalian rocks of the Derby district showing the principal coals, sandstone horizons (stippled), marine bands and proved horizons of tonsteins and related rocks. (The percentages are calculated from the text figures showing the representative sections of the measures in the 'Details of Stratigraphy')

The diagenetic process, according to Moore (1964) working on the Erda Tonstein from the Westphalian B rocks of Germany, involves microbiological degradation of organic matter in a soil or upon a surface below shallow and temporary water. This produces the necessary conditions, varying from anaerobic to aerobic, for the formation of kaolinite. The German theory of the pyroclastic origin of tonsteins in which ash fall-out into a peat bog environment is broken down into kaolinite, is supported in Scotland by the lateral passage from tonstein to tuff (Francis, 1961, pp. 199–201) and nearer this district by the presence of pumice within the First Piper Coal in the Nottingham district (Eden and others, 1963, p. 53). In support of a pyroclastic origin for the East Midlands tonsteins Francis (1969) emphasised the presence of suitable sources of material from vents within the Westphalian volcanic field proved in oil and coal boreholes east of Nottingham which, for example, gave rise to the closely related rock known as the Black Rake Tuffaceous Siltstone (see below).

Rocks related to tonsteins include Kaolin 'oolith' bands, Fragmental Clayrocks and the Black Rake Tuffaceous Siltstone.

Kaolin 'oolith' bands are widespread but discontinuous, thin (up to 2 in) beds of red-brown siderite with tiny spherules or irregular masses of pale kaolinite. They have been described in detail by Deans (1935) and Strauss (1971), the latter listing their horizons within the East Midlands Westphalian rocks. Hemingway (*in* Murchison and Westoll, 1968, p. 63) interprets them as late replacements of chamosite oolites. The most widespread band within this district is in the roof of the First Ell Coal, and others occur in the roofs of the First Piper and Clown coals.

Fragmental Clayrocks (Richardson and Francis, 1971), consisting of pale brown-grey disrupted and wispy kaolinitic mudstones up to about ½ inch in thickness, are known within the district in the roofs of the Brown Rake and High Main coals. They differ from true tonsteins in their disrupted nature and restricted distribution (although the above authors describe one extending over 500 square miles in northeast England). Mineralogically they contain the clay mineral illite which is missing from true tonsteins and they may also contain sporadic ostracods.

The Black Rake Tuffaceous Siltstone, some 23 to 50 ft below the Clay Cross Marine Band, has been described by Francis and others (1968). It is a hard, banded and graded siltstone containing basaltic glass shards and pumice. There is a high proportion of carbonates, mainly dolomite and kaolinite, the latter replacing the basaltic glass. It was formed from fall-out dust from at least three volcanic vents situated to the east and north-east of Nottingham and takes its name from the associated Black Rake Ironstone.

Ironstones

Ironstones are found in all parts of the cycle but are most common in the mudstones as nodules, lenses and bands, some of which are very fossiliferous. Some nodules are septarian, others occur in the grey seatearths, often occupying the irregular cavities left by decomposed root structures. Sphaerosiderite is also present in many seatearths. Some

iron is present as pyrite in the coal or overlying dark grey marine shales, but it occurs most commonly in coal as siderite and to a lesser extent as ankerite, particularly as an infilling in the cleat.

The source of iron in the Westphalian is considered by Hemingway (*in* Murchison and Westoll, 1968, p. 63) to be the pre-Carboniferous land masses surrounding the depositional basins. The manner in which the transported oxides were concentrated into their present forms is not clearly understood, but the early diagenesis is considered to be a reduction process caused by the presence of faunal and floral matter in shallow water and subsequent concentration of the products.

SEDIMENTATION

The cyclic repetition of coal measure lithologies and its mechanism of formation has been the subject of a voluminous literature, which has been summarised by Westoll (*in* Murchison and Westoll, 1968, pp. 71–103).

Thickness variations of Westphalian A strata (Figure 11) show marked similarities to those of the Namurian rocks presented by Howitt (*in* Kent, 1966, p. 334) and reveals the continued influence of the Widmerpool and Edale gulfs during the Westphalian, as noted by Kent (1966, p. 338). That influence is also apparent from thickness variations shown by some individual cycles, but the pattern is not consistent. Many cycles show variations apparently unrelated to the gulfs, though there is some evidence that gulf influence might still be detectable in Westphalian B rocks.

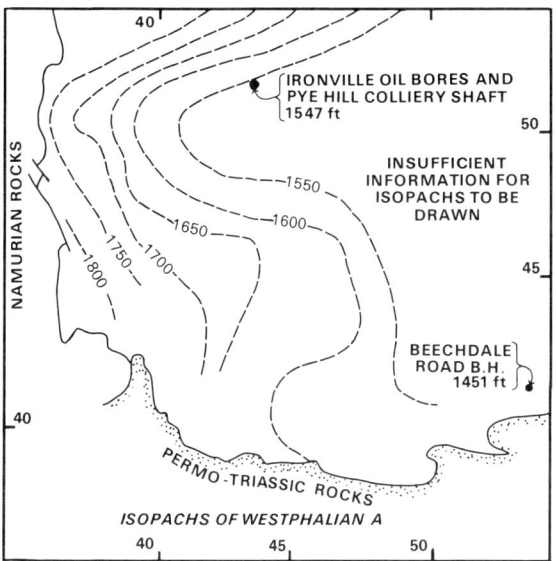

Figure 11   Isopachs (in feet) of Westphalian A

Sedimentary processes affect the thickness of the measures between the Deep Hard and Deep Soft coals which, over most of the district, form a single cycle. These measures show marked thickening towards a major sandstone 'washout' in the Deep Hard Coal, which was probably a main 'feeder' channel for the supply of sediment.

The coal isopach plans presented in the 'Details of Stratigraphy' show, with the notable exception of the Yard/ Blackshale Coal, that many seams are thicker in the west and south-west than in the east and north-east. Other coals deteriorate in quality or split to the east or north-east.

The isopach plans of five seams, the Belperlawn, Alton, Kilburn, Low Main and Brown Rake, reveal a relationship between coal thicknesses and the main axes of folding within the district. The relationship is by no means perfect, but these seams are thicker in the Shipley and Ripley synclines than on the axes of the Breadsall, Erewash, Crich and Ironville anticlines. Five more seams, the Deep Hard, Deep Soft, Bottom and Top Second Waterloo and Dunsil, show less well defined thickening in the synclines. It is thus evident that at times coal accumulation was influenced by precursory movements along Hercynian fold axes.

STRATIGRAPHY

Westphalian A (Lower Coal Measures)

BASE OF WESTPHALIAN STRATA TO KILBURN COAL

The measures between the Pot Clay Marine Band and Kilburn Coal range in thickness from about 590 ft near Bondland Colliery to about 800 ft S of Denby Colliery (Figure 12). In the east they are 625 ft thick in Beechdale Road Borehole.

Of the twelve component cycles within the district only that above the Belperlawn Coal and that capped by the Kilburn Coal lack marine or near-marine horizons at their bases. The highest marine horizon known to occur in these measures to the north and near Nottingham (Smith and others, 1967, p. 116)—the Burton Joyce Marine Band—remains unproved.

Within these measures are two principal sandstones—the Crawshaw Sandstone and Wingfield Flags—and three coals that have been worked underground—the Belperlawn, Alton and to a lesser extent, the Norton. Ganisters, present within the seatearths of many of the lower coals, were formerly worked in the north of the district.

The distribution of the marine and near-marine horizons was given by Eden (1954) and the faunal sequence is similar to that described by Smith and others (1967, 1973) in the Chesterfield and East Retford districts. The mussel beds between the Belperlawn and Alton coals contain the distinctive *Carbonicola fallax/C. protea* faunas described by Eagar (1947, 1952) from above the Soft Bed Coal in Yorkshire. Another distinctive horizon is the Norton Musselband with *Carbonicola proxima* which overlies the Norton Marine Band.

The Pot Clay Marine Band   The dark grey marine mudstones at the base of the Westphalian have been proved in Nether Heage, Park Brook and Beechdale Road boreholes. The characteristic goniatite, *Gastrioceras subcrenatum*, is present and the associated fauna consists of a further species of *Gastrioceras*, together with *Dunbarella papyracea*, *Lingula mytilloides*, *Orbiculoidea sp.*, conodonts including *Hindeodella sp.* and fish debris (see also Eden, 1954, p. 106).

The measures between the Pot Clay Marine Band and the Belperlawn Coal range from about 100 to 145 ft. Below the Crawshaw Sandstone they comprise largely argillaceous strata with a few feet of dark grey mudstones containing fish debris overlying the marine band. Wedd (*in* Gibson and others, 1908, p. 50) noted these strata to be '12 or 14 yards thick and composed of blue sandy micaceous shale' in a well [3634 4363][1] near Coxbench. In the Ambergate area a thin layer of penecontemporaneously contorted mudstone, similar to those described by Cope (1946), has been noted a few inches above the marine band.

The *Crawshaw Sandstone* ranges up to 112 ft in thickness and reflects a resumption of Namurian conditions of sedimentation. Its cross-bedding suggests derivation from the south. The sandstone is mainly medium to coarse-grained, gritty and arkosic; its colour is buff at outcrop, but grey to white when encountered in boreholes; pink and red staining is common both at outcrop and in boreholes and a green coloration has been sporadically recorded. Except at Ridgeway Quarry and in Beechdale Road Borehole the base of the sandstone is gradational into the underlying argillaceous strata: the lowest beds are either fine-grained for a thickness of up to 24 ft or interbedded with siltstone and mudstone. Within the main mass of sandstone, beds of siltstone and mudstone have been proved, and similar beds reach mappable proportions in the outcrop near Holbrook. Towards the top of the sequence the rock usually becomes finer grained and contains beds of siltstone and mudstone. At the top, below the Belperlawn Coal, there is a ganisteroid sandstone about 2 ft thick and mudstone seatearth up to about 1½ ft thick.

The Belperlawn Coal   (Figure 13) has been proved to be 66 in thick at Nodinhill Site and is 38 to 50 inches in the type area north-east of Belper, where it has been extensively worked, mainly by opencast methods.

Away from the western outcrop the thickness of coal only exceeds 30 in within the 'take' of Denby (Drury-Lowe)[1] Colliery. Near Heanor it is less than 20 in, but at Ilkeston the 27-in seam[2] is split by 9 in of dirt; at Oakwell Staple Pit the dirt parting is 18 in thick, separating two 9-in leaves of coal; the Ironville oil bores did not prove coal at this horizon.

The measures between the Belperlawn and Alton coals   (Figure 14), which have a maximum thickness of 78 ft at Denby (D-L) Colliery, thin northwards to 30 ft at Ridgeway Site. They comprise up to four cycles with three thin and impersistent coals, in ascending order the Holbrook, Second and First Smalley coals (Eden, 1954). The lowest

---

[1] National Grid references in this account are quoted at the first mention of the locality in each section, thereafter only when considered necessary to locate the section in context. For National Grid references of Boreholes and Shafts see Appendix 1.

[1] The name is condensed in this account to Denby (D-L) Colliery.

[2] Seam details in this account are based upon quality reports prepared by the Nottingham Coal Survey Laboratory, its successor the National Coal Board Scientific Department, the NCB Opencast Executive, and Geological Survey records.

Figure 12 Generalised section of the measures between the Pot Clay Marine Band and the Kilburn Coal with inset map showing the outcrop and isopachs of these measures and the locations of the principal provings

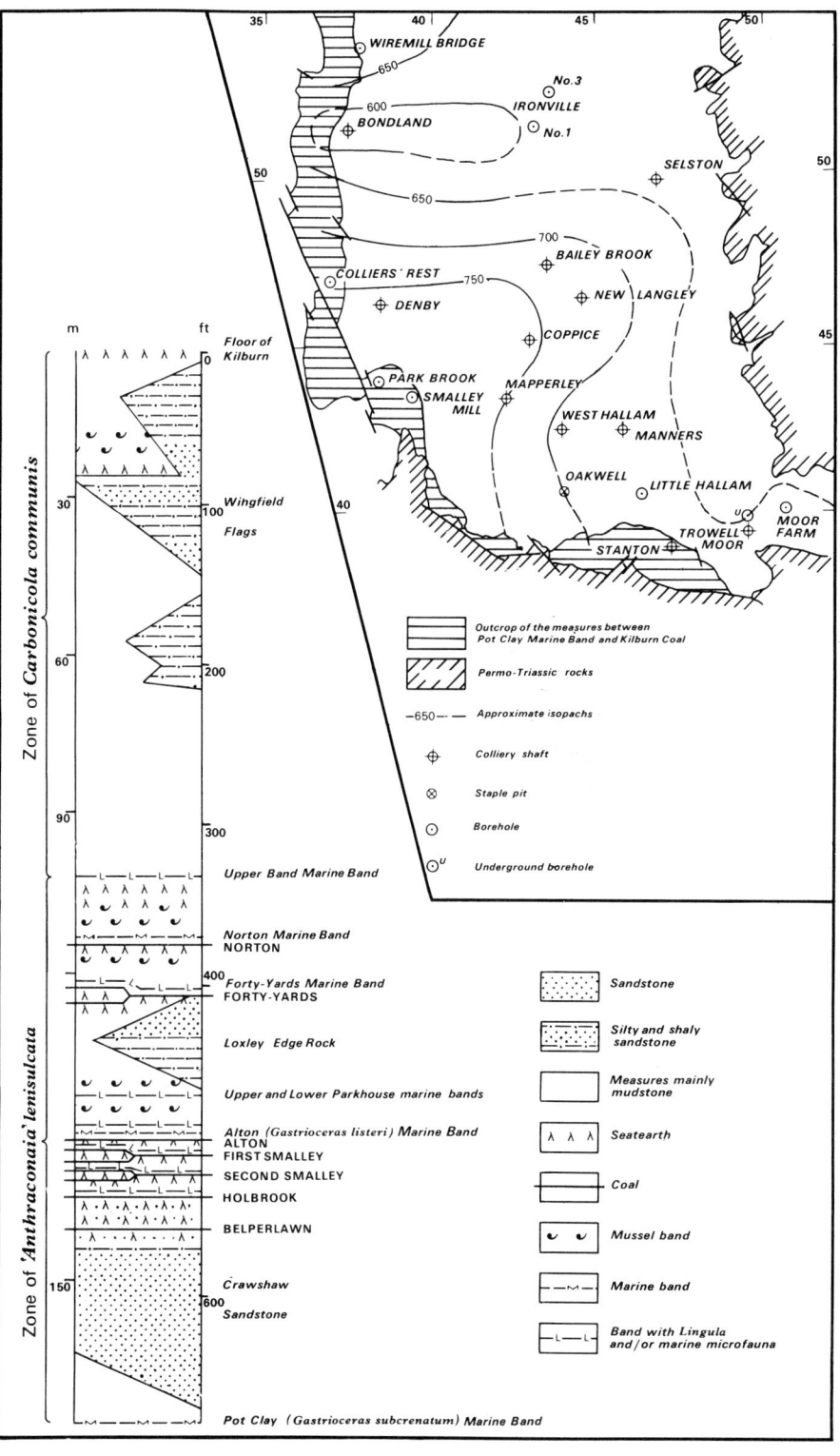

beds of each cycle, except those immediately overlying the Belperlawn Coal, are normally mudstones with *Lingula mytilloides* and, commonly, mussels. Fish debris, which does occur in the roof measures of the Belperlawn Coal, is intermittently recorded from the remaining cycles. It tends to be associated with, or replaces, the *Lingula* bands. Mussels are very rare in the cycle above the Second Smalley Coal and there is no record within this district of the mussels described

from the Yorkshire equivalent of the roof of the Belperlawn Seam. The distinctive mussel fauna of this sequence has been described by Eagar (1952, 1954) and by Smith and others (1967, 1973). In Denby (D-L) Colliery Underground Borehole it includes *Carbonicola haberghamensis* above the Holbrook Coal and *C. declinata*, *C. fallax*, *C. limax* and *Curvirimula sp.*, together with *Geisina arcuata* above the First Smalley Coal.

Figure 13 Plan of the Belperlawn Coal and representative sections of the seam. Localities mentioned in the text concerning the measures between the Belperlawn and Alton coals are also shown

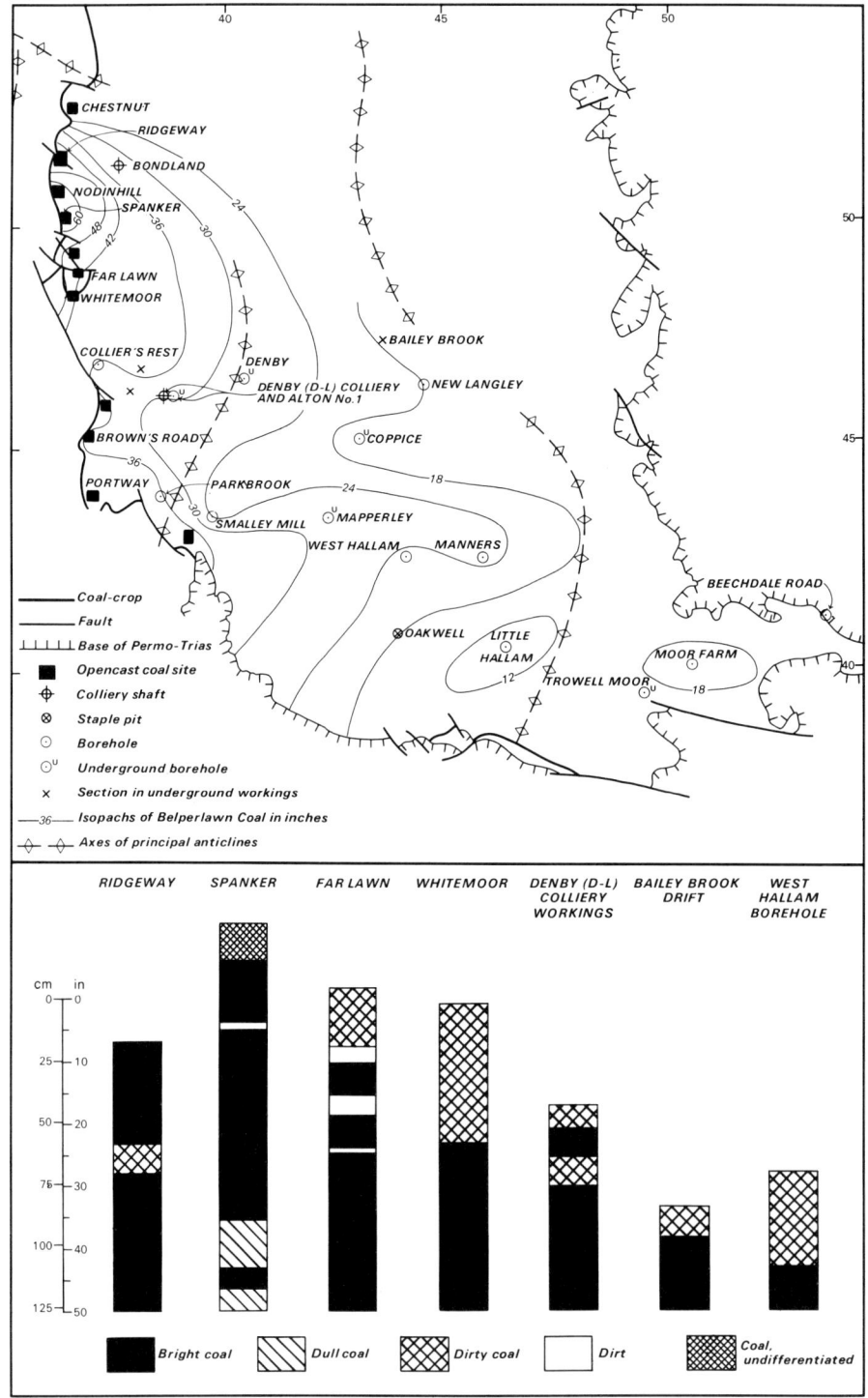

Each cycle contains variable amounts of sandstone and siltstone, and ganisteroid sandstone seatearths frequently underlie the coals. Sandstone in the uppermost cycle has been called the Sub-Alton Sandstone by Smith and others (1967, p. 105). The lithologies and faunas observed at most localities within the district are shown diagrammatically in Figure 14 and by Eden (1954, pl. 2A).

The *Holbrook Coal* is impersistent, ranging in thickness from 4 to 14 in of shaly coal, and is represented only by seatearth in places.

The *Second Smalley Coal* is a variable seam up to 18 in thick and composed of coal and batt in Smalley Mill Well (Gibson and others, 1908, p. 74), more usually it is less than 10 in thick (Figure 14). It may also be composed of split thin coals or represented by seatearth only.

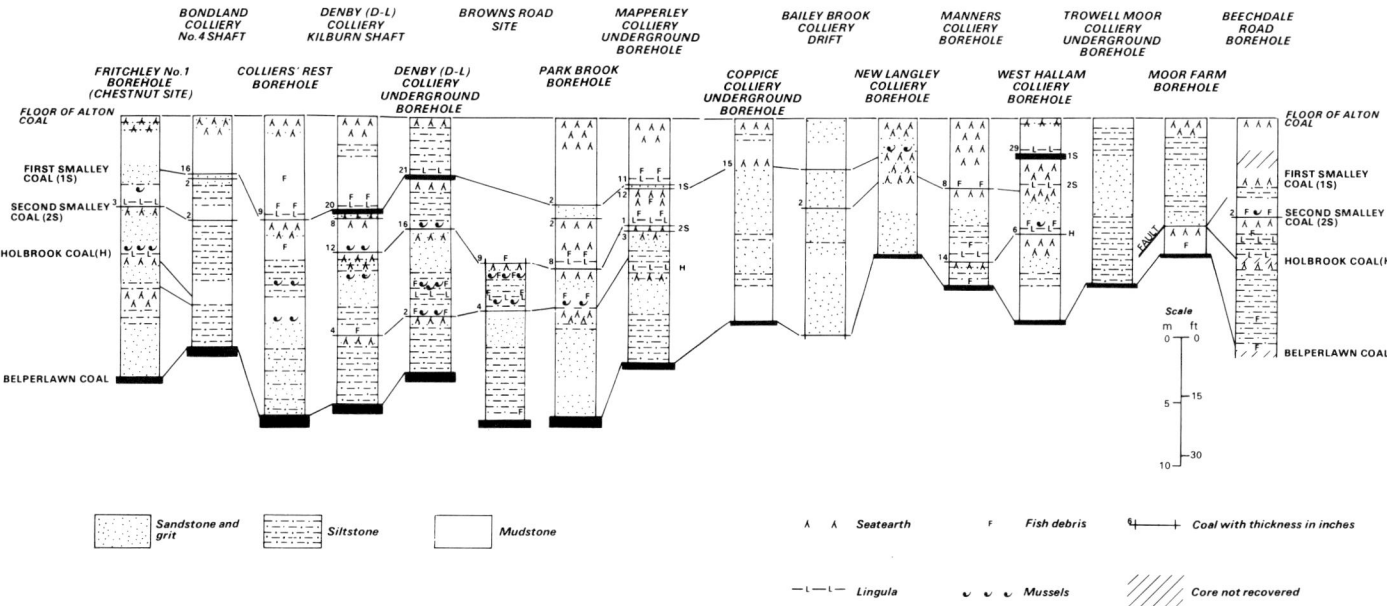

Figure 14 Sections of the measures between the Belperlawn and Alton coals. For locations see Figure 13

The *First Smalley Coal* ranges in thickness up to 29 in of shaly coal at West Hallam Colliery Borehole. Over parts of the district the seam consists of two coal leaves separated by sandstone (Figures 14, 67). At Smalley Mill Well (Gibson and others, 1908, p. 74), for instance, the whole was 3 ft 11 in thick and included a 1-ft rock parting.

The *Sub-Alton Sandstone* is an impersistent sandstone lying in the uppermost cycle of these measures.

The Alton Coal ranges from 10 in on Chestnut Site and up to 54 in on Ridgeway Site, both near Ambergate. The isopachs shown in Figure 15 represent only a regional tendency and exclude abrupt local variations in thickness; for instance other recordings on Chestnut Site range up to 31 in. The quality of the coal is good to medium with variable ash and high to very high sulphur contents; the latter originate from the many pyritised fusain partings.

The measures between the Alton and Forty-Yards coals (Figure 16) vary between 59 and 112 ft in thickness and, as described by Eden (1954), are composed of two cycles. The lower, 16 to 39 ft thick, commences with the Alton Marine Band and is largely argillaceous with significant sandstone or siltstone beds only locally recorded. A 2-in seatearth at the top of the lower cycle has been recorded in Wiremill Bridge Borehole; elsewhere the base of the upper cycle can be recognised only by the presence of the Parkhouse Marine Bands. This upper cycle, which is 43 to 96 ft thick, with variable thicknesses of sandstone and siltstone comprising the Loxley Edge Rock near the top, is capped by the seatearth or ganister below the Forty-Yards Coal.

The *Alton Marine Band* directly overlies the Alton Coal and consists of fossiliferous dark grey pyritous shales with 'bullions' of ironstone containing uncrushed goniatites. It is only a few inches thick in the Denby (D-L) Colliery district, but reaches 7 ft 10 in in Beechdale Road Borehole. In the

thicker sequences the marine band is split (Eden, 1954), the upper part consisting of dark mudstones with *Lingula sp.* and microfossils which may be separated from the lower part by up to 3 ft 8 in of mudstone without marine fossils, as in Beechdale Road Borehole.

The marine band was first noted in the Ambergate area by Wedd (1903, p. 12) and described, but not named, by Gibson and others (1908, pp. 64–67, 100, 185). The fauna includes *Dunbarella papyracea*, *Posidonia gibsoni*, *Caneyella multirugata*, turreted gastropods, *Anthracoceratites sp.*, *Gastrioceras listeri*, orthocone nautiloid, *Hindeodella sp.* and fish remains.

Fish debris and mussels have been recorded in the mudstones above the marine band.

The *Lower and Upper Parkhouse Marine Bands* occur at the base of the uppermost cycle below the Forty-Yards Coal and yield *Ammodiscus sp.*, *Glomospira sp.*, *Lingula mytilloides* and fish remains. Over parts of the district only one band is present.

The *Loxley Edge Rock* is a fine micaceous sandstone with siltstone and mudstone partings, lying in the upper part of the cycle which commences with the Parkhouse Marine Bands (Eden, 1954). It rarely exceeds 30 ft and is locally absent.

The Forty-Yards Coal commonly comprises two thin seams on average below 12 in thick separated by up to 15 ft of measures, which consist largely of seatearth, but may include mudstone, siltstone, sandstone and ganister. Most records of ganister come from below the lower, least persistent, leaf of the seam.

The measures between the Forty-Yards and Norton coals comprise one cycle normally about 30 ft thick, but ranging from 18 to 39 ft. The cycle consists largely of mudstone with ironstone bands and nodules (the 'Dale Moor

Figure 15 Plan of the Alton Coal and representative sections of the seam

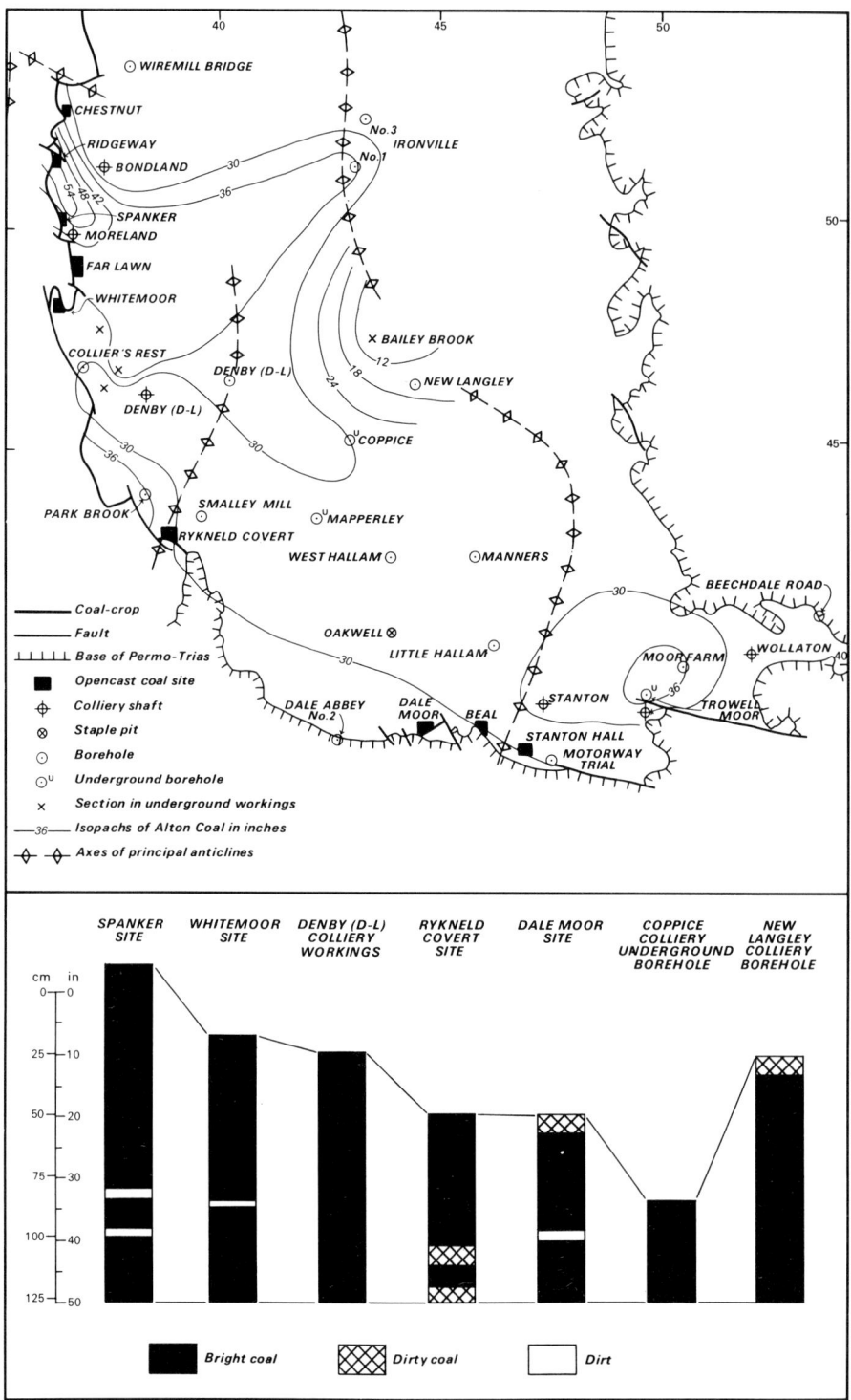

Rake' of Smyth, 1856, p. 59) and is now marked by extensive old excavations at outcrop near Stanton Ironworks [4630 3880]. Fish remains are plentiful (see Gibson and others, 1908, p. 101) and include *Diplodus sp.*, *Elonichthys sp.*, *Rhabdoderma sp.* and *Rhizodopsis sp.* The arenaceous part of the cycle is thin and impersistent, but the seatearth at the top is up to 14 ft thick.

The *Forty-Yards Marine Band* lies at the base of the cycle (Smith and others, 1967, p. 114) and contains *Ammodiscus sp.*, *Lingula mytilloides*, *Hindeodella sp.*, *Elonichthys aitkeni* and an acanthodian spine. The non-marine mudstones of this cycle contain fish and plant debris.

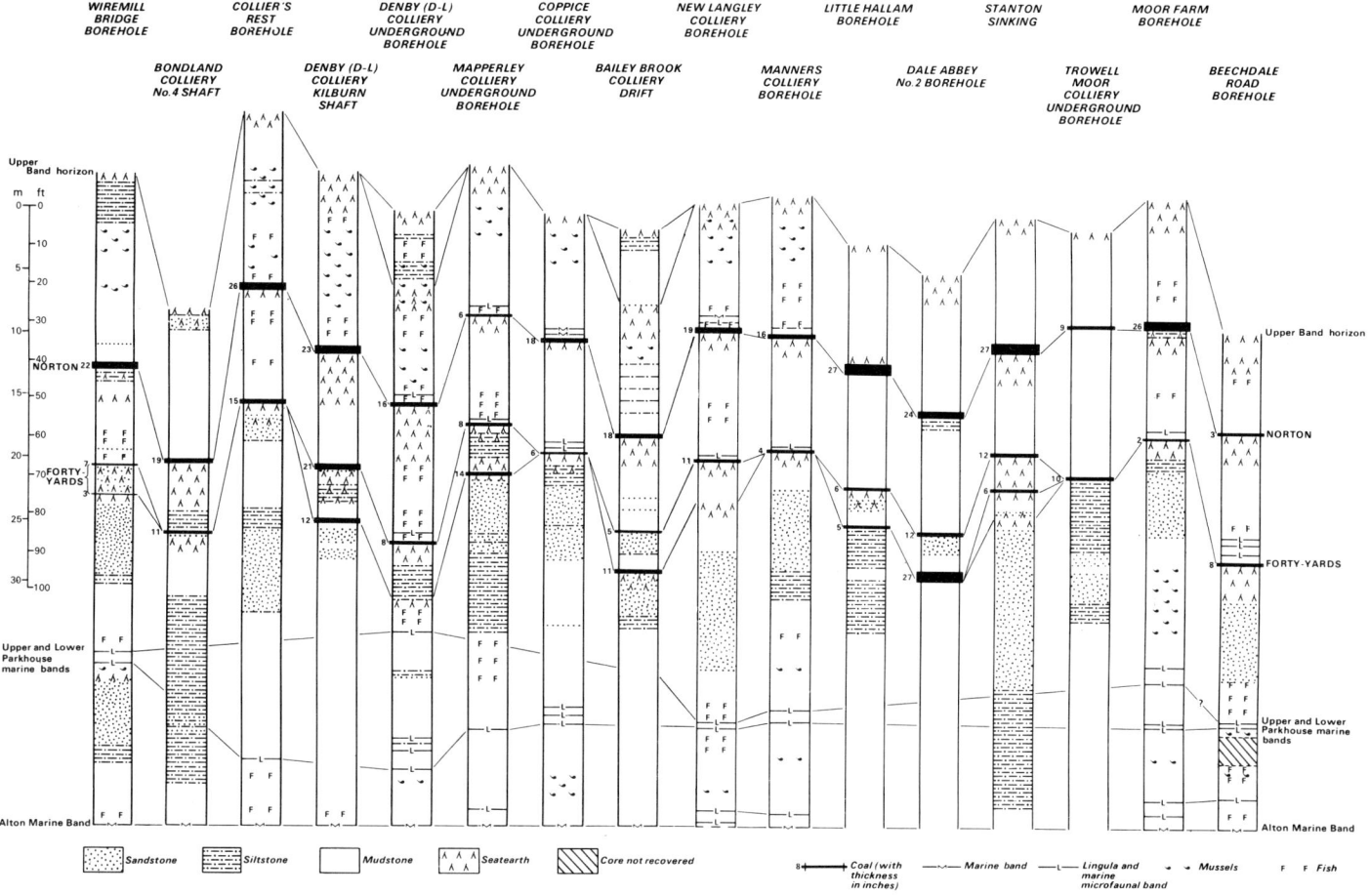

Figure 16 Representative sections of the measures between the Alton Coal and Upper Band horizon. For locations see Figure 15

The Norton Coal, which varies in thickness from 3 to 27 in, has been worked only in the south of the coalfield, near Stanton Gate (Gibson and others, 1908, p. 68).

The measures between the Norton Coal and the Upper Band horizon form an argillaceous sequence ranging from 25 to 53 ft in thickness (Beechdale Road and Wiremill Bridge boreholes). The sequence is composed of up to two cycles, although the evidence for this division has been found only in three sections (Figure 16).

The *Norton Marine Band* at, or very close to, the base of the sequence, consists of a thin bed of pyritous shale containing *Ammodiscus sp.*, *Glomospira sp.*, *Lingula mytilloides*, *Caneyella* aff. *multirugata*, *Dunbarella* aff. *papyracea*, and a conodont assemblage including *Hindeodella sp.* and *Ozarkodina sp.* The overlying mudstone and silty mudstones contain fish debris, plants and the Norton Mussel-band (Eden, 1954, p. 97; Eagar, 1956). The mussels are commonly preserved in calcitic material over a thickness of 5 to 20 ft, and comprise *Carbonicola crispa*, *C. proxima*, *Curvirimula belgica* and *Naiadites sp.*, in association with *Geisina arcuata*, *Rhabdoderma sp.*, *Rhadinichthys sp.* and *Rhizodopsis sp.*

The Measures between the Upper Band horizon and Kilburn Coal range in thickness from 306 to 395 ft, and consist of two cycles. The lower, thicker, cycle, much of it sandstone, extends up to a seatearth lying some 20 to 60 ft below the Kilburn Coal. The upper cycle, largely argillaceous, extends from the top of this seatearth to the top of the Kilburn Coal.

Arenaceous measures of both cycles are assigned to the Wingfield Flags, although their main development lies in the upper part of the lower cycle.

The *Upper Band Marine Band* is known only from New Langley Colliery Borehole, the type locality (Godwin, 1960, p. 33; Calver, 1968b, p. 34). It consists of dark mudstones with *Ammodiscus sp.*, *Lingula mytilloides*, and abundant fish remains including *Elonichthys sp.* The principal horizon with *Lingula* was 4 ft 6 in above the Upper Band seatearth. Fucoids only were recorded from this horizon in Wiremill Bridge Borehole. Up to 195 ft of mudstones with fish remains and mussels separate the Upper Band seatearth and the Wingfield Flags.

The *Burton Joyce Marine Band* (Smith and others, 1967, p. 116), which is present about 25 ft above the Upper Band seatearth in the Chesterfield and Nottingham districts, has not been proved within this district.

The *Wingfield Flags* are a sequence of pale grey flaggy sandstones, siltstones and silty mudstones in variable proportions up to about 200 ft in thickness. The top of the lower cycle (see above) is marked by a thin seatearth which, although not proved at outcrop, is present in most boreholes. A 3-in 'batt' in Bailey Brook Colliery Drift and a 3-in coal in Annesley Colliery (Deep Hard) Underground Bore-hole are the only records of coal at this horizon.

The upper cycle, normally 45 to 80 ft thick, consists largely of dark grey to black mudstones with ironstones; fish debris and ostracods have been recorded and mussels are common. The presence of the fauna serves to identify the cycle in sections such as Moor Farm Borehole where the underlying seatearth is missing; its sandstones tend to be generally thin.

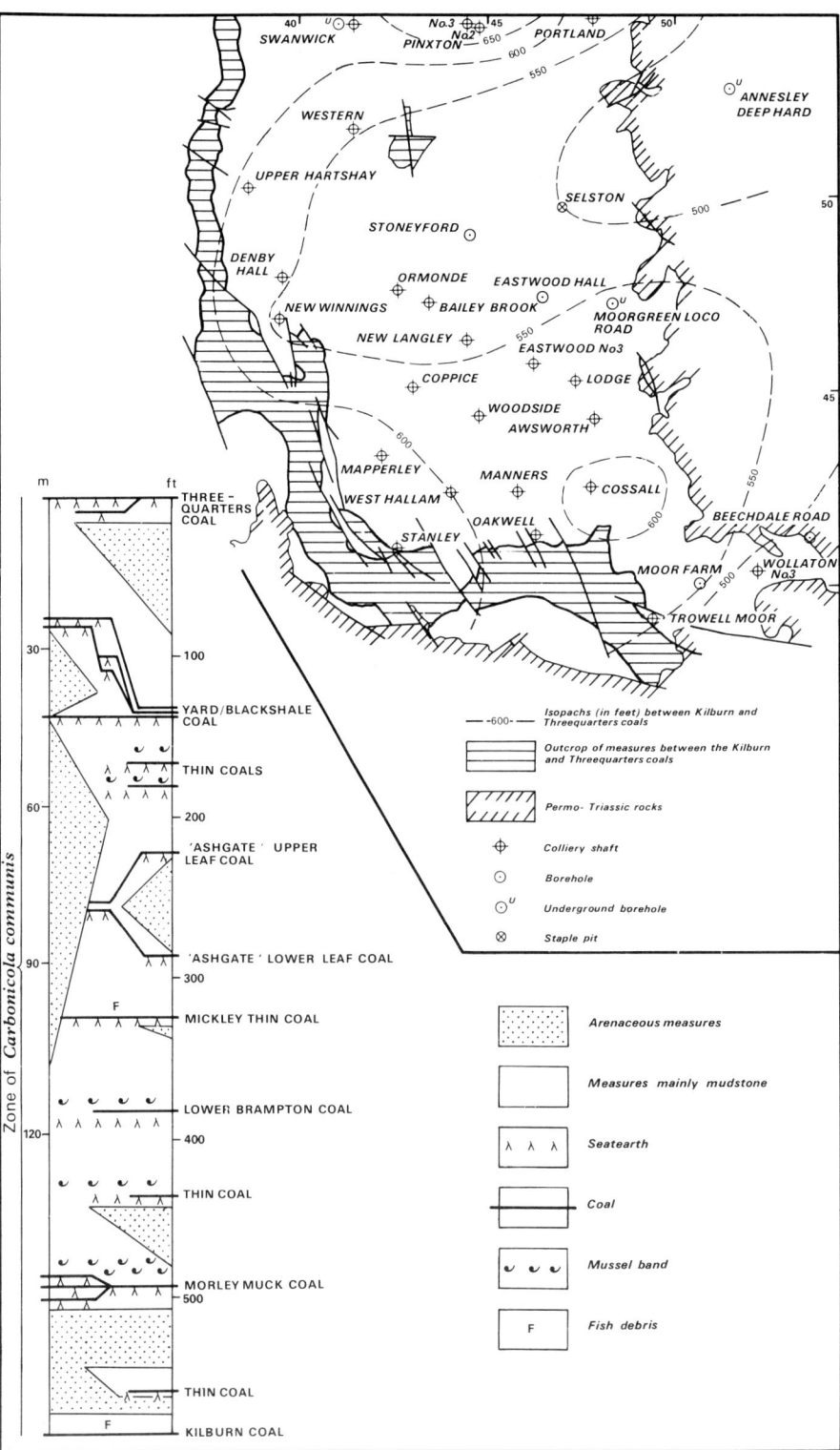

Figure 17   Generalised vertical section of the measures between the Kilburn and Threequarters coals with map showing their outcrop, locations of the principal provings and isopachs of the measures

KILBURN COAL TO THREEQUARTERS COAL

The measures between the Kilburn and Threequarters coals (Figures 17 and 18) are thickest in the north, where they reach a maximum of 661 ft. They exceed 600 ft in the south-west of the district. Eastwards and south-eastwards they thin to less than 500 ft. The sequence includes the Kilburn Coal, which was extensively worked for house coal; other important seams are the 'Ashgate', Blackshale, Yard and Threequarters. With the exception of the Yard and Blackshale, the coals tend to thin in an easterly direction.

The fauna of these measures consists largely of mussels of the *Carbonicola pseudorobusta* group, with fish debris present in the roofs of the coals up to and including the Mickley Thin; marine fossils are unknown in this district.

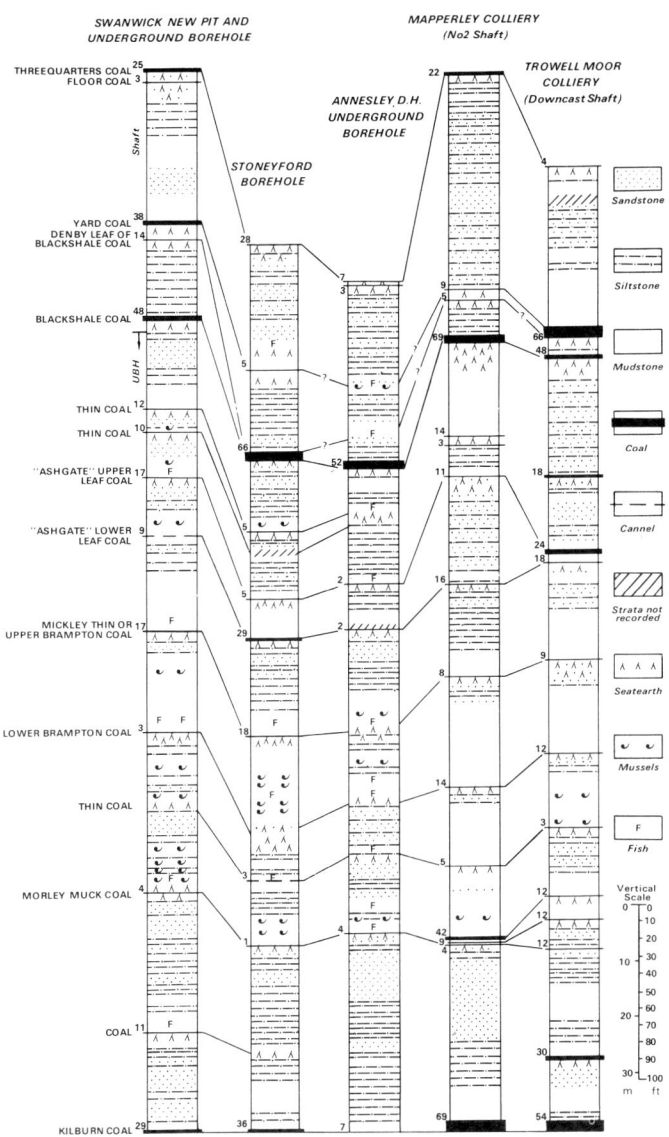

Figure 18 Representative sections of the measures between the Kilburn and Threequarters coals. Figures left of columns are thicknesses of coal or cannel in inches. There is an alternative version of the section of Trowell Moor Colliery Shaft—see Appendix 1

The Kilburn Coal ranges from 30 to over 70 inches in thickness in the south-western part of the exposed field, but in the north-east it is thin (Figure 19). The maximum proved thicknesses are at outcrop in the south-west, where $76\frac{1}{2}$ in of coal (including $5\frac{1}{2}$ in of dirt) and 74 in of coal were recorded in Hollies Farm and Mary sites respectively. The seam is composed predominantly of bright coal with subordinate dull bands, and is of good to medium quality with a low to moderate ash content.

The measures between the Kilburn and Mickley Thin coals range in thickness from 198 to 288 ft (Figure 20) and are composed of five main cycles. To the south-east of an irregular line from Swanwick to Stapleford the lowest cycle can be identified only by the presence of arenaceous measures at its top. Each cycle locally includes coal, but the seams are thin or of poor quality and have little economic value. The cyclic sequence of these measures is relatively uniform throughout the district except in the two Ironville oil bores and Moorgreen Loco Road Underground Borehole. The latter encountered additional seatearths, interpreted as splits of the two cycles above the Morley Muck Coal.

The lowest cycle, found only in the north-east of the district, ranges from 25 to 80 ft in thickness, its top commonly formed by a seatearth and with coal sporadically recorded. The upper cycle at Eastwood Hall Borehole is predominately arenaceous but *Spirorbis sp.*, *Carbonicola* cf. *browni* and *C.* cf. *pseudorobusta*, *Curvirimula subovata*, *Carbonita humilis*, *C. pungens* and fish remains were present in interbedded mudstone and siltstone.

The *Morley Muck Coal* may contain over 50 in of coal in up to three leaves of poor to very inferior quality; the seam is therefore of no economic value. The coal or its seatearth lies between 50 and 130 ft above the Kilburn Coal.

The cycle overlying the Morley Muck Coal varies from 36 to 56 ft. The coal, cannel or seatearth at the top can be recognised in many provings throughout the district; the coal is unnamed and, discounting a dubious record of 24 in in the chippings from the Ironville No. 1 Oil Bore, has a maximum thickness of 10 in. It is absent over wide areas in the east and north. The cycle includes up to about 50 ft of arenaceous measures, with *Spirorbis sp.*, *Carbonicola* cf. *polmontensis*, *C. pseudorobusta*, *Curvirimula sp.*, *Geisina arcuata* and fish remains including *Rhabdoderma sp.* in the underlying argillaceous beds above the Morley Muck Coal.

The succeeding cycle, ranging in thickness from 25 to 50 ft, is also widely recognised and has the Lower Brampton Coal at its top. Sandstones are less prominent than in the underlying cycle and the mussels less abundant.

The *Lower Brampton Coal* (Smith and others, 1967, p. 121) has a maximum thickness of 21 in including 12 in of cannel and is of very poor quality. Provings between Mapperley and West Hallam collieries show this seam to be split into two very thin leaves, each under 9 in thick, up to 63 in apart. The upper leaf is locally canneloid. In the north of the district the Lower Brampton is under 4 in thick; it is absent in the east and 9 to 12 in thick in the south-east.

The remaining cycle, that with the Mickley Thin Coal at the top, ranges in thickness from 46 to 67 ft and is almost wholly composed of argillaceous strata, locally bearing a mussel fauna; at Eastwood Hall Borehole *Carbonicola sp.* and

Figure 19 Plan of the Kilburn Coal with representative sections of the seam. Localities mentioned in the text concerning the measures between the Kilburn and 'Ashgate' coals are also shown

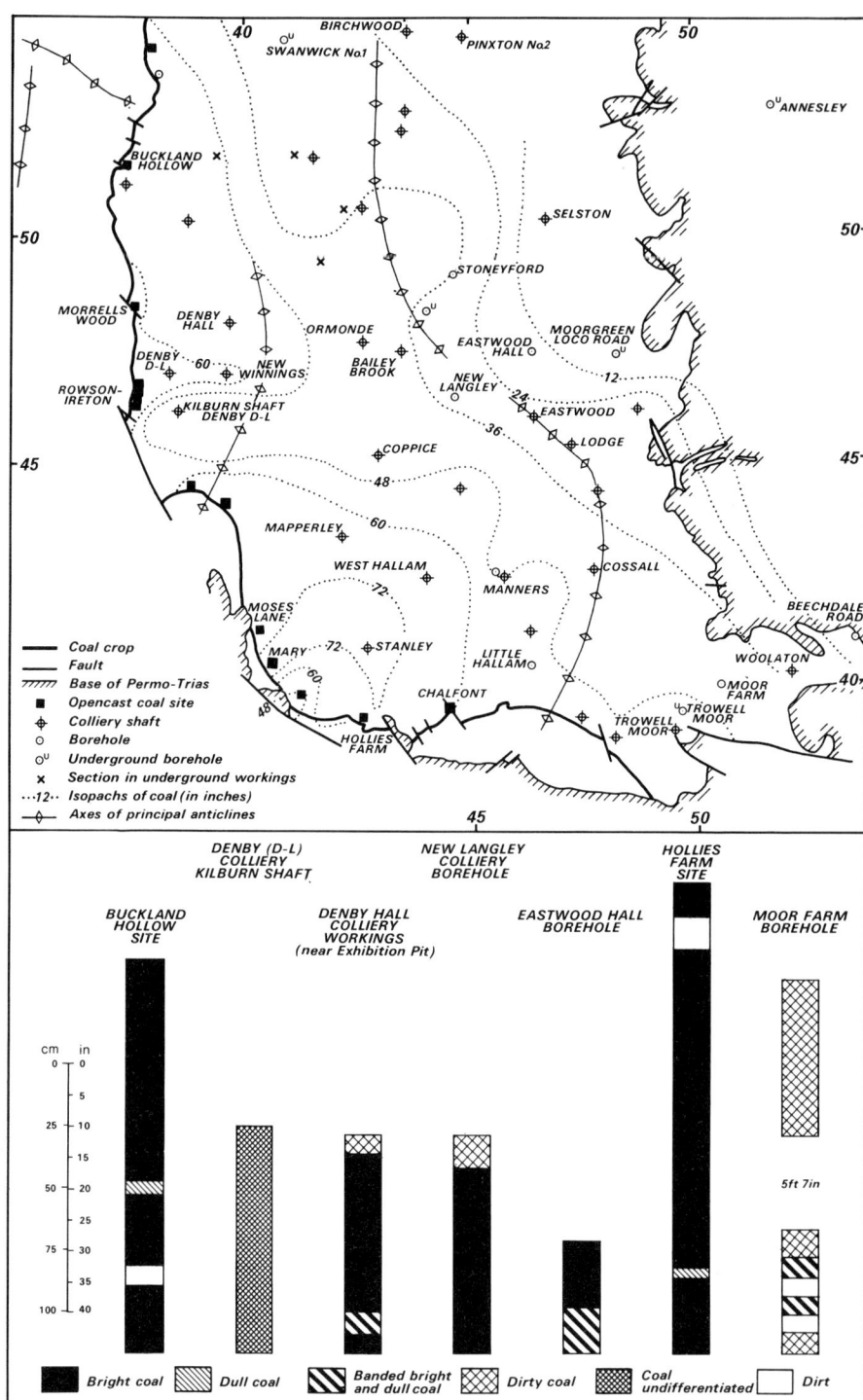

Curvirimula subovata were recorded. Sandstones are only a few feet thick in the floor measures of the coal.

The Mickley Thin (or Upper Brampton) Coal is thickest at Morrells Wood Site, where 20 in were recorded. As at Chalfont Site (19 in) and Beechdale Road Borehole (3 in) the coal is of good to moderate quality, but it deteriorates to poor in two analyses from Moses Lane Site (9 to 12 in). The thickness is 18 in or less in the remaining provings within the district, the thinnest recordings being in the east.

The measures between the Mickley Thin and 'Ashgate' coals form one cycle ranging from 41 to 68 ft. The roof measures of the Mickley Thin contain fish, ostracods and sporadic mussels.

The 'Ashgate' Coal consists of two leaves of coal each showing rapid variations in thickness up to a maximum of 54 in and lying up to 62 ft apart. Washouts are common. The upper and lower leaves were originally called the Ashgate and Mickley coals respectively and they have been ex-

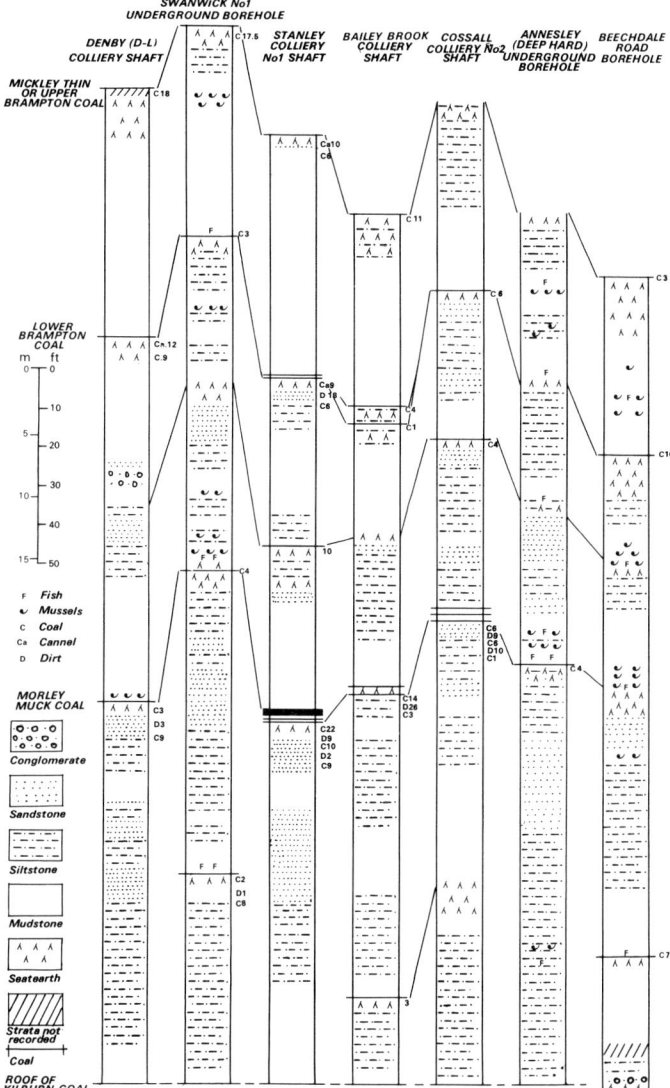

Figure 20 Detailed sections of the measures between the Kilburn and Mickley Thin coals

others (1967, p. 124). Each coal is itself split into up to three leaves, all thin and of no economic importance; the maximum thickness recorded is 28 in of 'batty coal' in Ormonde Colliery Shaft.

The Blackshale and Yard coals (Figures 21 and 22) are a complex sequence of up to six main leaves of coal lying within strata varying in thickness from 8 to 110 ft. The Blackshale Coal consists of the lowest three main leaves of the sequence plus, in this account, the two overlying leaves, here called the upper and lower Denby leaves after their striking development in Denby Hall Colliery Shaft (Figure 22). The three lowest leaves are the Top Softs, Middle Dirt with Tinkers Coal and Bottom Softs of the old miners' terminology, which together form the Blackshale Coal of the Chesterfield district (Smith and others, 1967, p. 125). The Low *'Estheria'* Band, which to the north-east occurs above the Bottom Softs (or Low Silkstone) is absent in the Derby district.

Each of the six areas shown in Figures 21 and 22 contains a different combination of the seams. In Area A all the coal leaves are present in close proximity to form the Yard/Blackshale, a seam which is equivalent with or approximates to the Silkstone of Yorkshire and which can be 96 in thick, including 18 in of dirt and dirty coal in partings. In Area B a thick Yard Coal is separated from the Blackshale Coal, which includes a variable sequence of the Denby leaves and ranges from a total of 65 in including 13 in of dirt to 155 in including 67 in of dirt. The Denby leaves attain a maximum in a single 54-in coal at Salterwood Site. Measures some 10 to 76 ft divide the Denby leaves from the Yard Coal, which itself ranges up to 59 in thick.

In areas C and D the Yard Coal is widely separated from insignificant Denby leaves which in turn are remote from the Top Softs – Middle Dirt – Bottom Softs sequence. In area C the Top Softs – Middle Dirt – Bottom Softs sequence varies between 33 and 54 in of which up to 10 in may be dirt partings; in area D it is generally 40 to 60 in thick with up to 13 in of dirt, but thick sections of 115 in (including 11 in of dirt) and 99 in (23 in of dirt) have been recorded.

In area C the Top Softs are separated from the Denby leaves by mainly arenaceous strata ranging in thickness from 13 to 80 ft. The Denby leaves are represented by seatearths up to 17 ft apart. The Yard Coal, lying 7 to 13 ft above the Denby leaves, is thin.

In area D a single Denby leaf was encountered only in Swanwick Colliery, New Pit, where 14 in of coal and dirt were encountered 44 ft above the Top Softs. Elsewhere the leaves appear to be absent or unrecorded.

The Denby leaves horizon is separated from the Yard Coal by sandy measures ranging in thickness from 7 to 85 ft. The Yard Coal, variable in quality, ranges from 17 to 38 in and has been extensively worked.

The centre line of area E coincides with the greatest thickness of sandy measures, which lie between an impoverished Yard Coal and the Top Softs – Middle Dirt – Bottom Softs sequence throughout the area. The Denby leaves are not proved; they may be merged with the Yard Coal or washed out by the sandy measures. Where wholly represented in the area, the Top Softs – Middle Dirt – Bottom Softs sequence

tensively worked underground near Denby, Heanor and Trowell Moor under these names. The seam is not the correlative of the Ashgate Coal of the Chesterfield district, which lies slighty higher in the sequence; instead the two leaves of the 'Ashgate' Coal of the present district are equivalent to the two thin coals present above the Mickley Thin Coal in Cotespark No. 1 Underground Borehole (Smith and others, 1967, fig. 11, p. 119).

The lower leaf of the seam is washed out within a sinuous belt of sandstone which extends from north-east of Eastwood to Ilkeston and thence south-east towards Wollaton, the upper leaf fails on the crest of this sandstone.

The measures between the 'Ashgate' and Blackshale coals are largely arenaceous and range from 60 to 146 ft. They contain up to two main coal horizons, which are probably the representatives of the Ashgate Coal of Smith and

Figure 21   Plan showing the locations of provings of the Yard and Blackshale coals and the boundaries of the areas within which these coals are described. Isopachs between the base of the Blackshale Coal and top of the Yard Coal are also shown. Some locations of underground provings in the Yard Coal between Swanwick and Pinxton are omitted

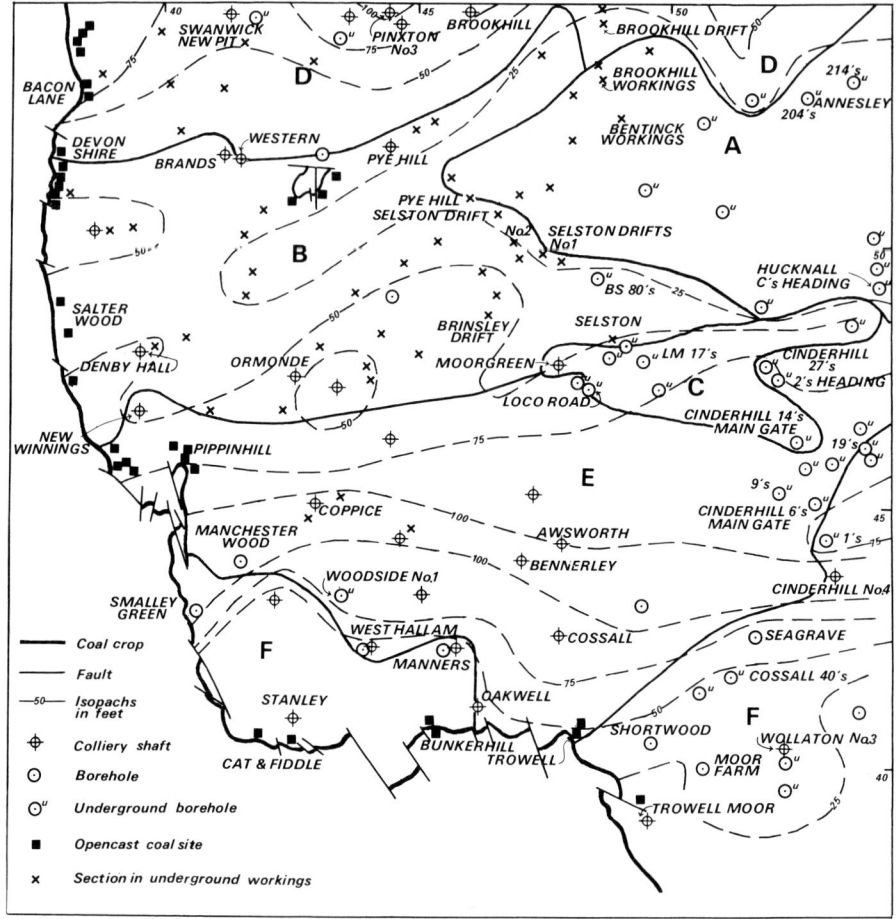

Figure 22   Sections showing typical developments of the Yard and Blackshale coals; the areas on Figure 21 which they represent are also indicated. For localities see also Figure 21

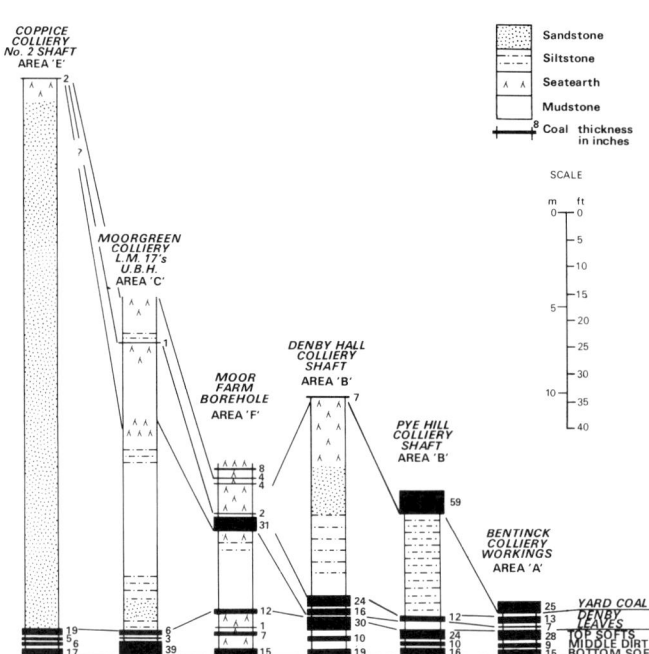

varies from 39 in, including $7\frac{1}{2}$ in of dirt to $78\frac{1}{2}$ in, including 26 in of dirt. The Denby leaves are only recorded with certainty from within this area at Pippinhill Site, where some provings show the Lower Denby leaf within 18 in of the Top Softs. The Yard Coal is represented mainly by a seatearth horizon.

Area F lies south of the thick belt of sandy measures and consequently a variable sequence of the Yard Coal and Denby leaves reappears. Moderately thick measures overlie a variable Top Softs – Middle Dirt – Bottom Softs sequence

—a feature which, apart from its location on the opposite side of the thick arenaceous measures, distinguishes this area from area C.

The Top Softs – Middle Dirt – Bottom Softs sequence, except in that part of the area near Cinderhill and Hucknall colliery provings, is characterised by an increase in the thickness of dirt partings compared with the rest of the district. The Top Softs of this area do not exceed 33 in and the Bottom Softs 20 inches in thickness. The thickest overall sequence of 101 inches in Stanley Colliery Shaft includes

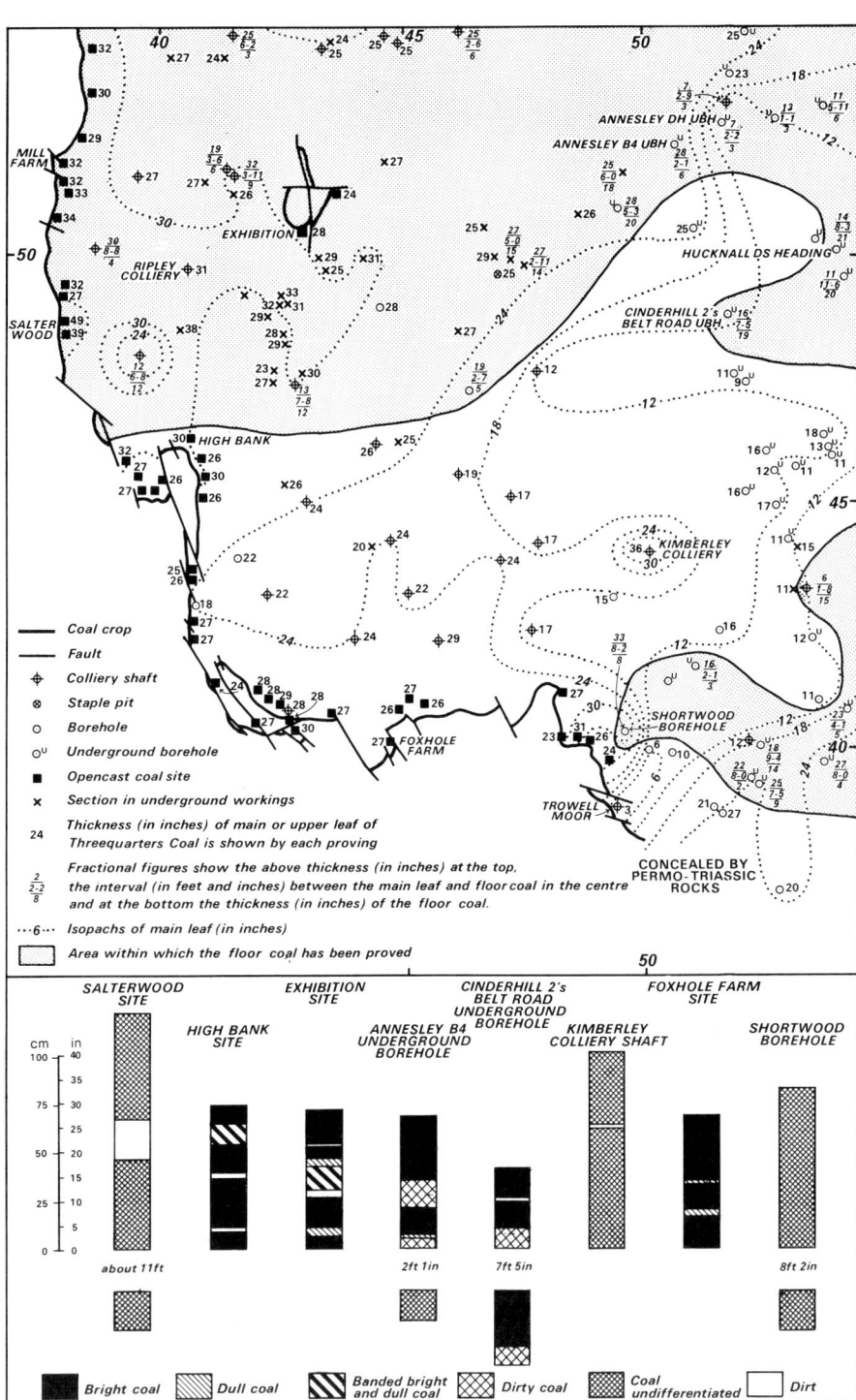

Figure 23 Plan of the Threequarters Coal showing provings and representative sections of the seam

51 in of dirt and dirty coal while the thinnest, of 48 inches in Cinderhill 19's Left Gate Underground Borehole, includes only 7 in of dirt.

The Denby leaves are 21 to 78 ft above the Top Softs but are represented only by a single seatearth or thin coal over much of the area. Measures ranging in thickness from 8 in to 30 ft separate the Denby leaves from the Yard Coal, which itself is mainly a single seatearth horizon or very thin coal.

The measures between the Yard and Threequarters coals range in thickness from 47 to 123 ft and normally form a single cycle containing variable arenaceous and argillace-

ous strata. The roof of the Yard Coal yields fish and plant debris and, more rarely, mussels; ironstones higher in the argillaceous measures form, in ascending order, the Striped Rake of Kirk Hallam, Blackshale Rake and Nodule Rake of the Morley Park district (Smyth, 1856, p. 38). The overlying sandstone is the 'Silkstone Rock' of the country to the north of this district.

The Threequarters (Tupton Threequarters or Dog-tooth) Coal ranges in thickness from 3 inches in Trowell Moor Colliery Shaft to 49 in at Salterwood Site. These figures exclude the thickness of a floor coal, the distribution of which is shown on Figure 23.

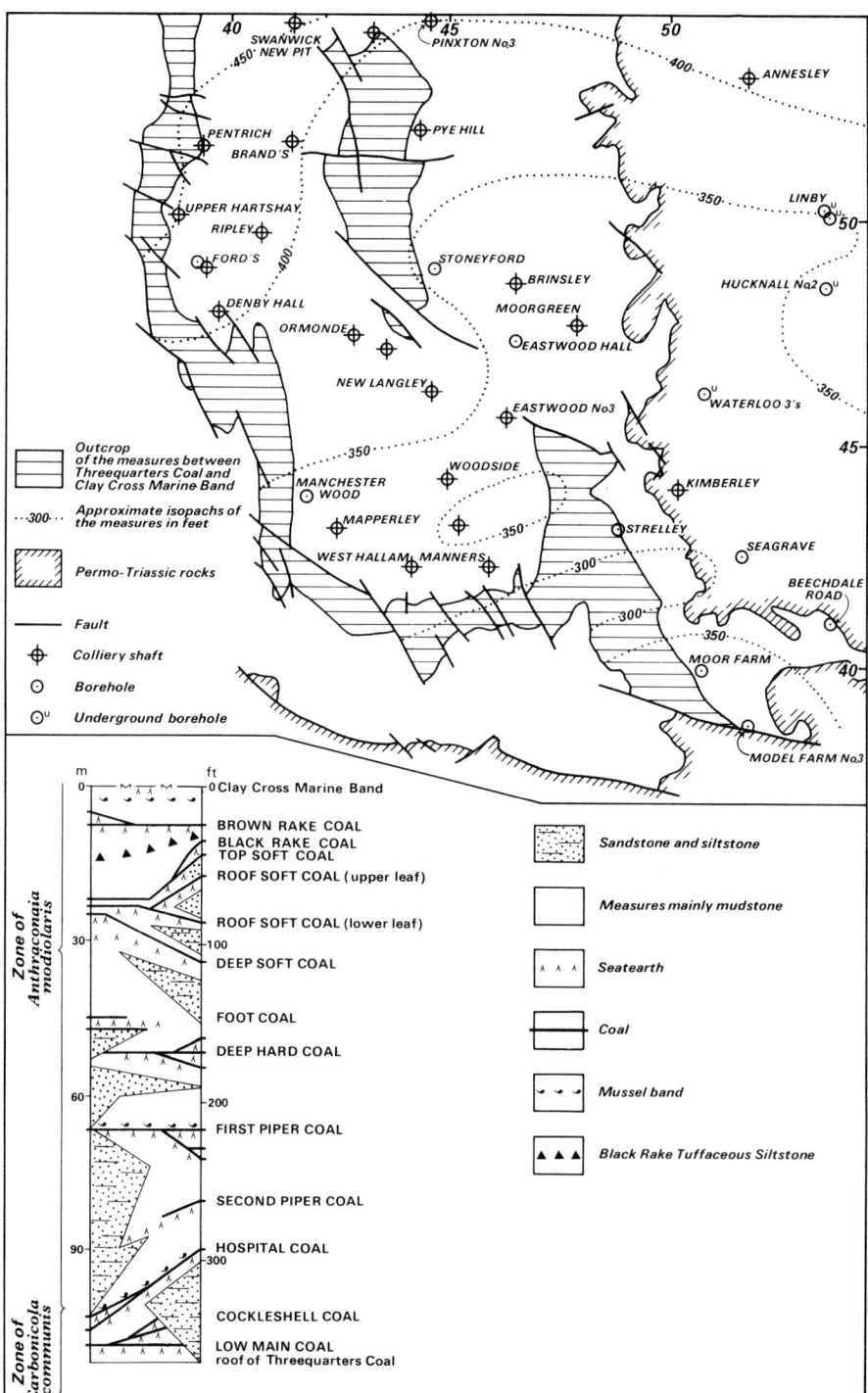

Figure 24 Generalised section of the measures between the Threequarters Coal and the Clay Cross Marine Band with inset map showing the principal provings, outcrop and isopachs of the measures

THREEQUARTERS COAL TO CLAY CROSS MARINE BAND

These measures range in thickness from 304 to 455 ft (Figures 24 and 25), and include the Low Main, First Piper, Deep Hard and Deep Soft coals, which were extensively mined as sources of household and steam coals throughout the district and have also been widely exploited by opencast methods. The Deep Soft group of coals is combined over a restricted area near Smalley to form the thickest coal sequence in the district. Other rock types in the sequence include the thin Black Rake Tuffaceous Siltstone which is an important lithological marker at the top of the Deep Soft group of coals. The two widespread and important sandstones of the measures are the 'Tupton Rock', found in the

west and also within a broad channel in the eastern part of the district, and the 'Deep Hard Rock' which occupies a narrow belt extending north-eastwards from its outcrop at Salterwood.

Boreholes in the east of the district illustrate the manner in which the Cockleshell and Hospital coals thin or disappear on the crests of thick sandstones. In contrast, however, the individual thicknesses of the coals of the Deep Soft group appear to be little related to the thickness of underlying sandstone, although the splitting of the coal sequence was controlled by the influx of sand at various times and locations. This group therefore resembles the Blackshale and Yard coal sequence (p. 35) though the area of maximum coal thicknesses is different.

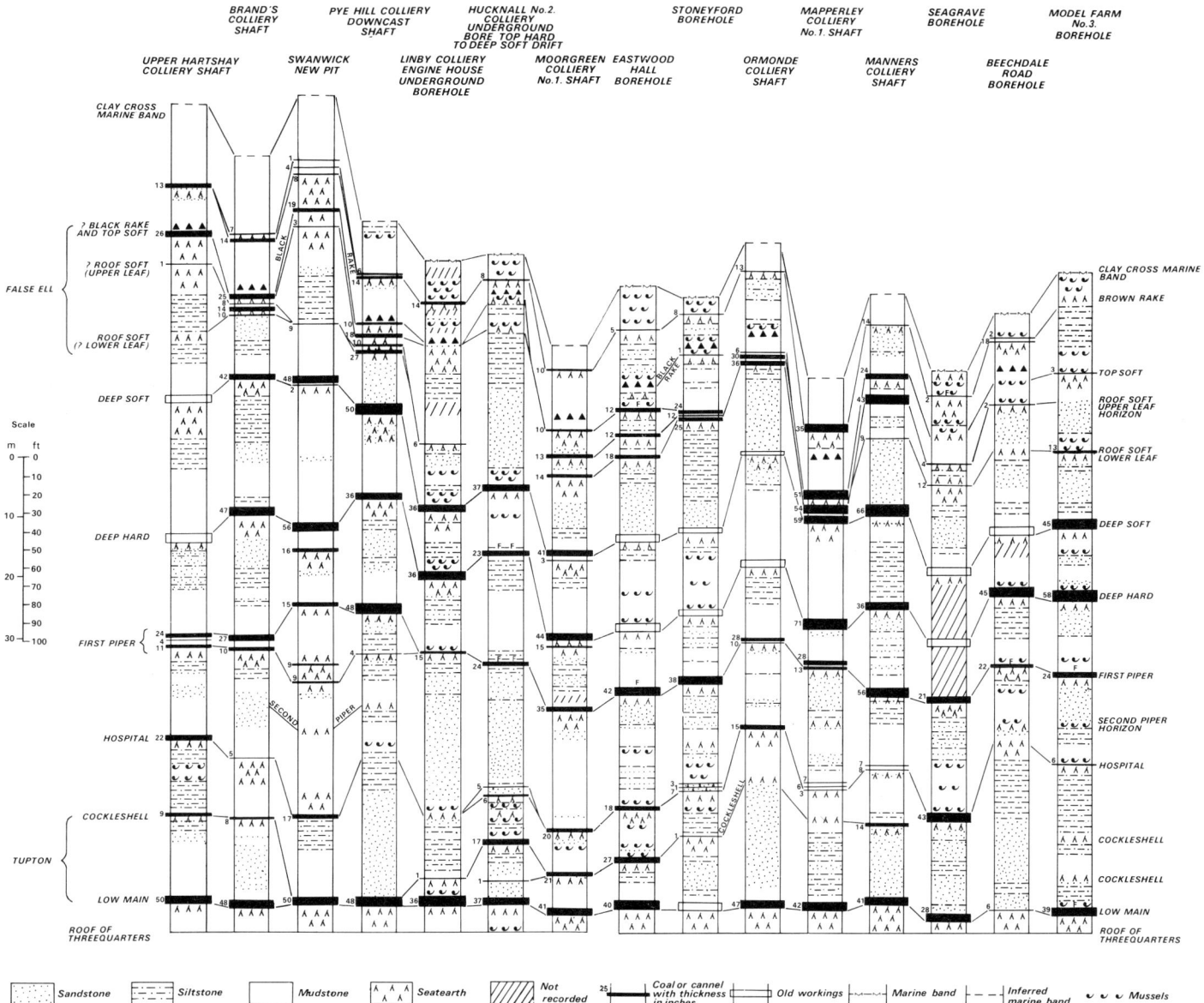

Figure 25 Representative sections of the measures between the Threequarters Coal and the Clay Cross Marine Band. For locations see Figure 24

The *Carbonicola cristagalli* mussel fauna (Smith and others, 1967, pp. 134–135) occurs above the Cockleshell Coal and passes upwards into the *Anthracosia regularis* fauna at about the horizon of the Deep Soft Coal (Smith and others, 1967, p. 96; 1973, p. 44).

The measures between the Threequarters and Low Main coals range in thickness from 6 to 23 ft and are mainly argillaceous with ironstone nodules; much of the sequence is composed of the seatearth of the Low Main Coal.

The Low Main (Low Tupton or Furnace) Coal has a maximum thickness of 66 in at Salterwood Site and is generally of good to very good quality. The seam thins irregularly and gradually from its outcrop to less than 36 inches in the south-east of the district. In the north-east (Figure 26) the top of the seam lies about 9 in below the Cockleshell Coal and together the two form the Tupton Coal. Within this area of union the Low Main element ranges in thickness from 36 to 42 in. A second and poorly defined area of union with the Cockleshell Coal exists near Swanwick, Birchwood and Pinxton collieries where the Tupton Coal is 45 to 50 in thick. The structure of the Low Main is frequently distinctive, particularly the basal 12 in, which are composed of bright coal with hard bands or interbedded hard and bright coal (Figure 28). The central portion of the seam in most records comprises soft bright coal, and the upper part is variable with hard coal, bright coal and cannel.

A dirt parting up to 2 in thick or a band of hard coal within this upper part of the seam can be correlated over wide areas. The ash content of the whole seam is low, the average of 138 analyses from the district being 4.4 per cent. Sulphur content is also low. The seam is frequently washed out within the eastern area of arenaceous roof measures (Figure 27).

The measures between the Low Main and First Piper coals range in thickness from 101 to 187 ft. The variation is directly related to the thickness of sandstones within the four major component cycles (Figure 29). The three lowest cycles are locally crowned by thin coals named, in ascending order, the Cockleshell, Hospital and Second Piper; the fourth is capped by the First Piper Lower Coal or First Piper Coal.

Washouts are present in these measures and, as both the Cockleshell and Hospital coals are split and the Second Piper Coal and seatearth are unrecorded in a number of the provings, the detailed correlation is not always certain. The sub-Hospital sandstone ('Tupton Rock') is thickest within the takes of Cinderhill and Wollaton collieries and crops out between Shortwood Site and Trowell Moor Colliery. Where this rock is thickest the Cockleshell Coal is absent as a result of erosion and/or non-deposition. The Hospital Coal also fails or is reduced on the 'crest' of this sandstone. In the west of the district, only the sub-Cockleshell sandstone is important, as displayed by Ormonde Colliery Shaft (Figures 25 and 29), where it is about 70 ft thick. This sandstone is

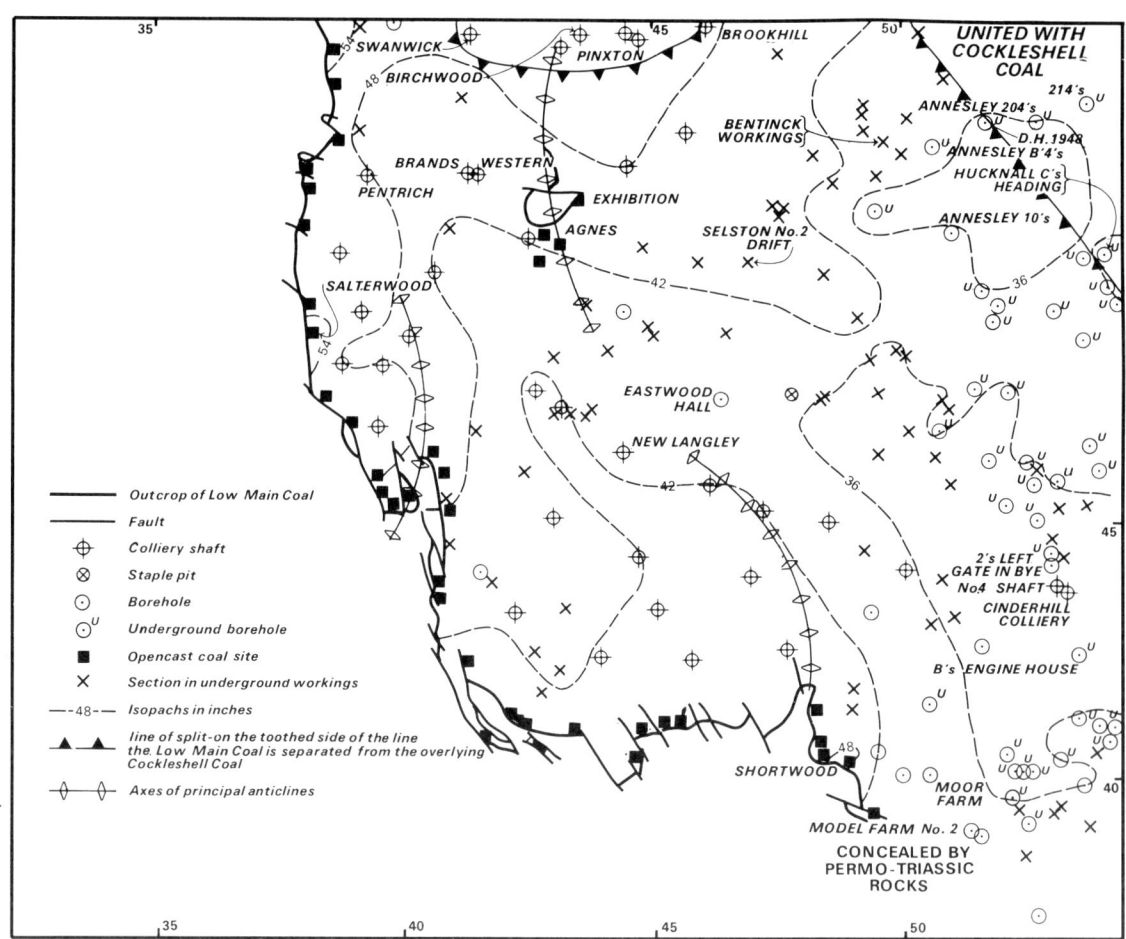

Figure 26 Plan of the Low Main Coal showing locations of provings, isopachs of the coal and localities mentioned in the text

Figure 27 Plan of the Cockleshell Coal showing provings, and isopachs of the measures between the Low Main and Cockleshell coals

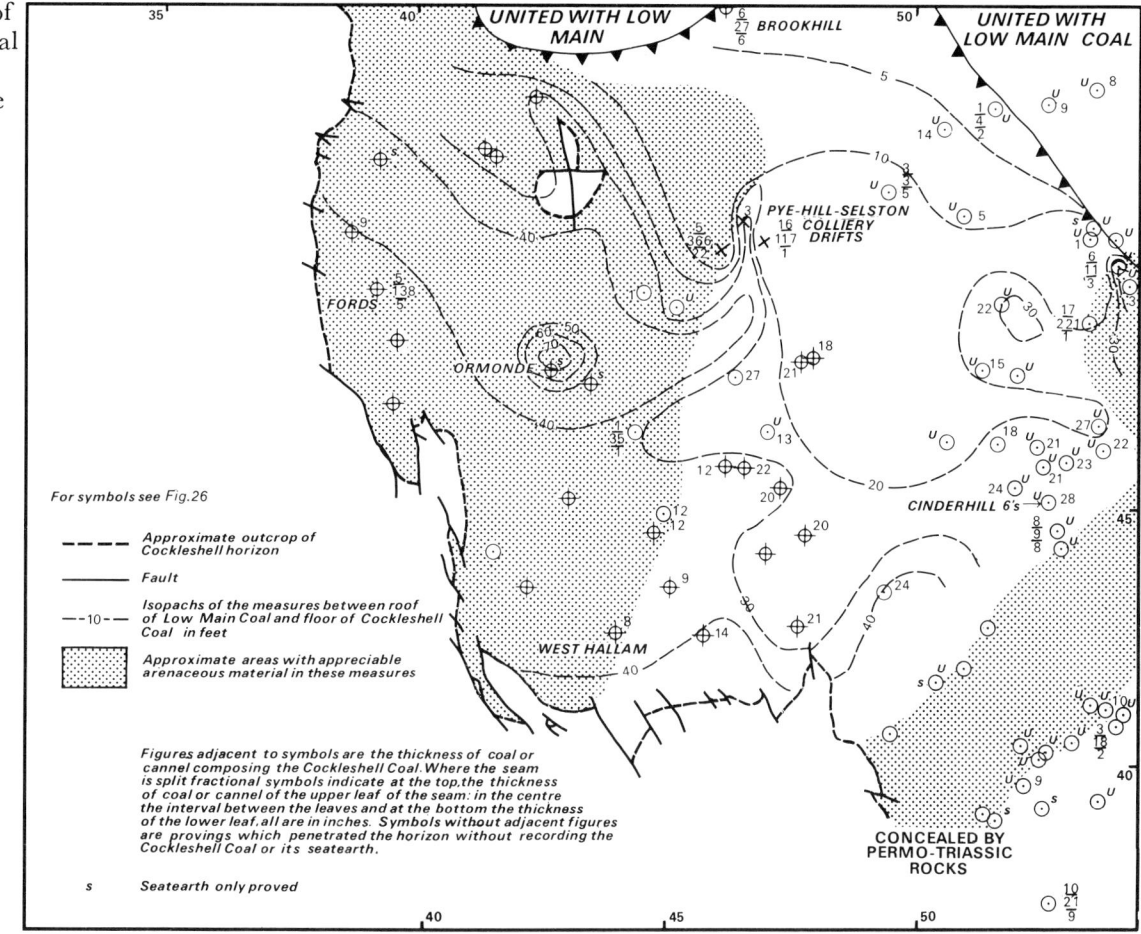

For symbols see Fig.26

- - - - Approximate outcrop of Cockleshell horizon

──── Fault

─ ─10─ ─ Isopachs of the measures between roof of Low Main Coal and floor of Cockleshell Coal in feet

▓ Approximate areas with appreciable arenaceous material in these measures

Figures adjacent to symbols are the thickness of coal or cannel composing the Cockleshell Coal. Where the seam is split fractional symbols indicate at the top, the thickness of coal or cannel of the upper leaf of the seam: in the centre the interval between the leaves and at the bottom the thickness of the lower leaf, all are in inches. Symbols without adjacent figures are provings which penetrated the horizon without recording the Cockleshell Coal or its seatearth.

s   Seatearth only proved

Figure 28 Representative sections of the Low Main and Cockleshell coals

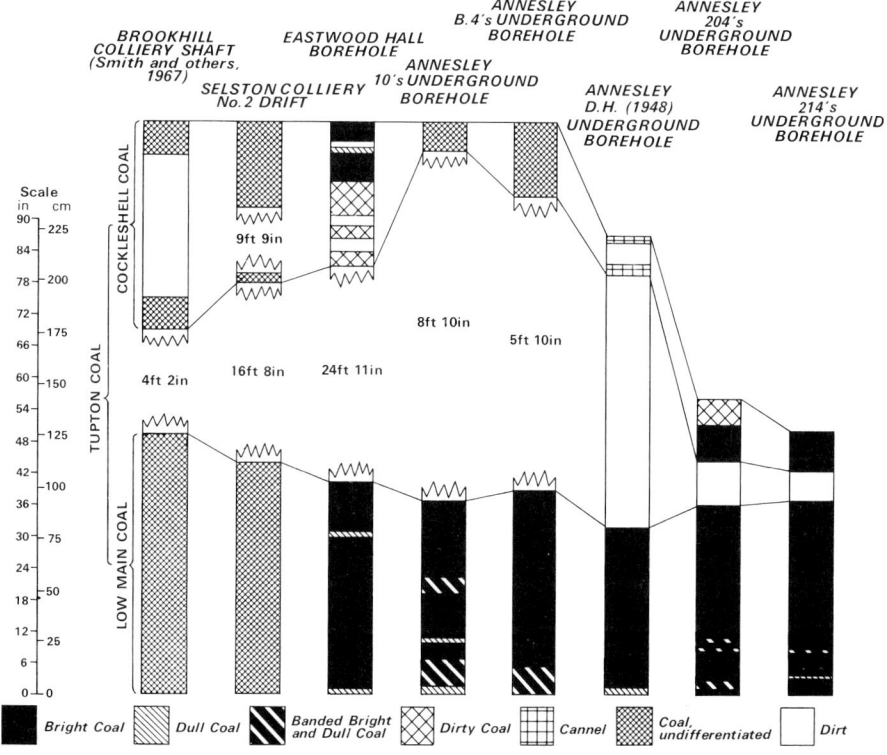

Bright Coal   Dull Coal   Banded Bright and Dull Coal   Dirty Coal   Cannel   Coal, undifferentiated   Dirt

marked by only minor ground features at outcrop except on the flanks of the Ironville Anticline. In the north-east of the district a thick sub-First Piper sandstone is the dominant sandstone; its sinuous line of maximum thickness enters and leaves the district within Annesley Colliery take. Its base falls to 14 ft above the Tupton Coal in Annesley Deep Soft G.4's Underground Borehole to the north of this district, and along part of the line of maximum thickness the Hospital Coal is reduced or washed out and the Second Piper Coal and seatearth are absent. Elsewhere in the district the sequence below the First Piper Coal includes only minor sandstones.

The *Cockleshell Coal* (Figure 27) lies up to 76 ft above the Low Main Coal and is united with it in the north and north-east of the district. Most of the intervening measures are sandy but 'cocklebeds' have frequently been recorded where mudstones form the roof of the Low Main Coal. The Cockleshell Coal may be composed of bright coal, dirty coal, cannel and/or interbedded coal and mudstone (batt) (Figure 28). Multiple sequences of coal and dirt, each a few inches thick, may range through several feet of measures. The thickest record of the unsplit seam, 28 inches in Cinderhill 6's Underground Borehole, includes laminae of dirt.

The measures between the Cockleshell and Hospital coals range in thickness from 8 to 75 ft. Widespread mussels in the roof of the Cockleshell Coal are frequently preserved in carbonate and include *Carbonicola cristagalli*, *C.* cf. *rhomboidalis* and *Naiadites flexuosus*. Sandstones are generally thin, except near Cinderhill and Wollaton collieries, where the top of the 'Tupton Rock' extends close to the base of the Hospital Coal (Figure 29) and the Cockleshell Coal is washed out along the line of maximum sandstone thickness.

The *Hospital Coal* is thickest, an exceptional 43 inches, in Seagrave Borehole. It ranges from 21 to 31 in at the outcrop in the west; elsewhere it is thinner or irregularly split into two or more leaves. Analyses of this coal are wholly from opencast provings; it is of good to medium quality in the west with low ash and sulphur contents, but the quality deteriorates in the south of the district.

The *measures between the Hospital and Second Piper coals* range in thickness from 12 to about 30 ft. *Carbonicola cristagalli*, *Geisina arcuata* and fish remains are present in the more argillaceous part of this cycle.

The *Second Piper Coal* attains a maximum thickness of 12 in of 'coaly shale' in Selston Colliery No. 2 Drift. There are only seven other records of a carbonaceous bed in deep provings; however the underlying seatearth is widespread, though locally impersistent.

The *measures between the Second Piper Coal and the Lower Coal of the First Piper or First Piper Coal* range in thickness from 16 to 84 ft, the increase being directly proportional to the

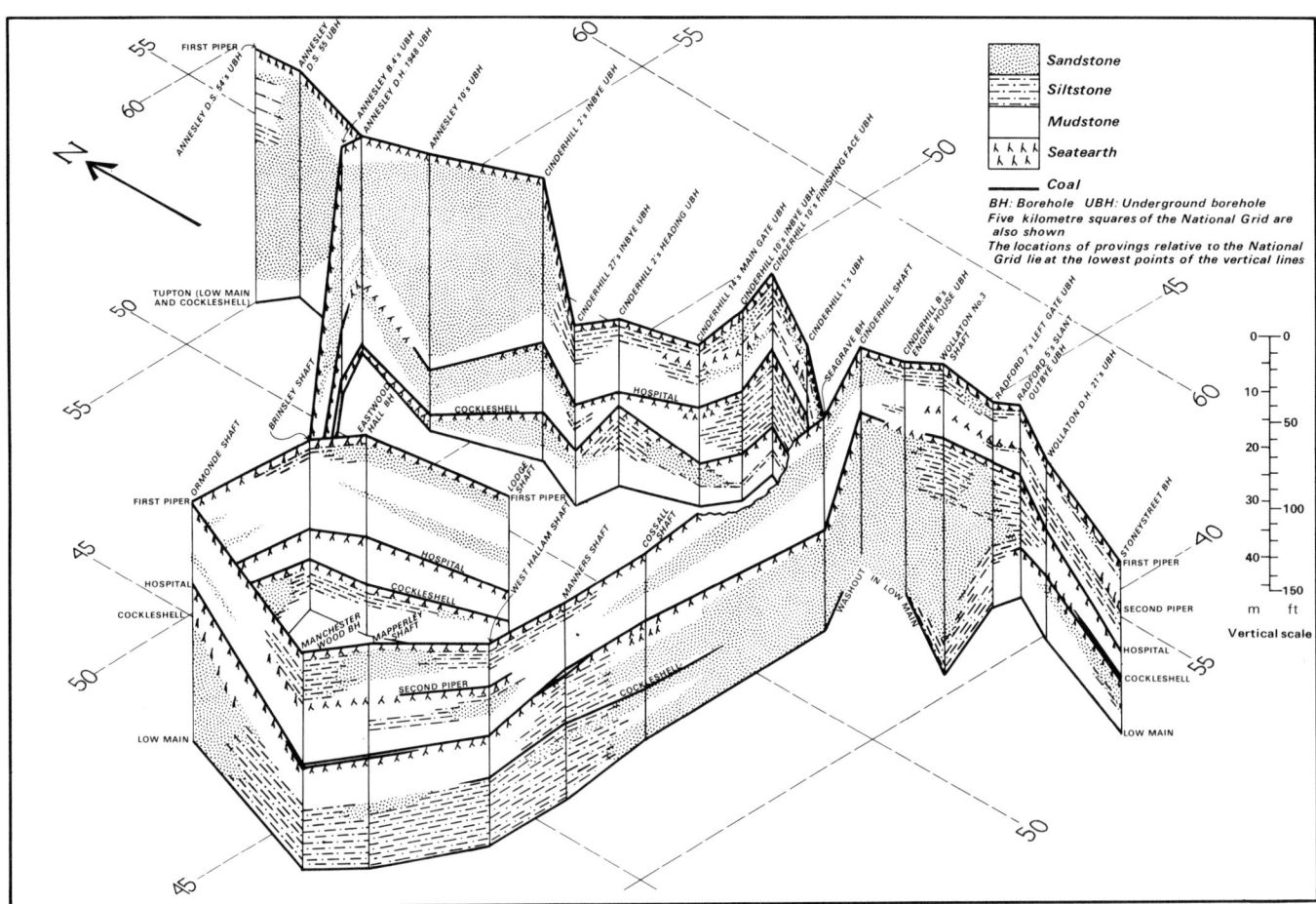

Figure 29 Isometric section showing the lateral variations of arenaceous measures between the Low Main and First Piper coals

thickness of the sub-First Piper sandstone in the north-east of the district (Figure 29).

The First Piper Coal (Figure 30) is composed of up to three main elements which comprise the lower and upper parts of the Lower Coal of the First Piper and the Upper Coal of the First Piper as figured by Smith and others (1967, fig. 18) for the Chesterfield district. Over a large part of the district they are combined in a single seam, up to 60 in thick, as in Bennerley Colliery Shaft. Many records of the combined seam show the tripartite subdivision, though a bipartite sequence of Upper and Lower coals is more widespread and important. The three elements of the seam divide irregularly so that they may be distributed through up to 43 ft of measures in two areas (Figure 30). In the north-easterly area the division is largely into two rather than three leaves. The Lower Coal of the First Piper is unimportant and usually of poor quality. The dirt parting between the lower and upper parts is normally a few inches in thickness, but increases rapidly to as much as 22 ft in the extreme north-west of the district (the 3-ft isopach is shown on Figure 30). The individual thicknesses of the two parts rarely exceed 12 in except west and south-west of Denby Hall Colliery, where analyses in opencast records show high ash contents.

Within the north-eastern area of split the Lower Coal is 10 in thick (including dirt partings) 6 in below the Upper Coal at Annesley 204's, 1 in (of batt) 3 ft 9 in below the Upper Coal at Annesley 10's, and 20 in at Cinderhill 2's Heading underground boreholes (Figure 30).

The Upper Coal of the First Piper within the north-western area of split varies from 15 to 40 in. Isopachs are shown on Figure 30. The coal is of variable quality, the thicker sequences being usually poor with high ash contents. In the north-eastern area of split it varies from 12 to 21 in and may contain up to $2\frac{1}{2}$ in of cannel at the top.

The combined First Piper seam ranges from 11 in (with 2-in cannel overlying) to 60 in.

The measures between the First Piper and Deep Hard coals are between 30 and 40 ft thick over much of the district but range from 29 to as much as 80 ft. They are thickest in the north and east. Eastwards there are two component cycles divided by the Deep Hard Floor Coal or its seatearth. The Floor Coal appears to separate from the Deep Hard Coal east of the sinuous north–south line shown on Figure 31.

The Deep Hard Coal is composed of three principal elements, the Floor Coal, the main seam and the Roof Coal, which may extend through up to 27 ft of strata. The main seam in turn can be divided into three parts.

The Floor Coal and the main Deep Hard seam are together in the west of the district, where their combined thickness ranges from 85 to 45 in. The Floor Coal never exceeds 12 in where the seam is combined. North of Openwood Site, the Floor Coal itself is split into two or three thin coals. The line of split between the Floor Coal and the main Deep Hard shown on Figure 31 is imprecise because most sections which record the separate seams lie well to the east of it. The Floor Coal is usually composed of dirty coal and may lie up to

19 ft below the main Deep Hard. East of the line of split most sections show only its seatearth.

The main Deep Hard Coal comprises the 'Bottom Brights', 'Hards' and 'Top Brights'—names used by the early miners (Anon., 1933). These terms are a simplification, for the composition of the seam is variable. The thickness of the main Deep Hard averages about 40 in but ranges from 14 to 76 in. The latter, in Hucknall Colliery No. 5 Shaft, is an abnormally thick swilley sequence.

The Roof Coal in the north-east of the district may lie up to 24 ft above the main Deep Hard and has a maximum thickness of 13 in of cannel.

A washout sandstone about 300 yd wide cuts out the main seam along a linear belt trending north-north-eastwards from the outcrop near Denby Hall Colliery.

The measures between the Deep Hard Coal or the Deep Hard Roof Coal and the Deep Soft Coal vary from 20 to 100 ft and are composed of up to three cycles; the lowest two are capped by thin, impersistent coals. The lowest cycle is composed largely of silty measures which may be up to 35 ft thick. The 'Deep Hard Rock' (Smith and others, 1967, p. 146) is only of local significance within this district, its most important outcrop lying near the washout described above. The Foot Coal at the top of this cycle is never thicker than 12 in. The second cycle either finishes at an impersistent thin coal or cannel lying up to about 2 ft below the Deep Soft Coal or extends upwards to include the Deep Soft Coal. The third cycle if present, therefore, includes these 2 ft of measures and the Deep Soft Coal. The measures are partly arenaceous but include thick seatearths of varying lithology. The underclay below the Deep Soft Coal was of considerable value as raw material for the ceramics industry.

The Deep Soft group of coals comprises in ascending order the Deep Soft, Roof Soft, Top Soft and Black Rake coals. Over a restricted area (A on Figure 32) near Smalley these coals lie within 25 ft of strata. Elsewhere they may be distributed in up to 96 ft of strata.

In area B of Figure 32 the Top Soft Coal splits off the combined seam. The Black Rake Coal, not recognised in area A, probably a split off the Top Soft, is present in this area. To the east, in area E, the sequence is further divided. There are two, or sporadically three, leaves of the Roof Soft, a thinner Top Soft and, locally, a Black Rake Coal. Only the Deep Soft remains as a persistent seam of importance.

In area C the Roof Soft and Top Soft seams, with or without the topmost Black Rake element, are close together, but are widely separated from the Deep Soft Coal. The combined Roof Soft and Top Soft coals, which may contain up to four separate leaves, was originally called the Ell Coal (Gibson and others, 1908, p. 69); latterly this name was altered to False Ell to avoid confusion with the Ell coals of Westphalian B. Type sections of the False Ell are provided by Grammer Street Borehole, where the sequence probably includes the Black Rake at the top (Figure 34), and Stoneyford Borehole, where the Black Rake is a separate seam.

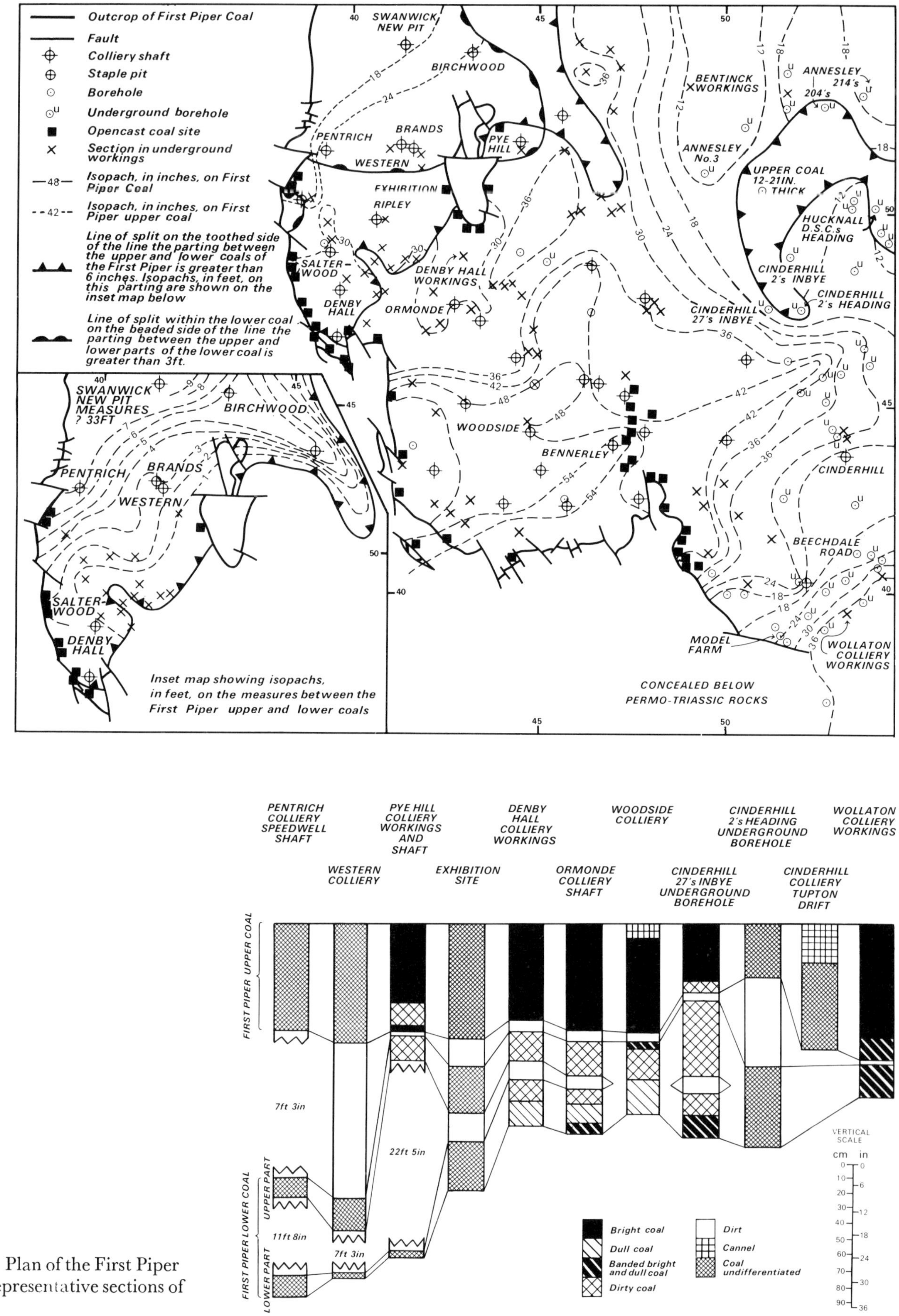

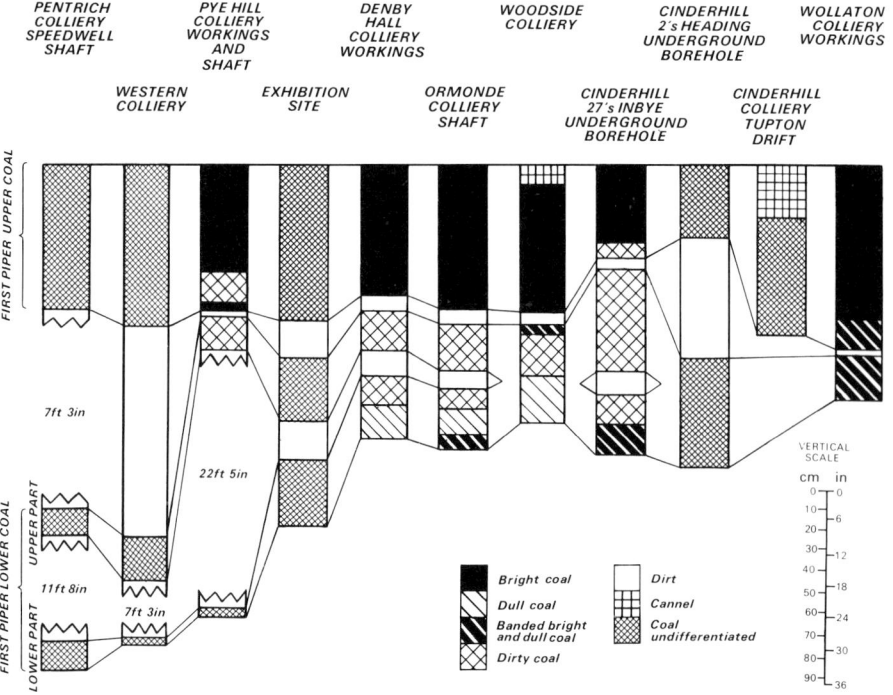

Figure 30   Plan of the First Piper Coal with representative sections of the seam

Figure 31 Plan of the Deep Hard Coal with representative sections of the seam. Localities mentioned in the text concerning the Deep Hard Coal and the overlying measures are also shown

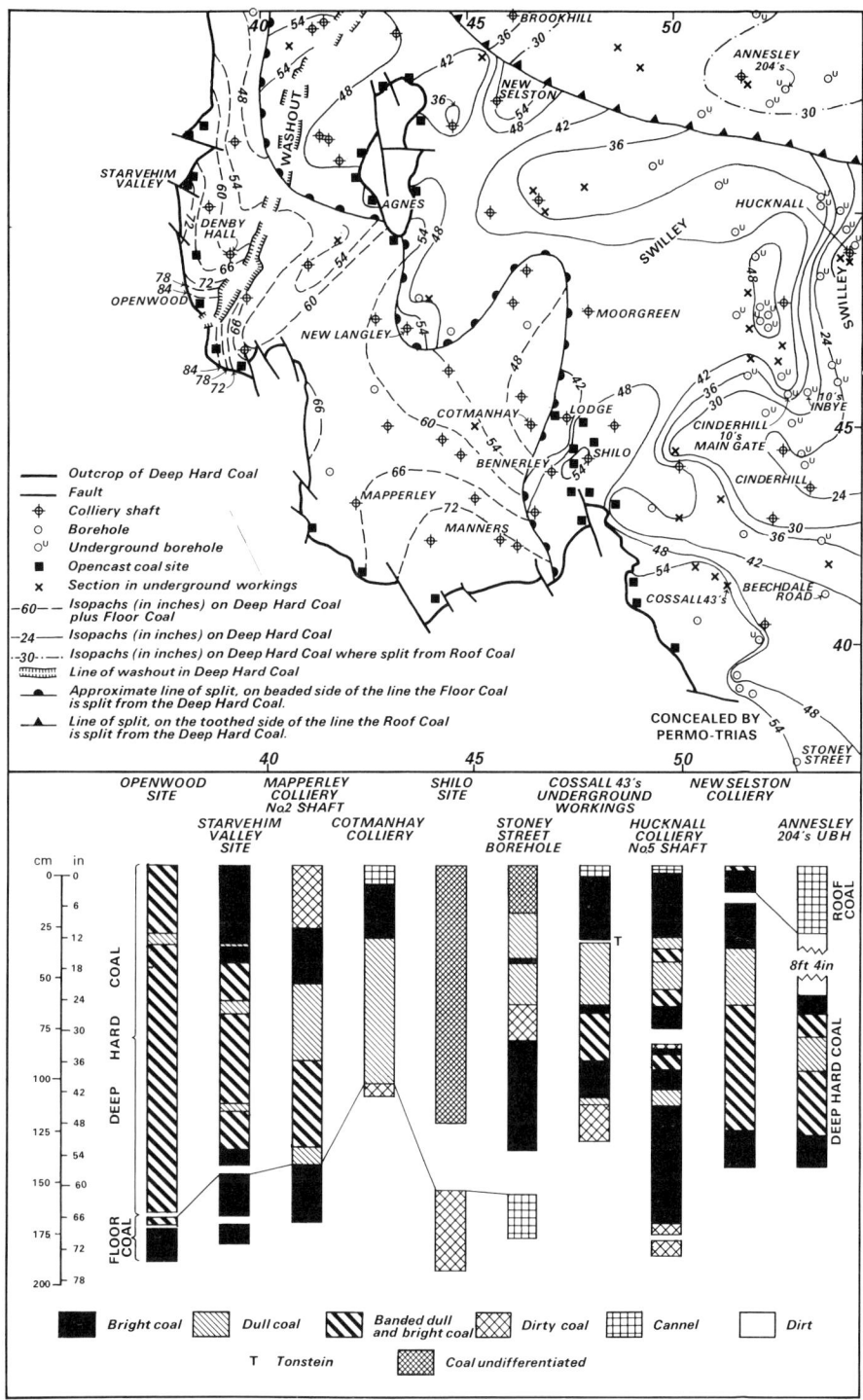

In area D, as in area E, the higher coals of the group are divided and remote from the Deep Soft. There is a thick Roof Soft (?lower leaf) in Birchwood Colliery Shady Shaft and Pinxton No. 3 Pit, but elsewhere the topmost coal is the principal seam. It is not certain if the latter is the Black Rake Coal or a combination of that seam with lower elements. It is however the Sitwell Coal (Smith and others, 1967, p. 149) of the south-west of the Chesterfield district.

The *Deep Soft Coal* itself (Figure 33) has a maximum thickness of 66 inches in Heanor Gate Borehole and thins irregularly towards the north-east. One or sometimes two thin floor coals or cannels, up to 7 in thick, lie as much as 2 ft below the main coal sequence.

The *measures between the Deep Soft and Roof Soft coals* (Figure 34) consist of up to 11 in of dirt over the major part of areas A and B; this increases in the north of these areas to between 24 and 57 in. Northwards into area C the sequence up to the base of the False Ell Coal consists of silty and sandy measures ranging in thickness from 27 to 71 ft. In the part of area D where the Roof Soft Coal or its seatearth is recognisable, these measures range in thickness from 28 to about 41 ft. Near the northern margin of the district, where the Roof Soft Coal is unrecognisable, the measures between the Deep Soft Coal and ?Black Rake Coal are as much as 91 ft

thick and include mudstone and siltstone with 24 ft of seatearth at their top. Near the south-west margin of area E, where they are largely argillaceous, these measures can be as thin as 12 ft, but over the rest of this area the incoming of arenaceous measures increases this interval to as much as 67 ft.

The *Roof Soft Coal* is thickest—up to 60 in—where combined with the Deep Soft Coal in areas A and B (Figure 32).

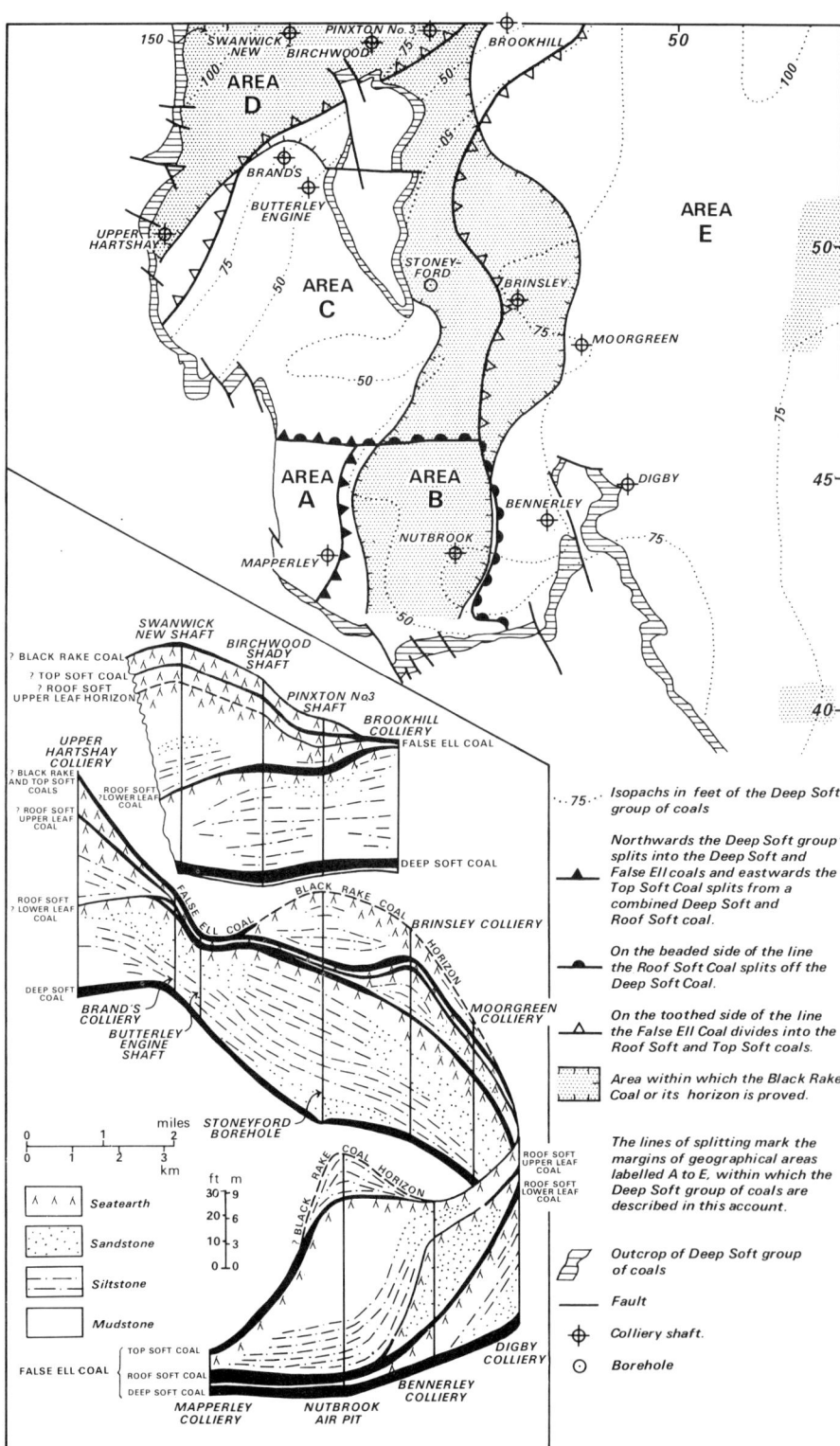

Figure 32 Diagram showing the areas of the various sequences of the Deep Soft group of coals described in this account together with a ribbon diagram illustrating these developments

Figure 33 Plan of the Deep Soft Coal with representative sections of the seam. Localities mentioned in the text concerning the Deep Soft group of coals are also shown

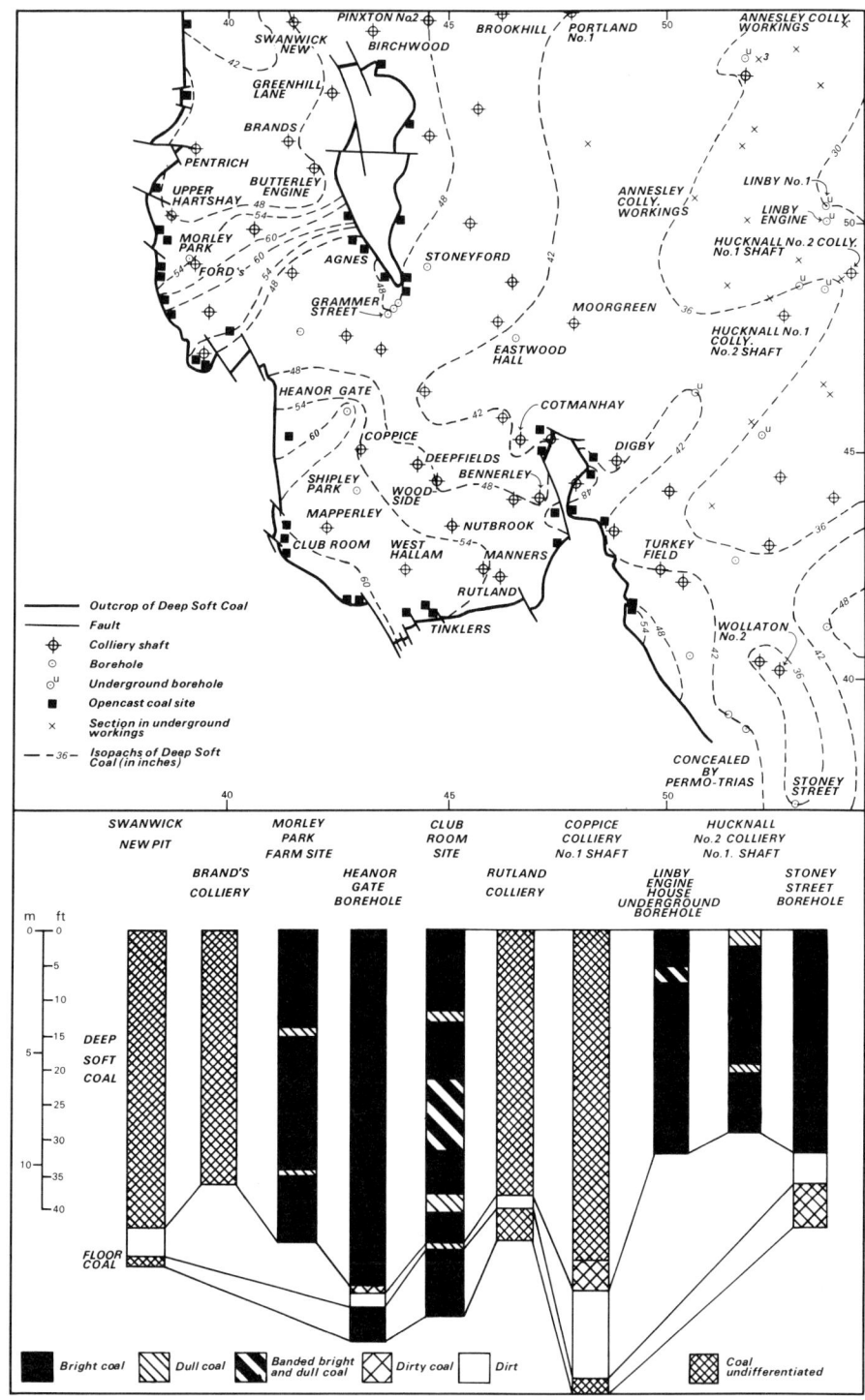

The Roof Soft is split in most places within area E into two leaves which are variable in thickness and distance apart.

Measures ranging in thickness from 8 ft at Turkey Field Colliery to 34 ft at Cotmanhay Colliery separate the lower and upper leaves of coal in area E, the thinner sequences being composed of argillaceous rocks and/or seatearth.

The Roof Soft Coal in area C is described below as part of the False Ell Coal. In area D, where much of the correlation is tentative, it is thought to be split in Birchwood Shady Shaft, where the lower 36-in leaf is separated by 21 ft of measures from the upper 6-in leaf and also in Swanwick Colliery

New Pit (Figure 34). Where the horizon has been recognised elsewhere within this area, the Roof Soft is either of insignificant thickness or represented only by seatearth.

The *False Ell Coal* in area C comprises the Roof Soft and Top Soft coals in three or four leaves in up to 16 ft of strata. Over part of area C it probably contains the Black Rake Coal at the top. The lowest leaf of the seam is everywhere the thickest and reaches a maximum of 37 inches in Grammer Street Borehole. The second leaf (the upper leaf of the Roof Soft) is never more than 14 in thick and lies up to 16 in above the lowest leaf. Exceptionally, as at Brands and

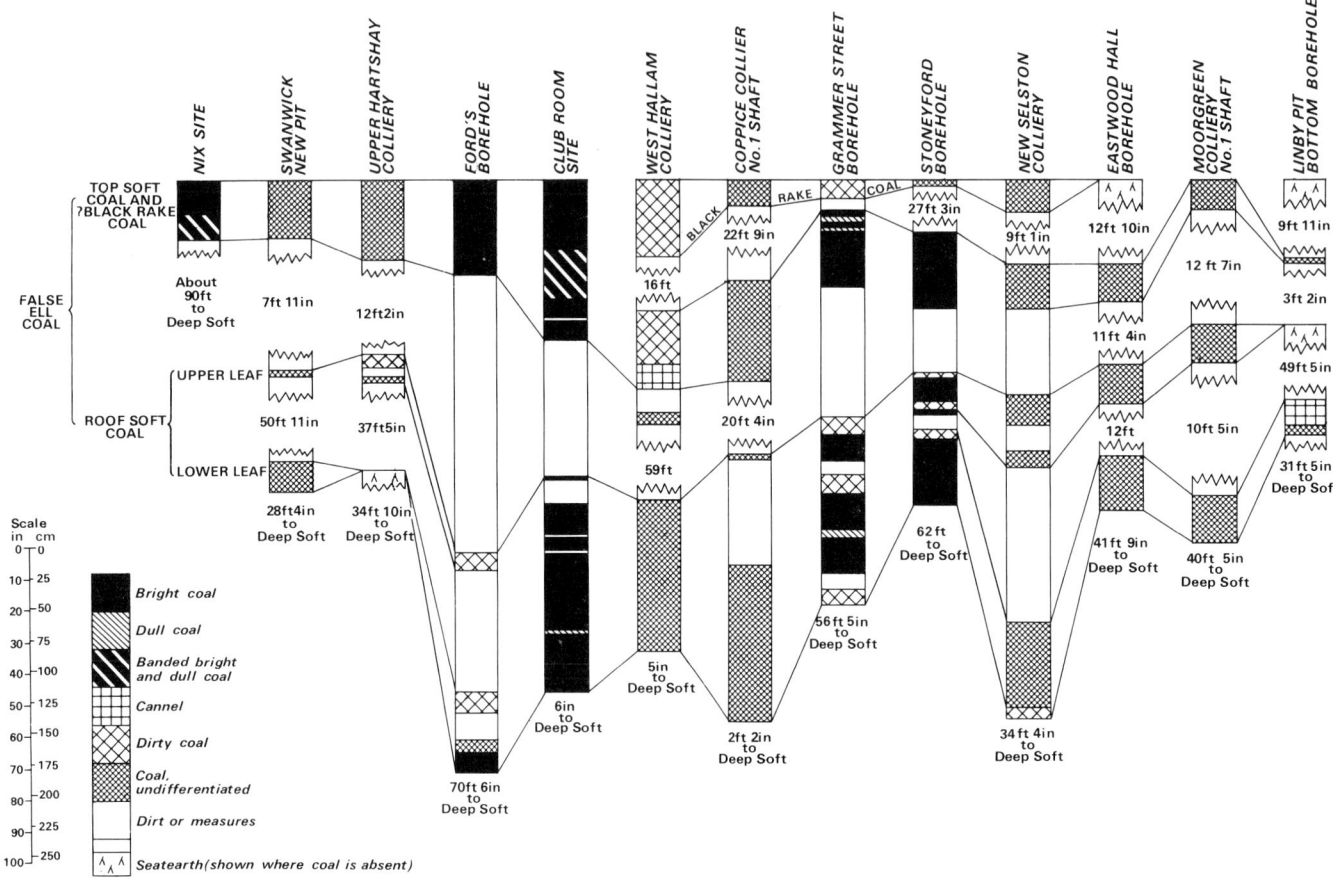

Figure 34 Sections showing variations in the Roof Soft and Top Soft coals. For locations
see Figure 33

Portland No. 1 collieries, there are three thin leaves to the Roof Soft component of the seam.

The *Top Soft Coal* is thickest, 32 to 52 in, in area A. Elsewhere within the Derby district the seam is of importance only as part of the False Ell Coal described above. In area B it is 8 to 39 in thick and lies up to 47 ft above the Roof Soft Coal, though over most of the district this interval is less than 15 ft. In area E the seam is usually thinner than 14 in but it is locally as much as 30 in. In the north-west (area D, Figure 32) the Top Soft is unrecognisable within the thick seatearths recorded below the Black Rake Coal, or it is incorporated within that coal.

The *Black Rake Coal* is the least important member of the Deep Soft group and is recognisable, mainly in a narrow central belt which widens towards the north-west (Figure 32), up to 27 ft above the Top Soft Coal. The coal varies from 6 to 24 in and is thickest in the north-west of the district. It is known locally in the eastern part of the district.

The measures between the Top Soft or Black Rake coals and the Brown Rake Coal range in thickness from 13 to 43 ft. Dark grey mudstones near the base of the cycle contain ironstone nodules of the Black Rake Ironstone (Smyth, 1856, p. 37) which have been worked both opencast and underground as a source of ore. Within these mudstones and forming a roof to underground workings is a widespread bed called the Black Rake Tuffaceous Siltstone described by Francis and others (1968).

A few feet of mudstone with ironstone nodules normally overlie the tuffaceous siltstone, the remainder of the cycle comprising arenaceous measures and seatearth. In Manchester Wood Borehole *Spirorbis sp.*, *Anthraconaia sp.*, *Anthracosia regularis*, *Carbonicola* cf. *venusta*, *Naiadites sp.* and *Geisina arcuata* were present in the mudstones of this cycle.

The Brown Rake Coal, which is up to 38 in thick at Club Room Site, thins rapidly towards the east where, over wide areas, it is under 12 in. In the north the seam is divided into two and sometimes three leaves, with up to 7 ft of measures between. The thickest single leaf is 18 in.

The measures between the Brown Rake Coal and the Clay Cross Marine Band vary in thickness from 8 to 35 ft and are wholly argillaceous in the district apart from a 3-ft sandstone in the vicinity of Linby Colliery.

The measures, including some of the ironstones, contain abundant *Anthracosia regularis*, *Naiadites sp.* and *Geisina arcuata*; *Spirorbis sp.* is common and *Planolites montanus* also occurs. Near the top, the fossils tend to be pyritised.

Westphalian B

CLAY CROSS MARINE BAND TO TOP HARD COAL

The measures between the Clay Cross Marine Band and Top Hard Coal (Plate 5) are about 460 ft thick at Heanor Gate Borehole and about 450 ft near Swanwick Colliery; they thin eastwards to about 330 ft at Hucknall No. 2 and Broxtowe collieries, though an incongruous record from Hempshill Heading in the east shows about 406 ft. Approximate isopachs are shown on the inset map of Plate 5. The measures comprise twelve or more cycles, most of which contain both arenaceous and argillaceous strata. There are ten named coals. Marine fossils are confined to the Clay Cross Marine Band. Mussels are present in the roof measures of all the coals except the Bottom First and Bottom Second Waterloo coals; the *Anthracosia aquilina/ovum* fauna lying below the Second Ell Coal grades upwards into the *A. phrygiana* fauna, which continues to the Top Hard Coal.

The named coals have been exploited mainly by opencast means, but the Second Waterloo Coal has also been extensively worked underground. The measures associated with the Second Waterloo Coal are worked for brickmaking.

The Clay Cross Marine Band consists of up to about 9 ft of dark grey mudstone with thin ironstone lenses. The marine fossils and their distribution within the marine band have been described by Edwards and Stubblefield (1948, p. 215) and Calver (*in* Smith and others, 1967, pp. 157–159; and 1968a, pp. 163–173). The 1 ft 10 in of mudstone between the Joan Coal seatearth and the *Dunbarella*/goniatite-bearing mudstones in Eastwood Hall Borehole contain *Lioestheria* at their base with *Lingula* and foraminifera above. They are indicative of an increasing marine influence on the conditions of deposition, but there is no clear separation of the foraminifera and *Lingula* phases. The *Dunbarella* and goniatite-rich horizons near or at the base of the sequence occur in dark mudstone, tougher and more fissile than the overlying *Lingula*-rich horizons. As noted by Gibson and Wedd (1913, p. 71) and Edwards and Stubblefield (1948, p. 216) the upper *Lingula*-bearing horizons alternate with mudstones containing stunted mussels.

The measures between the Clay Cross Marine Band and Second Ell Coal form part of a single cycle and range in thickness from 48 ft in the east to 95 ft, the thicker sequences being found mainly in the western and south-eastern parts of the district. Sandstones occur in the uppermost third of these measures; the most persistent is found to the west of Heanor. The mudstones are rich in mussels, which may be pyritised within 40 ft of the marine band, but may also be preserved in calcareous and ferruginous material. Many of the mudstones display worm tracks and burrows, and there are plants at the higher levels. Ironstones are common and frequently exhibit cone-in-cone structure.

The Second Ell Coal is extremely variable both in thickness and composition. Near outcrop in the north-west, and as far south as Upper Hartshay Colliery, it is a split seam similar to that described by Smith and others (1967, p. 158) in the Alfreton area to the north. The upper leaf (Tanyard Coal) is of medium quality and ranges in thickness from 18 to 28 in; it is separated from the 1-in to 8-in lower leaf of dirty coal by up to 26 in of seatearth.

Within a wide area in the north and west of the district the seam is composed wholly or largely of cannel. Two leaves are frequently present, up to 14 ft apart near Heanor. Cannel is also locally present in the upper part of the seam in the eastern part of the district. Elsewhere, the Second Ell is a single or split coal in which the lower leaf or the lower part of the single seam is commonly composed of dirty coal. The thickest known section is 19 in of coal underlying 48 in of dirty coal, and the maximum proved separation of the two leaves is 6 ft.

The measures between the Second and First Ell coals average about 50 ft over much of the district, with a minimum of 31 ft and a maximum of 65 ft.

The First Ell Coal shows great variations both in quality and thickness. It is thickest to the south-west of Heanor and in the Erewash Valley, east of Ilkeston. South-west of Heanor it is mainly 30 to 42 in thick, although locally only 23 in. Analyses from opencast sites show that it is rather poor to very poor in quality in this area.

North and north-west of Heanor the First Ell is of medium to very poor quality. It is 18 to 27 in thick at the northern boundary of the district, thickening to 36 in at Bailey Brook Colliery Shaft. Dirt partings, 5 in or less in thickness, are recorded from several localities.

The measures between the First Ell and Fourth Waterloo coals are about 40 ft thick over most of the district. They are thickest near Heanor, where they total 56 ft, and thinnest in the south-east of the district, where they are only 10½ to 16 ft thick. In most sections they comprise only one cycle with, in the immediate roof of the First Ell, a lenticular ironstone, usually up to about 2 in thick with kaolinite ooliths and pyrite, which forms a useful lithological marker.

The Fourth Waterloo Coal is variable both in thickness and in quality and may comprise up to three leaves of coal of which the middle leaf is the most persistent and important. It is 15 to 29 in thick in the north-west of the district, where it is medium to rather poor in quality. Elsewhere it ranges from 6 to 34 in.

The measures between the Fourth and Third Waterloo coals are usually 30 to 45 ft thick, but vary from 27 to 76 ft. Thicknesses are related to the proportion of sandstone present. The two component cycles cannot be separately distinguished, though 19 widely distributed provings in the district record a thin coal or seatearth 12 to 20 ft (exceptionally 38 ft) below the Third Waterloo Coal.

The argillaceous roof measures of the Fourth Waterloo Coal contain fish debris, ostracods, cf. *Planolites* and plants. *Anthracosia* cf. *beaniana* and *A. disjuncta* are also recorded from mudstones within a sequence of interbedded sandstone, siltstone and mudstone near the top of these measures.

The Third Waterloo Coal ranges from 12 in to a multiple sequence of coal and dirt, 48 in thick. In the north and north-east the seam is about 2 ft thick over wide areas, and dirt partings are recorded only locally. Well south of Heanor, at outcrop near Mapperley Colliery, there is a dirt parting near the top of the seam.

The measures between the Third and Second Waterloo coals vary in thickness from 16 to 46 ft and locally include a thin coal which is usually about 3 in thick, but may be thicker when close to the Second Waterloo Coal. The measures include a variable amount of sandy strata and the mudstones contain many mussels, some of which may be pyritic, together with fish debris, *Spirorbis sp.*, *Planolites montanus*, worm tracks and borings.

The Second Waterloo Coal consists of two principal elements which, where they constitute separate seams, are named the Bottom Second Waterloo and Top Second Waterloo coals (Figure 35). This split is irregular and the two leaves are locally re-united in the south-west of the district. The Top Second Waterloo Coal itself divides in the south-east of the district. The interpretation of the Second Waterloo Coal sequence is based upon a report by the former Nottingham Coal Survey Laboratory (Turner, 1961b).

Away from the line of split, the Second Waterloo Coal is 49 to 57 in thick with two dirt partings totalling up to 10 in. The lower parting divides the Bottom and Top Second Waterloo elements, but the upper parting is also conspicuous and can be widely traced in sections of both the Second Waterloo and Top Second Waterloo.

The *Bottom Second Waterloo*, up to 31 ft below the Top Second Waterloo in the west of the district, is generally between 12 and 18 in thick (Figure 35), though it ranges from 10 to 25 in. The few analyses show it to be of medium quality, but to the north-west of Ripley Colliery it is composed wholly of cannel up to 18 in thick. Deep mine provings at the eastern margin and in the north-east of the district show the seam to be not more than 17 ft below the Top Second Waterloo and under 13 in thick, but it varies from 8 to 27½ inches in opencast sites on the eastern side of the Erewash Valley to the east of Ilkeston.

The *measures between the Bottom and Top Second Waterloo coals* are largely argillaceous, but include appreciable amounts of sandy strata near Heanor. Mussels occur at the base of the cycle.

The *Top Second Waterloo Coal* is thickest, up to 64 in, in the extreme north-west of the district, where it may include a basal cannel up to 42 in thick. Elsewhere in the west the thickness ranges from 24 to 57 in. The detailed sequence of the seam is variable, including many dull coal bands and sometimes two dirt partings some 8 to 12 in above the base. Its quality is medium with a moderately low sulphur content. It has been worked underground from Coppice and Woodside collieries and has extensive opencast workings.

At the eastern margin of the district, at Hucknall and Linby collieries, the Top Second Waterloo Coal is 22 to 33 in thick; and over an area in the south-east it divides into two leaves along a line slightly north and west of the split between the Top and Bottom Second Waterloo coals. The line of split on Figure 35 is drawn approximately at the 18-in isopach of the separating measures and, as emphasised by Turner (1961b), it occurs at a higher horizon in the seam than the prominent dirt parting which can be widely traced in the district (Figure 74). The maximum proved separation of the two leaves is 15 ft in an outline record of Cinderhill Colliery No. 2 Shaft.

The measures between the Second and First Waterloo coals are reduced in thickness from 65 ft in the north to only 23 ft in the south, thereby continuing a trend mentioned by Smith and others (1967, p. 164). The attenuation is irregular. The measures comprise two main cycles divided by the Waterloo Marker Coal, plus, locally, a third thin cycle caused by splitting of that coal. There are mussels at the base of both the main cycles, and sineoids have been frequently recorded, mainly, however, from the lower cycle. *Anthracosia phrygiana* and *Spirorbis sp.* have been recorded from the roof of the Second Waterloo and *A. aquilina, A. beaniana, A. phrygiana, Naiadites quadratus* and *Carbonita humilis* from the mudstones overlying the Waterloo Marker Coal.

The *Waterloo Marker Coal* is a persistent thin coal or cannel which serves as an indicator horizon between the First and Second Waterloo coals. It rarely exceeds 12 inches in thickness and is of no economic importance.

The First Waterloo Coal forms a single seam over two large areas in the east of the district (Figure 36) as described by Turner (1961a). These two areas are separated by a narrow corridor within which the seam is split into the Bottom and Top Waterloo coals and which extends north-eastwards from Greasley Castle Borehole to Linby Colliery. The corridor opens at both ends into wide areas covering the northern and western parts of the district where the seam is similarly split.

In the northern area of undivided coal the minimum thickness of the First Waterloo is 34 in with a prominent dirt parting, normally 1½ to 6 in thick, but reaching a maximum of 12 in. The lower leaf varies from 16 to 25 in and the upper leaf from 14 to 27 in.

In the southern area of undivided coal, the First Waterloo reaches a maximum of 59 in, thinning eastwards to 21 in.

Over the remainder of the district, that is in the north and west, the seam is divided into Bottom and Top First Waterloo coals; the line of splitting shown on Figure 36 is drawn at the 12-in isopach of the dirt parting at which the split occurs. The maximum division of the leaves is in the north-west of the district, where it amounts to 39 ft. The dividing measures are composed of a single cycle.

The *Bottom First Waterloo Coal* is recorded as comprising 30 in of cannel in Swanwick Colliery, Common Pit, although this maximum thickness must be regarded as doubtful because only 12 to 15 in of coal were recorded in nearby shafts of the same colliery. The seam is mainly 18 to 24 in thick, but thicknesses of 27 and 28 in have been recorded locally.

The *Top First Waterloo Coal* is thickest adjacent to the lines of split in the First Waterloo Coal. It is 30 in thick in Moorgreen Colliery 10's workings, of which 15 in are dirty coal or dirt and 19 inches in Bentinck W1 Underground Borehole, of which 5½ in are dirty coal. Away from the lines of split, the

Figure 37 Plan of the Dunsil Coal with representative sections of the seam

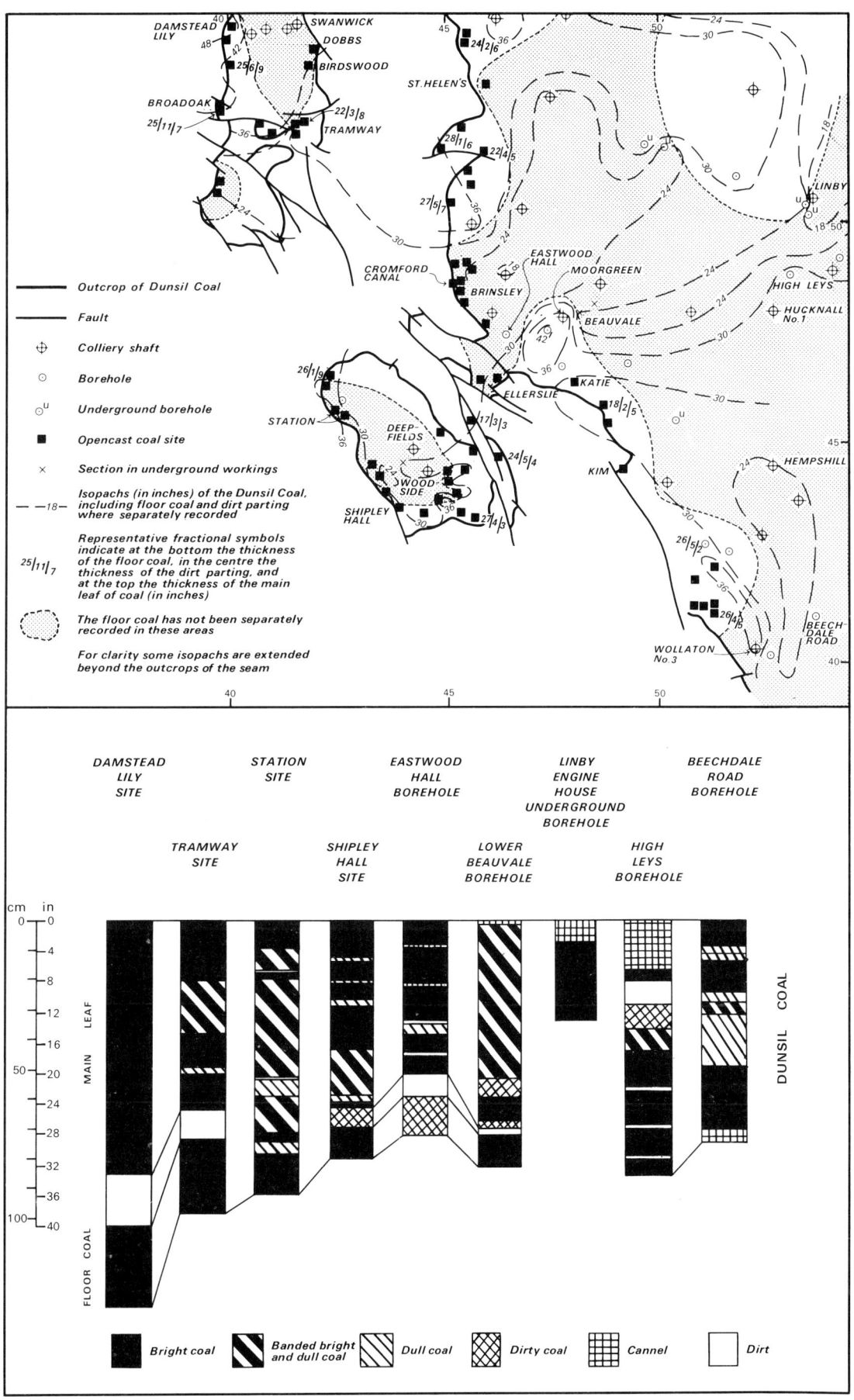

seam is frequently recorded as cannel, usually about 12 in thick, but reaching 23 inches in Brinsley Colliery shafts.

The measures between the First Waterloo and Dunsil coals form a single cycle normally about 30 ft in thickness, but ranging up to 57 ft. The measures are variable in lithology and many sections show an unusually thick seat-earth below the Dunsil Coal. Small specimens of *Anthracosia* cf. *phrygiana*, *A. disjuncta* and *Naiadites sp.* have been recorded from the roof measures of the First Waterloo Coal. Plant debris is present throughout the measures; sineoids and worm tracks have also been noted.

The Dunsil Coal usually consists of a main leaf 18 to 34 in thick and a thin floor coal a few inches below. Over large parts of the district (Figure 37) the floor coal is not recorded separately and it appears from detailed sections that it is locally united with the main leaf. Elsewhere, particularly at Cromford Canal Site and over most of the eastern part of the district, less reliable evidence suggests that it has died out.

The quality of the Dunsil is good to medium, with low to medium ash content and low to moderately low sulphur content. The seam has been mined locally, but it has mostly been worked by opencast means.

The measures between the Dunsil and Top Hard (or Top Hard Floor) coals form a single cycle about 40 ft in thickness, but ranging from 24 to 54 ft. Mussels and ostracods are recorded (p. 144d) from the roof of the Dunsil Coal and sineoids and worm tracks are also known. Ironstones are prominent in the mudstones, though in places, the cycle is mainly composed of sandstone.

TOP HARD COAL TO MANSFIELD MARINE BAND

The general thickness and lithological characteristics of this group of strata are shown on Figures 38 and 46. Thicknesses range from a maximum of about 590 ft in the High Park Colliery area to a minimum of 520 ft at Hempshill Colliery. The chief coals mined underground are the Top Hard–Comb at the base, and the Cinderhill, High Hazles, Lowbright and Mainbright higher in the sequence. Coals of lesser importance are the Mainsmut, Brinsley Thin and Two-Foot. Sandstones are common and locally prominent above the Cinderhill coals and between the Lowbright Coal and Haughton Marine Band. Washouts occur in the Top Hard and Clown seams.

The Two-Foot Marine Band is found throughout the area but the Manton '*Estheria*' Band and Haughton Marine Band are impersistent.

The Top Hard is a convenient boundary to take between the *A. modiolaris* and Lower *similis-pulchra* zones for it separates the characteristic *A. phrygiana* fauna below, from the equally distinctive *Anthraconaia pulchella* fauna above. The lowest distinctive fauna of the Lower *similis-pulchra* Zone occurs above the ?Second St John's Coal and includes *Anthraconaia pulchella*, *Anthracosia* cf. *planitumida* and *A. simulans*. Slightly higher in the sequence, above the Cinderhill Coal, representatives of the *Anthracosia caledonica* fauna were found. In addition to this species the assemblages included *A. sp.* cf. *fulva*. The mussel-bands between the High Hazles

and Two-Foot coals are notable for the first appearance of the *Anthracosia atra* group. The associated species include *Anthracosia simulans* and *Naiadites* cf. *obliquus*. The *A. atra* group attains its acme in the beds overlying the Clown Coal. In the higher beds up to the Mansfield Marine Band mussels are not common but include *Naiadites* aff. *angustus*.

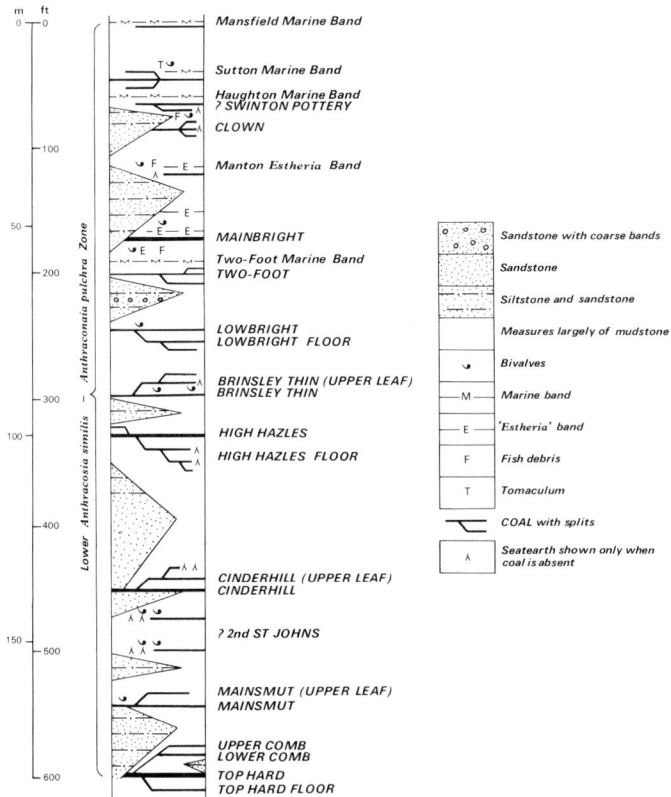

Figure 38 Generalised section of the strata between the Top Hard Coal and the Mansfield Marine Band

The Top Hard–Comb Coal consists of four elements, namely, the Top Hard Floor Coal, the main part of the Top Hard Coal and the Lower and Upper Comb coals, which occur in a variety of combinations in strata ranging in thickness from 12 to 87 ft (Figures 39 to 41).

Over most of the outlier at Heanor all four seams are united and the dirt partings in the sequence may total as little as 10 in. The various lines of split in the composite seam are shown on Figure 39A.

The *Top Hard Floor Coal* is separately identified north-east of Eastwood (Figure 39A). It is derived from the Top Hard seam by a split off the Bottom Brights (see below) and is of variable quality.

The *Top Hard Coal* itself consists of bright coal at the base and top of the seam, the Bottom Brights and Top Brights, separated by a variable thickness of hard coal, the Hards (Figure 40). The hard coal provided an excellent locomotive and steam fuel and the 'brights' a house coal. The overall high quality made it one of the principal economic seams in the coalfield and it was the first to be exhausted in the present district. In many early workings only the 'hards' were

Figure 39 Ribbon diagram showing the stratigraphical variations of the Top Hard Floor, Top Hard and Comb coals with inset plans showing: A their geographical distribution; B isopachs of the measures between the Top Hard Floor and Top Hard coals; C isopachs of the measures between the Top Hard and Comb coals

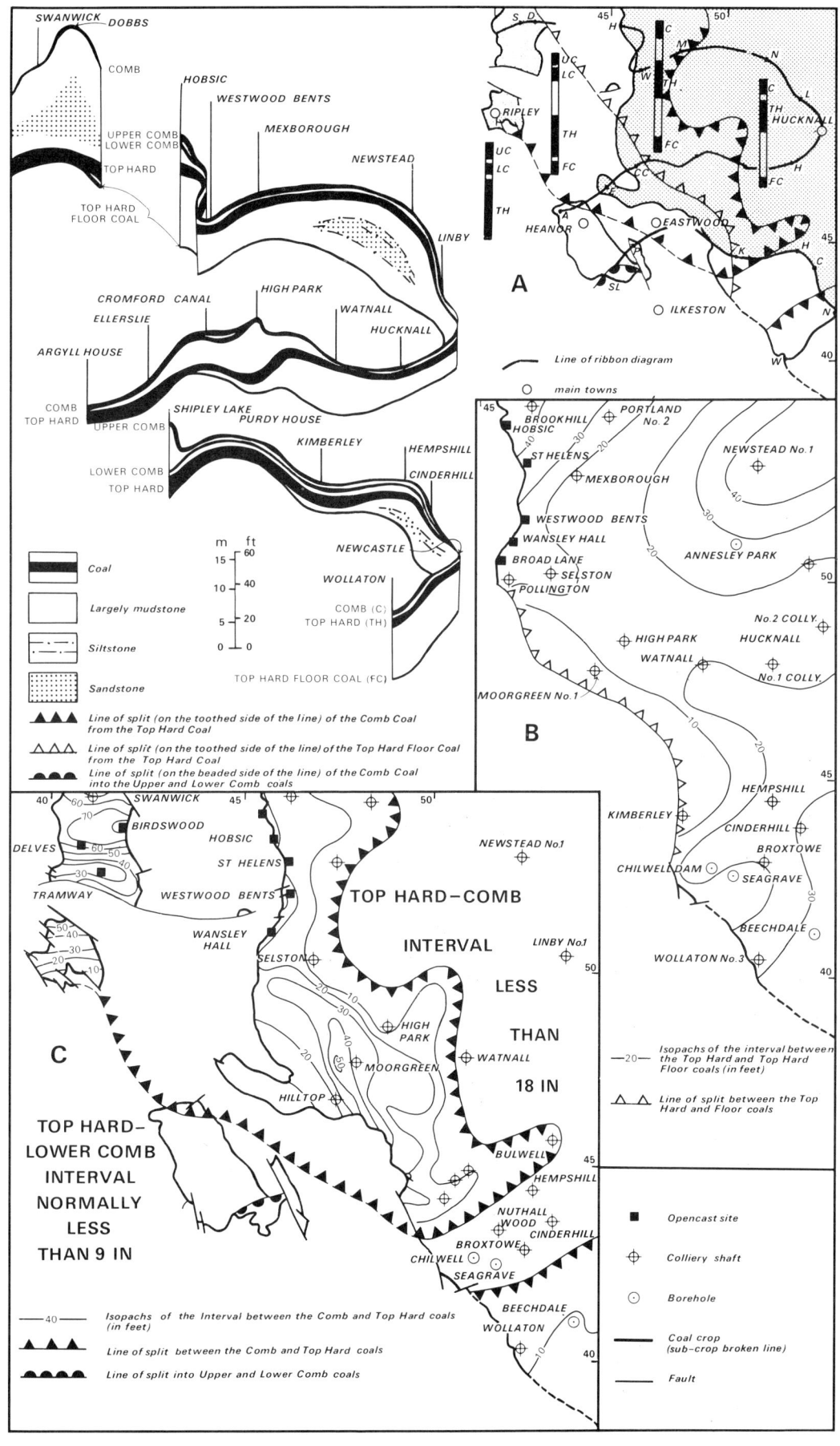

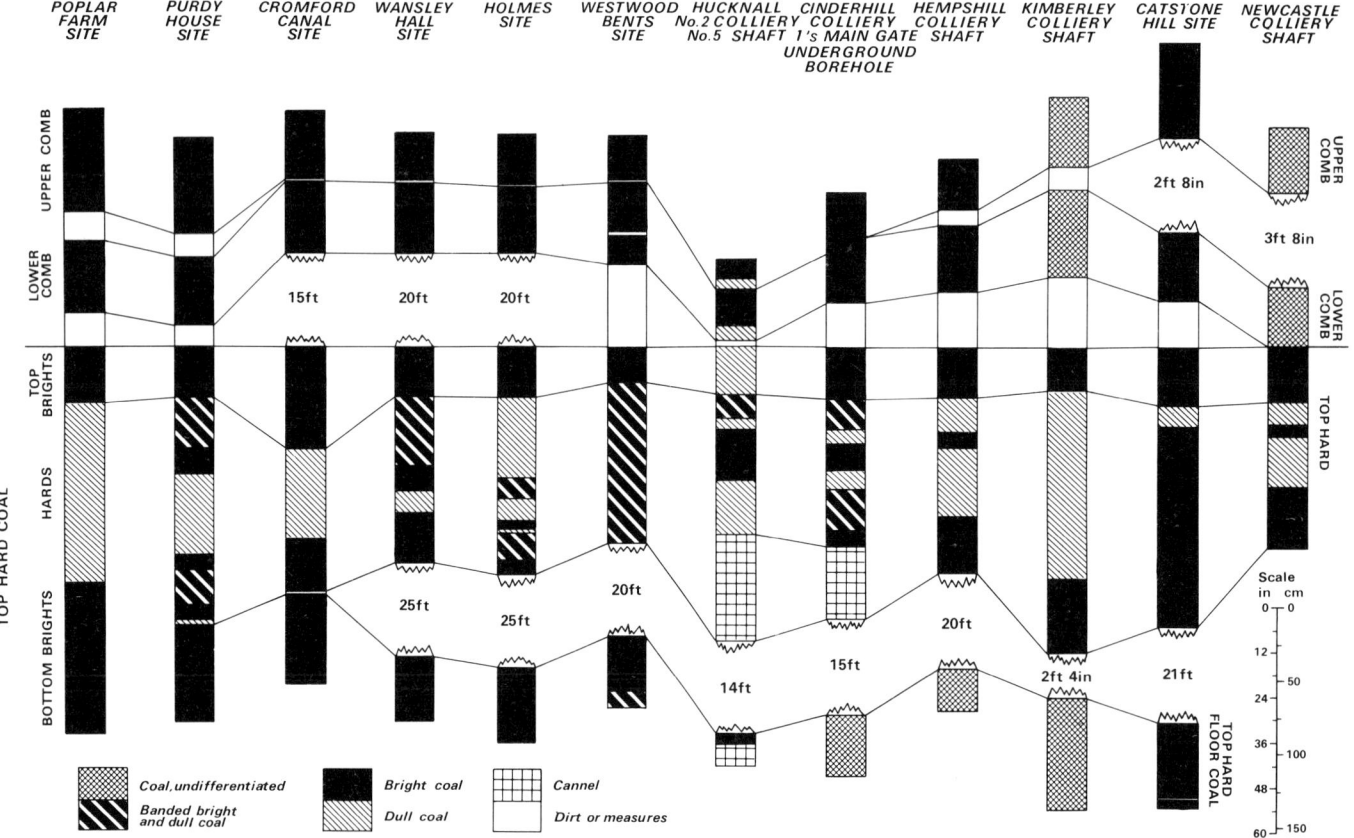

Figure 40 Sections showing the developments of the Top Hard Floor, Top Hard and Comb coals

removed leaving the 'brights' as a roof and floor; these remnants have been exploited by opencast means in recent years and are still eagerly sought.

Variations in thickness of the Top Hard are shown in Figure 41. In Hucknall Nos. 1 and 2 collieries there is cannel at the base of the seam, 27 in thick in No. 2 Colliery, No. 5 Shaft. According to Elliott (1965, pp. 133–141) the cannel was formed in a pool flanking a 'swilley' which is present to the east of this district.

The *Comb Coal* in the Heanor outlier is usually less than 9 in above the Top Hard, but locally as much as 15 in. The seam contains a dirt parting which expands gradually southwards and is 3 ft thick at the line of split shown on Figure 41. In Shipley Lake Site the dirt has passed into argillaceous measures about 40 ft thick. Cromford Canal Site exposed a 40-ft trunk of *Lepidodendron* in position of growth, with its base in the Top Hard Coal and its top close to the base of the Comb seam (Plate 6).

The measures between the Comb and Mainsmut coals range in thickness from 21 to 71 ft (Figure 75). The Comb Coal is closely succeeded over wide areas by sandstone or interbedded sandstone, siltstone and mudstone, locally referred to as the 'Top Hard Rock'. This is thickest (over 50 ft) in the Hucknall area, and over much of the exposed coalfield forms a pronounced feature or is part of the

a composite feature upon which Swanwick, Ripley and Heanor are sited.

The Mainsmut Coal in the type area around Shipley, Kimberley and Watnall, is generally 24 to 36 in thick (Figure 75). A maximum of 39 in was recorded at Shipley Park Borehole, where the seam was composed of two 17-in bands of bright coal separated by 5 in of 'hards'. North-eastwards the seam splits into two leaves, up to 27 ft apart, and deteriorates both in quality and thickness. The ash content of the seam in the south and west is usually moderately high (7 to 10 per cent), but in the north and east it is very high (29.8 per cent at Delves Site near Swanwick). The sulphur content is typically low.

The Mainsmut Coal is unusual because of the presence of inclusions within the upper part of the seam. They vary from sandy laminae to lenses of quartzitic sandstone.

The measures between the Mainsmut and Cinderhill coals range in thickness from 41 to 106 ft and contain, in the areas of thickest sedimentation, several cycles and two thin coals (up to 9 in thick) or seatearths. These coals may equate with the Second St John's Coal of the Chesterfield district to the north.

Sandstone, up to 50 ft thick, is common and forms good features in the Pinxton – Brinsley area. Mussels are recorded

Figure 41 Plans of the Top Hard and Comb coals

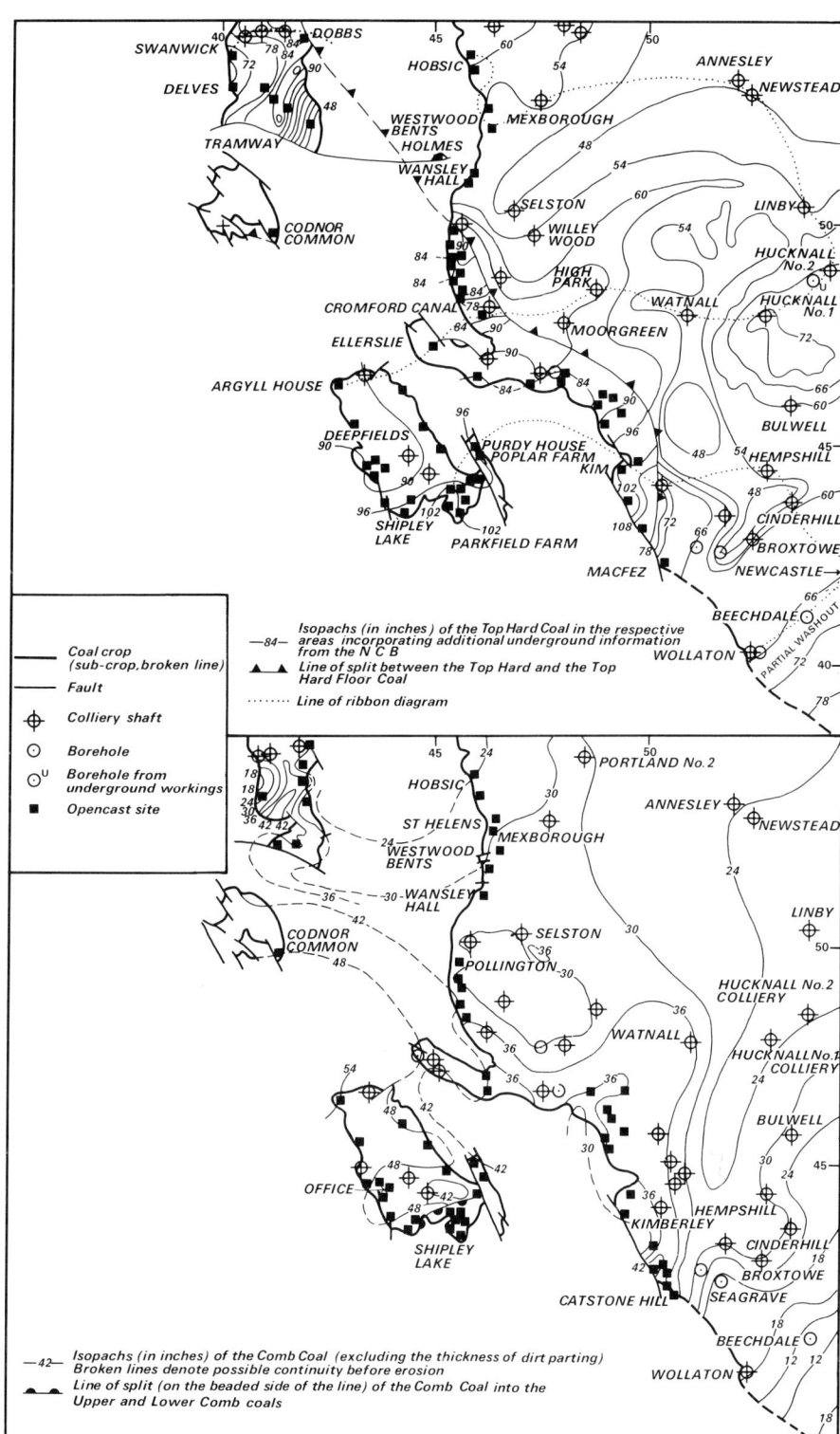

at various horizons, notably some 30 and 60 ft above the Mainsmut Coal. They are usually found either in black shale or preserved in ironstone nodules or bands.

The Cinderhill Coal, sometimes referred to as the Cinderhill Main, is 44 in thick in Cinderhill Colliery No. 4 Shaft and consistently a little under 4 ft in the surrounding area. It was mined in the 1920's on both sides of the Cinderhill Fault.

This belt of thick coal trends west-north-westwards through Nuthall and Kimberley and has been worked opencast in synclinal outliers south of Heanor, at Poplar Farm and Johnson Farm sites (Figure 76). In these thick sections the coal is composed essentially of 'brights' with banded dull and bright coal forming a thin parting in the middle of the seam. Ash content is low to moderate and sulphur percentages are moderate to moderately high.

Plate 6
Fossil tree at Cromford Canal
Opencast Site

Lepidodendron trunk 26 ft high,
standing nearly upright upon the
Top Hard Coal and truncated at or
near the base of the Comb Coal.
(L 923)

North-east of a line through Strelley, Kimberley and Eastwood, the Cinderhill splits and may be represented to the north and east by up to three coals (Figure 76B) i.e. an upward sequence of Cinderhill Coal, an Upper Leaf and a split from the Upper Leaf.

Smith and others (1967, p. 174) state that the Cinderhill Coal cannot be recognised with certainty in the shaft sections east of Alfreton, but infer from its position relative to the High Hazles that it may be the correlative of the First St John's Coal of districts to the north.

The measures between the Cinderhill and High Hazles coals vary between 105 ft in the south to 48 ft in the north, although leaves of the two coals are only 27 ft apart at Moorgreen Colliery. The succession in the areas of maximum deposition is dominated by sandstone, which is over 80 ft thick at Hempshill Colliery. At Newstead this sandstone was recorded as 'exuding grease'.

Figure 42 Plan of the High Hazles Coal with representative sections of the seam

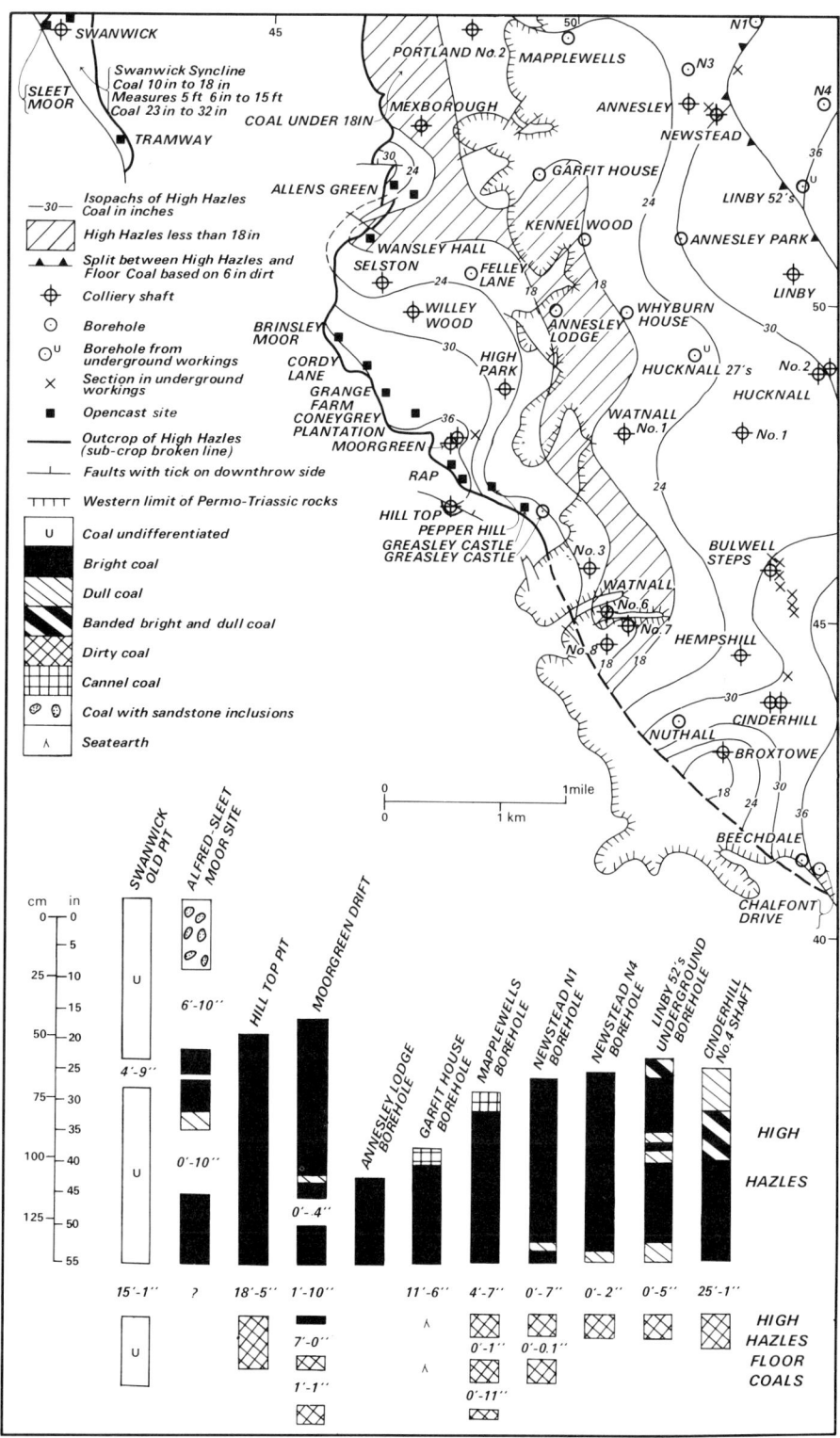

The High Hazles Coal is an economically important seam which has been worked in the past from Linby, Bulwell and Cinderhill (Babbington) collieries. It has promising reserves in the north-east of the district, where it is well documented from over 50 provings. The main seam has a maximum thickness of 36 inches in the north-east, but it splits and thins to the south and west beyond Hucknall and Annesley (Figure 42).

The High Hazles Coal consists largely of bright coal, but both the top and bottom of the seam are dirty. The ash content varies considerably; it is lowest in the areas of thickest coal at Linby (3.7 per cent) and Allens Green Site (2.9 per cent) and high to very high in the Swanwick Syncline. Sulphur content is moderately low.

The Brinsley Thin Coal is separated from the High Hazles Coal by largely argillaceous measures ranging in thickness from 20 ft at Hempshill Colliery to 45 ft at Portland Colliery No. 2 Shaft. In the type area near Brinsley village, 2 miles N of Eastwood, the Brinsley Thin consists of up to 48 in of coal and dirt, but its average thickness is 2 ft at outcrop between Selston and Watnall and it decreases to under 6 inches in the north-east of the district (Figure 77). In this north-eastern area a thin coal, the Brinsley Thin (Upper Leaf), is present up to 16 ft above the main seam.

The Brinsley Thin Coal is variable in quality with ash content ranging from good to very inferior (3.6 to 27.0 per cent). Sulphur content varies from low to moderately high (0.5 to 2.4 per cent). It has never been an economic proposition for deep mining.

The measures between the Brinsley Thin (Upper Leaf) and Lowbright Floor coals range in thickness from 50 ft at Selston Colliery in the north-west to 25 ft at Hempshill Colliery in the south-east. They comprise largely argillaceous strata containing sporadic mussels including *Naiadites* cf. *obliquus*, particularly common in the roof of the Brinsley Thin, together with plant fragments and *Planolites*. The mudstones contain ironstone bands up to 8 in thick.

The Lowbright (Furnace or Abdy) Coal is over 30 in thick in much of the district, reaching a maximum of 46 inches in the south-west at Lambclose House Site (Figure 43). It has been worked underground to a limited extent at Newstead, Watnall and Cinderhill (Babbington) collieries, but has been extensively exploited by opencast workings between Pinxton and Kimberley.

The Lowbright is composed largely of 'brights' with a persistent band of 'hards' about the middle of the seam, where the splits tend to occur.

The ash content varies from very low (2.5 per cent) in the Hucknall–Watnall area to high (over 15 per cent) in the Portland area. Sulphur percentages range from low to moderate.

The *Floor Coal* usually occurs less than 10 ft beneath the Lowbright. In the north-east of the district it closely approaches the main seam and in Newstead N1 Underground Borehole a separation of only 5 in is recorded.

The measures between the Lowbright and Two-Foot coals consist predominantly of interlaminated and interbedded sandstone and siltstone and are thickest, up to 44 ft, in Kennel Wood Borehole. The sandstone occurring at this locality is unusual. It contains two thin breccia horizons made up of siltstone fragments and ferruginous nodules, and shows many sedimentary structures such as current ripples, disturbed laminae, crumpled bedding and rib and furrow development. The sequence is thinnest (26 ft) in the vicinity of Selston Colliery, where the roof measures of the Lowbright Coal contain mussels.

The Two-Foot (Sough or Middlebright) Coal exceeds 18 in over much of the district, but varies from 29 in at Rap Site in the south-west to about 6 inches in the extreme north-east (Figure 78). In the Swanwick Syncline it averages 24 in. The seam is of little economic value for deep mining, but its close proximity to the Mainbright Coal along most of the outcrop has led to the extraction of both seams from the same opencast workings. The seam consists largely of 'brights', but many analyses show the upper half to be rather dirty. The ash content varies widely, from low to very high in under two miles, and sulphur percentages similarly range from low to very high.

A thin floor coal, 2 to 9 in thick, is present some 2 to 10 ft below the Two-Foot between Cinderhill and Selston.

The measures between the Two-Foot and Mainbright coals range between 19 ft in the east and 10 ft in the west. The Two-Foot Coal is invariably overlain by dark grey mudstone—'black bind' in the old shaft records. It is slightly silty, rather fissile and contains ferruginous laminae, some with a speckled appearance, possibly due to kaolinisation, whilst others are oolitic. The mudstone is richly fossiliferous, containing fish debris, '*Estheria*' *sp.*, mussels and ostracods, together with pyritised and carbonised plant fragments.

The *Two-Foot Marine Band* forms a reliable marker horizon and has been proved to be of wide lateral extent throughout the district. It comprises that part of the above-mentioned fossiliferous mudstones which contains *Lingula* and foraminifera, commencing immediately or close above the Two-Foot Coal and extending up to 5 ft 2 in above.

The dark grey mudstones in the lower half of this cycle pass up into silty mudstones with sporadic siltstone laminae containing a few mussels including *Anthraconaia cymbula* and scattered plant fragments.

The Mainbright Coal is an economically important seam, which has been worked from Hucknall, Watnall, Bulwell, Hempshill and Cinderhill collieries, but its best reserves lie beyond the eastern limits of the present district. The type section of the seam is at Hucknall No. 2 Colliery, No. 5 Shaft, where it consists of 34 in of bright coal containing a 5½-in parting of banded 'dull' and 'brights'.

In the main area of outcrop, the Mainbright Coal is thickest in the south and west with a maximum of 49 in recorded in a temporary section near Broxtowe Wood (Figure 44). It thins northwards to 36 inches in the Hucknall–Watnall–Selston area and to 24 in near Annesley Park and Mexborough. The coal deteriorates in the north-east corner of the district, where it splits into thin seams with dirt partings or is represented by cannel and inferior coal as at Newstead.

The seam generally is of good quality with low ash and sulphur contents. It is composed largely of 'brights', with subsidiary partings of dull or banded bright and dull coal throughout (Figure 44).

The measures between the Mainbright Coal and the Mansfield Marine Band vary from 246 ft at Annesley in the north to 185 ft at Cinderhill in the south. They include several thin coals of no economic value such as the Clown and ?Swinton Pottery. Marine bands and marker horizons proved in these measures in the Chesterfield and Ollerton districts to the north tend to be less persistent and only the Haughton Marine Band has been proved with certainty in

Figure 43   Plan of the Lowbright Coal with representative sections of the seam

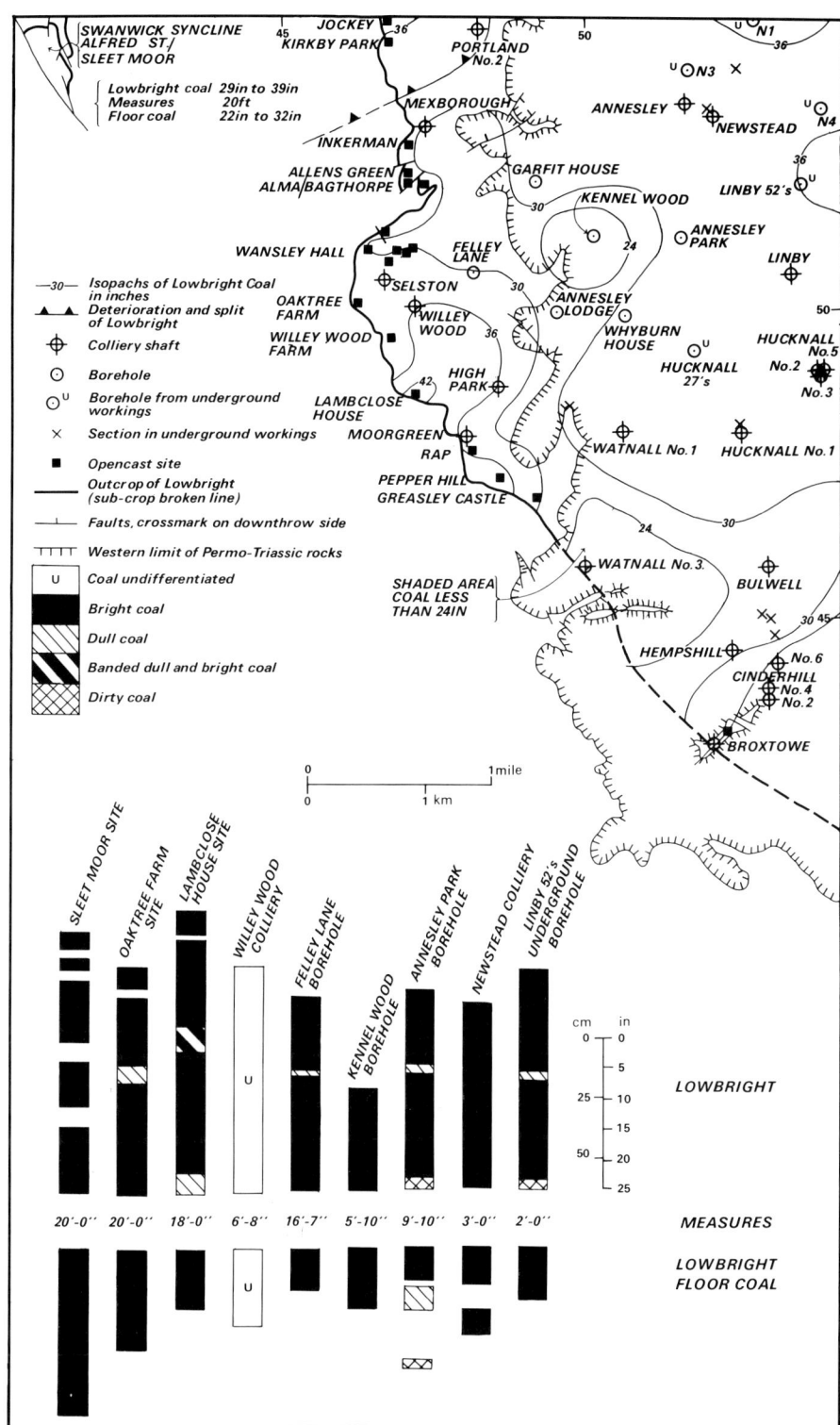

the present district. The measures are rich in ironstone nodules and ferruginous patches; sphaerosiderite is common locally, as in the seatearth of the Clown Coal. Worm burrows are much in evidence and are often lined by ferruginous residues; root-structures in the seatearths are replaced by ironstone.

'*Estheria*' has been found in the roof of the Mainbright Coal in many recent boreholes. The fossil is preserved mostly as irridescent films in a grey mudstone, and is associated with ostracods, small mussels and fish remains.

Figure 44 Plan of the Mainbright Coal with representative sections of the seam

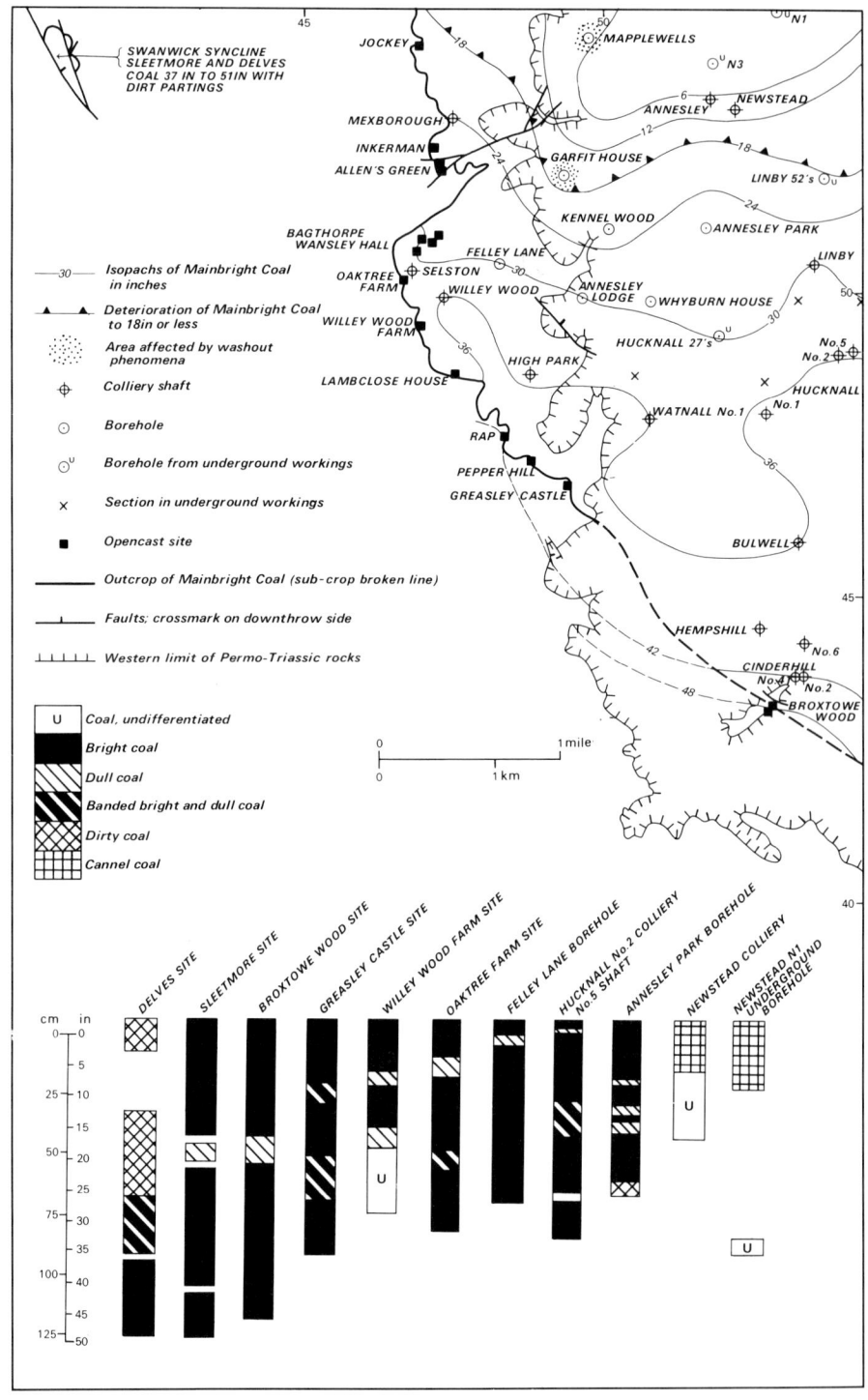

Distorted roof measures have been observed on the Mainbright Coal. Shirley (1955, p. 274) recorded near-vertical 'dykes' of hard sandstone, 2 to 5 ft thick, in roof measures at Bagthorpe Site. He attributed this phenomenon to earthquake tremors, the effects of which he has traced in other parts of the coalfield at this horizon (Shirley *in* Eden and others, 1957, p. 112).

The *Manton 'Estheria' Band* horizon occurs in Annesley Park Borehole, where '*Estheria*' associated with fish scales is present in 8 in of black mudstone.

The strata between the ?Manton 'Estheria' Band and the Clown Coal range from 78 ft in the west at High Park Colliery to 12 ft in the east at Annesley Park Borehole. South of Hucknall the absence of the Manton 'Estheria' Band and presence of washout sandstones at the Clown horizon make correlation uncertain.

The *Clown Coal* is thickest in the west, amounting to 78 inches in Wansley Hall Site, but thinning rapidly eastwards. Beyond Watnall and Newstead it is either washed-out, thin or represented by a seatearth (Figure 45).

The coal commonly contains many dirt partings which render the seam economically worthless.

The dark grey to black mudstone intermittently forming the roof of the Clown Coal contains variable amounts of fish and carbonaceous plant debris in the lowest few inches. In the Chesterfield district to the north, the mudstone also contains large specimens of *Lingula* and forms the Clown Marine Band.

The mudstones above the horizon of the Clown Marine Band contain numerous mussels typical of the *Anthracosia atra* fauna, together with pyritic plant fragments and ferruginous bands showing evidence of boring organisms. Some 20 ft above the Clown Coal in Mapplewells Borehole, such mudstones and siltstones are of a distinctive pale greenish grey colour. This colouring may be penecontemporaneous, for the Permo-Triassic unconformity is some 440 ft above.

There are a number of coals or seatearths between the Clown Coal and the Haughton Marine Band. The thickest seam is 29 in at Cinderhill Colliery No. 4 Shaft. At Hucknall Colliery No. 1 Shaft, a group of coals and dirts has an overall thickness of 139 in. It is not clear whether the first coal or seatearth below the Haughton Marine Band is the Swinton Pottery of districts to the north, or whether several or all of the coal horizons between the Marine Band and the Clown represent a split Swinton Pottery.

The *Haughton Marine Band* consists of dark grey mudstone with marine fossils. It is characterised by gastropods and *Tomaculum sp.* which occur together with *Lingula* and fish debris. The marine band is not seen at outcrop. In Felley Lane Borehole the assemblage included *Euphemites anthracinus*. Bellerophontoid gastropods were also recorded in Kennel Wood Borehole, where *Lingula mytilloides*, *Serpuloides stubblefieldi* and fish debris were found near the base of the marine band. The mudstones are particularly rich in ironstone and pyrite, and listric surfaces, some of which are associated with the zone of compression around ironstone nodules, are common.

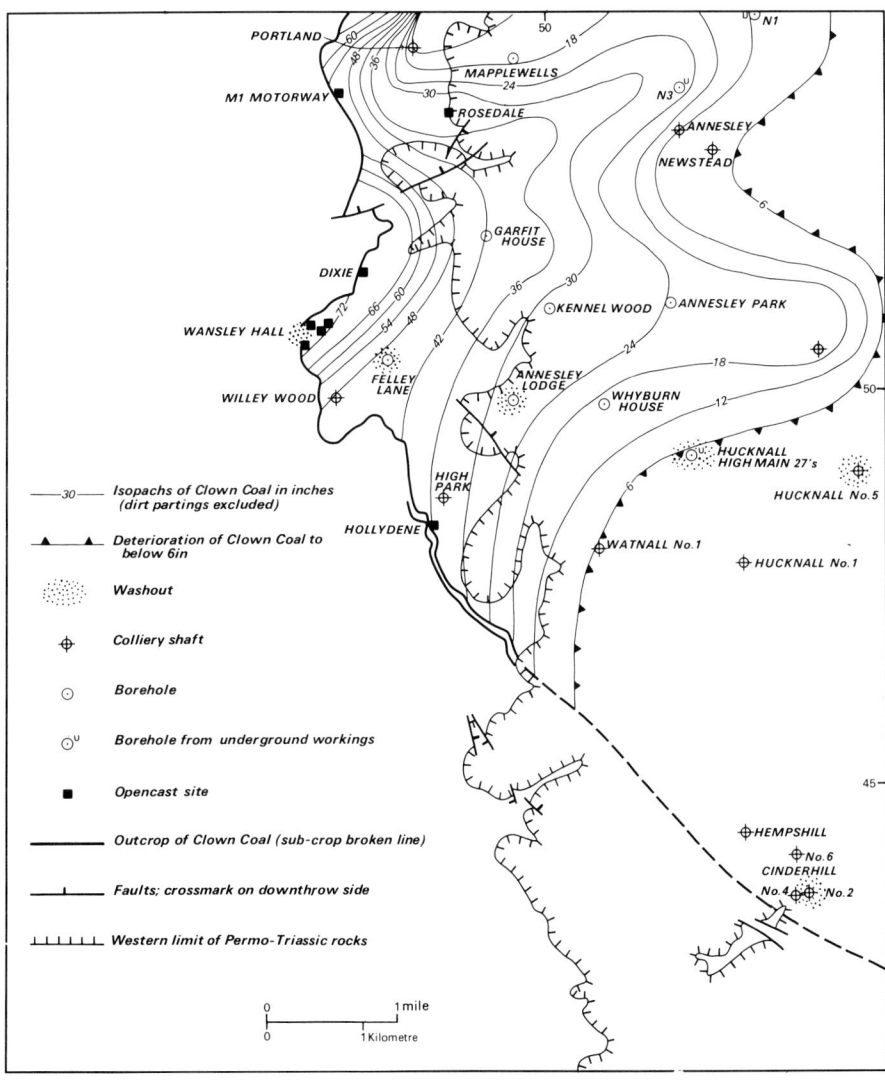

Figure 45 Plan of the Clown Coal

The measures between the Haughton Marine Band and the Mansfield Marine Band comprise two main cycles totalling 57 to 78 ft in thickness. They are largely mudstones with subordinate arenaceous strata; sphaerosiderite is common. The top of the first cycle is marked by a thin, but persistent, coal which has a maximum thickness of 15 in at Cinderhill Colliery No. 4 Shaft.

This coal is overlain by the *?Sutton Marine Band* in Mapplewells Borehole. The marine band may also be present in Hucknall Colliery High Main 27's Underground Borehole, but it is apparently missing over the greater part of the Derby district. At Mapplewells Borehole the possible marine band consists of 9 in of dark mudstone with pyrite and phosphatic nodules, its fauna consisting largely of fish debris.

The top of the cycle containing the Sutton Marine Band is invariably marked by a coal and/or seatearth. The coal is thickest (7 in) in Cinderhill Colliery No. 2 Shaft. This horizon is in places separated from the overlying Mansfield Marine Band by a few inches of mudstone or siltstone with pyritised and coaly plant fragments and non-marine bivalves.

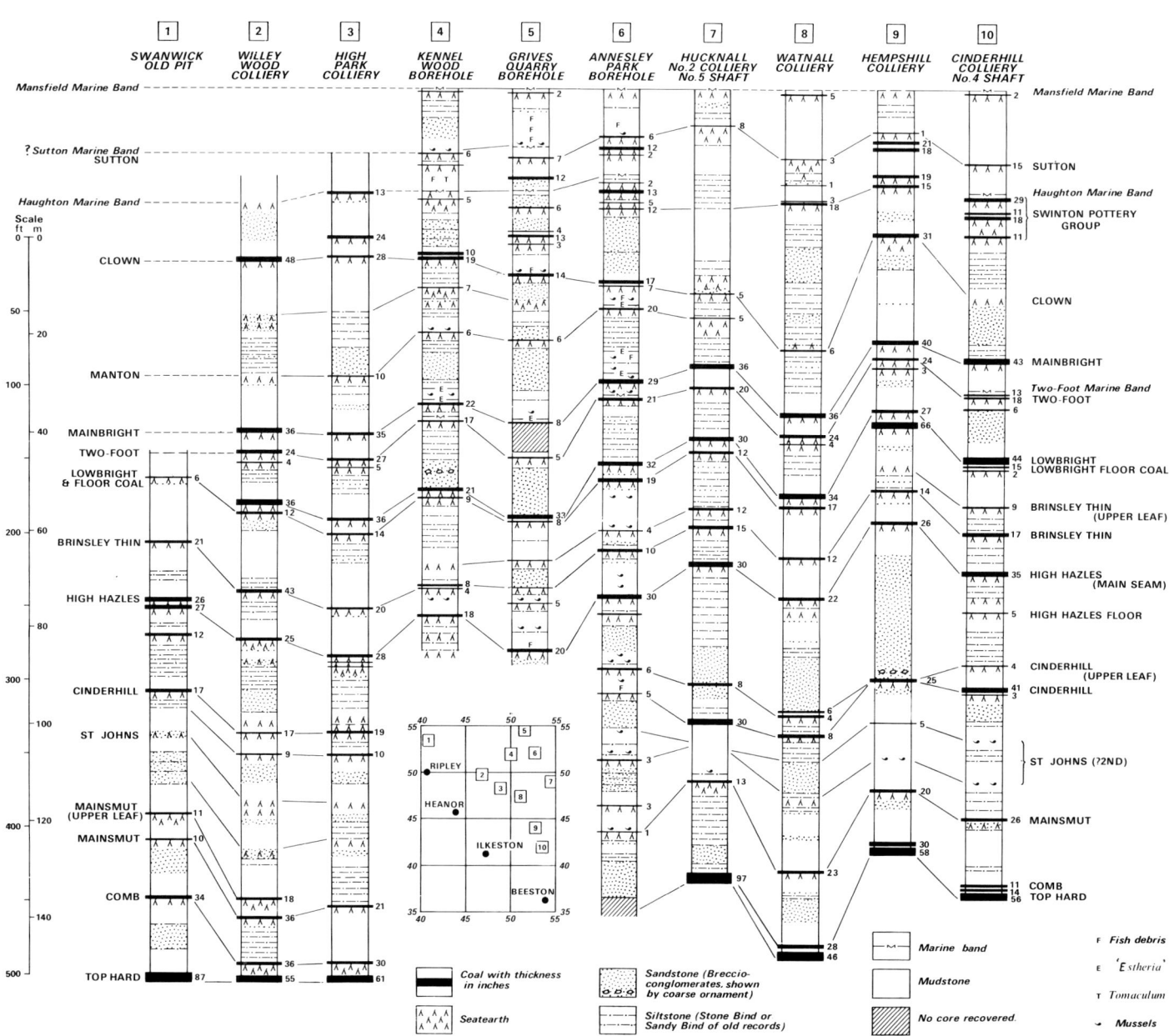

Figure 46 Representative sections of the measures between the Top Hard Coal and the Mansfield Marine Band (Grives Quarry Borehole lies just beyond the northern margin of the district)

Plate 7 Arthropods from the Westphalian

1, 2 *Lepidoderma moyseyi* (H. Woodward). Upper *A. modiolaris* Zone. Clay-ironstones a short distance below the Top Hard Coal, Shipley Clay Pit, 1¼ miles NNW of Ilkeston, Derbyshire.

1, Lectotype, dorsal view, with part of dorsal surface absent revealing metastoma. GSM 30192, × 1 (Figured H. Woodward, 1907, *Geol. Mag.*, pl. xiii, fig. 1).

2, Topotype, dorsal view with ventral surface exposed anterior to third mesosomal segment. GSM 30249, × 1 (Figured L. R. Moore, 1936, *Proc. Geol. Assoc.*, fig. 64).

3 *Belinurus bellulus* König. ?Upper *A. modiolaris* Zone, north-west of Ilkeston. GSM 16203, × 1.

4 *Euproops rotundata* (Prestwich). *A. modiolaris* Zone, ?Bretby Clay Pit, Ilkeston. Za 2832, × 1.

5 *Cyclus sp.* Between Silkstone and Deep Hard coals, Bondsmain Colliery, Temple Normanton, Derbyshire. [1″ Sheet 112, Chesterfield.] GSM 26061, × 2.

6, 7 *Palaeocaris praecursor* (H. Woodward). Upper *A. modiolaris* Zone. Clay-ironstones below Top Hard Coal, Shipley Clay Pit, 1¼ miles NNW of Ilkeston.

6, Lectotype, view of right hand side. GSM 30213, × 2 (Figured H. Woodward, 1908, *Geol. Mag.*, p. 386, fig. 1.).

7, Syntype, dorsal aspect. GSM 30216, × 2 (Figured H. Woodward, ibid, fig. 2).

8 *Pygocephalus [Anthrapalaemon] dubius* (Milne Edwards). Lower *A. modiolaris* Zone, above Piper Coal, Cossall Clay Pit, Nottingham. Za 3056, × 2.

9 *Prothelyphonus britannicus* (Pocock). Upper *A. modiolaris* Zone. Clay-ironstones below Top Hard Coal, Shipley Clay Pit, 1¼ miles NNW of Ilkeston. GSM 30220, × 2. (Figured R. L. Pocock, 1911, *Palaeontogr. Soc. [Monogr.]*, pl. 2, fig. 3).

All specimens in the collections of the Institute of Geological Sciences, Leeds.

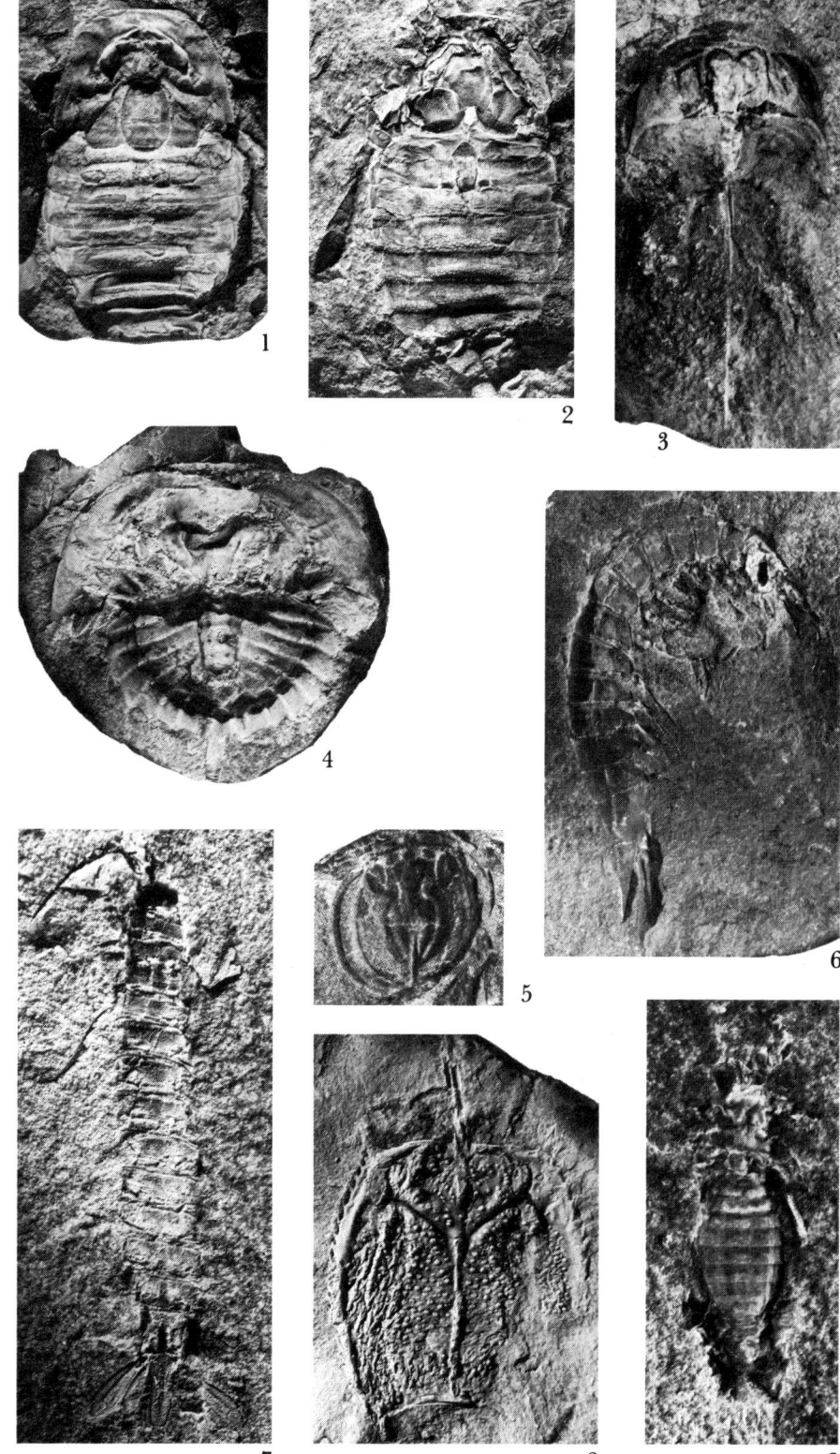

Westphalian C

MANSFIELD MARINE BAND TO TOP MARINE BAND

These strata, some 350 ft thick, contain only one economically important coal, the High Main (Figure 49). They are concealed by Permo-Triassic rocks except for a narrow strip cropping out between Greasley and Pinxton. The general characters of the strata are shown in Figure 47.

Above the Mansfield Marine Band, the mussel fauna is less varied than in the lower beds, with *Naiadites* as the dominant genus. Between the Mansfield Marine Band and the *Edmondia* Band the species include *Naiadites melvillei* and *N.* cf. *productus*. Above the *Edmondia* Band, *N.* cf. *hindi* is common. The lowest occurrence of *Lioestheria vinti* in this part of the sequence is in the roof measures of the first coal above the *Edmondia* Band, and this fossil recurs in several bands up to the Top Marine Band. Ostracods in these measures include *Geisina subarcuata* and *Carbonita pungens*.

The Mansfield Marine Band consists of up to 15 ft of dark grey mudstone with an abundant and varied fauna. A bed of fossiliferous 'cank' (hard ankeritic siltstone) occurs in the lower part and provides a reliable and persistent lithological marker horizon which is recognisable even in the records of old shafts and sinkings. The 'cank', which is 6 to 32 in thick, has a distinctive conchoidal fracture and is irregularly veined by calcite. Its petrography was described by Dunham (*in* Edwards and Stubblefield, 1948, pp. 249–252). Chonetoid brachiopods and small *Lingula* are commonly present (see also p. 147c).

The measures between the Mansfield Marine Band and High Main Coal vary from 158 ft in the north-west to 115 ft in the south-east, and contain up to four coals. The measures crop out just to the west of the Permo-Triassic escarpment in the Annesley Woodhouse – Selston Underwood areas, but are concealed by the younger rocks south of Watnall.

The variations of thickness in the measures are related to the presence of sandstones, particularly in the lowest and highest cycles.

Eden and others (1963, p. 54) record the sporadic occurrence of a kaolinite-rich band below the High Main Coal in various East Midlands localities. Boreholes indicate that this bed is widespread in the present district also, but because of the difficulty of differentiating such tonstein-like bands from normal ironstones and ferruginous mudstones it may have been missed in many of the records.

The measures between the Mansfield Marine Band and the High Main Coal are rich in ferruginous concretions, and sphaerosiderite is particularly common in the seatearths. Ironstone bands are locally oolitic and often have kaolin-filled centres. Kaolin patches and spots have been recorded below all the minor coal seams within these measures. The lowest of these minor coals is considered to be the probable equivalent of the Wales Coal of the East Retford district (Smith and others, 1973, p. 88). It is not known whether the overlying coals are splits from the Wales or represent higher seams including the Sharlston Low of Yorkshire.

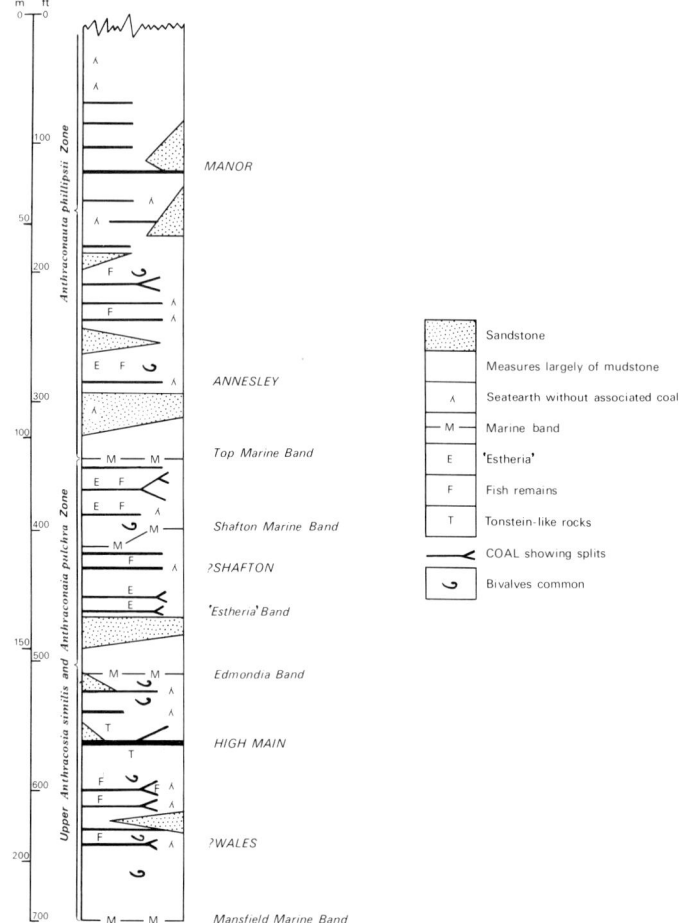

Figure 47 Generalised section of the strata above the Mansfield Marine Band (Westphalian C)

The *?Wales Coal* is about 12 in thick in the north-west of the district, where it lies some 85 ft above the Mansfield Marine Band. Traced south-eastwards this interval is halved and the seam deteriorates until, in the Cinderhill area, it is represented solely by seatearth. Analysis of this coal obtained from Newstead N1 Underground Borehole, just north of the district boundary, shows a medium ash content and low sulphur percentage.

The thickest single seam recorded in the measures up to the High Main Coal is 23 in at Hucknall Colliery No. 2 Pit, No. 3 Shaft, in the second cycle above the Mansfield Marine Band. Analysis of this seam from Babbington Pit Yard Boring showed poor quality (13 per cent ash) and high sulphur content (3.6 per cent).

The highest coal in these measures occupies a fairly persistent position some 40 ft below the High Main. An analysis from Linby 36's Slane Underground Borehole showed it to consist of bright coal which becomes inferior towards the middle and base of the seam. Overall quality is poor, with an ash content of 7.7 per cent and 0.93 per cent of sulphur.

Fish scales and associated debris have been recorded above this coal.

Sporadic mussels occur in the measures between the Mansfield Marine Band and the High Main including *Naiadites melvillei* and *N.* cf. *productus*.

The seatearth below the High Main Coal and most of the seatearths above this horizon have a distinctive grey-green, creamy grey or pale brown coloration. Similar colours commonly occur in the stained measures below the Permo-Triassic; those found in the High Main seatearth are some 300 ft below the stained zone in this district and are considered to be a reflection of the changing conditions of deposition in the upper part of the Westphalian sequence.

The High Main Coal is commonly 4 ft thick, but ranges from 43 to 62 in (Figure 48). It is mined in Newstead and Linby collieries at depths ranging from the 150-ft cover line in the west to 500 ft in the south and east. The seam consists of two major divisions (Figure 48)—a thin upper 'brights' and a thick lower 'brights' separated by a variable but generally thin dull coal or 'bluestone'. Between Annesley Woodhouse and Greasley the seam is split above the 'bluestone' over a zone some 600 to 1200 yd wide, comparable in plan with a meandering north–south channel (Figure 48). A detailed section across the split zone is afforded by Brookshill Farm Site, where some 200 trial boreholes were drilled. Figure 79 shows isopachs of the measures, largely seatearth and mudstone, separating the two leaves. A longitudinal section A–B illustrates the present configuration of the coals. The lower 'brights' have been partially washed out in only one borehole (Whyburn South) and the upper 'brights' are fairly uniform in thickness.

If it is assumed that the configuration of the coals in section A–B is largely of penecontemporaneous origin, then at the end of the formation of the High Main Coal, when the upper leaf should have been nearly horizontal, the lower 'brights' seam would have the profile a–b. Such a profile suggests a NE–SW belt of rapid subsidence along a pre-existing channel and/or tectonic line.

The quality of the coal varies from good to rather poor with ash content, less dirt, ranging between 4.6 and 9.6 per cent. Sulphur content is low.

Eden and others (1963) recorded a tonstein within the High Main Coal at several widely spaced localities in the East Midlands Coalfield. In the present district no definite occurrences are known.

The measures between the High Main Coal and Edmondia Band are some 30 to 40 ft thick and consist largely of mudstone though a washout siltstone, some 23 ft thick, overlies the High Main Coal in the Whyburn South Borehole and a 10-ft sandstone overlies the same seam at crop in the Beauvale Abbey area [490 490]. They contain up to two thin coals. The lower coal, present only in the south-east of the district, where the two cycles are clearly defined in the Linby, Hucknall and Watnall areas, is possibly a split from the High Main Coal. This coal lies 10 to 20 ft below the higher coal, which is up to 11 in thick. Mussels occur locally above both coals, and they lie generally 5 ft, but up to 17 ft, below the *Edmondia* Band.

The Edmondia Band has been found in 33 boreholes and shafts in the present district. It ranges in thickness from 2 to 24 ft, being thickest in a restricted belt of country which corresponds closely to the zone of split in the High Main Coal (Figure 48). This belt seems to have been liable to increased subsidence intermittently throughout this short period of geological time. Edwards (1967, p. 110) noted that the Band included a distinctive pale grey mudstone in the Ollerton area. Whilst pale mudstones are present within the Band in this district they are variable in both thickness and position. Foraminifera are found in both pale and dark grey mudstones and were particularly well preserved in the Whyburn House Borehole.

In Annesley Lodge Borehole the *Edmondia* Band contained abundant foraminifera including *Glomospira*, *Glomospirella* and *Tolypammina*. In the middle of the Band *Myalina* and *Hollinella* occur, an assemblage typical of the *Myalina* facies which is the usual development shown by this Band throughout the Yorkshire and East Midlands Coalfield. In Kennel Wood Borehole *Curvirimula* was found in association with *Myalina* cf. *compressa*. *Curvirimula* does not occur above the top of the *C. communis* Zone except in the retreat phases of some of the Westphalian B marine bands (Calver, 1968a, p. 151). The basal part of the Band contains cf. *Planolites ophthalmoides* and pyritised strap-like markings (fucoids). Thin bands and small nodules of ironstone occur throughout and pyrite is common as grains, clusters, concretions, coatings and infillings and replaces many of the plant fragments, worm burrows and fossils. The *Edmondia* Band contains galena in Misk Farm, Whyburn South and Kitty's Wood boreholes.

The measures between the Edmondia Band and the Shafton Marine Band are normally about 100 ft thick, though a maximum of 156 ft was recorded at Cavendish Crescent Borehole. They include up to six variable coals. The measures are particularly sandy and are broadly equivalent to the Mexborough Rock of south Yorkshire. They are nowhere exposed, being almost wholly concealed beneath the Permo-Triassic rocks.

The lowest cycle, commonly arenaceous and averaging 50 ft in thickness, is normally topped by a thin coal. The two succeeding thin cycles contain thin coals which are overlain by mudstones with *Lioestheria*.

The lower of these bands, which may be the "Main '*Estheria*' Band" of Edwards and Stubblefield (1948, p. 231), consists of 1 in to several feet of grey mudstone. It is rich in *Lioestheria* sp. together with fish debris and ostracods. Pyrite is common, often replacing plant and shell fragments and infilling worm burrows. The carapaces are distinctive, appearing as abundant irridescent blotches in the mudstone. The size of the carapaces both within the band and throughout the district is variable. The coal underlying this "Main '*Estheria*' Band" has been appropriately called the 'Bug Coal' by miners in the Ollerton area (Edwards, 1951, p. 110).

The higher of the *Lioestheria*-bearing bands up to 23 ft 5 in above the lower, is from 3 in to a few feet thick, and is restricted in its distribution to the areas where the High Main Coal is split and the *Edmondia* Band is thick. Whatever the cause of this localised increase in thickness of sediment (p. 148c), it affects over 100 ft of strata.

Figure 48 Plan of the High Main Coal with representative sections of the seam

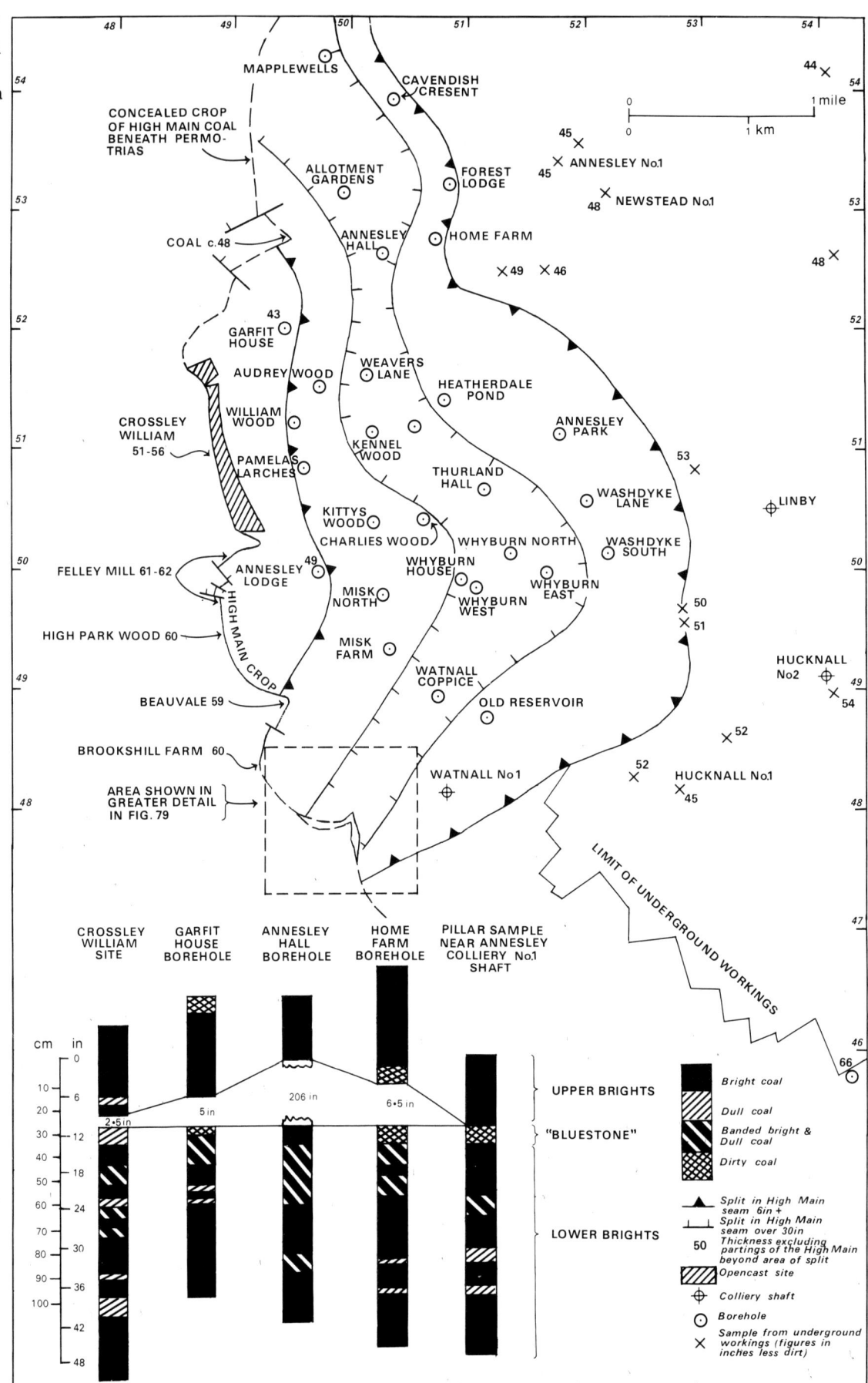

The first cycle above the higher *Lioestheria* band is usually between 20 and 30 ft thick with a seatearth and/or coal, up to 25 in thick, at the top. This coal, unnamed in the Nottingham area, may be the Shafton of south Yorkshire, though Edwards (1967, p. 111) correlates the coal close below the Shafton Marine Band in the Ollerton district with the Shafton Coal. In this memoir the seam is referred to as the ?Shafton Coal (Figure 80).

The cycle succeeding the ?Shafton ranges from 5 ft 8 inches in Annesley Hall Borehole to 25 ft 10 inches in Kitty's Wood Borehole. The cycle is topped by a coal which underlies the Shafton Marine Band and is apparently the equivalent of the seam in a comparable position in south Yorkshire, though evidence from the Ollerton district (see above) raises the possibility that it is the Shafton or a leaf of that seam. This coal is thickest in the Annesley Park area (Figure 80) and thins northwards, passing into batt and cannel in Mapplewells Borehole. The overall quality is poor with high ash and sulphur contents.

The Shafton Marine Band, recorded at 15 localities within the district, ranges in thickness from 8 in to 11 ft 7 in. The fauna includes *Anthraconaia spathulata*, ostracods such as *Paraparchites*, fish fragments, fucoids and worm burrows. *Lingula* has been recorded in three boreholes. The band usually closely overlies a coal or seatearth (see above), but may be separated by up to 14 ft 8 in of measures.

The measures between the Shafton Marine Band and Top Marine Band vary from 46 to 64 ft and contain up to three cycles with locally associated coals. Mussels occur sporadically in all three cycles, but are most common in the lowest, where they include *Naiadites daviesi*.

The Top Marine Band is a valuable and persistent marker horizon which has been proved throughout the district from Annesley Woodhouse in the north to Bulwell in the south. It usually comprises 2 to 7 ft of grey to dark grey mudstone containing a varied fauna of foraminifera, goniatites, brachiopods including *Lingula*, productoids, and *Dunbarella* gastropods, ostracods, fish, fucoids, *Planolites* and *Lioestheria*. Pyrite is common, replacing and coating plant and organic debris.

MEASURES ABOVE THE TOP MARINE BAND (UPPER COAL MEASURES)

Some 500 ft of measures above the Top Marine Band are calculated to have been preserved in the extreme north-eastern part of the district, but westwards they are overstepped by the Permo-Triassic rocks. The maximum thickness proved is 350 ft at Linby Colliery. In the past 10 years many boreholes have been drilled by the National Coal Board in the east of the district, but these measures are rarely cored because they lack workable coals.

The measures comprise two thick and persistent cycles near the base, overlain by many thinner and variable cycles. One of the highest cycles, some 220 ft above the Top Marine Band, contains a coal about 2 ft thick known as the Manor Coal (Edwards, 1951, p. 75).

The measures are commonly arenaceous and mussels occur only sporadically in the argillaceous beds. A change in the mussel fauna takes place near the base of the Upper Coal Measures with the loss of *Naiadites* and the incoming of *Anthraconauta*. The dominant form is *A. phillipsii*.

The Top Marine Band cycle reaches a maximum thickness of 78 ft in Washdyke North Borehole and is the thickest single cycle in Westphalian C. Sandstone and siltstone, containing scattered plant fragments, form a large proportion of the total, but a 15-in seatearth was recorded in Washdyke North Borehole some 19 ft below the top of the cycle. A coal, here named the *Annesley Coal*, is commonly present at the top of the cycle. It is 9 to 16 in thick in a localised east–west belt between Annesley and Watnall, but elsewhere variations are considerable. It is represented by a ganister at Linby and a cannel at Hucknall.

The Annesley Coal is overlain by dark grey mudstones containing *Lioestheria* and fish debris. Ostracods and phosphatic fragments have also been recorded. Mussels, including *Anthraconauta phillipsii*, are also present in the lower part of this cycle. It may well be that the Annesley Coal is the correlative of the Scofton Coal of the East Retford district, which is similarly overlain by mudstones with '*Estheria*', mussels and ostracods (Smith and others, 1973, p. 94).

The second cycle above the Top Marine Band ranges in thickness from 42 ft at Springfield Hosiery Borehole to 56 ft at Annesley Park Borehole. Above it there is a complex series of seatearths, coals, batts and carbonaceous shales spread over up to 40 ft of measures. One of these coals is 30 in thick in Hucknall No. 2 Colliery, No. 5 Shaft.

Fish fragments were recorded at two horizons between these coals at Annesley Park Borehole and at one horizon in Washdyke North Borehole. Ostracods are also present in both boreholes but at slightly different levels. Sphaerosiderite is a common constituent of the seatearths in Springfield Hosiery, Washdyke North and Whyburn East boreholes.

Above these coals are siltstones and sandstones 30 to 90 ft thick. They are succeeded by a group of eight minor impersistent cycles, which in Linby Colliery No. 1 Shaft comprise 120 ft of measures.

The *Manor Coal* at the base of this group, reaches a maximum of 27 inches in Hucknall No. 2 Colliery, No. 5 Shaft, where it is overlain by 36 in of cannel. In the Hucknall–Linby area the coal is generally over 2 ft thick but it thins northwards to 21 in at Newstead.

At Linby Colliery the Manor Coal is overlain by 110 ft of largely argillaceous strata, extending up to the unconformable Permo-Triassic cover. Sandstones are common in the Newstead and Hucknall areas. Primary red measures have not been recorded in the district.

STAINED MEASURES

The Westphalian strata underlying the Permo-Triassic rocks (Figure 49) are normally stained to about 50 ft below the unconformity, but range from 10 ft at Annesley and

Hucknall collieries to 115 ft at Whyburn East Borehole. Red, brown, pink, purple and green coloration is common in the mudstones and sandstones.

Seatearths show purple mottling, although in the measures above the Mansfield Marine Band they are commonly a pale grey or cream colour, contrasting with the darker seatearths of lower horizons (p. 24). Ironstone nodules, normally pale brown or buff, are markedly reddened. Sandstones often show staining to greater depths than do other lithologies. Staining is considered to be due to subaerial weathering and oxidation of the Westphalian land surface in late Carboniferous and possibly Lower Permian times (Anderson and Dunham, 1953; Trotter, 1953; Smith and others, 1967). Because argillaceous rocks are as highly coloured as other

lithologies, percolating solutions are probably less important than weathering, though the increased depths of staining in areas of porous strata would suggest that the former process took place penecontemporaneously with the denudation. Where the Permo-Triassic rocks are present, staining affects measures above the Mansfield Marine Band, but evidence from the west of the present district (p. 26) shows that the lowest sandstone in the Westphalian, i.e. the Crawshaw Sandstone, is similarly stained although the Permo-Triassic strata have long since been eroded.

The measures immediately below the unconformity are often leached to a depth of a few inches.

The approximate positions of the coal crops beneath the Permo-Triassic rocks are shown in Figure 50.

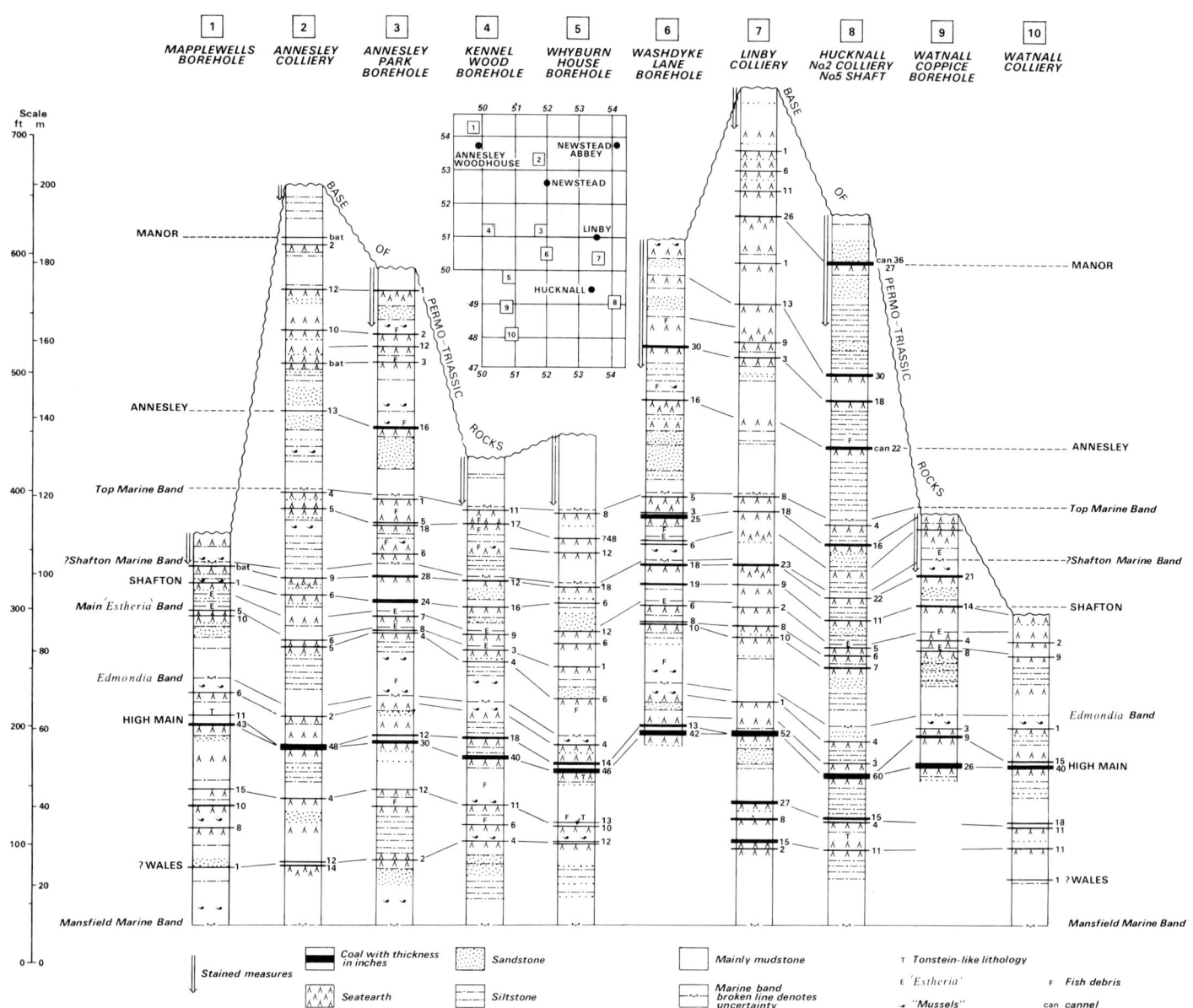

Figure 49   Representative sections of the measures above the Mansfield Marine Band

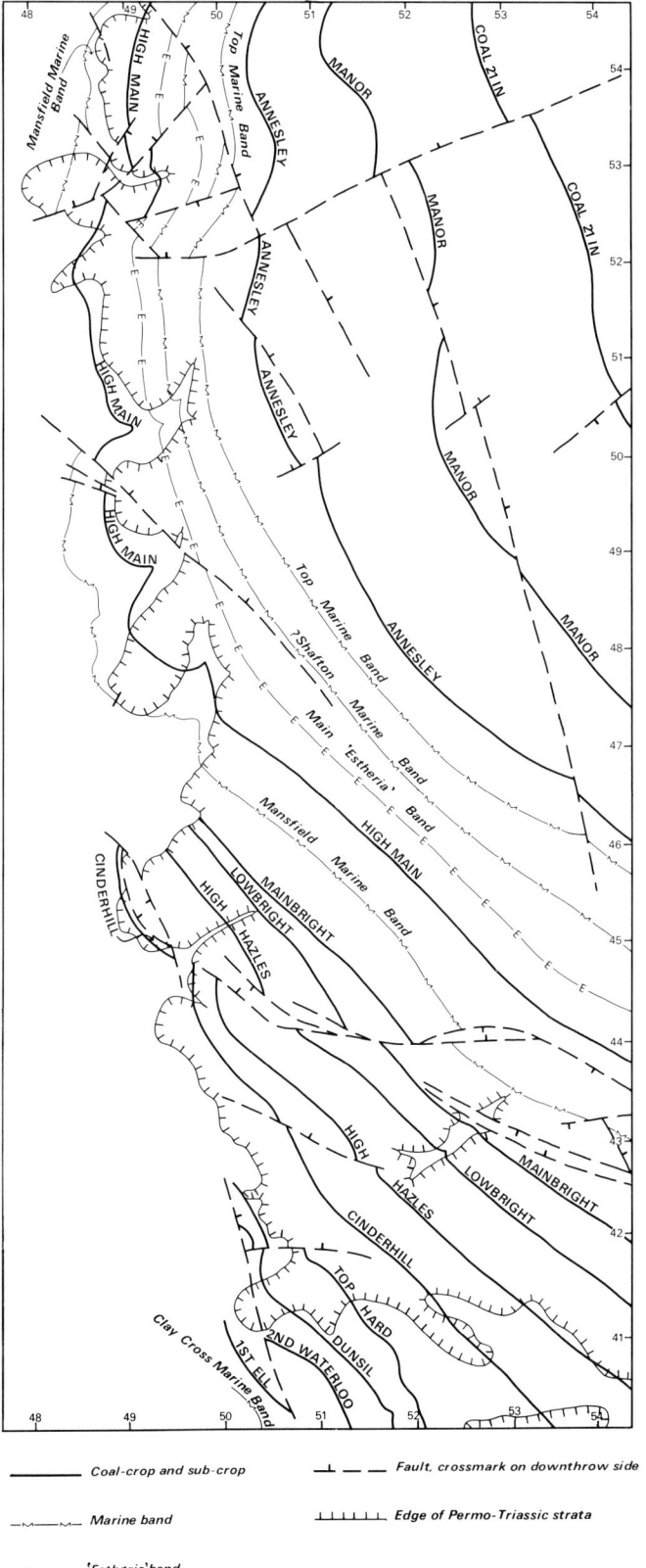

Figure 50 Plan showing the approximate position of the coal crops beneath the Permo-Triassic rocks

CHAPTER 5

Permo-Triassic

INTRODUCTION

Rocks of the Permian and Triassic systems, originally classified by Sedgwick (1829), were considered by early workers such as Aveline (1877, 1879) and Irving (1882) and later by Trechmann (1930) to be separated by an unconformity. This was questioned by Wilson (1876, 1881), who described the main divisions of the succession in north-east England, and Sherlock (1911, 1926, 1947), who suggested that there was a lateral passage between the Permian and the lower part of the Triassic. The usual practice of the Geological Survey since has been to group the two systems together.

Apart from the early papers of Wilson (1876, 1881) and Shipman (1889), Hickling (1906), Vernon (1910), Swinnerton (1910) and Smith (1910, 1912, 1913), the first comprehensive account of the present district was by Sherlock (*in* Gibson and others, 1908) in the memoir accompanying the primary 6-in survey of Sheet 125 (Derby). Further details have been added by Swinnerton (1914, 1918, 1948), Edwards (1951), Elliott (1961), Taylor (1964a and b, 1965, 1966, 1968, 1973), Taylor and Elliott (1971) and Smith (1968).

The strata unconformably overlie the Carboniferous rocks in the south-western, southern and eastern parts of the Derby district. Because of erosion and marked lateral thickness variations the thickest recorded sequence at any one locality is 450 ft near Chilwell, west of Nottingham. Thickness variations are exemplified by the Upper Permian strata, which are 140 ft thick in the north-east of the district and absent 8 miles to the south, near Nottingham. The lithostratigraphical subdivision of the Permo-Triassic sequence is shown in Figure 51, and the relationship between this sequence and the underlying Carboniferous rocks is depicted in Figure 52. Boreholes near Newstead in the north-east of the district proved Lower Marl overlying Westphalian C rocks whereas at Strelley the Lower Marl rests on Westphalian B rocks; at Breadsall, Pebble Beds rest on the Namurian Ashover Grit, and at Weston Underwood in the south-west of the district Waterstones overlie the Widmerpool Formation of Dinantian age.

CLASSIFICATION AND CORRELATION

The Permian rocks above the Basal Breccia were proved to be Upper Permian on the basis of miospore taxa in the 'Lower Permian Marl' (Clarke, 1965). Thus the use of 'Lower' and 'Middle' in the nomenclature of the various divisions is illogical and though they are retained on the published maps for the sake of continuity with adjoining areas, the modified terms, Permian Basal Breccia, Lower Marl and Middle Marl are used in this memoir.

As compared with over 1000 ft of Permian strata proved in County Durham (Smith and Francis, 1967), the maxi-

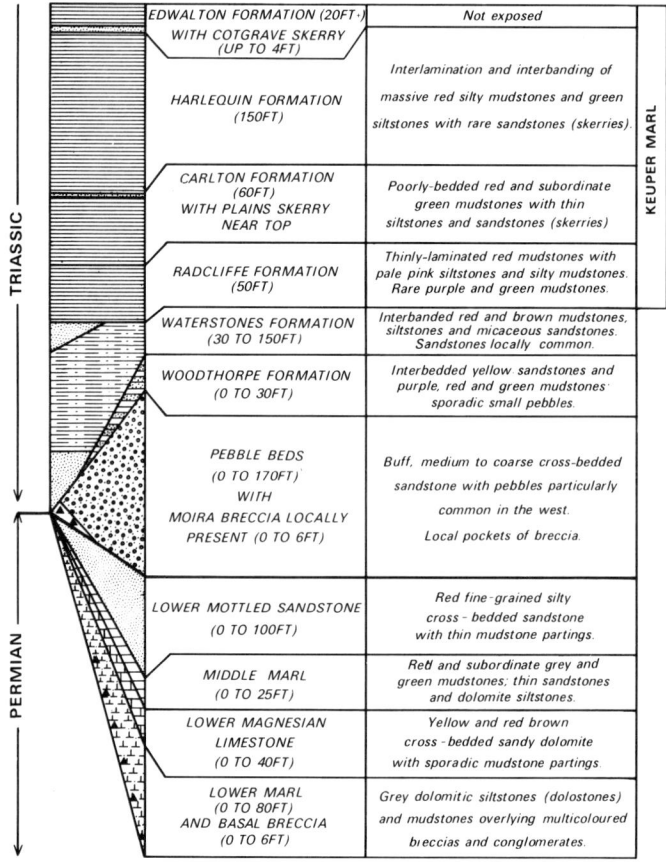

Figure 51   Generalised section of the Permo-Triassic rocks

mum thickness in the present district is only some 200 ft, and in the south and west they are absent.

In north-east England, Smith (1970a) has described five evaporite cycles, but only the lowest (EZ1) is represented with certainty in the Derby district. Thin bands of dolomite in the Middle Marl may be local representatives of the second cycle carbonates (EZ2).

Smith (1968) recognised the Hampole Beds in the Lower Magnesian Limestone of the present district. The beds form a distinctive marker horizon between the lower and upper subdivisions of the formation.

The Middle Marl is overlain by 'Bunter Sandstone', which Sherlock (1911) divided in the Nottingham area into a fine-grained, predominantly red marly lower division — the Lower Mottled Sandstone — and a coarser, pebbly, predominantly yellowish brown upper division — the Bunter Pebble Beds. His conclusion that there is a lateral passage of the Lower Magnesian Limestone into Lower Mottled Sandstone implied that the latter is partly of Permian age. This

74

Plate 8
Polished sections of basal
Permo-Triassic breccias
(about $\frac{4}{5}$ th natural size)

1 Permian Basal Breccia, Catstone Hill
Opencast Site; a petrographical
description of this specimen is on p. 150b

2 Permian Basal Breccia, Mappelwells
Borehole; described on p. 150b

3 Moira Breccia, Rose and Crown Opencast
Site, Morley; described on p. 152c, d

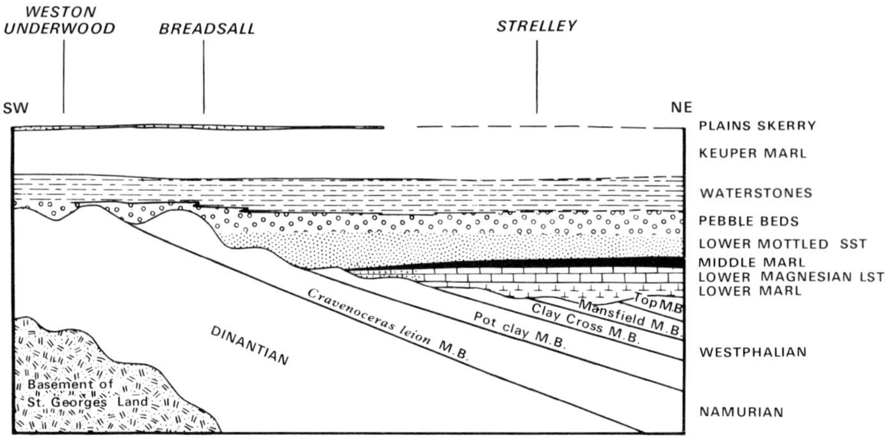

Figure 52 Diagram showing the general relationships of the Carboniferous, Permian and Triassic strata at the end of Triassic times

lateral passage was supported by the resurvey. The Lower Mottled Sandstone is therefore a sandy facies of the Upper Permian continental red beds and represents fluviatile activity on an increasingly important scale.

Smith and others (1973, p. 105) in the East Retford district have taken the base of the Trias at the base of the Pebble Beds — an arbitrary lithological boundary. This boundary in the south and east of the present district is gradational. In the west of the district a basal conglomeratic formation assigned to the Pebble Beds, but of uncertain age, transgresses on to an old land surface of Carboniferous rocks.

Recent palynological investigations (Warrington, 1967; Geiger and Hopping, 1968) have shown that the European stages of the Trias can be extended into the British Isles, and Smith and Warrington (1971) have thus been able to demonstrate that the Green Beds, Waterstones and Keuper Marl are diachronous formations to the north-east of the present district. No palynomorphs have been obtained from surface exposures in the Derby district, but in Trusley Borehole Dr Warrington reports Anisian miospore assemblages from the Waterstones and Anisian and probable Ladinian assemblages from the Radcliffe Formation; probable late Ladinian to early Carnian miospores occur some 20 ft above the Plains Skerry.

No distinct evaporite horizons are known in the present district. The well known Tutbury gypsum which is worked south of Derby occurs in the Trent Formation — above the highest rocks at outcrop in the present district. Gypsum is rarely seen in surface exposures. Elliott (1961, p. 228) has shown that gypsum is removed by solution in a zone which is up to 100 ft deep in the vicinity of faults. Fibrous gypsum was common below 62 ft in the Trusley Borehole in the south-west of the district. It occurs in veins as well as along bedding planes and is considered to be of a secondary origin.

Because of the repetitive nature of the Keuper Marl succession, detailed correlation of small exposures is particularly difficult, even with the aid of the various diagnostic sedimentary structures. Numerous old clay and skerry pits worked in the last century are flooded, grassed over, or infilled. The resurvey was fortunate therefore to have the results of a series of major roadworks and pipe-line schemes, the cuttings and excavations of which provided almost continuous sections through the Keuper rocks.

CONDITIONS OF DEPOSITION

The Permian sediments accumulated in a broad basin which extended eastwards to the Baltic and which on paleomagnetic and paleoclimatological evidence is considered to have lain in tropical and sub-tropical latitudes.

Following the Armorican Orogeny and associated uplift, the Carboniferous rocks in the east of the district were eroded into an eastward-sloping peneplain interrupted by minor irregularities of up to 10 ft amplitude. The sub-Permian peneplain was reddened by surface oxidation under desert-like conditions, usually to depths of 30 ft or so, depending on the porosity and fracture of the underlying beds. In many areas such reddening is the only evidence of a former cover of Permo-Triassic rocks. North-west of Wirksworth, a mantle of younger, probably Triassic rocks, overlay the Dinantian limestone and remnants of this cover are preserved in old pot-holes in the limestone surface despite a long history of re-working. The dolomitisation of the limestone has been attributed to downward percolation of magnesium-rich solutions from a once extensive Triassic sea (p. 12; Figure 56).

The Permian and younger rocks rest therefore with pronounced unconformity and progressive overlap on a planed surface of negligible relief in the east. To the south and west they were deposited on a land surface of considerable relief (Figure 56).

The Permian Basal Breccia was laid down on the Carboniferous peneplain under continental conditions as a piedmont breccia by intermittent fluviatile activity and was derived from local as well as distant sources. There followed a rapid southward transgression of the Zechstein Sea reaching almost to Nottingham (Figure 71).

It is convenient to think of Permian sedimentation in terms of desiccation cycles (Dunham, 1948; Stewart, 1954; Smith and Francis, 1967), each ideally consisting of a basal clastic member followed by carbonate and sulphate precipitates, and ending with salt and potash deposits. In marginal areas such as the present district these cycles are rarely complete, the bulk of the Permian succession being formed by the carbonate phase of the lowest (EZ1) cycle (Lower Magnesian Limestone and Lower Marl) (Figure 51; Smith, 1970b).

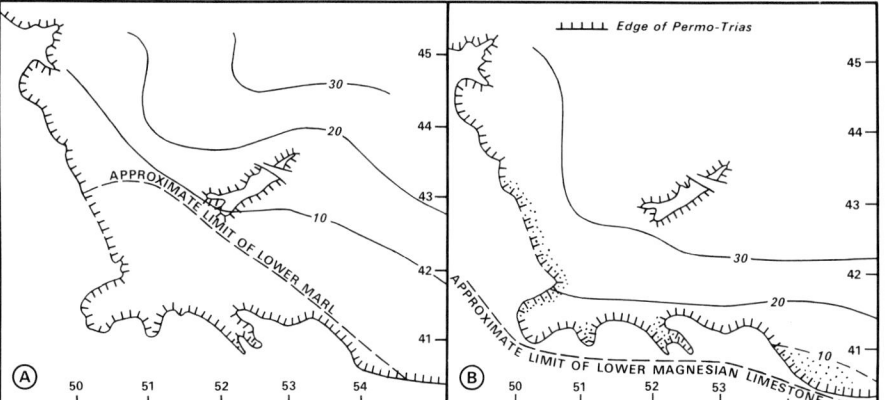

Figure 53 Sketch-maps of the Lower Marl and Lower Magnesian Limestone (isopachs in feet). A the southern depositional limit of the Lower Marl; B the southern depositional limit of the Lower Magnesian Limestone; sandy facies stippled

In this district much of the Lower Magnesian Limestone is recrystallised and dolomitised and the original textures have been destroyed. By analogy with the areas to the north, the rocks were deposited along the margins of a shallow sea which, judging from the restricted faunas, must have been highly saline. The Hampole Beds are considered to have accumulated intermittently on an extensive intertidal or supratidal coastal flat (Smith, 1968).

The overlying Middle Marl represents the silting up of the Zechstein basin and was deposited in playa-like depressions on a wide coastal plain. It indicates the establishment of arid continental conditions which were to persist into Triassic times.

In the rocks above the Middle Marl, differences between eastern and western lithologies reflect the palaeogeography of the period. In the east, the Zechstein basin persisted and the Bunter reveals a change in the energy environment from slow deposition in a shallow saline shelf sea to rapid accumulation in delta fans fed by fast-flowing rivers from the south. In the west, delta fans were deposited on uneven topography at the north-eastern margin of the Staffordshire or Needwood Basin (Figure 56). These two depositional basins were divided by an upland region formed by the outcrops of Namurian and Westphalian sandstone of the Morley area and called the Pennine–Charnwood Sill by Wills (1970). Moira Breccia formed here as a scree deposit, possibly from early Permian times. The upland area north of Morley village was not transgressed until the Waterstones were laid down; the underlying Pebble Beds were banked against the scarp face of Ashover Grit over much of its length (Wedd *in* Gibson and others, 1908, p. 120).

In the eastern basin, the Lower Mottled Sandstone thins westwards towards the Morley barrier and is probably banked against the dip-slope of the Crawshaw Sandstone (Westphalian A). The later Waterstones rocks overlap the Pebble Beds (Gibson and others, 1908, p. 124) and north of Morley rest either upon Moira Breccia or directly upon Carboniferous. They overlie a broad plateau feature in the older rocks, which appear to continue north-westwards as a planation surface linking the summits of the hills.

Wills (1970, p. 229) has divided the Bunter into cycles of sedimentation of a flood deposit type, in each of which there is an upward decrease of grain size.

During the deposition of the Keuper rocks the district was part of an area of relatively slow sedimentation lying across

the Pennine – Charnwood ridge (Wills, 1970, p. 274). The dominant lithology is a massive red silty mudstone. Sedimentary structures (Elliott, 1961; Klein, 1962) show that deposition took place in water, although the sporadic silt and sand grains may have been wind-borne from nearby land. Some authors consider transportation of the 'marls' to have been largely aeolian. At times climatic changes promoted an inflow of coarser detritus into the region, resulting in the formation of skerries. At other times aridity prevailed and evaporation led to the precipitation of soluble salts. Since the work of Shearman (1966) these evaporite sequences have been compared with the present-day sabkha environment of the Trucial Coast.

The origin of the red coloration of the Keuper rocks has provoked much discussion and controversy. The red colour is considered by Dunham (*in* Stevenson and Mitchell, 1955, p. 61) to be due to the even dissemination of minute granules of hematite through the rock. The green colour he considered is due to a magnesian clay mineral of the chlorite or montmorillonite group. The green rocks contain far less ferric oxide than red rocks, but the difference in the ferrous oxide content is not significant, so he ruled out the reduction of hematite *in situ*. Hematite is virtually absent in green rocks and Dunham (op. cit.) concluded that either it was not deposited or has been removed in solution; the latter possibility appears unlikely.

Clark (1962) has emphasised that the origin of red beds may be complex and result from the influence of more than one environment. Recent work by T. R. Walker (1967) and Norris (1969) provides evidence that the hematite pigment in many red beds, particularly those associated with evaporites and aeolian sandstones, has been formed *in situ* by alteration of non-red sediments in hot arid or semi-arid climates. The role of subsequent alteration however is considered to be a minor factor in the Keuper rocks of this district. In the finely laminated beds of the Harlequin Formation, for example, there is no diffusion of colour from one layer to another as might be expected from *in situ* alteration. Instead each colour shows lateral persistence and restriction to one layer. It is therefore concluded that the Keuper red beds are derived from sediments reddened at source during periods of aridity and the green beds are formed in less arid periods from green muds, which possibly owe their colour to reduction by bacteria either at source or during transport. The reducing role of bacteria has been

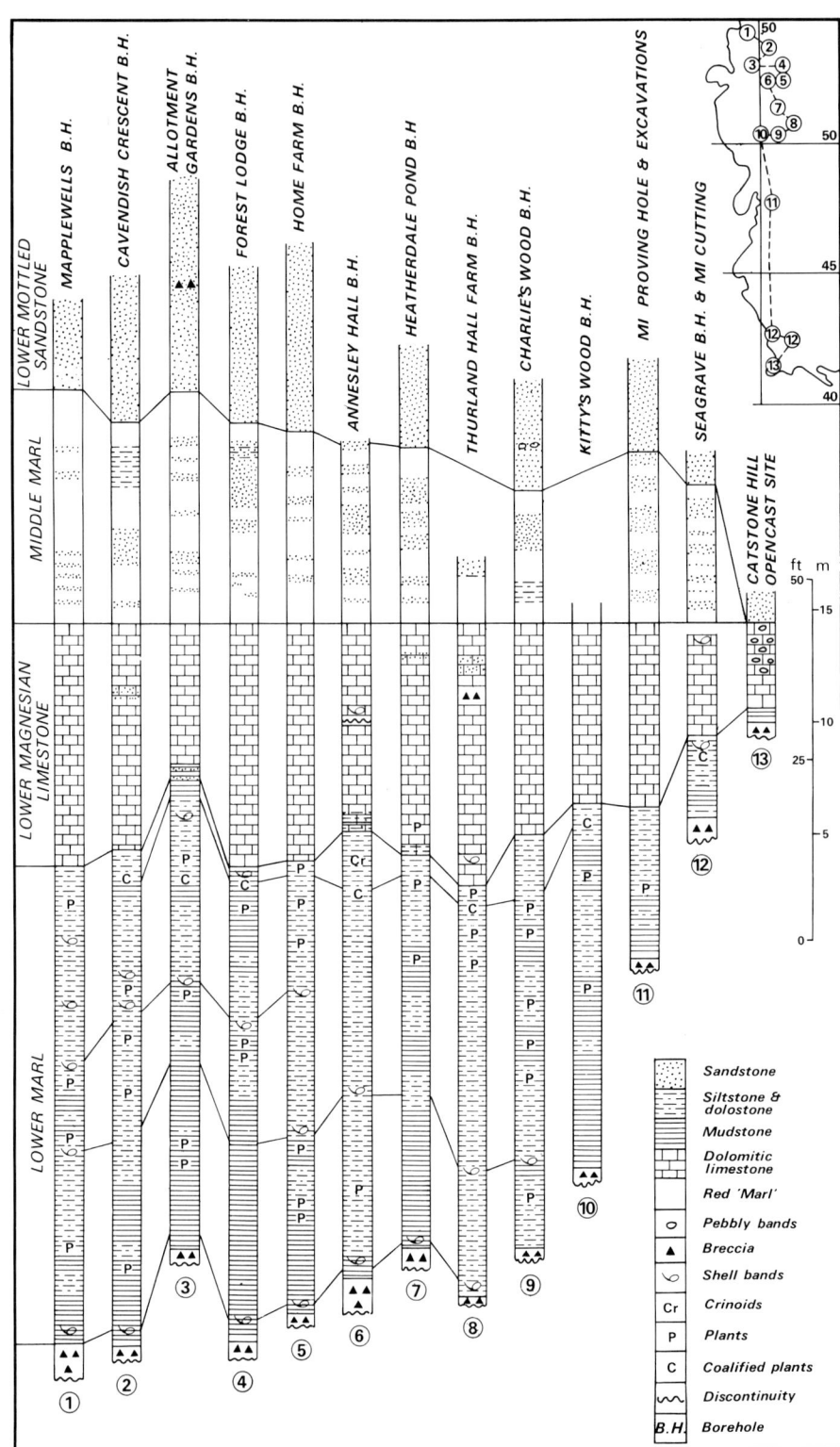

Figure 54 Sections showing variations of Permian strata in the eastern part of the Derby district

emphasised in recent studies of Keuper-type beds in the Persian Gulf.

PERMIAN

Permian Basal Breccia

The basal breccias, probably of Lower Permian age, represent the rapid accumulation and resorting of the products of erosion of the post-Carboniferous land surface. They include breccio-conglomerates which are rich in ironstone nodules derived from the Westphalian. The breccias are usually present throughout the district, though represented locally by only a few quartzite pebbles at the base of the Lower Marl or Lower Magnesian Limestone. In a large area south of Newstead, roughly coinciding with sub-Permian crops of thick Westphalian sandstones which were resistant to erosion, the breccia is less than 6 in thick (Figure 81). In the Annesley Woodhouse district and between Kimberley and Strelley accumulations of 48 in or more are not uncommon, the maximum thickness for the district of 5 ft 9 in being recorded in the Mapplewells Borehole. In Whyburn West Borehole, west of Hucknall, there is a 5-in sandstone at the horizon of the breccia.

The breccia is composed of rounded fragments of quartz and quartzite, with Westphalian ironstones, mudstones, siltstones and sandstones, and with angular fragments of igneous and metamorphic rocks of probable Charnian age, set in a grey-green sandy calcareous and/or dolomitic matrix. The fragments vary in size up to 12 in and the predominance of red-brown stained quartzites and dark red ironstones gives rise to local names such as 'plumcake rock' in the Kimberley area. The larger fragments occur in vaguely defined bands which occasionally show cross-lamination.

Lower Marl

The rocks now referred to as Lower Marl were described by Wilson (1876, p. 534) in a railway cutting near Hempshill as 'a series of thin-bedded slate-coloured sandstones and shales' which he considered representative of the Marl Slate of County Durham. They were noted by Farey (1811, p. 157) as a possible source of lime. These beds were mapped by Sherlock during the original survey of 1908 and included in the Lower Magnesian Limestone. He subsequently referred to them as 'Grey Beds' in his more extensive work (1926,

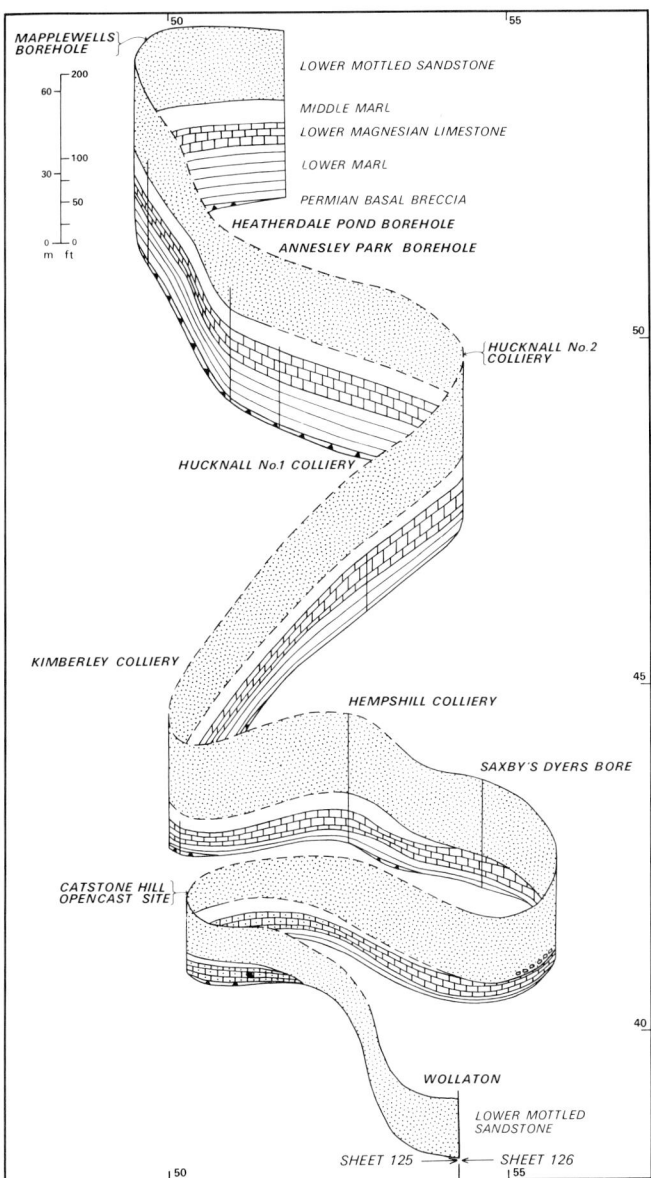

Figure 55 Ribbon section showing southerly thinning of the Zechstein deposits and their lateral passage into the Lower Mottled Sandstone. Broken line denotes present surface now eroded

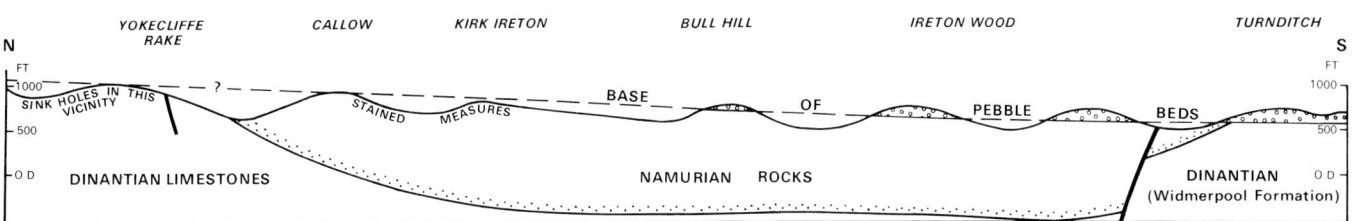

Figure 56 Section showing the relationship of the Triassic Pebble Beds to the underlying Carboniferous rocks in the western part of the district

p. 14) on the Permo-Triassic deposits of Britain. The present resurvey has shown that the Lower Marl forms an outcrop some 100 yd wide at the bottom of the Lower Magnesian Limestone escarpment from Annesley Woodhouse in the north to Strelley in the south with a narrow extension for over one mile to the east up the deep railway cuttings of the Kimberley–Nuthall area. There is an inlier at Broxtowe and a small outlier near Beauvale. In the north the formation is up to 80 ft thick but is only 50 ft at Hucknall and thins out rapidly in the five miles between Hucknall and Strelley, being absent south of a line approximately through Strelley, Bilborough and Radford (Figure 53). The beds are composed essentially of mudstones with siltstone, limestone and dolostone (dolomite of siltstone grade) bands. In general, dolostone bands are more abundant in the north where the overlying Lower Magnesian Limestone is thickest and decrease in importance to the south and east (see below). The more resistant carbonate lithologies are most common in the top part of the formation and mudstones increase in importance towards the base, but the mudstones comprise less than 50 per cent of the rock in most areas.

The beds are grey to pale grey when fresh but weather to a yellow-brown colour. In the top few feet they are irregularly red-brown in colour, and there is an upward passage into similarly coloured dolomitic Lower Magnesian Limestone by the intercalation of dolomite bands. Carbonaceous plant fragments are common, including leaves of cf. *Ullmannia bronni*, *U. frumentaria*, *Callipteris martinsi*, a young shoot of *?Pseudovoltzia liebeana* and undetermined conifer leaves and cone scales. Spores from the Lower Marl were listed by Clarke (1965). *Lingula* is rare in the present district except in the extreme north-east near Newstead.

Calcareous bands within the formation are commonly shelly but have an impoverished fauna, with *Bakevellia ceratophaga*, *Schizodus obscurus*, and other (indeterminate) bivalves and crinoids.

The Lower Marl was deposited in a shallow sea—a low energy environment but with a nearby land source of plant material. When traced north-eastwards there is an increase in thickness of the Lower Marl accompanied by a greater abundance and variety of fauna, indicating an approach to the main area of Zechstein deposition.

Lower Magnesian Limestone

In the north, the Lower Magnesian Limestone forms a marked westward-facing scarp some 100 ft high, forming the watershed between the Rivers Leen and Erewash. This feature decreases in prominence southwards and at Strelley is only some 30 ft high. The eastward-inclined dip-slope is about three miles wide and covers a large area (14 square miles) of outcrop which is disproportionate to the thickness of the formation. It forms gently rolling country modified by elongate dry hollows leading to narrow river valleys and covered by a rather thin red-brown sandy and marly loam suited to arable farming. The formation has been extensively quarried in the past and has been penetrated by some 40 boreholes and shafts extending into the Westphalian strata beneath. There is, in addition such a wealth of railway and motorway cuttings, together with natural exposures, opencast sites, and numerous temporary sections provided by new

development schemes in the outskirts of Nottingham that this chapter necessarily contains only a selection and condensation of the available information. Most of the surface exposures are shown on published six-inch maps SK 44 NE, SE; 45 SW; 54 NW, SW; and 55 SW.

The thickness of the formation varies between 20 and 40 ft over most of the outcrop, but generally thins southwards, and the southern limit of outcrop extending between Catstone Hill, Old Radford and Nottingham approximates to the depositional margin. That margin lies approximately one mile to the south of the depositional limit of the underlying Lower Marl. The southward thinning is accompanied by a change of colour from the pale greys and yellows of the northern area, through yellow-browns, to reddish browns and purples along the southern outcrop.

The formation consists of a flaggy, granular-textured dolomite or calcareous dolomite resembling a sandstone in appearance. The term 'limestone', used for convenience throughout the chapter, denotes dolomitic rock or dolomitic limestone.

Numerous red-brown silty micaceous mudstone bands, normally less than 1 in thick, accentuate the bedding-planes, which are commonly wavy, and at times, stylolitic. The Hampole discontinuity (Smith, 1968), noted in quarries and railway cuttings between Annesley and Bulwell, was located in Annesley Hall Borehole some 14 ft from the bottom of the limestone.

Because of the rather open granular texture produced by the dolomite rhombs, the rock tends to weather into dolomitic sand along the major joint planes, and isolated pockets of softened brown limonitic dolomite occur at depth. Hard finely crystalline calcareous dolomite commonly forms thin pinkish massive layers, particularly towards the base of the formation, and also occurs elsewhere as irregular nodules. Sherlock (*in* Gibson and others, 1908, p. 107) noted this concretionary phenomenon at the top of the Lower Magnesian Limestone at Cinderhill. Sporadic veinlets, cavity fillings up to 1 inch in diameter and small irregular patches of calcite occur within the formation. Pyrite associated with calcareous nodules and galena, both disseminated and in thin 'stringers', have also been recorded. Dendritic patterns and irregular speckling of black manganese dioxide are present along bedding and joint planes.

The transition from Lower Marl to Lower Magnesian Limestone involves changes in lithology and texture. The proportion of mudstone decreases upwards, and the grain size increases from fine silt grade in the Lower Marl to the coarse granular and saccharoidal textures of the Lower Magnesian Limestone. Accurate demarcation of the two formations is often difficult, particularly in the records of old borings and sinkings. The transition beds commonly contain calcite-filled cavities and veining.

The boundary of the Lower Magnesian Limestone with the overlying Middle Marl is usually a distinct lithological break. The top few inches of the limestone are commonly silicified, forming a hard pinkish finely crystalline dolomite. At many localities this top bed contains irregular concentrations of galena, and it was at this horizon that Deans (1961) proved the presence of wulfenite and uraniferous hydrocarbons including asphaltite. These deposits may have been carried in solution and trapped by the Middle Marl, acting as a cap-rock.

Sporadic sand grains occur throughout the Limestone; two distinct sandy bands, both about 2 in thick, occur in the Hucknall area, one 15 ft from the top and the other 8 ft from the base of the limestone. They thicken southwards until at Strelley the top half of the formation is largely a breccio-conglomeratic sandstone, representing a shoreline facies of the Lower Magnesian Limestone. Breccias have been recorded in Whyburn House Borehole, north-west of Hucknall, where a grey 'marly' band up to 2 in thick contained Magnesian Limestone pebbles approximately at the base of the Lower Magnesian Limestone. Gibson and others (1908, p. 106) recorded breccias in old quarries near Beauvale Abbey, and noted that the underlying 'marl slates' (Lower Marl) appear to be locally absent. Waring (1966, p. 207) came to similar conclusions and postulated an approach to a western shoreline in this northern area west of Hucknall. The present survey produced no evidence to support Waring's view.

The Lower Magnesian Limestone commonly shows cross-laminated units with foreset dips, normally some 10°, suggesting palaeocurrents from all quadrants except the north-east. Large areas of bedding planes showing assymmetrical and oscillatory ripple marks are common in the Linby area (Plate 9). The steepest slopes of the former variety are to the west, suggesting that the direction of wave movement was mainly from the east, i.e. the prevailing Permian wind direction.

Dome structures described by Smith and others (1967, p. 201) in the Pleasley–Mansfield area to the north of the present district occur towards the top of the Lower Magnesian Limestone, and are considered to be depositional in origin.

Fossils are not common in the formation as a whole and are usually present as solitary valves of *Bakevellia*, *Liebea* and *Schizodus*. At certain horizons they are concentrated into distinct bands as at Bulwell, where a 12-in shell band near the top of the Limestone contained *Bakevellia* cf. *binneyi*, *B.* cf. *ceratophaga*, *Liebea squamosa* and *Schizodus obscurus*. In addition *Permophorus costatus* has been found at Nuthall. Finely comminuted carbonaceous plant debris is sporadically present.

The Lower Magnesian Limestone was laid down in shallow saline water, largely as an oolite. Subsequent recrystallisation has destroyed most of its original structure in the present district. There is evidence in the Durham and East Retford areas that some of the beds formed under supratidal-sabkha conditions (Smith and Francis, 1967; Smith and others, 1973). In the Derby district the presence of large scale cross-bedding suggests fairly deep water. Preferred dip orientations of the cross-bedded units are from the north-west and south, confirming the likelihood of shorelines in those directions. Concentration of shell debris occurs at various horizons within the Lower Magnesian Limestone. Whilst numbers of individuals are large, the number of species is small (see p. 80), and it is inferred that conditions were not conducive to active marine growth in this marginal area of the Zechstein Sea. The present district was rarely beyond the reach of detrital sand grains which were either blown in from the nearby shorelines, or transported by

currents. Where the Lower Magnesian Limestone is the basal unit of the Permo-Trias the lowest few inches often contain conglomeratic fragments and coarse sand grains which probably represent a period of local erosion and re-working in Upper Permian times, which was distinct from the erosion producing the main Permian Basal Breccia, a bed largely consolidated and compacted in Lower Permian times.

At Strelley the top half of the Lower Magnesian Limestone is composed entirely of sand and small pebbles washed in from land adjacent to this shoreline. Onlap of the Lower Magnesian Limestone is about 1 mile to the south of the Lower Marl.

The Lower Magnesian Limestone is well jointed (see p. 159d), and cambering on valley sides has resulted in considerable widening of the joints which have been infilled subsequently with glacial debris. The jointing facilitates quarrying and throughout the Nottingham–Mansfield area the Limestone has been used extensively in the past for building stone, e.g. at Linby, Hucknall and Bulwell, and has also been burnt for lime. Its use for building has declined in recent years, but it is still in demand as an ornamental and rockery stone.

Middle Marl

The Middle Marl is confined to irregular outliers on the Lower Magnesian Limestone dip-slope west of the River Leen. A protective cap of Lower Mottled Sandstone usually overlies the Marl. Many small pockets of marl preserved in depressions in the dip-slope are too small to show on the map. Joints, widened by cambering, and glacial frost wedges in the top part of the Lower Magnesian Limestone contain a mixture of sand, derived Middle Marl and quartzite pebbles.

The formation weathers into a heavy, poorly drained, red-brown clay which is locally modified by a surface layer of sand and pebbles of Bunter origin to produce a light clay loam soil. The top soil is invariably rich in rounded quartzite pebbles and the Middle Marl outcrop contains Pleistocene erratics (see p. 158a), commonly attaining a length of between 2 and 3 ft.

The Middle Marl base marks the end of continuous carbonate deposition. Mud and silt are the dominant lithologies, but thin impersistent sandstone bands also occur, probably representing periods of excessive rainfall. Sporadic dolomitic layers indicate quiescent periods with a reversion to carbonate deposition. The limit of Middle Marl deposition lay to the north of the Lower Magnesian Limestone shoreline reflecting contraction, due to silting up, of the Zechstein depositional area.

In the present district the base of the Middle Marl is invariably sharply defined but there is upward as well as lateral passage into the overlying Lower Mottled Sandstone. The thickness of marl is therefore open to interpretation, but generally it does not exceed 25 ft in the north; a maximum of 34 ft was recorded in Mapplewells Borehole. The Middle Marl thins rather irregularly southwards, in which direction there is also a decrease in the clay content.

The Middle Marl consists of red clay or mudstone in beds up to 5 ft thick, interbedded with red-brown and green

Plate 9

1 Quarry Banks, Linby

Ripple-marked bedding plane within the Lower Magnesian Limestone (L 862)

2 Catstone Hill Opencast Site, Strelley

Prominent band of the Permian Basal Breccia resting upon Westphalian B strata and overlain by flaggy Lower Magnesian Limestone (L 866)

silty, sandy or sandstone bands. The harder sandstone bands have a dolomitic cement and are variable in thickness and extent. Thin beds of buff and yellow dolomitic limestone, lithologically comparable with the underlying Lower Magnesian Limestone, occur less commonly. The basal few inches of the marl overlying the Magnesian Limestone are invariably pale grey-green in colour, which is presumably due to the reducing action of migrating fluids in the limestone.

Deposition of the Middle Marl is considered to have occurred in the shallow land-locked remnants of the Zechstein Sea. The supply of clastic sediment, water borne as well as possibly wind borne, was normally too great to allow the formation of limestone. The thicker sandstone bands show current structures and probably represent rapidly accumulated sediments, following 'flash' floods. Such sandstones are not ideal for correlation, but in the absence of other markers have been so used with related breccia beds at Hempshill (Wilson, 1876) and beyond the limits of the district at Harworth (Mackintosh, 1937) and Calverton (Edwards, 1967, p.121).

The Middle Marl was formerly extracted for brick-making at Linby, Watnall, Hucknall, Bulwell and Cinderhill. The outcrop supports numerous artificial lakes and fish ponds in the grounds of Annesley, Papplewick, Bulwell, Basford and Strelley Halls and Newstead Abbey. The lake at Nuthall is only partly floored by the Marl and the clay was puddled to provide a watertight seal where permeable strata such as the Lower Magnesian Limestone form the base.

The marl has been exploited for the manufacture of plant pots at Bulwell but material suitable for firing is limited and production is tending to decrease in the face of competition from plastic pots.

TRIASSIC

In the south and east of the district the Middle Marl passes up into the Lower Mottled Sandstone, which may be wholly or in part of Permian age (p. 74). Where present, Pebble Beds are overlain by Waterstones, which, beyond the northern limits of the district, have been proved to be diachronous ranging in age from late Scythian in the north to Anisian in the south (Smith and Warrington, 1971). In the south-west of the present district beds of Waterstones lithology have yielded Anisian miospores.

Moira Breccia

According to Mitchell and Stubblefield (1948, p. 24) 'the Moira or Hopwas Breccia at the base of the New Red Sandstone is a basal breccia which may equally well occur at the base of the Keuper Marl (where that has overlapped the lower beds) or at the base of the Bunter Pebble Beds'. This adequately describes its restricted occurrences in the Derby district.

Only the large outcrop at Morley underlying transgressive sandy rocks of Waterstones lithology has been shown on the map; the occurrences of breccia near Breadsall, originally noted by Wedd (in Gibson and others, 1908,

pp. 110–112), are too thin or have too patchy a distribution to be shown separately from the overlying Pebble Beds. As Wedd pointed out, the Moira Breccia shows distinct affinities with the Permian Basal Breccia (pp. 150b and 152c).

In the field, the matrix of the breccia is normally a red or ochreous calcareous sand, but may be clayey in some sections; the fragments, which are up to about 6 in across, are of quartzitic rocks, hard and soft sandstones, siltstones, reddened ironstones, igneous rocks and soft mudstones. The rock is usually softer at outcrop than the Permian Basal Breccia.

The Moira Breccia is probably a scree deposit laid down in an area of high relief, but its age, whether Permian or Triassic, is uncertain.

Lower Mottled Sandstone

The largest area of outcrop of the Lower Mottled Sandstone, some 3 square miles, is a down-faulted block in the north-east of the district near Annesley. In the south-east the outcrop is a narrow belt marginal to the more extensive outcrop of the thicker overlying Pebble Beds. In the intervening ground the Lower Mottled Sandstone occurs as scattered small outliers on the Lower Magnesian Limestone dip-slope west of the River Leen. The sandstone gives rise to undulating, easily worked but not particularly fertile agricultural land, which readily reverts to birch scrub. Many acres have now been planted with conifers by the Forestry Commission.

A total thickness of about 100 ft is present in the north-east near Newstead Abbey and Robin Hood Hills. It thins southwards at crop to 50 ft in the Strelley area, but boreholes near Bulwell and Nottingham have proved red sandstone up to 100 ft thick. Farther south, between Wollaton and Dale Abbey, the thickness of the Lower Mottled Sandstone decreases to about 25 ft; farther west it appears to be overlapped by yellow Pebble Beds. At Morley, dolomitic Lower Mottled Sandstone is exposed near the western limits of deposition.

Despite its name, the sandstone is not obviously mottled in the present district except at the top, where there is an upward passage into the yellow to buff Pebble Beds. The base of the sandstone is usually gradational with the Middle Marl, though locally there is a basal breccia as in the Hempshill–Kimberley railway cutting, where a band up to 5 ft thick persists for at least 50 yd (Wilson, 1876, p. 533).

In the south-east near Bramcote, the Lower Mottled Sandstone is used as a moulding sand and in the past many pits were dug to supply Stanton Iron Works. To the north, in the Mansfield area (Smith and others, 1967, p. 209), the highest beds of the formation are the most important source of moulding sands, but in the present district it is the lowest 50 ft or so that contains the necessary 10 per cent clay fraction. Here the highest beds are used as building sand.

The formation is a red fine-grained friable sandstone, commonly argillaceous and locally pebbly. It contains red, grey and pale green mudstone partings and bands up to 4 ft thick together with angular fragments, inclusions and pellets of red, yellow and grey mudstone. The presence of sun cracks, contemporaneous clay-flake breccias, rounded and etched sand grains, rare dolomitic bands and pebbly horizons indicates semi-arid continental conditions of deposition. The Lower Mottled Sandstone appears to be a basin-edge facies of the Zechstein.

Petrographical analysis shows a bimodal grain size pattern, one of fine sand, the other of a clay fraction, which would suggest a reworking of the underlying marls. Cross-bedding is common and preferred orientations of the foreset beds suggest a northerly and westerly provenance for beds north of the Bulwell area, whilst southerly origins are indicated in the south of the district. From the study of numerous thin sections Dr N. Berridge reports that the Lower Mottled Sandstone consists of fine-grained, iron-stained sandstone with dominantly sub-angular, rather loosely packed, poorly sorted and weakly cemented grains. The feldspar content is variable, so that the sandstones include subarkoses and protoquartzites, of which the former are probably predominant. Many feldspar grains are highly altered to microcrystalline aggregates of other minerals and are then difficult to distinguish from compact siltstone. Other grains include true siltstone, chert, quartzite, gneiss, schist and hornfels and, rarely, igneous rocks. Accessory detrital minerals include opaque oxide ores, zircon, tourmaline, and rutile often intergrown with quartz; more rarely present are apatite, staurolite, andalusite, garnet and epidote.

Grains are characteristically coated by pellicles of limonite and locally ferric oxide is sufficiently abundant to act as a cement. Commonly the interstices of the sandstone are partly filled by calcite and dolomite, the latter typically in rhombic form with growth zones delineated by limonite inclusions. Scattered accessory patches of well crystallised kaolinite also occupy interstitial spaces in a majority of the thin sections examined. These are distinct from the small proportions of probably primary interstitial clay mineral aggregates that coat grains in many specimens and fill minor intergranular spaces.

The evidence provided by regional variation in composition, though limited, suggests provenance from a metamorphic terrain lying to the south and west, i.e., probably the Charnwood Forest area, or a former extension of it, but local sources were also available as shown by the presence of chert, siltstone and Magnesian Limestone particles. The upper part of the formation is not significantly richer in clay than the lower as is the case farther north (Smith and others, 1967, pp. 209–210).

The occurrence of baryte cementation at both Stapleford and Dale Abbey coincides with the presence of the same mineral in the overlying Pebble Beds. This supports the conclusions of Taylor and Houldsworth (1973) that the baryte is related to fault-controlled mineralisation and that stratigraphical control is limited to the availability of suitable host rocks.

Pebble Beds

The Pebble Beds crop out over some 8 square miles, largely in a belt between Nottingham in the south-east and Turnditch in the west. North of Nottingham they are present only as outliers or westerly extensions of the main crop which lies to the east of the district. They occupy some of the

highest ground, for example to the north-west of Nottingham, and in the Dale Hills area, where they form northward-facing scarps capped by Keuper rocks.

The Pebble Beds are thickest in the south-east, where they average 150 ft, but they thin westwards to half that figure at Dale Abbey. Farther west near Turnditch, however, the thickness ranges up to 130 ft, though they are absent south of Weston Underwood. The sandstone consists predominantly of subangular, rather poorly sorted and loosely packed grains of quartz and feldspar mainly medium-grained but with bands of fine-grained argillaceous sandstone as well as coarser pebbly horizons. Lithic clasts are abundant and predominantly composed of metamorphic rocks, especially quartzite and mica-schist, as well as fine-grained, mainly acidic, igneous rocks. Provenance from a metamorphic terrain is further suggested by the occurrence of poorly rounded grains of accessory garnet (E 36489) and staurolite (E 36488). The sandstones are generally weakly cemented by illite and limonite, but some contain spots of more concentrated ferruginous welding of clasts and the accessory occurrence of poikiloblastic baryte (e.g. Stapleford Hill and the Hemlock Stone.) At the top of the formation the sandstone is partially cemented by coarsely crystalline calcite that has poikiloblastically engulfed and partly replaced the affected sand particles. Irregular weathering at the top of the Pebble Beds is partly due to this calcite mineralisation and partly to analogous cementation and replacement by dolomite. The dolomite occurs as aggregates of small rhombohedra (0.02 mm diameter) which are engulfed by large crystals of calcite where both carbonate minerals are present together.

Mudstone is common as lenses, inclusions, and as a coating to the pebbles. There is much lateral variation, particularly in the west, where the beds are generally coarser, more pebbly and contain a greater proportion of argillaceous material. The highest beds are usually coarser than those low in the sequence. Cross-lamination is common, but foreset orientation is very variable. Derivation of sediments was predominantly from the north and north-west, but there is also an important element derived from the south.

Some pebbles came from the Charnwood area but the rest are mostly quartzites which seem to have been derived from sources to the south-west (Bonney, 1900; Matley, 1914; Campbell Smith, 1963). There is a northward and eastward decrease in size and abundance of pebbles, but the variations in the cross-bedding directions indicate considerable reworking and channelling in a complex deltaic area. The sandstones are coarser than the underlying Lower Mottled Sandstone and contain many more mudstone fragments, features indicating increased erosive activity. The Pebble Beds also have a wider distribution than the Lower Mottled Sandstone, which they overlap in the western half of the district.

Patches of pebbly sand between Turnditch and Wirksworth, originally mapped as glacial deposits (Gibson and others, 1908, p. 152), are now considered to be outliers of an original widespread Pebble Beds cover, e.g. at Blackwall [2540 4980]. The Triassic beds rest on a peneplained surface of Carboniferous rocks which rises northwards at some 3°. The extrapolation of this surface would carry the Pebble

Beds outcrop over the Dinantian rocks north-west of Wirksworth—an area noted for secondary dolomitisation of the limestone (p. 12) and the Brassington pocket deposits (Walsh and others, 1972). The present authors therefore maintain (p. 90) that these pocket deposits are of Triassic origin although partially reworked in Tertiary times.

Waterstones and Keuper Marl

The outcrop of these rocks, forming the upper part of the Trias, is two to four miles wide and covers 30 square miles. It is confined to the southern part of the district where, outside Derby city, it gives rise to pleasant undulating country suited to mixed farming.

In the Burton district to the south-west, these rocks are about 1000 ft thick (Stevenson and Mitchell, 1955), but only the lowest part of the succession, 400 ft around Derby and 270 ft in the Nottingham area, are preserved in the present district. The classification erected by Elliott (1961) is applicable in the ground between Derby and Nottingham but difficulties were found to the west of Derby as the Midland basin of Triassic deposition is approached. The Waterstones contain a preponderance of laminated, silty and sandy beds; the Keuper Marl comprises predominantly blocky red-brown mudstones, with sporadic green beds and patches and interlaminated and interbanded pale grey siltstones. Some of the thicker more persistent siltstones coalesce to form 'skerries' which are more resistant to erosion and form mappable features. Some 'skerries' are of sand grade and cemented by dolomite. Superimposed upon this overall pattern are repetitive cycles of sedimentation of various scales discerned by Elliott (1961, p. 225) and Wills (1970, p. 225). During the deposition of the Waterstones and Keuper Marl influxes of water became progressively less common and evaporation correspondingly more effective. In the present district, however, evaporites (gypsum and anhydrite) in the Keuper Marl are confined to secondary veins, stringers and nodules.

Sulphate solution in the surface beds leaves numerous small cavities and collapse-structures so that it is difficult to obtain good borehole cores from this part of the sequence. Elliott (1961, p. 228) has shown that the depth of the solution zone increases in the vicinity of faults.

Trusley Borehole (Appendix 1, p. 174a) was drilled through some 330 ft of Keuper rocks. Miospores of Anisian age were obtained from the lowest 98 ft, and probable Ladinian miospores to some 98 ft higher (Plate 10). The uppermost beds are inferred to be of Carnian age though no miospores have been found in them (see p. 156c). Dr Warrington reports that the borehole has provided the most abundant assemblage of mircroplankton so far recovered from the British Keuper. The occurrence in the Trusley assemblages of *Dictyotidium* with small *Micrhystridium*, abundant *Tasmanites* and some *Veryhachium* may indicate deposition in a somewhat turbulent environment near to a shoreline or in an area of shallow water.

In the east of the district the Waterstones overlie the Woodthorpe Formation (Elliott, 1961). In the west, the Woodthorpe Formation has not been recognised and the Waterstones are represented by an arenaceous facies. Trusley Borehole proved some 90 ft of Waterstones, the basal 50 ft of which are composed essentially of red-brown massive sandstone with a few laminae of siltstone and silty mudstone.

The thickness of the Waterstones in the east can only be estimated from the logs of old, inadequately described boreholes. The Waterstones and Woodthorpe formations together are considered to attain a maximum thickness of 150 ft in the Beeston area (Ericsson's Borehole) and to average 100 ft over much of the south-eastern part of the district. Sections [474 376] near Stanton show the Waterstones Formation to rest with angular unconformity upon the Woodthorpe Formation. A minor unconformity is also present within the Waterstones Formation.

Thicknesses of Waterstones are irregular in the west, with 70 ft proved at Weston Underwood, 90 ft at Trusley Borehole, 30 ft at outcrop near Kirk Langley and Mackworth and 90 ft inferred in boreholes in Derby city.

On the map the Keuper Marl is not subdivided, but this account follows the subdivisions of Elliott (1961), which are, in upward sequence, the Radcliffe, Carlton, Harlequin and Edwalton formations. The overlying Trent and Parva formations crop out beyond the margins of the district.

WOODTHORPE FORMATION

This formation is the southern Nottinghamshire equivalent of the Keuper Basement Beds—an omnibus term used by Lamplugh and Gibson (1910) for the variable coarse-grained basal Keuper rocks of the English Midland Province. Swinnerton (1918) considered these basal beds to be detritus derived from the erosion of the underlying Bunter rocks and redeposited in shallow water nearshore. Northwards from Nottingham the formation passes into the Green Beds (Smith, 1912; Smith and Warrington, 1971); westwards it thins and has not been recognised west of the Derwent Valley.

At the base of the formation there is a pebble bed apparently consisting of recycled quartzite pebbles from the underlying Pebble Beds. The bulk of the formation, as Elliott (1961) has shown, comprises rhythmic alternations of sandstone and mudstone in units up to 8 ft thick. Six such units were recognised in cuttings for the M1 Motorway near Stanton. Channelled sandstones also occur.

WATERSTONES FORMATION

This formation consists typically of fine-grained sandstone with bands and laminae of mudstone and siltstone. The mudstones are predominantly red-brown in colour with micaceous partings, and the sandstones range from pale grey through buff to red-brown. The formation is generally well laminated throughout in contrast with the underlying Woodthorpe Formation and the overlying Keuper Marl. Ripple-marks, sun cracks, salt pseudomorphs, mudstone pellets and clay-flake breccias, are common.

The top of the Waterstones Formation is gradational and particularly difficult to determine without detailed knowledge of the overlying strata. Elliott (1961) defines the junction as lying below the lowest skerry typical of the Keuper Marl and above the highest pale brown micaceous sandstone commonly found in the Waterstones. In the Sandiacre–Stapleford area a prominent sandstone, some 30 ft thick, occurs near this junction, the top of which acts as a useful marker and mapping horizon for delineating the top of the formation. In the west of the district the sequence is sandier throughout and the term 'Waterstones Formation' as defined by Elliott (1961) is not directly applicable.

The Waterstones Formation has been shown by palynological means to be of Anisian age in the present district, where it is therefore synchronous with Muschelkalk deposits of the North Sea Basin and Europe. The formation has yielded more macrofossils than any other unit in the Trias (except the Rhaetic) in the Nottingham–Derby area. Inarticulate brachiopods (*Lingula*) are recorded from the Waterstones near Eakring to the north (Rose and Kent, 1955) and numerous fish were discovered at Nottingham (Newton, 1887; Swinnerton, 1925, 1928). The record of plant remains (Wilson *in* Newton, 1887) from the fish-bearing horizon is doubtful however (Elliott, 1961). The Waterstones have yielded an extensive ichnofauna in the Nottingham–Derby area (Swinnerton, 1912; Sarjeant, 1967, 1970); the majority of these traces (of land vertebrates) occurred at Mapperley, Nottingham, but one (*Deuterotetrapous plancus*) was discovered within the present district at Dale Abbey, Stanton-by-Dale, Derbyshire (Sarjeant, 1967).

RADCLIFFE FORMATION

The Radcliffe Formation has a gradational base drawn below the lowest skerry but above the micaceous thinly bedded sandstones of the underlying Waterstones (Elliott, 1961, p. 216). It comprises evenly bedded, thinly laminated, red mudstones containing pale pink siltstone and silty mudstone layers with clearly defined salt pseudomorphs and hopper crystal outlines. Dolomitic slip layers were not observed at outcrop, but were recorded in Trusley Borehole and are reported to be common in the formation to the south and east of Nottingham. Purple mudstones are also a characteristic feature. The formation marks a change in deposition from dominantly quartzose and current-deposited rocks to beds which are primarily fine-grained and partly evaporitic.

The average thickness of the Radcliffe Formation in the type area is 40 ft (Elliott, 1961, p. 210). Some 45 ft were measured in exposures [471 371] on the M1 motorway near Sandiacre, west of Nottingham. In the south and west of the district it is about 30 ft thick in Trusley Borehole and in a pipe trench [2910 4142] near Weston Underwood.

CARLTON FORMATION

This formation, which comprises massive or poorly bedded marly red mudstones with flow breccias and several skerries, including the Plains Skerry towards the top, is well displayed in Chilwell Brick Pit [5130 3585] west of Nottingham (Plate 11), where its thickness of some 73 ft corresponds closely with that of the type area to the south-east of Nottingham. Farther west in cuttings of the M1 Motorway [471 365] and in Trusley Borehole it is about 60 ft thick. The purple ramifying structures which occur some 9 ft beneath the Plains Skerry (Elliott, 1961, p. 218) appear to be of considerable lateral extent for they have been detected in the present district at Chilwell Brick Pit, Sandiacre motorway cuttings and in Trusley Borehole.

Penecontemporaneous breccias similar to those in the formation have been found in Keuper rocks of the Cheshire Basin (Dr A. A. Wilson, personal communication 1977), where they are attributed to the former presence of salt, the solution of which disrupted the original laminae.

HARLEQUIN FORMATION

The Harlequin Formation is defined by Elliott (1961, pp. 218–219) as overlying all beds with flow-type brecciation but underlying beds containing a high proportion of even thin laminae and especially paper-thin laminae. The lower boundary lies some 15 ft above the Plains Skerry; the upper boundary is the base of the Cotgrave Skerry. The formation is over 130 ft thick in the M1 Motorway sections, and some 180 ft thick in the Trusley area.

The formation is characterised by repetitive cycles, each comprising red massive silty mudstone below and green laminated siltstone above. Interlamination and interbanding of the two lithologies in the middle of cycles is common. Red, brown and pink siltstone, as well as green mudstones, also occur in the sequence. A typical cycle recorded from the Harlequin Formation exposed in a cutting [4715 3585] for the M1 Motorway near Sandiacre is as follows:

	Thickness	
Top	ft	in
Siltstone, pale grey-green, mainly laminated but with a massive section near the centre	0	7
Mudstone, red, silty, and siltstone pale grey-green, finely interlaminated	0	11½
Mudstone, red to dark red, silty, predominantly massive; sporadic thin and discontinuous siltstone laminae, poorly sorted with sand grains; gypsum nodules	0	10½
Base		

The proportion of green and laminated beds increases upwards in each cycle, representing an increasing energy environment and/or nearness to a shoreline. The laminated part of the cycle may be missing, when the massive mudstone is capped only by a siltstone.

EDWALTON FORMATION

The Edwalton Formation extends from the base of the Cotgrave Skerry to the top of the Hollygate Skerries and is 150 ft thick in the type area (Elliott, 1961, p. 210). In the present district only the lowest 20 ft of the formation are represented, of which some 6 ft are exposed. The Trent and Parva formations are not preserved in the district.

CHAPTER 6

Tertiary (Neogene)

POCKET DEPOSITS

Around Brassington, immediately to the north-west of the district, is an area of Dinantian limestone in which solution hollows are filled with silica sands and clays.

The area extends into this district to the north of Hopton and Wirksworth, where small scattered hollows [2600 5470 and 2627 5453] are filled with sand and pebbles (Early and Dyer, 1964). The pocket deposits, first noted by Green and others (1887, p. 163) and Howe (1896), were thought by Yorke (1961, p. 22) to represent Triassic sediments originally spread over the Derbyshire Dome and trapped in pre-existing limestone hollows—in part solution cavities and in part surface watercourses. Kent (1957) also assumed a Triassic age for the sand, but considered that intermittent solution subsidence persisted through Mesozoic and Tertiary times. Ford and King (1969) thought that the distribution of the pockets was controlled by the form and position of the base of the dolomitised zone of the Dinantian limestone and that their age is post-mineralisation, in part Triassic, for they contain Pliocene plant remains and are post-faulting—probably late Tertiary. The sands are overlain by boulder clay. Boulter and others (1971) proposed the name Brassington Formation for these pocket deposits. Walsh and others (1972) thought that the formation represents a single cycle of sedimentation of Neogene age and noted that detached blocks of Namurian E_2 shales, in contact with the lowest sands found in the pockets, are stained lilac and thus represent a pre-subsidence land surface. Similar lilac staining was recorded during the present resurvey in the Namurian sediments that crop out south of Wirksworth and are, in places, overlain by Triassic sands. The present authors believe therefore that the origin of the pockets was by solution and subsequent collapse into limestone caverns in post-Triassic–pre-Pliocene times. The minor differences in grade and heavy mineral content between the Triassic Pebble Beds and the Brassington Formation are largely a result of reworking and resorting.

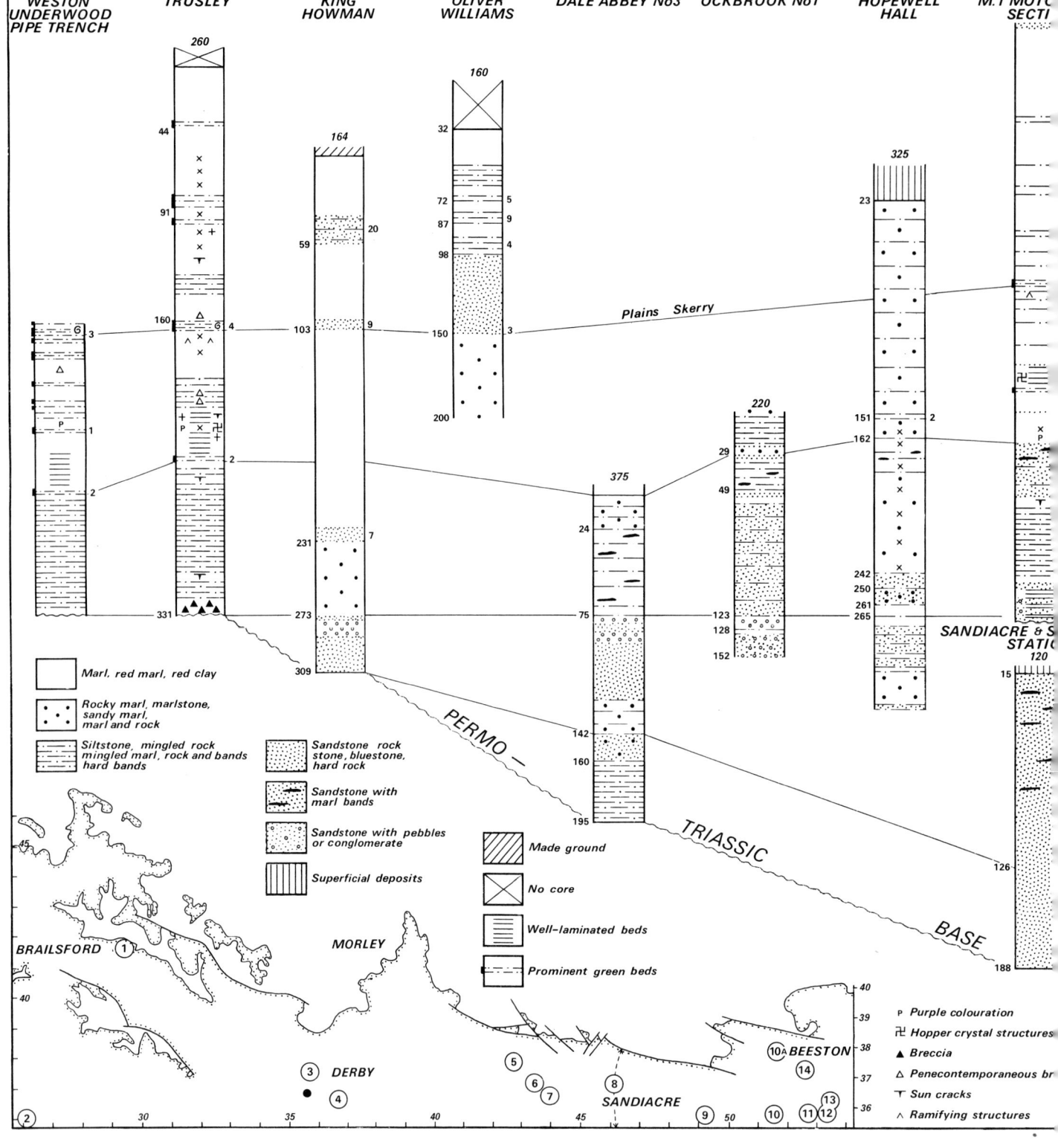

Plate 10 Sections and boreholes in the Permo-Triassic rocks in the south and west of the district.
Nos. 1, 9 and 11 are measured sections, the remainder are boreholes

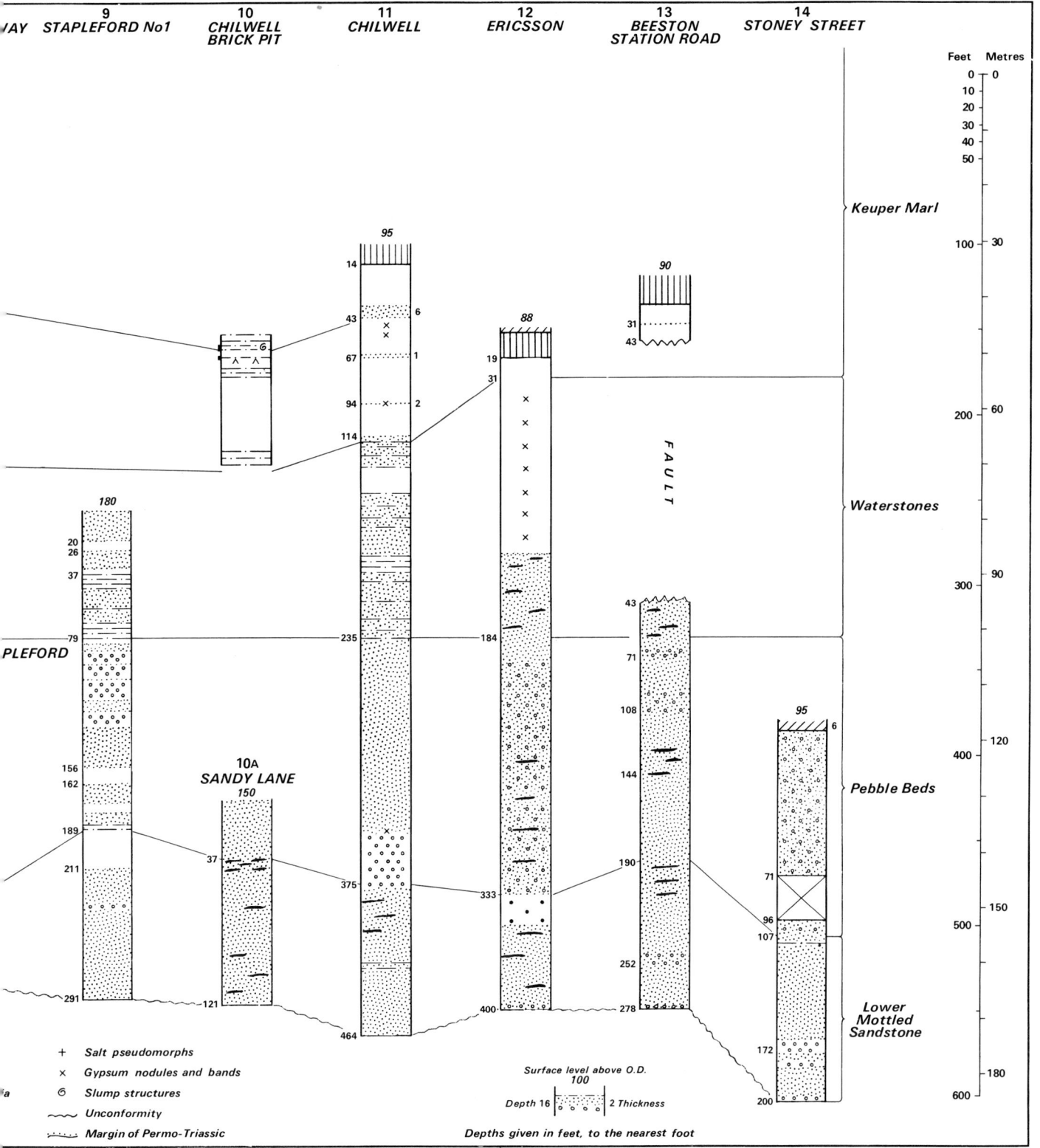

STAPLEFORD No1

9

10
CHILWELL
BRICK PIT

11
CHILWELL

12
ERICSSON

13
BEESTON
STATION ROAD

14
STONEY STREET

Keuper Marl

Waterstones

Pebble Beds

Lower
Mottled
Sandstone

Feet   Metres

+   Salt pseudomorphs
×   Gypsum nodules and bands
ᘔ   Slump structures
⌒⌒ Unconformity
⋯⋯ Margin of Permo-Triassic

10A
SANDY LANE

Surface level above O.D.
100
Depth 16 [○○] 2 Thickness

Depths given in feet, to the nearest foot

Plate 11

1 Keuper Marl, Carlton Formation with Plains Skerry at Chilwell Brick Pit

The Plains Skerry forms the topmost beds. The underlying mudstones are well laminated at the top and mainly massive below (L 860)

2 Triassic Pebble Beds near Mercaston

Cross-bedded pebbly sandstone in the Hilton Gravel Pit (L 860)

CHAPTER 7

Quaternary

INTRODUCTION

The distribution of the Quaternary deposits of the Derby district is shown on Figure 57. The glacial deposits, consisting of Boulder Clay and Glacial Sand and Gravel occur mainly on the higher ground, and are the eroded remnants of once larger sheets. The Post-Glacial deposits consist of: Head, mantling slopes and filling valleys cut in the solid rocks and earlier drift deposits; River Terraces, present in the valleys of major rivers and streams; Alluvium and its underlying Flood-plain Terrace, which occur along the water courses; and Peat, locally present in the west of the district. Also described under this heading are landslips, periglacial features, including cambering and the formation of dolomite tors, and the development of the drainage system.

The ages of the Quaternary deposits are summarised as follows:

Stage	Deposits
Flandrian	Alluvium, Peat
Devensian	Flood-plain Terrace, Head
Ipswichian	Beeston (First) Terrace
Wolstonian	Boulder Clay, Glacial Sand and Gravel, Hilton (Second) Terrace

There are no deposits in the district known to have been laid down between the Neogene (see p. 90) and the Wolstonian boulder clay. High-level gravels previously assigned to the Anglian glaciation (called Berrocian or Lowestoft Gravels) are now included within the Permo-Triassic rocks (p. 84).

During the Wolstonian, ice crossed the district from the north-west, but later in that glaciation it impinged against ice advancing from the east along the southern margin of the district. Very little evidence exists of the waning phases of this composite sheet of Wolstonian ice; there are no marginal drainage features within the district and it is assumed that the ice melted *in situ* in a stagnant condition.

The high undifferentiated terraces of the Derwent and the Hilton Terrace (p. 92) are assigned to this waning period and represent the re-establishment of the pre-Wolsaniton drainage system. According to Straw (1963) the high base level of the Trent, and hence that of the Derwent, during Hilton Terrace times was controlled by ice blocking the Humber mouth.

Posnansky (1960), on the evidence of warm-climate mammalian remains from Allenton (p. 301) to the south of the district, referred the Beeston or First Terrace to the Ipswichian (Eemian) Interglacial. This has been confirmed by Jones and Stanley (1974, 1975) who found an Ipswichian mammalian fauna in this terrace at Boulton Moor to the south of the district.

During the Devensian (Weichsel) glaciation ice failed to reach the district, but lay in the head-waters of the Trent to the west and also dammed the natural drainage outlet through the Humber to the north-east. The Flood-plain Terrace was probably deposited during this period. The periglacial climate during the Devensian Stage gave rise to a period, probably the main period, of Head formation and of landslipping.

During the succeeding Flandrian Stage alluvium was deposited.

BOULDER CLAY AND GLACIAL SAND AND GRAVEL

The boulder clay consists of stiff to sandy clay containing numerous boulders and fragments of local Dinantian, Namurian, Westphalian and Permian rocks and the almost ubiquitous quartzitic pebbles from the Trias. Coal fragments are common; far-travelled erratics include flints from the Chalk and volcanic rocks from the Lake District. When unweathered, the clay matrix is predominantly a similar grey colour to the Carboniferous mudstones from which it is largely derived. Red clays, probably derived from the Trias, are also present locally. The sands and gravels underlie, overlie and are interbedded with the boulder clay, and may be interpreted either as outwash deposits during local advances and retreats of the ice or as englacial deposits.

The boulder clay is discontinuously spread throughout the district. As in the Chesterfield district to the north (Smith and others, 1967), the spreads are the eroded remnants of more extensive deposits which were formed mainly as ground moraine of a composite ice sheet which covered the district. Because boulder clay drapes many of the valley sides it is likely that the main features of the present-day drainage pattern existed in pre-glacial times. The deposit is nevertheless so eroded as to indicate that it is of Older Drift (i.e. pre-Devensian) age. In the south-east Pennine uplands such deposits are assigned to the Wolstonian (Saalian) glaciation.

It has been known since early in the century (Wedd *in* Gibson and others, 1908, pp. 153–154) that the movement of ice near Crich (p. 157b) was parallel to the Derwent Valley and the presence of Lake District erratics indicate that the ice must have crossed the Derbyshire limestone massif. This ice flowed southwards into the Midlands (Posnansky, 1960, p. 289), where it deposited boulder clay with Pennine erratics (Pennine Drift). In the Midlands this drift is overlain by more widespread, easterly derived boulder clay containing Mesozoic erratics, principally chalk and flint. On the evidence of these erratics, Clayton (1953), Posnansky (1960), King (1966) and Rice (1968) show that the eastern ice reached the southern margin of the Derby district, where it overrode or impinged against the Pennine-derived ice somewhat north of the valley of the Trent. The limits of the Eastern Ice depend upon the recognition of flints in the glacial

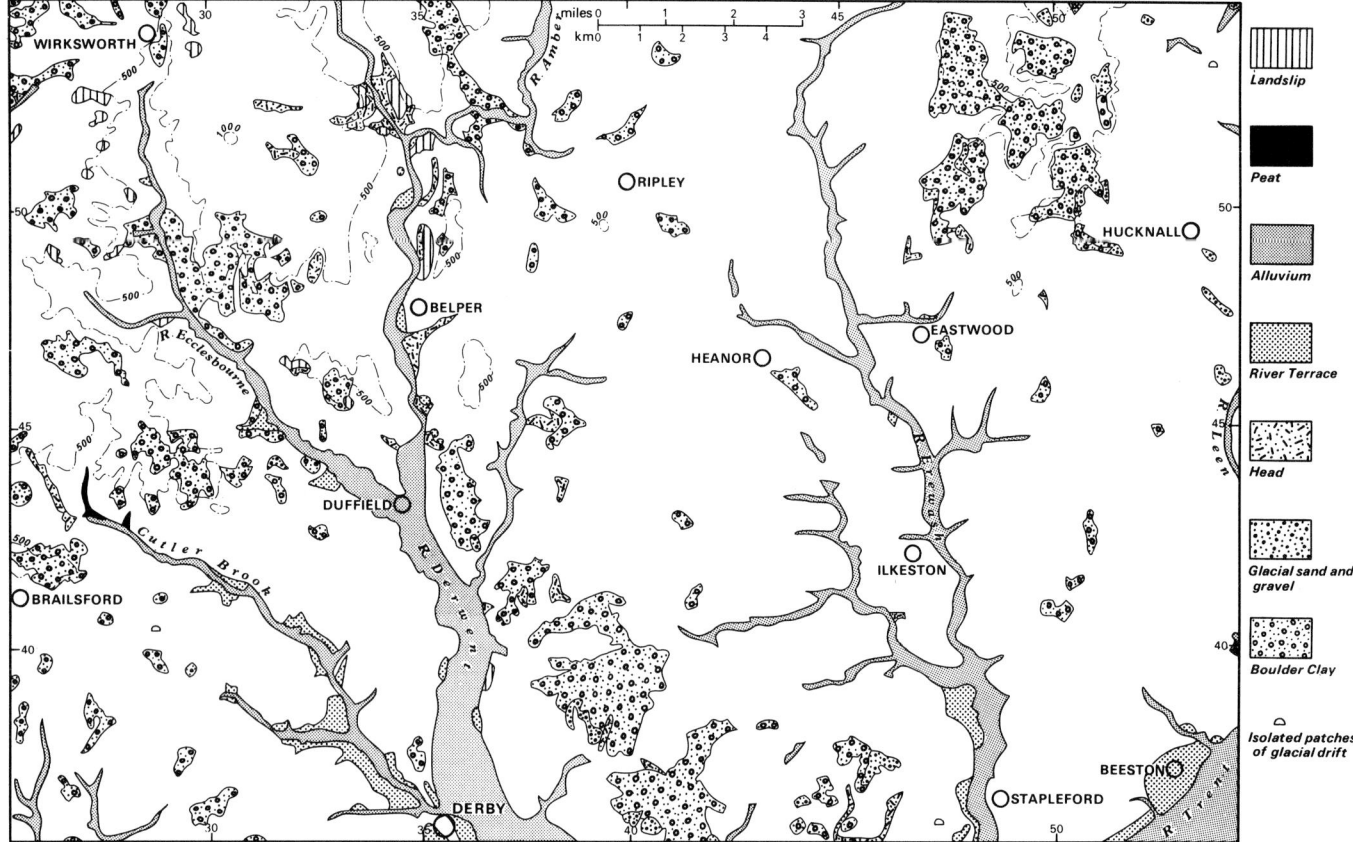

Figure 57 Sketch-map showing the Quaternary deposits

deposits and should perhaps now include Turnditch (see Figure 84), Breadsall [3691 4052] (p. 157c) and Heanor (Farey, 1811, p. 138). In all probability there was considerable fluctuation and intermingling of the two ice sheets.

In the north-eastern part of the district the glacial deposits contain Permian limestone erratics and thicken eastwards, indicating probable derivation from the north-east. They were deposited from ice sweeping south-westwards along the broad Trent Valley, and mounting the eastern flank of the Pennines until it encountered the Pennine Ice.

RIVER TERRACES

Terrace deposits have been preserved above many of the flood-plains of the rivers of the district, but are most extensive in the valleys of the Trent, the Derwent and its right-bank tributaries, and the lower reaches of the Erewash.

The highest deposits, lying up to 140 ft above the flood-plain of the Derwent, usually form well marked terrace features, but their heights relative to the flood-plain are too irregular for classification simply as a Third Terrace and they are shown on the maps as Terrace, undifferentiated. Two lower terraces have been identified and are shown as Second and First terraces on the maps. Their bases lie about 35 ft and up to about 10 ft respectively above the flood-

plain. Each terrace, however, appears to be an aggradation of sand and/or gravel.

The Derwent terrace system is analogous to that of the River Trent, of which only a very small tract lies within the district. The Trent terraces have been extensively studied (Deeley, 1886; Pocock, 1929; Swinnerton, 1935; Clayton, 1953; Stevenson and Mitchell, 1955; Posnansky, 1960; Straw, 1963 and extensive review accounts by King, 1966 and Rice, 1968). A tripartite sequence, named in descending order the Hilton, Beeston and Flood-plain Terraces, has received near general agreement, at least in respect of the nomenclature. There is disagreement in respect of the chronology of the sequence and of the mode of origin of the Hilton Terrace, but neither dispute can be resolved by a study restricted to the present district.

The flood-plain terrace is included with the alluvium on the maps of the Derby district. The Beeston Terrace, the type area of which lies within the district, is the First Terrace of the Derby sequence, and the Hilton the Second.

The chronology of the Trent terraces is controversial. Clayton (1953) named the Hilton Terrace and assigned it to the Ipswichian Interglacial. The terrace includes gravels mapped by Stevenson and Mitchell (1955, p. 92) as fluvio-glacial and which they considered to be younger than the Chalky Boulder Clay (of probable Wolstonian age). Posnansky (1960, p. 300) agreed that part of the Hilton Terrace was fluvioglacial and dated it on the evidence of included artifacts to the waning phase of the Wolstonian Glaciation. Straw (1963, pp. 187–188) also referred the Hilton Terrace to this phase with the Trent ponded against ice in its lower reaches.

The Beeston Terrace was assigned both by Clayton (1953) and Posnansky (1960, p. 301) to the Eemian (Ipswichian) Interglacial' (Ipswichian) and Straw (1963) to a late remains (Plant, 1859; Arnold-Bemrose and Deeley, 1896). Straw (1963) considered the Beeston Terrace aggradation to be of early Devensian age but Jones and Stanley (1974, 1975), on the basis of mammalian remains, place it in the Ipswichian.

There is agreement that the Flood-plain Terrace is Devensian but Clayton (1953) assigned it to the 'Acheulian Interglacial, (Ipswichian) and Straw (1963) to a late Devensian glacial episode when an ice readvance blocked the Humber.

The present authors accept the chronology presented by Shotton (1973), i.e. Hilton Terrace, Wolstonian; Beeston Terrace, Ipswichian; Flood-plain Terrace, Devensian.

Some doubt exists on the correlation of low terraces in the middle reaches of the River Ecclesbourne and the Markeaton (Cutler, Mercaston) Brook because of the presence of two knickpoints which can be traced in the long profiles of these two streams. Knickpoints occur at 500 ft OD near Wirksworth and 325 ft OD near Idridgehay in the Ecclesbourne and at 400 ft OD near Mercaston and about 250 to 275 ft OD (within the ornamental lakes of Kedleston Park) in the Markeaton Brook. The terraces in question lie between the knickpoints in both streams and may therefore be older than their relationship to the adjacent flood plains would suggest; they are shown therefore as undifferentiated.

The age of the high undifferentiated terraces of the Derwent are problematical. They rest on well-marked bench features below the level of adjacent Wolstonian boulder clay, which they therefore presumably post-date. They are higher than the Hilton Terrace, but may also date from the waning period of the Wolstonian ice sheet.

HEAD DEPOSITS

The Head deposits (Andersson, 1906; Dines and others, 1940) of the district consist of accumulations on slopes or valley floors of material derived by solifluction from drift deposits or solid rock. Their lithology therefore varies according to the source of the material. They were largely formed under Devensian periglacial conditions, but recent downwash deposits are included locally.

ALLUVIUM (including FLOOD-PLAIN TERRACE)

Alluvial deposits consisting of sandy loam, clay, sand and gravel border the rivers and streams of the district. Gravels at depth in these deposits are assigned to the flood-plain terrace. The near-surface deposits tend to reflect the parent rock of the rivers, for instance the Ecclesbourne alluvium tends to be more clayey than that in the Derwent Valley.

PEAT

Mappable peat is found only near Ravensdale Park in the west of the district; it has formed where drainage of springs from the Pebble Beds is impeded, and locally [262 437] overlies head deposits. Bridges (1966, p. 45) noted a patchy thin surface peat on some poorly drained head deposits in the upland regions.

LANDSLIPS

The commonest landslips of the district occur where thick Namurian sandstone overlies mudstone on steep valley sides. Water from the sandstone lubricates the underlying mudstones, facilitating the development of shear planes along which the slip moves. The displacement may be rotational and, where the dip is into the valley, down the bedding planes. Faults and joints are additional factors in the origin of slips. Other slips are triggered off by undercutting of valley sides by rivers. Related mudflows are initiated in exceptionally wet periods on slopes underlain by mudstones and superficial deposits.

PERIGLACIAL AND SOLUTION PHENOMENA

These are best displayed in the Permo-Triassic rocks, in particular in the Lower Magnesian Limestone and the Pebble Beds. Cryoturbation structures are preserved in the Pebble Beds in the western part of the district between Hulland and Brailsford. They vary from crenulations, convolutions and small faults, well depicted by thin marl partings and laminae, to frost-wedge structures, a small example of which near Muggington has been described by Bridges (1964, p. 262).

An isolated structure (Figure 84) about 1 mile S of Turnditch [2814 4549] in the Pebble Beds is of such proportions that its formation solely by ice is considered unlikely. The structure, lying at about 700 ft OD, is a vertical parallel-sided fissure some 60 ft deep, and 15 ft wide, which is 'V' shaped in the basal 5 ft. It has an east-north-easterly trend and a mapped length of over 200 yd. It contains red-brown clayey sand on the north side and friable fine buff sand on the south, and is traversed by red mudstone laminae which depict a synclinal or collapse structure in the centre. Scattered irregularly throughout are erratics of sandstone (red fine-grained and yellow coarse-grained, of Ashover Grit type), Dinantian limestone and flint. Near the walls of the fissure the deposit contains numerous 'Bunter' quartzite pebbles commonly coated with black manganese oxide, with their long axes aligned along the margin. The regular geometrical form of the fissure suggests initiation along a plane of structural weakness. The internal structure is comparable with that produced by subsidence simulation experiments (Walsh and others, 1972), used to explain the formation of the Brassington 'pocket deposits' in the Dinantian limestone. Its abrupt termination downwards suggests that it originated as a 'gull' due to the gravitational movement to the north-west of a large area of Pebble Beds. Despite the extensive sand and gravel workings in this area, only one such structure is known and it seems to be the result of a rare set of circumstances.

VALLEY BULGE AND CAMBER STRUCTURES

Most of the stream exposures in the Namurian rocks show anomalous disturbed and disrupted strata. These structures, also found in the Permo-Triassic rocks, are believed to result from local pressure of overlying competent beds on the valley sides squeezing the incompetent argillaceous strata in the load-free valley bottom. These bulge structures—compressional folds, faults and thrusts—were described by Jones and Weaver (1975) in Namurian strata near Derby and by Hollingworth and others (1944) in the Mesozoic rocks of the Midlands. They are known to decrease in magnitude downwards and to die out at depths of 100 ft or so (see, for instance, Stevenson and Gaunt, 1971, p. 338). Ideal conditions for such superficial movement occurred during the Pleistocene Period when the argillaceous strata were softened at surface during the thawing of ground ice. However, in order to account for the association of large bulges with low relief on certain valley sides, even when allowing for unloading of the bulge by erosion, Kellaway (1972) has suggested that the movements were assisted by the weight of ice on the valley sides. These conditions he considered would be effective during deglaciation, where erosion of the bulge occurs within a subglacial drainage channel along the valley floor.

Cambering (Hollingworth and others, 1944) consists of a lowering of the competent rocks on interfluves and valley sides associated with valley bulging. In the competent rocks, joints approximately at right-angles to the direction of slope are widened to form fissures or 'gulls'.

The formation of the valley bulges and the cambering of the district cannot be dated accurately, but probably occurred under periglacial conditions as well as during deglaciation.

GENERAL GEOMORPHOLOGY AND DRAINAGE

Fearnsides (1932, p. 177) recognised a vestigial peneplain in the Peak District—a relic of the post-Hercynian sub-Permo-Triassic surface. The present survey confirmed the existence of this surface, which is an important geomorphological feature of the Derby district. The base of the Pebble Beds rises westwards from some 400 ft near Weston Underwood and Brailsford to 800 ft at Kirk Ireton and once lay close above the existing dolomitic limestone surface near Brassington, Hopton and Wirksworth (Figure 56).

Linton (1951, p. 452) considered that the higher reaches of the Welsh Dee were the headwaters of a proto-Trent which flowed eastwards across the south of the district. Supporting evidence for this is present in the southern Pennines where deeply incised upland streams flow in a south-south-easterly direction. He noted Fearnsides' comment that the Derwent is a subsequent or strike stream, but emphasised the discordant nature of the river, which cuts indiscriminately across different rocks and structures. Since they are clearly superimposed, Linton was of the opinion that the

Derwent and Erewash were tributaries of a proto-Trent established on a pre-Eocene down-fold of the Chalk, all traces of which have since been removed by erosion.

Clayton (1968, p. 324) suggested the presence of a 1000-ft subaerial surface cut wholly in Carboniferous rocks, maintaining that many relics of it are still present. Assuming that this 1000-ft surface existed, its attempted reconstruction suggests that the Derwent then followed a syncline in the Westphalian rocks. However at Ridgeway Quarry [3587/5150] near Ambergate the Crawshaw Sandstone, at between 300 and 500 ft OD, shows red and purple staining, pointing to the proximity of a Permo-Triassic cover.

In the west of the Derby district the drainage trends south-south-eastwards towards the River Ecclesbourne. Many of the streams were initiated on Triassic rocks and have subsequently been incised into Carboniferous strata, e.g. Cutler, Sherbourne and Alton brooks. There is also a close parallelism in this western area between the stream pattern and the trend of the faulting.

Farther north an east-south-easterly drainage trend is preserved on the surface of the Dinantian limestone north and west of Wirksworth. The limestone has been eroded into a series of valleys which are now largely without streams for most if not all of the year. Smith and others (1967, p. 233) assumed that the main valleys were in existence before the glaciation.

Figure 58 shows the probable sequence of events leading up to the present drainage system in this north-western part of the district. The earliest drainage is thought to have been by the Dale, Warmbrook and Eniscloud valleys, which formerly linked up with three valleys to the east of Wirksworth now occupied by Pendleton Brook, Peatpits Brook and a channel north of Alport Hill.

The linear nature of these channels suggests that they are part of a superimposed drainage system originally formed on a Triassic sandstone dip-slope, prior to the development of the Ecclesbourne Valley.

The depth and length of these channels suggest that large quantities of water flowing from the limestone uplands rapidly eroded the softer Namurian shales below the Ashover Grit of the area south of Wirksworth.

The Derwent Valley was deepened and the Ecclesbourne Valley enlarged and cut back, resulting in the capture of the Dale, Warmbrook and Eniscloud streams. Minor modifications to these channels also occurred. The Eniscloud stream cut a deep gorge known as Stone Dene and its flow was diverted southwards through Hopton and into Scow Brook assisted possibly by the relative softness of the weathered Hopton Agglomerate. Similarly a southern tributary of the Dale channel was captured and a major valley created trending southwards through Godfreyhole into Scow Brook. The south-eastern part of the Eniscloud channel was therefore beheaded and the lower part of the valley is now much too large for the brook flowing in it.

As the ice sheets diminished and the water table was lowered by progressive downcutting, the drainage of the limestone area moved underground. Subsequent attempts to de-water the lead mines of the Wirksworth basin by soughs have re-established the link with the Derwent. The River Ecclesbourne, however, is still deprived of much of its source

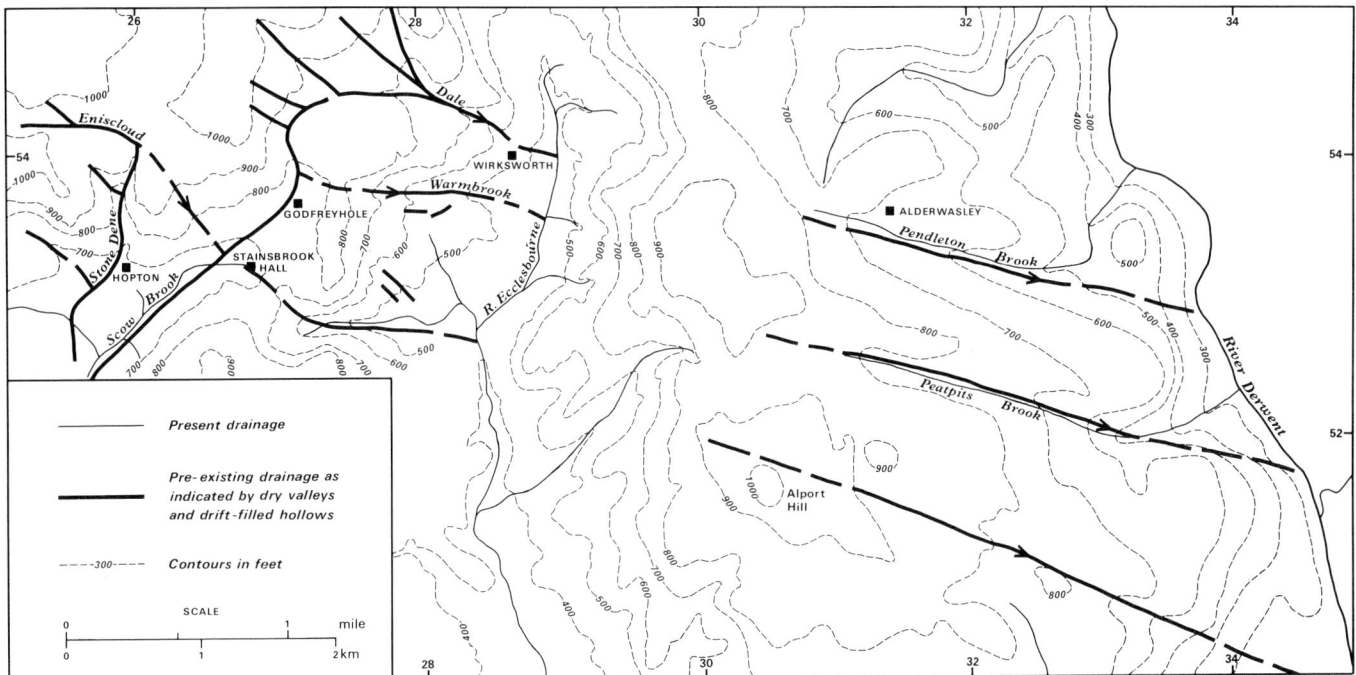

Figure 58 Sketch-map of the Wirksworth–Alport Hill area showing the pre-existing east-south-east drainage to the River Derwent and its eventual capture by the Scow Brook and River Ecclesbourne

water, which drains to the north and east, and it appears a paltry river for the size of the valley.

Clayton (1968) recognised subaerial surfaces at 560 to 580 ft, 460 to 595 ft and 280 to 360 ft OD, which he considered were stages in the excavation of the river valleys formed before the principal glaciation, which left a thin sheet of drift upon them. The present survey has shown that some of these surfaces are planar because of a thick cover of drift masking irregularities at a lower level, and others can be explained by near horizontal outcrops of sandstone.

CHAPTER 8

Structure

INTRODUCTION

The Derby district lies at the southern end of the Pennines. The main period of folding is of Hercynian age, but the district was subsequently uplifted during the Alpine Orogeny during the Tertiary period. The Alpine movements resulted in an easterly dip of 1° to 2° over most of the East Pennine coalfield, but in the Derby district the dip swings to the south as part of the southward plunge of the main Pennine upfold. Not all the movements within the district occurred during these two orogenies; there were precursory movements during deposition of the Carboniferous rocks and there is also evidence of contemporaneous tilting within the Permo-Triassic rocks.

The main structural elements of the Derby district are shown on Figure 59 and structural contours on selected horizons in the Carboniferous rocks are shown on Plate 12.

INTRA-CARBONIFEROUS MOVEMENTS

Direct evidence of earth movements during deposition of the Carboniferous rocks is found in the Dinantian limestones and Namurian shales of the Wirksworth area, where the order of events may be summarised as follows:

Zone

D_1 Local uplift and concomitant formation of knoll reefs. Folding and probable early movements along the Yokecliffe Rake. Outpouring of the Matlock Lower Lava from the Hopton vent. Erosion and reworking of the D_1 limestone top as evidenced by rolled and rounded corals of D_1 age in the lowest beds of D_2.

D_2 Strong local movements in D_2 times, giving unconformities (Plate 2.1). Complete erosion and/or non-deposition of D_2 limestones in the west; progressive erosion in the south and east. Reefs were formed in the south and subsequently eroded with only patches remaining. Breccia formation in the highest beds north of Wirksworth. Silicification and chert formation.

P_2 Deposition of Cawdor Limestone on an irregular surface of Hoptonwood (D_1) and Matlock (D_2) limestones. Formation of numerous patch reefs. Penecontemporaneous folding and thrusting (Dale Quarry). Continued erosion or non-deposition in the west.

Late P_2 to early E_1 Renewed subsidence with transgression of shales on to eroded limestone. Reactivated movement along Yokecliffe and Gulf faults.

Elsewhere within the district, insofar as is known, there is a conformable sequence throughout the Carboniferous rocks, but thickness variations show the effects of contemporaneous earth movements. For instance precursory movements of the anticlines, particularly that at Breadsall, during deposition of Namurian sediments and Westphalian coals resulted in slight thickness variations. Much more significant, however, was the differential subsidence in the Edale and Widmerpool gulfs, which dominated Dinantian and Namurian sediment-

ation in the Derby district and continued at least into Westphalian B times. This subsidence represents a downfolding of greater magnitude than the subsequent Hercynian folding (Kent, 1966, p. 339). The Edale Gulf (*ibid.*) lies to the north of the Derby district and the centre line of the south-eastward-trending Widmerpool Gulf (Falcon and Kent, 1960, p. 18) crosses the district north of Derby.

The Carboniferous succession in the central part of the Widmerpool Gulf has been proved by Duffield Borehole, but the only indications of the positions of the margins are the gravity and magnetic observations (p. 117), certain sedimentological observations discussed on p. 5 and structural evidence. The last depends upon the recognition of adjustment faults in overlying strata (Kent, 1966, p. 325) and suggests that the north-eastern margin of the gulf corresponds with the Ridgeway–Pentrich–Porter Barn–Parkgate–Smalley–Hagg fault belt or the Ridgeway–Pentrich–Godkin–Aldercar–Cinderhill fault belt (Figure 59). Because the first-mentioned belt lies south of Little Hallam Water Borehole, which proved a thick gulf sequence of Namurian rocks (Falcon and Kent, 1960, figs. 8 and 9, pp. 18–19), it is suggested tentatively that it marks the gulf margin during the Dinantian, and that the Ridgeway–Cinderhill belt delineates the margin during the Namurian and Westphalian. The position of the south-western margin of the gulf has not been recognised and may lie to the south of the Derby district.

On the block some 200 ft of limestone overlying tuff in Ironville No. 3 Oil Bore are equated with 1740 ft of D_2 and P strata in Duffield Borehole. Such an eight-fold increase in thickness into the 'gulf' is unlikely to apply to the whole of the Dinantian; probably subsidence was very irregular, and overall figures for thickness increases from 'block' to 'gulf' are speculative. In the construction of the sketch-sections in Figure 61 the overall thickness of Dinantian rocks proved in the Eyam Borehole (Dunham, 1973) has been applied (perhaps somewhat generously) to the Ironville area. The two-fold increase in thickness of D_2 Zone sediments from the Eyam area[1] to the Duffield area has been applied to the remainder of the Dinantian at Duffield.

The variation in thickness of the Namurian rocks is based on the borehole evidence at Duffield and Ironville together with the field mapping; the thicknesses are tabulated on p. 18 and indicate a three-fold increase into the 'gulf'. The thickness variations in the Westphalian rocks are based on an extension of thickness variations in Westphalian A strata (Figure 11) to the remainder of the sequence below the Top Marine Band; they may err somewhat upon the conservative side (see Howitt *in* Kent, 1966, p. 334). The sections (Figure 60) show a total subsidence at Duffield of about 19 000 ft from the beginning of the Carboniferous until the

---

[1] Mr I. P. Stevenson has provided evidence of 100 ft at the top of the D_2 Zone which were not proved in the Eyam Borehole.

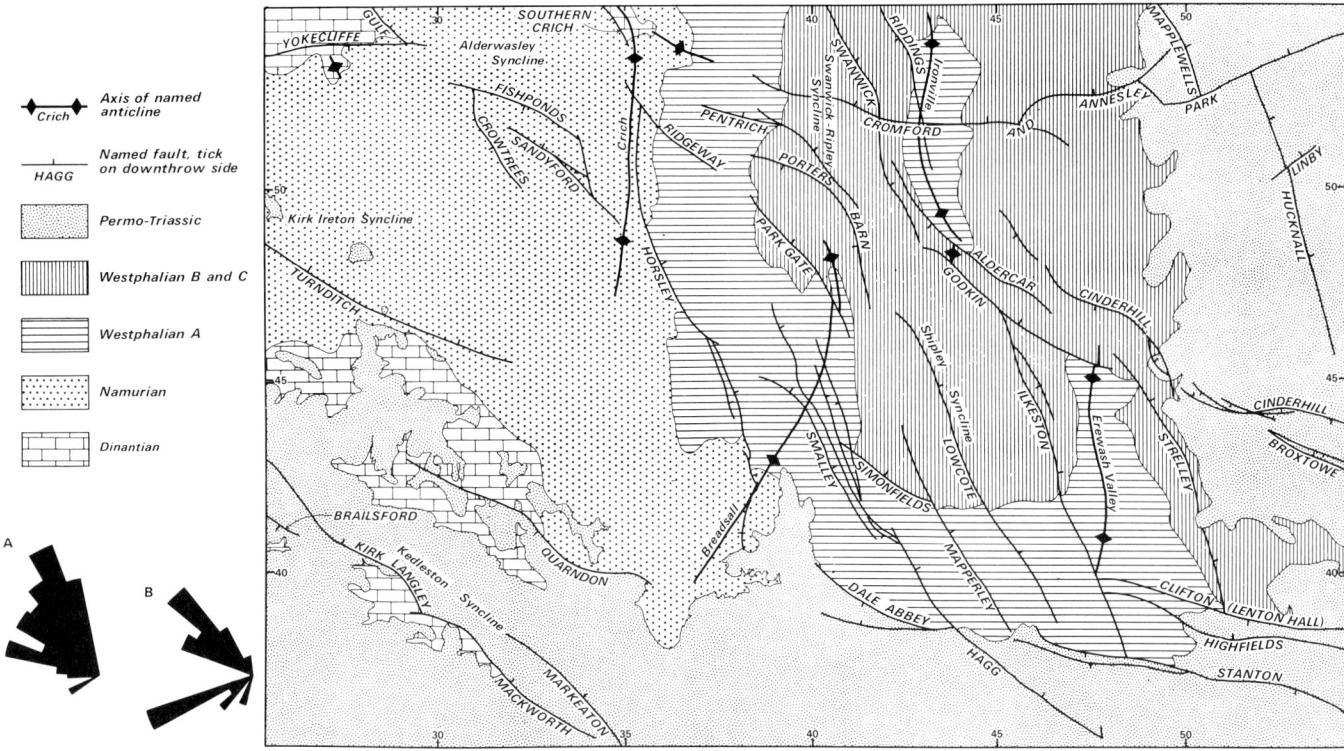

Figure 59   Sketch-map showing the main structural elements of the Derby district with rose diagrams based: A upon the lengths of the faults proved in the district and; B (from Evans, 1968, p. 3) upon the lengths of faults in the Charnwood area

deposition of the Top Marine Band; the equivalent figure for Ironville is 9500 ft. The amounts of subsequent uplift, including the effects of the Hercynian and Alpine orogenies, are a minimum of 6500 and 2000 ft respectively.

THE HERCYNIAN OROGENY AND LATER MOVEMENTS

The Hercynian Orogeny affected the rocks of the district from late in the Westphalian until early in the Permian (Edwards, 1951, pp. 76–77). Early folding and faulting in response to east–west compressional forces was followed by normal faulting. The main trends of the faulting (Figure 59, inset A) lie between north and west with a minor trend in a south-west direction. The throws of these faults, including later Alpine movements, are generally 100 to 200 ft but the maximum known is more than 800 ft (p. 160c). The folding and the major faults of the district are described in the detailed account (p. 160b). Towards the close of the Hercynian Orogeny the district was subjected to erosion, which removed 7000 ft more strata from the uplifted south-west of the district than from the north-east prior to the deposition of the Permo-Triassic rocks.

Differential movements apparently took place during the deposition of the Permo-Triassic rocks and continued up to the Waterstones; thereafter, i.e. from Anisian times, subsidence was more or less equal throughout the district.

Mesozoic rocks younger than the Trias are unrepresented in the district but the tectonic history is likely to have been one of continued subsidence until the end of the epoch. There followed uplift and erosion, and in the Miocene the district was subjected to renewed faulting and gentle folding representing the Alpine Orogeny.

THE ALPINE OROGENY AND LATER MOVEMENTS

The main Alpine fold is a south-easterly trending anticline or monocline crossing the south-east of the district; it gives rise to different regional dips of the Permo-Triassic rocks in the east and south.

Structural contours on the base of the Permo-Triassic strata are shown on Figure 60. In the western and south-central parts of the district these contours reflect irregularities of the original dissected surface upon which the Triassic rocks were deposited. In the east the sub-Permian surface was peneplained.

In the west of the district there is additional gentle folding along axes aligned north-west–south-east and north–south. Re-activation of Hercynian faults is displayed by lesser displacements in the Permo-Triassic rocks than in the Carboniferous. Displacements were greatest in the south-east of the district. In the east a few weak anticlinal structures were formed in the Permo-Triassic rocks along lines of faults in the underlying Carboniferous.

The Alpine Orogeny was followed by subsidence and the deposition of the Neogene sediments now found in pockets to the north-west of the district (p. 90). The district was subsequently uplifted to its present form.

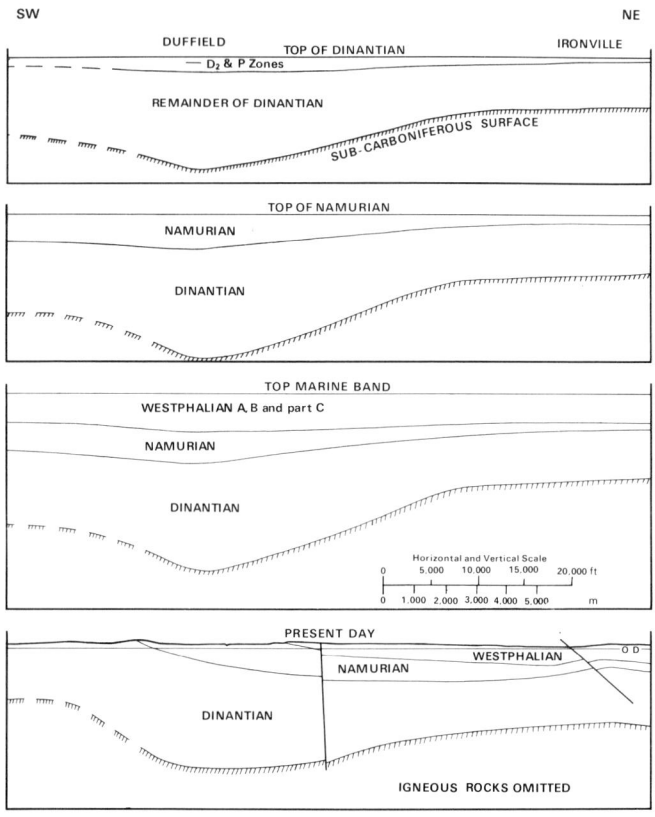

Figure 60 Natural scale sketch-sections between Duffield and Ironville to illustrate the relative earth movements during deposition of the Carboniferous rocks and the orogenic movements of post-Carboniferous age

ORIGIN OF THE FAULTS

The folding in the Carboniferous rocks appears to have resulted from compressional forces acting in an east–west field. It was followed by faulting with a dominant north-north-west (Charnian) trend and subsidiary trends which are west-north-west and south (see insets, Figure 59). The majority of these faults are normal, with dips of their planes averaging 60° although showing considerable variation, and are apparently of tensional origin. The mechanics of vertical displacement following horizonal stress, a general association in northern England, has been investigated by Moseley and Ahmed (1967). They considered that the faults were controlled by earlier shear fractures, and described their formation as commencing with folding and wrench faulting under stress, followed first by a diminution of stress during which joints were formed and secondly by

renewed compression and folding, passing finally into a regime suitable for normal faulting. Moseley and Ahmed (*ibid.*, p. 80) also stated that the pattern of faulting in the Carboniferous rocks in the north of England is inherited from that in underlying Lower Palaeozoic strata. The north-north-west and south-west fault trends of the Derby district agree closely with the fault pattern of Charnwood Forest (Figure 59) and, although some of the steeply dipping faults may have resulted from shearing stress associated with the folding, the majority of the normal faults are believed to have been controlled by pre-existing faults in the basement rocks of the district. Similarly the Permo-Triassic structures reflect the Carboniferous faults.

EARTHQUAKES

There have been twelve earthquakes of an intensity of 5 or more (Davison Scale) in central England since 1900 (Hains and Horton, 1969, p.119), of which several were located near Derby. A map by De Rance (1896) which showed the occurrence of earthquakes since 1248 also emphasised the importance of the Derby area in British seismic activity.

Known seismic foci in the vicinity of Derby fall into three groups. There is a north-western group with epicentres 1 mile E of Ashbourne and 3 miles W of Wirksworth. The Wirksworth focus corresponds closely with the westerly surface termination of the Yokecliffe Rake Fault—an important east–west boundary fault on the southern end of the Derbyshire limestone block (see p. 160c). A second group of foci occur south-west of Nottingham, near Beeston, again associated with long and important westerly trending faults affecting Carboniferous and Permo-Triassic strata. The third group lies south-east of Derby with some foci in north-west Leicestershire (Dollar, 1957). Here structural lines are obscured by Permo-Triassic rocks, but strong north-west trending faults are known.

According to Davison (1924) the strongest of all recorded earthquakes in Derbyshire occurred in 1795. Other earthquakes in the district were recorded in 1084, 1180, 1678, 1734, 1738, 1784, 1857, 1865, 1872, 1873, 1903–6, 1956 and 1957.

The foci of the earthquakes of 1903–6 were in the Ashbourne–Wirksworth area. Slight tremors were felt on 2nd March 1903. The shock of 3rd May 1903 with a maximum intensity of 5 was widely felt north-west of Derby. The next reported seismic event was on 3rd July 1904 with an intensity of 7. On 27th August 1906 a shock of intensity 5 was centred around Ashbourne. Tremors at Winster 17 miles NNW of Derby in 1952 and 1954 are possibly associated with the focus near Wirksworth.

The most recent and important earthquake in the district was on 11th February 1957 and has been studied in detail by Dollar (1957) and Lees (1958). Its epicentral region included Derby, Ilkeston and Nottingham and the damage to chimneys, masonry and brickwork indicated an intensity of 8 (Davison Scale). The initial quake was felt as a double shock, which may indicate two points of origin. The area of country disturbed was at least 30 000 square miles, stretching from Doncaster and Preston to the north, Oakham to the east, Nuneaton to the south and Shrewsbury

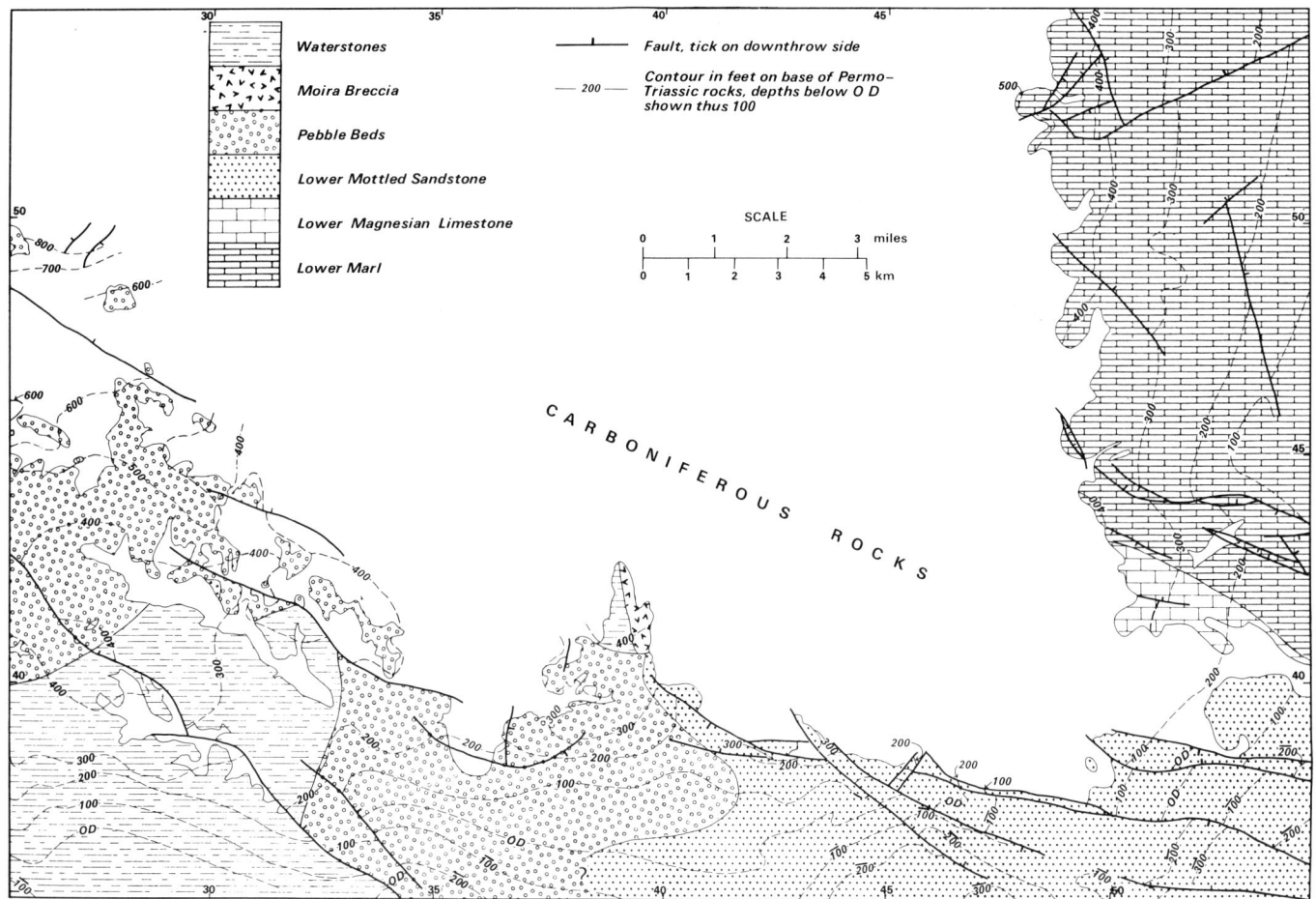

Figure 61 Structure contours of the base of the Permo-Triassic rocks and map showing the distribution of Permo-Triassic rocks resting unconformably upon the Carboniferous surface (basal Permian breccias excluded)

and Chester to the west. Its focus is considered to be near Diseworth, a small village south of the Trent and roughly mid-way between Derby and Nottingham.

Davison (1924) suggested that there had been a westerly migration of the foci of earthquakes from Northampton (1750 and 1768), through Leicester (1893 and 1904) to Derby 1905–6. It appears that such foci lie on a north-westerly trending fault—a dominant structural direction in the area south of Derby. As noted by Dollar (1957), however, the 1954 earthquake with its epicentre at Diseworth does not fit this progression. An investigation by Lees (1958) into the effects of this earthquake on vibration-sensitive relays at local power stations in the Trent and Erewash valleys showed that the relays which tripped were always those sited on the river side of the power station and consequently over the maximum thickness of alluvial deposits. This factor appeared more significant than the orientation of the relays. Most of the macroseismic evidence concerning the direction of fall of chimneys, etc. suggested that although the epicentre was south-east of Derby the shock waves travelled through the town from a more easterly direction, i.e. along the Derwent Valley. A similar rough parallelism

between the shock waves and the trend of the Erewash Valley deposits was also noted.

The importance of the configuration of the Widmerpool Gulf (see p. 5) has, however, not been considered in relation to the epicentral areas. There was considerable crustal distortion and volcanic activity along the margins of the Widmerpool Gulf during Carboniferous times and evidence of downwarping in post-Liassic strata. The southern boundary of the gulf is particularly sensitive to changes of stress since it is the northern limit of the old resistant block of St George's Land in the Charnwood area. Kent (1966) has drawn attention to the presence of large and important faults as indicators of the margins of such gulfs. Whilst the limits of the Widmerpool Gulf are known only in outline, the three groups of seismic foci known in the present district correspond to its estimated margins. Thus Diseworth, situated near Breedon Cloud, is on the southern margin of the gulf, Beeston near Nottingham lies on the northern margin and the Ashbourne–Wirksworth area forms the north-eastern edge. The influence of valley deposits, causing refraction of the earthquake waves, is therefore considered to be a surface modification of deep-seated faulting along the margins of the Widmerpool Gulf.

The district has evidence of past instability in the form of tuffs and sills in the Dinantian of the Duffield area, turbidites perhaps triggered off by earthquakes, in the Dinantian and Namurian rocks north of Derby, disturbed coal seams and roof strata in the Westphalian north of Nottingham and tuffaceous siltstones interpreted as 'fall-out' of volcanic ash in the Westphalian north-east of Nottingham. These phenomena probably represent response to the imbalance brought about in this part of the earth's crust by the differential sedimentation rates between Wirksworth and Derby. The sulphurous and warm waters of Wirksworth and Matlock are indications of the continuing links between the deep-seated structures and the surface.

Table 4 Properties of coal seams and reserves

Seam	Ash Range %	Sulphur %	Chlorine %	Coal Rank Code (Coke Type)	Volatile Matter (d.m.m.f.)	Calorific Value (d.a.f.)
High Main	6– 8	0.5 –1.0	Trace–0.20	902(B)	38–40	13 900–14 150
First Waterloo	5– 9	1.0 –1.5	0.30–0.55	802(C)	39–40	14 250–14 450
Second Waterloo	8–11	1.25–2.0	0.30–0.45	802(D)	38–39	14 250–14 450
Deep Soft	3– 5	0.5 –1.0	0.45–0.60	702(F)–602(G2)	37–40	14 650–14 900
Deep Hard	4– 7	0.5	0.40–0.60	802(D)	36–37	14 600–14 750
Tupton	3– 6	0.75–1.5	0.50–0.70	802/1(C)–702/1(G)	35–39	14 500–14 900
Blackshale	6–14	2.0 –5.0	0.30–0.70	802(C)–602(G3)	36–42	14 350–14 850

have been subsequently revived, profitable working having been made possible by improved economic circumstances and the introduction of sophisticated plant.

Sites have varied in size from those yielding a few thousand tons to some yielding a million tons or more, but in some cases adjacent prospecting sites have been combined to form single working sites. Some areas have been especially prolific, such as that between Loscoe and Denby, where 12 sites produced 2.5 million tons from seams from the Waterloo group to the Threequarters. There are substantial quantities of proved coal still unworked near Denby.

South of the Loscoe–Denby area lies the Shipley area, where opencast production has been spectacular. Of all the seams worked, from the Mainsmut to the Second Ell, the most rewarding has been the Top Hard, which with the two Combs, forms a thick seam with partings. In spite of old workings in the Top Hard, and the considerable excavation depths reached, opencast production has been profitable and ratios of overburden to coal have been relatively low. At the time of writing production in the Shipley area is still in progress, but from 23 sites completely worked and restored the total production was 5.4 million tons.

North-east of the Shipley area there is a complex of sites, now restored, occupying land formerly owned by the Butterley Company. Here, all seams from the Waterloo group to the Blackshale were worked, the most productive being the Deep Soft in combination with the Top Soft and Roof Soft. Six sites (including combined sites) produced 1.9 million tons. One site, Corfield, was notable in consisting mainly of a nine-hole golf course which after excavation was satisfactorily returned to its former use. North of these sites, at Swanwick, there was Tramway Site, where 946 000 tons were recovered from the High Hazles, Comb, Top Hard and Dunsil seams.

In recent years there has been a general policy, where practicable, of associating opencast coal production with reclamation or landscaping of derelict or unsightly areas of land. Obvious cases for treatment are disused collieries, with their waste heaps, slurry ponds, abandoned buildings and railway sidings, and both the Opencast Executive and licensed operators have carried out projects in the Region in consultation with the County Councils. In the area of the Derby Sheet one of the first schemes was the grading of two high colliery tips in association with the working of Cromford Canal Site and adjacent sites, the seams involved being all those from the Mainsmut to the Second Ell. A somewhat more ambitious project embraces the derelict area around Woodside and Coppice collieries at Shipley, where the principal seams are the Combs and Top Hard. Large waste heaps and many disused buildings were involved, and an extensive country park is taking shape. The scheme includes the recovery of over 1¼ million tons of coal, some of which lay under a lake which has been drained and will be reinstated. A smaller reclamation scheme has been completed by a licensed operator at Stanley Colliery, West Hallam, where extraction of the

Low Main and Threequarter seams enabled a massive waste heap to be partly removed and reshaped. Shilo Site in the Awsworth–Eastwood area includes a number of derelict features, and the estimated production of about 500 000 tons will be from the Deep Soft, Deep Hard, Piper and associated seams.

In the route-planning of the M1 Motorway cooperation between the consulting engineers and the Opencast Executive Prospecting Department was essential. The contractors were warned of hazards such as old workings, and where it was known that excavations would cut coal seams recovery contracts could be arranged in advance. In cases where the motorway was to affect sites proposed for contract, liaison was organised between the motorway contractors and the Opencast Executive's contractors. Such a case was Catstone Hill Site between Trowell and Nuthall, where working of the Top Hard, Waterloo and Ell seams, in conjunction with the motorway construction, provided 466 000 tons. An important example of motorway liaison was provided at Trowell. Here, the proposed service area roughly coincided with the Opencast Executive's Shortwood Site, where the Deep Soft and Deep Hard seams had been extensively worked but there was enough coal left for economic production. The site was duly excavated, the replaced overburden was compacted in 4-ft layers, and the resultant surface formed a firm foundation for the service area.

In conclusion, it can be stated that coal resources suitable for economic opencast working, even within the well-worked area of this district, are by no means exhausted. Considerable tonnages have been proved which are not yet worked, and there are probably workable sites still to be discovered and assessed.

The writer is obliged to the National Coal Board Opencast Executive for permission to publish this contribution, and in particular he is grateful to Mr G. Jago for his constructive interest.'

LEAD

The earliest records of lead mining in Derbyshire date from 81 AD based on Roman inscriptions upon pigs of smelted lead. Mining at Wirksworth was active in Saxon times, in the 9th century, but there are few records until many centuries later when the district became an important lead producing and administrative centre with Wirksworth as 'the Westminster of the Peak'. Detailed records have been kept for many hundreds of years by the Barmaster of the King's Field in the Soak and Wapentake of Wirksworth. The Barmote Court for the regulation of trade sat in the Moot Hall, where the famous Miners' Standard Dish made in the reign of Henry VIII is still in use today.

Plate 12
Structure map
of the
Carboniferous
rocks

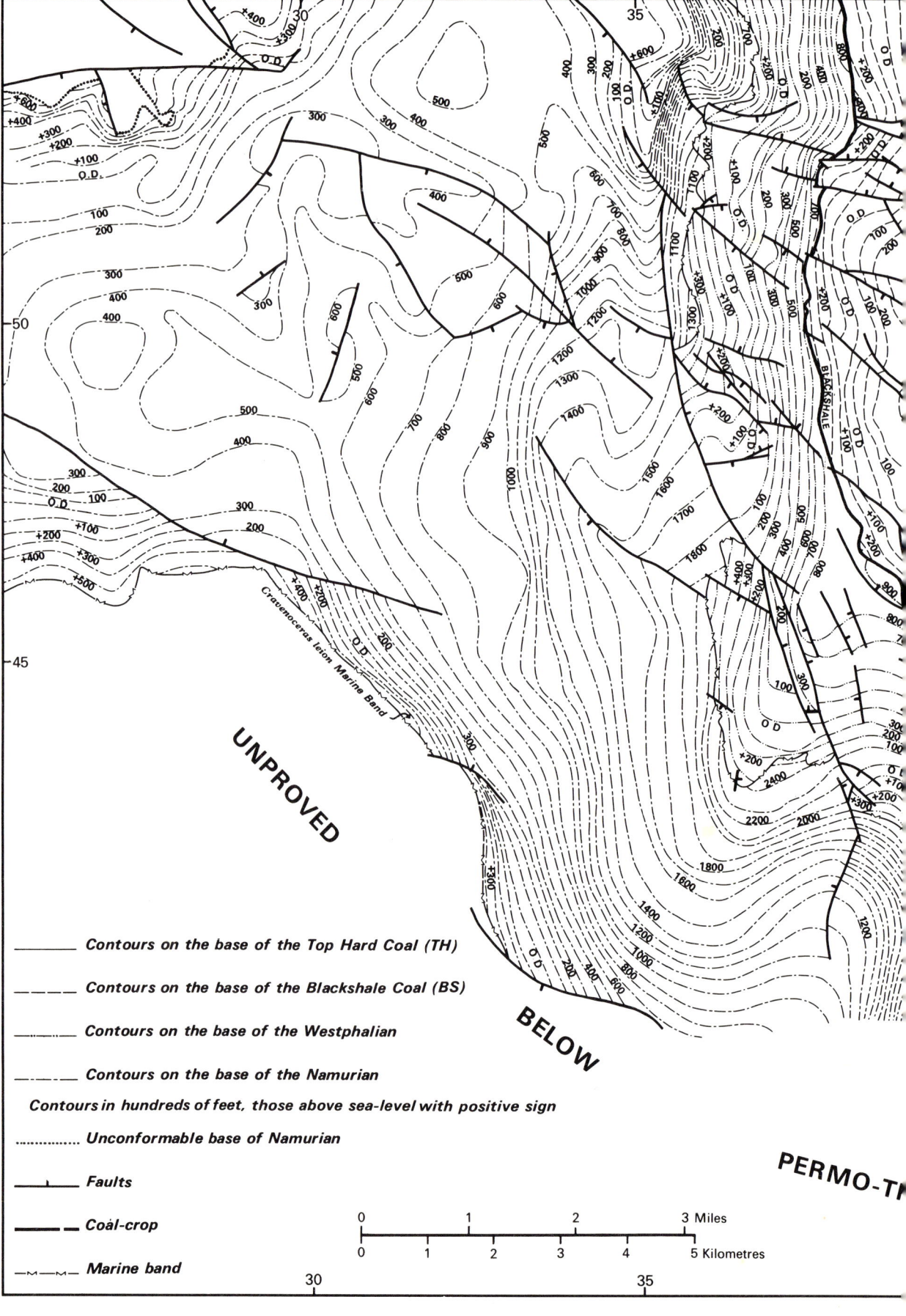

UNPROVED

BELOW

PERMO-TR

_____ *Contours on the base of the Top Hard Coal (TH)*

_ _ _ _ _ *Contours on the base of the Blackshale Coal (BS)*

_ · _ · _ *Contours on the base of the Westphalian*

_ ·· _ ·· _ *Contours on the base of the Namurian*

Contours in hundreds of feet, those above sea-level with positive sign

············ *Unconformable base of Namurian*

_|_|_ *Faults*

_ ▬ _ *Coal-crop*

_–M–M_ *Marine band*

0		1		2		3 Miles

0	1	2	3	4	5 Kilometres

30   35

SSIC

ROCKS

CHAPTER 9

Economic products and water supply

INTRODUCTION

The present wealth and importance of the district owes much to the underlying rocks and over the years every formation represented on the map has provided something of economic use. As early as Roman times the Dinantian limestones were being exploited for the lead deposits they contained. Recently other minerals associated with the lead, such as fluorite, have attained importance. The limestones have been quarried for roadstone and for lime for agricultural and smelting purposes. The Namurian sandstones have supplied millstones and grindstones and have been an abundant source of excellent building stone. Every rock type in the Westphalian sequence has been used, coal, for heating and a multitude of by-products, the mudstones for brick-making, the sandstones for building, roofing and paving, the ironstones for smelting and the seatearths and ganisters for heat-resistant bricks for use in the steel industry.

In the Permian, the Lower Magnesian Limestone has been quarried for lime and for building and ornamental stone, the Middle Marl was dug for manufacturing bricks and plant pots, and the Lower Mottled Sandstone provides a valuable source of moulding sand. The Pebble Beds yield building sands and a pebbly fraction which is ideal for aggregates. The Keuper Marl provides material for brick-making as do the boulder clays. Glacial, fluvioglacial and riverine sands and gravels are now of great importance in the Trent Valley.

The economic products both past and present are described below, followed by an account of the water supply of the district.

COAL

When the resurvey of the Derby district began in 1961 there were some 30 working collieries. Today, as a result of closure and amalgamations, there are only seven, of which five are on the concealed part of the coalfield. In addition the district has been extensively exploited for opencast coal and, although the scale has recently been reduced, two major sites were producing in 1973 and the present trend is one of expansion.

A description of the physical and chemical properties of the coals of the district was given by Dawe and Coles (1948), and the following up-to-date account, including Table 4, has been provided by Mr M. Lock of the National Coal Board.

'The south-western part of the Nottinghamshire and North Derbyshire Coalfield, covered by Sheet 125, contains Annesley, Newstead, Linby, Hucknall, Cinderhill (Babbington), Pye Hill and Moorgreen collieries. The seams currently being deep-mined are the High Main, First Waterloo, Second Waterloo, Deep Soft, Deep Hard, Tupton and Blackshale. The following notes on the properties of these seams are based on analyses of *in situ* seam less dirt samples.

The Deep Soft, Deep Hard and Tupton are the cleanest of the above coals, with an ash range of 3 to 7 per cent. Of slightly poorer quality, with an ash range of 5 to 9 per cent, are the High Main and First Waterloo seams. The Second Waterloo has 8 to 11 per cent ash and the Blackshale a range of 6 to 14 per cent.

The amount of sulphur in the seams worked is generally low to moderately low (0.5 to 1.5 per cent) with the exception of the Blackshale, which has high values ranging from 2 to 5 per cent. Chlorine, an important minor constituent, shows a general easterly increase in this part of the coalfield, with values ranging from a trace to 0.70 per cent and normally 0.30 to 0.60 per cent.

The rank of the seams being worked shows a general increase towards the north-east; seams currently mined are non-caking (rank code 900) to medium caking (rank code 600), the High Main having the lowest rank and the Blackshale the highest. Calorific values (d.a.f.) of the coals are of the order of 13 900 to 14 900 B.T.U./lb. and volatile matter contents (d.m.m.f.) 35 to 42 per cent, but normally 36 to 40 per cent.

The coals being worked are suitable for household and industrial purposes, depending on the size of the product. Larger sizes are utilised for house coal and domestic purposes; the smaller sizes are used mainly for electricity generation, but also for industrial steam raising.

Seams worked at collieries now exhausted produced good, medium and variable quality coals. The Mainbright, Lowbright, Top Hard, Deep Soft, Deep Hard, Tupton, Mickley and Kilburn seams gave ash values ranging from 2 to 6 per cent, the First Waterloo, Brown Rake, Top Soft and Alton seams values ranging from 5 to 10 per cent and the High Hazles, First Piper, Blackshale and Ashgate seams, values ranging from 4 to 18 per cent. Sulphur in the Mainbright, Lowbright, High Hazles, Top Hard, Top Soft, Deep Soft, Deep Hard and Tupton seams was generally under 1.5 per cent, but higher values—up to 5 per cent—were encountered in the First Piper, Blackshale and Alton seams. The extremely high sulphur content of the Alton is associated with the presence of the overlying marine band.'

The following account concerning opencast coal operations, with special reference to prospecting has been provided by Mr J. E. Metcalfe, formerly Regional Opencast Manager (Development), National Coal Board.

'Official opencast coal activity in Great Britain was instituted in 1942, when the Directorate of Opencast Coal Production (DOCP), was established. Control passed to the National Coal Board in 1952, when the responsible body became the Opencast Executive. In the Derbyshire and Nottinghamshire exposed coalfield the whole sequence of workable seams from the Belperlawn to the High Main has been explored.

At the time of publication, there are about 300 prospected sites within the confines of the Derby Sheet. Of these, slightly over half have been worked and restored, and of the remainder, many have been proved to be economically workable but are not yet worked. Some sites which were prospected and abandoned in earlier years

In 1842 the local Barmaster informed a Royal Commission that there were 70 to 80 small mines in the Wirksworth–Cromford area, nearly all of which were worked by their owners.

The mediaeval mining laws, administered by the Barmote Court at Wirksworth, gave the Derbyshire miners legal rights over the landowners in the King's Field to win the lead. This system, established to promote a pioneer expansion of the industry, was not conducive to 19th century capitalist methods. Parliamentary bills in 1857 did little to rectify the situation and the old pattern of mining lingered on. Floatations of lead mining companies were often unsuccessful despite the speculation of thousands of pounds on underground drainage and sough construction (see p. 114.)

The peak years of lead production for Derbyshire were between 1850 and 1870 when the county produced some 9 per cent of the country's total (i.e. 7000 to 11 000 tons of lead). The southern area of the Peak, south of the Wye Valley, proved to be the most important half of the county's lead producing area. Up till 1861 the number of men in Derbyshire lead mines was fairly stable at around 2000, but it declined rapidly to only 281 in 1901. In 1851 there were only twice as many Derbyshire coal miners as lead miners, but ten years later the coal miners were four times as numerous. The technical advances of the Industrial Revolution made little impact on the Derbyshire lead mining industry and output per man remained very low. After 1900 the industry was maintained against all economic odds by miners having part-time jobs in agriculture and in the rapidly expanding textile industries at Matlock, Belper and Duffield.

The industry survives today mainly by virtue of the associated minerals such as fluorite, with lead ore produced only as a by-product. There is little mining activity within the district at the present time.

The history and development of the lead mining industry has attracted a voluminous literature. Contributions by Pilkington (1789), Farey (1811), Green and others (1887), Varvill (1959) and Fuller (1965) provide valuable details and list references for further reading and research into an industry second only to coal in past importance.

Future prospects

Varvill (1959, p. 200) noted that the future working of lead-zinc ores is likely to prove economic only when carried out in conjunction with gangue mineral mining. The example of Millclose Mine (Smith and others, 1967, p. 244), where exploration of the deeper levels proved so fruitful, was not followed by the deepening of any of the old mines in the Wirksworth area. Ford and Ineson (1971, p. 205) suggest that attention be paid to the areas of limestone at shallow depth surrounding the Wirksworth area.

The present survey has revealed a gentle anticline trending south-south-eastwards from the Wirksworth limestone massif with the base of the Namurian lying at an estimated depth of 600 ft, some 1½ miles south of Wirksworth (Plate 12). Geophysical prospecting has confirmed this estimate and geochemical sampling across mapped faults nearby on the sides of the Ecclesbourne Valley has indicated statistically significant enrichment of lead and zinc at surface.

OREBODIES

The Wirksworth area is a continuation of the Matlock mineralised belt, the mining and mineral deposits of which have been described by Smith and others (1967, p. 39).

The primary metallic ores are galena (lead sulphide) and sphalerite or blende (zinc sulphide); the principal gangue minerals are calcite, baryte and fluorite. Secondary alteration of the ores resulted in the formation of small pockets of ochre (limonite), wad (manganese oxide) and calamine (zinc carbonate or silicate). Maximum exploitation of lead ore has been concentrated along the veins with an approximate east–west trend (see Figure 62), but veins of another set trending north-west–south-east have been worked on a smaller scale.

There are three main areas of mineralisation and past mining in the Wirksworth district:
1 A main east–west vein—The Yokecliffe Rake
2 The 'Gulf' area north-east of Wirksworth
3 The Dream Mine area, south-west of Wirksworth.

The Yokecliffe Rake trends westwards from Wirksworth for 2 miles, passing north of Hopton and Carsington before petering out some 700 yd W of St Margaret's Church outside the present district. Its course is clearly marked on the ground by a series of spoil heaps which are distinctly larger than those associated with the north-east trending veins. The Rake is exposed sporadically along its length and occupies the plane of a fault having a downthrow to the south of between 50 and 150 ft. It is best exposed in Yokecliffe Wood [2760 5382] ½ mile W of Wirksworth, where it produces a spectacular fault scarp of Matlock Limestone overlooking the Namurian shales to the south. The fault hades at about 30° to the south and the vein rock, some 5 in thick, consists of radiating aggregates of crystalline calcite, together with disseminated galena. Farther west, in an adit entrance [2697 5383] near Boulderflats Mine, the calcite crystals are arranged normal to the fault plane and the vein is up to 10 ft wide. The Rake was reported to be some 30 ft wide in excavations beneath the dining room of the Gatehouse [2853 5387], Wirksworth (Gibson and others, 1908, p. 175). The calcite-baryte-galena mineralisation of the Yokecliffe Rake accords with the statement by Dunham (1952, p. 83) that the present area lies within the baryte-calcite zone, Wirksworth marking the western limit of the zone with predominant fluorite.

Gibson and others (1908, p. 176) noted that wad (manganese oxide) was dug from a hollow in the 'toadstone' on the Yokecliffe Rake north of Hopton. The Ochre Mine, to which it is assumed they were referring, is north of the Rake and walled by dolomitic limestone, and appears to be the weathered top of a north-trending vein.

The 'Gulf' area north-east of Wirksworth comprises a topographical hollow of down-faulted Namurian shale and Dinantian limestone between the Gulf Fault and the Ashover Grit scarp. The limestones contain numerous veins trending north-west parallel with the Gulf Fault; some are also fault-planes, e.g. the Rantor Vein. These veins were worked in the 19th century but the combination of a high

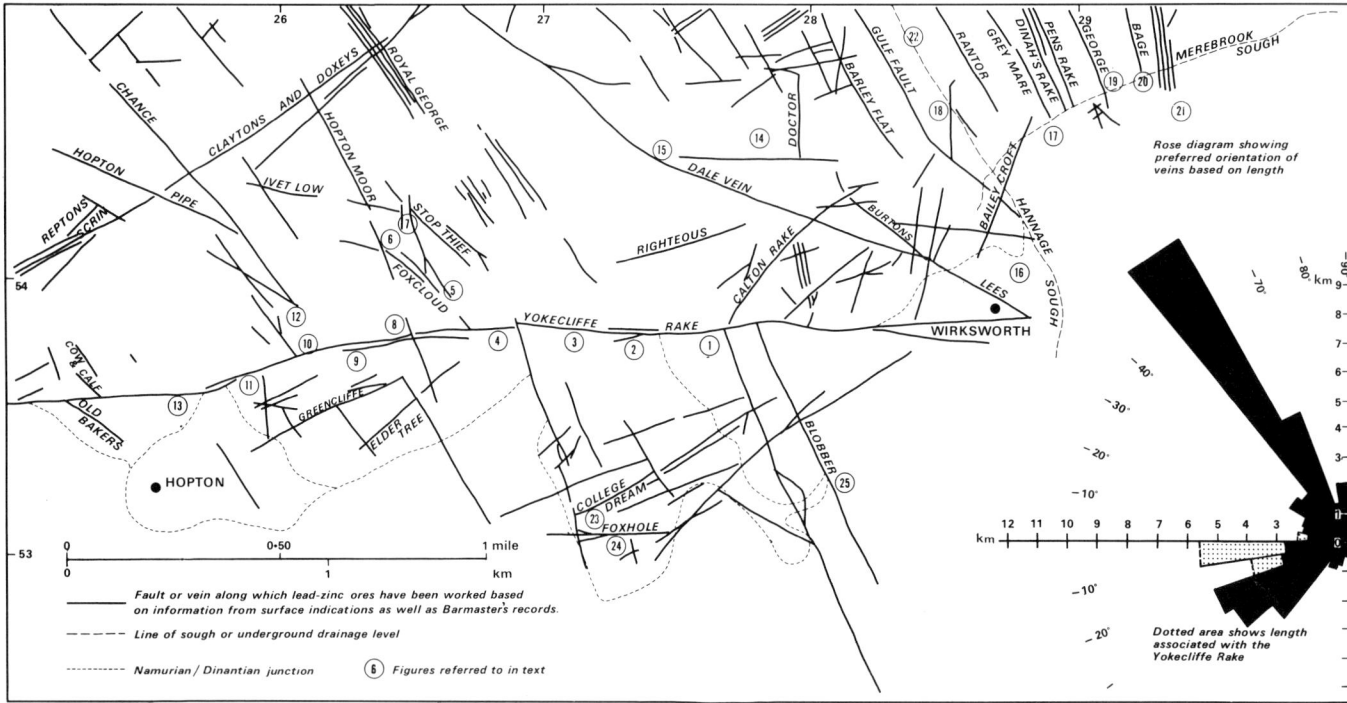

Figure 62 Vein trends in the Wirksworth area

water table and low structural position gave rise to exceptional problems of underground drainage in the mines. To overcome these a network of soughs, or underground drainage channels, were made in the area and its extension towards Matlock. The Hannage Sough, which was started about 1663, was probably the earliest construction. Others followed, including the Cromford Sough and finally Meerbrook Sough (Figure 63), which empties into the River Derwent near Whatstandwell. According to Farey (1811, p. 330), the Meerbrook Sough was started in 1773 and not completed until 1889. It is over $2\frac{1}{2}$ miles long, up to 600 ft below surface and was large enough for the use of boats (*idem*). The cost was over £45 000.

Many of the springs encountered in the mines in this region were of warm water obviously derived from considerable depth.

The most important mines, e.g. Bage and Ratchwood, in the 'Gulf' are just north of the present district. Smaller mines such as Sitch, George, Merebrooksough and Twentylands were less economic owing to smaller size of veins and greater working depths. A zone of mineralisation often occurred below the limestone/shale junction, where the shale had acted as a cap rock to the rising mineralising solutions.

Two veins trending north-west have been worked under Wirksworth itself—the Bailey Croft, and the Blackman Croft—the latter passing under the west end of the church (Figure 62).

The 'promontory' of limestone south of the Yokecliffe Rake contains numerous old shafts and workings, the most successful of which was the Dream Mine. Here an important vein trending WSW–ENE was worked for 140 yd

to the west 'before the shale comes on' (Gibson and others, 1908, p. 175). The present survey suggests that the shale is brought down by a fault and it is possible that the vein continues at a greater depth in the underlying limestone.

The Sandhole Vein runs parallel to the Yokecliffe Rake through the Foxhole—an open chasm from which calamine (smithsonite, $ZnCO_3$) and ochre have been extracted. Buckland found a rhinoceros skeleton in the early 19th century in a cavern discovered in the mine workings referred to as the Dream Cave by Arnold-Bemrose (1910a, p. 48). The intensity of mining on the 'promontory' seems to have been greatest at the southern end, where the structure is anticlinal. Approximately on the axis of the fold in Sprink Wood [2725 5295] fluorite was extracted on a small scale in about 1960. This locality is one mile west of the suggested periphery of fluorite mineralisation. Small fluorite crystals also occur in parts of the pale grey reefy Hoptonwood Limestone on the crest of the 'promontory'. As suggested by Shirley (1959, p. 420), the maximum mineralisation is often associated with the anticlinal areas.

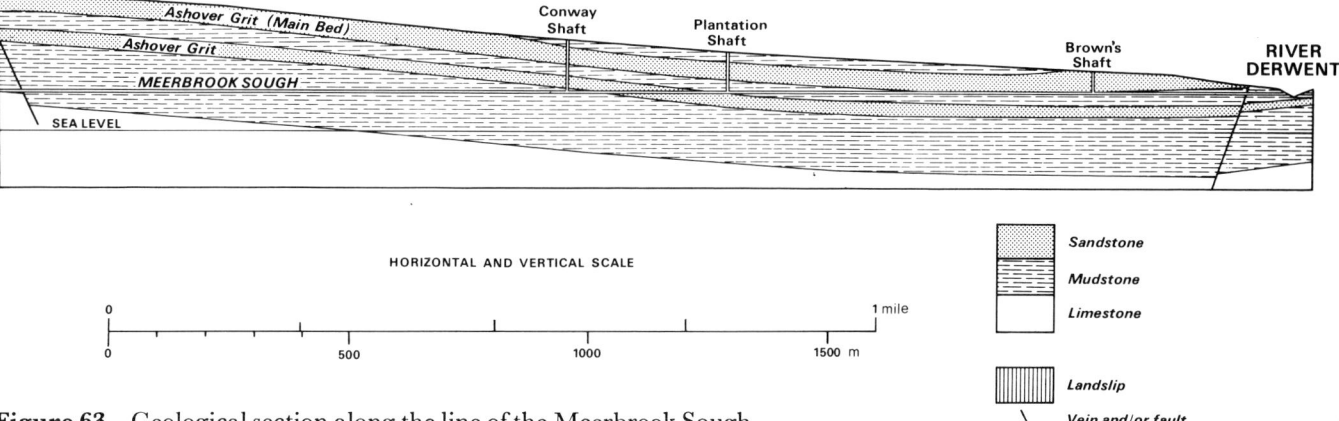

Figure 63 Geological section along the line of the Meerbrook Sough

Details

MINES ASSOCIATED WITH THE YOKECLIFFE RAKE

This rake follows the line of a major fault from about 70 yd S of Wirksworth Church westwards for over 2 miles to Carsington beyond the western margin of the sheet.

The vein has been extensively worked throughout its length and has yielded galena and much smithsonite. In one place the ore was found in the top of the vein, scattered under the soil, over a field known as California, 750 yd west by south of Wirksworth Church (Carruthers and Strahan, 1923, p. 7).

Yokecliffe Mine [2757 5381] (1)[1] was situated at the foot of the Yokecliffe Rake scarp. Vein calcite and galena are still attached to the scarp face, which dips southwards at 70°. Several shafts were sunk through Namurian shales to intersect the vein at depth. The main Engine Shaft [2757 5381] was 342 ft deep. The mine yielded galena and smithsonite, and the main gangue mineral was calcite.

Windmill Lane [2733 5380] (2) Two shafts were sunk on the south side of Yokecliffe Rake to intersect it at depth; no details are known.

Boulderflats Mine [2715 5382] (3) No details known.

Oldgell's Mines [2682 5381] (4) No underground details are known, but extensive surface tips and the presence of several shafts in addition to the main one suggest that the mine was as large and important as Yokecliffe Mine. It differs from mines farther east (see above) in that the mineralisation was associated with dolomitised limestone. An old level [2701 5384] near the Godfreyhole–Middleton Road still shows the Yokecliffe Rake some 10 ft wide, hading at 20° to the south. Calcite commonly occurs as gangue. The main vein is crossed by a north-westerly vein which is of importance

south of Godfreyhole (see Figure 62). A possible extension of this vein north-westwards was worked from Foxcloud Mine [2661 5393] (5) and Nile Mine [2650 5422] (7).

Several scrins leave the Rake in a north-westerly direction, but have usually been worked for only a few yards.

Smithycove Mine [2651 5383] (9) No underground details are known, but many old shafts and a large area of spoil are present. Between Oldgell's Mine and Smithycove Mine a north-westerly vein of some importance has been worked for up to 700 yd in Foxcloud Mine and Shiningcloud Mine.

Shiningcloud Mine [2645 5414] (6), 360 ft deep, is situated on a pipe at the junction of three veins. It produced galena, baryte, and hemimorphite. The mine was dry.

Quickset Mine [2635 5377] (9) No details are available. Boulders of toadstone incorporated in nearby dry-stone walls may be erratics but it is more likely that they came from a lava associated with the fault plane of the Yokecliffe Rake.

Newclose Mine [2613 5376] (10) No underground information is available but the surface tips are extensive. Where the Rake is partially exposed [2595 5369] west of the mine, it consists of calcite and baryte.

Old Jacob's Mine [2590 5367] (11) No details are known. North of Old Jacob's and Newclose mines a small north-west-trending vein 400 yd long was worked from Hewitstone Mine [2604 5394] (12).

Doglow Mine [2564 5360] (13) No details are known. Some 100 yd to the north [2563 5370] ochre was worked at surface from a north-trending excavation.

There are many old shafts to the north of the Yokecliffe Rake which worked small veins and scrins trending north-westwards.

---

[1] Numbers refer to locations on Figure 62.

Unnamed mines and shafts to the south of the Rake appear to have worked veins more or less parallel to it.

Mines north-west of Wirksworth

Much of this area has now been quarried for limestone on such a large scale that the old pattern of rakes and mine workings has been obscured. The two main north-westerly trending veins, Burton Vein and Blackman Croft or Dale Veins were worked from the *Shovel and Toby Mines* [2761 5463] (14) and *Dalefield Mine* [2740 5446] (15) respectively. The Blackman Croft Mine was situated at the junction of these two veins. Dale Quarry was restricted in its working by the presence of Burton's Vein which passes through the middle of the quarry; the adjacent limestone was left as a pillar owing to the unstable and irregular nature of the workings. A southerly extension of this mineralisation, the Lees Vein, passes under Wirksworth Church [2875 5395]. The Little Glory Vein is parallel to but some 150 yd to the south of the Lee's Vein.

A group of small veins approximately at right angles to the main veins were also worked such as the Pinfold Vein from Hawleys Garden Shaft [2847 5399], Pulpit Scrin (beneath the church) from Bright's Mine [2884 5406] (16), and Good Luck Vein from Meerbrooksough Mine [2885 5457] (17) east of the Gulf Fault.

Mines north-east of Wirksworth

Most of the mines in this area on the downthrow side of the Gulf Fault are north of the Derby district and were described by Eden (*in* Smith and others, 1967, p. 45). The veins have a dominant north-westerly trend parallel to the Gulf Fault and were worked from the following mines:

Jenny Shaft [2865 5438] on the Northcliffe Vein.
Twentylands Mine [2852 5462] (18) and *Captain Mine* [2862 5456] on numerous veins between the Gulf Fault and the Rantor Vein.
Meerbrooksough Mine [2885 5457] on the Rantor and Greymare Veins and the Goodluck Vein (north-easterly trend). The Goodluck Shaft was some 258 ft deep; the water level fluctuated from 192 to 242 ft below surface.
George Mine [2908 5477] (19) on George Vein; the mine, linked to the Cromford Sough, was dry.
Hallam's Mine [2925 5475] (20) on Bage Vein; Hallam's Shaft was 357 ft deep with the water level varying from 309 to 367 ft.
Sitch Mine [2935 5464] (21) on veins east of Bage Vein; lead and fluorite are present in its tips.

Mines south of Godfreyhole

The southerly promontory of limestone is anticlinal and surrounded on three sides by shale. A number of veins have been worked in the limestone at crop and for some distance beneath the shale cover, particularly to the east. The veins trend north-north-west or east-north-east and contain galena, baryte, calcite and fluorite.
Dream Mine (or Stafford's Dream) [2726 5322] (23) according to Carruthers and Strahan (1923, p. 80) was very profitable to a depth of 360 ft. Farey (1811, p. 267) noted that the mine yielded much lead and ochre. The vein runs east-north-eastwards and passes under shale.
Sandhole Mine [2735 5307]   The Sandhole Vein runs east–west and has been confused with the Dream Mine on some Ordnance Maps. The mine was drained by Jewlows Sough, driven 900 yd from the stream near Speedwell Mill to the workings. An openwork known as the Foxhole [2728 5308] 75 yd W of the shaft yielded lead, smithsonite and ochre (Farey, 1811, p. 258). From its name it can be surmised that the Foxhole was a pot-hole in the limestone containing sand like that farther north at Brassington (see p. 90).
Bells Scrin is an east-north-east vein some 100 yd to the south of

Dream Vein. Sweenclose Vein is a north-east-trending vein and fault extending from Sandhole Vein 130 yd E of Sandhole Mine to Blobber Vein, a distance of some 700 yd.
Blobber Mine [2809 5327] worked lead from two shafts, each said to be 135 ft deep. An unsuccessful attempt was made to reopen the mine in 1924–25. An old plan by W. M. Else, dated 1917, shows two north-easterly trending veins, both called Blobber. The workings however run north-west–south-east between the two shafts, and plans dated 1925 show the main Blobber Shaft to reach 176 ft. A 90-ft level, now silted up, which ran into the Mill Pond at Millers Green, was used by the miners as a sough. The water in summer stood at about 144 ft below the surface without pumping but in winter rose to 90 ft. Lead was found in lumps weighing 4 cwt in 1925 at the 144-ft level some 60 yd north of the shaft.

In Namurian and Westphalian rocks

Lead was worked at Alport Hill Spout (Farey, 1811, p. 252) in the Alport Mine.

Minor occurrences   Mineralisation of a minor nature has also been noted in the Westphalian rocks. In 1954 Mr A. V. Priest of the National Coal Board noted galena in a small fault [4210 4270] $\frac{1}{2}$ mile S of Mapperley Colliery (Craven, in discussion and contributions to Schnellmann, 1955, p. 626). Galena was also said to occur in masses up to 12 in thick in the floor of the First Piper Coal at the same locality. This is the locality 'a few miles west of Ilkeston' mentioned by Varvill (1959, p. 202). A specimen (M.I.29837) from this locality is on display in the Geological Museum, Exhibition Road. Another specimen which cannot be located more accurately than from 'Mapperley Colliery (brought out by the manager)' was described by Dr J. Phemister in 1943 (E 20377) as 'an argillaceous sandstone cut by veins of zinc blende and calcite'. The horizon of the specimen is not known but if it came from the floor of the First Piper Coal then the locality must have been close to the shafts of the colliery on account of the known dates of workings in that coal.

Varvill (*ibid.*) also recorded lead ore at an opencast site near Ambergate and Mr J. Scarborough of the National Coal Board Opencast Executive noted traces of lead ore in the roof of the First Ell Coal at Godbers Lum Site [405 486] near Denby and in the Second Waterloo Coal at Cromford Canal Site [452 491] in the Erewash Valley.

OTHER MINERALS ASSOCIATED WITH LEAD

The production of lead has in the past overshadowed that of associated minerals, but recently the position has changed.

Zinc

In 1720 the production of zinc in Derbyshire was only 1500 tons. Between 1850 and 1949 Derbyshire produced a total of 80000 tons, nearly half of which came from Millclose Mine to the north of the present district (Smith and others, 1967, p. 243). Zinc is not mined in the district at present, but is the main product of the separation process carried out by C. E. Giulini (Derbyshire) Ltd at Hopton, who are re-working the old tailing dumps from Millclose Mine.

Silver

The percentage of silver in Derbyshire lead averages only 2 to 4 oz a ton, which makes recovery uneconomic.

Baryte

In Derbyshire the mineral is mostly discoloured and of low grade. It has been produced at Upper Golconda Mine just beyond the north-west corner of the district.

Fluorite

As a result of introduction of the basic open-hearth process for steel manufacture, fluorite has become an important mineral.

Wirksworth lies on the western edge of the zone of pyrite-fluorite concentrations (Dunham, 1952), but a small deposit in Sprink Wood [2726 5295] was dug about 1960. It occurs in pale grey limestone of D_1 age in the crest of the anticline forming the promontory. Fluorspar veining was recorded in Dinantian limestone at depths of 2300 to 2400 ft in Ironville No. 3 Oil Bore some 5 miles from the nearest outcrop.

Wad

Manganese ores, known locally under the collective term 'wad', have been extracted in the past at Doglow Wood [2563 5369] near Hopton. Wad was used as a black pigment and delivered to paint mills at Bonsall and Matlock. Wad associated with ochre was also dug in a shallow working [2741 5309] at Foxhole, near Godfreyhole.

LIMESTONE

In recent years owing to increased demand for roadstone and aggregate the output of Dinantian limestone from the Wirksworth area has increased rapidly. The low value of raw limestone restricts the market area. However, the shortage of competing aggregates in the areas to the south and east makes road transport economic even as far as London, 140 miles from Wirksworth. Where there is a steady demand for large quantities, as at Corby, it is economic to use rail transport. A new bunker is proposed at Wirksworth to allow liner trains to carry the products farther afield.

The principal quarries are Stoneycroft [285 543] now closed, and Middlepeak [282 547] which has been described in detail in the Quarry Managers' Journal for December, 1967 (Vol. 51, No. 12). The latter quarry originally produced lime and stone for use as flux in the steel industry, but it is now extensively modernised and produces road aggregate with a capacity well in excess of 400000 tons annually. The quarry is now operated by Tarmac Roadstone Holdings Ltd. Dale Quarry [284 541] has closed since the completion of the survey due to the increasing height of the faces and the lack of space within the quarry for tipping. Analysis of the white limestone from the bottom of the quarry (Hoptonwood Limestone, D_1) in per cent was as follows: SiO_2 0.36; Fe_2O_3 0.07; Al_2O_3 0.05; CaO 55.00; MgO 0.43. Percentages for the dark blue limestone from the top of the quarry (Matlock Limestone, D_2) were SiO_2 2.04; Fe_2O_3 0.26; Al_2O_3 2.18; CaO 52.70; MgO 0.72 (see also Smith and others, 1967, p. 241).

In the Crich area, some quarries are now abandoned and used for tipping. Crich Cliff Quarry, north of Crich village (Smith and others, 1967, p.35), supplied the limeworks at Ambergate until 1959. The works were erected over 100 years ago by George Stephenson's Clay Cross Company and used a steeply graded tramway to transport the limestone from the quarry at 750 ft OD to the works almost 2 miles due south at 300 ft OD. After 1959 the works obtained limestone by road from Cromford but increasing costs and falling markets caused their closure in 1965. The Ambergate site has now been taken over by the East Midlands Gas Board.

Quarrying of dolomitic Dinantian limestone was carried out in recent years by Magnesium Elektron Ltd at Hopton. The quarry [2570 5465] aimed to provide some 80000 tons of dolomite annually and this was expected to yield 5000 tons of magnesium by a thermal process using over 20000 tons of fuel oil, 5000 tons of ferrosilicon and a small quantity of fluorspar. Unfortunately the dolomitisation of the limestone was too variable and the works are now closed. Parsons (1922, p. 114) gives two analyses of the dolomitic limestone at Hopton. The percentages of $CaCO_3$ were 56.8 and 62.2 and of $MgCO_3$ 40.5 and 34.3 respectively. Iron content was 0.9 and 1.1.

The Lower Magnesian Limestone has been extensively worked in the past, and quarries near Linby [534 522] are said to have been used by monks from Newstead Abbey in the 12th century. It has been quarried for building stone, road metal and agricultural lime. It is a particularly attractive building stone, with a yellow-brown colour and coarse texture, which has withstood the years of industrial atmosphere particularly well. Many of the older houses and terrace rows in Hucknall and Bulwell are constructed of the stone. Linby village was built almost entirely from the local stones. The rather friable nature and open texture of the Lower Magnesian Limestone makes it less suitable for motorway construction, for which Dinantian limestone has been preferred. It is however still used locally as hard-core. The stone is also in demand as rockery, walling and ornamental material. The principal working quarries are at

Quarry Banks [537 525], Linby, and at Bulwell [533 455]. The following chemical analysis (percentage figures) from the quarry at Linby is of limestone about 12 ft up in the quarry face (E15986) (analyst, Mr I. G. Evans of the Building Research Station): SiO_2 6.38; Al_2O_3 0.93; Fe_2O_3 1.28; MgO 18.78; CaO 28.40; Na_2O 0.07; K_2O 0.12; TiO_2 0.04; Mn_2O_4 0.13; SO_3 0.07; loss on ignition 43,86—total 100.06.

IRONSTONE

Ironstone in the form of nodules, either dispersed or concentrated in bands or ('rakes'), are present in the Westphalian mudstones. Most of the horizons that have been worked lie between the Top Hard and Kilburn coals. They were mined in the past by means of closely spaced bell-pits or by opencast pits; at least one, the Black Rake (p. 48), was also worked extensively underground.

Yields were up to 6000 tons of ore per acre with up to 33 per cent metallic iron content. The annual output in 1873 from the Derbyshire–Nottinghamshire Coalfield reached almost 500 000 tons, but after that date the industry went rapidly into decline.

Smyth (1856) and Gibson and others (1908) list the 'rakes' of the district and their positions in the Westphalian sequence. There is however confusion of the naming in some working plans, for ironstones of similar character but at different horizons have been given the same name. Some of the more important ironstone horizons are indicated in the detailed account of the Westphalian rocks. Smyth and Gibson and others quote analyses of several 'rakes' from within the district. The following are the percentage range of the results of the seven available analyses from six 'rakes': 'Protoxide of iron' 33.31 to 40.01; 'Peroxide of iron' 1.04 to 2.71; 'Protoxide of manganese' 0.96 to 2.18; 'Alumina' 0.41 to 1.14; 'Lime' 2.32 to 4.53; 'Magnesia' 2.44 to 5.43; 'Carbonic acid' 24.83 to 29.92; 'Phosphoric acid' 0.34 to 1.12; 'Bisulphide of iron' 0.05 to 0.26; 'Water hygroscopic' 0.45 to 0.74; 'Water in combination' 0.87 to 1.87; 'Organic matter' 0.36 to 1.85; 'insoluble residue' 15.8 to 27.42. Of the insoluble residues, 'Silica' ranged from 10.04 to 17.24; 'Alumina' 4.51 to 7.9; 'Peroxide of iron' 0.45 to 1.22; 'Lime' trace to 0.07; 'Magnesia' 0.03 to 0.27; 'Potash' 0.34 to 0.74. The total amount of iron ranged from 26.79 to 33.20 per cent.

SANDSTONE

All the major Namurian and Westphalian sandstones have been worked at various times for building, walling and paving stones, but none are exploited on a large scale at present. Gibson and others (1908, pp. 177–179) described the quarrying of the Namurian sandstones of the district, an industry which is today virtually defunct.

The Ashover Grit is the thickest Carboniferous sandstone and has been quarried in the past at Alport Hill [304 516] and Crich [346 530]. The quarries at Shiningcliff Wood and Crich Chase produced millstones.

The largest quarries have worked the Main Bed of the Ashover Grit particularly between Milford and Little Eaton (p. 134c). Small quarries also worked the thinner sandstones beneath this Main Bed, which in the Callow area are stained pink and purple due to a former cover of Triassic rocks. The village of Kirk Ireton is notable for buildings made from this coloured stone.

The Rough Rock is quarried at Ridgeway [3585 5150] for building and ornamental stones. The topmost beds are pale coloured, fine-grained and particularly hard and well cemented. The same quarry has yielded greater quantities of the coarser grained Crawshaw Sandstone, largely for walling purposes. This is the only quarry in the district regularly worked at the present time for the supply of building stone.

The Wingfield Flags are, as their name implies, too thinly-bedded to be of much use in building. They were used in the past however to build Wingfield Manor [374 547] on the northern margin of the district. On the outcrop are many small quarries once worked for paving and walling stones.

Sandstone in the measures above the Morley Muck Coal near Rowson Green [376 470] is sporadically worked for ornamental purposes. The sandstone overlying the Blackshale Coal was once worked for house building from Pentrichcommon Quarry [3830 8385]. The 'Tupton Rock' used in the construction of Codnor Castle [4335 4995] came from thin flaggy beds in the nearby quarries [4333 5180], but the stone has weathered badly and the castle walls are now in ruins.

Numerous sandstones higher in the Westphalian, notably those above the Deep Soft, Top Hard and Cinderhill and those below the Clown and High Main seams, were once worked in small quarries.

MOULDING AND BUILDING SANDS

An old quarry in the Lower Mottled Sandstone at Catstone Hill [5075 4145] was once worked to provide moulding sand for Stanton Ironworks but has now been filled in. The same beds have been quarried at Bramcote Hills [503 389] and are now worked by British Industrial Sand Ltd for both moulding and building sand. The company has kindly provided the following analysis of saleable sand for foundry use (percentages figures): SiO_2, 87.5; Al_2O_3, 4.41; Fe_2O_3, 2.51; CaO, 1.00; Na_2O, 0.08; K_2O, 1.41; loss on ignition 2.16. Grading data for the two types of sand are as follows:

Grading on washed Moulding sand		Grading on washed Building sand	
Mesh	Percentage retained	Mesh	Percentage retained
16	Nil	$\frac{3}{8}$ in	Nil
30	Nil	$\frac{3}{16}$ in	0.25
44	2	7	2.85
60	21	14	2.60
100	38	25	7.73
150	24	52	43.69
200	8	100	33.19
−200	7	−100	9.95
Clay content	8	Clay content	5.0

Many other quarries in the district have worked the Lower Mottled Sandstone for building and moulding sand, e.g. Stanton by Dale [460 382 and 452 366] Dale Abbey [4355 3882 and 4305 3930], Nottingham [546 388] and Nuthall [5275 4430].

The Pebble Beds have also been extensively quarried in the past for building sand. The principal quarries were Sandiacre [478 372], Dale Abbey [4355 3882], the Flourish [4285 3870] and Dunnshill [4220 3846]. Sands from the Pebble Beds near Muggington worked by Blue Circle Aggregates Ltd are extensively quarried (see below) and utilised for brick-making, plastering and for concrete, including spun concrete pipes.

AGGREGATES

The use of Dinantian limestones for aggregate has already been noted (p. 109).

In the west of the district the Pebble Beds are variable in lithology but in general pebbles are abundant and provide a valuable supply of aggregate. The beds thin and show an increase in the clay content southwards. The largest quarries are north of Mugginton [2785 4520, 2670 4440 and 2920 4840], where up to 130 ft of friable pebbly sandstones are present. They are worked by Blue Circle Aggregates Ltd, who produce about 8000 tons per week of aggregates and sands for the building industry after multiple crushing and screening processes. The aggregates range from $\frac{1}{16}$ to $\frac{7}{8}$ in and the small sizes are sold to firebrick manufacturers and for potting compost; $\frac{1}{4}$-in aggregate is used locally for concrete products and, with $\frac{3}{8}$-in, as road chippings. The larger sizes are utilised as concrete aggregates.

A large outlier of Pebble Beds has been removed almost entirely by quarrying at Hulland Ward [2600 4570]. The site is now occupied by Charcon Products Ltd, who make concrete slabs and kerbstones using imported raw materials.

Aglite (Midlands) Ltd of Denby [478 390] produce a low-density cellular aggregate from local Westphalian mudstones by means of a sintering process to 1250°C. The end product after crushing and grading is utilised in the construction industry for lightweight concrete with good thermal insulation and fire resistant properties. It is also made up into concrete blocks and slabs.

REFRACTORY MATERIALS

Ganister

A mine [3583 5202] belonging to Mr H. Glossop at Ridgeway, Ambergate, worked in the 1920's a very fine, hard, siliceous and ferruginous cemented sandstone about 3 ft 9 in thick (Wedd *in* Strahan, 1920, p. 75) immediately beneath the Pot Clay Marine Band. It was also worked by opencast methods. The stone was sold for making ordinary silica bricks for Sheffield.

Farther south a trial excavation was made in ganister at the same horizon and Wedd (op. cit., p. 76) quotes the following upward sequence: ganister, 2 ft+; coal smut, 1 in; ganister 6 in. The stone dries to almost white and is com-

parable in appearance with the best Durham ganisters, but not so compact and quartzitic as the best Sheffield ganister. Wedd (*idem.*) gave a petrographic description of the lower ganister (E 11789), and an analysis is shown in Table 5.

In 1920 Litchfield's Ganister Mine [359 505] (Wedd, op. cit., p. 77) was no more than a trial shaft some 24 ft deep through the base of the Crawshaw Sandstone into the underlying shale and the ganister at the top of the Rough Rock. The ganister was similar to that at the previous two localities but was apparently unworked.

Sabine, Guppy and Sergeant (1969) have petrographically described the Glossop's and Litchfield's ganisters as follows:

Glossop's Ganister (E 11792) A grey sandstone impregnated with hematite along joint faces and containing some plant remains. The rock is composed of angular quartz grains mainly 0.06 to 0.15 mm across with ferruginous and quartzose cement. There are scattered muscovite flakes and plentiful heavy mineral grains.

Litchfield's Ganister (E 11791) A greyish brown fine-grained sandstone with very small plant remains. The rock is composed of angular quartz grains commonly 0.05 to 0.1 mm across, scattered muscovite flakes and some heavy mineral grains, with interstitial iron-stained clay minerals, iron ore granules, carbonaceous material and a siliceous cement.

Chemical analyses of these rocks from the same source are given in Table 5.

A ganister said to have been recognised by Mr F. Russell as Glossop's Ganister in the railway cutting north of Wingfield Park (Wedd *in* Strahan, 1920, p. 70) is in fact a higher bed below the Norton Coal. A ganister beneath the Forty-Yards Coal was worked in two pits nearby [3705 5305 and 3665 5255] by the Derby Silica Brick Company in 1942. The ganister, some 3 ft thick, underlay a 7-in coal and rested on fireclay.

Other old ganister pits [3698 5372 and 3713 5352] in the Wingfield Park area worked the ganister, 30 to 33 in thick, beneath the Alton Coal. This ganister was also worked in the Bullbridge Pit [362 519] near Ambergate (Wedd *in* Strahan, 1920, p. 74).

Chemical analyses of the 'White Clay' and 'Bottom Clay' presumably respectively overlying and underlying the ganister and taken from Ennos and Scott (1924, p. 50) are tabulated below (Table 5). These authors quote one further analysis of fireclay from the Pyebridge area but its horizon is not known.

Wedd (*in* Strahan, 1920, p. 69) noted the presence of workings in fireclay below the Deep Hard Coal and in silica rock in a sandstone above the Tupton Coal in mines near Riddings. The silica rock was mixed with the fireclay to make bricks for boilers, coke ovens, cement and pipe kilns. The company, James Oakes and Company (Riddings) Ltd, have kindly provided the following information on their activities together with analyses of their clays. 'Brick-making continued until about five years ago (i.e. 1968), but the main development has been in the manufacture of vitrified clay pipes and fittings. Originally the clay used was won in the deep mine and adjacent to the pipeworks (Pye Hill) but since the 1930's the source of clays has been from opencast sites in the area.

Table 5 Chemical analyses of ganisters and fireclays

	1	2	3	4	5	6
	%	%	%	%	%	%
SiO_2	96.10	94.36	60.20	80.20	87.83	61.75
Al_2O_3	0.37	1.45	22.46	11.32	7.50	23.11
Fe_2O_3	1.90	1.46	3.43	1.26	0.84	1.69
MgO	0.12	0.18	0.64	0.19	0.11	0.81
CaO	0.15	0.11	0.24	0.09	0.14	0.12
Na_2O	0.15	0.15	1.24	0.49	0.73	0.29
K_2O	0.10	0.38	2.06	0.86		3.27
$H_2O > 105°C$	0.46	0.84	—	—	—	1.55
$H_2O < 105°C$	0.14	0.22	—	—	—	5.61
TiO_2	0.22	0.42	0.71	0.88	0.38	1.10
P_2O_5	0.02	0.02	—	—	—	0.05
MnO	0.09	0.13	—	—	—	tr.
CO_2	0.03	0.05	—	—	—	0.03
ZrO_2	0.19	0.14	—	—	—	—
FeS_2	n.d.	n.d.	—	—	—	0.04
Fe_2S_3	—	—	—	—	—	0.06
H (organic)	—	—	—	—	—	0.02
Li_2O	n.d.	tr.	—	—	—	tr.
C	0.09	0.12	—	—	—	0.56
Loss on ignition	—	—	9.40	4.85	2.68	7.94
Total	100.13	100.03	100.38	100.14	100.21	100.06

Saturated density	2.50	2.51
Dry density	2.41	2.38
Grain density	2.66	2.64
Effective porosity	9	13

1 Glossop's Ganister.
2 Litchfield's Ganister.
3 Bullbridge Pit, white clay overlying Alton Ganister.
4 Bullbridge Pit, bottom clay below Alton Ganister.
5 Fireclay below Glossop's ganister.
6 Riddings fireclay below Deep Hard Coal.

The analyses of Glossop's and Litchfield's ganisters are from Sabine, Guppy and Sergeant (1969) and the remainder from Ennos and Scott (1924).

Table 6 Analyses of fireclays

	Deep Hard Fireclay	Main Soft (Deep Soft) Fireclay	Dunsil Fireclay
	%	%	%
SiO_2	63.00	56.10	65.06
Al_2O_3			
TiO_2	25.90	30.74	21.52
Fe_2O_3			
MgO	0.87	1.05	0.45
CaO	0.40	0.68	0.40
K_2O	3.28	2.23	5.30
Na_2O			
Loss on ignition	6.55	9.20	7.04

The clays used are low-grade fireclays, principally those below the Deep Hard and Main [Deep] Soft coals but clays below the Piper, Yard, Threequarters and Dunsil seams have also been used.

In addition to the fireclays the shale overlying the coal seams has been used for brickmaking and to a lesser extent for pipemaking.

The clays are dry ground after blending and are tempered to about 15 per cent moisture content to make them suitable for extrusion from vertical auger type presses. At this works both traditional intermittent kilns and tunnel kilns are used to fire the product to over 1100°C to achieve the low porosity needed for drainage purposes. The consumption of clay is currently about 75 000 tons per annum, but a considerable increase in usage is envisaged in the near future.'

Typical analyses of three of the fireclays mentioned are given in Table 6.

Other ganisters were noted in the Kilburn Shaft deepening, some 22 and 50 ft below the Alton Coal and 52 ft below the Norton Coal (Wedd *in* Strahan, 1920).

Joseph Bourne and Co. Ltd of the Denby Pottery, who make high quality ceramic ware for kitchen use, originally utilised fireclays within the Deep Soft group of coals from the Denby area, but now obtain their raw material from elsewhere.

BRICK CLAY

Carboniferous rocks

Large quarries exist near the Ambergate Brick Works [3600 5182], where a variety of house bricks are made from the shales and associated strata above the Pot Clay Marine Band. The quarries were formerly worked for ganister (p. 176a).

Large areas of excavated ground are present in the Riddings–Lower Somercotes area, where James Oakes and Co. have removed Westphalian A strata for brick-making. A large field [4315 5225], now largely composed of made ground, was once worked for clay, the measures extracted being between the First and Second Piper coals. At Lower Somercotes [434 533] the measures between the Deep Soft and Roof Soft coals have been extracted. New trial excavations [4375 5328] made to the south during the resurvey proved that suitable mudstones continue along the strike.

Measures which include the Waterloo group of coals are worked at Waingroves [408 488] near Ripley by the Butterley Co. Ltd for class B engineering and facing bricks; production is about 960 000 bricks per week. The Manners Brick Co. Ltd near Eastwood [469 458] also work measures at the same horizon for building bricks.

The Loscoe Brickworks Ltd [426 471] near Heanor utilises the measures between the Bottom Second Waterloo and Top First Waterloo coals for the production of bricks for most types of building work including good quality facing bricks. Production in 1972 was of the order of 1 200 000 bricks.

Messrs G. S. Langton and Co. excavate [374 470], near Rowson Green, up to 2000 tons per week of the measures overlying the Morley Muck Coal for use in local pipe and brickworks, and the Denby Mining Co. at Ryefields [393 470] (New Winnings Colliery) intermittently work the measures between the Deep Hard and Deep Soft coals.

An analysis of the brickclay from Birchwood Quarry [435 533], Lower Somercotes, in the Deep Soft group of coals has been provided by James Oakes and Co. (percentage figures): SiO_2 62.80; Al_2O_3 20.00; Fe_2O_3 3.75; MgO 1.20; CaO 0.45; Na_2O 1.60; K_2O 2.14; TiO_2 1.36; loss on ignition 6.42.

Permo-Triassic rocks

The largest brickworks in operation at the time of the resurvey was at Chilwell Brick Pit [513 357]. The works, which have since closed, used the Keuper Marl for its raw material. The Plains Skerry formed the highest beds of the pit and the marl below was massive and relatively free from the siltstone bands and laminae detrimental to the colour and strength of the bricks. Keuper Marl at a similar horizon was formerly used extensively in Derby [334 358] for the manufacture of bricks.

Good quality bricks were made from the Middle Marl at Watnall [510 480]. The Watnall Brick Yard is now used by the National Coal Board who import clay from elsewhere. At Bulwell the main use for the clay was for the manufacture of flower pots by Richard Sankey and Son Ltd. As the maximum workable thickness of marl was about 20 ft the old clay workings have left large areas of waste ground. Recently plastic pots have been manufactured at the same site.

OIL AND GAS

Attention was first drawn to the Alfreton area in 1847 by seepage of oil into colliery workings in the New Deep Pit near Riddings (Marshall, 1875). The workings were in the Kilburn Coal some 720 ft below surface and the 'spring' produced 300 gallons of oil daily for over 12 months before drying up. It is perhaps significant that the explosion which occurred during the sinking of the New Deep Pit was probably a result of gas seepage from the Kilburn Sandstone some 120 ft above the Kilburn Coal.

Ironville No. 1 Oil Bore was sited on the southern end of the Ironville Anticline, near the region of previously known seepage and close to the Riddings Fault which, it was assumed, had acted as a feeder tapping oil from the Dinantian limestone. Drilling commenced in March 1919. At 2031 ft there was a show of residual heavy oil and at 2405 ft an artesian head of water was encountered. The Dinantian limestone was drilled for some 1600 ft and further traces of oil were found at 2500 and at 3600 ft, 30 ft from the bottom of the borehole. Large quantities of salt water at 3500 ft could not be shut-off; this made attempts to 'shoot' the well very difficult. A 180-lb dynamite charge failed to initiate an oil flow and the hole was abandoned in September 1920. Ironville No. 2 Oil Bore, drilled to the north in the Chesterfield district, also found traces of oil (Smith and others, 1967, p. 38).

Ironville No. 3 Oil Bore commenced drilling in October 1956 at a site on the eastern flank of the anticline, well clear of known surface faults. The top of the Dinantian limestone was reached at 2220 ft. Despite penetration of 522 ft of limestone, no oil was tapped and the hole was abandoned at a depth of 2742 ft. It was cemented back to about 1800 ft to preserve a head of gas originating from the Ashover Grit. In 1958 Ironville No. 4 Oil Bore was drilled about 450 yd SSW of No. 3 in a final but unsuccessful attempt to locate an oil accumulation.

The results of the Carboniferous oil bores were disappointing in view of the promising structures. The presence of oil traces at many horizons suggests that most of the original oil had escaped upwards along faults with only the heavy fraction remaining. The Hardstoft Borehole farther north (Smith and others, 1967, p. 249) was fortunate in penetrating a small reservoir which had an effective seal.

There still appears to be an escape of gas in the Riddings area, for an inflammable vapour was recorded in cellars and manholes in 1966.

During the resurvey a seepage of oil was observed from sandstone at the top of the Waterstones near Stapleford [493 361]. Some 30 ft of the sandstone had been excavated for the new Sandiacre–Stapleford By-pass, revealing a gentle anticlinal structure which had trapped this local pocket of oil. British Petroleum drilled an exploratory hole, Stapleford No. 1 [4905 3596], nearby in 1966 through some 300 ft of Triassic strata (p. 173c) into the Westphalian but without tapping any further oil.

Other oil 'shows' have been encountered, as in Newstead Colliery shaft some 1180 ft below the surface in a sandstone below the High Hazles Coal. There may be many pockets of residual oil in the district, but the chances that an effective trap containing high grade oil of economic dimensions is present are remote.

MISCELLANEOUS

The 'smudge' of the Belperlawn Coal was worked [3595 5120] near Ambergate in 1943–44 for sale to the Via Gellia Colour Co. for making the dye known as vandyke brown. Under an overburden some 3 ft thick the weathered seam was between 24 and 36 inches in thickness.

WATER SUPPLY

Until the late 19th century the upper reaches of the Derwent and its tributaries were only of local importance and were used mainly as sources of power and soft water for the textile industries. In 1899 the Derwent Valley Water Board was created to develop the water resources of this highly favourable area and to supply Derby, Nottingham, Leicester and Sheffield. With effect from April 1974 the Severn–Trent Water Authority is responsible for the public water supplies of the district.

The rainfall of the district varies from 35 inches in the north-west to 25 inches in the south-east. There are no large reservoirs but supplies are obtained from wells and boreholes in the Carboniferous, Permian and Triassic strata and from old mine workings in the Dinantian limestone and Westphalian rocks. There is also abstraction from shallow wells, large diameter boreholes and tunnels in the river deposits of the main valleys of the Trent, Derwent and Erewash.

A comprehensive account of the groundwater hydrology of Derbyshire was given by Stephens (1929) and of the Trent River Basin by Downing and others (1970). More detailed work on specific aquifers has been undertaken by Land (1966) on the Bunter Sandstone of Nottinghamshire and by Downing and Howitt (1969), who reported on the saline groundwaters in the Carboniferous rocks of the East Midlands.

The following account considers sequentially each geological formation of the district. Acknowledgement is made for the help of the former Trent River Authority and the South Derbyshire Water Board in providing the modern statistical data upon which much of the account is based.

Dinantian

The Dinantian limestone has a low intergranular porosity and permeability contrasting with a high fissure flow. Because of the presence of numerous joints, infiltration is rapid and normally surface run-off is nil. Impervious beds within the limestone such as clay bands, tuffs and lavas cause perched water tables. Mining has interfered with the natural flow of underground water and much of the limestone in the district is artificially drained by the Meerbrook Sough (p. 106 and Figure 86) which takes the limestone groundwater eastwards into the Derwent Valley.

Well yields in limestone aquifers tend to show considerable variation depending on the abundance of fissures (Downing and others, 1970, p. 36). Groundwater temperature is normally between 9° and 13°C but warmer temperatures are known locally, e.g. Matlock Bath (20°C) north of the present district and Warmbrook Sough [286 537] south of Wirksworth. The temperature of the water at the outfall of the Meerbrook Sough varies between 13.5°C and 15°C. Such occurrences of thermal water are thought to be due to local meteoric waters circulating to considerable depths before emergence (Edmunds, 1971).

In the early part of the century there was little correspondence between the flow of Meerbrook Sough and the rainfall amounts. Recently however there has been a 3 months' interval between maximum rainfall and maximum outfall. Water abstracted from Meerbrook Sough during the year ending March 1973 totalled 3611 m.g.

Water from the disused Ladyflatts Mine is pumped from Blobber Shaft (72 in diam.) [2811 5322] at a licensed annual rate of 183 m.g. for public use. Attempts were made to pump the workings dry preparatory to re-opening the mine but powerful pumps had little effect on the water levels. Analysis of the water compares closely with that of water from the Meerbrook Sough (Table 8).

There are two chemical types of groundwater, i.e. a calcium bicarbonate type and a calcium sulphate/bicarbonate type, which show no regular regional pattern. Total hardness of shallow groundwater varies between 148 and 400 mg/l as $CaCo_3$.

The Widmerpool Formation, composed largely of impermeable strata, is an unimportant aquifer. Annual licensed abstraction from springs for public supply is 10 m.g. whilst that from shallow boreholes is 2 m.g.

Namurian

The two principal aquifers of the Namurian are the Ashover and Chatsworth grits. They are separated by impervious shales and normally there is no hydraulic continuity between them. The Ashover Grit is coarse-grained and some intergranular flow occurs near the surface, but the largest yields are obtained from fissured strata. Springs commonly occur at or near the base of sandstones and the water is collected and stored in tanks such as those at Kirk Ireton which supply the needs of the village. Farther south at Millington Green [2647 4775] a spa 'well' provides a constant flow of cold sulphurous water formerly sought after as a cure for rheumatism.

The natural flow from springs in the district is subject to considerable seasonal variations and ranges between 7000 and 240 000 gallons per day (g.p.d.). Yields from boreholes in the Namurian range from a few thousand to 40 000 g.p.d. The most reliable supplies come from boreholes penetrating more than one sandstone and some have an artesian flow. Total hardness varies between 23 and 236 mg/l.

The total annual licensed abstraction from Namurian strata is 752 m.g. Most of the supply is from boreholes and 727 m.g. are from public undertakings. Agriculture uses 9 m.g. and industry only some 4 m.g.

Analysis of water from a spring at Whatstandwell [332 543] is shown in Table 8.

Westphalian

The thicker beds of sandstone in the Westphalian sequence, such as the Crawshaw Sandstone, 'Tupton Rock' and 'Top Hard Rock', provide aquifers of local importance. They are commonly well jointed but faulting breaks them into separate blocks between which there may or may not be hydraulic continuity. Most blocks therefore possess only limited outcrop areas over which surface recharge can take place and for this reason yields from boreholes are poorly sustained.

Mining, by making artificial connections between aquifers has disturbed many of the natural rest water levels. Large areas of backfilled opencast sites have effectively sealed off many square miles of permeable outcrop from possible surface recharge. Thus underground workings originally troubled by water have sometimes become dry after the same seam, up-dip, has been removed by opencast methods. The Derbyshire pits are some of the wettest in the country, largely due to their location in synclinal belts. Large quantities of water are impounded in old workings of the Top Hard, Dunsil, Second Waterloo, Deep Soft, Deep Hard, Tupton and Blackshale coals, which are overlain by fissured water-bearing sandstones. The concealed part of the coalfield is drier because the impervious Lower Marl of the Permian seals off the Westphalian 'crops' and the underground workings from the Permo-Triassic water. Workings also tend to be deeper with less associated fissuring.

The chemical nature of groundwater in abandoned mine workings is very variable. Considerable treatment is necessary in most cases to render the water usable, with frequent checks on chemical and bacterial quality.

The typical hardness of water from colliery workings ranges from 397 to 578 mg/l. High figures obtained from

Cinderhill (Babbington) No. 6 Shaft and Annesley Colliery may reflect seepage of Permo-Triassic water.

Table 7 shows the volume of water drained annually from mines in the district since 1970.

Table 7 Volume of water (in millions of gallons) abstracted from mines in 1970–72

Colliery	1970	1971	1972
Woodside	531	598	585
Pyehill (Soft Coal Shaft)	80	67	73
Pyehill (No. 1 Shaft)	125	25	24
Selston (Pyehill No. 1)	15	17	16
Annesley (No. 1 Shaft)	3	4	6
Cinderhill (Nos. 4 and 6 Shafts)	255	192	186
Cinderhill (No. 7 Shaft)	7	7	15
Hucknall No. 2 (No. 5 Shaft)	10	12	16
Linby (No. 1 Shaft)	3	4	1
Moorgreen	24	25	35
Moorgreen (Watnall No. 1 Borehole)	65	48	46
Newstead (No. 1 Upcast Shaft)	8	11	15
Total	1126	1010	1018

As in the Namurian, borehole yields in Westphalian rocks are more reliable where several sandstone aquifers are penetrated. In general 200 ft × 12 in diameter boreholes in the Westphalian rocks yield up to 6000 g.p.h. Slack Lane Waterworks [3948 5028], Ripley, produced some 20 000 g.p.d. from an 8-ft diameter 180-ft water bore in 1923; nearby mining was blamed for a decrease in supply and eventual abandonment of the borehole. Comparison of the analyses (Table 8) shows that the borehole water from the Westphalian rocks is softer than that from the old workings.

Permo-Triassic

The Lower Magnesian Limestone is confined both above and below by impervious 'marls' and where the Lower Magnesian Limestone/Lower Marl junction is exposed in the old railway cuttings at Kimberley (p. 150d) there is a continual seepage of water. Although the Lower Magnesian Limestone is a coarsely granular rock composed largely of dolomite rhombs, its porosity and intrinsic permeability are low and its transmissivity is related to the number of open discontinuities. Yields for boreholes up to 8 in diameter rarely exceed 800 g.p.h. The water is excessively hard with a range of 400 to 1000 mg/l. Sulphate values are also particularly high. The total annual licensed abstraction of water from the Lower Magnesian Limestone is some 127 m.g., most of which is for industrial use.

The Lower Mottled Sandstone and Pebble Beds together form the most important aquifer, generally known as the Bunter Sandstone of the East Midlands. In the Derby district the water table reaches a maximum elevation of about 550 ft near Annesley Woodhouse in the north-east and near Turnditch to the west. The annual range of water levels is only of the order of a few feet (Land, 1966). Excessive abstraction of groundwater around Nottingham has reversed the natural water flowage from the Bunter into the River Trent, with resultant groundwater contamination. The Lower Mottled Sandstone is finer in grain size and a 10 per cent clay fraction reduces its porosity compared with the coarser, 'cleaner' Pebble Beds.

The Pebble Beds are highly permeable and bores with diameters of 24 to 36 in commonly yield between 1 and 2 m.g.d. Variations in yield are due largely to differing degrees of cementation and fissuring of the sandstone. The isolated outcrops of Pebble Beds in the west of the district yield up to 3000 g.p.d. from boreholes of 100 ft or so. Total hardness is usually above 100 mg/l. Replenishment of the aquifer is restricted by a cover of boulder clay to the west.

The total annual licensed abstraction from Triassic sandstones is 350 m.g., of which 40 m.g. are derived from the Pebble Beds west of the River Derwent. Some 300 m.g. are used by industry, particularly the National Coal Board. Table 8 shows a typical analysis of water from the Sandiacre and Stapleford Station boreholes [4821 3658; 4820 3659]; twin boreholes, 30 in in diameter and 203 ft deep, yield 25 m.g. per annum.

Groundwater from the Keuper Marl and Waterstones, usually extremely hard, owes its non-carbonate hardness to solution of gypsum bands and nodules. Total annual abstraction is 11 m.g., over half of which is derived from springs and used for agricultural purposes.

Quaternary deposits

River terraces and alluvium are important sources of water, particularly where they are in hydraulic continuity with the river. In the case of the Trent alluvium there is only limited connection between the river and the groundwater of the alluvium. This is probably due to a high proportion of clay in the alluvium and some degree of puddling of the river bed.

The iron content is very high in water from gravels in the Trent Valley. Molyneux (1869) suggested that the iron is associated with organic deposits in old channels of the Trent. Such gravels are however rich in ironstone (p. 158c) which may be largely responsible for the concentration.

Twin boreholes [5372 3670] of 48 in diameter and 16 and 19 ft deep in the Trent alluvium at Beeston Lane have an annual licensed abstraction of 40 m.g. for industrial use. Similar quantities are taken from shallow boreholes in the River Derwent alluvium, but only 8 m.g. per annum are used by industry from the Erewash valley deposits. The South Derbyshire Water Board extracted 2471 m.g. in the year ending March 1973 from an extensive tunnel system in the Derwent alluvium between Little Eaton and Allestree.

Excavations [4935 5166] for the M1 Motorway showed that where glacial sand and gravel overlie, or are interbedded with boulder clay, springs and seepages are common.

Table 8 Analyses of some important groundwater sources of the district

	1	2	3	4	5	6	7	8	9	10
PH	7.6	7.4	—	7.6	7.4	—	7.2	7.4	7.5	7.7
Electrical conductivity	428	470	—	—	—	—	2700	930	1200	603
Free and Saline Ammonia (as N)	0.01	0.01	0.005	—	0.019	0.10	—	—	—	0.08
Albuminoid Ammonia (as N)	0.01	0.01	0.02	—	—	0.05	—	—	—	0.06
Nitrite Nitrogen (as N)	—	0.001	—	—	—	—	0.6	0.2	—	0.024
Nitrate Nitrogen (as N)	1.78	1.81	2.75	—	—	—	79.4	12.2	—	2.05
Oxygen absorbed from Permangamate (in 4 hrs)	0.10	0.10	0.24	—	0.92	—	—	—	—	1.96
Dissolved Oxygen	—	6.9	—	—	—	0.25	—	—	—	8.4
Free CO_2	—	22.0	—	—	—	—	—	—	—	5.0
Hardness as $CaCo_3$										
Temporary	213	218	140	135	290	159	680	250	120	157
Permanent	44	70	125	422	288	231	280	210	100	76
Total	257	288	265	435	578	390	960	460	220	233
Total Solids	—	335	225	—	663	415	—	—	—	416
Calcium as Ca	85.2	88.4	—	121.6	—	—	—	—	—	72.8
Magnesium as Mg	—	16.3	—	321.4	—	—	—	—	—	12.5
Sodium as Na	5.6	9.9	—	—	—	—	261	25.6	261	52.4
Potassium as K	0.9	0.9	—	—	—	—	11.2	7.6	4.5	4.5
Iron as Fe	None	None	—	—	—	—	—	0.02	0.09	0.28
Manganese as Mn	0.01	None	—	—	—	—	0.03	0.04	0.04	0.09
Copper as Cu	—	0.05	—	—	—	—	—	—	—	0.05
Lead as Pb	—	0.25	—	—	—	—	—	—	—	0.05
Zinc as Zn	—	0.25	—	—	—	—	—	—	—	0.05
Aluminium as Al	—	—	—	—	—	—	—	—	—	—
Silica as SiO_2	—	8.3	—	—	—	—	—	—	—	7.6
Sulphates as SO_4	27.7	47.5	—	11.8	—	—	720	146	42	75.8
Chlorides as Cl	17.0	21.0	—	941.0	150	—	205	59	216	72.0
Fluorides as F	0.31	0.72	—	—	—	—	0.30	0.10	0.38	0.40
Phosphates as PO_4	—	—	—	—	—	—	—	—	—	—
Nitrate (NO_3) calc.	—	8.02	—	—	—	—	—	—	—	9.08
B.O.D.	None	—	—	—	—	—	3.6	0.3	0.8	5.2
Detergents	—	—	—	—	—	—	—	—	—	0.047

1 Ladyflatts Mine, Wirksworth . . . . 1973 Dinantian
2 Meerbrook Sough. . . . . 1973 Dinantian
3 Whatstandwell Spring. . . . . 1902 Namurian
4 Cinderhill Colliery No. 6 Shaft, mine drainage . . . . . — Westphalian
5 Annesley Colliery. . . . . 1952 Westphalian
6 Slack Lane Waterworks Borehole, Ripley. . . . . — Westphalian
7 Hucknall No. 1 Colliery, mine drainage . . . . — Permian
8 Kimberley Brewery Borehole. . . . . — Permian
9 Sandiacre and Stapleford Station Borehole. . . . . . . — 'Bunter'
10 Little Eaton Pumping Station. . . . . — Quaternary

The chemistry of the groundwaters of superficial deposits is very variable and supplies of water from all such deposits are liable to chemical and bacterial pollution.

Future development

Downing and others (1970) concluded that the development of groundwater resources in the district will be ultimately at the expense of river flow. Further extraction from the Bunter is possible, except in the Nottingham area. The other aquifers should only be developed locally and left to maintain base flows, to dilute sewage effluents and as a source of water for artificial recharge of the Bunter Sandstone.

The main prospect of meeting the future demands of the district appears to be by improving the quality of the River Trent water, one of the recommendations in the Report on the Water Resources in England and Wales (Anon., 1973).

CHAPTER 10

Geophysical investigations

The association in the Derby district of low density Westphalian and Permo-Triassic sediments with older, higher density limestones of Dinantian age would be expected to produce pronounced Bouguer anomalies due to near-surface density contrasts. While this is generally true the Bouguer anomaly map also reflects more deeply buried structures such as the thick Namurian sequence in the Widmerpool Gulf and a largely unproven ridge of Lower Carboniferous, or older, rocks which extends southwards from the Derbyshire Dome.

Magnetic surveys reveal extensive zones of magnetic material at depth beneath a cover of non-magnetic sediments. The magnetic zones bear little overall relationship to the structures revealed by the gravity data, although there is a certain correspondence in minor features; they probably represent magnetic bodies at a deeper level in the crust.

PHYSICAL PROPERTIES OF THE MAIN ROCK TYPES

Table 9 contains a summary of the physical properties of the main rocks in the Derby district, using both published information and the results of surveys carried out by the Institute. The data are based on sample measurements, ground surveys and, in the case of resistivity results, geophysical borehole logs.

Susceptibility determinations on samples indicate values of about 4×10^{-4} emu for the dolerite intrusion at Ible, some 3 miles NW of Wirksworth, and 1×10^{-4} emu for the Hopton agglomerate, some 2 miles W of Wirksworth. Sedimentary rocks in the area would be expected to be practically non-magnetic, with susceptibilities less than 1×10^{-5} emu.

Sonic velocity data for the main rock types in the Nottingham area have been obtained from well velocity surveys, mostly in oil boreholes, including those at Eakring and Widmerpool (Wyrobeck, 1959).

GRAVITY SURVEYS

The Bouguer anomaly map of the Derby district is dominated by two elongated highs which merge near Wirksworth where the maximum value reaches +23 mGal (Figure 64). The north–south high passing through Trusley has an anomaly level only about 3 mGal lower than the value over the main area of Dinantian rocks in Derbyshire and it would be most convenient to explain the feature as being due to a buried limestone ridge, forming the western boundary of the Widmerpool Gulf. However the northern part of the high north of Trusley coincides with an area of shaly limestones of the Widmerpool Formation, and a borehole at Mickleover Hospital, $4\frac{1}{2}$ miles SW of Derby,

encountered similar strata below a depth of −69 m OD (Stephens, 1929). This shaly limestone facies should be less dense than the main massive limestone sequence and it seems likely that it is underlain by denser limestone or older rocks responsible for the Bouguer anomaly high.

South of the Derby district this elongated Bouguer anomaly changes direction to the south-south-east and passes between the Leicestershire and South Derbyshire coalfields, coincident with the Ashby Anticline. The north–south line can be traced southwards as a Bouguer anomaly feature in the Burton upon Trent area.

In the northern part of the Derby district the regional decrease in Bouguer anomaly values to the east (Figure 64) is largely due to the increasing thickness of low density Westphalian rocks and, farther east, to the appearance of Permo-Triassic sediments.

South of a line from Wirksworth to Nottingham, Westphalian rocks are largely absent and the Bouguer anomaly low in the Derby–Long Eaton area is thought to be due to the thick Namurian sequence in the Widmerpool Gulf (Falcon and Kent, 1960; Kent, 1966) for the thickness of Trias is inadequate to explain the anomaly and the Bouguer anomaly gradient values exclude the possibility of any deep-seated origin. The full extent of this anomaly is difficult to judge but it would seem that the amplitude must be at least 11 mGal (Figure 89), assuming that the regional anomaly continues southwards to a level of about +10 mGal near Charnwood Forest. With a density contrast against the Dinantian limestone of -0.25 g/cm³, an additional thickness of at least 1.05 km (3450 ft) of Namurian sediments is necessary to explain this anomaly. As the Namurian sequence is about 0.3 km (1000 ft) thick at Ironville (Kent, 1966) away from the gulf development, the total depth to the base of the Namurian would have to be at least 1.4 km (4600 ft), allowing for the comparatively thin Trias. The estimate of 1.4 km (4600 ft) exceeds the Namurian thickness of 0.8 km (2600 ft) at Widmerpool and 0.8 km (2600 ft) in the Duffield area. The southward decrease in Bouguer anomaly values in this area is therefore probably not due solely to the thickening of lower density sediments in the Namurian but probably also to a thickening of the shaly facies in the upper part of the Dinantian. In both Widmerpool (Kent, 1966) and Duffield boreholes the Widmerpool Formation is considerably thicker than equivalent strata in the Ironville area, containing mudstones and sandstones which are less dense than the massive limestone and therefore tending to cause Bouguer anomaly lows. Any gravitational effect of the igneous intrusion which is presumably responsible for the magnetic anomaly west of Nottingham (Figure 65) is obscured by the steep Bouguer anomaly gradient.

The Wirksworth–Nottingham Bouguer anomaly high can be traced from near Wirksworth through the limestone inlier of Crich (Figure 64) and the Ironville Anticline. In the extreme south-east the high coincides with an abrupt change

Table 9 Summary of the physical property data for the main rock types of the district

		Saturated bulk density	Porosity	Velocity	Resistivity
		g/cm³	%	km/s	Ωm
Trias	Keuper Marl	—	—	2.44[1]	25–75
	'Bunter' (Pebble Beds and Lower Mottled Sandstone)	—	—	4.20[1]	50–100
Permian	Lower Magnesian Limestone	2.52[3]	—	—	—
Carboniferous	Westphalian	2.47–2.59[3]	—	3.05[1]	30–50
	Namurian sandstone	2.38(3)	16.6(3)	2.45(3)	500–700[2]
	Namurian sandstone	—	—	2.7	—
	Namurian sandstone	—	—	5.04[1]	—
	Namurian shale	2.42(1)	—	—	—
	Namurian shale	—	—	2.83(1)	—
	Namurian shale	—	—	2.5	—
	Dinantian limestone	2.70(4)	2.0(4)	5.95[1] 6.18(4)	500–1000
	Igneous rocks				
	Olivine-basalt	—	—	—	200–400[2]
	Basaltic breccia	—	—	—	200[2]
	Tuff	2.51(1)	—	—	240[2]
	Dolerite	2.88(1)	1.6(1)	5.73(1)	—
	Agglomerate	2.43(1)	13.7(1)	2.53(1)	—

Unless referred to by suffixes, data were obtained by IGS surveys. Results based on sample measurements are followed by the number of sites (in brackets) sampled.

[1] Bullerwell (*in* Stevenson and Mitchell, 1955).
[2] Bullerwell (*in* Ramsbottom and others, 1962).
[3] Whetton and others, 1961.

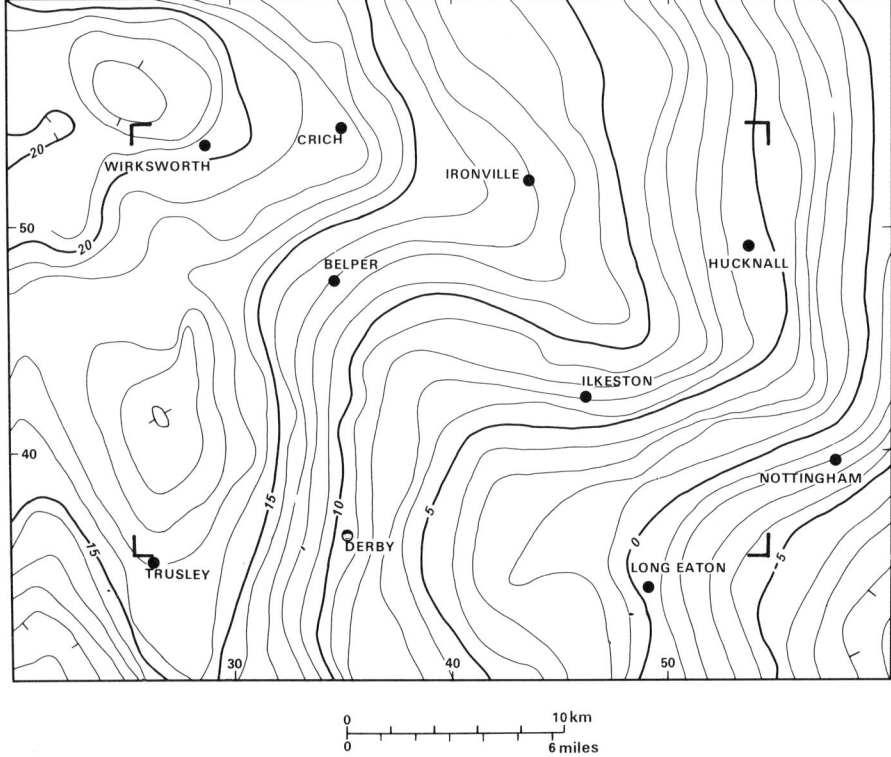

Figure 64 Bouguer anomaly map of the region around Derby. Contours at 1 mGal intervals. Corners of Sheet 125 (Derby) are indicated

in the direction of the Bouguer anomaly contours from north–south (reflecting the increasing thickness of Westphalian strata) to east–west through Ilkeston (largely reflecting the increased thickness of Namurian strata in the Widmerpool Gulf). The small Bouguer anomaly peak along the crest of the high of about 1 or 2 mGal can be explained in places by structures such as the Ironville Anticline. The belt of faults in the Westphalian (Plate 12) also coinciding with the Bouguer anomaly high may reflect the presence of a buried limestone ridge.

The highest Bouguer anomaly value (+23 mGal) in the area is located over the main outcrop of limestone north-west of Wirksworth at the intersection point of the trends of the two elongated highs described above. The amplitude of the closure in this area may be exaggerated by local features such as a thicker development of dolomite in the Dinantian sequence or high density basic intrusions similar to that at Ible [252 575].

Superimposed on the broad, major anomalies in Figure 64 are local small-amplitude Bouguer anomalies which can usually be related directly to mapped features at the surface. These are not easily recognisable in Figure 64 but more detailed measurements across the Dinantian–Namurian boundary in the Wirksworth area, for example, revealed step-like anomalies of 2 to 3 mGal which are due to the density contrast between the shales and limestones. Interpretations of these profiles generally show that the limestone surface dips at angles of up to 20° near the mapped contact but then levels off to a more gentle slope.

MAGNETIC SURVEYS

The Derby district was included in the aeromagnetic survey flown in 1955 with a mean terrain clearance of 1000 ft (308 m). The map shown in Figure 65 is taken from the 1:625 000 scale aeromagnetic map (Sheet 2) of Great Britain (Geological Survey, 1965). The total field measurements were made with a fluxgate magnetometer along east to west flight lines 1.61 km (1 mile) apart and north to south tie lines 9.65 km (6 miles) apart.

The magnetic field in the Derby district is dominated by a broad magnetic high, about 30 km wide and at least 80 km long, trending south-eastwards from Wirksworth through Nottingham. The smooth, widely spaced magnetic contours in the south-west suggest that the structure causing the anomaly probably lies at a considerable depth.

Superimposed on the main magnetic high are several sharply defined anomalies suggesting smaller magnetic bodies at comparatively shallow depths of 1 or 2 km (3300–6600 ft). The pronounced circular anomaly at Hucknall is the most obvious of these, but other examples are the circular feature some 6 miles E of Belper and the ridge trending north-west from near the Hucknall anomaly. Broader magnetic highs west of Nottingham and north-west of Wirksworth suggest magnetic bodies at intermediate depths (3 to 4 km or 10 000–13 000 ft).

The main north-west-trending magnetic zone in Figure 65 can be interpreted as a broad horizontal slab extending from a depth of −3 km (10 000 ft) down to −6 km (20 000 ft) with a susceptibility of 2×10^{-4} emu. The depth extent however can be varied without invalidating the interpretation and the susceptibility changed accordingly, but it seems

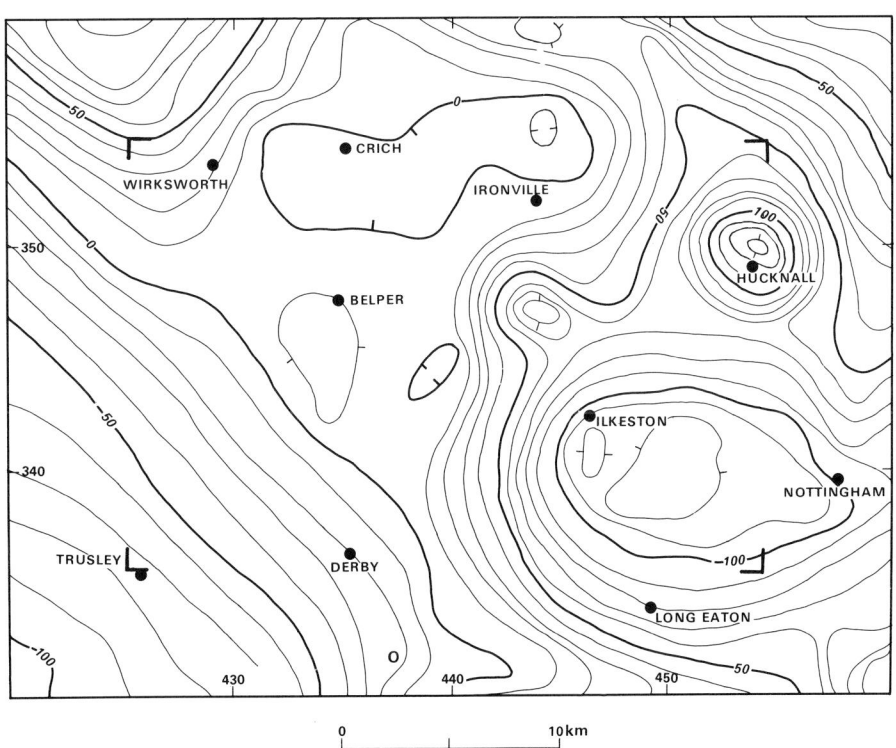

Figure 65 Aeromagnetic map of the region around Derby. Contours at 10 gamma intervals. Corners of Sheet 125 (Derby) are indicated

that the south-west margin must be inclined in all cases to account for the low gradient west of Derby. The steeper north-eastern margin crosses the corner of the area shown in Figure 65 north of Hucknall but it is necessary to postulate the existence of some magnetic material at depth northeast of this line to reproduce the higher background value in this area. The depth and shape of this model, its elongate form and the lack of corresponding structures in the Carboniferous rocks suggest that it could be due to a belt of volcanic or metamorphic rocks without an unusually high density in Lower Palaeozoic or Precambrian basement rocks.

Superimposed on the broad magnetic belt are more localised anomalies of shallower origin which may be due to intrusions, some of which may rise through the Lower Carboniferous strata. The shallowest of these is the anomaly 9 km E of Belper (Figure 65) which is probably due to a plug-like intrusion rising to less than −0.7 km (−2300 ft) OD. The anomaly lies on an extended belt of faulting on the northern flank of the south-east-trending Bouguer anomaly high, but the slightly larger and deeper source of the anomaly at Hucknall lies beneath Permian sediments and has no surface indication. The east–west anomaly west of Nottingham coincides in position and direction with the Bouguer anomaly gradients associated with the northern margin of the Widmerpool Gulf (Figure 64). The positive magnetic anomaly near the north-west corner of the area shown in Figure 65 coincides with the outcrop of Dinantian limestones on the south-east corner of the Derbyshire Dome and represents a broad rise in the magnetic material.

A comparison of the structures suggested by the magnetic map with those deduced from the main Bouguer anomalies (Figure 64) reveals little in common. The broad magnetic zone appears to lie below the Carboniferous rocks responsible for almost all the Bouguer anomaly features, but its north-west to south-east trend is repeated by the Bouguer anomaly high interpreted as a ridge in the Lower Carboniferous.

The broad magnetic zone has a Charnian trend and a Precambrian or Lower Palaeozoic age is suggested. Farther to the south, outside the Derby district, a distinct group of magnetic anomalies, probably due to intrusions related to the Caledonian Mountsorrel granodiorite, occur within the broad magnetic zone. From this evidence it would appear that the zone has been the scene of repeated igneous activity extending from Lower Palaeozoic or even Precambrian times through to the late Carboniferous.

North–south-trending structures of Hercynian age are prominent on the Bouguer anomaly map, which mainly reflects density contrasts within the Carboniferous. There are only two places where this trend can be seen on the magnetic map. One is the slight bulge in the contours west of Wirksworth, which is the only magnetic feature correlating with the pronounced north–south Bouguer anomaly high. The other is the north–south-trending western margin of the shallower magnetic anomalies west and north-west of Nottingham which coincides at the surface with the eastern flank of the anticline beneath Denby [398 462]. On the Bouguer anomaly map this feature is shown only as a weak local anomaly of −1 mGal.

SEISMIC SURVEYS

A seismic reflection and refraction survey was carried out by the Applied Geophysics Unit to trace the continuation of the Dinantian limestones beneath the Namurian shales in the Ecclesbourne Valley area south of Wirksworth. The contact of the shale with the limestone was expected to be a good reflecting horizon, but it was found that several, poorly defined, reflectors existed, making correlation difficult in an area without any borehole control. The survey produced some evidence to support the existence of the small anticlinal structure trending south-south-east from near Wirksworth (Plate 12), but it appeared that the mean velocity of the Namurian shales could not be greater than 2.5 km/s if depths consistent with the gravity and geological evidence were to be produced.

From refraction data obtained in the same survey, drift thicknesses were found to be mostly less than 12 m in the area and the bedrock velocities varied between 1.5 and 3.5 km/s with an average of 2.46 km/s for the Namurian shales. Sandstone (Ashover Grit) horizons in the sequence gave a slightly higher mean velocity of 2.71 km/s.

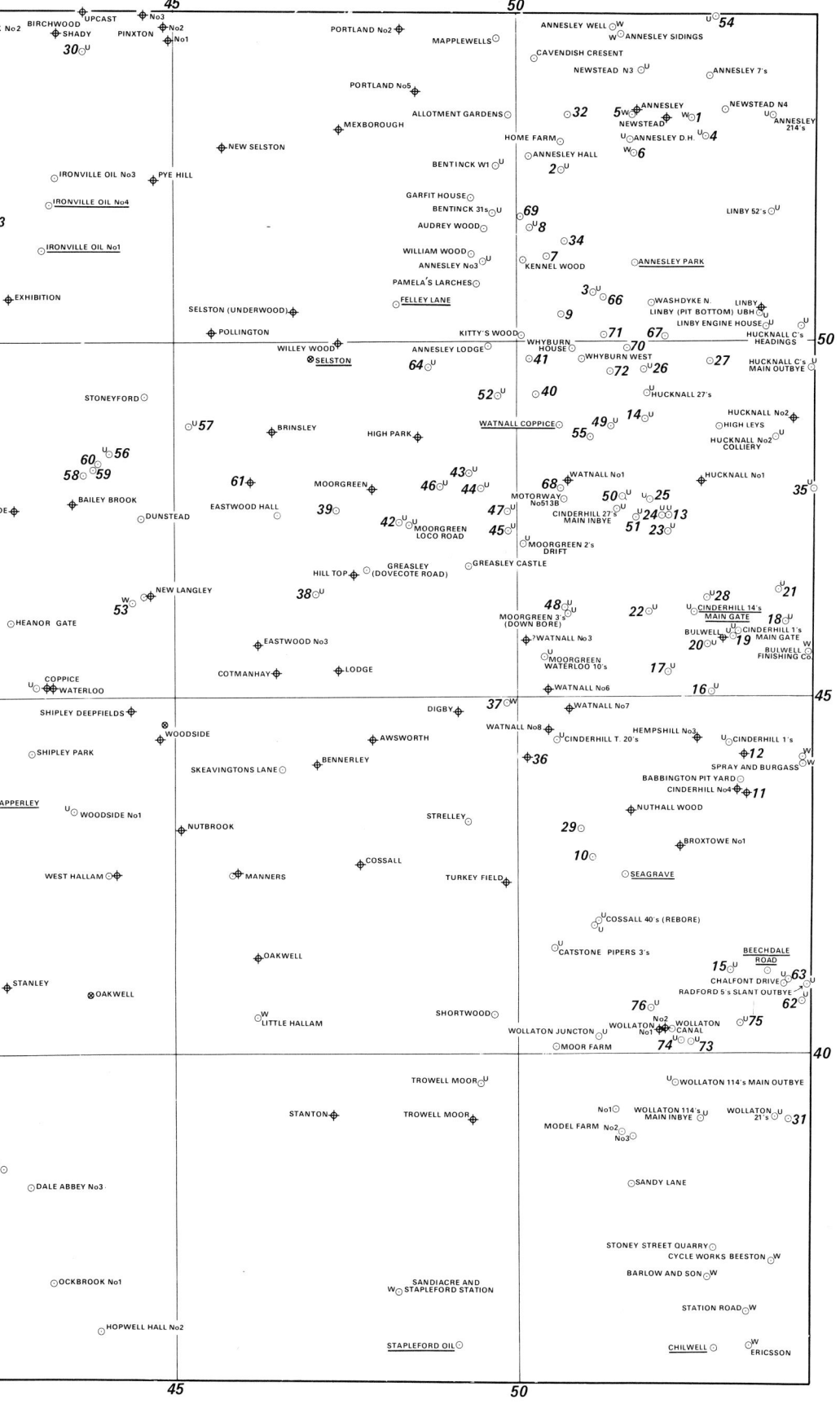

LIST OF NUMBERED PROVINGS.

1 Annesley British Railways Water BH
2 Annesley Colliery B4 UBH
3 Annesley Colliery 10's UBH
4 Annesley Colliery 204's UBH
5 Annesley Colliery Water BH
6 Annesley Hooper Water BH
7 Balaclava Wood BH
8 Bentinck Colliery 27's UBH
9 Charlie's Wood Cottages BH
10 Chilwell Dam Farm BH
11 Cinderhill Colliery No's 1 and 2 Shafts
12 Cinderhill Colliery No 6 Shaft
13 Cinderhill 2's Heading UBH
14 Cinderhill 2's Main Inbye Belt Road UBH
15 Cinderhill 2's Left Gate Inbye UBH
16 Cinderhill 6's Main Gate UBH
17 Cinderhill 9's Left Gate Heading UBH
18 Cinderhill 10's Finishing Face UBH
19 Cinderhill 10's Inbye UBH
20 Cinderhill 10's Main Gate UBH
21 Cinderhill 14's Finishing Face UBH
22 Cinderhill 19's Left Gate UBH
23 Cinderhill 25's Left UBH
24 Cinderhill 25's Right Gate Slant UBH
25 Cinderhill 27's Main Outbye UBH
26 Cinderhill S. 32's UBH
27 Cinderhill H. 68's UBH
28 Cinderhill T. 10's UBH
29 Cossall 3's UBH
30 Cotespark Colliery No3 UBH
31 Crow Wood BH
32 Forest Lodge BH
33 Greenhill Lane BH
34 Heatherdale Pond BH
35 Hucknall South UBH
36 Kimberley Colliery No1 Shaft
37 Kimberley Water BH
38 Lodge Colliery A1 UBH
39 Lower Beauvale (Greasley) BH
40 Misk Farm BH
41 Misk North BH
42 Moorgreen Colliery Low Main 2's UBH
43 Moorgreen Colliery Low Main 16's UBH
44 Moorgreen Colliery Low Main 17's UBH
45 Moorgreen Colliery Low Main 25's UBH
46 Moorgreen Colliery Main Dips UBH
47 Moorgreen Colliery 2's /10's UBH
48 Moorgreen Colliery 3's UBH (Upbore)
49 Moorgreen Colliery 66's /21's UBH
50 Moorgreen Colliery 66's /28's UBH
51 Moorgreen Colliery 66's /33's UBH
52 Moorgreen Colliery Waterloo 71's UBH
53 New Langley Water BH
54 Newstead Colliery NI UBH
55 Old Reservoir BH
56 Ormonde Colliery Blackshale 70's UBH
57 Ormonde Colliery 6's RHG UBH
58 Plastic No1 BH
59 Plastic No2 BH
60 Plastic No3 BH
61 Plumtree Colliery Shafts
62 Radford 5's Slant UBH
63 Radford 7's Left Gate UBH
64 Selston Colliery Blackshale 80's RH UBH
65 Swanwick Colliery Engine Shaft
66 Thurland Hall Farm BH
67 Wash Dyke Lane (South) BH
68 Watnall Colliery Inset UBH
69 Weavers Lane BH
70 Whyburn East BH
71 Whyburn North BH
72 Whyburn South BH
73 Wollaton Colliery South UBH
74 Wollaton Colliery South Dips UBH
75 Wollaton Colliery 15's Main UBH
76 Wollaton Colliery 30's N UBH

BH......Borehole

UBH....Underground borehole

References

AITKENHEAD, N. 1966. P. 52 in INSTITUTE OF GEOLOGICAL SCIENCES, *Annual Report for 1965*, Part I. (London: HMSO.)

— 1968. P. 82 in INSTITUTE OF GEOLOGICAL SCIENCES, *Annual Report for 1967*. (London: HMSO.)

— 1977. Institute of Geological Sciences Borehole at Duffield, Derbyshire. *Bull. Geol. Surv. G.B.*, No. 59.

ALSOP, J. 1845. On the toadstones of Derbyshire. *Rep. Br. Assoc. Adv. Sci. (for 1844)*, pp. 51–52.

ANDERSON, W. and DUNHAM, K. C. 1953. Reddened beds in the Coal Measures beneath the Permian of Durham and south Northumberland. *Proc. Yorkshire Geol. Soc.*, Vol. 29, pp. 21–32.

ANDERSSON, J. G. 1906. Solifluction: a component of subaerial denudation. *J. Geol.*, Vol. 14, pp. 91–112.

ANON. 1973. P. 58 in *Report on the Water Resources of England and Wales*, Vol. 1. (London: HMSO for Water Resources Board.)

— 1942. The Top Hard Seam. *In* The Yorkshire, Nottinghamshire and Derbyshire Coalfield. *Fuel Research Board, Physical and Chemical Survey of the national coal resources*, No. 53.

ARNOLD-BEMROSE, H. H. 1894. On the microscopical structure of the Carboniferous dolerites and tuffs of Derbyshire. *Q. J. Geol. Soc. London*, Vol. 50, pp. 603–643.

— 1899. A sketch of the geology of the Lower Carboniferous rocks of Derbyshire. *Proc. Geol. Assoc.*, Vol. 77, pp. 55–64.

— 1907. The toadstones of Derbyshire: their field relations and petrography. *Q. J. Geol. Soc. London*, Vol. 63, pp. 241–281.

— 1910a. *Derbyshire*. (Cambridge.)

— 1910b. The Lower Carboniferous rocks of Derbyshire. Pp. 540–563 in *Geology in the Field*, Part 3. MONCKTON, H. W. and HERRIES, R. (Editors). (London: Geological Association; Jubilee Vol.)

— and DEELEY, R. M. 1896. Discovery of mammalian remains in the Old River-Gravels of the Derwent near Derby. *Q. J. Geol. Soc. London*, Vol. 52, pp. 497–510.

AVELINE, W. T. 1877. The Magnesian Limestone and New Red Sandstone in the neighbourhood of Nottingham. *Geol. Mag.*, Vol. 24, pp. 155–156, 380.

— 1879. The geology of Nottinghamshire and Derbyshire. *Mem. Geol. Surv. G.B.*

— 1880. The geology of parts of Nottinghamshire, Yorkshire and Derbyshire. *Mem. Geol. Surv. G.B.*

BISAT, W. S. 1924. The Carboniferous goniatites of the north of England and their zones. *Proc. Yorkshire Geol. Soc.*, Vol. 20, pp. 40–124.

BONNEY, T. G. 1900. The Bunter Pebble Beds of the Midlands and the source of their material. *Q. J. Geol. Soc. London*, Vol. 120, pp. 369–396.

BOULTER, M. C., FORD, T. D., IJTABA, M. and WALSH, P. T. 1971. Brassington Formation: a newly recognised Tertiary Formation in the southern Pennines. *Nature Phys. Sci., London*, Vol. 231, pp. 134–136.

BOUMA, A. H. 1962. *Sedimentology of some flysch deposits: a graphic approach to facies interpretation*. (Amsterdam: Elsevier.)

BRIDGES, E. M. 1964. Examples of periglacial phenomena in Derbyshire. *East Midland Geogr.*, Vol. 3, pp. 262–266.

— 1966. The Soils and Land Use of the District North of Derby (Sheet 125). *Mem. Geol. Surv. G.B.*

BROADHURST, F. M. and SIMPSON, I. M. 1967. Sedimentary infillings of fossils and cavities in limestone at Treak Cliff, Derbyshire. *Geol. Mag.*, Vol. 104, pp. 443–448.

CALVER, M. A. 1956. Die stratigraphische Verbreitung des nicht-marinen Muscheln in den penninischen Kohlenfeldern Englands. *Z. Dtsch. Geol. Ges.*, Vol. 107, pp. 23–39.

— 1968a. Coal Measures invertebrate faunas. Pp. 147–177 in *Coal and Coal bearing strata*. MURCHISON, D. G. and WESTOLL, T. S. (Editors). (Edinburgh: Oliver and Boyd.)

— 1968b. Distribution of Westphalian Marine faunas in northern England and adjoining areas. *Proc. Yorkshire Geol. Soc.*, Vol. 37, pp. 1–72.

CARRUTHERS, R. G. and STRAHAN, A. 1923. Lead and Zinc ores of Durham, Yorkshire and Derbyshire. *Mem. Geol. Surv. Spec. Rep. Miner. Resour. G.B.*, Vol. 26.

CHISHOLM, J. I. 1977. Growth faulting and sandstone deposition in the Namurian of the Stanton Syncline, Derbyshire. *Proc. Yorkshire Geol. Soc.*, Vol. 41, pp. 305–323.

CLARK, J. 1962. Field interpretation of Red Beds. *Bull. Geol. Soc. Am.*, Vol. 73, pp. 423–428.

CLARKE, R. F. A. 1965. British Permian saccate and monosulcate miospores. *Palaeontology*, Vol. 8, pp. 322–354.

CLAYTON, K. M. 1953. The glacial chronology of part of the Middle Trent Basin. *Proc. Geol. Assoc.*, Vol. 64, pp. 198–207.

— 1955. The geomorphology of the area around Nottingham and Derby. *East Midland Geogr.*, Vol. 1, pp. 16–20.

— 1968. Structure surface relationships in the middle part of the Derwent Basin. *East Midland Geogr.*, Vol. 4, pp. 321–328.

COPE, F. W. 1946. Intraformational contorted rocks in the Upper Carboniferous of the southern Pennines. *Q. J. Geol. Soc. London*, Vol. 101, pp. 139–176.

DAVISON, C. 1905. The Derby earthquake of 1904. *Q. J. Geol. Soc. London*, Vol. 61, pp. 8–17.

— 1924. *A history of British earthquakes*. (Cambridge: Cambridge University Press.)

DAWE, A. and COLES, G. 1948. The Coal Seams of Derbyshire, Nottinghamshire and Lincolnshire. *J. Inst. Fuel*, Vol. 22, pp. 12–23.

DEANS, T. 1935. Some Oolitic Ironstones from the Coal Measures of Yorkshire. *Trans. Leeds Geol. Assoc.*, Vol. 5, pp. 161–187.

— 1961. A galena-wolframite-uraniferous-asphaltic horizon in the Magnesian Limestone of Nottinghamshire. *Mineral. Mag.*, Vol. 32, pp. 705–715.

DEELEY, R. M. 1886. The Pleistocene succession in the Trent Basin. *Q. J. Geol. Soc. London*, Vol. 42, pp. 437–480.

DE RANCE, C. E. 1896. The earthquake of Dec. 17th 1896. *Trans. North Staffordshire Nat. Field Club*, Vol. 31, pp. 159–173.

DINES, H. G., HOLLINGWORTH, S. E., EDWARDS, W., BUCHAN, S. and WELCH, F. B. A. 1940. The mapping of head deposits. *Geol. Mag.*, Vol. 77, pp. 198–226.

DOLLAR, A. T. J. 1957. The Midlands earthquake of February 11th 1957. *Nature, London*, Vol. 179, pp. 507–510.

DOWNING, R. A. and HOWITT, F. 1969. Saline ground-waters in the Carboniferous rocks of the English East Midlands in relation to the geology. *Q. J. Eng. Geol. London*, Vol. 1, pp. 241–269.

— LAND, D. H., ALLENDER, R., LOVELOCK, P. E. R. and BRIDGE, L. R. 1970. The Hydrology of the Trent River Basin. *Hydrogeol. Rep. Inst. Geol. Sci.*, No. 5.

DUNHAM, K. C. 1948. A contribution to the petrology of the Permian evaporite deposits of north-eastern England. *Proc. Yorkshire Geol. Soc.*, Vol. 27, pp. 217–227.

— 1952. Fluorspar. (4th Edition.) *Mem. Geol. Surv. Spec. Rep. Miner. Resour. G.B.*, Vol. 4.

DUNHAM, K. C. 1973. A recent deep borehole near Eyam, Derbyshire. *Nature Phys. Sci., London*, Vol. 241, pp. 84–85.

EAGAR, R. M. C. 1947. A study of a non-marine lamellibranch succession in the *Anthraconaia lenisulcata* Zone of the Yorkshire Coal Measures. *Philos. Trans. R. Soc.*, Vol. 233, pp. 1–54.

— 1952. The succession above the Soft Bed and Bassy Mine in the Pennine region. *Liverpool Manchester Geol. J.*, Vol. 1, pp. 23–56.

— 1954. New species of Anthracosiidae in the Lower Coal Measures of the Pennine region. *Mem. Proc. Manchester Lit. Philos. Soc.*, Vol. 95, pp. 40–65.

— 1956. Additions to the non-marine fauna of the Lower Coal Measures of the North Midlands coalfields. *Liverpool Manchester Geol. J.*, Vol. 2, pp. 328–369.

— 1962. New Upper Carboniferous non-marine lamellibranchs. *Palaeontology*, Vol. 5, pp. 307–339.

EARLY, K. R. and DYER, K. R. 1964. The use of a resistivity survey on a foundation site underlain by karst dolomite. *Géotechnique*, Vol. 14, pp. 341–348.

EDEN, R. A. 1954. The Coal Measures of the *Anthraconaia lenisulcata* Zone in the East Midlands Coalfield. *Bull. Geol. Surv. G.B.*, No. 5, pp. 81–106.

— ELLIOT, R. W., ELLIOTT, R. E. and YOUNG, B. R. 1963. Tonstein Bands in the Coalfields of the East Midlands. *Geol. Mag.*, Vol. 100, pp. 47–58.

— MITCHELL, M., ORME, G. R. and SHIRLEY, J. 1964. A study of part of the margin of the Carboniferous Limestone Massif in the Pin Dale area, Derbyshire. *Bull. Geol. Surv. G.B.*, No. 21, pp. 73–118.

— STEVENSON, I. P. and EDWARDS, W. 1957. Geology of the country around Sheffield. *Mem. Geol. Surv. G.B.*

— RHYS, G. H. and SMITH, E. G. 1959. P. 33 in *Summ. Prog. Geol. Surv. G.B. for 1958*.

EDMUNDS, W. M. 1971. Hydrogeochemistry of groundwaters in the Derbyshire Dome with special reference to trace constituents. *Rep. Inst. Geol. Sci.*, No. 71/7.

EDWARDS, W. N. 1951. The Concealed Coalfield of Yorkshire and Nottinghamshire. (3rd Edition.) *Mem. Geol. Surv. G.B.*

— 1967. Geology of the Country around Ollerton. *Mem. Geol. Surv. G.B.*

— and STUBBLEFIELD, C. J. 1948. Marine bands and other faunal marker horizons in relation to the sedimentary cycles of the Middle Coal Measures of Nottinghamshire and Derbyshire. *Q. J. Geol. Soc. London*, Vol. 103 (for 1947), pp. 209–260.

— and TROTTER, F. M. 1954. The Pennines and adjacent areas. (3rd Edition.) *Br. Reg. Geol., Inst. Geol. Sci.*

ELLIOTT, R. E. 1961. The stratigraphy of the Keuper Series in southern Nottinghamshire. *Proc. Yorkshire Geol. Soc.*, Vol. 33, pp. 197–234.

— 1965. Swilleys in the Coal Measures of Nottinghamshire interpreted as palaeo-river courses. *Mercian Geol.*, Vol. 2, pp. 133–142.

— 1968. Facies, sedimentation successions and cyclothems in productive Coal Measures in the East Midlands, Great Britain. *Mercian Geol.*, Vol. 2, pp. 351–372.

— 1969. Deltaic processes and episodes: the interpretation of productive Coal Measures occurring in the East Midlands, Great Britain. *Mercian Geol.*, Vol. 3, pp. 111–135.

ENNOS, F. R. and SCOTT, A. 1924. Refractory Materials: Fireclays. Analyses and Physical Tests. *Mem. Geol. Surv. Spec. Rep. Miner. Resour. G.B.*, Vol. 28.

EVANS, A. M. 1968. Charnwood Forest *in* Precambrian Rocks in *The Geology of the East Midlands*. SYLVESTER-BRADLEY, P. C. and FORD, T. D. (Editors). (Leicester: Leicester University Press).

— FORD, T. D. and ALLEN, J. R. L. 1968. Precambrian Rocks in *The Geology of the East Midlands*. SYLVESTER-BRADLEY, P. C. and FORD, T. D. (Editors). (Leicester: Leicester University Press.)

FALCON, N. L. and KENT, P. E. 1960. Geological results of petroleum exploration in Britain 1945–1957. *Mem. Geol. Soc. London*, No. 2.

FAREY, J. 1811. *A general view of the agriculture and minerals of Derbyshire*. Vol. 1. (London.)

FEARNSIDES, W. G. 1932. The geology of the eastern part of the Peak District. *Proc. Geol. Assoc.*, Vol. 43, pp. 152–191.

FITCH, F. J., MILLER, J. A. and WILLIAMS, S. C. 1970. Isotopic ages of British Carboniferous Rocks. *C. R. 6me Congr. Av. Etud. Stratigr. Géol. Carbonif.*, Part 2, pp. 771–790.

FORD, T. D. 1963. The dolomite tors of Derbyshire. *East Midland Geogr.*, Vol. 3, pp. 148–153.

— 1967. Some mineral deposits of the Carboniferous Limestone of Derbyshire. Pp. 53–75 in *Geological Excursions in the Sheffield Region and the Peak District National Park*. NEVES, R. and DOWNIE, C. (Editors). (Sheffield: University of Sheffield Press.)

— 1968a. The Carboniferous Limestone. Pp. 59–82 in *The Geology of the East Midlands*. SYLVESTER-BRADLEY, P. C. and FORD, T. D. (Editors). (Leicester: Leicester University Press.)

— 1968b. The Millstone Grit. Pp. 83–94 in *The Geology of the East Midlands*. SYLVESTER-BRADLEY, P. C. and FORD, T. D. (Editors). (Leicester: Leicester University Press.)

— and INESON, P. R. 1971. The fluorspar mining potential of the Derbyshire ore field. *Trans. Inst. Min. Metall., Sect. B: Appl. Earth Sci.*, Vol. 80, pp. 186–210.

— and KING, R. J. 1969. The origin of the silica sand pockets in the Derbyshire limestone. *Mercian Geol.*, Vol. 3, pp. 51–69.

FRANCIS, E. H. 1961. Thin beds of graded kaolinized tuff and tuffaceous siltstone in the Carboniferous of Fife. *Bull. Geol. Surv. G.B.*, No. 17, pp. 191–215.

— 1969. Les tonsteins du Royaume-Uni. *Ann. Soc. Géol. Nord*, Vol. 89, pp. 209–214.

— SMART, J. G. O. and RAISBECK, D. E. 1968. Westphalian volcanism at the horizon of the Black Rake in Derbyshire and Nottinghamshire. *Proc. Yorkshire Geol. Soc.*, Vol. 36, pp. 395–416.

FROST, D. V., EDEN, R. A. and RHYS, G. H. 1968. Six-inch geological maps SK 25 SE. *Inst. Geol. Sci.*

FULLER, G. J. 1965. Lead mining in Derbyshire in the mid-thirteenth century. *East Midland Geogr.*, Vol. 3, pp. 373–393.

GEIGER, M. E. and HOPPING, C. A. 1968. Triassic stratigraphy of the southern North Sea basin. *Philos. Trans. R. Soc. London*, Ser. B, Vol. 254, pp. 1–36.

GEIKIE, A. 1897. *The ancient volcanoes of Great Britain*, Vol. 2. (London.)

GEOLOGICAL SURVEY. 1965. Aeromagnetic map of Great Britain. Sheet 2.

GEORGE, T. N., JOHNSON, G. A. L., MITCHELL, M., PRENTICE, J. E., RAMSBOTTOM, W. H. C., SEVASTOPULO, G. D. and WILSON, R. B. 1976. A correlation of Dinantian rocks in the British Isles. *Spec. Rep. Geol. Soc. London*, No. 7.

GIBSON, W., POCOCK, T. I., WEDD, C. B. and SHERLOCK, R. L. 1908. The geology of the southern part of the Derbyshire and Nottinghamshire Coalfield. *Mem. Geol. Surv. G.B.*

121

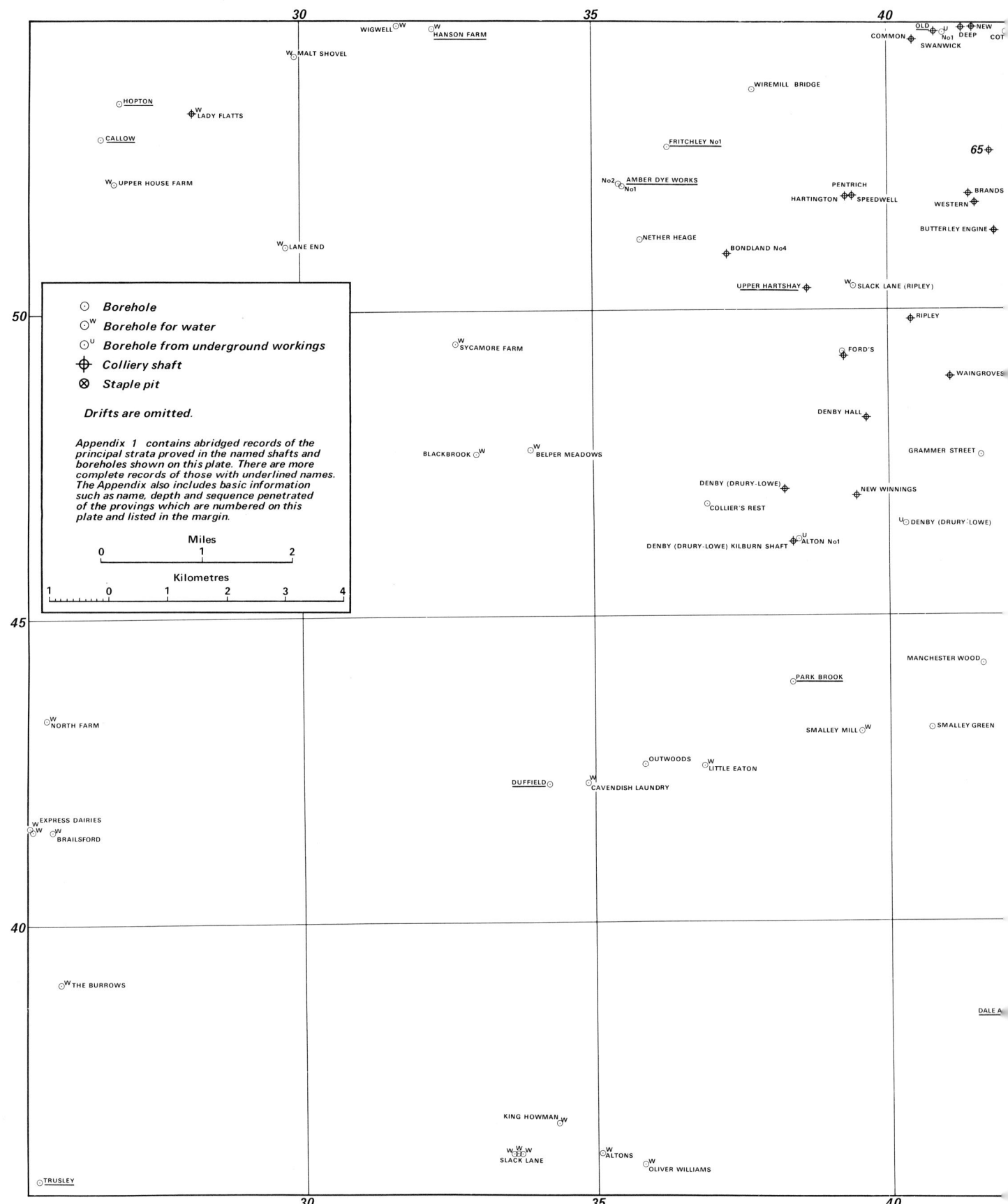

Plate 13   Site map of boreholes and shafts in the Derby district

— and WEDD, C. B. 1913. The geology of the northern part of the Derbyshire Coalfield and bordering tracts. *Mem. Geol. Surv. G.B.*

GODWIN, C. G. 1960. P. 33 in *Summ. Prog. Geol. Surv. G.B. for 1959.*

GREEN, A. H., FOSTER, C. LE NEVE and DAKYNS, J. R. 1869. The geology of the Carboniferous Limestone, Yoredale Rocks and Millstone Grit of North Derbyshire and the adjoining parts of Yorkshire. *Mem. Geol. Surv. G.B.*

— — — — and STRAHAN, A. 1887. Carboniferous Limestone, Yoredale rocks and Millstone Grit of north Derbyshire. (2nd Edition, with additions by A. H. Green and A. Strahan.) *Mem. Geol. Surv. G.B.*

GRIFFIN, A. R. 1971. *Mining in the East Midlands, 1550–1947.* (London: Frank Cass.)

HAINS, B. A. and HORTON, A. 1969. Central England. *Br. Reg. Geol., Inst. Geol. Sci.*

HICKLING, G. 1906. On footprints from the Permian of Mansfield (Nottinghamshire). *Q. J. Geol. Soc. London*, Vol. 62, pp. 125–131.

HIND, W. 1909. The present state of our knowledge of Carboniferous geology. *The Naturalist*, pp. 149–156.

HOLDSWORTH, B. K. 1963. Pre-fluvial, autogeosynclinal sedimentation in the Namurian of the southern Central Province. *Nature, London*, Vol. 199, pp. 133–135.

— 1966. A preliminary study of the palaeontology and palaeoenvironment of some Namurian limestone 'bullions'. *Mercian Geol.*, Vol. 1, pp. 315–337.

HOLLINGWORTH, S. E., TAYLOR, J. H. and KELLAWAY, G. H. 1944. Large-scale superficial structures in the Northampton Ironstone field. *Q. J. Geol. Soc. London*, Vol. 100, pp. 1–14.

HOWE, J. A. 1896. Notes on the pockets of clay and sand in the limestones of Staffordshire and Derbyshire. *Trans. North Staffordshire Field Club*, Vol. 31, pp. 143–149.

HULL, E. 1869. The Triassic and Permian Rocks of the Midland Counties of England. *Mem. Geol. Surv. G.B.*

INSTITUTE OF GEOLOGICAL SCIENCES. 1968. *Annual Report for 1967.* (London: HMSO.)

IRVING, A. 1882. On the classification of the European rocks known as Permian and Trias. *Geol. Mag.*, Vol. 29, pp. 158–164, 219–223, 272–278, and 316–322.

JONES, P. F. and STANLEY, M. F. 1974. Ipswichian mammalian fauna from the Beeston Terrace at Boulton Moor, near Derby. *Geol. Mag.*, Vol. 111, pp. 515–520.

— — 1975. Description of *Hippopotamus* and other mammalian remains from the Allenton Terrace of the lower Derwent valley, South Derbyshire. *Mercian Geol.*, Vol. 5, pp. 259–271.

— and WEAVER, J. D. 1975. Superficial valley folds of late Pleistocene age in the Breadsall area of south Derbyshire. *Mercian Geol.*, Vol. 5, pp. 279–290.

JOWETT, A. and CHARLESWORTH, J. K. 1929. The glacial geology of the Derbyshire Dome and the western slopes of the southern Pennines. *Q. J. Geol. Soc. London*, Vol. 85, pp. 307–334.

KELLAWAY, G. A. 1972. Development of non-diastrophic Pleistocene structures in relation to climate and physical relief in Britain. *24th Int. Congr.*, Sect. 12, pp. 136–146.

KENT, P. E. 1957. Triassic relics and the 1000-foot surface in the southern Pennines. *East Midland Geogr.*, Vol. 1, Part 8, pp. 3–10.

— 1966. The structure of the concealed Carboniferous rocks of north-eastern England. *Proc. Yorkshire Geol. Soc.*, Vol. 35, pp. 323–352.

KENT, P. E. 1967. A contour map of the sub-Carboniferous surface in the north-east Midlands. *Proc. Yorkshire Geol. Soc.*, Vol. 36, pp. 127–133.

— 1968. The buried floor of eastern England. Pp. 138–148 in *The Geology of the East Midlands.* SYLVESTER BRADLEY, P. C. and FORD, T. D. (Editors). (Leicester: Leicester University Press.)

KING, C. A. M. 1966. Geomorphology. Pp. 41–59 in *Nottingham and its Region.* EDWARDS, K. C. (Editor). (Nottingham: Br. Assoc. Adv. Sci.)

KLEIN, G. DE V. 1962. Sedimentary structures in the Keuper Marl (Upper Triassic). *Geol. Mag.*, Vol. 99, No. 2, pp. 137–144.

LAMPLUGH, G. W. and GIBSON, W. 1910. The geology of the country around Nottingham. *Mem. Geol. Surv. G.B.*

— and SMITH, B. 1914. The Water Supply of Nottinghamshire from Underground Sources. *Mem. Geol. Surv. G.B.*

LAND, D. H. 1966. Hydrogeology of the Bunter Sandstone in Nottinghamshire. *Hydrogeol. Rep. Inst. Geol. Sci.*, No. 1.

LEES, G. 1958. An investigation into the effects of the East Midlands earthquake of Feb. 11th on vibration sensitive relays at local power stations. *East Midland. Geogr.*, Vol. 10, pp. 41–45.

LINTON, D. L. 1951. Midland drainage: some considerations bearing on its origin. *Br. Assoc. Adv. Sci.*, Vol. 7, pp. 449–456.

LLEWELLYN, P. G., BACKHOUSE, J. and HOSKIN, I. R. 1969. Lower-Middle Tournaisian miospores from the Hathern Anhydrite Series, Carboniferous Limestone, Leicestershire. *Proc. Geol. Soc.*, Vol. 23, pp. 282–301.

MACKINTOSH, A. H. G. 1937. Notable features in the Middle Permian Marls of south Yorkshire and north Nottinghamshire. *Proc. Yorkshire Geol. Soc.*, Vol. 23, pp. 282–301.

MÄDLER, K. 1964. Die geologische Verbreitung von Sporen und Pollen in der Deutschen Trias. *Beih. Geol. Jahrb.*, No. 65, pp. 1–147.

MARSHALL, R. A. 1875. Mineral oil as found at the Deep Main Pits, Riddings, Derbyshire. *Trans. North Staffordshire Inst. Min. Mech. Eng.*, Vol. 1, pp. 126–134.

MATLEY, C. A. 1914. Note on the source of the pebbles of the Bunter Pebble Beds of the East Midlands. *Geol. Mag.*, Vol. 52, pp. 211–215.

MAYHEW, R. W. 1967. The Ashover and Chatsworth Grits in north-east Derbyshire. Pp. 94–103 in *Geological Excursions in the Sheffield Region and the Peak District National Park.* NEVES, R. and DOWNIE, C. (Editors). (Sheffield: University of Sheffield.)

McCULLAGH, M. J. 1968. A note on the composition of the terraces of the Middle Trent. *East Midland Geogr.*, Vol. 4, pp. 303–313.

MITCHELL, G. H. and STUBBLEFIELD, C. J. 1948. The geology of Leicestershire and the South Derbyshire Coalfield. (2nd Edition.) *Wartime Pam. Geol. Surv. G.B.*, No. 22.

MOLYNEUX, W. 1869. *Burton-on-Trent, its history, its waters and its breweries.* (London.)

MOORE, L. R. 1964. The microbiology, minerology and genesis of a tonstein. *Proc. Yorkshire Geol. Soc.*, Vol. 34, pp. 235–291.

MOSELEY, F. and AHMED, S. M. 1967. Carboniferous joints in the north of England and their relation to earlier and later structures. *Proc. Yorkshire Geol. Soc.*, Vol. 36, pp. 61–90.

MURCHISON, D. and WESTOLL, T. S. 1968. *Coal and coal-bearing strata.* (Edinburgh and London: Oliver and Boyd.)

NEVES, R. 1967. The Crich Inlier. Pp. 42–46 in *Geological Excursions in the Sheffield Region and the Peak District National Park.* NEVES, R. and DOWNIE, C. (Editors). (Sheffield: University of Sheffield.)

NEWTON, E. T. 1887. On the remains of fishes from the Keuper of Warwick and Nottingham. (With notes on their mode of occurrence by the Rev. P. B. Brodie and E. Wilson.) *Q. J. Geol. Soc. London*, Vol. 43, pp. 537–543.

NORRIS, R. B. 1969. Dune reddening and time. *J. Sediment. Petrol.*, Vol. 39, pp. 7–11.

ORME, G. R. 1973. Silica in the Viséan limestones of Derbyshire, England. *Proc. Yorkshire Geol. Soc.*, Vol. 40, pp. 63–104.

PARSONS, L. M. 1922. Dolomitization in the Carboniferous Limestone of the Midlands. *Geol. Mag.*, Vol. 59, pp. 104–117.

PENDLETON, J. 1886. *A history of Derbyshire.* (London.)

PETTIJOHN, F. J. 1957. *Sedimentary Rocks.* (2nd Edition.) (New York: Harper and Row.)

PILKINGTON, J. 1789. *A view of the Present State of Derbyshire: with an account of its most remarkable Antiquities.* Vol. I. *Derby.* (London.)

PLANT, J. 1859. Notice of the Occurrence of Mammalian Remains in the Valley of the Soar. *The Geologist*, Vol. 2, p. 174.

POCOCK, T. I. 1929. The Trent valley in the glacial period. *Z. Gletscherkd. Glazialgeol.*, Vol. 17, pp. 302–318.

POSNANSKY, M. 1960. The Pleistocene succession in the Middle Trent basin. *Proc. Geol. Assoc.*, Vol. 71, pp. 285–311.

RAISTRICK, A. and MARSHALL, C. E. 1939. *The nature and origin of coal and coal seams.* (London: English Universities Press.)

RAMSBOTTOM, W. H. C. 1970. The Namurian of Britain. Pp. 219–232 in *C. R. 6me Congr. Int. Stratigr. Géol. Carbonif.* (*Sheffield, 1967*).

— 1973. Transgressions and regressions in the Dinantian: a new synthesis of British Dinantian stratigraphy. *Proc. Yorkshire Geol. Soc.*, Vol. 39, pp. 567–607.

— RHYS, G. H. and SMITH, E. G. 1962. Boreholes in the Carboniferous rocks of the Ashover district, Derbyshire. *Bull. Geol. Surv. G.B.*, No. 19, pp. 75–168.

READ, J. I., MURPHY, J. M., SERGEANT, G. A. and POWIS, D. R. 1973. P. 20 in INSTITUTE OF GEOLOGICAL SCIENCES, *Annual Report for 1972.* (London: Institute of Geological Sciences.)

RICE, A. J. 1968. The Quaternary Era. Pp. 332–355 in *The Geology of the East Midlands.* SYLVESTER-BRADLEY, P. C. and FORD, T. D. (Editors). (Leicester: Leicester University Press.)

RICHARDSON, G. and FRANCIS, E. H. 1971. Fragmental Clayrock (FCR) in coal-bearing sequences in Scotland and north-east England. *Proc. Yorkshire Geol. Soc.*, Vol. 38, pp. 229–260.

RIEUWERTS, J. H. 1972. *Derbyshire old lead mines and miners.* (Leek: Moorland.)

ROSE, G. N. and KENT, P. E. 1955. A *Lingula*-Bed in the Keuper of Nottinghamshire. *Geol. Mag.*, Vol. 92, pp. 476–480.

SABINE, P. A., GUPPY, E. M. and SERGEANT, G. A. 1969. Geochemistry of Sedimentary Rocks. 1. Petrography and Chemistry of Arenaceous Rocks. *Rep. Inst. Geol. Sci.*, No. 69/1.

SARGENT, H. C. 1912. On the origin of certain Clay-bands in the Limestone of the Crich Inlier. *Geol. Mag.*, Vol. 9, pp. 406–412.

— 1921. The Lower Carboniferous chert formations of Derbyshire. *Geol. Mag.*, Vol. 58, pp. 265–278.

SARJEANT, W. A. S. 1967. Fossil footprints from the Middle Triassic of Nottinghamshire and Derbyshire. *Mercian Geol.*, Vol. 2, pp. 327–341.

— 1970. Fossil footprints from the Middle Triassic of Nottinghamshire and the Middle Jurassic of Yorkshire. *Mercian Geol.*, Vol. 3, pp. 269–382.

SCHNELLMANN, G. A. 1955. Concealed lead-zinc fields in England. *Trans. Inst. Min. Metall.*, Vol. 64, pp. 617–636.

SEDGWICK, A. 1829. On the geological relation and internal structure of the Magnesian Limestone and the lower portions of the New Red Sandstone. *Trans. Geol. Soc. London*, Vol. 3, pp. 37–124.

SHEARMAN, D. J. 1966. Origin of marine evaporites by diagenesis. *Trans. Inst. Min. Metall.*, Vol. 75, pp. 208–215.

SHERLOCK, R. L. 1911. The relationship of the Permian to the Trias in Nottinghamshire. *Q. J. Geol. Soc. London*, Vol. 67, pp. 75–117.

— 1926. A correlation of the British Permo-Triassic rocks. *Proc. Geol. Assoc.*, Vol. 37, pp. 1–72.

— 1947. *The Permo-Triassic Formations: a world review.* (London: Hutchinson.)

SHIPMAN, J. 1889. Notes on the geology of Nottingham. *Trans. Nottingham Nat. Soc.*, No. 37, pp. 26–36.

SHIRLEY, J. 1955. The disturbed strata on the Fox Earth Coal and its equivalents in the East Pennine Coalfield. *Q. J. Geol. Soc. London*, Vol. 111, pp. 265–282.

— 1959. The Carboniferous Limestone of the Monyash–Wirksworth area, Derbyshire. *Q. J. Geol. Soc. London*, Vol. 114, pp. 411–429.

SHORT, T. 1734. *The Natural Experimental and Medicinal History of the Mineral Waters of Derbyshire, together with a Natural History of the Earths, Minerals and Fossils through which the chief of them pass.* (London.)

SHOTTON, F. W. 1973. English Midlands. Pp. 18–22 in MITCHELL, G. F., PENNY, L. F., SHOTTON, F. W. and WEST, R. G. A correlation of Quaternary deposits in the British Isles. *Spec. Rep. Geol. Soc. London*, No. 4.

SMILES, S. 1871. *The life of George Stephenson.* (London.)

SMITH, A. V. H. and BUTTERWORTH, M. A. 1967. Miospores in the coal seams of the Carboniferous of Great Britain. *Spec. Pap. Palaeontol.*, No. 1. (London: Palaeontological Association.)

SMITH, B. 1910. The Upper Keuper Sandstones of east Nottinghamshire. *Geol. Mag.*, Vol. 7, pp. 302–311.

— 1912. The green Keuper Basement Beds in Nottinghamshire and Lincolnshire. *Geol. Mag.*, Vol. 9, pp. 252–257.

— 1913. The geology of the Nottingham district. *Proc. Geol. Assoc.*, Vol. 24, pp. 205–240.

SMITH, D. B. 1968. The Hampole Beds – a significant marker in the Lower Magnesian Limestone of Yorkshire, Derbyshire and Nottinghamshire. *Proc. Yorkshire Geol. Soc.*, Vol. 36, pp. 463–477.

— 1970a. Permian and Trias *in* The geology of Durham County. JOHNSON, G. A. L. (Compiler). *Trans. Nat. Hist. Soc. Northumberland, Durham and Newcastle-upon-Tyne*, Vol. 41, pp. 66–91.

— 1970b. The Palaeogeography of the British Zechstein. Pp. 20–23 in *Third Symposium on Salt.* Vol. 1. DELWIG, L. F. and RAU, J. L. (Editors). (Cleveland: Northern Ohio Geological Society.)

— 1972. *In* Geology of Saline Deposits. *Proc. Hanover Symp., 1968, Earth Sciences*, Vol. 7. (Unesco.)

— and FRANCIS, E. A. 1967. Geology of the country between Durham and West Hartlepool. *Mem. Geol. Surv. G.B.*

SMITH, E. G., RHYS, G. H. and EDEN, R. A. 1967. The geology of the country around Chesterfield, Matlock and Mansfield. *Mem. Geol. Surv. G.B.*

— — and GOOSSENS, R. F. 1973. Geology of the country around East Retford, Worksop and Gainsborough. *Mem. Geol. Surv. G.B.*

— and WARRINGTON, G. 1971. The age and relationships of the Triassic rocks assigned to the lower part of the Keuper in north Nottinghamshire, north-west Lincolnshire and south Yorkshire. *Proc. Yorkshire Geol. Soc.*, Vol. 38, pp. 201–227.

SMITH, W. CAMPBELL. 1963. Description of the igneous rocks represented among pebbles from the Bunter Pebble Beds of the Midlands of England. *Bull. Br. Mus. (Nat. Hist.), Mineralogy*, Vol. 2, pp. 9–17.

SMYTH, W. W. 1856. The Iron Ores of Great Britain. Part 1. *Mem. Geol. Surv. G.B.*

STEPHENS, E. A. 1952. On the 'Rough Rock' and Lower Coal Measures, near Crich, Derbyshire. *Proc. Yorkshire Geol. Soc.*, Vol. 28, pp. 221–227.

STEPHENS, J. V. 1929. Wells and Springs of Derbyshire. *Mem. Geol. Surv. G.B.*

STEVENSON, I. P. and GAUNT, G. D. 1971. Geology of the country around Chapel en le Frith. *Mem. Geol. Surv. G.B.*

— and MITCHELL, G. H. 1955. Geology of the country between Burton upon Trent, Rugeley and Uttoxeter. *Mem. Geol. Surv. G.B.*

STEWART, F. H. 1954. Permian evaporites and associated rocks in Texas and New Mexico compared with those of northern England. *Proc. Yorkshire Geol. Soc.*, Vol. 29, pp. 185–235.

STRAHAN, A. 1920. Refractory Materials: Ganister and Silica-rock, Sand for open-hearth steel furnaces, Dolomite. Resources and Geology. (2nd Edition.) *Mem. Geol. Surv. Spec. Rep. Miner. Resour. G.B.*, Vol. 6.

STRAUSS, P. G. 1971. Kaolin-rich rocks in the East Midlands Coalfields of England. Pp. 1519–1532 in *C.R. 6me Congr. Av. Etud. Stratigr. Géol. Carbonif.*, Vol. 4.

STRAW, A. 1963. The Quaternary evolution of the Lower and Middle Trent. *East Midland Geogr.*, Vol. 3, pp. 171–189.

STUBBLEFIELD, C. J. and TROTTER, F. M. 1957. Divisions of the Coal Measures on Geological Survey Maps of England and Wales. *Bull. Geol. Surv. G.B.*, No. 13, pp. 1–5.

SWINNERTON, H. H. 1910. The Bunter Sandstone of Nottinghamshire. *Rep. Trans. Nottingham Nat. Soc.*, No. 58, pp. 17–28.

— 1912. The Palmistry of the Rocks. *Rep. Trans. Nottingham Nat. Soc.*, No. 60, pp. 65–68.

— 1914. Periods of dreikanter formation in south Notts. *Geol. Mag.*, Decade 6, Vol. 1, pp. 208–211.

— 1918. The Keuper Basement Beds near Nottingham. *Q. J. Geol. Soc. London*, Vol. 81, pp. 87–99.

— 1925. A new catopterid fish from the Keuper of Nottingham. *Q. J. Geol. Soc. London*, Vol. 81, pp. 87–99.

— 1928. On a new species of *Semionotus* from the Keuper of Nottingham. *Geol. Mag.*, Vol. 65, pp. 406–409.

— 1935. The denudation of the East Midlands. *Rep. Br. Assoc. Adv. Sci.*, p. 375.

— 1946. The Middle Grits of Derbyshire. *Geol. Mag.*, Vol. 83, pp. 118–120.

— 1948. The Permo-Trias. Pp. 53–59 in *Guide to the Geology of the East Midlands*. MARSHALL, C. E. and others (Editors). (Nottingham: University of Nottingham Press.)

TAYLOR, F. M. 1964a. An oil seepage near Toton Lane, Stapleford, Notts. *Mercian Geol.*, Vol. 1, pp. 23–30.

— 1964b. The geology of the M1 Motorway in north Leicestershire and south Nottinghamshire. *Mercian Geol.*, Vol. 1, pp. 221–229.

— 1965. The Upper Permian and Lower Triassic formations in southern Nottinghamshire. *Mercian Geol.*, Vol. 1, pp. 181–196.

— 1966. Geology. Pp. 11–40 in *Nottingham and its Region*. EDWARDS, H. C. (Editor). (Nottingham: British Association for the Advancement of Science.)

— 1968. Permian and Triassic Formations. Pp. 149–173 in *The Geology of the East Midlands*. SYLVESTER-BRADLEY, P. C. and FORD, T. D. (Editors). (Leicester: Leicester University Press.)

— 1973. The distribution of baryte in Permo-Triassic sandstones at Bramcote, Stapleford, Trowell and Sandiacre, Nottinghamshire. *Mercian Geol.*, Vol. 4, pp. 171–178.

— and ELLIOTT, R. E. 1971. Permo-Triassic stratigraphy of the Great Central Railway cutting, north-west of Annesley Tunnel SK 505 550), Nottinghamshire. *Mercian Geol.*, Vol. 4, pp. 23–28.

— and HOULDSWORTH, A. R. E. 1973. The Permo-Triassic/ Upper Carboniferous unconformity at Swancar Farm, Trowell Moor, Nottinghamshire. *Mercian Geol.*, Vol. 4, pp. 165–170.

TRECHMANN, C. T. 1930. The relation of the Permian and Trias in north-east England. *Proc. Geol. Assoc.*, Vol. 41, pp. 323–325.

TREWIN, N. H. 1968. Potassium Bentonites in the Namurian of Staffordshire and Derbyshire. *Proc. Yorkshire Geol. Soc.*, Vol. 37, pp. 73–91.

— and HOLDSWORTH, B. K. 1973. Sedimentation in the Lower Namurian rocks of the North Staffordshire Basin. *Proc. Yorkshire Geol. Soc.*, Vol. 39, pp. 371–408.

TROTTER, F. M. 1953. Reddened beds of Carboniferous age in North-West England and their origin. *Proc. Yorkshire Geol. Soc.*, Vol. 29, pp. 1–20.

TRUEMAN, A. E. 1954. *The Coalfields of Great Britain.* (London: Edward Arnold.)

— and WEIR, J. 1946–56. A monograph of British Carboniferous non-marine Lamellibranchiata. Parts 1–9. *Palaeontogr. Soc.* [Monogr.]

TURNER, D. W. 1961a. Variations in the First Waterloo Seam. *Rep. National Coal Board Scientific Dept., Coal Survey, East Midlands Division*, No. 3290.

— 1961b. Variations in the Second Waterloo Seam. *Rep. National Coal Board Scientific Dept., Coal Survey, East Midlands Division*, No. 3291.

VARVILL, W. W. 1959. The future of lead-zinc and fluorspar mining in Derbyshire. Pp. 175–203 in *The future of non-ferrous mining in Great Britain and Ireland*. (London: Institute of Mining and Metallurgy.)

VERNON, R. D. 1909. The Geology of the Lower Coal Measures of the Derbyshire and Nottinghamshire portion of the Yorkshire Coalfield. *Geol. Mag.*, Vol. 6, pp. 289–299.

— 1910. On the occurrence of *Schizoneura paradoxa* Schimper and Mougeot in the Bunter of Nottingham. *Proc. Cambridge Philos. Soc.*, Vol. 15, pp. 401–405.

VISSCHER, H. and COMMISSARIS, A. L. T. M. 1968. Middle Triassic pollen and spores from the Lower Muschelkalk of Winterswijk (The Netherlands). *Pollen et Spores*, Vol. 10, pp. 161–176.

WALKER, R. G. 1967. Turbidite sedimentary structures and their relationship to proximal and distal depositional environments. *J. Sediment. Petrol.*, Vol. 37, pp. 25–43.

WALKER, T. R. 1967. Formation of red beds in modern and ancient deserts. *Bull. Geol. Soc. Am.*, Vol. 78, pp. 353–368.

WALSH, P. T., BOULTER, M. C., IJTABA, M. and URBANI, D. M. 1972. The preservation of the Neogene Brassington Formation of the southern Pennines and its bearing on the evolution of upland Britain. *Q. J. Geol. Soc. London*, Vol. 128, pp. 519–559.

WARING, L. H. 1966. The basal Permian beds north of Kimberley, Nottinghamshire. *Mercian Geol.*, Vol. 1, pp. 201–211.

WARRINGTON, G. 1967. Correlation of the Keuper Series of the Triassic by miospores. *Nature, London*, Vol. 214, pp. 1323–1324.

— 1970. The stratigraphy and palaeontology of the 'Keuper' Series of the central Midlands of England. *Q. J. Geol. Soc. London*, Vol. 126, pp. 183–223.

— 1974. Studies in the palynological biostratigraphy of the British Trias. I. Reference sections in west Lancashire and north Somerset. *Rev. Palaeobot. Palynol.*, Vol. 17, pp. 133–147.

WEDD, C. B. 1903. In *Summ. Prog. Geol. Surv. G.B. for 1902.*

WEIR, J. 1960–68. A monograph of British Carboniferous non-marine lamellibranchia. Parts 10–13. *Palaeontogr. Soc.* [Monogr.]

WHETTON, J. T., MYERS, J. O. and BURKE, K. B. S. 1961. Tracing the boundary of the concealed coalfield of Yorkshire using the gravity method. *Min. Engineer*, Vol. 120, pp. 657–674.

WHITEHURST, J. 1778. *An inquiry into the original state and formation of the Earth.* (London.)

WILLIAMSON, I. A. 1970. Tonsteins—their nature, origins and uses. *Mineral. Mag.*, Vol. 122, pp. 119–126, 203–211.

WILLS, L. J. 1970. The Triassic succession in the central Midlands in its regional setting. *Q. J. Geol. Soc. London*, Vol. 126, pp. 225–283.

WILSON, E. 1876. On the Permian of the north-east of England (southern margin) and their relationship to the under and overlying formations. *Q. J. Geol. Soc. London*, Vol. 32, pp. 533–537.

— 1881. The Permian formations in the north-east of England. *Midland Nat.*, Vol. 4, pp. 97–101, 121–124, 187–191 and 201–208.

WYROBEK, S. M. 1959. Well velocity determinations in the English Trias, Permian and Carboniferous. *Geophys. Prospect.*, Vol. 7, pp. 218–230.

YORKE, G. 1961. *The Pocket Deposits of Derbyshire.* (Birkenhead.)

CHAPTER 2

Dinantian (Carboniferous Limestone Series)

DETAILS

a

b

HOPTONWOOD LIMESTONE

NORTH OF THE YOKECLIFFE RAKE.

The Hoptonwood Limestone crops out on high ground between Carsington and Hopton in the south, and the Ible valley to the north. Exposures are common, often occurring as isolated crags or 'tors' of flat-lying, pale yellow and brown dolomitic limestone, e.g. at Foxcloud Plantation [2665 5394]. Despite dolomitization, destruction of fossils is not always complete and lithostrotionoids, gigantoproductoids and crinoid ossicles are recognisable in places.

In the type section at Hoptonwood Quarries (Smith and others 1967, p. 15), on the southern edge of the Chesterfield district, the Hoptonwood Limestone is 250 ft thick and is unaltered, but dolomitization in the Hopton—Carsington area precludes detailed correlation. Sargent (1912) quotes an analysis by E. Sinkinson of the clay from the Matlock Lower Lava in Hoptonwood Quarry as follows: SiO_2 47.83%, Al_2O_3 22.74%, Fe_2O_3 and FeO 4.76%, FeS_2 5.02%, MgO 2.50%, CaO 1.19% Na_2O and K_2O 3.30%, H_2O 12.66%. Boreholes three miles east of Hoptonwood Quarries, and just beyond the north-western corner of the Derby district, proved limestones with up to seven clay horizons and with oolite bands which were particularly common towards the base. One bore penetrated a greenish grey siliceous rock at least 24 ft thick which may be tuffaceous and related to the Hopton Agglomerate and the Matlock Lower Lava.

In the now disused Magnesium-Electron quarry [2560 5480] the formation consists of up to 50 ft of massive grey-brown dolomite and dolomitic limestones which are strongly but irregularly jointed. Bedding is vague but one variable plane contains up to 6 in of pale green, possibly tuffaceous clay. The top surface of the dolomite is very irregular with joints widened to form pockets up to 10 ft at the top and containing brown sand stratified with clay, pebbles and chert. This area is on the fringe of the well-known Brassington pocket deposits (Early and Dyer 1964; Ford 1967, p. 70), the origin of which is discussed on p. 90.

QUARRIES NORTH-WEST OF WIRKSWORTH.

The Hoptonwood Limestone is exposed as inliers in the lower benches of the Middlepeak, Stonycroft, Baileycroft and Dale quarries. It is thickest in Stonycroft Quarry (locs. 1 and 2)* [286 543], where 120 ft or so of limestones have been excavated in close proximity to the Gulf Fault. The limestones are massive, pale grey and grey-brown, weathering to off-white. Fossils are rare and largely comprise fine crinoid debris.

*These localities are listed on p. 132 and a full list of fossils with the authors' names for the species is given on p. 131. Only the more important fossil records are given in the text.

The limestone is largely inaccessible, but displays a uniformity of bedding and colour. The top is marked by a 6-in dark grey shale band [2850 5431] with rolled brachiopods and corals, including Dibunophyllum bipartitum bipartitum and Carcinophyllum vaughani, which have probably been derived from the Hoptonwood Limestone. In the Middlepeak (loc. 3) [282 546] and Dale [283 542] quarries the massive nature and uniform bedding of the Hoptonwood Limestone is again well displayed. The maximum thickness of the limestone in Dale Quarry is estimated at a little over 100 ft.

The Baileycroft Quarry [287 542] is now disused and has been partially obscured by tipping. The generalized succession is:

	ft	in
Cawdor Limestone and Shale (P) loc. 38 (p. 132b)		
Dark shale overlying thinly bedded dark grey limestone with shale lenses	12	0
Matlock Limestone (D_2) loc. 23 (p. 132a)		
Limestone, grey, crinoidal	12	0
Limestone, grey, massive, brecciated; shelly fragments and brachiopods at base (wedges out southwards)	4	0
Limestone, grey to dark grey, with corals	3	0
Hoptonwood Limestone (D_1) loc. 4 (p. 132a)		
Limestone, pale grey, crinoidal, partly oolitic	3	6
Limestone, grey to pale grey; shell debris	3	6
Limestone, grey	1	0
Limestone, grey to pale grey	2	0

The Hoptonwood Limestone is overlain by 19 ft of Matlock Limestone (loc. 23). Both limestones are truncated by thinly bedded dark grey Cawdor Limestone (loc. 38) showing an undulating base (See Plate 2 and Fig. 5).

In the access tunnel [2870 5410] to Dale Quarry some 60 yd S of Baileycroft Quarry, typical pale grey massive limestones lying some 80 ft below the top of the Hoptonwood Limestone are exposed.

SOUTH OF THE YOKECLIFFE RAKE.

The Hoptonwood Limestone is in the main undolomitized, comprising pale grey to off-white fine-grained massive limestones. The topmost bed, some 4 ft thick, is particularly rich in gigantoproductoids and colonial corals and is exposed in the small quarries (loc. 5) [2709 5372] south-east of Godfreyhole. The limestones bordering the Hopton Agglomerate are richly fossiliferous and of a reefy nature (locs. 6 and 7).

c

A distinct reef-knoll north of the Almshouses at Hopton (loc. 8) [2608 5333] yielded a rich assemblage, typical of D_1 reef limestones, including: Fenestella sp., Acanthoplecta mesoloba, Aliteria cf. panderi, Antiquatonia cf. antiquata, A. hindi, A. insculpta, Avonia youngiana, Dielasma hastatum, Echinoconchus punctatus, Eomarginifera sp. lobata group, Fluctuaria sp., Gigantoproductus sp., G.? sp. nov. [wrinkled concentric ornament], Martinothyris cf. lineata, Overtonia fimbriata, Pleuropugnoides pleurodon, Plicatifera plicatilis, Plicochonetes cf. buchianus, Pugnax acuminatus platylobatus, P. pugnus [small form], Pugnoides triplex, Spirifer cf. grandicostatus, S. trigonalis, orthocone nautiloids, Cyclus sp. and ostracods.

Similar assemglages were obtained from limestones on the western side of the Hopton Agglomerate e.g. at Carsington Wood (loc. 9) [2542 5340]. An old quarry [2625 5340] east of Hopton is largely inaccessible, but appears to comprise about 50 ft of massive, pale grey, fine-grained to porcellaneous limestone without obvious fossil concentrations.

A site investigation borehole (loc. 10) [2652 5340] drilled on the extreme edge of the limestone outcrop some 200 yd east of the above quarry proved 131 ft of pale grey to grey-brown fine-grained massive Hoptonwood Limestone. Thin shelly bands in bioclastic limestone contained a D_1 fossil assemblage including Dibunophyllum bourtonense (at 129 ft 7 in). The limestones were intersected by numerous fault planes and associated minor fractures.

The most southerly outcrops of limestone [2723 5302], in the vicinity of Dream Mine (loc. 11) and Sprink Wood (loc. 12), form an elevated promontory surrounded on three sides by Namurian shales. They lie within the top 50 ft or so of the Hoptonwood Limestone and are of reef facies. They are richly fossiliferous, and brachiopods often with red coatings of algal material are particularly common. Shirley (1959, p. 420) noted the reef-like fauna from this locality and recorded the characteristic D_1 fossil Davidsonina septosa. A collection during the present resurvey from near Dream Mine (loc. 8) yielded a D_1 reef assemblage including the following: Koninckopora inflata, Lithostrotion martini, L. pauciradiale, Athyris cf. expansa, Chonetipustula carringtoniana, D. hastatum, Echinoconchus punctatus, Eomarginifera sp. lobata group, Gigantoproductus? sp. nov. [wrinkled concentric ornament], Leptagonia analoga, Overtonia sp. [juv.], Pugnax pugnus [small form], Schizophoria resupinata, Spirifer cf. grandicostatus, S. trigonalis, S. bisulcatus, Straparella fallax, Straparollus dionysii, Turbonitella biserialis, Conocardium alaeforme, Girtypecten? tessellata, Griffithides? and ostracods.

The margin of the promontory is complicated by faulting and mineralization, but the structure appears anticlinal near Dream Mine. On the nose of the 'promontory' southerly dips up to 65° carry the beds down the slope, at the base of which they pass below Namurian shales of estimated E_1 age.

This folding possibly accentuates already existing depositional dips.

MATLOCK LIMESTONE

NORTH OF THE YOKECLIFFE RAKE.

The outcrop of the Matlock Limestone is only about half the area of that of the Hoptonwood Limestone. North-west of Wirksworth at the base of the Matlock Limestone, a ledge-like feature is considered to be the outcrop of a thin representative of the Matlock Lower Lava, the only evidence of which was noted [2646 5475] 140 yd W of the Hopton Tunnel entrance where it is marked by fragments of greenish weathered clay. The succession in the cuttings in the approaches to the Tunnel is:

	ft	in
Limestone, grey, crinoidal, with large irregular chert masses	10	0
Limestone, grey, crinoidal, partly dolomitized	9	0
Limestone, massive, dolomitized, cherty; shelly with colonial coral	10	0
Limestone, grey, fine-grained, cherty, partly dolomitized	11	0
Limestone, dolomitic, with clay partings	1	0
Limestone, grey, fine-grained to porcellanous, cherty; brachiopods and corals in basal 3 ft	8	6
Limestone, irregularly bedded, dolomitic, cherty; rare gigantoproductoids	16	0
Dolomite, cherty, poorly bedded; rare shells	5	0
Dolomite, well-bedded	19	0
Dolomite, thinly-bedded, shelly	5	0
Dolomite, massive	4	0
Dolomite, nodular, shelly	2	0
Dolomite, massive; shelly bands	12	0
Dolomite, thinly and irregularly bedded	25	0
Dolomite, level-bedded; sporadic corals	15	0
	152	6

The passage from dolomite to limestone occurs approximately in the middle of the tunnel.

The Yokecliffe Rake Fault scarp (locs. 13-15), some 60 ft high, provides an excellent section of Matlock Limestone, which is massive, grey and crinoidal with sporadic shelly bands. At the top of the scarp [2807 5391] the junction with the overlying Cawdor Limestone can be seen.

QUARRIES BETWEEN MIDDLETON AND WIRKSWORTH.

In Middlepeak Quarry (locs. 16-19) [282 546] a maximum thickness of 140 ft of Matlock Limestone was recorded in 1956. Subsequent quarrying has extended the face by some 100 yd south-westwards and by 300 yd north-westwards. Eden (in Smith and others 1967, p. 22) noted that, when traced southwards, the Matlock Limestone becomes progressively darker, less cherty and more thinly bedded. It includes a prominent unconformity; lenticular bedding and beds of breccia are

c

d

particularly common at the top.

The following succession was measured in the south-western corner of the quarry [2815 5441]:

		ft	in
Limestone, grey, massive, coarsely crinoidal, with pockets of Gigantoproductus sp.; breccias common; uneven top surface		6	0
Limestone, dark grey, fine-grained, well-bedded		1	5
Limestone, grey-brown, massive		2	6
Limestone, dark grey, fine-grained, rubbly bedded; small shell debris and a 6-in brachiopod band near top		8	0
Limestone, grey, with concentrations of 'Girvanella' sp. (Upper 'Girvanella' Band)		2	0
Limestone, grey-brown, massive, with sporadic shelly lenses towards base		5	0
Limestone, dark grey, thinly bedded, with chert		18	0
Gap (horizon of unconformity)	abt	14	0
Limestone, pale grey, massive		6	0
Limestone, dark grey, well-bedded		13	0
Limestone, dark grey, pale weathering		5	0
Limestone, dark grey, well-bedded, with 6 in of dark grey shale at base with rolled corals and brachiopods		2	6
		83	5

In Stonycroft Quarry (locs 20-22), 300 yd SE of the Middlepeak section, the thickness of the Matlock Limestone is reduced to 62 ft. The overlying Cawdor Limestone has overlapped the uppermost part and rests on limestones below the Upper 'Girvanella' Band. The Matlock Limestone is thinly bedded, individual beds lensing out south-eastwards.

On the north-eastern face of Baileycroft Quarry, the Matlock Limestone comprises some 19 ft of mostly massive, grey to dark grey beds which contain breccias towards the base. In the basal 7 ft, beds wedge out southwards but contain a rich fauna (loc. 23) including Aulophyllum fungites, Dibunophyllum bipartitum, Diphyphyllum lateseptatum, Lithostrotion junceum, L. martini, L. pauciradiale, Lonsdaleia floriformis, Palaeosmilia murchisoni and P. regia. The beds continue to thin southwards (p. 129b) and 120 yd along the quarry face they are overlapped completely by the Cawdor Limestone. Some 80 ft of Matlock Limestone are therefore cut out in 200 yd. Between Middleton and Wirksworth some 140 ft of Matlock Limestone are removed in a distance of 1000 yd.

When traced south-westwards from Stonycroft Quarry to Dale Quarry (locs 24, 25) [282 540] the Matlock Limestone again shows a decrease in thickness. The succession in the southern end of Dale Quarry is:

	ft	in
Limestone, dark grey, thinly bedded, fine-grained; sporadic chert lenses at top	18	0
Clay, black	0	1
Limestone, dark grey, thinly bedded fine-grained shaly limestone partings; shelly in upper part	11	0
Limestone, dark grey, fine-grained, rare small shells	12	0
Clay, brown	0	1
Limestone, grey, fine-grained, crinoidal; abundant small shells	6	0
Clay, brown	0	1
Limestone, dark grey; sporadic shelly fauna	6	0
Shale, black with abundant decomposed shells	0	2
	53	5

In this southern part of the district the Matlock Limestone is generally dark grey and thinly bedded and difficult to differentiate from the overlying Cawdor Limestone where angular discordance is not pronounced.

SOUTH OF THE YOKECLIFFE RAKE.

The eastern side of the Godfreyhole promontory (locs 26-29) comprises a maximum of about 50 ft of Matlock Limestone, though on the southern part of the promontory the thickness is estimated to be less than 25 ft. This may be due to overlap by the Namurian or to thinning across a penecontemporaneous anticline (p. 96).

The basal beds of Matlock Limestone exposed south-east of Godfreyhole (p. 129b) are massive, grey and partly oolitic. In a small quarry, some 100 yd to the north-east (loc. 30) [2729 5363], 21 ft of massive, grey, evenly bedded, crinoidal limestone are exposed. The discovery of 'Girvanella' nodules in these beds may indicate an equivalence with the Lower 'Girvanella' Band of the Castleton area which there lies some 6 ft above the boundary between the D_1 and D_2 zones (Stevenson and Gaunt 1971, p. 73).

The fauna includes: 'Girvanella' nodules, foraminifera, Aulophyllum fungites, Dibunophyllum bipartitum, Lithostrotion pauciradiale, Palaeosmilia murchisoni, Gigantoproductus sp. [latissimoid], Spirifer bisulcatus, Sulcatopinna flabelliformis and ostracods.

Some 150 yd E of Dream Mine (loc. 31) [2741 5308] grey thinly bedded limestone, about 10 ft above the base of the Matlock Limestone, is exposed. The northern face of the quarry is rather rubbly bedded and coarsely crinoidal; it yielded a fauna including: Dibunophyllum bipartitum, Eomarginifera sp. lobata group, Gigantoproductus sp., Megachonetes sp. papilionaceus group and M. siblyi.

The dip increases eastwards and the Matlock Limestone plunges beneath the overlapping presumed Namurian shales at an angle of 33°.

CAWDOR GROUP

The Cawdor Limestone is mainly exposed north-east of Wirksworth. In Middlepeak Quarry most of it has been worked out, but the remaining isolated patches are readily accessible. The limestone is predominantly grey and dark grey, thinly bedded and cherty with shaly partings. Crinoidal debris is common, associated with rich reefy patches of brachiopods. The basal beds are invariably irregular and reflect the undulations of the eroded and channelled top surface of the underlying Matlock Limestone. Some pockets are coarsely crinoidal and often contain broken brachiopod shells and brecciated fragments of limestone.

The following fauna was obtained from the lowest 15 ft of Cawdor Limestone in Middlepeak Quarry (locs. 34 and 35): Caninia juddi, zaphrentoid, Aliteria panderi, Antiquatonia insculpta, Avonia davidsoni, A. youngiana, 'Brachythyris' planicostata, Dielasma hastatum, Eomarginifera lobata aff. laqueata, E. longispina, Gigantoproductus crassiventer, orthotetoids, Productus concinnus, Pugilis pugilis, Rugosochonetes sp., Schellwienella sp., Schizophoria sp., smooth spiriferoids, Spirifer trigonalis, S. bisulcatus, Naticopsis sp., Aviculopecten?, Edmondia?, Leiopteria sp. [juv.], Griffithides sp., Weberides sp., ostracods, fish tooth and plate.

In the east of Baileycroft Quarry (see p. 129b) for section) thinly bedded dark grey limestones with shales show considerable undulation and are overlain in the south by dark laminated shales which contain abundant Caneyella membranacea (McCoy), characteristic of the highest beds of P_2 (See locs. 36, 37 and 38 for Baileycroft Quarry, west side). In Dale Quarry (loc. 39) similar shales of Cawdor age rest unconformably on the Matlock Limestone. This is the most southerly outcrop of the Cawdor Shales; farther south [2800 5300] they are overlapped by the Namurian shales of low E_2 or E_1 age which rest on successively lower horizons within the Cawdor Limestone.

South of the Yokecliffe Rake, the Sweenclose Vein separates the main outcrop of Hoptonwood and Matlock limestones to the north-west from an irregular patch of poorly exposed and variable limestone, near Pittywood Farm (locs. 40 to 45). This limestone, partly exhumed from beneath the Namurian shale cover, has an irregular surface with a relief of up to 25 ft. Its western boundary, straight over a distance of 300 yd, suggests a fault. The beds vary from pale grey, partly oolitic limestones to grey and dark grey limestones. Shirley (1959) refers them to the Cawdor Limestone. A 'knoll' of massive pale grey limestone is bordered on the south by thinly bedded fine-grained dark grey limestones which may be remnants of a once more extensive Cawdor Limestone outcrop. The present survey has confirmed that the faunas are of D_2 age and include typical Cawdor reef forms from the vicinity of the knoll. The fauna recorded from limestones between Sweenclose Vein and Pittywood Farm includes: Dibunophyllum bipartitum bipartitum, Syringopora cf. ramulosa, Antiquatonia antiquata, cf. 'Brachthyris' planicostata, Echinoconchus punctatus, Gigantoproductus cf. edelburgensis, orthotetoid, Phricodothyris cf. insolita, Pleuropugnoides pleurodon, Plicochonetes sp., Productus sp., Schizophoria sp., Spirifer trigonalis, S. bisulcatus, Striatifera striata, Bellerophon sp., Straparollus sp. and Weberides sp.

CRICH

The best exposures are at Hilts Quarry [3530 5425], south-east of the village. The general succession is:

	ft	in
CAWDOR LIMESTONE (locs. 47-49)		
Limestone, predominantly dark grey medium- to coarse-grained; pale and dark chert lenses; Gigantoproductus sp. common, sporadic corals	57	10
MATLOCK LIMESTONE (loc. 32 a-c)		
Limestone, grey fine- to medium-grained; cherty lenses at base; sporadic brachiopods and corals	98	9

A 6-in brown clay band (First Clay Band - see p. 130b) with limestone partings occurs some 11 ft below the top of the Matlock Limestone, and may be the lateral extension of a bed of toadstone (Alsop 1845). Farther north, a 2-ft clay band (Second Clay Band, see below) some 70 ft below the top of the Matlock Limestone, has been regarded as the lateral equivalent of the Matlock Upper Lava (Smith and others 1967, p. 36), but this horizon is unfortunately now obscured in Hilts Quarry. Recent excavation in Cliff Quarry [343 557], north of Crich, exposed a 2-ft mudstone some 70 ft below the top of the Matlock Limestone. The mudstone was grey silty and pyritic and contained bryozoa and shell fragments. It extended throughout the quarry, which is on the axis of the Crich Anticline (p. 10). The basal few inches comprised green and purple-stained clay and the top 3 in or so were altered to a dark grey and ferruginous brown colour. This description confirms that of Arnold-Bemrose (1899, p. 77) and modifies the observations of Sargent (1912) who considered such a lithology to be limited to one small exposure.

Such evidence indicates that not all clay wayboards are decomposed lavas or tuffs and that in the Crich area, limestone sedimentation was interrupted towards the end of D_2 times by an influx of argillaceous material which compares with that found in the overlying Cawdor Limestone and Shale. The shale rests on a weathered limestone surface which together with the altered shale margins indicates local periods of emergence.

a

b

Sargent (1912, p. 412) quotes the following analyses by E. Sinkinson of two clay bands:

	First Clay-Band (blue central portion), Hilt's Quarry	First Clay-Band (ochreous margin), Hilt's Quarry	Second Clay-Band Hilt's Quarry
SiO_2	46.66	48.60	47.50
Al_2O_3	24.76	24.23	29.62
F_2O_3 and FeO	8.82	21.99	4.48
FeS_2	-	-	-
FeS	-	-	-
MgO	2.72	trace	1.35
CaO	1.97	0.42	0.96
Na_2 and K_2O	2.02	0.71	3.00
H_2O	12.52	3.72	12.73
CO_2	0.53	0.33	0.36
	100.00	100.00	100.00

BASIN FACIES

In the following account it is convenient first to summarize the borehole results and then to pass on to the surface exposures of the district.

DUFFIELD BOREHOLE [3428 4217]

This proved upper Viséan rocks belonging to the Upper Beyrichoceras (B_2), Lower Posidonia (P_1) and Upper Posidonia (P_2) zones, the succession being broken by faults which probably cut out only a small thickness of strata. Apart from igneous rocks and two relatively thin calcareous mudstone sequences of 13 ft and 13 ft 6 in at depths of 1406 ft 4 in and 1443 ft 11 in respectively, the succession consists largely of graded silty mudstone/sandstone and muddy limestone units with dark grey mudstone intercalations, comprising the Widmerpool Formation.

Graded Units

The graded units range in thickness from less than an inch to 4 ft but are normally under 2 ft thick. They all have sharp contacts with the underlying mudstone and the soles often show linear structures such as groove and bounce casts or small, irregular, knobbly structures, some of which are probably the sand-filled burrows of bottom-dwelling animals.

In general, the graded units show two distinct divisions, a lower grey to pale grey arenaceous division and an upper dark grey calcareous silty mudstone or pelitic division. An intervening poorly delineated, relatively thin third division of silty or sandy calcareous mudstone with faintly developed parallel lamination is also present in most units.

The arenaceous division consists mainly of quartz sand and silt or of bioclastic debris. Graded units with quartz sand in the lower divisions predominate in the top 34 ft of P_2 immediately below the Cravenoceras leion Band, between 2633 ft and 3030 ft in the lowest 400 ft of P_1 and between 3353 ft 3 in and 3436 ft 10 in in B_2. In this last sequence sandstone makes up about 57 per cent of the total thickness and 65 per cent of the turbidite beds with the sandstone divisions averaging 4.5 in. These sandstones usually show thin sets of ripple lamination. In a few thicker graded units one or two additional lower divisions show strong parallel laminations or massive sandstone, but sedimentary structures are often masked by diagenetic calcium carbonate recrystallization. The arenaceous divisions in which bioclastic debris predominates usually show parallel lamination with alternating pale and dark laminae. In a significant proportion of units the arenaceous division is absent.

The pelitic divisions form the major proportion of the succession except between 3353 ft 3 in and 3436 ft 10 in, in the lowest 84 ft of the B_2 sequence (see above). They are usually massive with a slight upwards decrease in grain size, and the calcium carbonate content (qualitatively estimated from the amount of effervescence with dilute HCl) is generally higher where the arenaceous division of the unit is rich in bioclastic debris. One exceptional unit at 1717 ft 3 in is 7 ft 8 in thick and grades unevenly upwards from limestone conglomerate to dark grey calcareous mudstone. The conglomerate contains angular phenoclasts of muddy limestone up to 4 in across, set in a dark grey muddy limestone matrix with coarse crinoidal and broken shelly debris.

Fossils in the graded units are restricted to small shells and bioclastic debris and to finely divided plant debris which mainly occurs in the laminated silty mudstone and sandstone divisions.

Mudstone Beds

The mudstones between the graded units are normally less than 1 in thick; some are merely films which adhere to the soles of the overlying graded units. In a few cases they are absent, presumably removed by the current depositing the succeeding graded unit. In the strongly calcareous sequences their presence may be masked by recrystallization. They are then distinguished by sharp contacts with the pelitic top of the underlying graded units, their darker colour (often nearly black), their low silt content and their fissility. Fossils, particularly goniatites and bivalves, are common in the mudstone beds.

Igneous rocks

Six tuff bands and two dolerite sills were encountered. The tuffs range from fine- to coarse-grained and vary in colour from pale blue to greenish-grey (p. 167d). They usually contain pumice fragments and in some beds lapilli are present. Some intermixing of tuff with calcareous mud has been noted, particularly in the thickest band.

The two dolerite sills are 27 ft 9 in and 220 ft 3 in thick, their bases occurring at 1779 ft 7 in (P_1) and 3337 ft 3 in (B_2) respectively. Both have chilled upper and lower margins and have baked the adjacent sediments.

TRUSLEY BOREHOLE [2548 3588]

The Dinantian rocks in this borehole comprise a turbidite sequence of red and purple sandstones, siltstones and silty mudstones which show grading, slumping and sole markings. The lower part of the sequence is calcareous. Interbedded with the turbidite units are red, brown and purple mudstones interlaminated with pale grey-green mudstones.

Detailed examination of representative lengths of Trusley core (330 ft 9 in to 340 ft 8 in, 368 ft 11 in to 380 ft 7 in, 410 ft to 420 ft, 455 ft 2 in to 465 ft 6 in and 496 ft 7 in to 507 ft 2 in) indicates that graded sandstone - mudstone (turbidite) units comprise about 84 per cent of the core and interturbidite or pelagic shale about 16 per cent. Sandstone makes up about 49 per cent of the turbidites or 42 per cent of the total core with individual sandstone divisions averaging 2.8 inches in thickness. The sandstone-bearing turbidites have a proximality index (R.G. Walker 1967) of 20 per cent compared with 96 per cent in the B_2 Zone sandstone-bearing turbidites of similar age in Duffield Borehole. Thus both sequences are of distal turbidites, but that in the Duffield Borehole appears to be more distal than that in Trusely. However, there is no evidence that the two sequences are stratigraphically equivalent and sedimentological comparisons are of doubtful validity.

The Trusley Borehole sequence is sparsely fossiliferous but contains enough to justify its classification as the Widmerpool Formation. Fragments of orthotetoid brachiopods have been identified from 503 ft. In thin sections made from the lowest and most calcareous parts of the sequence (504 ft) Dr W. H. C. Ramsbottom identified bryozoa and crinoid fragments and the following D_1 Zone algae and foraminifera — Koninckopora inflata, Archaediscus krestovnikovi, A. moelleri, Endothyra spp. and Tetrataxis sp.

A typical section with turbidite units from the Trusley Borehole is shown on the next page.

AT OUTCROP

The Widmerpool Formation outcrop borders that of the lower Namurian in the Turnditch, Quarndon and Weston Underwood districts and forms inliers in the valley of Mackworth Brook near Kirk Langley, at Wildpark Wood, one mile east of Brailsford, and in a valley immediately south of Brailsford (Fig. 3).

In general, the formation consists of thin beds of dark grey argillaceous limestone and grey calcareous siltstone exposed sporadically in streams such as Blind Brook, Kedleston and its tributaries [3088 4199 and 3159 4194], Cutler Brook, Weston Underwood [2908 4188] Black Brook, Ravensdale Park [2730 4454], Waterlaggs Brook, Hulland Ward [2765 4597], Flagshaw Brook (see below) and a stream near Windley [3058 4438]. Similar hard beds have also been noted in old pits, temporary sections and road banks. They commonly contain plant fragments, with occasional poorly preserved bivalves (see also Gibson and others 1908, p. 24). A few of the coarser limestones contain conspicuous crinoid and brachiopod debris). An extensive microflora obtained from a sequence near Mackworth [3137 3768] included no diagnostic species. In the details listed below the numbers correspond to the locations shown on Fig. 3.

TURNDITCH (4)

In an old limestone quarry at Flower Lilies [2946 4550], south-west of Turnditch, 30 ft of alternating dark grey limestones, calcareous siltstones and grey fossiliferous mudstones are poorly exposed and largely inaccessible. The limestones contain crinoid debris and brachiopods including Buxtonia sp., Dielasma sp., productoid fragments [costate] and Spirifer sp. The shale tips [2933 4563] near the quarry entrance yielded specimens of Posidonia corrugata indicative of a P_2 age. The same age is indicated by Neoglyphioceras spirale (Phillips), found near this locality by Ford (1968a).

At the only exposure of the basal Namurian Cravenoceras leion Band in the Derby district — a stream section and its nearby tributary near Champion Farm [3243 4328] — the underlying Dinantian rocks are thinly bedded quartzitic sandstones and siltstones. Similar sandstones, probably belonging to the same sequence, have been worked in shallow pits in Park Nook Wood [3254 4142].

White tuffaceous calcareous clay bands, 3 ft and 2½ ft thick in Blind Brook [3097 4233] and a temporary exposure [3037 4352] respectively, are the only exposed representatives of the upper P_1 - lower P_2 tuffaceous sequence recorded in Duffield Borehole. Microscopic examination by Mr R.K. Harrison shows a distinct pyroclastic assemblage altered to clay consisting predominantly of a mixed-layer mineral (E 34597). Differential thermal analysis confirmed the presence of a mixed montmorillonite-illite clay mineral but with no detectable sepiolite. The clay bands are therefore of volcanic origin, though some reworking as well as mineral replacement seems to have occurred.

WILD PARK INLIER [2726 4127] (2)

Some 4 ft of purple and buff-stained calcareous sandstone and siltstone are poorly exposed in the banks of Wild Park Brook. The colouration here and in other exposures (see above) is due to secondary alteration close beneath a former cover of Triassic beds. Farey (1811, p. 159) described patches of yellow limestone from this locality and mistakenly thought them to be Magnesian Limestone overlying Westphalian rocks.

KIRK LANGLEY INLIER (1)

Exposures in an old quarry [2919 3891] comprise about 12 ft of red and red-brown sandstone in beds

c

d

up to about 2 ft thick, with sole structures aligned east-west and east-north-east—west-south-west, and with thin intercalations of red and green clays and silty clays (see also Green and others, 1887, p. 88). Crinoids and Fenestella sp. have been recorded. This sequence is more arenaceous than any other recorded either at outcrop or in the Duffield and Trusley boreholes, but its stratigraphic level is uncertain.

Mr K. S. Siddiqui described a thin section [E 38065] from this locality as a 'well-consolidated feldspathic sandstone lightly stained with red-brown iron oxide. The clastics are coarse-to medium-grained (0.8 x 0.6 mm - 0.4 x 0.3 mm), mainly angular to subangular though a few are subrounded, very poorly sorted, much fragmented and scattered with aggregated patchy cement. Polygranular quartz (average 0.3 mm), angular in places is the predominant minerals, pitted with numerous inclusions and cracks and showing secondary overgrowths, Orthoclase (average 0.3 mm), plagioclase (0.2 mm), and a few grains of microcline (0.2 mm) make up the bulk of the section. Rarer scales of muscovite and chlorite, mostly orientated along the bedding, occur and dispersed aggregates of illite, kaolinite, dusty carbonates and well-crystallized dolomite (in places etching the silicates) are common. Cherty grains of cherty sillica (0.6 mm), particles of siltstone and sandstone and quartzite (0.6 mm), grains of pyrite, and leucoxene and a few granules of red-brown iron oxide and red organic matter are also present. A microchemical test confirmed the presence of ferric radical. A heavy mineral analysis showed the predominance of opaque ferric grains followed by perfect to broken prisms of zircon (some with zonal growth), tourmaline (O = dark brown, E = colourless to light yellow) anhedral grains of garnet (some step-etched) and rounded an prismatic grains of apatite'.

The presence of plagioclase suggests a B_2 age for the beds, the mineral have only been recorded from the B_2 Zone in the Duffield and Trusley boreholes.

In Flagshaw Brook, interbedded mudstones, siltstones and sandstones are exposed sporadically for nearly a mile, and at one locality [2977 3894] Sudeticeras sp. and Posidonia corrugata, indicative of the P_2 Zone, have been collected from 6 ft of grey silty mudstones containing bands up to 6 in thick, of hard grey siltstone with sole structures.

In a road cutting [3137 3768] at Mackworth, faulted against Waterstones, are 3 ft 7 in of laminated grey and dark-grey mudstone containing bands, up to 4 in thick, of grey-brown slightly calcareous siltstone and one band, 8 in thick, of brown and red-tinged fine calcareous sandstone. Dr B. Owens refers the extensive microflora from this sequence (Sample SAL 1247) to the upper Dinantian or Namurian A. Since siltstone and fine sandstones are absent from the lowest 150 ft of the Namurian in Duffield Borehole, the sequence at Mackworth is assigned to the top part of the Widmerpool Formation.

There are no exposures in the Brailsford Inlier (3).

A typical section with turbidite units from the Trusley Borehole is as follows:

Lithology	Types (following Bouma 1962)	Thickness ft in	Depth ft in	Sample Nos
Sandstone, massive, slight load casting	A - E		368 - 0	DF (R) 1181
Mudstone, red-brown and purple, inter-laminated with pale grey-green; ? sphaerosiderite and numerous plant fragments	pelagic	0 - 4½	368 - 4½	DF (R) 1182
Siltstone, red-brown, massive		0 - 2	368 - 6½	
Sandstone/siltstone, evenly laminated, load cast base	B - E	0 - 0½	368 - 7	DF (R) 1183
Mudstone, interlaminated as above	pelagic	0 - 4	368 - 11	DF (R) 1184
Siltstone, purple, massive		0 - 2½	369 - 1½	
Mudstone, laminated		0 - 1	369 - 2½	
Siltstone, massive, red-brown	B - E	0 - 2½	369 - 5	
Siltstone/sandstone, interlaminated, showing complex internal grading; slumped beds in top half		0 - 10½	370 - 3½	DF (R) 1185, 1186 and 1189
Mudstone, purple	pelagic	0 - 0½	370 - 4	DF (R) 1190
Sandstone	A - E	0 - 2	370 - 6	DF (R) 1191
Mudstone, red-purple, interlaminated	pelagic	0 - 2	370 - 8	DF (R) 1192
Siltstone, massive	C - E	0 - 2	370 - 10	DF (R) 1193
Sandstone, laminated, rippled		0 - 2	371 - 0	DF (R) 1194

a

b

PALAEONTOLOGY OF THE DINANTIAN LIMESTONES

The stratigraphical palaeontology of the Dinantian limestones is similar to that already described by Ramsbottom (in Smith and others 1967, pp. 47-57) from the adjacent Matlock district and is therefore not described here. The available palaeontological data for the various limestone divisions in the district is summarized below in two lists, the first giving the fossil determinations in biological order with cross-references to the numbered localities where they were collected, the second giving the complementary list of localities. Brief notes on the fauna and age of the limestones is given in the general account (p.5).

FOSSILS COLLECTED FROM THE LIMESTONES

The names of the fossils are followed by the numbers of the localities from which each has been recorded. The use of '?', 'cf.' or 'aff.' before a locality number in these lists respectively indicates doubt as to the identification of, similarity to, or departure from the genus or species named.

HOPTONWOOD LIMESTONE

Koninckopora inflata (de Koninck) 11
Algal nodules 11
Foraminifera 4
Aulophyllum fungites (Fleming) 4
Caninia cf. densa Lewis [of Hudson & Cotton 1945, p. 306] 4
C. sp. ?3, 4, 75, 7
Carcinophyllum vaughani Salée 4
Chaetetes depressus (Fleming) 4
C. septosus (Fleming) 4
Dibunophyllum bipartitum bipartitum (McCoy) 4
D. bourtonense Garwood & Goodyear 4, 10
Diphyphyllum sp. 4
Koninckophyllum cf. proprium Sibly 4
Lithostrotion martini Milne Edwards & Haime ?3, 11
L. pauciradiale (McCoy) 4, 11
Palaeosmilia murchisoni Milne Edwards & Haime 4, 7
Syringopora sp. 4, 5
Zaphrentoid 4
Serpula sp. 11
Fenestella sp. 6, 8
Acanthoplecta mesoloba (Phillips) 8
Aliteria cf. panderi (Muir-Wood & Cooper) 8
Antiquatonia antiquata (J. Sowerby) 8, cf. 8, 9
A. hindi (Muir-Wood) 8, 12
A. insculpta (Muir-Wood) 8
A. sp. 7
Athyria cf. expansa (Phillips) 11
Avonia youngiana (Davidson) 8
A. sp. [juv.] 11
Brachythyris sp. 11
Buxtonia sp. 8, 9
Choneipustula carringtoniana (Davidson) 11
C. sp. ?8, 11
Davidsonina septosa (Phillips) 7
D.? sp. 2b
Dictyoclostus pinguis (Muir-Wood) 8
Dielasma hastatum (J. de C. Sowerby) 8, 11
D. sp. 7, 9
Echinoconchus punctatus (J. Sowerby) 8, 11, cf. 11
E. sp. 7
Eomarginifera sp. lobata (J. Sowerby) group, 5, 8, 11
E. sp. 11
Fluctuaria sp. 8
Georgethyris cf. acutiloba (George) 7
Gigantoproductus spp. 1, 2a, 2b, 3, 7, 8, 9, 11

Gigantoproductus? sp. nov. [wrinkled concentric ornament: Mitchell in Stevenson and Gaunt 1971, pl. 14, fig. 2] 8, 11
G. sp. [latissimoid] 8
Krotovia? [juv.] 8
Leptagonia analoga (Phillips) 11
Linoprotonia sp. ?1, 2a, 8, 9
Martinia sp. 6
Martinothyris cf. lineata (J. Sowerby) 8
Megachonetes sp. 5
Orthotetoid 5, 6, 7, 9
Ovatia sp. 6
Overtonia fimbriata (J. de C. Sowerby) 8
O. sp. [juv.] 11
Phricodothyris sp. 8
Pleuropugnoides pleurodon (Phillips) 6, 8
P. sp. [juv.] 8
Plicatifera plicatilis (J. de C. Sowerby) 8
P. sp. 8, 11, ?12
Plicochonetes cf. buchianus (de Koninck) 8
P. sp. [juv.] 11
Proboscidella proboscidea (de Verneuil) 6, 9
Productus? 8
Pugnax acuminatus (J. Sowerby) platylobatus (J. de C. Sowerby) 8
P. pugnus (Martin) [small form] 8, 11
Pugnoides triplex (McCoy) 8
Reticularia cf. mesoloba (Phillips) 9
Schizophoria connivens (Phillips) 7
S. resupinata (Martin) 7, 11
S. sp. 6, 8, 12
Smooth spiriferoid 5, 6, 7, 8, 9, 11
Spirifer bisulcatus J. de C. Sowerby 11
S. cf. granicostatus McCoy 8, 11
S. trigonalis (Martin) 8, 11, 12
S. sp. 4
Bellerophon sp. [juv.] 8
Euomphalus? 2b, 11
Naticopsis sp. [juv.] 11
Straparella fallax (de Koninck) 11
Straparollus dionysii de Montfort 11
S. sp. 8, 11
Turbonitella biserialis (Phillips) 11
Conocardium alaeformis (J. de C. Sowerby) 11
Edmondia sp. 6, ?8, 11
Girtypecten tessellatus (Phillips) 11
Pectinoid 8
Sanguinolites sp. [juv.] 8
Streblopteria laevigata (McCoy) 6
Bivalve indet. 11

Orthocone nautiloid 8
Cyclus sp. 8
Griffithides? 11
Ostracods 8, 11

MATLOCK LIMESTONE

'Girvanella' nodules 30c
Saccamminopsis sp. 27
Foraminifera 19b, 20, 23, 24b, 30a, 31, 32a
Amplexizaphrentis enniskilleni (Milne Edwards & Haime) 19b
Aulophyllum fungites (Fleming) 20, 23, 24b, 30c
A. fungites pachyendothecum Thomson 19a
Caninia benburbensis Lewis 32a
C. juddi (Thomson) ?18a, 18b, ?19d, 20
C. sp. ?19b, 23, ?24a
Carcinophyllum vaughani Salée 17, 21
Chaetetes depressus (Fleming) 23
Clisiophyllum keyserlingi McCoy 19a
C. sp. 32a
Dibunophyllum bipartitum bipartitum (McCoy) 13, 18a, 18b, 21, 23, 30a, 30c, 31
Diphyphyllum lateseptatum McCoy 16, 18b, 23, 24b, aff. 33
D. sp. 13, ?31
Koninckophyllum proprium Sibly 32a
K. sp. 13
Lithostrotion junceum (Fleming) 23
L. martini Milne Edwards & Haime 19a, 20, 23
L. pauciradiale (McCoy) 14, 18b, 23, 30a, 30c, 32a
Lonsdaleia floriformis floriformis (Martin) 23
Palaeosmilia murchisoni Milne Edwards & Haime 23, 30c
P. regia (Phillips) 17, 23
Syringopora cf. geniculata Phillips 23
S. cf. ramulosa Goldfuss 18b, 20
Zaphrentites sp. 33
Fistulipora incrustans (Phillips) 23
Antiquatonia cf. muricata (Phillips) 25
A. sp. 13, ?14, 19c, 28
Avonia davidsoni (Jerosz) 31
A. youngiana (Davidson) 25
A. sp. [juv.] 30a
'Brachythyris' planicostata McCoy 31
B. sp. 31
Dielasma sp. 31, 32b
Echinoconchus sp. 26, 31
Eomarginifera sp. lobata (J. Sowerby) group 31
E. sp. 24c, 26
Gigantoproductus crassiventer (Prentice) 24c
G. cf. edelburgensis (Phillips) 15, 27
G. aff. gigantoides (Paeckelmann)
G. inflatus (Sarycheva) 20
G. sp. [latissimoid] 18a, 19b, 30a
G. spp. 13, 14, 16, 18b, 19a, 19d, 20, 21, 24a, 24b, 25, 27, 28, 30a, 30c, 31, 32a, 32c
Linoprotonia sp. 19c, 28
Martinia sp. 19b, 24b, 26
Megachonetes cf. papilionaceus (Phillips) 26
M. siblyi (I. Thomas) 20, 22, 31
M. sp. 31
Orbiculoidea? 19c
Orthotetoid 13, 20, 31
Phricodothyris sp. 20
Pleuropugnoides cf. greenleightonensis Ferguson 32b

Plicatifera? 25
Productus productus (Martin) 19c, 28, 32b
P. sp. 19b, 24b, 24c, 30a, 31, 32a
Pugilis pugilis (Phillips) 15, 20, ?25
Pustula sp. 26
Rhynchonelloid 27
Rugosochonetes sp. 30c
Schizophoria resupinata (Martin) 20
S. sp. 13, 19b, 24b, 25, 26, 30c, 31
Smooth spiriferoid 13, 19a, 19b, 19c, 20, 25, 26, 27, 30a, 30c
Spirifer bisulcatus J. de C. Sowerby 14, 18a, 19b, 24b, 24c, 25, 30a, 30c, 32c
S. trigonalis? (Martin) 27
Spiriferoid 27
Striatifera striata (Phillips) 27
Bellerophon sp. 29
Eoptychia? 31
Euphemites sp. 32b
Straparollus sp. [juv.] 20
Pectinoid 26
Sanguinolites? 20
Sulcatopinna flabelliformis (Martin) 30a
Griffithides sp. 25
Ostracods 20, 30c
Fish tooth and plate

CAWDOR LIMESTONE

Foraminifera 47a, 47b
Amplexizaphrentis enniskilleni (Milne Edwards & Haime) 37a
Aulophyllum fungites (Fleming) 47a
Caninia benburbensis Lewis 47a
C. aff. cornucopiae Michelin 37a
C. juddi (Thomson) 34b, cf. 35
C. sp. 37a, ?46, 48a, 48b, 49
Chaetetes depressus (Fleming) 47a
Clisiophyllum keyserlingi McCoy 48c
C. sp. ?46, 48b
Clisiophylloid 37a, 48d
Cyathaxonia cornu Michelin/rushiana Vaughan group 37a, 38, 46, 48d
Dibunophyllum bipartitum bipartitum (McCoy) 38, 45
D. bipartitum craigianum (Thomson) 47a, 47b
Diphyphyllum sp. 47a, 48c, 48d
Fasciculophyllum densum (Carruthers) 38, 46
F. sp. 47a, 48c
Koninckophyllum interruptum Thomson & Nicholson 47a
K. sp. 48b
Lithostrotion junceum (Fleming) 47b, 48c, 48d
L. pauciradiale (McCoy) 47b, 48a
Rhopalolosma aff. bradbournense (Wilmore) 46
Rotiphyllum costatum (McCoy) 37a, 47a
R. sp. 38
Syringopora cf. geniculata Phillips 48b
S. cf. ramulosa Goldfuss 45, 47a, 47b
Zaphrentites? sp. [juv.] 37
Zaphrentoid 34b
Fenestella sp. 36, 47b
Alitaria panderi (Muir-Wood & Cooper) 34b
Antiquatonia antiquata (J. Sowerby) 36, 38, 40
A. hindi (Muir-Wood) cf. 35, 38
A. insculpta (Muir-Wood) 34b, 36
A. cf. muricata (Phillips) 48d
A. sp. 34a

c

d

Athyris sp. 37a
Avonia davidsoni (Jarosz) 34a
A. youngiana (Davidson) 34a, 34b, 38
'Brachthyris' planicostata McCoy 34b, 36, cf. 40
B. sp. 38, 45
Buxtonia sp. 34b, 35, 37a, 38
'Camarotoechia' sp. 36
Chonetoid 35, 36, 37a
Dictyoclostus sp. 34b
Dielasma hastatum (J. de C. Sowerby) 34b
D. sp. 34a, 36, 38
Echinoconchus elegans (McCoy) cf. 37b, 48d
E. punctatus (J. Sowerby) 38, 41
E. sp. 37a
Eomarginifera lobata (J. Sowerby) aff. laqueata (Muir-Wood) 34b, 38
E. cf. lobata 34a, 34b
E. cf. longispina (J. Sowerby) 34a
E. cf. setosa (Phillips) 35
E. cf. 36, 47b, 48d
Gigantoproductus crassiventer (Prentice) 34b
G. crassus (Fleming) 47b
G. cf. edelburgensis (Phillips) 44
G. giganteus (J. Sowerby) 35, aff. 35
G. cf. tulensis (Bolkhovitinova) 35
G. sp. [latissimoid] 34b
G. spp. 34a, 34b, 36, 40, 41, 42, 45, 46, 47a, 48c, 48d
Krotovia spinulosa (J. Sowerby) 35
Linoprotonia sp. 43, 45
Martinia sp. 35, 37a, 47b
Orthotetoid 34a, 34b, 38, 42, 44, 46, 47b
Phricodothyris cf. insolita George 41, 46
P. sp. 38
Pleuropugnoides pleurodon (Phillips) 44
Plicochonetes cf. crassistria (McCoy) 40, 46
P. sp. 38, 40, 45, ?46
Productus concinnus J. Sowerby 34b
P. productus (Martin) 48c, 48d
P. sp. ?34a, 38, 42, 48a
Pugilis pugilis (Phillips) 34a, 34b
P.? 36
Pustula sp. 38
Rugicostella? 46
Rugosochonetes sp. 34a, 34b
Schizophoria 34b, 36, 40, 42, 45
Smooth spiriferoid 34a, 34b, 35, 36, 37a, 38, 40, 42, 43, 47b, 48d
Spirifer bisulcatus J. de C. Sowerby 34b, 35, 36, 38, 40, 45, 47b, 48c, 48d
S. trigonalis (Martin) 34b, 37a, cf. 37a, ?38, 41
S. sp. 46, 47b
Striatifera striata (Phillips) 42, 43
Tylothyris sp. 38
Bellerophon sp. 42, 43, 47b
Naticopsis sp. 34b, 47b
Pleurotomarian 43
Straparollus sp. 40
Turreted gastropod 46
Aviculopecten? 34b
Caneyella membranacea (McCoy) 37b
Edmondia? 34b
Leiopteria squamosa (Phillips) 47b
L. sp. [juv.] 34b
Streblopteria 47b

Bivalve indet. 36, 48d
Griffithides sp. 34b
Weberides sp. 34a, 34b, 37a, 39, 41
Trilobite fragments 36, 38
Ostracods 34b
Archaeocidaris sp. [plate] 47b
Fish teeth and plates 34b, 35, 36, 37a, 47b

LIST OF LOCALITIES

Where more than one face in a quarry is listed, the first-mentioned yielded the most comprehensive assemblage. Each item includes the IGS registered numbers of the specimens. The fossil assemblage at a particular locality is found by searching the faunal lists (p. 131) for entries bearing the locality number.

HOPTONWOOD LIMESTONE (HL)

Locality

1   Stonycroft Quarry [2861 5441], Wirksworth, 510 yd N 14° W of Wirksworth church near top of HL: PJ 6601-3.

2   Stonycroft Quarry [2863 5438], Wirksworth, 470 yd N 13° W of Wirksworth church; (SE face of quarry).
Bed b: near top of HL: PJ 6604-6
Bed a: about 80 ft below top of HL; PJ 6607-10

3   Middlepeak Quarry [2825 5476], Wirksworth, 1020 yd N 31° W of Wirksworth church; near entrance to main working quarry, about 8 ft below top of HL: PJ 6617-20.

4   Baileycroft Quarry [2872 5420], Wirksworth, 270 yd, N 18° W of Wirksworth church; section on E side of road, below unconformity loc. 3 of Smith and others 1967, top of HL: WTD 1-23, LL 134-52.

5   Old quarry [2709 5372] in Godfreyhole, 720 yd N 41° E of Stainsbro' Hall, near Hopton; PJ 5720-7.

6   Exposure [2540 5351] near Hopton, 300 yd N 35° W of Hopton Hall: reef facies limestone: DF 1515-25.

7   Exposure [2547 5342] near Hopton, 200 yd N 34° W of Hopton Hall; reef facies limestone: DF 1526-44.

8   Exposure [2608 5333] in field, 660 yd W 13° N of Stainsbro' Hall, near Hopton; reef facies limestone: PJ 5837-5941, DF 1576-88.

9   Exposure [2542 5340] in Carsington Wood, 200 yd W 44° N of Hopton Hall, reef facies limestone: DF 1561-74.

10   Borehole (Site Investigation No. C1) [2652 5340] 270 yd N 37° W of Stainsbro' Hall, near Hopton; depth 129 ft 7 in: DF 2401

11   Old quarry [2723 5302] near Dream Mine, 580 yd E 15° S of Stainsbro' Hall, near Hopton; reef facies limestone: PJ 5581-5680.

12   Exposure [2725 5286] in Sprink Wood, 665 yd E 28° S of Stainsbro' Hall, near Hopton; reef facies limestone: DF 1557-60.

MATLOCK LIMESTONE (ML)

13   Yokecliffe Rake [2760 5382] at bottom of cliff, 1230 yd W 5° S of Wirksworth church: PJ 6472-80.

14   Yokecliffe Rake [2775 5385], middle

Locality

section of cliff, 1060 yd W 6° S of Wirksworth church: PJ 6481-7.

15   Yokecliffe Rake [2794 5390], top of cliff section, 850 yd W 3° S of Wirksworth church: PJ 6488-96.

16   Middlepeak Quarry [2820 5483], Wirksworth, 1110 yd N 31° W of Wirksworth church; N end of main working quarry, near top of ML: PJ 6611-2.

17   Middlepeak Quarry [2818 5472], Wirksworth, 1020 yd N 35° W of Wirksworth church; middle part of main working quarry, near top of ML: PJ 6613-6.

18   Middlepeak Quarry [2812 5474], Wirksworth, 1080 yd N 38° W of Wirksworth church; section along roadway next to lower working face;
b: beds in basal 20 ft of ML: PJ 6628-35.
a: beds in basal 1 ft 2 in of ML: PJ 6621-7.

19   Middlepeak Quarry [2802 5460], Wirksworth, 1040 yd W 44° N of Wirksworth church: main working face
Bed d: 0 to 8 ft above Upper Girvanella Band: PJ 6683-5
Bed c: associated with Upper Girvanella Band: PJ 6667-82
Bed b: 0 to 40 ft below Upper Girvanella Band: PJ 6646-66
Bed a: basal 40 ft of ML: PJ 6636-45.

20   Stonycroft Quarry [2863 5433], Wirksworth, 430 yd N 14° W of Wirksworth church; top of S face near top of ML: PJ 6566-96.

21   Stonycroft Quarry [2849 5430], Wirksworth, 460 yd N 33° W of Wirksworth church; W face of quarry, 6 in shale parting at junction of HL and ML: PJ 6597-6600.

22   Stonycroft Quarry [2860 5428], Wirksworth, 380 yd N 18° W of Wirksworth church; S side of quarry above ascending footpath, 40 ft below top of quarry: CBW 1755-7.

23   Baileycroft Quarry [2875 5423], Wirksworth, 290 yd north of Wirksworth church; section on E side of road, basal 7 ft of ML: PJ 8740-59

24   Dale Quarry [2835 5407], Wirksworth, 420 yd N 20° N of Wirksworth church; section at SE end of quarry, collected from roadway leading from quarry entrance to quarry floor

Locality

Bed c: at top of ML: PJ 6527-40
Bed b: in middle of ML: PJ 6541-56
Bed a: near base of ML: PJ 6557-60

25   Dale Quarry [2820 5420], Wirksworth, 630 yd W 26° N of Wirksworth church; section along roadway at W end of quarry above main working face, from top 3' of ML: PJ 6512-26.

26   Hillside exposure [2750 5314] near Dream Mine, 870 yd E 4° S of Stainsbro' Hall: PJ 5681-8.

27   Old quarry [2725 5351] near Godfreyhole, 690 yd E 29° N of Stainsbro' Hall; PJ 5700-19, DF 1575.

28   Old quarry [2731 5397] 390 yd E 43° N of Godfreyhole: PJ 5942-56.

29   Exposure [2725 5348] near Godfreyhole, 670 yd E 27° N of Stainsbro' Hall: DF 1545.

30   Old quarry [2729 5363] near Godfreyhole, 800 yd E 36° N of Stainsbro' Hall
Bed c: Limestone, dark grey: PJ 5728-38.............7'
Bed b: Limestone, dark grey, massive..................8'
Bed a: Limestone, dark grey, massive: PJ 5739-65.....6'

31   Old quarry [2741 5308], 860 yd E 6° S of Stainsbro' Hall: PJ 5689-99, 8730-39.

32   Hilt's Quarry [3515 5440], Crich, 140 yd E 42° N of Methodist church, Crich (See also locality 47)
Bed c: horizon uncertain: CBW 661, 748-9
Bed b: 1 ft to 5 ft below lowest chert band on E side of quarry: JM 100-5
Bed a: 1 ft to 15 ft below lowest chert band on N side of quarry: JM 106-16

33   Ironville No. 3 Oil Borehole [4318 5192] 950 yd S 19° E of St James Church, Riddings; depth 2249 ft 6 in to 2250 ft 3 in. Zi 7710-2

CAWDOR LIMESTONE (CL)

34   Middlepeak Quarry [2802 5460], Wirksworth, 1040 yd W 44° N of Wirksworth church, main working face (See also locality 19)
Bed b: pocket of reef limestone about 10 ft above base of CL: PJ 6708-6800
Bed a: 3-in shell bed about 7 ft above base of CL: PJ 6686-6707

35   Middlepeak Quarry [2823 5451], Wirksworth; top beds from southern bay of quarry (loc. 47 of Smith and others 1967): WTD 69-88.

36   Baileycroft Quarry [2863 5425], Wirksworth, 340 yd N 20° W of Wirksworth church; W side of road

Locality

cutting, beds immediately above base of CL: PJ 6561-5, 8760-76.

37   Baileycroft Quarry [2868 5416], Wirksworth, W side of road
Bed b: shale, W of road near footbridge: LL 153-63
Bed a: limestone, S end of main face: WTD 27-68, 255-7

38   Baileycroft Quarry [287 542], Wirksworth, 330 yd N 15° W of Wirksworth church; shell bed, dark grey limestone within Cawdor Shales, 6 ft above top of CL: Bu 5873-5939.

39   Dale Quarry [2820 5418], Wirksworth, 620 yd W 23° N of Wirksworth church base of CL: DF 1546.

40   Old quarry [2794 5299] 90 yd E 28° N of Pittywood Farm, about 1 mile SW of Wirksworth: PJ 5766-80.

41   Exposure on hillside [2788 5310] 160 yd N 14° E of Pittywood Farm about 1 mile SW of Wirksworth; reef facies limestone: PJ 5781-9.

42   Exposure in field [2775 5325] 330 yd N 18° W of Pittywood Farm, about 1 mile SW of Wirksworth: PJ 5790-5803.

43   Exposure in field [2780 5318] 240 yd N 10° W of Pittywood Farm, about 1 mile SW of Wirksworth: PJ 5804-16.

44   Exposure in field [2783 5316] 210 yd N 4° W of Pittywood Farm, about 1 mile SW of Wirksworth: PJ 5817-23.

45   Exposure in field [2784 5314] 200 yd due N of Pittywood Farm, about 1 mile SW of Wirksworth: PJ 5824-36.

46   Old quarry [2808 5390] on top of Yokecliffe, 700 yd W 2° S of Wirksworth church: PJ 6497-6511, DF 1555-6.

47   Hilts Quarry [3515 5440], Crich, 140 yd E 42° N of Crich Methodist church
Bed b: top beds of CL: JP 2244-88
Bed a: beds from 1 to 15 ft above lowest chert band [3530 5431] on E side of quarry: JM 77-99.

48   Old quarry [357 542] E of Hilts Quarry, Crich, about 700 yd E 12° S of Crich Methodist church
Bed d: topmost beds forming dip-slope [3560 5427] on NW side: CBW 656-660, JM 29-54.
Bed c: 3 to 10 ft below bed d, N end of W side: JM 55-63
Bed b: 25 ft below bed d, N end of W side: JM 64-8
Bed a: below lowest chert band at S end of quarry: JM 74-6

49   Old quarry [3499 5472] at Crich, 180 yd E 16° N of St. Michael's Church CBW 666

CHAPTER 3

Namurian (Millstone Grit Series)

DETAILS

PENDLEIAN (E₁)

The maximum thickness of this stage where complete over the 'block' areas is estimated at about 50 ft. In the Widmerpool Gulf some 566 ft were proved in the Duffield Borehole.

The Cravenoceras leion band, marking the base of the stage, is exposed at only one locality in the district, in a stream [3243 4328] near Farnah House, Duffield, where it occurs in about 4 ft of dark grey weathered shale with a decalcified limestone band 1 ft above the base. Both lithologies have yielded Cravenoceras cf. leion (p. 14). Since the mapping was completed, C. leion together with Posidonia cf. membranacea have been found in fragments of brown decalcified limestone ploughed up at Windley [3057 4547]. The conjectural position of the Namurian-Dinantian boundary shown on the first published editions of 1 : 10 560 sheet SK 34 NW and 1 : 50,000 sheet therefore requires correction. Shale tips [2710 4435] on the Dinantian outcrop near Black Burn yielded a single specimen of Eumorphoceras tornquisti and an orthocone nautiloid (E₁ₐ), but the source of the debris is not known.

South-west of Duffield, a series of scarp and dip-slope features trending north-north-west to south-south-east are the surface expression of unexposed sequences of graded mudstone-sandstone turbidite beds of E₁ᵦ to H₁ₐ age proved in the Duffield Borehole.

ARNSBERGIAN (E₂)

The lowest E₂ fossiliferous horizon seen during the resurvey was poorly exposed in a shale bank [2802 5305] ¾ mile SW of Wirksworth Church and yielded Eumorphoceras bisulcatum (E₂ₐ). The locality is some 10 yards east of the Dinantian limestone/Namurian shale contact. The E₁ stage is therefore thin or missing, with E₂ shales resting directly on Cawdor Limestone (p. 000). Steep dips and poor core recovery in the Hopton Borehole [2677 5334] indicate disturbances in these lowest shales of the Namurian.

Cravenoceratoides sp. (E₂ᵦ) was identified at 43 ft in Hopton Borehole, where the sequence consisted predominantly of grey silty mudstone with subordinate calcareous siltstone. The fossils, which were confined to certain horizons, proved that the E₂ stage continues to the bottom of the borehole at 149 ft. This thickness is double that in the Ashover area (Ramsbottom and others 1962) and suggests that the margin of the limestone 'massif' was close to the margin of the Widmerpool Gulf in early Namurian times. There were no indications from the cores of a true 'gulf' facies.

In a temporary excavation [2872 5340] 1/3 mile S of Wirksworth, finely laminated shales and ferruginous orange-coloured rottenstone contained Posidonia corrugata, suggesting an E₂ or older age. The same species was recorded some 1000 yd SW of Wirksworth in temporary exposures [2812 5326] made for house foundations. Ford (1968b) recorded fossils of E₂ age in shales near Wirksworth station [? 2894 5442]. The over-all thickness of the Arnsbergian Stage around and south of Wirksworth is therefore probably between 150 and 200 ft.

In sporadic stream exposures near Kirk Ireton, grey silty mudstones and dark grey calcareous shales predominate. Thin limestone (calcareous siltstone) bands up to 12 in thick contain goniatites, bivalves and gastropods. Ferruginous, fine-grained, hard sandstone bands and laminae often exhibit groove, flute and tool marks, typical of turbiditic sedimentation.

In the Ridgeway Brook [2924 4694] north-east of Turnditch dark grey shales with calcareous siltstone bands yielded P. corrugata together with indeterminate goniatites and gastropods considered to be E₂ or older.

In Biggin Brook [2576 4777] the following section was recorded:

	ft	in
Sandstone, brown, fine-grained	4	0
Mudstone with siltstone bands	27	0
Mudstone, blue-grey, laminated, fossiliferous	8	0
Siltstone, dark grey, calcareous, with bivalves and goniatites	0	9

The lower calcareous beds yielded Nuculoceras nuculum (E₂c). Sporadic exposures in the stream [2572 4779 to 2634 4766] yielded the following E₂c fauna: Eumorphoceras bisulcatum, Dimorphoceras sp. [s. l.] orthocone nautiloid, P. corrugata, Posidoniella variablis, Actinopteria regularis.

Nearby, at Millington Green [2646 4779], tough calcareous mudstone contained Cravenoceratoides fragilis, palaeoniscid fish scales and carbonised plant fragments probably from the N. nuculum Zone (E₂c).

A 10-ft scar [3178 4508] on the left bank of the River Ecclesbourne shows thin alternations of brown and grey mudstone containing Ct. edalensis (E₂ᵦ) and Posidonia corrugata in the lowest 4 ft. A band with Ct. edalensis occurring some distance upstream [3165 4521], is probably at the same horizon.

The next scar upstream [3137 4555] consists of variegated shale with N. stellarum and Posidonella aff. vetusta 8 ft 5 in above the bottom of the section, marking the base of the E₂c Zone. These shales overly 5 ft 7 in of tough cherty interlaminated dark mudstones and pale siltstones containing Ct. nititoides, E. rostratum, crinoid columnals and a pectenoid. The benthonic fauna and associated cherty lithology of these beds constitute a distinctive marker band in the Namurian of the Central Province (Stevenson and Gaunt 1971, pp. 189-190).

There are few exposures in the Franker Brook of the beds between the N. stellarum band and the goniatite band with N. nuculum and Eumorphoceras sp. [3082 4729]. However, the sequences with alternations of mudstone and thin quartzose sandstones (formerly known as Crowstones) proved in the E₂ succession in the Duffield Borehole are represented by minor scarps, aligned sub-parallel to the Ecclesbourne Valley, north-north-west and south-south-east of Postern Lodge. Temporary trench exposures across these features showed the beds to be folded with axial trends in the same direction as the scarps. A small exposure of mudstone with thin quartzose siltstones, 50 yd downstream from the N. nuculum band noted above, probably belongs to this sequence.

The Franker Brook section provides exposures of several of the many goniatite bands in the E₂c to R₁c sequence (Fig. 66). The dip, generally to the north-east, varies from 0 to 19° with a few minor reversals, which, together with the many gaps between exposures, allow only approximate estimation of thickness. The N. nuculum band occurs in a 10-in grey carbonate siltstone and probably correlates with a bed with a similar lithology and fauna at a depth of 123 ft 2 in in Duffield Borehole which forms part of the highest of three N. nuculum bands occurring extensively in the Central Province.

N. nuculum was recovered from bullions in the bed [3738 4053] of the Boosemoor Brook, Breadsall (Fig. 66). It could not be established whether or not the bullions were in place and since H. subglobosum Zone sediments are present both in the adjacent bank and downstream (p.133c). it is inferred that the top of the Arnsbergian is here raised to stream level by a superficial structure.

In Ferriby (or Dam) Brook, Breadsall (Fig. 66), N. nuculum was found [3758 3954] in a 10-in siltstone band underlain by about 15 ft of soft grey mudstone with darker more shaly bands exposed low in the stream bank; the horizon is probably low in the H. subglobosum Zone.

CHOKIERIAN (H₁)

Homoceras cf. beyrichianum and H. beyrichianum (H₁ᵦ) together with Anthracoceras or Dimorphoceras have been discovered in the Scow Brook [2537 5238] some 2 miles SW of Wirksworth. H. beyrichianum was also recorded [2618 4763] in Biggin Brook nearly 2 miles south of Kirk Ireton, but is not considered to be in situ for surrounding strata are of the Arnsbergian Stage (p.133b).

A gap in Franker Brook between the N. nuculum band (p. 00) and the H. beyrichianum band [3059 4761], representing about 116 ft of strata, is referred to the H. subglobosum (H₁ₐ) Zone; it contains only a few small exposures of black carbonaceous shaly mudstone. As well as H. beyrichianum present in a bullion, the latter band has yielded Caneyella semisulcata, an orthocone nautiloid and H. cf. subglobosum. A badly preserved fauna exposed higher up the brook [3060 4778] includes H. cf. subglobosum and H. sp. nov. and possibly represents an horizon slightly lower than the H. beyrichianum band noted above. The stream section may therefore contain unexposed faults and minor folding to cause this reversal in the succession. A higher H₁ᵦ band occurs in a 6-ft exposure [3057 4774] of mudstone containing H. sp. aff. beyrichianum and H. sp. of the subglobosum group.

A temporary section [3725 3970] for school playing-fields at Breadsall exposed about 30 ft of decalcified dark grey mudstone with weathered (limonitic) bullions and, near the top, sporadic thin (up to 3 in) sandstone ribs. H. subglobosum was plentiful.

In Boosemoor Brook north-east of Breadsall H. subglobosum was collected from a few feet of weathered grey mudstone in the bank [3738 4053] adjacent to the bullions with N. nuculum in the stream bed (p.133c). H. subglobosum also occurs in a large bullion about 35 yd downstream. No other diagnostic fossils were collected from this stream section which includes a 20-ft section [3748 4060] showing iron-stained and weathered grey mudstone.

In Mill Plantation [3759 3955] near Breadsall the following section is exposed low in the bank of Ferriby (Dam) Brook over a distance of about 20 yd:

	Thickness ft	in
Mudstone, grey and dark grey, with coaly plant fragments		
Bullion, 6 in long	0	3
Mudstone, dark grey, contorted	1	6
Mudstone, soft, dark grey	4	0
Bullion, 5 ft long, with H. subglobosum	0	10
Mudstone, dark grey, soft at top (goniatites, see below, 4 ft down)	10	0
Arnsbergian strata (see p. 133b)	about 16	0

The section has yielded H. subglobosum in quantity. Mr M. J. Reynolds has examined the bullions from this section (including those in the Arnsbergian strata) for their conodont fauna and found the following species to be present: Euprioniodina microdenta, Gnathodus girtyi Hibbardella acuta, H. ortha, Hindeodella ibergensis, H. subtilis, H. uncata, Idiognathoides minutus, Idiognathoides noduliferus, Neoprioniodus cf. armatus, Ozarkodina cf. macer, Prioniodina subaequalis. Adjacent exposures upstream show poor sections, up to about 17 ft, of mudstone with interbedded sandstones up to 2½ ft thick, dipping towards the east. These are probably the representatives of the highest turbidite sequence in the Duffield Borehole (Aitkenhead, in press). Near the eastern end of Mill Plantation sections of unfossiliferous mudstones are referable to either this stage or the overlying Alportian.

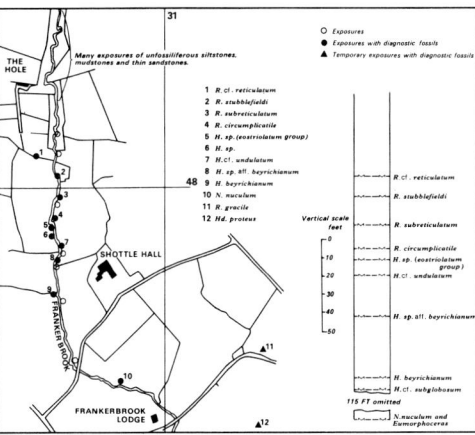

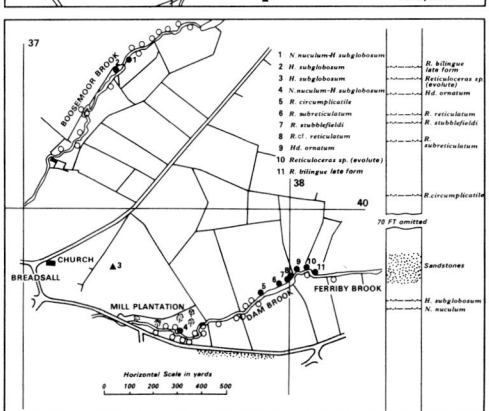

Figure 66 Sketch-maps showing Namurian exposures and fossiliferous localities in the Franker Brook near Breadsall

a

ALPORTIAN (H_2)

Shales of the Hudsonoceras proteus zone (H_{2a}) are exposed in Scow Brook [2540 5240] ½ mile S of Hopton Hall. The goniatites are concentrated in large calcareous bullions. The basal H_{2a} band of H. proteus is not exposed in Franker Brook, but a bullion containing this fossil was found in a temporary exposure [3135 4714], 830 yd SE of Shottle Hall. The Homoceras undulatum Zone (H_2b) is probably represented in Franker Brook by an exposure [3055 4782] containing badly preserved fossils, including H. cf. undulatum obtained from 2 ft of grey mudstone containing a bullion. Another exposure [3058 4785], some 6 ft higher in the sequence yielded H. sp. (eostriolatum group), probably from the Ht. prereticulatus Zone (H_{2c}).

The stage has not been established in the Ferriby Brook, but is probably represented by the scattered small sections showing mudstones and thin grey sandstones, the latter with sole-marks, in the banks for 200 yd E of Mill Plantation. Caneyella sp. is the only fossil recovered from this part of the stream [3781 3962].

KINDERSCOUTIAN (R_1)

Shales containing goniatites of the Reticuloceras nodosum group have been seen at two localities [2898 5354 and 2755 5209] near Wirksworth in a small tributary stream of the River Ecclesbourne. The fauna included Homoceratoides aff. divaricatus, Dimorphoceras sp., Homoceras sp. as well as Caneyella sp.

R. coreticulatum has been identified in Callow Borehole [2665 5282] at depths between 61 and 90 ft, together with the following fauna: C. rugata, Dunbarella sp., Posidonia sp. juv., Anthracoceratites sp., Reticuloceras spp.

The Franker Brook exposes two goniatite bands in the R. circumplicatile (R_{1a}) Zone. The lowest [3056 4788], probably about 9 ft above the H_{2c} band noted above, consists of 5 ft 9 in of grey mudstone with a bullion band. The fauna obtained from the mudstone includes H. henkei, Homoceratoides sp. and R. circumplicatile. About 9 ft of unexposed strata separate this band from a small exposure of mudstone [3059 4796] which has yielded Dimorphoceras sp. and R. subreticulatum. About 15 ft of mainly unexposed mudstone separates this band from the only exposure [3060 4805] of the R. nodosum (R_{1b}) Zone band in the section, comprising poorly exposed mudstone with a bullion band. The fauna from the mudstone includes H. cf. striolatum and R. stubblefieldi. Above are about 10 ft of poorly exposed mudstone overlain by grey mudstone containing Dimorphoceras sp., and R. cf. reticulatum, the only exposure in the R. reticulatum (R_{1c}) Zone [3061 4812].

In the Ferriby Brook section Kinderscoutian rocks, in disturbed condition, are intermittently exposed between localities [3787 3966 and 3808 3978] in the 5-ft banks of the stream. Over the western 50 yd of this outcrop R. circumplicatile and H. henkei have been collected from up to 5 ft of dark grey and ochreous mudstones containing a 4 to 5-in band of siltstone and a bullion. R. cf. subreticulatum (R_{1a}) is present in westerly dipping grey mudstone [3797 3972] containing coaly laminae, plants and ferruginous bands; at one locality [3800 3973] mudstones which are overturned both to the east and west contain a hard grey septarian siltstone which yielded R. cf. stubblefieldi and R. sp. (nodosum group) (R_{1b}). At least two horizons of bullions (up to 3 ft x 1 ft in size) within 5 ft of gently dipping dark grey mudstones [3802 3975], are of R_{1b} or R_{1c} age since they contain R. cf. reticulatum, Dimorphoceras sp. Posidonia sp. and a fragmentary orthocone nautiloid. The remaining exposures [3804 3977 and 3808 3978] of Kinderscoutian age include thin sandstone and siltstone ribs and lenses. Hudsonoceras ornatum, and Reticuloceras sp., collected from bullions, are indicative of the R. nodosum Zone. The presence of the highest zone (R_{1c}) is not definitely established on the available material.

The total thickness of Kinderscoutian rocks in this brook section is of the order of 70 ft.

Mr M. J. Reynolds found the following conodonts in the bullions from the Kinderscoutian sections in this stream. In one section [3800 3973] he records Hindeodella sp. fragments, Idiognathoides corrugatus, I. sulcatus, Lonchodina furnishi, Neoprioniodus sp. and Ozarkodina roundyi; at another [3802 3975] Hindeodella sp. fragments, I. noduliferus, I. sinuatus, I. sulcatus, Ligonodina sp. fragments Lonchodina cf. furnishi and Streptognathodus lateralis; and at another [3808 3978] Euprioniodina microdenta, Neognathodus bassleri, H. ibergensis, H. subtilis, H. uncata, I. noduliferus, O. hindei, Prioniodina stipans, S. elegantulus and S. lateralis.

NAMURIAN BEDS BELOW THE MARSDENIAN IN IRONVILLE NOS. 1 AND 3 OIL BORES

Downing and Howitt (1969, p. 246, fig. 3) have recognised the R. gracile horizon in Ironville No. 3 Oil Bore at about 1894 ft depth on the gamma log. It lies immediately above a light brown sandy limestone overlying a thin sideritic sandstone. The whole sequence, about 15 ft thick, is about 380 ft above the Dinantian limestone. No sandy limestone or sideritic sandstone was recorded in the Ashover boreholes at this level, though their gamma logs through the R. gracile sequence are similar to that of No. 3 Bore.

The late C. E. N. Bromehead noted 'fine brown micaceous calcareous grit' probably representing the same horizon, in chipping samples from 1715 to 1735 ft in Ironville No. 1 Oil Bore, again lying about 300 ft above the Dinantian limestone. The approximate depths of 1814 ft in No. 3 Bore and 1715 ft in No. 1 are therefore taken to mark the base of the Marsdenian.

The underlying sequence down to the top of the Dinantian limestone is recorded in No. 3 Bore as largely black, pyritic mudstone, slightly silty near the top with thin 'fine-grained white sandstones' noted at about 2010, 2020, 2030, 2118 and 2165 ft, the lowest being shelly. The top of the Dinantian limestone has been recognised from the gamma

b

log at about 2220 ft (Howitt in litt.). In No. 1 Bore predominantly argillaceous rocks are recorded. The top of the Dinantian limestone is at 2035 ft.

The similarities of the gamma log of No. 3 Bore to those of the Ashover boreholes (Cosgrove in Ramsbottom and others 1962) is the evidence on which the stage boundaries shown in Plate 4 are based. The base of R_1 is thought to lie at about 2057 ft, the base of H at about 2108 ft and the base of E_2 probably at about 2184 ft.

MARSDENIAN

BEDS BELOW THE ASHOVER GRIT

In Callow Borehole a ferruginous band in fragmentary cores at 20 ft depth yielded Reticuloceras gracile. At outcrop this horizon is some 30 ft below the lowest mapped sandstone in the Wirksworth district, proving the absence of the Kinderscout Grit in south Derbyshire.

The basal goniatite band of the R. gracile Zone was found in a temporary excavation [3137 4741] near Shottlegate, where weathered grey mudstone yielded Dunbarella sp. and R. gracile.

In Franker Brook, the R. gracile and R. bilingue bands are probably cut out by a fault [3058 4833] crossing the stream. Upstream to a locality [3063 4931] near Carrbrook Farm, there are numerous exposures of grey silty mudstone and siltstone with bands of ironstone nodules and a few hard grey or grey-brown sandstone beds up to 5 ft thick. These sporadically exposed and variably dipping beds, which are about 250 ft thick, probably represent an early pro-delta intermittent arenaceous phase preceding the major sand influxes of the Ashover Grit.

Similar rocks are exposed in streams west and south of Blackbrook where the Blackbrook Water Borehole [3308 4775], sunk towards the end of the last century (Wedd in Gibson and others 1908, p. 37), proved alternations of shale and sandstone to a depth of 231 ft 2 in. The deeply incised streams near Lumb Grange below the Ashover Grit escarpment are mainly in grey silty mudstones, siltstones and fine sandstone, but 2 ft above the base of one 10-ft section [3314 4675] is a 2-in rottenstone band from which R. bilingue was collected. The Blackbrook Water Borehole probably started near the crop of this band. A lower band is poorly exposed in a roadside bank at Hazelwood [3327 4549], where weathered mudstone has yielded R. cf. bilingue.

A specimen of R. bilingue late form was collected during the first 6-in survey (Wedd in Gibson and others 1908, p. 25) from the north side of the railway cutting at Duffield [3429 4364]. The fossil was found in a 'thin band of pyritous shale' at the top of the cutting and was then identified as Glyphioceras bilingue (Salter). A fossiliferous band, 4 in thick, was excavated here during the present survey, but no diagnostic fauna was found. The R. bilingue late form band, the defined base of strata containing the Ashover Grit (Smith and others 1967, p. 68) is about 50 ft below the lowest mappable Ashover Grit sandstone at Duffield.

R. bilingue late form occurring with Caneyella sp., Posidonia sp. nov. and fragmentary Dunbarella sp., in small sections of dark grey mudstones with silty bands in the river bank [3808 3978], marks the lowest Marsdenian horizon proved in the Ferriby Brook. A few yards upstream are intermittent exposures of micaceous mudstones with siltstone bands, some containing plant remains and one, 2½ in thick, with a vertical dip [3811 3976]. Farther upstream, dips become more orderly towards the east in grey micaceous silty mudstone with siltstone bands, and near the top of the sequence are thin ferruginous sandstone bands indicative of a passage upwards into the lowest sandstone of the Ashover Grit. Calculations of thickness of these deposits are unreliable because of the variations in dip; it is possible that the R. bilingue band noted above may lie some 300 ft below the lowest mapped leaf of the Ashover Grit in this area.

The base of the Marsdenian in Ironville No. 3 Oil Bore is drawn at a gamma peak at about 1894 ft and above a "sandy limestone" (Downing and Howitt 1969, p. 246). The overlying strata up to the base of the Ashover Grit (1738 ft on the gamma log) is recorded as 'dark grey-black micaceous silty mudstone with white silt and tight very fine-grained sandstone'.

ASHOVER GRIT BELOW THE MAIN BED

Wirksworth and Ecclesbourne Valley

North of Wirksworth and near Crich, the Ashover Grit Main Bed is underlain by silty laminated mudstones with subordinate siltstone and sandstone horizons. To the south of Wirksworth, between Shottle Hall and Duffield, however, these beds give way to sandstones, which vary from isolated lenses, up to 20 ft thick, to continuous fine-grained, cross-laminated beds up to 140 ft thick, forming good features on the sides of the Ecclesbourne Valley.

The lowest thick sandstone, some 500 ft below the base of the Main Bed, is exposed in the disused Stainsbro' Quarry [2660 5265] ¼ mile N of Callow. It comprises some 20 ft of light brown fine-grained sandstone with mudstone inclusions in the basal 8 ft and with shaly micaceous bands up to 2 in thick containing carbonaceous plant fragments.

Some 20 ft of sandstone, about 150 ft below the Main Bed, are exposed in Callow Quarry [2660 5120], ½ mile S of Callow. Shaly micaceous bands within the sandstone show purple and pink colorations which suggest post-Carboniferous staining, related to a Triassic cover since removed.

These sandstones and shales are exposed in many places along the sides of the Ecclesbourne Valley (See 6-in geological map SK 25 SE). The shales are predominantly silty with ironstone nodules and bands; apart from scattered plant fragments, fossils are rare.

Calculations based on information obtained from the old plans of the Meerbrook Sough together with recent details from the Mines Research and Exploration Group led by Mr Nash, indicate that the base of the Main Bed is some 600 ft above the top of the Dinantian limestone in the area east of Wirksworth. A further sandstone of the Ashover

c

Grit some 100 ft thick occurs about 100 ft below the Main Bed (Fig. 8b).

West of the River Derwent

In an old quarry [3343 4471] in a garden at Hazelbrow, Duffield, 10 ft of pale brown fine-grained cross-bedded sandstone with poorly delineated soft-weathering ferruginous concretions are exposed. Similar cross-bedded sandstone, but without concretions, occurs in the railway cutting at Duffield.

The overlying sandstones are thicker, coarser and more continuous at outcrop, particularly to the north-west and east of Shottle. Here an old quarry [3121 4988] exposes about 30 ft of coarse, pale brown sandstone with lens-shaped cross-bedded units up to 3 ft thick which commonly contain bands with pebbles up to ½ in diameter. A few incomplete casts of fossil trees are present on bedding planes. Smaller exposures of similar sandstone in old quarries near Hollyseat [3200 4876] and Holly House [3302 4823] represent the most important leaf of the Ashover Grit below the Main Bed. In Belper Meadows Water Borehole [3399 4776] 112 ft of grit were recorded below the Main Bed (Wedd in Gibson and others 1908, p. 40) the only proved thickness of this lower leaf. The leaf is best exposed in the base of an old quarry near Milford [3466 4501] where 30 ft of pale brown medium-grained sandstone with cross-bedded sets 1 in to 3 ft thick are seen. At the northern end of the Milford railway tunnel 25 ft of the same sandstone are exposed.

Derwent Valley

A lower leaf of the Ashover Grit forms an inlier in the valley of the Derwent between Whatstandwell and Ambergate. A temporary exposure at the works near Oak Hurst [3404 5230] revealed 25½ ft of pale brown fine to coarse-grained cross-bedded sandstone, in beds up to 2 ft thick, interbedded with fine ripple-laminated sandstone.

The strata, undivided on the maps, which lie between the mapped Ashover Grit sandstones, are generally poorly exposed. They probably consist mainly of grey to grey-brown laminated shaley siltstones, very silty mudstones and fine micaceous sandstones. They are best seen in the cutting at the southern end of the Milford railway tunnel [3446 4468] and are sporadically exposed in the stream west of Oak Hurst [336 521]. Fossils other than small plant fragments have been found only near Milford [3438 4525], where a specimen of the non-marine bivalve Anthraconaia angulosa was collected from an 8-ft section of purplish grey very silty mudstones with carbonaceous partings (Aitkenhead 1966).

Strata underlying the Main Bed were temporarily exposed on the roadside [3544 4251] near Outwoods, where below 4 ft of soil and wash (p. 158c) they comprised some 5 ft of grey-brown, ochreous-stained and banded sandy micaceous mudstones with thin (up to 2 in) sandstone ribs.

Little Eaton to Breadsall

A section within the lower leaves of the Ashover Grit seen in an old quarry [3607 4166] near the Vicarage at Little Eaton comprises 5 ft of coarse laminated red and red-brown micaceous clayey sand with 6-in bands of sandstone, overlain by 6 ft of brown and pale brown speckled coarse sandstone. East of Little Eaton equivalent beds are seen in a roadside section [3673 4163], where they consist at the base of 25 ft of medium to coarse cross-bedded khaki and red speckled sandstone with large brown and red friable ovoid patches up to 6 ft in diameter. This is overlain by 18 ft of thin and wedge-bedded, medium to coarse, red, red-brown and khaki micaceous sandstone locally containing fragments of angular red silty mudstone in the basal few feet.

In Camp Wood, south of Little Eaton, a section [3668 4115] in the lowest mapped sandstone shows 25 ft of cross-bedded (towards the south-west) medium to coarse-grained brown and red-stained micaceous sandstone with friable patches. This sandstone was not encountered, according to Wedd (in Gibson and others 1908, pp. 44-45) in the well at the Paper Mill [about 354 423] and has probably died out below the cover of the Derwent alluvium. To the east of Camp Wood it dies out, or is cut out by the Main Bed (see p. 134d) before Breadsall Priory is reached. However, it may be represented by the lowest mapped leaf of Ashover Grit in the Ferriby Brook (see below).

There is little doubt that the sandstone forming Burley Hill on the west side of the Derwent near Little Eaton is the Ashover Grit. Wedd (in Gibson and others 1908, p. 32), although calling it Shale Grit, implied correctly that it is a continuation of the sandstone cropping out in Duffield and correlated by the presence of R. bilingue late form about 50 ft below. Debris of black mudstone with Dunbarella sp. from a roadside excavation [3500 4135] near Burley is assigned to this marine horizon. Exposures at the Hill itself are poor; a few feet of khaki and pale grey speckled sandstone are seen on the north side, and in the bank of the River Derwent about 20 ft of steeply dipping brown micaceous medium-grained sandstone with harder nodules containing plant remains are underlain by 3 to 4 ft of coarser sandstone with similar nodules and overlain by 3 ft of grey silty micaceous mudstones. This sandstone dies away to the east of Burley Hill, but it is perhaps represented by siltstones proved in trial boreholes for roadworks on the hill at the Water Works [367 407], south of Little Eaton. It is absent in the stream section at Breadsall where the lowest sandstone is some 300 ft above the R. bilingue Marine Band.

There are poor sections of the lowest mapped leaf of the Ashover Grit in the Ferriby Brook [3838 3978] where 10 ft+ of soft bedded micaceous silty sandstone with 1-ft harder ferruginous bands are exposed near the base of the mapped sandstone; 15 ft of similar beds with thin sandstone bands are exposed some 10 to 15 yd upstream. This sandstone, which may correlate with the sandstone leaf in Camp Wood, near Little Eaton, is about 75 ft thick and its top lies a calculated 300 ft below the base of the Main Bed; only a few feet of the intervening mudstones are exposed in the stream banks.

d

ASHOVER GRIT (MAIN BED)

East of Wirksworth to the Derwent Valley

The Ashover Grit (Main Bed) forms the main part of the interfluve between the Ecclesbourne and Derwent valleys. At levels above 900 ft it is often stained a purplish red colour by percolation of ground waters and by weathering below a Permian land surface which was eventually overlain by Triassic rocks. There are small sections [2974 5313 and 2980 5355] on the scarp overlooking Wirksworth.

The full thickness of the Main Bed has been proved in boreholes at Wigwell [3170 5465] and Hanson Farm [3233 5459], where 163 ft and 170 ft 6 in respectively were recorded. A similar thickness (171 ft) was proved at Belper Meadows Water Borehole [3399 4776], but this borehole started in a shaly sandstone slack within the Main Bed and did not prove the full thickness.

Exposures of the Main Bed are small and scattered in the extensive outcrops west of the Derwent valley, the best being in the old quarry at Alport Hill [3038 5157]. Here, a column of sandstone, known as the Alport Stone, left standing in the middle of the quarry, shows 17 ft of coarse-grained sandstone with dispersed pebbles up to ¾ in diameter, resting discordantly on a single set of cross-bedded coarse pebbly sandstone (Plate 3). The upper bed here probably represents a restricted channel fill similar to those described by Mayhew (1967, pp. 96-97) at Cocking Tor [348 540] in the Chesterfield district.

Crich Area

The Main Bed forms the lowest mappable feature east of the Dinantian limestone inlier. Where completely exposed in a temporary trench [3564 5334], near Mill Green, its thickness totalled some 160 ft, the basal 15 ft being coarse-grained, and in places conglomeratic. Coaly partings were common and cross-bedding was well displayed in the central portion. It was immediately overlain by a 20-in coal.

Sporadic exposures of massive, buff, cross-laminated sandstone occur along the outcrop at Fritchley [3555 5303]. In the Amber Valley near Bullbridge some 500 ft of predominantly arenaceous measures were seen in excavations [3546 5207] below the top of the Ashover Grit.

The extensively quarried east-facing escarpment extending south from The Tors near Crich also has good exposures, though in general the cross-bedding is not well weathered out. At the highest (50-ft) quarry face Mayhew (1967, p. 97) has recognised the upper pebbly planar cross-stratified facies lying with a transgressive base on the lower non-pebbly trough cross-stratified facies.

The Derwent Valley

The escarpments flanking the Derwent valley from Whatstandwell to Ambergate show numerous craggy exposures. The best sections are found in disused quarries on the west flank of the valley [3291 5465] near Hankin Farm, at Shining Cliff Woods [3332 5227 to 3332 5285] and on the east flank at Dukes Quarries [3344 5467] and Chase Cliff [3415 5370], where thicknesses of up to 70 ft, 45 ft, 40 ft and 60 ft respectively are seen. The sandstone is buff to pale purple-brown, cross-bedded and coarse-grained except in the Dukes Quarries where it is fine- to medium-grained and more micaceous, probably representing a fining-upwards facies near the top of the Main Bed similar to that exposed at the Ambergate railway cutting (see below). Small pebbles, up to ¾ in across, are more common in the lower beds seen in the south-facing 45-ft exposure at Shining Cliff Woods [3350 5230]. The planar cross-bedding which appears to characterise most of the lower two thirds of the Main Bed is here particularly well displayed with individual sets up to 6 ft thick truncated by thin interbedded units showing indistinct current-ripple lamination. There are small sections in the Main Bed near Thackers Wood [352 513] and extensive exposures of the pebbly cross-bedded facies by the main A6 road at Ambergate [3481 5115 and 3504 5160], near Belper [3884 4837], in the old quarry above Blackbrook [3363 4768], in the railway cutting at Chevinside [3468 4594] and at the old quarry west of Milford [3480 4525].

The uppermost part of the Main Bed is well exposed only in the railway cutting at Ambergate [3467 5080], where about 50 ft of sandstone are seen with sporadic exposures of the overlying sequence (p. 135a). About 20 ft below the top of the Grit the cross-bedding appears to pass upwards from planar to broad lenticular sets. The top 6 ft are flaggy bedded and micaceous with parallel and current-ripple lamination. There is an overall fining upwards.

Two local exceptions to the prevailing planar cross-bedded facies are exposed at quarries on Firestone Hill. The first [3359 4650] shows 18 ft of coarse cross-bedded sandstone, soft and micaceous in the basal 2 to 3 ft, overlying 10 ft of coarse massive sandstone. The second, in a disused quarry 200 yd to the south [3368 4587], shows 70 ft of coarse sandstone with unusually thick cross-bedded sets, some exceeding 10 ft. The cross-bedding dip directions indicate current supply from the north-west, which is quite contrary to those for the Main Bed as a whole (p. 18).

A shale slack has been mapped within the Main Bed at Crossroads Farm, where a small temporary exposure revealed grey-brown silty mudstone with carbonaceous micaceous partings. Other small temporary excavations suggest that this slack contains a thin coal seam.

Duffield to Breadsall

The Main Bed is almost continuously exposed in old quarries on the escarpment between Milford and Breadsall. Most of the quarries lie today within private gardens. The sandstone is largely pebbly, and medium- to coarse-grained, its basic pale grey colour frequently brown-speckled and reddened; soft friable limonitic patches are commonly present and the planar cross-bedding, which is inclined largely towards the north-west (Fig. 9) at angles of 15 to 20°, may reach inclinations as high as 33° [3563 4243] when coincident with the dip of the beds. Cross-bedding units up to 30 ft thick have been observed. The following is a list of the principal quarries in the Main Bed and the estimated thicknesses exposed:-

Roadside at Milford [352 452], 85 ft; quarry near Makeney Lodge [3534 4440], 60 ft; quarries at Turpins [353 437], three faces totalling about 70 ft; Bank Wood Quarry [3532 4354], 25 ft; Manor Quarry [3538 4329], 80 ft; quarry near Manor Farm [3530 4317], 32 ft; quarry [3536 4292], 60 ft with, near the base, 12 ft of lenticular sandstones, interbedded with red micaceous flaggy sandstone with clay laminae - probably the lower facies of Mayhew (1967) - underlain by 10 ft of brown medium to coarse-grained cross-bedded sandstone; quarry near Outwoods [3551 4260], 25 ft; Blue Mountains Quarry [3563 4243], 50 ft; Outwoods Quarry [358 421], 70 ft; Hathering Wood Quarry [360 420], 75 ft; quarry at Eaton Hill [3637 4215], 75 ft; Little Eaton Quarries [366 420], 100 ft (Mayhew (1967) records his lower facies here); old quarries in Horsley Carr [3708 4241], 12 ft.

The Main Bed, comprising brown coarse sandstone, is exposed for 11 ft below Head (p.158c) in an old quarry [3843 4139] near Breadsall Priory and includes 2½ ft of laminated brown and purple sandstone. In Ferriby Brook, east of Breadsall, the poorly exposed Main Bed is softened and reddened due to close proximity to the unconformity at the base of the Permo-Triassic.

Some discussion is necessary on the total thickness and sequence of the Ashover Grit in the vicinity of Little Eaton and Breadsall. Wedd (*in* Gibson and others 1908, pp. 29-30) gives a figure of 'quite 500 feet thick' of which the now-named Main Bed would account for '300 feet or more'. Outwoods Borehole (p. 172a) entered the Main Bed sandstone a calculated 40 ft below the top and continued to the total depth of 320 ft without encountering shale. It is estimated that it reached the bottom of the Main Bed which must therefore be about 360 ft thick. Wedd also noted that a water well at the Paper Mills [about 354 423] was sunk 80 ft without proving any sandstone, although some of this thickness is in alluvium. This thickness together with a further 200 ft of shale and sandstone mapped as lying below the Main Bed near the Paper Mills brings the total to about 640 ft and there is still the unproved thickness of the sandstone on Burley Hill to be added. This suggests an overall thickness of about 700 ft for the Ashover Grit, and therefore a minimum of 800 ft for the interval between the *R. bilingue* and *R. superbilingue* marine bands is credible.

There is confirmation of the thickness of the Main Bed from the Little Eaton Water Bores [3684 4249] which penetrated Ashover Grit from 73 ft to 427 ft. The overall thickness of Ashover Grit here therefore appears to be similar to that of the Outwoods-Paper Mills area.

Only the Main Bed appears to be represented at outcrop east of the faults near Breadsall Priory.

However a lower leaf of sandstone is present in the Ferriby Brook section about one mile to the south. The low leaves of sandstone appear to be missing over the axial line of the Breadsall Anticline, but it is not known if this is due to non-deposition or erosion.

In the north of the district most of the Ashover Grit was cored in Ironville No. 4 Oil Bore (Appendix 1, p. 169c and Plate 4), where 116 ft 2 in of fine- to medium-grained sandstones with two mudstone beds were proved. The sequence in the Ironville Nos. 1 and 3 bores are shown graphically on Plate 4.

Little Hallam Water Bore is interpreted as entering Ashover Grit at 1406 ft and continuing in the Main Bed to 1792 ft, 9 ft above the bottom of the bore. The succession, recorded by Stephens (1929, pp. 97-98) includes a 'trace of coal' at about 1504 ft.

STRATA BETWEEN THE ASHOVER GRIT AND THE CHATSWORTH GRIT

West of Hankin Farm [3214 5448] *R. superbilingue* occurs only a few yards upstream from the exposed flaggy top of the Ashover Grit (Smith and others 1967, p. 72), whereas in the Ambergate railway cutting the following composite section was measured:

	ft	in
Mudstone, dark grey, shaly, ferruginous, with *R. superbilingue*	1	0
Mudstone, grey shaly, silty, with a few thin sandstone beds; *Lingula* sp. 8 ft below top	50	0
Coal	2	0
Seatearth, mudstone, silty with rootlets	1	4
Mudstone, shaly, with carbonaceous and coaly partings	0	4
Coal	0	10
Seatearth, mudstone, silty	0	3
Siltstone, grey-brown micaceous	1	0
Sandstone (Ashover Grit, Main Bed) see p. 134d	50	0

A soft weathered coal up to 33 in occurred 10 ft or so above the top of the Main Bed of the Ashover Grit in an excavation [3546 5207] in the Amber Valley near Bullbridge.

A sequence of coal-bearing strata overlying the Ashover Grit was recorded by Wedd (*in* Gibson and others 1908, p. 48) from a trial borehole and quarry section north of Crich Carr, just beyond the boundary of the district.

In the same part of the succession, 9 in of weathered coal are exposed in a ditch [3092 5367] near Lanehead, 2 ft of coal and dirt on clayey seatearth were recorded in a temporary exposure at Wiggonlee Farm and 2 ft of coal overlying 2 ft of seatearth, mudstone and siltstone are exposed in the right bank of the River Derwent [3428 4802] near Belper. Coal was formerly worked along the foot of the Ashover Grit (Main Bed) dip slope near Belperlane End. *R. superbilingue* was found in shale dug out of a house foundation [3428 4830] on the hillside about 60 ft above the coal.

a

b

A higher coal and its underlying fireclay were formerly mined from an adit [3210 5425] beneath the outlier of Chatsworth Grit north of Alderwasley village. The coal appears to lie about 45 ft above the *R. superbilingue* Marine Band at [3214 5448] (see p. 00), but *R. superbilingue* has also been found in old shale tips on the slope below the old shaft [3151 5412]. A borehole record nearby, not precisely located, gives the following section from the base of the Chatsworth Grit (Wedd *in* Gibson and others 1908, p. 42):

	ft	in
Blue shale	60	0
Coaly shale	0	3½
Coal	2	0
Coaly shale	0	2
Seatearth clay	3	0
Sandstone	0	3

A thin bed full of *Lingula mytilloides* formerly exposed in an old brickpit [3226 5431] about 20 ft above the top of the Ashover Grit is also recorded by Wedd.

The coal seam in the above section was proved recently in a trench [3170 5345] near Alderwasley where it was 2 ft 0½ in thick overlying 3 ft of seatearth clay.

The *Reticuloceras superbilingue* Marine Band was proved some 85 ft above the Main Bed of the Ashover Grit in Amber Dye Works Nos. 1 and 2 boreholes (Plate 4) and named by Swinnerton (1946) the Belper Grit Marine Band. He recorded *Carbonicola* cf. *recta* and *C.* aff. *lenicurvata* some 30 ft below. The latter species was then considered to be the lowest record of a non-marine bivalve from the Namurian. The *R. superbilingue* Marine Band was discovered in a temporary exposure [3551 5201] for a gas holder at Bullbridge, where it also contained numerous specimens of *Lingula*, and was recorded by Mr R. A. Eden in a temporary exposure [3548 5184] near Hag Wood, in black carbonaceous shale.

Between Milford and Little Eaton there are no significant exposures. The thickness of the sequence appears to increase from about 120 to 160 ft between these two localities. Wedd (*in* Gibson and others 1908, p. 45) recorded an unsuccessful attempt to prove the coal overlying the Ashover Grit near Milford. A section is provided by water bores at Little Eaton (p.170b

The 148 ft 8-in sequence lying between the Ashover and Chatsworth Grits and consisting mainly of mudstone was cored in Ironville No. 4 Oil Bore (p.169c). The base of the *R. superbilingue* Marine Band is taken at 1321 ft, with *Lingula* and poorly preserved goniatites present, sporadically, in the cores for about 40 ft above this level. The sequences in Ironville No. 1 and No. 3 Oil bores are shown on Plate 4. In the latter bore, gamma peaks indicate the assumed position of the *R. superbilingue* Marine Band (Downing and Howitt 1969, p. 246) at about 1578 ft.

Little Hallam Water Bore proved 'sandy shales' 38 ft, on 95 ft of 'blue shale containing a few very small *Lingulae*' (Stephens 1929, p. 97) overlying 11 ft of sandy shale at 1406 ft 0½ in. The '*Lingulae*' are assumed to represent the *R. superbilingue* Marine Band.

THE CHATSWORTH GRIT

West of its main outcrop the Chatsworth Grit forms several outliers flanking the Derwent Valley near Whatstandwell and Alderwasley. The full thickness is not proved and the only significant exposures are in two old quarries [3157 5415 and 3163 5420] near Knob Farm, where up to 15 ft of pale brown and partially red-stained fine- to coarse-grained sandstone are seen.

In another outlier, on the west side of the Derwent Valley north-west of Belper, the grit is estimated to be at least 90 ft thick. An old quarry [3400 4936] in the north-east corner of this outlier shows 30 ft of thin-bedded fine- to medium-grained pale and purple-brown sandstone with numerous ripple-laminated micaceous partings; it is cross-bedded in sets up to 6 in thick.

Similar sandstone is exposed in faces up to 25 ft high in old quarries above the main A6 road about 830 yd N of Broadholm. A shale slack in the outcrop of the Chatsworth Grit indicates a split into upper and lower leaves about 70 ft, and over 100 ft thick respectively.

The grit forms a pronounced feature in the Crich - Fritchley - Ambergate area and comprises some 65 ft of fine- to medium-grained sandstone with shaly partings and lenses, the base of which lies some 40 ft above the *R. superbilingue* Marine Band.

In temporary sections in the gas pipe trench at Mill Green [3576 5336] the highest beds of the grit were flaggy, micaceous and purple-stained.

Sporadic exposures in the Ambergate - Toadmoor area show a brown, fine- to medium-grained sandstone up to 20 ft thick. It is cross-laminated and commonly exhibits ripple structures. Pink weathering affects the rock along its joints. The best exposures are at Hag Wood Railway cutting [3557 5194] and at Thackers Wood [3559 5152].

The upper leaf of the Chatsworth Grit is about 60 ft thick and is exposed in old quarries north of Belper. One quarry [3536 4889] shows 25 ft of pale brown, ochreous and pink mottled sandstone with friable patches and with cross-bedding dipping towards the south-east. The same leaf is seen in old quarries high on the scarp face at Belper [3531 4810 to 3538 4761], where up to 20 ft of cross-bedded units (up to 5 ft thick) of pale brown sandstone with softer pink micaceous sandstones (as bottom-set beds) are exposed. In the southernmost quarry there are pipe-like masses of incoherent sand 5 ft wide in which only vague traces of the original bedding are preserved.

Near Milford the division of the grit into upper and lower leaves is well displayed by ground features; the only section [3567 4506] shows 4 ft of flaggy fine-grained sandstone. South-eastwards the grit is nowhere exposed. It is calculated to be about 60 ft thick north of Little Eaton.

In Ironville No. 4 Oil Bore the Chatsworth Grit is only 20 ft 9 in thick composed of sandstone with interbedded mudstones (p. 169c). The sequences in Nos. 1 and 3 oil bores are illustrated graphically on Plate 4.

In Little Hallam Water Bore strata between 1158 ft 4½ in and 1262 ft 0½ in are assigned to the Chatsworth Grit (Stephens 1929, p. 97). There is an upper sandstone 25 ft 8 in thick overlying 'sandy shale', below which is a lower leaf comprising 3 ft of 'hard close-grained sandstone'.

Trowell Moor Colliery Underground Borehole proved 66 ft of hard, grey or variegated sandstone without reaching the base of the Chatsworth Grit.

BEDS ABOVE THE CHATSWORTH GRIT

There are no significant exposures of this sequence, but it is well recorded in boreholes. In Nether Heage Borehole (p.171c) the Chatsworth Grit is overlain successively by 13 ft 4 in of mudstone, 3 ft 2 in of siltstone and 7 ft 6 in of ganisteroid sandstone. This sandstone is taken to represent the Redmires Flags (Smith and others 1967, p. 75) at the top of the Marsdenian. In Ironville No. 4 Oil Bore the cores of this sequence (p.169c) 19 ft 7 in thick, included 4 ft 10 in of 'gritstone' at 1202 ft 10 in, also thought to represent the Redmires Flags.

Sandstones above the Chatsworth Grit mapped locally near Ambergate and Milford may represent these flags at outcrop, but in the absence of palaeontological control they are not named on the geological maps.

Overlying the Chatsworth Grit in Trowell Moor Colliery Underground Borehole are some 40 ft of 'grey sandy shales' overlain by 15 ft 6 in of grey sandstone (including 2 ft of sandy shale) assigned to the Redmires Flags. At the inferred top of the Marsdenian is 6 in of 'Simmondley Coal'.

In Little Hallam Water Bore, Stephens (1929, p. 97) recorded some 8 in of coal 4 ft 9 in above the Chatsworth Grit. This coal is thought to be the representative of the Baslow Coal (Smith and others 1967, p. 54) or Ringinglow Coal (Stevenson and Gaunt 1971). A 4-in coal 55 ft 8 in higher in the borehole marks the assumed top of the Marsdenian and probably represents the Simmondley Coal.

Coaly debris above the Chatsworth Grit outcrop was noted on the northern edge of the district east of Culland Farm [3657 5384].

YEADONIAN

BEDS BELOW THE ROUGH ROCK

Nether Heage Borehole and Ironville No. 4 Oil Bore (pp.171c and 169c) provide the only significant records of the Yeadonian beds below the Rough Rock in the district. Their respective sequences are 140 ft 2 in and 111 ft 2 in thick and each proved the *G. cumbriense* Marine Band. In both boreholes the base of the Yeadonian is marked by *Lingula* bands which are taken to represent the *G. cancellatum* Marine Band. This interpretation is supported by a comparison of the gamma logs with that of Calow No. 1 Oil Bore in the Chesterfield district, in which both the *G. cumbriense* and *G. cancellatum* marine bands were proved (Smith and others 1967, fig. 7, p. 62). Bivalve and *Lingula* beds are known elsewhere in the Central Province both between the marine bands and also below the *G. cancellatum* Marine Band (Smith and others 1967, pp. 74, 77).

The lithological correlation of the Trowell Moor Colliery Underground and Little Hallam Water boreholes [4621 4048] is shown on Plate 4. The Simmondley Coal is respectively 6 in and 4 in thick.

THE ROUGH ROCK AND OVERLYING BEDS

The Rough Rock was exposed in a trench [3650 5326] north-west of Beech Hill Farm, where it consisted of brown cross-bedded fine- to medium-grained sandstone some 60 ft thick.

At the top of the Rough Rock, in an old quarry [3590 5251] near Bullbridge, a 4-in ganister overlies some 18 in of pale grey fireclay. Exposures in the railway cutting nearby [3595 5213] show the following section:

	ft	in
Pot Clay Marine Band (p. 136b)		
Ganister, white	2	6
Seatearth; siltstone with carbonaceous rootlets	1	0
Siltstone, pale grey	1	0
Rough Rock		
Sandstone, brownish yellow, fine-grained; shaly partings	5	0
Sandstone, massive	20	0
Sandstone with shaly laminae	6	0
Sandstone, massive, cross-laminated	12	0

The Rough Rock has a gradational base down into mudstone. The section does not accord with the report by Neves (1967, p. 44) that in the Crich area the Rough Rock is reduced to a few feet of siltstone, but agrees with that proved in Nether Heage Borehole.

Ten feet of flaggy fine-grained sandstone forming part of the Rough Rock were seen in temporary exposures for extensions to the Brick Works [3584 5196], Ambergate.

The top of the Rough Rock together with overlying beds are exposed in Ridgeway Quarry [3585 5145]:

	ft	in
Shale, black fissile (Pot Clay Marine Band, see p.136b)	0	1
Ganister, pale grey - white, very hard	2	6
Mudstone, sandy	0	6
Rough Rock		
Sandstone, fine-grained pale grey	1	3
Mudstone, laminated silty	0	3
Sandstone, white, weathering yellow-brown (in 12-in beds)	8	0+

In Ironville No. 4 Oil Bore the Rough Rock is 44 ft thick with its base at about 1086 ft; the lowest 14 ft is recorded as sandy micaceous siltstone, the main mass of the rock as fine-grained whitish sandstone, finer and micaceous at the base, and the top as a 3-ft bed of ganister. The Cromford Fault (p.160d) intersects the borehole at approximately the top of the Rough Rock. In No. 3 Bore the Rough Rock (about 1325 to about 1378 ft) was cored over the greater part of its thickness and is recorded as silty fine-

c

d

grained cemented sandstone, micaceous and laminated, increasingly silty towards the base.

The Rough Rock at Pinchom's Hill, to the south of the Horsley Fault, has a mapped thickness of about 110 ft. Discontinuous exposures in the quarries [3604 4693 and 3595 4689] show up to 30 ft of cross-bedded sandstones in units or 'posts' about 3 ft thick; rippled and laminated micaceous sandstones, probably formed as bottom-set beds, are also present. Towards the summit of the hill, a separate low feature with ganister fragments in the soil is taken to represent the Pot Clay seat-earth. There is no palaeontological evidence of the age of these rocks.

At Shawlane Quarry [3592 4559] the uppermost 11 ft of the Rough Rock there exposed are composed of laminated and flaggy sandstone in which weak cross-bedding is apparent in places; the cross-bedding is truncated by the flagstone bedding. Underlying are some 20 ft of medium to coarse-grained, brown to red-brown cross-bedded (towards the north-west) sandstone. A large ferruginous mass, 8 ft long and 1½ ft thick, was noted.

The Rough Rock is well exposed in quarries east and west of Coxbench in the Bottle Brook valley and is about 100 ft thick. West of the old Coxbench - Little Eaton road [3701 4320] about 40 ft of medium-grained micaceous sandstone coloured in various shades of brown are exposed, parts of the face being in a very friable condition. Immediately south-west of this pit a quarry-face nearly 300 yd long and up to 75 ft high exhibits similar but harder sandstone with irregular and often curved fractures. Clear bedding-planes are rarely seen and although westerly dipping cross-bedding was observed in one place, the few parts of the face which are suitably weathered exhibit extreme contortions of the bedding with amplitudes of up to 10 ft (Plate 3). It is apparent that the irregular fractures are the result of quarrying the contorted or slumped sandstone which comprises the entire face. A further quarry [3657 4297] 300 yd SW of the above face shows about 60 ft of cross-bedded pale grey to brown medium- to fine-grained sandstone; the cross-bedding dips slightly west of south.

On the west side of the valley the lower 30 ft in the Coxbench Quarry [374 433] is cross-bedded (towards the west-north-west) sandstone in units up to 15 ft thick; there is some red staining mainly on the bottomset beds and ferruginous concretions up to 5 ft by 8 ft occur. In the uppermost 20 ft of the quarry the rock is brown and red-stained and contains small hard nodules, small-scale cross-bedding and slump structures.

The base of the Rough Rock is exposed among quarry waste in the adjacent yard of Castle Farm [3738 4317]. It is underlain by 1 ft 3 in of purple mudstone on grey micaceous mudstone.

Cross-bedding in the sections in old quarries near Horsley Castle dip mainly towards the west-north-west, but units dipping west and west-south-west were also measured; up to 60 ft of the rock is exposed.

On the east side of the Horsley Fault system the Rough Rock has a thickness of little more than 50 ft in old quarries near Morley Moor Farm. At Morley Moor Farm [388 425] the lower part of the north face of the quarry shows 10 to 18 ft of brown and purple mottled sandstone with slump structures; cross-bedding, some of which dips towards the south-east, is best seen in the lowest 5 ft exposed. At the west end of this face there is a passage between slump and cross-bedding and convolutions are present in the laminated foreset beds of the latter. The upper part of the face shows 12 ft of cross-bedded medium-grained brown and purple mottled sandstone with softer purple micaceous bands. At the top of the quarry 4 to 6 ft of inaccessible, red and grey mottled clays are assumed to include the Pot Clay horizon.

No detailed sedimentological study of the Rough Rock has been attempted, but it is concluded that in the Coxbench - Morley Moor Farm area, instability during deposition was responsible for the slump structures. These and the differential subsidence revealed by local thickness changes are interpreted as evidence of precursory movements associated with the formation of the Breadsall Anticline (Figure 59).

In Park Brook Borehole [3830 4385] the following sequence was proved below the Pot Clay Marine Band base at 505 ft 8 in:

	ft	in
Mudstone, dark grey, with bivalves, fish debris	1	6
Seatearth, siltstone, grey, with pink patches and small orange-red nodules	2	0
Sandstone, (Rough Rock) light grey and pink-stained; massive to bottom of borehole at 520 ft	10	10

The Pot Clay Coal, 2 in thick, was encountered in Trowell Moor Colliery Underground Borehole and was separated by 7 in of 'grey sandy clunch' from the top of the Rough Rock. The latter, 79 ft 6 in thick, is recorded as follows:

	ft	in
Sandstone, hard grey laminated	7	6
Sandstone, hard grey calcareous	17	0
Sandstone, hard grey laminated, with shale bands	27	0
Sandstone, dark grey laminated	28	0

Mr R. A. Eden placed the base of the Westphalian in Little Hallam Water Bore at 939 ft 7½ in. The Rough Rock is 112 ft thick with base at 1051 ft 7½ in. The detailed section, recorded by Stephens (1929, pp. 95-98), is from the top: sandy shale, 4 ft; grey mudstone, 16 ft; sandy shale, 3 ft 6 in; variegated sandstone, 11 ft; grey sandstone ('dark streak'), 67 ft 6 in; strong sandy shale, 10 ft.

In Beechdale Road (Robin's Wood) Borehole, the base of the Pot Clay Marine Band is at 2068 ft 8 in. The underlying downward sequence is dark grey silty mudstone with fish debris, 4 in; Pot Clay Coal, 2 in; ganister, 4 in. This rested on Rough Rock composed of pale greenish grey fine-grained sandstone with dark greenish grey coarse siltstone bands at base, 8 ft 6 in; on greyish white micaceous sandstone, 12 ft.

a

CHAPTER 4

Westphalian (Coal Measures)

DETAILS

b

WESTPHALIAN A (LOWER COAL MEASURES)

BASE OF WESTPHALIAN TO KILBURN COAL

The measures between the Pot Clay Marine Band and Kilburn Coal range in thickness from about 590 ft near Bondland Colliery to about 800 ft south of Denby Colliery (Fig. 12). In the east they are 625 ft thick in Beechdale Road Borehole.

Of the twelve component cycles within the district only that above the Belperlawn Coal and that capped by the Kilburn Coal lack marine or near-marine horizons at their bases. The highest marine horizon known to occur in these measures to the north and near Nottingham (Smith and others 1967, p. 116) - the Burton Joyce Marine Band - remains unproved.

Within these measures are two principal sandstones - the Crawshaw Sandstone and Wingfield Flags - and three coals that have been worked underground - the Belperlawn, Alton and, to a lesser extent, the Norton. Ganisters, present within the seatearths of many of the lower coals, were formerly worked in the north of the district.

The distribution of the marine and near-marine horizons was given by Eden (1954) and the faunal sequence is similar to that described by Smith and others (1967, 1973) in the Chesterfield and East Retford districts. The mussel beds between the Belperlawn and Alton coals contain the distinctive Carbonicola fallax/C. protea faunas described by Eagar (1947, 1952) from above the Soft Bed Coal in Yorkshire. Another distinctive horizon is the Norton Mussel-band with Carbonicola proxima which overlies the Norton Marine Band.

The quality of coal mentioned in this section is assessed according to a scale used by the Coal Survey Laboratory for the ash and sulphur contents as follows:

ASH

Ash %	Comment: Quality	Comment: Ash Content
2.5 & under	Very good	Very low
2.6 - 5.0	Good	Low
5.1 - 7.5	Medium	Moderate
7.6 - 10.0	Rather poor	Moderately high
10.1 - 15.0	Poor	High
15.1 - 25.0	Very poor)	Very high
25.1 - 40.0	Very inferior)	
Over 40.0	Dirt	-

SULPHUR

Sulphur %	Comment
0.5 & under	Very low
0.6 - 1.0	Low
1.1 - 1.5	Moderately low
1.6 - 2.0	Moderate
2.1 - 2.5	Moderately high
2.6 - 4.0	High
4.1 & over	Very high

THE POT CLAY MARINE BAND. The dark grey marine mudstones at the base of the Westphalian have been proved in three boreholes. The characteristic goniatite, Gastrioceras subcrenatum, is present and the associated fauna consists of a further species of Gastrioceras, together with Dunbarella papyracea, Lingula mytilloides, Orbiculoidea sp., conodonts including Hindeodella sp. and fish debris (see also Eden 1954, p. 106).

Strata inferred to represent the marine band are exposed near Ambergate at Bullbridge Brickworks Quarry [3602 5203], and at Ridgeway Quarry [3583 5146]; they may occur also as reddened mudstones at Brackley Gate. At Ridgeway Quarry, the ganister at the top of the Namurian is overlain by some 12 in of laminated grey silty and listric mudstones containing fish scales and foraminifera; within the basal 1 in are rare specimens of Lingula sp. and Orbiculoidea sp. This thin sequence of the marine band is unusual in view of the sparse fauna and the presence of foraminifera (see Calver 1968, p. 26; Smith and others 1967, p. 102). In the nearby Nether Heage Borehole both goniatites and Lingula were present in 1 ft 2 in of dark grey mudstone.

Posidonia sp., Anthracoceras sp. and G. subcrenatum were collected from 4 ft 8 in of dark grey mudstone in Park Brook Borehole; in Beechdale Road (Robin's Wood) Borehole the marine band was 2 ft 4 in thick (Eden 1954, p. 84). The position of the marine band is marked by a 'high' between 1304 and 1308 ft depth in the gamma log of Ironville No. 3 Oil Bore. Smalley Mill Well, Little Hallam Borehole and Trowell Moor Colliery Underground Borehole also penetrated the horizon, but yielded insufficient evidence to establish its precise position.

THE MEASURES BETWEEN THE POT CLAY MARINE BAND AND THE BELPERLAWN COAL range from about 100 ft in Beechdale Road and Nether Heage boreholes to 145 ft in Park Brook Borehole. Below the Crawshaw Sandstone they comprise largely argillaceous strata with a few

c

feet of dark grey mudstones containing fish debris overlying the marine band. Wedd (in Gibson and others 1908, p. 50) noted these strata to be '12 or 14 yards thick and composed of blue sandy micaceous shale' in a well [3634 4363] near Coxbench. In the Ambergate area a thin layer of penecontemporaneously contorted mudstone similar to those described by Cope (1946) has been noted a few inches above the marine band. These argillaceous measures are exposed in the Bullbridge Brickworks Quarry and adjoining rail cutting and in Ridgeway Quarry (see Appendix 2).

The Crawshaw Sandstone ranges up to 112 ft in thickness in Little Hallam Borehole and reflects a resumption of Namurian conditions of sedimentation. Its cross-bedding suggests derivation from the south. The sandstone is mainly medium to coarse, gritty and arkosic; its colour is buff at outcrop, but grey to white when encountered in bores; pink and red staining is common both at outcrop and in boreholes and a green coloration has been sporadically recorded. Except at Ridgeway Quarry (Appendix II) and in Beechdale Road Borehole the base of the sandstone is gradational into the underlying argillaceous strata: the lowest beds are either fine-grained for a thickness of up to 24 ft as at Beechdale Road Borehole or interbedded with siltstone and mudstone, as in the vicinity of Ambergate. Within the main mass of sandstone, beds of siltstone and mudstone (about 3 ft at Coppice Colliery Underground Borehole and 7 ft at Stapleford No. 1 Oil Bore have been proved, and similar beds reach mappable proportions in the outcrop near Holbrook. Towards the top of the sequence the rock usually becomes finer grained and contains beds of siltstone and mudstone. At the top, below the Belperlawn Coal, there is a ganisteroid sandstone about 2 ft thick and mudstone seatearth up to about 1½ ft thick.

North of Ridgeway, silty bands occur in both the highest and lowest parts of the sandstone, and at the southern end of the Brick Works Quarry the sandstone is entirely fine-grained and silty.

Southwards from Ambergate the bold escarpment of the Crawshaw Sandstone is broken by faulting. Between Rowson Green (south-east of Belper) and Holbrook there are several small exposures of coarse pebbly sandstone. Subangular brown ferruginous sandstone fragments and quartz pebbles (up to 1 in diameter) were seen in 5 ft of coarse brown feldspathic sandstone near Holbrook Moor [3622 4580] and 18 ft of cross-bedded and shattered coarse sandstone are exposed south of the roadway to Rowson Green [3678 4606].

From Holbrook to Coxbench, and also near Brackley Gate, the outcrop is divided by an argillaceous bed near the base of the sandstone. At Portway* Site, it is believed that the Crawshaw Sandstone is locally united with a sandstone lying below the Alton Coal (see p. 136d); similarly a coal, presumably the Belperlawn, was proved up to 24 in thick within sandstone at Beechhill Farm Site, ½ mile east of Fritchley.

The Crawshaw Sandstone is exposed locally between Stanton (or Stanton by Dale) and Dale in the south-east of the district. It has been quarried [4715 3783] immediately east of Stanton, where 50 ft of massive brown to off-white micaceous medium- to coarse-grained sandstones are seen. The quartz grains are angular and well cemented, and cross-lamination suggests a southerly origin. West of Stanton, in an old sand pit [4604 3833], a section of a fault plane in the Crawshaw Sandstone shows grain variations from fine to medium sandstone to coarse grit.

THE BELPERLAWN COAL (Fig. 13) has been proved to be 66 in thick at Nodinhill Site and is 38 to 50 in in the type area north-east of Belper, where it has been extensively worked, mainly by opencast methods.

The following section, recorded by R.A. Eden at Far Lawn Site [3665 4901] prior to working, typifies the seam at outcrop near Belper:

	ft	in
Soil	3	0
Siltstone, buff, micaceous, carbonaceous; ferruginous lenses	2	0
Siltstone, soft clayey; streaks of coal and clay	0	7
Coal	0	10
Siltstone, soft, grey, micaceous		0 to 1
Coal	3	0
Silt, black, soft, carbonaceous	0	2
Fireclay, white, micaceous	1	0+

Away from the western outcrop the thickness of coal only exceeds 30 in within the 'take' of Denby (Drury-Lowe)* Colliery. Near Heanor it is less than 20 in, but at Ilkeston (Manners Colliery Borehole) the 27-in seam is split by 9 in of dirt; at Oakwell Staple Pit the dirt parting is 18 in thick, separating two 9-in leaves of coal; the Ironville oil bores did not prove coal at this horizon.

In many sections† the seam is divided by a bed of dirt or dirty coal. Above this parting the seam is variable and frequently of poor quality; below the parting it is of medium to poor quality with, on average, a moderately high ash content.

A number of bell pits have been sunk to the seam between Ridgeway and Heage, including Bowman's Colliery [3590 5087], where the Belperlawn was recorded as 5 ft thick at 18 ft depth. A quarter of a mile to the north [3594 5122] the weathered crop ('smudge') of the seam was once worked for dye-making.

A small area of Belperlawn Coal up to 32 in thick was mined near Sawmill from Bullbridge Colliery. The seam, 2 ft 10 in thick, is exposed in the adjacent railway cutting [3621 5212], where it is overlain by unfossiliferous shale and underlain by 8 in of grey seatearth.

*Locations of Opencast Coal Sites mentioned in the text are recorded in Appendix 3

*The name is condensed in this account to Denby (D-L) Colliery.

†Seam details in this account are based upon reports prepared by the Nottingham Coal Survey Laboratory, its successor the National Coal Board Scientific Department, the N.C.B. Opencast Executive, and Geological Survey records.

d

THE MEASURES BETWEEN THE BELPERLAWN AND ALTON COALS (Fig. 14), which have a maximum thickness of 78 ft in Denby (D-L) Colliery Alton No. 1 Underground Borehole, thin northwards to 30 ft at Ridgeway Site and eastwards to 42 ft in Trowell Moor Colliery Underground Borehole. They comprise up to four cycles with three thin and impersistent coals, in ascending order the Holbrook, Second and First Smalley coals (Eden 1954). The lowest beds of each cycle, except those immediately overlying the Belperlawn Coal, are normally mudstones with Lingula mytilloides and, frequently, mussels. Fish debris which occurs in the roof measures of the Belperlawn Coal and is intermittently recorded from the remaining cycles tends to be associated with, or replaces, the Lingula bands. Mussels are very rare in the cycle above the Second Smalley Coal and there is no record within this district of the mussels described from the Yorkshire equivalent of the roof of the Belperlawn seam. The distinctive mussel fauna of this sequence has been described by Eagar (1952, 1954) and by Smith and others (1967, 1973). In Denby (D-L) Colliery Underground Borehole it includes Carbonicola haberghamensis above the Holbrook Coal and C. declinata, C. fallax, C. limax and Curvirimula sp. together with Geisina arcuata above the First Smalley Coal.

Each cycle contains variable amounts of sandstone and siltstone, and ganisteroid sandstone seatearths frequently underlie the coals. Sandstone in the uppermost cycle has been called the Sub-Alton Sandstone by Smith and others (1967, p. 105). The lithologies and faunas observed at most localities within the district are shown diagrammatically in Fig. 14 and by Eden (1954, pl. 2A) and sections near Ambergate and Holbrook are described in Appendix 2.

The Holbrook Coal is impersistent, ranging in thickness up to 14 in (of shaly coal) in Manners Colliery Borehole. It is only 4 in thick at Browns Road Site near Holbrook in the west (Eden 1954, p. 89) and is represented only by seatearth at, for instance, Moor Farm Borehole. Twin seatearths however separated by 6 ft of sandstone and siltstone in Fritchley No. 1 Borehole and 'sandstone with dark markings' 18 ft above the Belperlawn Coal in Bailey Brook Colliery Drift further illustrate the variable nature of this horizon. Eden (1954, p. 87) also observed a non-sequence cutting out a 2-in coal (probably the Holbrook) lying 8 ft above the Belperlawn Coal on part of Ridgeway Site.

The Second Smalley Coal is a variable seam up to 18 in thick and composed of coal and batt in Smalley Mill Well (Gibson and others 1908, p. 74); more usually it is less than 10 in thick (Fig. 14). It may also be composed of split thin coals such as in Mapperley Colliery Underground Borehole where the downwards sequence is coal ½ in, seatearth and mudstone 17½ in, coal 3 in or it may be represented by seatearth only.

The First Smalley Coal where developed as a single seam ranges in thickness up to 29 in (of shaly coal at West Hallam Colliery Borehole), but over parts of the district the seam consists of two coal leaves separated by sandstone (Figs. 14, 67). In Smalley Mill Well (Gibson and others 1908, p. 74) for instance, the whole was 3 ft 11 in thick and included a 1-ft rock parting.

The Sub-Alton Sandstone is an impersistent sandstone lying in the uppermost cycle of these measures. In the west it becomes important on part of Portway Site where, on the evidence of a few boreholes, it is united with or rests upon the Crawshaw Sandstone with none of the intervening coals proved. The sandstone - siltstone sequence between the Belperlawn and Alton coals at Trowell Moor Colliery Underground Borehole and in Moor Farm Borehole where it is faulted against the Holbrook horizon are considered to be this sandstone (Fig. 14). It is less certainly recognised at Foreclose Farm and Farlawn sites near Belper where, as in other opencast sites, the thin coals are difficult to correlate because of splits and non-sequences.

THE ALTON COAL is only 10 in thick in Fritchley No. 1 Borehole (on Chestnut Site) and also near Heanor (Bailey Brook Colliery Drift) but ranges up to 54 in at Ridgeway and Spanker sites near Ambergate. The isopachs shown in Fig. 15 represent only a regional tendency and exclude abrupt local variations in thickness; for instance other recordings on Chestnut Site range up to 31 in. The quality of the coal is good to medium with variable ash and high to very high sulphur contents; the latter originate from the many pyritised fusain partings.

The seatearth has been worked for ganister at Wingfield Park [3700 5370] and for fireclay at Sawmill [3625 5195]. The coal, ranging from 30 to 45 in in thickness, has been extensively worked by underground and opencast means in the area east of Belper. It has also been worked at outcrop at Rykneld Covert Site [387 430] near Brackley Gate and in the south-east of the district at Beal Site, where the thickness varied between 33 and 36 in.

Away from the outcrop the Alton Coal was recorded as 27 and 36 in in the uncored Ironville Nos. 3 and 1 oil bores respectively. It is thin near Heanor, where 16 in of coal were encountered in New Langley Colliery Borehole; in Beechdale Road Borehole only a trace of coal was recorded, but a maximum thickness of 41 in was proved in Trowell Moor Colliery Underground Borehole. A 10-in 'smut or coal' at 1721 ft 1 in in Wollaton Colliery No. 3 Shaft Boring is only doubtfully correlated with the Alton. In concealed measures along the southern margin of the district in Dale Abbey No. 2 Borehole, the coal was 39 in thick at 227 ft.

A motorway proving hole [4764 3789] showed 27 in of coal underlain by light grey mudstone seatearth with ganister at the base.

THE MEASURES BETWEEN THE ALTON AND FORTY-YARDS COALS (Fig. 16) vary between 59 and 112 ft in thickness and, as described by Eden (1954), are composed of two cycles. The lower, 16 to 39 ft thick, commences with the Alton Marine Band and is largely argillaceous with significant sandstone or siltstone beds only in Bondland Colliery Shaft, Wiremill Bridge Borehole and also, probably, at Stanton Sinking, where all but 5 ft of the 87 ft of these measures are recorded as 'rock' or 'stone bind'. A 2-in seatearth at the top of the lower cycle has been recorded in Wiremill Bridge Borehole; elsewhere the base of the upper cycle can be recognised only by the presence of the Parkhouse Marine Band. The upper cycle, which is 43 to 96 ft thick, with variable thicknesses of sandstone and siltstone comprising the Loxley Edge Rock near the top, is capped by the seatearth or ganister below the Forty-Yards Coal.

The Alton Marine Band directly overlies the Alton Coal except in New Langley Colliery Borehole (see below) and consists of fossiliferous dark grey pyritous shales with 'bullions' of ironstone containing uncrushed goniatites. It is only a few inches thick in the Denby (D-L) Colliery district, but reaches 7 ft 10 in in Beechdale Road Borehole. In the thicker sequences the marine band is split (Eden 1954), the upper part consisting of dark mudstones with Lingula sp. and microfossils which may be separated from the lower part by up to 3 ft 8 in of mudstone without marine fossils, as in Beechdale Road Borehole.

The marine band was first noted in the Ambergate area by Wedd (1903, p. 12) and described, but not named, by Gibson and others (1908, pp. 64-67, 100, 185). The fauna includes Dunbarella papyracea, Posidonia gibsoni, Caneyella multirugata, turreted gastropods, Anthracoceratites sp., Gastrioceras listeri, orthocone nautiloid, Hindeodella sp. and fish remains.

There are no in situ exposures, but tips near Sawmill [3630 5195], Ridgeway [3617 5142] and Denby (D-L) Colliery Engine Pit [3836 4734] and slipped debris on Rykneld Covert Site near Brackley Gate [3915 4294] contain fossiliferous shales and bullions. Detailed provings of the marine band are as follows: Wiremill Bridge Borehole, 5 in; Fritchley No. 1 Borehole, 6 ft 2 in; Bondland Colliery Shaft, 2 ft 1 in; Ridgeway Site and vicinity, 5 ft; Colliers' Rest Borehole, 4 in; Denby (D-L) Colliery Kilburn Shaft, thin; Denby (D-L) Colliery Underground Borehole, 8 in (see Eden 1954, p. 93 for details); Rykneld Covert Site near Brackley Gate, 4 ft (see Appendix 2); Park Brook Borehole, 6 ft 10 in; Mapperley Colliery Underground Borehole, 4 ft 3 in including 2 ft of shale between the two parts (see Eden 1954, p. 93 for details); Coppice Colliery Underground Borehole, 3 ft; New Langley Colliery Borehole, 4 ft 3 in including 9 in of siltstone between the marine band and the coal; Bailey Brook Colliery Drift, 1 in; West Hallam Colliery Borehole, 1 ft 5 in; Manners Colliery Borehole, 2 ft 7 in; Oakwell Colliery Staple Pit, 1 ft 4 in (Vernon 1909, p. 295); Moor Farm Borehole, 7 ft 8 in, including 2 ft 5 in between the upper and lower parts; Beechdale Road Borehole, 7 ft 10 in, including 3 ft 8 in between the upper and lower parts; and a motorway proving east of Stanton [4764 3789], 9 in. Twin peaks on the gamma log of Stapleford No. 1 Oil Bore represent the bipartite nature of the marine band and spot cores yielded Lingula sp. and fish debris. A gamma 'high' at about 1180 ft in Ironville No. 3 Oil Bore is also correlated with this marine band.

Fish debris occurs in the mudstones above the marine band and mussels (Eden 1954, pp. 94, 100) have been recorded in Denby (D-L) Colliery Underground Borehole, New Langley Colliery Borehole, Beechdale Road Borehole, Manners Colliery Borehole and, abundantly, in Coppice Colliery Underground Borehole. At Denby Carbonicola prisca and palaeoniscid scales occurred and at Coppice Curvirimula belgica, Geisina arcuata and Rhadinichthys sp. were present with Carbonicola prisca. Examples of the latter species were described and figured by Eagar (1954, pp. 60-61; 1956, p. 345) who noted its value as a horizon marker.

A sandstone 25 ft thick, lying close above the Alton Coal, forms a feature in Wingfield Park and in the Nether Heage area [3620 5100].

The Upper and Lower Parkhouse marine bands occur at the base of the uppermost cycle below the Forty-Yards Coal and yield Ammodiscus sp., Glomospira sp., Lingula mytilloides and fish remains. Over parts of the Derby district only one band is present. At Wiremill Bridge Borehole in the north of the district 4 ft of dark silty mudstone, with fish and mussels, separate the seatearth at the top of the lower cycle from further mudstones containing two thin Lingula bands, 1 ft apart; presumably these are the Upper and Lower Parkhouse marine bands.

A gamma 'high' at 1147 ft in Ironville No. 3 Oil Bore is correlated with this marine band. In Collier's Rest Borehole there was only one 6-in band with Lingula and 'Estheria' (note also Eden 1954, p. 95), some 16 ft above the Alton Coal. In Denby (D-L) Colliery Underground Borehole the Lower Parkhouse Marine Band lies 17 ft above the Alton Coal and contains Lingula at three levels within 7 ft 6 in of shale and siltstone (Eden 1954, p. 94): the Upper Parkhouse Marine Band, consisting of 2 ft 9 in of carbonaceous shale with Lingula, lies 49 ft 4 in above the Alton Coal.

Mr R. A. Eden (1954, p. 95) found a microfauna 16 ft above the Alton Coal and 'Estheria' in a 2-in ironstone band some 18 in higher at Rykneld Covert Site. Lingula, was found at Mapperley Colliery Underground Borehole, in a 19-in band within dark shale, 25 ft 6 in above the Alton Coal and in 5 ft 11 in of pyritous mudstones, 27 ft 11 in above this coal in Coppice Colliery Underground Borehole. Two bands of black mudstone with Lingula, separated by 9 in of micaceous mudstone, lie 25 ft 2 in above the Alton Coal in New Langley Colliery Borehole, and two bands, 3 ft apart within grey mudstone, lie 26 ft 10 in above the Alton Coal in Manners Colliery Borehole.

In Moor Farm Borehole, Lingula is present through 1 ft 4 in of the Lower Parkhouse Marine Band 27 ft 5 in above the Alton Coal, and in two thin beds 4 ft 4 in apart, representing the Upper Parkhouse Marine Band 39 ft 5 in above the coal. Both bands lie within grey mudstone with ironstone bands and minute pyritic nodules and 'tubes'. In Beechdale Road Borehole the marine bands are apparently united to form a single 3 ft 8 in band within grey mudstone some 27 ft 4 in above the Alton Coal.

The mudstones between the Parkhouse Marine Band and the Loxley Edge Rock contain sporadic fish remains and mussels.

The Loxley Edge Rock comprises a fine micaceous sandstone with siltstone and mudstone partings, variable in thickness, lying in the upper part of the cycle which commences with the Parkhouse Marine Band (Eden 1954). The sandstone exceeds 30 ft only in New Langley Colliery Borehole and Stanton Sinking. The rock was absent in the Denby (D-L) Colliery Underground Borehole, Dale Abbey No. 2 Borehole and from one of the records of the Smalley Mill Well.

East of Belper the Loxley Edge Rock forms a topographic feature which appears to merge locally with that of a sandstone close above the Alton Coal within Whitemoor Site and near Holbrook; it forms a gentle feature between Dale and Stanton; in a motorway borehole [4779 3798] it consisted of 20 ft of massive, light grey, fine-grained micaceous sandstone with carbonaceous partings.

THE FORTY-YARDS COAL commonly comprises two thin seams separated by up to 15 ft of measures, which consist largely of seatearth, but may include mudstone, siltstone, sandstone and ganister. Most records of ganister, however, come from below the lower leaf of the seam. This lower leaf is the least persistent and is represented solely by seatearth or ganister in Denby (D-L) Colliery Underground Borehole, New Langley Colliery Borehole and Stanton Sinking. In those sequences where only one leaf of coal and seatearth are recorded, the rock succession and interval to the Alton Coal indicate that it is the lower leaf which is missing.

At Stanton Sinking the split is threefold with the lowest leaf represented by 1 ft 3 in of ganister; a 6-in coal lies 7 ft above and a 12-in coal a further 8 ft 9 in higher in the sequence.

The thickness of coal quoted for Stanton Sinking are near average for the district and the seam is therefore not of economic importance. The lower leaf has a maximum proved thickness of 27 in in Dale Abbey No. 2 Borehole. The upper leaf, contains dirt partings, 2½ in thick, in an over-all 21 in of coal and 'batt' in Denby (D-L) Colliery Kilburn Shaft and 6 in of dirt within an over-all 16-in in Dale Abbey No. 3 Borehole.

Old ganister pits [3710 5310 and 3660 5248], now overgrown, occur south and east of Beech Hill Farm where dip and slope coincide and shallow workings are extensive. Some 9 in of 'smut' were recorded in a trench near Beech Hill Farm [3721 5315], and 8 in of 'smutty' coal on 14 in of ganister were proved at Chestnut Site. Old opencast workings in Wingfield Park [3720 5366] proved 6 in of coal overlying 4 in of ganister, on fireclay.

Wedd (in Gibson and others 1908, p. 75 and ms. map) recorded 12 in of coal, probably the Forty-Yards, in a temporary section near Holbrook [3697 4489], and in a drive cutting east of Brackley Gate [3934 4279] the coal is 7 in thick, overlain by 3 in of black canneloid mudstone and underlain by 2 to 3 in of silty mudstone resting on ganisteroid sandstone.

THE MEASURES BETWEEN THE FORTY-YARDS AND NORTON COALS comprise one cycle normally about 30 ft thick, but ranging from 18 ft in Bondland Colliery Shaft to 39 ft in Trowell Moor Colliery Shaft. The cycle consists largely of mudstone with ironstone bands and nodules (the 'Dale Moor Rake' of Smyth 1856, p. 59) and is now marked by extensive old excavations at outcrop near Stanton Ironworks [4630 3880]. Fish remains are plentiful (see Gibson and others 1908, p. 101) and include Diplodus sp., Elonichthys sp., Rhabdoderma sp. and Rhizodopsis sp. The arenaceous part of the cycle is thin and impersistent, but the seatearth at the top is up to 14 ft thick.

At the base of the cycle the Forty-Yards Marine Band (Smith and others 1967, p. 114), containing Ammodiscus sp., Lingula mytilloides, Hindeodella sp., Elonichthys aitkeni and an acanthodian spine has been found in the following boreholes. Denby (D-L) Colliery Underground (1 in thick); Mapperley Colliery Underground (1 in); Coppice Colliery Underground (3 ft 1 in); New Langley Colliery (3 in); West Hallam Colliery (trace); Manners Colliery (trace); Moor Farm (½ in); and Beechdale Road (5 ft 4 in).

The non-marine mudstones of this cycle contain fish and plant debris.

THE NORTON COAL which varies in thickness from 27 in in Little Hallam Water Borehole and Stanton Sinking to 3 in in Beechdale Road Borehole, has only been worked in the south of the coalfield, near Stanton Gate (Gibson and others 1908, p. 68).

In the north, in Wiremill Bridge Borehole, the section consists of 8 in of cannel on 22 in of coal. Other records are: Bondland Colliery, 19 in of coal; Ridgeway Site, 22 in of bright coal containing dirt partings in the uppermost 3½ in and a 1-in fusain parting. A coal presumed to be the Norton, varying between 5½ and 10½ in is exposed in the cutting of the Ambergate-Ironville Railway on either side of the tunnel [3710 5300] south-east of Beech Hill Farm. Poorly exposed dark shales overlie the seam and it is underlain by fireclay with lenses of ganister up to 12 in thick. Trenches nearby [3728 5311] showed 13 in of coal underlain by pale grey ganister.

In Denby (D-L) Colliery Kilburn Shaft the Norton Coal is 20 to 22 in thick plus 1 to 3 in of 'batt' in the roof. Other records are Colliers' Rest Borehole, 26½ in; Denby (D-L) Colliery Underground Borehole, 16 in; Smalley Mill Well, 3 in or 22 in according to different records; Mapperley Colliery Underground Borehole, 6 in. In five provings near Heanor and Ilkeston the Norton Coal is 16 to 19 in thick and it is 21 to 27 in in deep mine provings in the south-east of the coalfield; with the exception of Wollaton Colliery No. 3 Shaft, where a seatearth at 1651 ft 4 in in depth is assigned to the Norton horizon. The coal has been proved to be up to 26 in thick in opencast boreholes on Dale Moor Site and 17 in along the line of the motorway [4799 3798], east of Stanton.

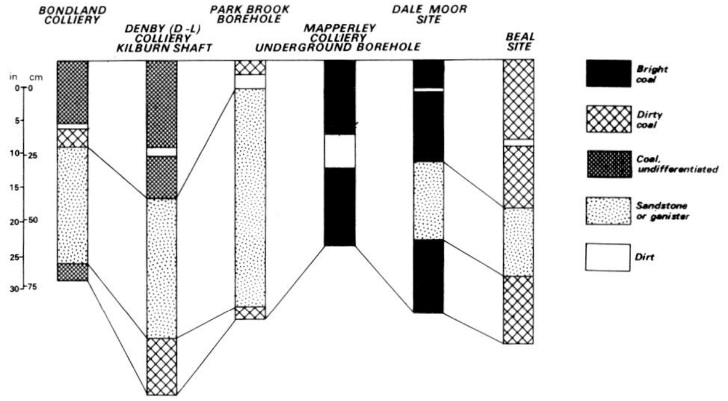

Figure 67 Representative sections of the First Smalley Coal

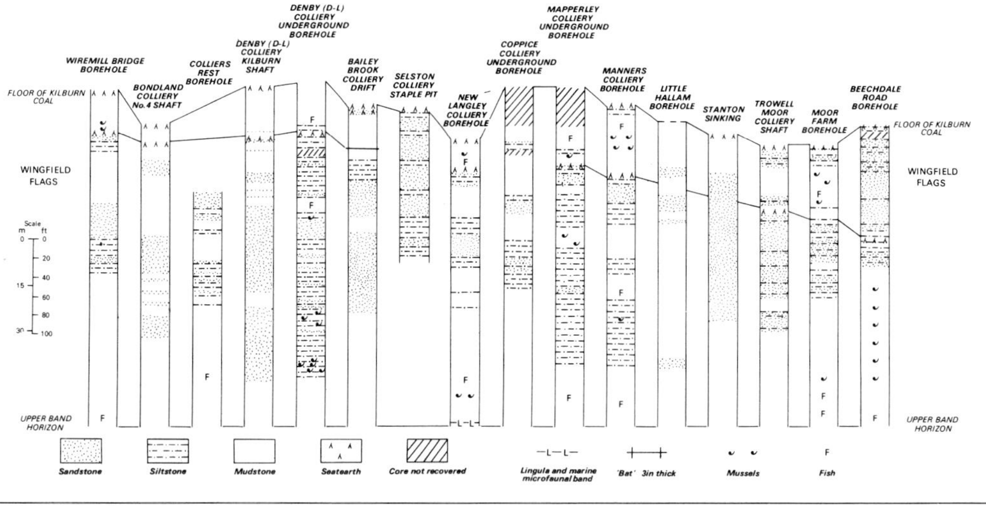

Figure 68 Sections of the measures between the Upper Band horizon and Kilburn Coal showing the development of the Wingfield Flags

THE MEASURES BETWEEN THE NORTON COAL AND THE UPPER BAND horizon form an argillaceous sequence ranging from 25 ft to 53 ft in thickness in Beechdale Road and Wiremill Bridge boreholes respectively. The sequence is composed of up to two cycles although the evidence for this division has been found only in Denby (D-L) Colliery Underground Borehole, Beechdale Road Borehole and Bailey Brook Colliery Drift (Fig. 16).

The Norton Marine Band, at or very close to the base of the sequence, consists of a thin bed containing Ammodiscus sp., Glomospira sp., Lingula mytilloides, and a conodont assemblage including Hindeodella sp. and platformed conodonts lying within pyritous shale in Denby (D-L) Colliery and Mapperley Colliery Underground boreholes. In Coppice Colliery Underground Borehole the band, 1 ft 11 in thick, includes terebelloid worm tubes, Dunbarella aff. papyracea, Caneyella aff. multirugata, the conodont Ozarkodina sp., and Elonichthys sp.

The overlying mudstone and silty mudstones contain fish debris, plants and the Norton Mussel-band (Eden 1954, p. 97; Eagar 1956, 1962). The mussels are commonly preserved in calcitic material over a thickness of 5 to 20 ft, and comprise Carbonicola crispa, C. proxima, Curvirimula belgica and Naiadites sp., in association with Geisina arcuata, Rhabdoderma sp., Rhadinichthys sp., and Rhizodopsis sp. The mussels were exceptionally well preserved in Mapperley Colliery Underground Borehole and the holotypes of both C. crispa Eagar and C. proxima Eagar came from this borehole. The mussel-band is absent in Moor Farm and Beechdale Road boreholes. A seatearth, its top 28 ft 3 in above the Norton Coal, divides these measures in Denby (D-L) Colliery Underground Borehole. The seatearth lay near the base of the mussel band and contained Carbonicola cf. extenuata, Curvirimula sp., Geisina sp., and Rhadinichthys sp. (Eden 1954, p. 97). The equivalent seatearth in Beechdale Road Borehole is separated from the seatearth at the summit of the upper cycle by only 2 ft 6 in of dark grey silty mudstone with green mottling.

Seatearth up to 12 ft thick marks the top of these measures, the Upper Band Coal being absent from every proving although coaly traces seen at the surface [3641 4745] east of Belper may have come from this horizon.

THE MEASURES BETWEEN THE UPPER BAND HORIZON AND KILBURN COAL range in thickness from 306 ft in Trowell Moor Colliery Shaft to 395 ft in Denby (D-L) Colliery Kilburn Shaft, and consist of two cycles. The lower thicker cycle, much of it sandstone, extends up to a seatearth lying some 20 to 60 ft below the Kilburn Coal. The upper cycle, largely argillaceous, extends from the top of this seatearth to the top of the Kilburn Coal.

Arenaceous measures of both cycles are assigned to the Wingfield Flags, although their main sequence lies in the upper part of the lower cycle.

The Burton Joyce Marine Band (Smith and others 1967, p. 116), which is present about 25 ft above the Upper Band seatearth in the Chesterfield and Nottingham districts, has not been proved within this district.

The Upper Band Marine Band is known only from New Langley Colliery Borehole, the type locality (Godwin 1960, p. 33; Calver 1968b, p. 34). It consists of dark mudstones with Ammodiscus sp., Lingula mytilloides, and abundant fish remains including Elonichthys sp. The principal horizon with Lingula was 4 ft 6 in above the Upper Band seatearth. Fucoids only were recorded from this horizon in Wiremill Bridge Borehole.

Up to 195 ft of mudstones separate the Upper Band seatearth and the Wingfield Flags. The mudstones are dark and pyritous in the lowest 30 to 40 ft, suggestive of near-marine conditions; fish debris is plentiful and includes Elonichthys sp., Megalichthys sp., Rhabdoderma sp., and Rhadinichthys sp., while Carbonicola sp., and Naiadites sp. are also present. The upper part of the mudstone (with Carbonicola cf. subconstricta in Denby (D-L) Colliery Underground Borehole) is interbedded with silty mudstones and passes gradually upwards into the Wingfield Flags. Ironstones are present which, together with those in the Norton-Upper Band cycle, form the Civilly Rake of Smyth (1856, p. 39). Fossils in the upper cycle include Carbonicola sp., Geisina arcuata and fish remains including Rhadinichthys sp., and an acanthodian spine.

The Wingfield Flags are a sequence of pale grey flaggy sandstones, siltstones and silty mudstones in variable proportions up to about 200 ft in thickness. This account is largely concerned with their outcrop while Fig. 68 depicts them in boreholes and shafts.

The Wingfield Flags form the first marked feature in the Westphalian rocks above the Crawshaw Sandstone. On the east bank of the River Amber, near Amberley Farm, a trench temporarily exposed 40 ft of brown fine-grained sandstone dipping to the south-east at 10 to 20°. Up to 20 ft of finely laminated sandstone are sporadically exposed near Lodge Hill Farm [3717 5274], where the Wingfield Flags feature is split into a main scarp below and a smaller scarp above.

In the railway cutting at Buckland Hollow Branch Junction [370 520] over 70 ft of brown fine-grained sandstone are excellently exposed. The overall massive bedding is split up into 2-foot 'posts' containing a finer lamination which weathers into typical 'flags'. The base of the main body of sandstone is irregular, and sandstone balls completely enclosed within the underlying silty shales suggest slumping.

The sandstone is commonly cross-bedded and an exposure in Graves Wood [3684 5151] suggests a current direction from the south-east.

North of Heage, 10-ft lenses of sandstone are common at the top of the Wingfield Flags and increase in importance northwards. South of Heage shale separates the Flags into two halves, but there is a single marked escarpment as far south as Morrel Wood Farm [3722 4830]. There, shale 'slacks' separate the upper leaf of sandstone from lower sandstones which increase in thickness southwards. Some 5 ft of laminated fine sandstone is exposed in the roadside at Openwoodgate [3682 4739].

In the Stanton district temporary excavations for the 'Ironworks' [4675 3909] exposed 11 ft of brown fine-grained sandstone, flaggy at the top, and showing sporadic purple staining.

The top of the lower cycle is marked by a thin seatearth which, although not proved at outcrop, is present in most boreholes. A 3-in 'batt' in Bailey Brook Colliery Drift and a 3-in coal in Annesley Colliery (Deep Hard) Underground Borehole are the only records of coal at this horizon.

The upper cycle, normally 45 to 80 ft thick, includes part of the Wingfield Flags, but consists largely of dark grey to black mudstones with ironstones; fish debris and ostracods have been recorded and mussels are common. The presence of the fauna serves to identify the cycle in sections such as Moor Farm Borehole where the underlying seatearth is missing. Sandstones tend to be thin except in Selston Colliery Staple Pit and Underground Borehole, where they lie close below the Kilburn Coal and apparently continue downwards into the main body of Wingfield Flags. In Beechdale Road Borehole the upper cycle is abnormally thick, comprising 125 ft of largely arenaceous strata.

KILBURN COAL TO THREEQUARTERS COAL

The measures between the Kilburn and Threequarters coals (Figs. 17 and 18) are thickest in the north, where they reach a maximum of 661 ft in Pinxton Colliery No. 2 Shaft. They exceed 600 ft near Stanley and Mapperley collieries in the south-west of the district and in an isolated area near Cossall Colliery. Eastwards and south-eastwards they thin to about 550 ft at Trowell Moor Colliery, 489 ft in Beechdale Road Borehole and 486 ft in Annesley (Deep Hard) Underground Borehole. The sequence includes the Kilburn Coal, which was extensively worked for house coal, and other important seams are the "Ashgate", Blackshale, Yard and Threequarters. With the exception of the Yard and Blackshale, the coals tend to thin in an easterly direction.

The fauna of the measures consists largely of mussels of the Carbonicola pseudorobusta group, with fish debris present in the roofs of the coals up to and including the Mickley Thin; marine fossils are unknown in this district.

THE KILBURN COAL ranges from 30 to over 70 in in thickness in the south-western part of the exposed field, but in the north-east it is thin (Fig. 19). The maximum proved thicknesses are at outcrop in the south-west, where 76½ in of coal (including 5½ in of dirt) and 74 in of coal were recorded in Hollies Farm and Mary sites respectively. The seam is composed predominantly of bright coal with subordinate dull bands, and is of good to medium quality with a low to moderate ash content. Much in demand for house coal, it was originally one of the more important worked seams of the coalfield and one of the first to be exhausted by underground workings. The top and bottom parts of the seam are frequently recorded as inferior coal (Fig. 19). In the north-west a floor coal included in the seam section has a distribution approximately complimentary to that of the uppermost cycle within the Wingfield Flags. In Moor Farm Borehole the seam was split.

THE MEASURES BETWEEN THE KILBURN AND MICKLEY THIN COALS range in thickness from 198 ft in Moorgreen Loco-Road Underground Borehole to 288 ft in Swanwick No. 1 Underground Borehole (Fig. 20) and are composed of five main cycles. To the south-east of an irregular line from Swanwick to Stapleford the lowest cycle can be identified only by the presence of arenaceous measures at its top. Each cycle locally includes coal, but the seams are thin or of poor quality and have little economic value. The cyclic sequence of these measures is relatively uniform throughout the district except in the two Ironville oil bores and Moorgreen Loco-Road Underground Borehole. The latter encountered additional seatearths, interpreted as splits of the two cycles above the Morley Muck coal.

Gibson and others (1908, p. 101) have described the fish remains in dark grey mudstones within the roof measures of the Kilburn Coal at Denby (D-L) Colliery, and similar remains have since been proved locally throughout the district with scales of Strepsodus sp. present in Eastwood Hall Borehole. The fish remains are sometimes associated with cf. Planolites, but Lingula, known to the north (Smith and others 1967, p. 121), has not been found.

Sandstones are present, in variable proportions and at various levels. They form the 'Kilburn Rock' of Gibson and others (1908, p. 70) in all recordings of the measures below the Morley Muck except New Langley Colliery Borehole, Little Hallam Water Borehole, Lodge Colliery Shaft, and Beechdale Road Borehole. The sandstones are generally thicker in the south-west of the coalfield, where the thin coal and seatearth of the lowest cycle is absent; even here, however, the sequence up to the Morley Muck Coal commonly displays sandstones at two distinct levels. These latter sandstones form a minor escarpment feature in the west, at the lower limits of the dip-slope of the Wingfield Flags, but they are less important in the south of the coalfield. Sections of the overburden of the Kilburn Coal near Rowson Green and Stanley are described in Appendix II.

The lowest cycle, found only in the north-east of the district, ranges from 25 to 80 ft in thickness, its top commonly formed by a seatearth; coal is sporadically recorded as follows: Pinxton Colliery No. 2 Shaft, 12 in; Swanwick Colliery No. 1 Underground Borehole, 11 in; Bailey Brook Colliery Shaft, 3 in batt; Lodge Colliery Shaft, 5 in; Cossall Colliery No. 2 Shaft, trace; Trowell Moor Colliery Shafts, 27 to 30 in; Moor Farm Borehole, 1 in; Beechdale Road Borehole, 7 in; Wollaton Colliery No. 3 Shaft, 2 in batt. In Annesley Colliery (Deep Hard) Underground Borehole, where the coal and seatearth are missing, the presence of two cycles up to the Morley Muck Coal is revealed only by the sandstones. The upper cycle at Eastwood Hall Borehole is predominately arenaceous but Spirorbis sp. Carbonicola cf. browni and C. cf. pseudorobusta, Curvirimula subovata, Carbonita humilis, C. pungens, and fish remains were present in interbedded mudstone and siltstone.

The Morley Muck Coal may contain over 50 in of coal in up to three leaves of poor to very inferior quality; the seam is therefore of no economic value. The coal or its seatearth lies between 50 and 130 ft above the Kilburn Coal. In the north and north-east of the district it is up to 11 in thick, including 2½ and 4 in of cannel in its roof in Cotespark No. 3 Underground Borehole and Birchwood Colliery Upcast Shaft respectively. Near Denby the 59 in of 'Batty Coal' recorded from the Denby Hall Colliery Shaft is the thickest single record of the district; a more typical sequence in Denby Colliery, New Winnings Shaft, comprises coal 3 in, dirt 3 in, coal 9 in on 'Batty Coal and Clunch' 19 in. At Rowson-Ireton Site the 19 in of coal included 4½ in of dirt. Two leaves of the seam are exposed in the clay pit at Rowson Green [3751 4688], where the roof measures contain fine specimens of Carbonicola pseudorobusta. The section is:

	ft	in
Clay	6	0
Mudstone, pale grey and dark grey; mussels at top	15 to 18	0
Mudstone, dark grey, with brown ferruginous bands and large mussels	2	0
Coal		10
Clay, ochreous and grey	1	0
Clay, grey and dark grey		7
'Batt'		4
Fireclay, pale grey		

The full three-leaf sequence is present between Mapperley and Trowell Moor collieries, but near Heanor this sequence is reduced.

TABLE 10

Provings in the Morley Muck Coal

	Top Coal	Dirt or Measures	Coal	Dirt or Measures	base Coal
Mapperley Colliery No. 1 Shaft	40	12	9	9	4
West Hallam Colliery Borehole	6½	1¾	10	6	8
West Hallam Colliery Shaft	16	15	7	7	4
Stanley Colliery Shaft	22 (Top 12 dirty)	9	10	2	9
Woodside Colliery Shaft	9	3 (batt)	12	12	6
*Manners Colliery Shaft	15	8½	9	2½	3
Manners Colliery Borehole	3	10 ft 10	24	-	-
Cossal Colliery	6	about 9	6	about 10	1
Oakwell Colliery	9	18 (cannel)	-	31	5 (batt)
Trowell Moor Colliery	12	12 ft	12	16 ft	12
Moor Farm Borehole	10	7	2	-	-
Coppice Colliery No. 2 Shaft	16	18	6	-	-
New Langley Colliery Borehole	8	32	2	-	-

*From an outline record of No. 2 Shaft. I. G. S. record is 38 in of cannel.

The cycle overlying the Morley Muck Coal varies from 36 ft in Stoneyford Borehole to 56 ft in Denby Colliery, New Winnings Shaft. The coal, cannel or seatearth at the top can be recognised in many provings throughout the district; the coal is unnamed and discounting a dubious record of 24 in in the chipped Ironville No. 1 Oil Bore, it has a maximum thickness of 10 in in Stanley Colliery Shaft. It is absent over wide areas in the east and north. The cycle includes up to about 50 ft of arenaceous measures, with Spirorbis sp., Carbonicola cf. polmontensis, C. pseudorobusta, Curvirimula sp., Geisina arcuata and fish remains including Rhabdoderma sp. in the underlying argillaceous beds above the Morley Muck Coal.

The succeeding cycle, ranging in thickness from 25 ft in New Langley Borehole to 50 ft in Little Hallam Water Bore, is also widely recognised and has the Lower Brampton Coal at its top. Sandstones are less prominent than in the underlying cycle and the mussels less abundant.

The Lower Brampton Coal (Smith and others 1967, p. 121) has a maximum thickness of 21 in including 12 in of cannel in Denby (D-L) Colliery Shaft and, on the evidence of an analysis taken in the nearby Morrells Wood Site, is of very poor quality. Provings between Mapperley and West Hallam collieries show this seam to be split into two very thin leaves, each under 9 in thick, up to 63 in apart. The upper leaf is locally canneloid. In the north of the district the Lower Brampton is under 4 in thick; it is absent in the east and 9 to 12 in thick in the south-east.

The remaining cycle, that with the Mickley Thin Coal at the top, ranges in thickness from 46 ft in Eastwood Hall Borehole to 67 ft in Denby (D-L) Colliery Shaft and is almost wholly composed of argillaceous strata, locally bearing a mussel fauna; at Eastwood Hall Borehole Carbonicola sp. and Curvirimula subovata were recorded. Sandstones are only a few feet thick in the floor measures of the coal.

THE MICKLEY THIN (OR UPPER BRAMPTON) COAL is thickest at Morrells Wood Site, where 20 in were recorded. As at Chalfont Site (19 in) and Beechdale Road Borehole (3 in) the coal is of good to moderate quality, but it deteriorates to poor in two analyses from Moses Lane Site (9 to 12 in). The thickness is 18 in or less in the remaining provings within the district, the thinnest recordings being in the east. The coal is exposed in the clay pit near Rowson Green [3783 4725] (see Appendix 2). Washouts, which also affect the Blackshale group of coals, were encountered in Eastwood Colliery and Lodge Colliery shafts. This coal is the same as the ? Mickley Thin Coal of the Chesterfield area which Smith and others (1967, p. 122) found difficult to correlate with the Mickley Thin of the Sheffield area.

THE MEASURES BETWEEN THE MICKLEY THIN AND 'ASHGATE' COALS form one cycle ranging from 41 ft in Cotespark No. 2 Underground Borehole to 68 ft in Pinxton Colliery No. 2 Shaft. The roof measures of the Mickley Thin contain fish, ostracods and sporadic mussels. A section near Rowson Green is described in Appendix 2.

THE 'ASHGATE' COAL (Fig. 69) consists of two leaves of coal throughout most of this district. The upper and lower leaves were originally called the Ashgate and Mickley coals respectively and they have been extensively worked underground near Denby, Heanor and Trowell Moor under these names. The seam is not the correlative of the Ashgate Coal of the Chesterfield district, which lies slightly higher in the sequence; instead the two leaves of the "Ashgate" Coal of the present district are equivalent to the two thin coals present above the ? Mickley Thin Coal in Cotespark No. 1 Underground Borehole (Smith and others 1967, fig. 111, p. 119).

At Starvehimvalley Site and in the area between Denby and Heanor the two leaves lie close together, while in Awsworth Colliery Shaft they are combined with a total thickness of 81 in, the thickest known section in the district. The two leaves divide rapidly from the Heanor area to the north, west and south, with a maximum known separation of 62 ft 2 in in Mapperley Colliery shafts. The leaves are close together again at outcrop between Horsley Woodhouse and Morley Hayes Site and near Trowell Moor Colliery and were exposed, some 3 ft apart, in motorway cuttings near Trowell, a separation which is maintained until their outcrop is concealed beneath Permo-Triassic rocks near Stapleford.

A sinuous belt of sandstone which affects the 'Ashgate' Coal extends from north-east of Eastwood to Ilkeston and thence south-east towards Wollaton. It probably represents a major watercourse along which coarse material was transported into the district and which widened and became increasingly active during deposition of the upper leaf of the seam. As a result the lower leaf of the coal is apparently cut out over a restricted area of this watercourse, and in Cossall Colliery Shaft, Moor Farm Borehole and Wollaton Colliery Shaft this erosion removed measures almost down to the horizon of the Mickley Thin Coal and has even removed this coal in Eastwood No. 3 and Lodge Colliery shafts. Deposition along the watercourse affected the formation of the upper leaf of the 'Ashgate' over a wider area than the lower leaf. Where the two leaves are separated the intervening measures are sandstone except in the area between Heage and Horsley Woodhouse. The coal of the upper leaf as it 'rides' the sandstone tends to die away to cannel or is represented only by its seatearth. Sandstone is also widely distributed in the roof measures of the coal and its deposition continued so as to affect the formation of the overlying Blackshale group of coals.

The provings of sandstone in the watercourse and the first underlying recognisable horizon at each are as follows: Eastwood Hall Borehole, lower leaf of 'Ashgate' Coal present; Moorgreen Loco Road Underground Borehole, seatearth of lower leaf preserved; Eastwood Colliery No. 3 Shaft and Lodge Colliery Shaft, extends to below Mickley Thin horizon; West Hallam Borehole and Colliery Shaft, lower leaf of 'Ashgate' Coal present (as cannel); Kimberley Colliery Shaft, lower leaf of 'Ashgate' Coal present; Cossall Colliery Shaft, Moor Farm Borehole and Wollaton Colliery Shaft, roof measures of Mickley Thin Coal.

The lower leaf of the seam (discounting Awsworth Colliery Shaft, where the distinction between the two leaves fails) is thickest at Avenue Site [401 439], where it includes 41 in of coal. It shows rapid variations in thickness; for instance it ranges from 24 to 36 in at Morley Hayes Site [402 418]. It is 20 to 30 in thick between Heanor and the outcrop in the west, but thins in the northern part of the district, where some provings (Cotespark No. 2 Underground Borehole and Birchwood Upcast Shaft) encountered cannel. Cannel was also present in Stanley Colliery Shaft (18 in), West Hallam Borehole (14 in) and West Hallam Colliery Shaft (17 in). To the east of the line of the watercourse the lower leaf was 19 in of coal in Kimberley Colliery Shaft, 4 in at Cinderhill Colliery No. 4 Shaft (where the correlation is doubtful) and at Cossall 40's Underground Rebore (including 2 in of cannel). In Beechdale Road Borehole it is represented by seatearth only and in Annesley (Deep Hard) Underground Borehole by 2 in of coal.

The thickness of the upper leaf is related to the presence of sandstone below and in the roof. Like the lower leaf it shows rapid local variation; at Morrells Wood Site it varies from 14 to 54 in, the latter being the thickest record in the district. It is 20 to 24 in thick near Heanor but thins north-east of that town to 4 in in New Langley Colliery Borehole and 10 in in Ormonde 70's Underground Borehole. In Annesley (Deep Hard) Underground Borehole it is 2 in thick. In the north of the district the upper leaf 'rides' a relatively constant thickness of sandstone and is consistently 9 to 12 in thick except at Cotespark No. 2 Underground Borehole, where it is represented only by seatearth.

Between Horsley Woodhouse and Heage the thickness of the upper leaf varies inversely with the thickness of the measures in the seam split, which in this area contain little sandstone. The great coal thickness at Morrells Wood Site is the notable exception to the above. On part of Bottle Site, between Kilburn and Horsley Woodhouse, the upper leaf is cut out by an overlying sandstone which is widely distributed in the vicinity. South of Horsley Woodhouse the upper leaf varies between 33 and 36 in Gypsy, Avenue and Rose and Crown sites. Farther south, on the evidence of Morley Hayes and Moses Lane sites, the split

a

b

of the two leaves is again abrupt and the upper leaf has only been proved and worked adjacent to the line of the split. This leaf again 'rides' arenaceous measures and dies away to leave only a seatearth horizon as shown by Stanley Colliery Shaft. It is also probably subject to washouts, which may only be local, at the outcrop from Moses Lane Site to near Trowell Moor Colliery, but it is not known whether the leaf has been eroded as in West Hallam Colliery Shaft and Borehole, or whether it was an area of non-deposition as in Stanley Colliery Shaft. Near Trowell Moor Colliery its thickness in the shafts and in opencast sites varies from 24 to 44 in. It was exposed in an M1 cutting [4857 3912] 800 yd S of Trowell Church, (see Appendix 2) where the washout sandstone cuts down to within 5 ft of the top of the coal.

To the east of the sinuous washout deposits (Fig. 69) the leaf is represented only by seatearths in Cossall 40's Underground Rebore and Beechdale Road Borehole, by 2 in of coal in Cinderhill Colliery No. 4 Shaft, where the correlation is doubtful, and by 5 in and 11 in in Hucknall C's Heading and Main Outbye underground boreholes respectively.

THE MEASURES BETWEEN THE 'ASHGATE' AND BLACKSHALE COALS are largely arenaceous and range from 60 ft in Pinxton No. 2 Shaft to 146 ft in Stanley Colliery Shaft. They contain up to two main coal horizons, which are probably the representatives of the Ashgate Coal of Smith and others (1967, p. 124). Each coal is itself split into up to three leaves, all thin and of no economic importance; the maximum thickness recorded is 28 in of 'batty coal' in Ormonde Colliery Shaft.

THE BLACKSHALE AND YARD COALS (Figs. 21 and 22) are a complex sequence of up to six main leaves of coal lying within strata varying in thickness from 8 ft in Annesley 204's Underground Borehole to 110 ft in Coppice Colliery No. 2 Shaft. The Blackshale Coal consists of the lowest three main leaves of the sequence plus, in this account, the two overlying leaves here called the upper and lower Denby leaves after their striking development in Denby Hall Colliery Shaft (Fig. 22). Where combined or not separately identified they are known as the Denby leaf. The three lowest leaves are the Top Softs, Middle Dirt with Tinkers Coal and Bottom Softs of the old miners' terminology, which together form the Blackshale Coal of the Chesterfield district (Smith and others 1967, p. 125). The Low '*Estheria*' Band, which to the north-east occurs above the Bottom Softs (or Low Silkstone) is absent in the Derby district. The sequence of coal leaves appears to be controlled by deposition of sandy and silty measures below, between and above the Denby leaves. As a result the Denby leaves may be associated with the Top Softs or the Yard or may lie in the measures between and widely separated from these seams.

The correlation of the Denby leaves is somewhat arbitrary and there is also the possibility of additional splitting of the base of the Yard Coal, although this appears to be restricted to the area of the underground boreholes at Hucknall and Cinderhill collieries. In this account, however, splitting of the Yard Coal refers to minor splitting of the uppermost of the six leaves of coal of the standard sequence.

The Top Softs - Middle Dirt - Bottom Softs sequence has been extensively mined and the Yard Coal has been worked in the north, where it is thickest. The quality of the coal in all the leaves is extremely variable, but it is rarely better than medium quality, and then only in the Top Softs.

To simplify description of this coal complex, six areas, each containing a different development of the seams, have been selected and are shown in Fig. 21.

Area A is known only from underground provings; all the coal leaves are present in close proximity to form the Yard/Blackshale, a seam which is equivalent or approximates to the Silkstone of Yorkshire. In the table below, Annesley 204's Underground Borehole is an example of a virtually dirt-free sequence while Brookhill 5's section in underground workings shows a more expanded sequence typical of the area margin.

In area B the Yard Coal is thick and separated from the Blackshale Coal, which includes a variable sequence of the Denby leaves and ranges from a total of 65 in including 13 in of dirt (Selston Colliery underground workings) to 155 in including 67 in dirt (Denby Hall Colliery underground workings). The Denby leaves were not recorded in Selston B 80's Underground Borehole and Pye Hill - Selston Colliery Drift, where they may be washed out; elsewhere they lie within 23 in of the Top Softs, but are variable. They are thickest in the west where both leaves are usually present; in the east there is a single thin coal, which is either the lower leaf or represents both leaves. The Denby leaves attain a maximum in a single 54-in coal at Salterwood Site. Measures some 10 ft thick at Selston Colliery No. 1 Drift to 76 ft at Selston Colliery, Brinsley Drift, divide the Denby leaves from the Yard Coal, which itself ranges up to 59 in in thickness at Pye Hill Colliery Shafts.

In areas C and D, the Yard Coal is widely separated from insignificant Denby leaves which in turn are remote from the Top Softs - Middle Dirt - Bottom Softs sequence. In area C the Top Softs - Middle Dirt - Bottom Softs sequence varies between 33 and 54 in of which up to 10 in may be dirt partings; in area D it is generally 40 to 60 in thick with up to 13 in of dirt, but thick sections of 115 in (including 11 dirt) and 99 in (23 dirt) were recorded at Bacon Lane [3835 5293] and Devonshire sites [3785 5182] respectively. These appear to be due to local thickening of the Bottom Softs at Bacon Lane Site and of the Top Softs at Devonshire Site.

In area C the Top Softs are separated from the Denby leaves by mainly arenaceous strata ranging in thickness from 13 ft at Moorgreen Colliery Underground Borehole to 80 ft at Cinderhill 14's Main Gate Underground Borehole. The Denby leaves are represented by seatearths up to 17 ft apart as at Moorgreen Colliery Underground Borehole, but there is only one seatearth in the Cinderhill underground borehole

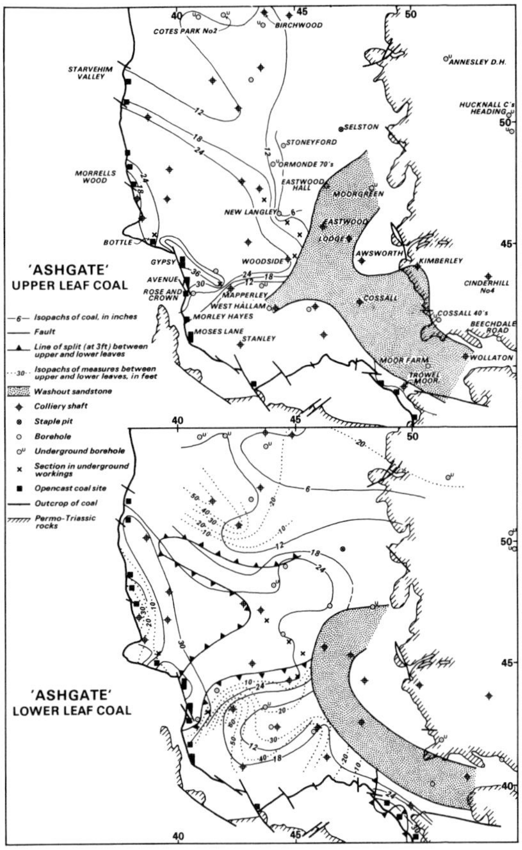

Figure 69 Provings of the 'Ashgate' Coal, showing isopachs of the upper and lower leaves and of measures dividing the leaves. Washout sandstones at the horizons of the two leaves are also shown

c

d

TABLE 11

Comparative provings in the Yard/Blackshale Coal

	Annesley 204's U.B.H.		Brookhill 5's [4803 5300]	
		in.		in.
Yard	Dirty coal	$9\frac{1}{2}$	Coal	28
	Coal	16		
	Dirty coal parting		Dirt	10
Denby leaf	Coal	$13\frac{1}{2}$	Coal, dirt and dirty coal	$24\frac{1}{2}$
			Dirt	14
			Dirty coal	$1\frac{1}{4}$
	Dirt	1	Dirt	$19\frac{3}{4}$
Top Softs	Coal	27	Dirty coal	$5\frac{1}{2}$
	Dirty ooal	4	Coal	$22\frac{1}{2}$
	Dirt	1	Dirt	$3\frac{3}{4}$
Tinkers	Coal	$11\frac{1}{2}$	Coal	9
	Dirt	$\frac{1}{2}$	Dirt	$1\frac{3}{4}$
Bottom Softs	Coal	10	Dirty coal	$\frac{3}{4}$
	Dirty coal	2	Coal	$15\frac{3}{4}$
			Dirt	$3\frac{1}{4}$
			Dirty coal	$1\frac{1}{2}$

mentioned above. The Yard Coal lying 7 to 13 ft above the Denby leaves is thin with the possible exception of the 45 in of 'Batty Shale' recorded in Moorgreen Main Dips Underground Borehole.

In area D a single Denby leaf was encountered only in Swanwick Colliery New Pit, where 14 in of coal and dirt were encountered 44 ft above the Top Softs. Elsewhere the leaves appear to be absent or unrecorded, but there is a seatearth at this horizon, 15 ft above the Top Softs, in Pinxton Colliery No. 3 Shaft.

The Denby leaves horizon is separated from the Yard Coal by sandy measures ranging in thickness from 7 ft at Swanwick New Pit to 85 ft at Pinxton No. 3 Shaft. The Yard Coal ranges from 17 in in Pinxton Colliery No. 3 Shaft to 38 in in Swanwick Colliery New Pit and has been extensively worked from Cotespark and Pinxton collieries, where analyses show it to be of very variable quality.

The splitting from Area A to Area D occurs between the Top Softs and Denby leaf, as demonstrated by the section in Bentinck Colliery Drift, just north of the district boundary, where the Denby leaf lies 32 ft above the Top Softs but is separated from the Yard Coal by only 2 in of dirt. This is followed by further splitting between the Denby leaves and the Yard Coal.

The centre line of area E coincides with the greatest thickness of sandy measures which lie between an impoverished Yard Coal and the Top Softs - Middle Dirt - Bottom Softs sequence throughout the area. The Denby leaves are not proved; they may be merged with the Yard Coal or washed out by the sandy measures.

These sandy measures, which probably follow a former watercourse, have cut out all the Top Softs - Middle Dirt - Bottom Softs sequence in Manchester Wood Borehole and parts of that sequence in Woodside No. 1 Underground Borehole, Oakwell Colliery Shaft, and Cinderhill 27's Main Inbye Underground Borehole and perhaps also at Denby Colliery, New Winnings Shaft and Cossall Colliery Shaft. Where wholly represented in the area the Top Softs - Middle Dirt - Bottom Softs sequence varies from 39 in, including $7\frac{1}{2}$ in dirt, in Cinderhill 6's Main Gate Underground Borehole to $78\frac{1}{2}$ in, including 26 in of dirt, in Coppice Colliery workings. The sequence was exposed in the motorway cutting near Trowell (Appendix 2). No certain Denby leaves are recorded from within this area except at Pippinhill Site, where some provings show the lower Denby leaf within 18 in of the Top Softs.

The Yard Coal is represented mainly by a seatearth horizon, but up to 12 in of coal have been encountered in West Hallam Colliery Shaft. The Yard horizon was not recorded in Bennerley, Awsworth and Kimberley colliery shafts.

Area F lies south of the thick belt of sandy measures and consequently a variable sequence of the Yard Coal and Denby leaves reappears, mainly in association. Moderately thick measures overlie a variable Top Softs - Middle Dirt - Bottom Softs sequence - a feature which, apart from its location on the opposite side of the thick

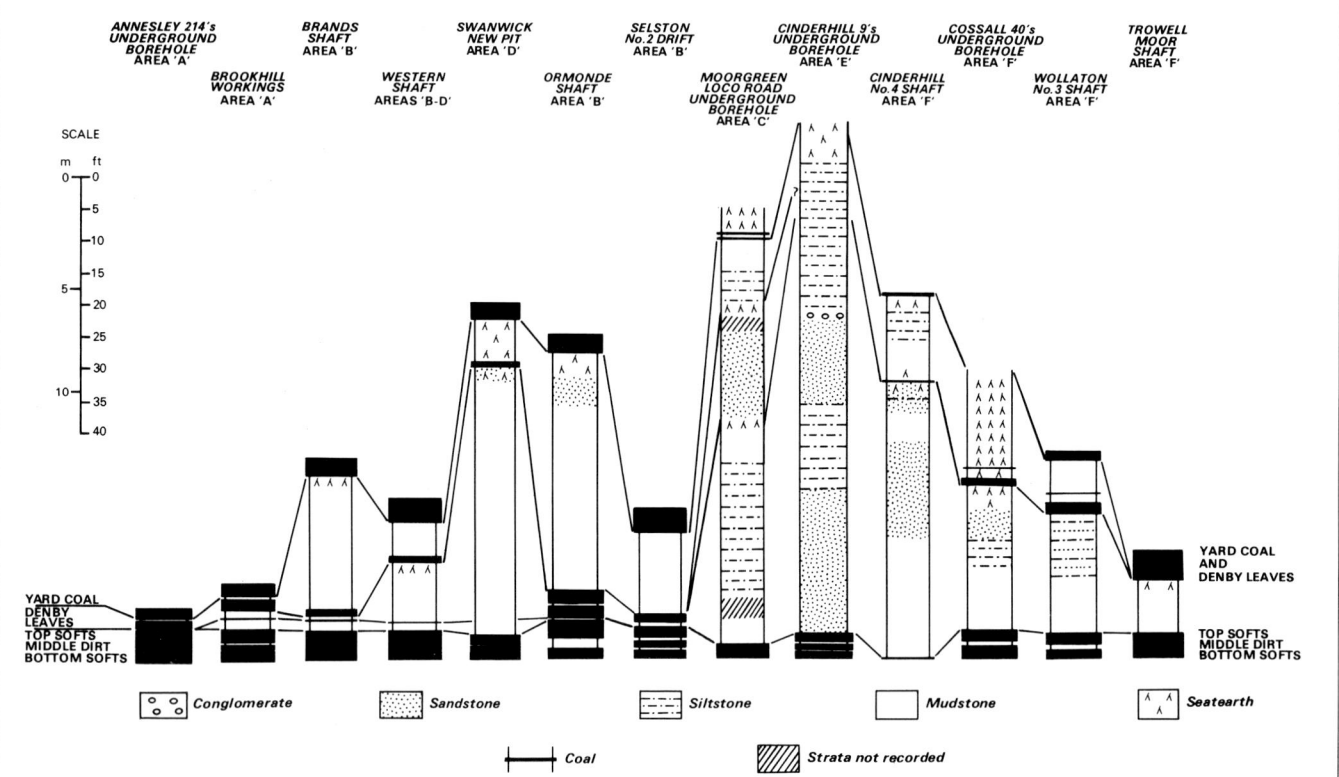

Figure 70 Comparative sections of the Yard and Black-shale coals and the intervening measures. For locations see Figure 21

a

b

arenaceous measures, distinguishes this area from area C.

The Top Softs - Middle Dirt - Bottom Softs sequence, except in that part of the area near Cinderhill and Hucknall colliery provings, is characterized by an increase in the thickness of dirt partings compared with the rest of the district. The Top Softs of this area do not exceed 33 in and the Bottom Softs 20 in in thickness. The thickest overall sequence of 101 in in Stanley Colliery Shaft includes 51 in of dirt and dirty coal while the thinnest, of 48 in in Cinderhill 19's Left Gate Underground Borehole, includes only 7 in of dirt. There is no coal (Tinkers) between the Bottom Softs and Top Softs in Smalley Green Borehole, Cat and Fiddle Site, Bunkerhill Site, and Wollaton Colliery No. 3 Shaft. The thickest dirt, 6 ft 9 in, between the Bottom Softs and Top Softs is found in Smalley Green Borehole.

At Cinderhill Colliery No. 4 Shaft Underground Borehole the Top Softs - Middle Dirt - Bottom Softs sequence is probably washed out and represented by seatearth only. In a small part of the area, near Shortwood and Moor Farm boreholes, this sequence contains up to four thin leaves of coal, none of which exceeds 18 in in thickness.

The Denby leaves are 21 ft and 78 ft above the Top Softs in Cossall 40's Underground Borehole (Rebore) and Smalley Green Borehole respectively. They are represented only by a single seatearth or thin coal over much of the area. Exceptions are in Manners Colliery Shaft and Borehole, where the lower of two leaves is 25 and 22 in thick respectively, and between Trowell Site and Wollaton Colliery, where each leaf may exceed 20 in of dirty coal. At Cinderhill 1's Underground Borehole the Denby leaf is only 3 ft 3 in above the Top Softs, an anomalous section for the area.

Measures ranging in thickness from 8 in at Shortwood Borehole to 30 ft in Manners Colliery Shaft separate the Denby leaves from the Yard Coal. The Yard Coal is mainly a single seat-earth horizon or very thin coal; however up to three thin leaves have been proved in Moor Farm Borehole.

THE MEASURES BETWEEN THE YARD AND THREEQUARTERS COAL range in thickness from 47 ft in Cinderhill 10's Main Gate Underground Borehole to 123 ft in Mapperley Colliery Shafts and normally form a single cycle containing variable arenaceous and argillaceous strata. Exceptionally, at Stoneyford Borehole, a weak rootlet bed 11 ft above the Yard Coal indicates the presence of a second cycle. This seatearth is perhaps the correlative of one recorded by Edwards (1967, p. 80) in the Ollerton district. The roof of the Yard Coal yields fish and plant debris and more rarely mussels; ironstones higher in the argillaceous measures form, in ascending order, the Striped Rake of John Hallam, Blackshale Rake and Nodule Rake of the Morley Park area (Smyth 1856, p. 38). The overlying sandstone is the 'Silkstone Rock' of the country to the north of this district, but it forms prominent ground features only near Smalley;

elsewhere it is subordinate to those of the unnamed rocks which lie between the 'Ashgate' and Blackshale coals and between the Blackshale and Yard coals.

THE THREEQUARTERS (TUPTON THREE-QUARTERS OR DOGTOOTH) COAL ranges in thickness from 3 to 4 in in Trowell Moor Colliery Shaft to 49 in at Salterwood Site. These figures exclude the thickness of a floor coal (see below). However, the section at Salterwood Site includes an 8-in dirt parting which itself is the thickest recorded within the main sequence of the seam. This parting is ubiquitous at outcrop between High Bank and Mill Farm sites and is present in many deep mine provings in the central part of the district. The coal above the parting is generally of good quality with a low ash content and is composed largely of bright coal with sporadic bands of hards. Below the parting, the quality of the coal is poor and variable; some analyses show dirty coal. This is epitomised by the section at Salterwood Site where the whole 19-in leaf contained over 20 per cent ash. In provings where the dirt parting is absent there is, in general, an upwards improvement in the quality of the coal, though the overall quality of the seam is medium to poor.

The Threequarters Coal has been worked underground in small areas near Ripley Colliery and near the northern margin of the district, its main attraction however lies in its close proximity to the Low Main/Tupton Coal and it has been extensively won with that coal by open-cast methods.

A floor coal is present some 2 to 11 ft below the main seam. The distribution of this floor coal is indicated on Fig. 23 and its maximum known thickness is 21 in, proved in Hucknall C's Heading Underground Borehole.

c

THREEQUARTERS COAL TO CLAY CROSS MARINE BAND

These measures range in thickness from 304 ft in Seagrave Borehole to about 455 ft in Swanwick Colliery New Pit (Figs. 24 and 25). They include the Low Main, First Piper, Deep Hard and Deep Soft coals, which were extensively mined as sources of household and steam coals throughout the district and have also been widely exploited by opencast methods. The Deep Soft group of coals is combined over a restricted area near Smalley to form the thickest coal sequence in the district. Other rock types in the sequence include the thin Black Rake Tuffaceous Siltstone which is an important lithological marker at the top of the Deep Soft group of coals. The two widespread and important sandstones of the measures are the 'Tupton Rock', found in the west and also within a broad channel in the eastern part of the district, and the 'Deep Hard Rock' which occupies a narrow belt extending north-eastwards from its outcrop at Salterwood.

Boreholes in the east of the district illustrate the manner in which the Cockleshell and Hospital coals thin or disappear on the crests of thick sandstones, although the splitting of the coal sequence appear to be little related to the thickness of underlying sandstone. In contrast, however, the individual thicknesses of the coals of the Deep Soft group appear to be little related to the thickness of the coal sequence was controlled by the influx of sand at various times and locations. This group therefore resembles the Blackshale and Yard coal sequence in that the area of maximum coal thicknesses is different.

The Carbonicola cristagalli mussel fauna (Smith and others 1967, p. 134-135) occurs above the Cockleshell Coal and passes upwards into the Anthracosia regularis fauna at about the horizon of the Deep Soft Coal (Smith and others 1967, p. 96 and 1973, p. 44).

THE MEASURES BETWEEN THE THREE-QUARTERS AND LOW MAIN COALS range in thickness from 6 ft in Seagrave Borehole to 23 ft in Pentrich Colliery, Hartington Shaft. They are mainly argillaceous with ironstone nodules; much of the sequence is composed of the seatearth of the Low Main Coal. The measures are well exposed in an old quarry at Little Matlock [4345 5137] near Ironville (Appendix 2).

THE LOW MAIN (LOW TUPTON OR FURNACE) COAL has a maximum thickness of 66 in at Salterwood Site and is generally of good to very good quality. From its outcrop the seam thins irregularly and gradually to less than 36 in in the south-east of the district. In the north-east (Fig. 26) the top of the seam lies about 9 in below the Cockleshell Coal and together the two form the Tupton Coal. Within this area of union the Low Main element ranges in thickness from 36 in in Annesley 204's Underground Borehole to 42 in in Hucknall C's Heading Underground Borehole. A second and poorly defined area of union with the Cockleshell Coal exists near Swanwick, Birchwood and Pinxton collieries where the Tupton Coal is 45 to 50 in thick. In the north-central part of the district the Cockleshell Coal is recorded only in Brookhill Colliery (Smith and others 1967, p. 308), where its lower leaf lies

d

50 in above the Low Main Coal, and in Agnes and Exhibition sites where it lies about 25 ft above that coal. To the west, in Brands, Western, Pentrich and Pye Hill collieries the Cockleshell horizon is inferred to lie at about the latter distance above the Low Main Coal.

The Low Main Coal is known to be split only in New Langley Colliery Shaft, where the section is coal 39 in, dirt 15 in, coal 16 in, and in Moor Farm Borehole where the respective thicknesses were 33, 9 and 16 in.

The structure of the Low Main Coal is distinctive, particularly the basal 12 in, which are composed of bright coal with hard bands or interbedded hard and bright coal. The central portion of the seam comprises soft bright coal, and the upper part is variable with hard coal, bright coal and cannel, the latter being 21 in thick in Bentinck Colliery workings. A dirt parting up to 2 in thick or a band of hard coal within this upper part of the seam can be correlated over wide areas. The ash content of the whole seam is low, the average of 138 analyses from the district being 4.4 per cent. Sulphur content is also low.

The seam is frequently washed out within the eastern area of arenaceous roof measures (Fig. 27), as exemplified by Cinderhill Colliery No. 4 Shaft and 2's Left Gate underground boreholes and Model Farm No.2 Borehole. The edge of the washout was observed in Shortwood Site (p.179c).

THE MEASURES BETWEEN THE LOW MAIN AND FIRST PIPER COALS range in thickness from 101 ft in Wollaton Colliery D.H. 21's Underground Borehole to 187 ft in Wollaton Colliery, junction of South Main and South Dips Underground Borehole. The variation is directly related to the thickness of sandstones within the four major component cycles (Fig. 29). The three lowest cycles are locally crowned by thin coals named, in ascending order, the Cockleshell, Hospital and Second Piper; the fourth is capped by the First Piper Lower Coal or First Piper Coal.

Washouts are present in these measures and, as both the Cockleshell and Hospital coals are split and the Second Piper Coal and seatearth are unrecorded in a number of the provings, the detailed correlation is not always certain. Additionally some of the coals recorded in boreholes have no underlying seatearths and are probably washout rafts. The sub-Hospital sandstone ('Tupton Rock' is thickest within the takes of Cinderhill and Wollaton collieries and crops out between Shortwood Site and Trowell Moor Colliery. Where this rock is thick the Cockleshell Coal is absent as a result of erosion and/or non-deposition. The Hospital Coal also fails or is reduced on the "crest" of this sandstone. In the west of the district, only the sub-Cockleshell sandstone is important, as displayed by Ormonde Colliery Shaft (Figs.25 and 29), where it is about 70 ft thick. This sandstone is marked by only minor ground features at outcrop except on the flanks of the Ironville Anticline. In the north-east of the district a thick sub-First Piper

a

sandstone is the dominant sandstone; its sinuous line of maximum thickness enters and leaves the district within Annesley Colliery take. Its base falls to 14 ft above the Tupton Coal in Annesley Deep Soft G4's Underground Borehole to the north of this district, and along part of the line of maximum thickness the Hospital Coal is reduced or washed out and the Second Piper Coal and seatearth are absent. Elsewhere in the district the sequence below the First Piper Coal includes only minor sandstones.

Nodular ironstones are a common feature of the more argillaceous part of these measures. They were once worked locally under names such as Wallis's Rake and Whetstone Rake (Smyth 1856, p. 41). Wedd (in Gibson and others 1908, pp. 79 and 171) records an oolitic ironstone in the roof of the Furnace (Low Main) Coal at Marehay and Salterwood collieries, near Denby. In Eastwood Hall Borehole *Carbonicola* sp. and *Naiadites* sp. were present about 19 ft above the Low Main Coal.

Sections in these measures near Denby, Codnor, Heanor and Smalley Common are described in Appendix 2.

The Cockleshell Coal (Fig. 27) lies up to 76 ft above the Low Main Coal in Ormonde Colliery Shaft and is united with that coal in the north and north-east of the district. Most of the intervening measures are sandy but 'cocklebeds' of the Low Main Coal have frequently been recorded where mudstones form the roof. The Cockleshell Coal may be composed of bright coal, dirty coal, cannel and/or interbedded coal and mudstone (batt). Multiple sequences of coal and dirt each a few inches thick may range through several feet of measures. The thickest record of the unsplit seam, 28 in in Cinderhill 6's Underground Borehole, includes laminae of dirt. There are two distinct elements to the seam at Ford's Pit, Brookhill Colliery, Agnes Site and the Pye Hill - Selston Colliery area, separated by up to 28 ft of measures. These two elements can be traced into the north-east of the district where they unite with the Low Main Coal, but it is not known how they are related to the single seam found between West Hallam and Cinderhill collieries and in boreholes in the east of the district. The seam is unproved over wide areas within the washout in the south-east and also near the western outcrop.

The measures between the Cockleshell and Hospital coals range in thickness from 8 ft in Cinderhill 2's Heading Underground Borehole to 75 ft in Wollaton 114's Main Outbye Underground Borehole. The widespread mussels in the roof of the Cockleshell Coal are frequently preserved in carbonate and include *Carbonicola cristagalli*, *C.* cf. *rhomboidalis* and *Naiadites flexuosus*. Sandstones are generally thin, except near Cinderhill and Wollaton collieries, where the top of the 'Tupton Rock' extends close to the base of the Hospital Coal (Fig. 29), and the Cockleshell Coal is washed out along the line of maximum sandstone thickness. This sandstone is interbanded and interbedded with siltstone and includes conglomeratic layers with pebbles and derived ironstones up to 6 in long. About 9 ft of cross-bedded ferruginous sandstone are exposed on the roadside near Shortwood [4913 4005]. The cross-bedding dips mainly to the east, but both north- and south-dipping units were also observed. In the nearby Shortwood Site the lowest 25 ft of the washout measures consisted of silty mudstones with poorly defined bedding features arranged in a channel, of which only the eastern margin was exposed. These mudstones cut across the Low Main Coal to rest upon the Threequarters Coal and contained large rafts and masses of derived seatearth which both transgressed and interfingered with the bedding of the mudstone. Within the seatearth were large ironstone nodules with kaolin. Fine khaki-coloured sandstones overlay the mudstones and, away from the channel margin itself, cut out the Low Main Coal in part of the site.

Mr R. E. Elliott (in litt.) recorded the washout in the adjacent Swancar Farm Site [491 397]; the Low Main was washed out and the irregular base of the 'Tupton Rock' contained much coal debris. The underlying 12 ft of measures down to the Threequarters Coal are disrupted siltstones, seatearths and ironstone; the siltstones appear to be rafts within derived seatearth and they interfinger with overlying mudstones.

The Hospital Coal is thickest, an exceptional 43 in, in Seagrave Borehole. It ranges from 21 to 31 in at the outcrop in the west; elsewhere it is thinner or irregularly split into two or more leaves. The split is most commonly recorded near Pentrich, Cinderhill and Wollaton collieries and near the Ironville Anticline. Individual leaves of the split seam may be up to 32 ft apart, as at Brands Colliery, but this is exceptional as are also the 15-ft split in Cinderhill 6's Main Gate Underground Borehole and the 17-ft split in Radford 5's Slant Underground Borehole. The seam thins away on the crest of the 'Tupton Rock' near Wollaton Colliery. Analyses of this coal are wholly from opencast provings; it is of good to medium quality in the west with low ash and sulphur contents, but the quality deteriorates in the south of the district.

The measures between the Hospital and Second Piper coals range in thickness from 12 ft in Beechdale Road Borehole to about 30 ft in Cinderhill 9's Left Gate Heading Underground Borehole. *Carbonicola cristagalli*, *Geisina arcuata* and fish remains are present in the more argillaceous part of this cycle.

The Second Piper Coal attains a maximum thickness of 12 in of 'coaly shale' in Selston Colliery No. 2 Drift. There are only seven other records of a carbonaceous bed in deep provings; however the underlying seatearth is widespread, though locally impersistent.

The measures between the Second Piper Coal and the Lower Coal of the First Piper or First Piper Coal range in thickness from 16 ft in Cinderhill 9's Left Gate Underground Borehole to 84 ft in Annesley No. 3 Underground Borehole, the increase being directly proportional to the thickness of the sub-First Piper sandstone in the north-east of the district (Fig. 29). The full thickness of this sandstone however may be as much as 120 ft in Annesley 10's Underground

b

Borehole, where it extends from a few feet above the Hospital Coal up to the seatearth of the First Piper. The sandstone contains bands of breccia and fine- and coarse-grained units of both sandstone and siltstone. Bands with a hard ferruginous cement (cank) up to about $1\frac{1}{2}$ ft thick are also present.

In the remainder of the district these measures are about 35 ft thick and contain a variable amount of arenaceous material. The roof of the Second Piper Coal may contain mussels of the *cristagalli* fauna and plants.

THE FIRST PIPER COAL (Fig. 30) is composed of up to three main elements which comprise the lower and upper parts of the Lower Coal of the First Piper and the Upper Coal of the First Piper as figured by Smith and others (1967, fig. 18) for the Chesterfield district. Over a large part of the district they are combined in a single seam, up to 60 in thick, as in Bennerley Colliery Shaft. Many records of the combined seam show the tripartite subdivision, though a bipartite sequence of Upper and Lower coals is more widespread and important. The three elements of the seam divide irregularly so that they may be distributed through up to 43 ft of measures in two areas (Fig. 30). One area in the north-west includes Denby Hall, Pentrich, Ripley, Brands, Western, Birchwood, Swanwick and Pye Hill collieries and the second in the north-east lies within the takes of Annesley and Hucknall collieries. In this latter area the division of the seam is largely into two rather than three leaves. The three leaves in Hucknall C's Heading Underground Borehole are distributed through 10 ft of measures.

The irregular nature of the splitting is most evident at Pye Hill Colliery, where the shaft and adjacent workings show a 22-ft division between the lower 4-in and upper 6-in parts of the Lower Coal, while only 2 in separate the Lower Coal from the 28-in Upper Coal (Fig. 30). The main value of this section and also that at Swanwick New Pit, however, lies in the presence of both the Second Piper and First Piper horizons, thus facilitating the correlation of the latter within the north-western area of split.

The Lower Coal of the First Piper is unimportant and usually of poor quality. The dirt parting between the lower and upper parts is normally a few inches in thickness, but increases rapidly in the extreme north-west of the district (the 3-ft isopach is shown on Fig. 30) to 12 ft in Pentrich Colliery Speedwell Shaft and 22 ft at Pye Hill Colliery. The individual thicknesses of the two parts rarely exceed 12 in except in opencast records west and south-west of Denby Hall Colliery, where analyses show high ash contents. The greatest thickness recorded is 27 in for the lower part in Denby Hall Colliery workings, though much of it was dirty coal.

Within the north-eastern area of split the Lower Coal is 10 in thick (including dirt partings) 6 in below the Upper Coal at Annesley 204's, 1 in of batt 3 ft 9 in below the Upper Coal at Annesley 10's, and 20 in thick and 21 in below the Upper Coal at Cinderhill 2's Heading underground boreholes. In the more divided sequence at Hucknall C's Heading Underground Borehole the two parts of the Lower Coal are 2 ft 11 in apart and each 4 in thick.

The Upper Coal of the First Piper within the north-western area of split varies from 15 in at Swanwick New Pit to 40 in at Salterwood Site. Isopachs are shown on Fig. 30. The coal is of variable quality, the thicker sequences being usually poor with high ash contents. In the north-eastern area of split it varies from 12 in at Hucknall C's Heading Underground Borehole to 21 in at Annesley 204's Underground Borehole and may contain up to $2\frac{1}{2}$ in of cannel at the top.

The combined First Piper seam ranges from 11 in (with 2-in cannel overlying) in Annesley No. 3 Underground Borehole to 60 in at Bennerley Colliery. Two dirt partings divide the seam in the northern part of the district. They are seldom recorded above the 'Tupton Rock' crest in the south, where the seam is reduced to 14 in at Cinderhill Colliery No. 4 Shaft and to 22 in at Model Farm No. 2 Borehole, and where a bipartite subdivision is apparent.

Apart from a few inches of cannel at the top in some analyses, the Upper Coal part of the combined seam consists of bright coal containing variable quantities of ash and sulphur ('Top Brights'), the basal few inches being mainly dirty coal. The elements composing the Lower Coal are variable and, as in the areas of split, are frequently composed of dirty coal, although some units of banded bright and dull coal are recorded. A rock less than 2 in thick and having a close affinity to tonstein occurs $3\frac{1}{2}$ in above the base of the seam at Lodge Colliery. Eden and others (1963, p. 52) described it as a quartzitic silty rock containing lenses of mudstone composed of variable amounts of kaolinite and illite. They also described a correlative of this rock from Clifton Colliery in the Nottingham district as tuffaceous.

THE MEASURES BETWEEN THE FIRST PIPER AND DEEP HARD COALS are between 30 and 40 ft thick over much of the district but range from 29 ft in Eastwood Hall Borehole to as much as 80 ft in Birchwood Shady Shaft. They are thickest in the north and east. There are one or two component cycles divided by the Deep Hard Floor Coal or its seatearth. The Floor Coal appears to separate from the Deep Hard Coal east of the sinuous north-south line shown on Fig. 31. There are few records of the overlying cycle up to the floor of the Deep Hard Coal, but it can be up to 19 ft thick as in Cinderhill 10's Main Gate and 10's Inbye Underground boreholes. This cycle and the Floor Coal are described with the Deep Hard Coal below.

The roof measures of the First Piper Coal contain mussels and some fish debris, but the fauna is mainly recorded in the eastern part of the district. Plant debris and nodular ironstones have been noted. Sandy measures are comparatively thin and are largely restricted to the uppermost 10 or 20 ft except near Kimberley and Cinderhill collieries, where they are dominant. Temporary sections near Codnor, Heanor and West Hallam are described in Appendix 2.

THE DEEP HARD COAL is composed of three principal elements, the Floor Coal, the main seam and the Roof Coal, which may extend through up to 27 ft of strata. The main seam in turn can be divided into three parts (see below).

The Floor Coal and the main Deep Hard seam are together in the west of the district, where their combined thickness ranges from 85 in at Openwood Site to 45 in in Cotmanhay Colliery Shaft and workings. Of these only $8\frac{1}{2}$ and 2 in respectively represent the Floor Coal, which never exceeds 12 in where the seam is combined. North of Openwood Site the Floor Coal itself is split into two or three thin coals. The line of split between the Floor Coal and the main Deep Hard shown on Fig. 31 is imprecise because most sections which record the separate seams lie well to the east of it. Exceptions are Moorgreen No. 1 and Bailey Brook Colliery shafts, where the interval is 1 ft 7 in and 2 ft respectively, and near Shilo Site, where it varies up to 6 ft. The Floor Coal is usually composed of dirty coal except where it is a persistent seam up to 18 in thick, as in Shilo Site and Bennerley Colliery Shaft, and 6 in to 6 ft below the main Deep Hard. East of the line of split most sections show only its seatearth. This is separated by up to 19 ft (Cinderhill 10's Main Gate and 10's Inbye Underground boreholes) of argillaceous and arenaceous measures from the main Deep Hard.

The main Deep Hard Coal comprises the 'Bottom Brights', 'Hards' and 'Top Brights' - names used by the early miners (Anon 1942). These terms are a simplification, for the composition of the seam is variable. The thickness of the main Deep Hard ranges from 14 in in Cinderhill Colliery No. 4 Shaft to 76 in in Hucknall Colliery No. 5 Shaft. The latter is an abnormally thick swilley sequence which varies in the shaft down to 25 in. The seam averages about 40 in for the district, but it thins rapidly in the extreme east. In the north-east thicknesses include the Roof Coal - recognised as a separate seam only in the area shown on Fig. 31 and two adjacent sections.

The detailed seam sections (Fig. 31) show the variable composition of the seam; the 'Bottom Brights' in the west include part or all of the Floor Coal element and are mainly of very poor quality. The 'Hards' have been the main unit worked underground and consist of a variable sequence of hard coal overlying interbanded hards and brights, all of good quality with low to very low ash and sulphur contents. Bright coal is common in the 'Top Brights' in many sections within the eastern third of the district, where it may be capped by a thin cannel (up to $2\frac{1}{2}$ in). Bright coal is however rare in the remainder of the district, where 'scuds' (an inferior coal) predominate, capped by up to 12 in of 'gees' (carbonaceous shale) or 'jays' (inferior cannel); the last two terms appear to have been used indiscriminately. The Roof Coal element is made up of all or part of this sequence of 'scuds' and 'jays'; a dirt parting between the Roof Coal and main Deep Hard is found only at New Selston Colliery and in one record from Agnes Site. Strauss (1971, pp. 1526-1527) has recorded and described a thin tonstein lying at or near the top of the 'Hards' in Cossall Colliery 43's workings. The same tonstein is noted in the log of Stoney Street Borehole and from Wollaton Colliery (Eden and others 1963, p. 53) as illustrated by Strauss (1971, pp. 1526-1527). It appears to extend only into the south-eastern part of this district.

The Roof Coal in the north-east of the district may lie up to 24 ft above the main Deep Hard, as in Brookhill Colliery (Smith and others 1967, p. 308). Here its thickness is 7 in, of which the uppermost 3 in are cannel. Its maximum thickness is 13 in of cannel, in Annesley 204's Underground Borehole. Cannel may also occur at the top of the 'Hards' even within the area of division from the Roof Coal.

A washout sandstone about 300 yd wide cuts out the Deep Hard Coal along a linear belt trending north-north-eastwards from the outcrop near Denby Hall Colliery. The northern continuation has been noted by Smith and others (1967, p. 144). According to Gibson and others (1908, pp. 79-80), the coal thickens as the washout is approached, but sections in clay pits at Salterwood [3888 4750] show that the abnormally thick coals are derived and lie within the sandstone; one such coal was 9 ft thick and died out within a few yards. This coal contained thin sandstone lenses.

Elliott (1965, p. 138) has described a swilley in the Deep Hard Coal in Hucknall Colliery shafts and mentions a second in workings at Moorgreen Colliery. The swilley at Hucknall enters this district from the north-east and has been proved in No. 5 Shaft (see above) and in No. 3 Shaft, whence it turns southwards and its course is lost.

THE MEASURES BETWEEN THE DEEP HARD COAL OR THE DEEP HARD ROOF COAL AND THE DEEP SOFT COAL vary from 20 ft in Lodge Colliery Shaft to about 100 ft in Denby Hall Colliery Shaft and are composed of up to three cycles; the lowest two are capped by thin, impersistent coals. The lowest cycle is composed largely of silty measures which may be up to 35 ft 1 in thick, as in Manners Colliery Shaft, where 7 in of batt at the top represent the Foot Coal. In Eastwood Hall Borehole *Naiadites* sp. and *Carbonita* sp. were present in the roof of the Deep Hard Coal. The 'Deep Hard Rock' (Smith and others 1967, p. 146) is only of local significance within this district, its most important outcrop lying near the washout described above. The Foot Coal is never thicker than the 12 in recorded in Bennerley Colliery Shaft. (It is sporadically represented in sections near Ilkeston and Hucknall Deep Soft C's junction Underground Borehole to the east of this district). In the absence of the Foot Coal, the top of the cycle is drawn at the summit of the silty measures. The second cycle either finishes at an impersistent thin coal or cannel lying up to about 2 ft below the Deep Soft Coal or extends upwards to include the Deep Soft Coal. The third cycle therefore, if present, includes these 2 ft of measures and the Deep Soft Coal. The measures are partly arenaceous but include thick seatearths of varying lithology. In Denby Hall Colliery Shaft for instance, 'clunch' and principally, 'stone clunch' was recorded for 35 ft below the Deep Soft Coal and comprised all but 3 ft of the second cycle. The underclay below the Deep Soft Coal was of

c

d

considerable value as raw material for ceramics.

THE DEEP SOFT GROUP OF COALS comprises in ascending order the Deep Soft, Roof Soft, Top Soft and Black Rake coals. Over a restricted area (A on Fig. 32) near Smalley these coals lie within 25 ft of strata, the most condensed sequence being at Club Room Site, where there were only 4 ft 6 in of dirt in 19 ft 5 in of coal-bearing strata. Elsewhere these coals may be distributed in up to 96 ft of strata, as in Pentrich Colliery, Speedwell Shaft.

In area B of Fig. 32 the Top Soft Coal splits off the combined seam. The Black Rake Coal, not recognised in area A, probably a split off the Top Soft, is present in this area.

To the east, in area E, the sequence is further divided. There are two, or sporadically three, leaves of the Roof Soft, a thinner Top Soft and, locally, a Black Rake Coal. Only the Deep Soft remains as a persistent seam of importance.

In area C the Roof Soft and Top Soft seams, with or without the topmost Black Rake element, are close together, but are widely separated from the Deep Soft Coal. The combined Roof Soft and Top Soft coals which may contain up to four separate leaves was originally called the Ell Coal (Gibson and others 1908, p. 69); latterly this name was altered to False Ell to avoid confusion with the Ell coals of Westphalian B. Type sections for the False Ell are provided by Grammer Street Borehole, where the sequence probably includes the Black Rake at the top (Fig. 34) and Stoneyford Borehole where the Black Rake is a separate seam.

In area D, as in area E, the higher coals of the group are divided and remote from the Deep Soft. There is a thick Roof Soft (? lower leaf) in Birchwood Colliery Shady Shaft and Pinxton No. 3 Pit, but elsewhere the topmost coal is the principal seam. It is not certain if the latter is the Black Rake Coal or a combination of that seam with lower elements. It is however the Sitwell Coal (Smith and others 1967, p. 149) in the south-west of the Chesterfield district.

The Deep Soft Coal itself (Fig. 33) has a maximum thickness of 66 in in Heanor Gate Borehole and thins irregularly towards the north-east, being only $30\frac{1}{2}$ in in Linby Colliery (Pit Bottom) Underground Borehole. One, or sometimes two, thin floor coals or cannels, up to 7 in thick, lie as much as 2 ft below the main coal sequence. The main coal consists largely of 'brights' with, in the middle of the thicker sequences, banded bright and dull coal. A thin hard coal parting has been widely recorded near the base of the seam. This parting is probably the representative in the undivided seam, of the dirt interval which separates off the floor coal elsewhere. A second hard coal parting near the top of the seam is also widely recorded. The seam is generally of good quality with low ash and sulphur contents.

The measures between the Deep Soft and Roof Soft coals consist of up to 11 in of dirt over the major part of areas A and B; this increases in the north of these areas to between 24 in in Woodside Colliery Shaft and 57 in in Shipley Colliery Deepfields Pit. See also the section in Top Dumbles Site near Heanor (Appendix 2). Northwards into area C the sequence up to the base of the False Ell Coal consists of silty and sandy measures ranging in thickness from 27 ft in Butterley Park Engine Shaft to about 71 ft in Ford's Borehole. In the part of area D where the Roof Soft Coal or its seatearth is recognisable, these measures range in thickness from 28 ft in Swanwick New Pit to about 41 ft in Birchwood Shady Shaft. In sections near the northern margin of the district, where the Roof Soft Coal is unrecognisable, the measures between the Deep Soft Coal and ? Black Rake Coal are as much as 91 ft thick in Pentrich Colliery Speedwell Shaft and include mudstone and siltstone with 24 ft of seatearth at their top. Near the south-west margin of area E, where they are largely argillaceous, these measures can be as thin as 12 ft (*Rutland Colliery in Ilkeston), but over the rest of this area, the incoming of arenaceous measures increases this interval to as much as 67 ft, as in Wollaton Colliery No. 2 Shaft.

The Roof Soft Coal is thickest where combined with the Deep Soft Coal in areas A and B (Fig. 32); it is 51 in at Tinklers Site and 60 in at Woodside Colliery Shaft. Analyses from opencast sites in these areas show the seam to be composed of bright coal at the base, with a banded bright and dull sequence in the centre. The top of the banded sequence is frequently marked by a thin dirt parting which separates off the uppermost 8 to 13 in of bright coal. One or two thin roof coals may be present close above the seam. The Roof Soft is split into two leaves in every record within area E with the exception of Hucknall No. 2 Colliery No. 1 Pit and Linby Colliery Engine House and Pit Bottom Underground boreholes, where it was represented by a few inches of coal or cannel. The leaves of the seam are very variable in thickness and in distance apart, with the upper leaf frequently near the Top Soft Coal. The lower leaf is thickest near the line of split which apparently takes place along the same line as that of the seam split from the Deep Soft and False Ell coals. The maximum known thickness of the lower leaf is 30 in in Cotmanhay Colliery Shaft, thinning northwards to 18 in at Pollington Colliery and eastwards and southwards to 12 in or less.

Measures ranging in thickness from 8 ft at Turkey Field Colliery to 34 ft at Cotmanhay Colliery separate the lower and upper leaves of coal in area E, the thinner sequences being composed of argillaceous rocks and/or seatearth. Except in the north-east of this area, the sandstones within this interval are concentrated in two belts, one along the line of splitting and the other along the eastern margin of this district. The sandstones in the region of the split appear to have been deposited from currents moving towards the north. The upper leaf of coal is normally thin, but reaches an exceptional thickness of 43 in of coal and dirt at Manners Colliery; the 18 in recorded at Turkey Field Colliery is also above average for the area.

*The precise site of this shaft is not known and the section is not included in Appendix 1; it penetrated measures from the Second Ell to Deep Hard coals.

a

The Roof Soft Coal in area C is described below as part of the False Ell Coal. In area D, where much of the correlation is tentative, it is thought to be split in Birchwood Shady Shaft, where the lower 36-in leaf is separated by 21 ft of measures from the upper 6-in leaf and in Swanwick Colliery New Pit (Fig. 34). Where the horizon has been recognised elsewhere within this area the Roof Soft is either of insignificant thickness or represented only by seatearth.

The False Ell Coal in area C comprises the Roof Soft and Top Soft coals in three or four leaves in up to 16 ft of strata. Over part of area C it probably contains the Black Rake Coal at the top (See Appendix II for such a section near Codnor). The lowest leaf of the seam is everywhere the thickest and reaches a maximum of 37 in in Grammer Street Borehole. The second leaf (the upper leaf of the Roof Soft) is never more than 14 in thick and lies up to 16 in above the lowest leaf. Exceptionally, as at Brands and Portland No. 1 collieries, there are three thin leaves to the Roof Soft component of the seam; in Ford's Borehole the lowest of the three is 24 in thick, all but $5\frac{1}{2}$ in being either dirty coal or dirt. The Top Soft component ranges from 18 to 30 in thick and is separated from the Roof Soft components by up to 44 in of seatearth. Analyses of deep borehole provings show the Roof Soft components to be poor in quality with high ash and moderately high to high sulphur contents and the Top Soft component to be of good quality with low ash and moderately low sulphur contents.

The Top Soft Coal is thickest, 32 to 52 in, in area A. In structure, this coal displays a banded bright and dull coal sequence in the middle of the seam and is of medium quality with moderate ash content and moderately low to low sulphur content. Elsewhere within the Derby district the seam is of importance only as part of the False Ell Coal described above. In area B it is 32 to 39 in thick in Shipley Colliery Deepfields Shaft, West Hallam Colliery No. 1 and Coppice Colliery shafts and Shipley Park Borehole, but thins to 8 in and 18 in in Nutbrook Air and Woodside Colliery shafts respectively. At the last locality the seam lies some 47 ft above the Roof Soft Coal, and this is the thickest interval between these coals in the district. Over most of the district this interval is less than 15 ft. The seam in Shipley Park Borehole was of poor quality. In area E the seam is usually thinner than 14 in but it thickens to 18 in in Turkey Field Colliery, 24 in at Manners Colliery and to a maximum of 30 in at Plumtree Colliery. In the north-west (area D, Fig. 32) the Top Soft is unrecognisable within the thick seatearths recorded below the Black Rake Coal, or it is incorporated within that coal.

The Black Rake Coal is the least important member of the Deep Soft group and is recognisable, mainly in a narrow central belt which widens towards the north-west (Fig. 32) up to 27 ft (Stoneyford Borehole) above the Top Soft Coal. The coal is thickest in the north-west of the district and varies from 6 in in Birchwood Shady Shaft to 24 in in Pentrich Colliery, Hartington Shaft. The seam is known locally in the east of the district, at Hucknall No. 1 and No. 2 collieries and Linby Colliery and also near Wollaton Colliery No. 2 Shaft. On Fig. 32 the Black Rake Coal is shown as a split off the Top Soft seam, but this is based upon little evidence and the seam could be part of an impersistent cycle distinct from that of the Top Soft. Apart from the records in the north-west of the district the coal is only a few inches thick and frequently the horizon is represented by a seatearth; there are however 24 in of 'batt' recorded in West Hallam Colliery. This seam has been worked opencast at the outcrop in the north-west of the district, where analyses show it to be of medium to rather poor quality.

THE MEASURES BETWEEN THE TOP SOFT OR BLACK RAKE COALS AND THE BROWN RAKE COAL vary from 13 ft at Greenhill Lane Colliery, Riddings to 43 ft in Ormonde Colliery Shaft. Dark grey mudstones near the base of the cycle contain ironstone nodules of the Black Rake Ironstone (Smyth 1856, p. 37) which have been worked both opencast and underground as a source of ore. Within these mudstones and forming a roof to underground workings is a widespread bed called the Black Rake Tuffaceous Siltstone by Francis and others (1968). Previously called the Black Rake Carbonate Rock (Edwards 1951, pp. 33, 147), it is a striped, hard bed of flinty rock up to 15 in thick with up to 50 successive pale grey graded layers which in places contain basaltic glass shards and pumice. At many localities the rock is dolomitic. It is formed from volcanic fall-out emanating from vents lying to the east of the present district; in Stoneyford Borehole it contains sufficient basaltic glass to merit the name vitric tuff while at Hucknall Colliery No. 2 Drift the bed includes layers of tuff with accretionary lapilli (Francis and others 1968, p. 399). The following petrographical description of this rock is by Dr E.H. Francis ... 'a graded, partly carbonated rock with layers 1.0 to 10.0 mm thick. The coarser elements include quartz grains up to 0.3 mm diameter, together with fragments of shale, coal and abundant basaltic debris. Most of the basaltic debris is elongate and sharp-shaped. It is mainly altered to turbid brown decomposition products or, less commonly, it has been kaolinized, but there remain a few fresh, yellow, isotropic vesicular fragments. The finer elements at the tops of graded units are silty mudstone without any recognisable volcanic debris'.

A few feet of mudstone with ironstone nodules normally overlie the tuffaceous siltstone, the remainder of the cycle comprises arenaceous measures and seatearth. In Manners Wood Borehole Spirorbis sp., Anthraconaia sp., Anthracosia regularis, Carbonicola cf. venusta, Naiadites sp. and Geisina arcuata were present in the mudstones of this cycle.

THE BROWN RAKE COAL (Fig. 71), which is up to 38 in thick at Club Room Site, thins rapidly towards the east where, over wide areas, it is under 12 in. In the north the seam is divided into two and sometimes three leaves, as in Swanwick New Pit, with up to 7 ft of measures between. The thickest single leaf is 18 in at Greenhill Lane Colliery.

b

The quality of the coal at outcrop is variable, but on average it is poor with high ash and moderate sulphur contents; in two analyses from deep-mine boreholes it was medium in quality with moderate ash and moderately high sulphur.

The seam has been extensively exploited by opencast means, usually in conjunction with the underlying coals of the Deep Soft group.

THE MEASURES BETWEEN THE BROWN RAKE COAL AND CLAY CROSS MARINE BAND vary in thickness from 8 ft in Stoneyford Borehole to 35 ft in Ford's Borehole, and are wholly argillaceous in the district apart from a 3-ft sandstone in the vicinity of Linby Colliery. Temporary sections near Smalley, Codnor (Agnes Site) and Strelley (Catstone Hill Site) are described in Appendix 2.

At the base of the measures a kaolinitic rock, now considered to be a Fragmental Clayrock (Richardson and Francis 1971), of doubtful persistence, has been noted in Ford's Borehole. This correlates with one of the tonstein horizons described by Strauss (1971, p. 1525). The grey mudstones everywhere contain nodules of the Brown Rake Ironstone (Smyth 1856). Some of these are calcareous and exhibit cone-in-cone structure. The measures, including some of the ironstones, contain abundant Anthracosia regularis, Naiadites sp. and Geisina arcuata; Spirorbis sp. is also common and Planolites montana is present. Near the top, the fossils tend to be pyritised. A sporadic seatearth, a few inches thick, at the summit of the measures has been recorded, mainly in the eastern part of the district. It is the only representative of the Joan Coal horizon; in Eastwood Hall Borehole the rootlets were pyritised.

Figure 71 Plan of the Brown Rake Coal showing provings and isopachs. Localities mentioned in the text are also shown

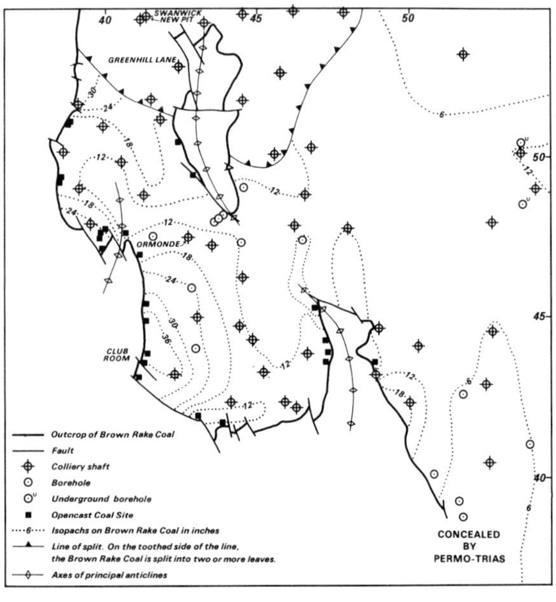
40
45
50
SWANWICK NEW PIT
GREENHILL LANE
ORMONDE
CLUB ROOM
30
24
18
12
36
6
50
45
40
Outcrop of Brown Rake Coal
Fault
Colliery shaft
Borehole
Underground borehole
Opencast Coal Site
Isopachs on Brown Rake Coal in inches
Line of split. On the toothed side of the line, the Brown Rake Coal is split into two or more leaves.
Axes of principal anticlines
CONCEALED BY PERMO-TRIAS

c

WESTPHALIAN B

CLAY CROSS MARINE BAND TO TOP HARD COAL

The measures between the Clay Cross Marine Band and Top Hard Coal (Plate 5) are about 460 ft thick at Heanor Gate Borehole and about 450 ft near Swanwick Colliery; they thin eastwards to about 330 ft at Hucknall No. 2 and Broxtowe collieries, though an incongruous record from Hempshill Heading in the east shows about 406 ft of these measures. Approximate isopachs are shown on the inset map of Plate 5. The measures comprise twelve or more cycles, most of which contain both arenaceous and argillaceous strata. There are ten named coals. Marine fossils are confined to the Clay Cross Marine Band. Mussels are present in the roof measures of all the coals except the Bottom First and Bottom Second Waterloo coals; the Anthracosia aquilina/ovum fauna lying below the Second Ell Coal grades upwards into the A. phrygiana fauna which continues to the Top Hard Coal.

The named coals have been exploited mainly by opencast means, but the Second Waterloo Coal has also been extensively worked underground. Extensive temporary sections in these measures at Catstone Hill Site near Strelley are described in Appendix 2. The measures associated with the Second Waterloo Coal are worked for brick-making.

THE CLAY CROSS MARINE BAND consists of up to about 9 ft of dark grey mudstone with thin ironstone lenses. The marine fossils and their distribution within the marine band have been described by Edwards and Stubblefield (1948, p. 215) and Calver (in Smith and others 1967, pp. 157-159 and 1968a, pp. 163-173). The 1 ft 10 in of mudstone between the Joan Coal Seatearth and the Dunbarella and goniatite-bearing mudstones in Eastwood Hall Borehole contain Lioestheria at their base with Lingula and foraminifera above. They are indicative of an increasing marine influence on the conditions of deposition, but there is no clear separation of the foraminifera and Lingula phases. The Dunbarella and goniatite-rich horizons near or at the base of the sequence occur in dark mudstone tougher and more fissile than the overlying Lingula-rich horizons. As noted by Gibson and Wedd (1913, p. 71) and Edwards and Stubblefield (1948, p. 216) the upper Lingula-bearing horizons alternate with mudstones containing stunted mussels. Paraparchites and abundant foraminifera including Glomospira and Glomospirella associated with stunted Anthracosia spp. were present at this level in the Eastwood Hall Borehole.

The 5 ft $11\frac{1}{2}$-in sequence of the Clay Cross Marine Band in Seagrave Borehole is shown on Fig. 72. Here some 2 ft $1\frac{1}{2}$ in of non-marine mudstone underlie the marine band. This is unusually thick for this district; normally there are only a few inches of such mudstone above the Joan Coal horizon. The mussels from this mudstone include Anthracosia regularis and Naiadites aff. quadratus. Anthracoceratites vanderbeckei and Dunbarella papyracea mut ♂ are common near the base of the marine band and are indicative of maximum marine influence. The mussels which are interbedded with the Lingula horizons in the retreat phase of the marine incursion include Anthracosia spp. of the nitida and ovum/aquilina groups.

There is no permanent section of the Clay Cross Marine Band in this district, except at the bottom of a shallow ditch [4975 4267] 450 yd SW of Strelley Park Farm, where dark mudstones containing Dunbarella sp. can be found. The marine band has been recorded in the following temporary and underground sections; where known the thicknesses over which marine fossils have been found are indicated:

Agnes Site [4250 5030], 2 ft 9 in and [4300 4949], 1 ft 3 in (Appendix II): Beechdale Road Borehole, 8 ft 7 in: Carrington Coppice Site [413 455], 3 ft 8 in: Catstone Hill Site [4974 4151], 2 ft (Appendix II): Denby Hall Colliery Drift [3992 4859], 2 ft: Digby Clay Pit [c. 485 450]: Eastwood Hall Borehole, 4 ft 7 in: Ford's Borehole, 6 ft 7 in: Godber's Lum Site [4053 4787]: Grammer Street Borehole, 2 ft 11 in: Heanor Gate Borehole, 6 ft 5 in: Hucknall No. 1 Colliery; No. 2 Shaft, 4 ft 3 in; Drift, 6 ft 3 in; Boring from Top Hard to Deep Soft Drift, 7 ft 10 in: Hucknall No. 2 Colliery; No. 3 Shaft, 3 ft; No. 5 Shaft: Linby Pit Bottom Underground Borehole, 7 ft 5 in: Linby Engine House Underground Borehole, 6 ft 11 in: Manchester Wood Borehole, 4 ft: Model Farm No. 1 Borehole, 5 ft 1 in; No. 3 Borehole, 5 ft 1 in: Moorgreen Waterloo 3's Borehole: Motorway Cutting [4959 4112]: Park Meadows Site [3977 4718], 1 ft 3 in: Seagrave Borehole, 5 ft $11\frac{1}{2}$ in: Shipley Park Borehole, 1 ft 8 in: Stoneyford Borehole, 4 in: Swanwick Colliery, Dyson's Heading [c. 404 532], 2 ft 6 in: Tavern Houses Site [4115 4683], 3 ft 3 in.

THE MEASURES BETWEEN THE CLAY CROSS MARINE BAND AND SECOND ELL COAL form part of a single cycle and range in thickness from 48 ft in Linby Pit Bottom Underground Borehole to $94\frac{1}{2}$ ft in Grammer Street Borehole. The thicker sequences are mainly in the western and south-eastern parts of the district. Sandstones occur in the uppermost third of these measures; the most persistent is found to the west of Heanor. The mudstones are rich in mussels, which may be pyritized within 40 ft of the marine band, but may also be preserved in calcareous and ferruginous material. Many of the mudstones display worm tracks and burrows, and there are plants at the higher levels. Ironstones are common and frequently exhibit cone-in-cone structure.

In Eastwood Hall Borehole the mussels are present in mudstones throughout the cycle and are commonly pyritized and of small size. Anthracosia aff. ovum, A. phrygiana, A. aquilina and variants, Anthracosphaerium exiguum, Naiadites fragments and indeterminate ostracods were present in the lowest 37 ft of the cycle underlying a 10-ft thick sandstone. In the 14 ft of

d

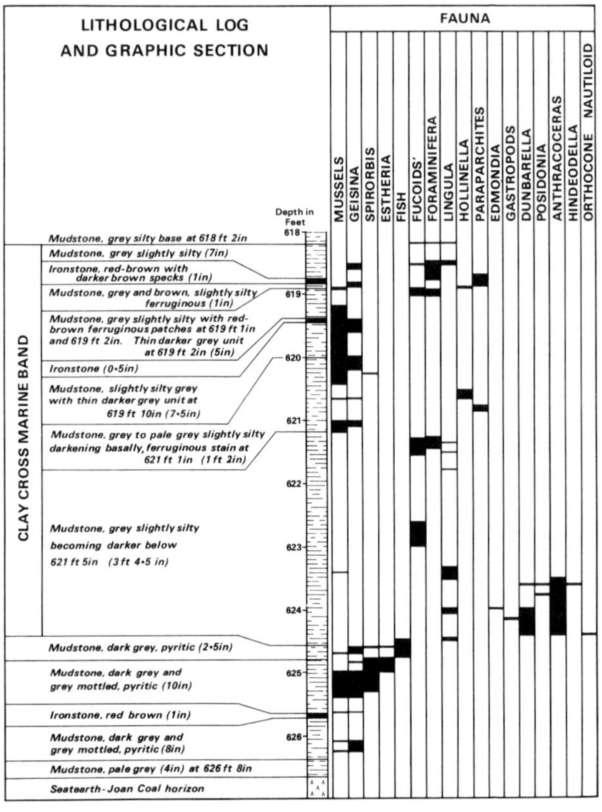
LITHOLOGICAL LOG
AND GRAPHIC SECTION
FAUNA
MUSSELS
GEISINA
SPIRORBIS
ESTHERIA
FISH
'FUCOIDS'
FORAMINIFERA
LINGULA
HOLLINELLA
PARAPARCHITES
EDMONDIA
GASTROPODS
DUNBARELLA
POSIDONIA
ANTHRACOCERAS
HINDEODELLA
ORTHOCONE NAUTILOID
Depth in Feet
618
619
620
621
622
623
624
625
626
CLAY CROSS MARINE BAND
Mudstone, grey silty base at 618 ft 2in
Mudstone, grey slightly silty (7in)
Ironstone, red-brown with darker brown specks (1in)
Mudstone, grey and brown, slightly silty ferruginous (1in)
Mudstone, grey slightly silty with red-brown ferruginous patches at 619 ft 1in and 619 ft 2in. Thin darker grey unit at 619 ft 2in (5in)
Ironstone (0•5in)
Mudstone, slightly silty grey with thin darker grey unit at 619 ft 10in (7•5in)
Mudstone, grey to pale grey slightly silty darkening basally, ferruginous stain at 621 ft 1in (1 ft 2in)
Mudstone, grey slightly silty becoming darker below 621 ft 5in (3 ft 4•5 in)
Mudstone, dark grey, pyritic (2•5in)
Mudstone, dark grey and grey mottled, pyritic (10in)
Ironstone, red brown (1in)
Mudstone, dark grey and grey mottled, pyritic (8in)
Mudstone, pale grey (4in) at 626 ft 8in
Seatearth-Joan Coal horizon

Figure 72 Lithological log and section of the strata overlying the Joan Coal horizon in Seagrave Borehole, showing the distribution of marine and non-marine fossils associated with the Clay Cross Marine Band

a

mudstone overlying this sandstone were Spirorbis sp. Anthracosia sp. nov. aff. phrygiana and Naiadites quadratus.

THE SECOND ELL COAL is extremely variable both in thickness and composition. Near outcrop in the north-west and as far south as Upper Hartshay Colliery it is a split seam similar to that described by Smith and others (1967, p. 158) in the Alfreton area to the north. The upper leaf (Tanyard Coal) is of medium quality and ranges in thickness from 18 in at Broad Oak Site to 28 in at Upper Hartshay Colliery; it is separated from the 1 to 8-in lower leaf of dirty coal by up to 26 in of seatearth.

Within a wide area in the north and west of the district the seam is composed wholly or largely of cannel. Two leaves are frequently present and the sequence reaches exceptional thicknesses at Ormonde Colliery (9 ft 4 in, see p. 172a) and the adjacent Bailey Brook Colliery (20 ft 1 in, see p. 165b). The 14-ft separation of the leaves at Bailey Brook Colliery is exceptional; normally it is 1 to 4 in. The individual cannel beds are normally 15 to 20 in thick in sections near Ilkeston, but are locally as much as 31 in, as at Coppice Colliery No. 2 Shaft.

The sequence at Shipley Park Borehole — dirty cannel 9½ in, ironstone 3 in, coal 5½ in, seatearth 4½ in on dirty coal 1½ in — appears to represent the passage from cannel to the 18 in of coal proved in Mapperley Colliery shafts.

Cannel is present in the upper part of the seam in two further, but restricted, areas. Firstly in part of Catstone Hill and Macfez sites, Turkey Field Colliery and Seagrave Borehole (p.172c) the two leaves include up to 16 in of cannel at the top. Secondly, in the shaft sections of Hucknall No. 1 and 2 collieries there are 7 to 9 in of cannel at the top of the seam. Elsewhere within the district the Second Ell is a single or split coal in which the lower leaf or the lower part of the single seam is commonly composed of dirty coal. The thickest known section is 19 in of coal underlying 48 in of dirty coal in Strelley Borehole, and the maximum proved separation of the two leaves is 6 ft at Moor Farm Borehole.

THE MEASURES BETWEEN THE SECOND AND FIRST ELL COALS average about 50 ft over much of the district, with a minimum of 31 ft at Hempshill Colliery Heading and a maximum of 65 ft at Turkey Field Colliery. They are composed of a single cycle which includes, over most of the district, only a small amount of arenaceous material; the 36 ft of sandstone underlying the First Ell seatearth at Turkey Field is exceptional. The mudstones are very fossiliferous, and the mussels may have a pyritous, calcareous or ferruginous mode of preservation. Ironstones are fairly common. The fauna in Eastwood Hall Borehole included Spirorbis sp., Anthracosia spp., Anthracosphaerium bellum, fragmentary Naiadites and fish remains, including Rhizodopis sp.

THE FIRST ELL COAL shows great variations both in quality and thickness. It is thickest to the south-west of Heanor and in the Erewash Valley, east of Ilkeston. South-west of Heanor it is mainly 30 to 42 in thick, although only 23 in were recorded in Coppice Colliery No. 2 Upcast Shaft. Analyses from opencast sites show that it is rather poor to very poor in quality in this area.

North and north-west of Heanor the First Ell is of medium to very poor quality. It is 18 to 27 in thick at the northern boundary of the district, thickening to 36 in at Bailey Brook Colliery Shaft. Dirt partings, 5 in or less in thickness, are recorded from several localities.

East of Ilkeston, analyses from opencast sites in the Erewash Valley show sequences of up to 48 in, as at Coventry Lane Site, but up to 17 in of the sequence may be dirt partings and cannel. The over-all quality of the seam is very poor to very inferior.

In the eastern part of the district the First Ell is absent in Hucknall No. 2 Colliery No. 2 Drift and Underground Borehole. Elsewhere it is a divided inferior seam.

THE MEASURES BETWEEN THE FIRST ELL AND FOURTH WATERLOO COALS are about 40 ft thick over most of the district. They are thickest near Heanor, where they total 56 ft in Coppice Colliery No. 2 Shaft, and thinnest in the south-east of the district, where they are only 10½ ft at High Leys Borehole, Hucknall, 12 ft thick at Broxtowe Colliery Shaft and 16 ft at Wollaton Colliery No. 1 Shaft. In most sections they comprise only one cycle, but at Bailey Brook, Ormonde and Mapperley collieries, Shipley Park Borehole and in the Woodlinkin Drift a thin coal or seatearth is recorded 6 to 17 ft below the Fourth Waterloo. This coal is probably a split from the Fourth Waterloo (see below) although direct evidence is lacking.

In the immediate roof of the First Ell a lenticular ironstone, usually up to about 2 in thick with kaolinite ooliths and pyrite, forms a useful lithological marker. It is exposed in a stream [4001 4852] near Marehay. Its discontinuous nature, observed in the nearby Godber's Lum Site (Appendix 2), accounts for its absence in many borehole sections, although it is very widely distributed. In this site the nodules were up to 8 in thick and protruded into the underlying coal. In Eastwood Hall Borehole the ironstone was 1 in thick and infilled Stigmaria.

THE FOURTH WATERLOO COAL is variable both in thickness and in quality and may comprise up to three leaves of coal of which the middle leaf is the most persistent and important. Only Linby Pit Bottom Underground Borehole (p.170b) and Hucknall No. 1 Colliery No. 2 Shaft Deepening (p.168d) encountered all three leaves and detailed correlation throughout the district is therefore subjective. The seam has been worked locally by opencast means.

A thin floor coal, up to 7 in thick and lying up to 15 in below the middle leaf of the seam, has been proved in New Selston Colliery Shaft, Linby Pit Bottom Underground Borehole, shafts at Hucknall No. 1 and No. 2 collieries, and in Codnor Common, Catstone Hill, Coventry Lane and Kim sites. This floor coal is probably the correlative of the thin coal in the measures underlying the seam elsewhere (see above).

The main or middle leaf of the seam is 15 to 29 in thick in the north-west of the district, where it is medium to rather poor in quality; at Agnes

b

Site it is 21 to 22 in thick with quality ranging from good to poor; it was 6 in thick at Coppice Colliery No. 2 Shaft, 10 in in Mapperley Colliery Shaft and 17 in in Manners Colliery Shaft. The leaf is thickest, 34 in, in a water bore at Wollaton and is 29 to 30 in thick in Wollaton Colliery; northwards it thins to 18 in or less in provings at Hucknall and Linby collieries. The quality of the 31-in leaf in Beechdale Road Borehole was good to medium, but in Coventry Lane and Catstone Hill sites, where the thickness varied from 24 to 32 in, the quality varied from good to poor. It consists of 11 to 15 in of cannel in some records from Shilo Site and Pye Hill Colliery Trial Borehole, and of 7 to 8 in of cannel in Moorgreen Colliery shafts; a few inches of cannel are also sporadically recorded at the top of the leaf elsewhere.

A thin roof coal or seatearth is present in Brookhill Colliery Shaft (Smith and others 1967, p. 308), Pye Hill Colliery Trial Bore, Selston Colliery (Underwood) Shaft, Eastwood Hall Borehole, Linby Pit Bottom Underground Borehole, Linby Engine House Underground Borehole, Hucknall No. 1 Colliery No. 2 Shaft and No. 2 Colliery No. 2 Drift, Kimberley Colliery Shaft, Hempshill Colliery Heading and Broxtowe Colliery Shaft. This roof coal is usually very thin and dirty and is as much as 18 ft above the middle leaf of the seam in Linby Engine House Underground Borehole.

THE MEASURES BETWEEN THE FOURTH AND THIRD WATERLOO COALS are usually 30 to 45 ft thick, but vary from 27 ft in Butterley Park Engine Shaft to 76 ft in Ormonde Colliery Shaft. Thicknesses are related to the proportion of sandstone present. The two component cycles cannot be separately distinguished, though 19 widely distributed provings in the district record a thin coal or seatearth 12 to 20 ft (exceptionally 38 ft in Woodside Colliery Shaft) below the Third Waterloo Coal. It is 1 in thick 16 ft 3 in below the Third Waterloo Coal in Brookhill Colliery Shaft, where Smith and others (1967, pp. 308 and 161) suggest that it may be a leaf of the Third Waterloo Coal. In Linby Engine House Underground Borehole and Kimberley Colliery Shaft (Plate 5) this coal lies some 20 ft above the roof coal or seatearth of the Fourth Waterloo, which has a more restricted distribution. Woodside Colliery Shaft encountered an anomalous 24-in coal at this horizon instead of the more usual 1 to 3 in.

The argillaceous roof measures of the Fourth Waterloo Coal contain fish debris, ostracods, burrows and plants. Anthracosia cf. beaniana occurred in Eastwood Hall Borehole in mudstones within a 16-ft sequence of interbedded sandstone, siltstone and mudstone near the top of these measures. A. disjuncta was present in these measures in Annesley Park Borehole.

THE THIRD WATERLOO COAL ranges from 12 in in Ripley Colliery Shaft to a multiple sequence of coal and dirt, 48 in thick at Catstone Hill Site. In the north and north-east the seam is about 2 ft thick over wide areas, and dirt partings are recorded only in Wansley Hall Site, where an overall thickness of 21½ in includes ½ in of dirt 2 in from the top, and in Dobbs Site, where a total thickness of 17½ in includes ¼ in of dirt 5 in from the top of the seam. A hard coal parting is present in the bright coal of the seam. Usually in this part of the district the leaf is of medium quality, but one analysis from William IV Site showed poor quality.

Near Heanor, in Agnes, Corfield and Plastic sites, detailed sections show a dirt or hard coal parting near the base of the seam. The dirt is thickest in Plastic Site, where the section is coal 29 in, dirt 4 in, on coal 6 in; the leaf is medium in quality. A section in Agnes Site showed that the lower 7 in of the seam was cannel with 21 in of overlying coal. It is possible that the cannel is the equivalent of the thin coal or seatearth which occurs 12 ft or more below the Third Waterloo in Shipley Park and Eastwood Hall boreholes, Selston Colliery Underwood Shaft, New Selston Colliery and Swanwick Colliery New Pit.

Well south of Heanor, at an outcrop near Mapperley Colliery, there is a dirt parting near the top of the seam; at Mapperley Colliery No. 2 Shaft the sequence is coal 2 in, batt 4 in, on coal 24 in. The seam is at a local maximum thickness in the adjacent Jople Site, where the sequence is coal 2½ in, dirt 3½ in, on coal 31 in; there is a hard coal band in the middle of the lower coal in this sequence. In Heanor Gate Borehole a 2-in dirt parting separates 3 in of dirty coal at the top of the seam from the underlying 25 in of coal, and in Shipley Park Borehole the upper 5 in of the 22-in sequence comprises dirty coal. The quality in this part of the district varies from good to poor with moderate to high sulphur content.

At Shilo Site the seam ranges from 17 to 31 in with only one of 15 recordings showing the upper dirt parting; the quality varies from good to poor and the sulphur content from low to high.

In the south-east of the district, although the overall thickness of the seam tends to be greater than elsewhere there are more dirt partings. The dirt parting near the top of the seam is present in most of the detailed records and shows a rapid and local expansion as indicated by the representative sections in Table 12.

In Beechdale Road Borehole the seam was 7 in thick. Its quality in opencast sites in the south-east of the district is medium to poor.

THE MEASURES BETWEEN THE THIRD AND SECOND WATERLOO COALS vary in thickness from 46 ft at Swanwick New Pit to 16 ft at Wollaton Colliery No. 1 Shaft. They locally contain a thin coal which is usually up to 3 in thick, but may be thicker when in close proximity to the Second Waterloo Coal.

The measures include a variable amount of sandy strata and the mudstones contain many mussels, some of which may be pyritic, together with fish debris, Spirorbis sp., Planolites montanus, worm tracks and borings.

THE SECOND WATERLOO COAL consists of two principal elements which, where they constitute separate seams, are named the Bottom Second Waterloo and Top Second Waterloo coals (Fig. 35). This split is irregular and is illustrated by borehole sections from Babbington

c

TABLE 12

Representative sections of the Third Waterloo Coal (in inches)

	Fez Site	Catstone Hill Site	Coventry Lane Site	Seagrave Borehole	Broxtowe Colliery Shaft	Hempshill Colliery Shaft	Wollaton Colliery No. 1 Shaft
Coal	4	4½	3	2	3	4½	1
Dirt and measures	0½	19	35	40	177	4	4
Coal	3	1½	2	1	-	0½	-
Dirt	0¾	3½	6	3	-	15	-
Coal	20	20	23	22	21	17	24

and Macfez Sites (Fig. 73). The two leaves are re-united in the south-west of the district only within a restricted area of Jople Site. The Top Second Waterloo Coal itself divides in the south-east of the district. Graphic sections of the seam are shown on Fig. 74. The interpretation of the Second Waterloo Coal sequence is based upon a report by the former Nottingham Coal Survey Laboratory (Turner 1961b).

Away from the line of split the Second Waterloo Coal is 49 to 57 in thick with two dirt partings totalling up to 10 in. The lower parting divides the Bottom and Top Second Waterloo elements, but the upper parting is also conspicuous and can be widely traced in sections of both the Second Waterloo and the Top Second Waterloo. The Second Waterloo Coal has been extensively worked from Moorgreen Colliery and is generally of medium quality with a moderately low sulphur content.

The Bottom Second Waterloo, up to 31 ft below the Top Second Waterloo in the west of the district, is generally between 12 and 18 in thick (Fig. 35), though in Shipley Park Borehole it is 21 in and in Shilo Site it ranges from 10 to 25 in. The few analyses show it to be of medium quality, but to the north-west of Ripley Colliery it is composed wholly of cannel up to 18 in thick.

Deep mine provings at the eastern margin and in the north-east of the district show the seam to be not more than 17 ft below the Top Second Waterloo and under 13 in thick, but it varies from 8 to 27½ in in opencast sites on the eastern side of the Erewash Valley to the east of Ilkeston.

The measures between the Bottom and Top Second Waterloo coals are largely argillaceous, but include appreciable amounts of sandy strata near Heanor. Mussels occur at the base of the cycle.

The Top Second Waterloo Coal is thickest in the extreme north-west of the district where it may include a basal cannel up to 42 in thick, as in Swanwick Colliery, Common Pit. The thickest proved sequence is 64 in at Swanwick Colliery, Old Pit; elsewhere in the west the thickness ranges from a doubtful 24 in at Shipley Park Borehole to 57 in at Codnor Common Site. The detailed sequence of the seam is variable, including many dull coal bands and, as at Brake House Site, one and sometimes two dirt partings some 8 to 12 in above the base. Its quality is medium with a moderately low sulphur content. It has been worked underground from Coppice and Woodside collieries and has extensive opencast workings.

At the eastern margin of the district, at Hucknall and Linby collieries, the Top Second Waterloo Coal is 22 to 33 in thick; and over an area in the south-east it divides into two leaves along a line slightly north and west of the split between the Top and Bottom Second Waterloo coals. The line of split on Fig. 35 is drawn approximately at the 18-in isopach of the separating measures and, as emphasized by Turner (1961b), it occurs at a higher horizon in the seam than the prominent dirt parting which can be widely traced in the district (Fig. 74). The maximum proved separation of the two leaves is 15 ft in an outline record of Cinderhill Colliery No. 2 Shaft. Very localized splits up to about 3 ft at the same horizon are apparent in some records from Shilo Site and Macfez Site (Fig. 73), but are not shown on Fig. 35. A typical section of the split seam is provided by Cinderhill Colliery No. 4 Shaft Deepening: coal 22 in, measures 14 ft 8 in, on coal 11 in (Turner 1961, fig. 1). Similar sequences were present in Seagrave and Chilwell Dam boreholes and Broxtowe and Wollaton No. 3 Colliery shafts (Fig. 74).

THE MEASURES BETWEEN THE SECOND AND FIRST WATERLOO COALS are thickest - 65 ft - in Swanwick Colliery New Pit in the north of the district and thinnest in Beechdale Road Borehole where they total only 23 ft. This southward attenuation continues a trend mentioned by Smith and others (1967, p. 164), but is irregular for the measures are only 31 ft thick at Ripley Colliery, 29 ft in Seagrave Borehole and 36 ft in Hucknall No. 2 Colliery No. 5 Shaft. Temporary sections near Ripley, Smalley, Heanor, Ilkeston, Brinsley and Strelley are described in Appendix 2. They comprise two main cycles divided by the Waterloo Marker Coal, plus, locally, a third

d

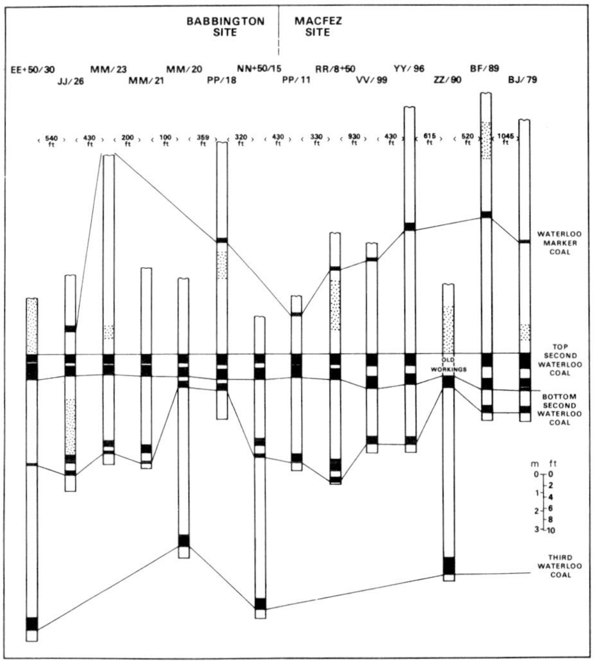

Figure 73 Variations in the development of the Second Waterloo Coal in selected boreholes from Babbington and adjacent Macfez opencast sites. The stippling denotes sandy measures

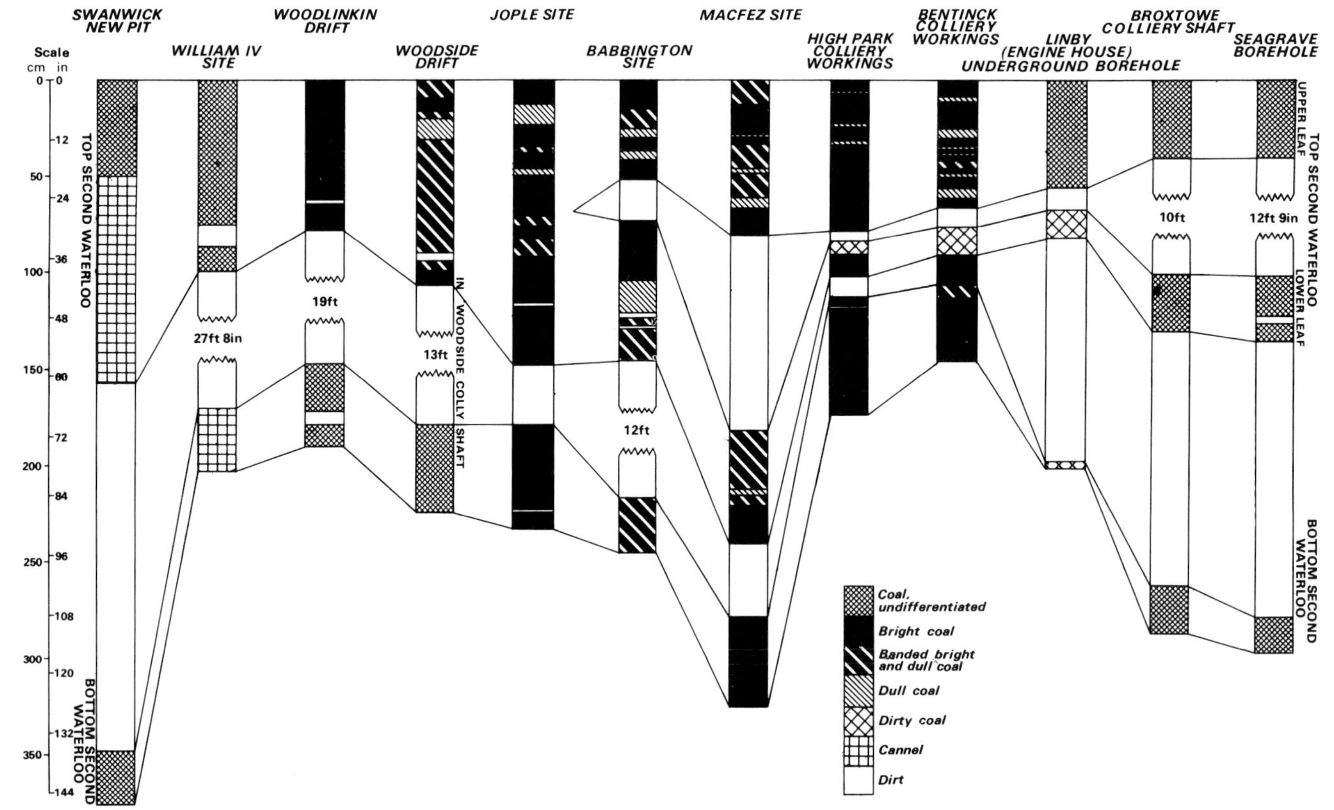

Figure 74 Representative sections of the Second Waterloo Coal. For localities see Figure 35

thin cycle caused by splitting of that coal. The thicknesses of the main cycles are variable, for instance the lower, which averages about 20 ft for the district, ranges from less than 1 ft at Gutter Slang and New Covert sites in the south-east to 30 ft in Swanwick Colliery New Pit (see also Fig. 73). Both cycles contain sandy measures, but these are absent from the lower where it is thin. There are mussels at the base of both the main cycles, and 'sineoids' have been frequently recorded, mainly, however, from the lower cycle. Anthracosia phrygiana occurs in the roof of the Second Waterloo in Eastwood Hall Borehole with Spirorbis sp., A. aquilina, A. beaniana, A. phrygiana, Naiadites quadratus and Carbonita humilis in the mudstones overlying the Waterloo Marker Coal.

The Waterloo Marker Coal is a persistent thin coal or cannel which serves as an indicator horizon between the First and Second Waterloo coals. It rarely exceeds 12 in in thickness and is of no economic importance. Seatearth alone marks its horizon at Greenhill Lane Colliery, Riddings, and Wollaton Colliery No. 3 Shaft; it is represented by 15 in of cannel in Brinsley Colliery Shaft and by two 6-in leaves of cannel, 18 in apart, in Annesley Colliery Shaft. Where split, the two thin leaves are usually a few inches apart, but are as much as 8 ft 2 in apart in Selston Colliery Underwood Shaft, where the leaves are each 1 in thick. Exceptional thicknesses have been noted in Swanwick Colliery, New Pit, 48 in of 'coal and clunch'; Nutbrook Air Shaft, 48 in of coal; Moorgreen Colliery No. 1 Shaft, cannel 46 in, underlain and overlain by 9 in and by 5 in of 'batt' respectively; Kimberley Colliery Shaft, coal 16 in, 'batt' 5 in, on coal 13 in. The coal is exposed in the Loscoe Brick Pit [426 471], Heanor (Appendix 2).

THE FIRST WATERLOO COAL forms a single seam over two large areas in the east of the district (Fig. 36). As described by Turner (1961a), these two areas are separated by a narrow corridor within which the seam is split into the Bottom and Top Waterloo coals and which extends north-eastwards from Greasley Castle Borehole to Linby Colliery. The corridor opens at both ends into wide areas covering the northern and western parts of the district where the seam is similarly split.

In the northern area of undivided coal the minimum thickness of the First Waterloo is 34 in, in Annesley Colliery Shaft. It contains a prominent dirt parting, normally $1\frac{1}{2}$ to 6 in thick, but reaching a maximum of 12 in, in Linby Colliery Shaft. The lower leaf of coal varies from 16 in at Annesley Colliery Shaft to 25 in at Moorgreen 1's Underground workings and the upper leaf from 14 in a drift [4851 4816] at Moorgreen Colliery to 27 in in Moorgreen 2s/10s Underground Borehole.

In the southern area of undivided coal the First Waterloo reaches a maximum of 59 in in Moorgreen 10's RH Underground Borehole; it thins eastwards to 21 in in Cinderhill 1's Main Gate Underground Borehole and 29 in in Beechdale Road Borehole. The dirt parting, 5 to 11 in thick near the line of split, is only $\frac{1}{2}$ to $\frac{3}{4}$ in thick in the sections of opencast sites on the east site of the Erewash Valley and is not recorded in many of the deep mine provings in this part of the district. Below the prominent central dirt parting, the seam (Fig. 36) consists largely of bright coal of medium ash and moderately high to high sulphur content; there is frequently a dirty coal or dirt parting 2 to 4 in above the floor of the seam, which is traceable into the areas of divided coal in the west and north of the district. Above the prominent dirt parting the seam is more variable with bands of hard and dirty coal; its quality is variable, again a thin dirt parting is present near the base in some sections. There are small isolated underground workings in the combined seam.

The corridor of split coal which divides the field of the First Waterloo was encountered in Moorgreen 2's Drift, Moorgreen Watnall Drift (for locations see Fig. 36) and in Linby Engine House and Pit Bottom Underground boreholes; its extension between Moorgreen Watnall Drift and Linby Colliery is not proven, but the interpretation shown in Fig. 36 follows Turner (1961a). Turner emphasised that the split in the corridor which takes place at the prominent dirt parting, is accompanied by partial or complete washout of the bottom leaf of the seam. The two leaves of the First Waterloo are 31 ft apart in Moorgreen 2's Drift with 6 in of cannel representing the lower leaf or Bottom First Waterloo. The dividing measures are largely siltstone, and the Top First Waterloo consists of dirty coal 8 in, coal 9 in, dirt $2\frac{1}{2}$ in, on coal $2\frac{1}{2}$ in; in a drift [5060 4795] near Watnall Colliery, the Bottom First Waterloo is $2\frac{1}{2}$ in of cannel lying 37 ft below the Top First Waterloo, here coal, 20 in, capped by 2 in of dirty coal. In Linby Pit Bottom Underground Borehole only the Top First Waterloo, 16 in thick, was proved, the Bottom First Waterloo Coal being washed out; in the Engine House Underground Borehole, only 270 yd to the south, the Bottom First Waterloo is $11\frac{1}{2}$ in thick separated by 33 ft of mainly silty measures from the 14-in Top First Waterloo. In workings 1200 yd NW of Linby Colliery Shaft the First Waterloo consisted of hard coal 12 in, bright coal 3 in, dirt 4 in, on bright coal 24 in.

Over the remainder of the district, that is, in the north and west, the seam is divided into Bottom and Top First Waterloo coals; the line of splitting shown on Fig. 36 is drawn at the 12-in isopach of the dirt parting at which the split occurs. The maximum division of the leaves is in the north-west of the district and amounts to 39 ft in Swanwick Colliery, Common Pit. The dividing measures are composed of a single cycle.

The Bottom First Waterloo Coal is recorded as comprising 30 in of cannel in Swanwick Colliery, Common Pit, although this maximum thickness must be regarded as doubtful because only 12 to 15 in of coal were recorded in nearby shafts of the same colliery. The seam is mainly 18 to 24 in thick, but reaches 27 in at Brinsley Wharf and Nutbrook sites and 28 in in Dovecote Road Borehole. The seam was not recorded in Nutbrook Air Pit, where presumably it was washed out amongst the 18 ft of sandy measures below the Top First Waterloo Coal.

Apart from cannel, recorded as being 24 in thick in Woodside Colliery Shafts, some 20 to 22 in at Shilo Site, 60 in at Whiteley's Plantation Site and 18 in at Codnor Common Site, the Bottom First Waterloo consists largely of bright coal of variable quality. An impersistent dirt parting near the base divides off a thin floor coal of poor quality. Some sections, however, show a band of hard coal within the brights and others, like Dovecote Road and Lower Beauvale boreholes, may have up to 11 in of dirty coal at or near the top of the seam. The sulphur content is variable, many records showing it to be moderately high. At Broad Lane, Wansley Hall and Hobsic sites, in the north-central part of the district, the seam improves to good or medium quality, but still has a variable but generally moderate sulphur content. There are sporadic opencast workings in the seam. It is exposed in the Loscoe Brick Pit.

The Top First Waterloo Coal is thickest adjacent to the lines of split in the First Waterloo Coal. It is 30 in thick in Moorgreen Colliery 10's workings of which 15 in is dirty coal or dirt; 22 in in a drift [5060 4795] near Watnall Colliery, of which 2 in is dirty coal; and 19 in in Bentinck W. 1 Underground Borehole, of which $5\frac{1}{2}$ in is dirty coal.

Away from the lines of split the seam is frequently recorded as cannel, usually about 12 in thick, but reaching 23 in in Brinsley Colliery shafts. Some 22 in of coal occur in Portland Colliery No. 7 Shaft immediately north of the district and 24 in at Headhouse Farm Site in the south-west. The coal is economically unimportant.

THE MEASURES BETWEEN THE FIRST WATERLOO AND DUNSIL COALS form a single cycle normally about 30 ft in thickness, but ranging up to 57 ft as in Bentinck W. 1 Underground Borehole. The measures are variable in lithology and many sections show an unusually thick seatearth below the Dunsil Coal amounting to, for instance, 12 ft in Selston Colliery Underwood Shaft and Wollaton Colliery No. 2 Shaft and 16 ft in Coppice Colliery shafts. Small Anthracosia cf. phrygiana were recorded from the roof measures of the First Waterloo Coal in Eastwood Hall Borehole and A. disjuncta and Naiadites sp. in Annesley Park Borehole. Plant debris is present throughout the measures; sineoids and worm tracks have also been noted.

THE DUNSIL COAL usually consists of a main leaf 18 to 34 in thick and a thin floor coal a few inches below. Over large parts of the district (Fig. 37) the floor coal is not recorded separately and it appears from detailed sections that it is locally united with the main leaf. Elsewhere, particularly at Cromford Canal Site and over most of the eastern part of the district, less reliable evidence suggests that it has died out. The isopachs of the seam are shown in Fig. 37.

The thickest overall sequence at Damstead Lily Site is 50 in, of which the floor coal is 10 in and the dirt parting 7 in; the remaining 33 in of coal represents the second thickest record of the main leaf in the district. Other records from the same site show a 7-in floor coal, a dirt parting varying from 5 to 9 in and a 29-in main leaf. The seam also exceeds 42 in in the adjacent Broad Oak Site (floor coal $7\frac{1}{2}$ in, dirt 11 in, main leaf $25\frac{1}{2}$ in) and in Beauvale Borehole (floor coal 1 in, dirt 10 in, main leaf 32 in). The thickest record of the main leaf was in Moorgreen Colliery No. 1 Shaft, adjacent to Beauvale Borehole, and totalled 34 in of which the basal 4 in was composed of dirt and 'batt'; the floor coal was 1 in of 'batt' 4 in below the main leaf. The record of the Dunsil Coal in No. 2 Shaft at Moorgreen Colliery is:- floor coal, 1 in; 'clunch' 26 in; 'batt' 4 in; coal ('contains stone and smut') 32 in. The dirt parting does not exceed 11 in elsewhere in the district. Throughout the remaining areas of split seam the total thickness is not less than 24 in and the main leaf not less than 18 in.

In those parts of the district where the floor coal has not been separately recorded, the seam is thickest, 40 in, in Swanwick Colliery New and Deep pits. A record of 38 in in Wollaton Colliery No. 3 Shaft is doubtful, for only 23 in were noted in No. 1 Shaft. More detailed records from opencast sites nearby suggest that the floor coal is united with the main leaf. The seam is 36 in thick at Birdswood Site and Shipley Colliery, Deepfields Shaft, 35 in at Dobbs Site, 34 in at Ellerslie Site, 35 in in Bentinck 31's Underground Borehole and 31 in at St. Helen's Site and Woodside Colliery Shaft. The minimum thicknesses are 18 in in Brinsley Colliery Shaft and 16 in, of which the top 3 in are cannel, at Linby (Engine House) Underground Borehole. In Deepfields Shaft and Woodside Colliery the floor coal appears to be united with the main leaf.

The floor coal normally comprises bright coal which, in some analyses, is dirty; the main leaf has variable thicknesses of bright coal at the base and top with dull or banded bright and dull coal in the middle. Cannel, up to 11 in thick, is present at the top of the seam near Hucknall No. 2 and Linby collieries and at Katie and Kim sites to the south-east of Eastwood. An analysis from Station Site and another from High Leys Borehole showed a thin dirt parting near the top of the seam. The quality of the Dunsil is good to medium, with low to medium ash content and low to moderately low sulphur content. The seam has been mined locally, but it has mostly been worked by opencast means.

THE MEASURES BETWEEN THE DUNSIL AND TOP HARD (OR TOP HARD FLOOR) COALS form a single cycle about 40 ft in thickness, but as much as 54 ft in Shipley Colliery Deepfields Shaft and as little as 24 ft in Selston Colliery Underwood Shaft. Anthracosphaerium exiguum, fragmentary Anthracosia sp. and Carbonita humilis were present in the roof of the Dunsil Coal in Eastwood Hall Borehole and A. disjuncta, A. aff. phrygiana, C. humilis, and C. cf. scalpellus in Annesley Park Borehole and sineoids and worm tracks are also known. Ironstones are prominent in the mudstones, though in places, as at Deepfields Shaft, the cycle is mainly composed of sandstone.

TOP HARD COAL TO MANSFIELD MARINE BAND

The general thickness and lithological characteristics of this group of strata are shown on Fig. 46. Thicknesses range from a maximum of about 590 ft in the High Park Colliery area to a minimum of 520 ft at Hempshill Colliery.

The chief coals mined underground are the Top Hard - Comb at the base, and the Cinderhill, High Hazles, Lowbright and Mainbright higher in the sequence. Coals of lesser importance are the Mainsmut, Brinsley Thin and Two-Foot.

Sandstones are common and are locally prominent above the Cinderhill coals and between the Lowbright Coal and Haughton Marine Band. Washouts occur in the Top Hard and Clown seams.

The Two-Foot Marine Band is found throughout the area but the Manton 'Estheria' Band and Haughton Marine Band are impersistent.

The Top Hard is a convenient boundary to take between the A. modiolaris and Lower similis-pulchra zones for it separates the characteristic A. phrygiana fauna below, from the equally distinctive Anthraconaia pulchella Broadhurst fauna above.

The lowest distinctive fauna of the Lower similis-pulchra Zone occurs above the ? Second St Johns Coal and includes Anthraconaia pulchella, Anthracosia cf. planitumida and A. simulans.

Slightly higher in the sequence, above the Cinderhill Coal, representatives of the Anthracosia caledonica fauna were proved. In addition to this species the assemblages included A. sp. cf. fulva.

The mussel-bands between the High Hazles and Two-Foot coals are notable for the first appearance of the Anthracosia atra group. The associated species include Anthracosia simulans and Naiadites cf. obliquus. The A. atra group attains its acme in the beds overlying the Clown Coal. In the higher beds up to the Mansfield Marine Band mussels are not common but include Naiadites aff. angustus.

THE TOP HARD - COMB COAL consists of four elements, namely, the Top Hard Floor Coal, the main part of the Top Hard Coal and the Lower and Upper Comb coals, which occur in a variety of combinations in strata ranging in thickness from 12 ft at Shipley Colliery Deepfields Shaft to 87 ft at Dobbs Site (Figs. 39 to 41).

Over most of the outlier at Heanor all four seams are united and the dirt partings in the sequence may total as little of 10 in, as at Shipley Colliery. The various lines of split in the composite seam are shown on Fig. 39a.

The Top Hard Floor Coal is separately identified north-east of Eastwood (Fig. 39a). It is derived from the Top Hard seam by a split off the Bottom Brights (see below) and is of variable quality. It is separated from the Top Hard by up to 48 ft of measures in Newstead Colliery No. 1 Shaft. It is thickest near the line of split, e.g. 20 in at Moorgreen Colliery shafts, 5 to 9 ft below the Top Hard, and a maximum of 28 in in Kimberley Colliery Shaft, 2 ft below the Top Hard. It averages 16 in in the Selston area and at the nearby Hobsic Site 17 in of bright coal are underlain by 4 in of cannel. In Annesley Park Borehole in the north-east, 3 in of cannel overlie $1\frac{1}{2}$ in of coal. A cannel is well known at this horizon in the south-west of the Chesterfield district (Smith and others 1967, p. 169).

At Hucknall No. 2 Colliery the Floor Coal is 9 in thick some 14 ft below the Top Hard and at Broxtowe Colliery 17 in thick, 20 ft below. It is thinner elsewhere in the south-east of the district and in Beechdale Road Borehole is absent. Sporadic records, e.g. at Parkfield Farm and Catstone Hill sites (Fig. 40) show a thin (up to 3 in) parting near the base of the Floor Coal, separating off the lowest 1 to 2 in of coal.

The Top Hard Coal itself consists of bright coal at the base and top of the seam, the Bottom Brights and Top Brights, separated by a variable thickness of hard coal, the Hards (Fig. 40). The hard coal provided an excellent locomotive and steam fuel and the 'brights' a house coal. The overall high quality made it one of the principal economic seams in the coalfield and it was the first to be exhausted in the present district. In many early workings only the 'hards' were removed leaving the 'brights' as a roof and floor; these remnants have been exploited by opencast means in recent years and are still eagerly sought.

Variations in thickness of the Top Hard are shown in Fig. 41. In Hucknall Nos. 1 and 2 collieries there is cannel at the base of the seam, 27 in thick in No. 2 Colliery No. 5 Shaft. According to Elliott (1965, pp. 133-141) the cannel was formed in a pool flanking a 'swilley' which is present to the east of this district. At Macfez Site one record shows an anomalous 63 in of cannel at the base of the 130-in seam section.

The Comb Coal in the Heanor outlier is usually less than 9 in above the Top Hard, but locally as much as 15 in. The following section of the Comb from Office Site (2 in above the Top Hard) may be taken as typical: Lower Comb Coal, 18 in; dirt, $8\frac{1}{2}$ in; Upper Comb Coal, 27 in. The dirt parting within the Comb expands gradually southwards and is 3 ft thick at the line of split shown on Fig. 41. In Shipley Lake Site the dirt has passed into argillaceous measures about 40 ft thick. The Upper Comb Coal, normally 24 to 30 in thick in this site, is 60 in in one proving, and the Lower Comb, 12 to 27 in thick, lies 4 in to 3 ft above the Top Hard. This split in the Comb Coal occurs sub-parallel to, and appears to be related to, a linear 'washout' of the Lower Comb and Top Hard coals which occurs in the southern part of the site.

At Codnor Common Site, in the outlier at Ripley, the Comb Coal totals 52 in, including the 4-in dirt parting, and is 5 ft above the Top Hard. The separation from the Top Hard is up to 87 ft in Dobbs Site, in the north of the district, where the overall thickness of the seam is 25 in, of which only $\frac{1}{4}$ in is dirt. Nearby, at Damstead Lily Site, the upward sequence was coal $19\frac{1}{2}$ in, dirt 2 in, coal 1 in, dirt 2 in, coal $13\frac{1}{2}$ in.

Cromford Canal Site [460 481] exposed a 40-ft trunk of Lepidodendron in position of growth, with its base in the Top Hard Coal and its top

a

b

close to the base of the Comb seam. (Plate 6).

In the remainder of the district the Comb Coal varies from 21 in at Linby Colliery to a maximum of 46 in, including 6 in of dirt, in the Kimberley Colliery area. Over a wide area it lies less than 18 in above the Top Hard Coal.

THE MEASURES BETWEEN THE COMB AND MAINSMUT COALS range in thickness from 21 ft at Moorgreen Colliery No. 1 Shaft to 71 ft at Annesley Park Borehole (Fig. 75). The Comb Coal is closely succeeded over wide areas by sandstone or interbedded sandstone, siltstone and mudstone, which is locally referred to as the 'Top Hard Rock'. This is thickest (over 50 ft) in the Hucknall area, and over much of the exposed coalfield forms a pronounced feature or is part of a composite feature upon which Swanwick, Ripley and Heanor are sited.

The following is a composite section of these measures built up from measurements along a face [4568 4823] at Cromford Canal Site, near Eastwood.

		ft	in
Coal, dirty (Mainsmut)	up to		5
Seatearth; sandstone		1	5
Sandstone, fine, brown		5	0
Siltstone		1	6
Mudstone, pale grey silty; plants and nodules		4	0
Sandstone, fine, brown, carbonaceous debris; massive beds up to 4 ft ('Top Hard Rock')		28	0
Siltstone		2	2
Mudstone, silty		4	0
Coal, bright (Comb)		3	6

THE MAINSMUT COAL has been traced from Pinxton in the north of the district to Wollaton in the south, but its crop is concealed beneath Permo-Triassic rocks for about 1 mile in the Strelley - Kimberley area. It also occurs to the west of the main crop in several faulted synclines near Swanwick, Ripley, Eastwood and Shipley.

In the type area around Shipley, Kimberley and Watnall, the seam is generally 24 to 36 in thick (Fig. 75). A maximum of 39 in was recorded at Shipley Park Borehole, where the seam was composed of two 17-in bands of bright coal separated by 5 in of 'hards'. North-eastwards the seam splits into two leaves, up to 27 ft apart, and deteriorates both in quality and thickness. The ash content of the seam in the south and west is usually moderately high (7 to 10 per cent) but in the north and east it is very high (29.8 per cent at Delves Site near Swanwick). The sulphur content is typically low.

The Mainsmut Coal is unusual because of the presence of inclusions within the upper part of the seam. They vary from sandy laminae to lenses of quartzitic sandstone. In City Site, the Coal Survey Laboratory recorded that sandstone replaced large parts of a tree stump. 'Stone' inclusions have also been recorded from Hobsic, Cromford Canal and Poplar Farm sites. Earthquake activity following deposition may be a contributory factor to the formation of the inclusions but the coal is typically overlain by silty mudstone, and only rarely by sandstone.

THE MEASURES BETWEEN THE MAINSMUT AND CINDERHILL COALS range in thickness from 106 ft in Moorgreen No. 1 Colliery Shaft to 42 ft at Hucknall No. 2 Colliery No. 5 Shaft. They contain, in the areas of thickest sedimentation, several cycles and two thin coals (up to 9 in thick) or seatearths. These coals may equate with the Second St John's Coal of the Chesterfield district to the north.

Sandstone, up to 50 ft thick, is common and forms good features in the Pinxton - Brinsley area. Bivalves are recorded at various horizons, notably some 30 and 60 ft above the Mainsmut Coal (Appendix 1, p. 165a). They are usually found either in black shale as at Beechdale Road Borehole (at 153 ft), or preserved in ironstone nodules or bands. Fish scales were recorded above a 1-in coal at Seagrave Borehole, some 45 ft above the Mainsmut.

THE CINDERHILL COAL sometimes referred to as the Cinderhill Main, is 44 in thick in Cinderhill Colliery No. 4 Shaft and consistently a little under 4 ft in the surrounding area. It was mined in the 1920's on both sides of the Cinderhill Fault. This belt of thick coal trends west-north-westwards through Nuthall and Kimberley and has been worked opencast in synclinal outliers south of Heanor, at Poplar Farm and Johnson Farm sites (Fig. 76). In these thick sections the coal is composed essentially of 'brights' with banded dull and bright coal forming a thin parting in the middle of the seam. Ash content is low to moderate and sulphur percentages are moderate to moderately high.

At Hempshill, less than a mile to the north of the Cinderhill area, the upper half of the coal contains many dirt partings. North-east of a line through Strelley, Kimberley and Eastwood the Cinderhill splits and may be represented to the north and east by up to three coals (Fig. 76b) i.e. an upwards sequence of a Cinderhill Coal, an Upper Leaf and a split from the Upper Leaf.

The interval between the Cinderhill and the Upper Leaf seam ranges from 9 in at Watnall Colliery No. 7 Shaft to 30 ft at Cromford Canal Site. The Upper Leaf is 18 in thick in Watnall Colliery No. 6 Shaft and varies between 10 in and 24 in in Cromford Canal Site. It thins to the north to less than 6 in. The Upper Leaf is split over a large area by a parting up to 22 ft thick at Hill Top Colliery. The highest leaf, however, is commonly absent and represented only by a seatearth. Smith and others (1967, p. 174) state that the Cinderhill Coal cannot be recognised with certainty in the shaft sections east of Alfreton, but infer from its position relative to the High Hazles that it may be the correlative of the First St John's Coal of districts to the north.

THE MEASURES BETWEEN THE CINDERHILL AND HIGH HAZLES COALS vary between 105 ft at Hempshill Colliery in the south and 48 ft at Newstead Colliery in the north, although leaves of the two coals are only 27 ft apart at Moorgreen Colliery. The succession in the areas of maximum deposition is dominated by sandstone, which is over 80 ft thick at Hempshill Colliery. At Newstead this sandstone was recorded as 'exuding grease'.

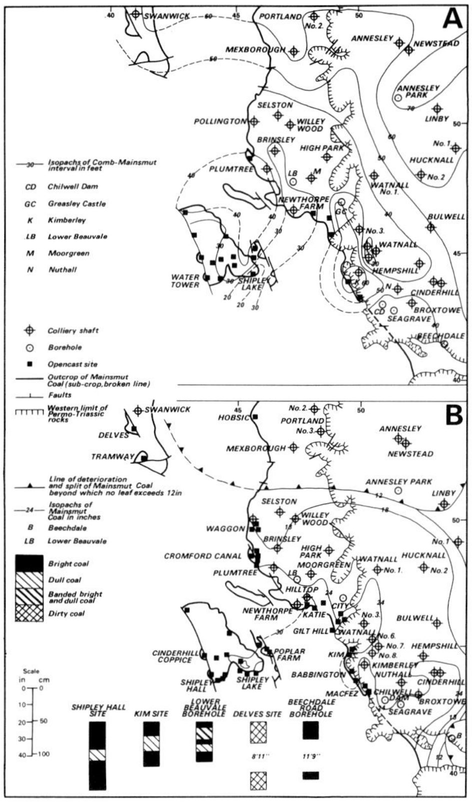

c

Figure 75 A Isopachs of the interval between the Comb Coal and the Mainsmut Coal (upper leaf where split)
B Isopachs of the Mainsmut Coal showing location of provings and representative sections of the seam. Borehole information from opencast sites not named on this plan has also been taken into consideration

d

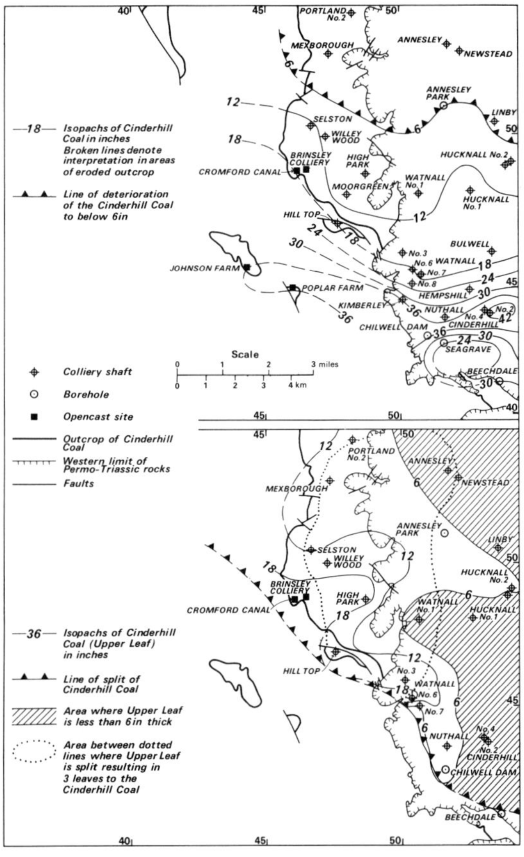

Figure 76 Isopachs of the Cinderhill Coal showing location of provings

a

THE HIGH HAZLES COAL is an economically important seam which has been worked in the past from Linby, Bulwell and Cinderhill (Babbington) collieries. It has promising reserves in the north-east of the district, where it is well documented from over 50 provings. The main seam is a maximum thickness of 36 in in Linby 52's Underground Borehole in the north-east, but it splits and thins to the south and west beyond Hucknall and Annesley (Fig. 42). There are three thin floor coals, all 12 in or less in thickness, within 27 ft of strata below the main seam in Cinderhill Colliery No. 4 Shaft. They are represented by seatearths in Garfit House Borehole and farther south are missing, eg at Annesley Lodge Borehole.

West of the main split of dirty coal from the bottom of the High Hazles the main seam is thin in a sinuous belt extending from Portland Colliery through Annesley Lodge to Kimberley (Fig. 42). This belt corresponds closely in position to underlying sandstones and siltstones and is attributed by Elliott (1969, p. 120, fig. 3) to a swamp drainage feature subsequently filled with distributory deposits. West of this belt the main seam is 42 in thick at Rap Site and 45 in thick in an isolated area at Allens Green Site. Farther west in the Swanwick Syncline, opencast sites such as Tramway and Sleet Moor/Alfred Sleet show up to four divisions in the main seam with dirt totalling up to 15 ft. The top part of the main seam commonly contains 'stone' inclusions over much of Sleet Moor Site.

The High Hazles Coal consists largely of bright coal, but both the top and bottom of the seam are dirty. A sinuous belt of inferior coal developed in the top of the seam has been traced by Elliott (1969, fig. 3) north-east of Hucknall, and is comparable in trend with the belt of thin coal referred to above.

Patches of cannel are developed locally at the top of the seam, as at Newstead. The ash content varies considerably; it is lowest in the areas of thickest coal at Linby (3.7 per cent) and Allens Green Site (2.9 per cent) and high to very high in the Swanwick Syncline. Sulphur content is moderately low.

THE BRINSLEY THIN COAL is separated from the High Hazles Coal by largely argillaceous measures ranging in thickness from 20 ft at Hempshill Colliery to 45 ft at Portland Colliery No. 2 Shaft. In the type area near Brinsley village, two miles north of Eastwood, the Brinsley Thin consists of up to 48 in of coal and dirt, but its average thickness is 2 ft at outcrop between Selston and Watnall and it decreases to under 6 in in the north-east of the district. In the Swanwick Syncline the coal is characterised by numerous splits and dirt partings. A belt of thin coal is present between Annesley Lodge Borehole and Cinderhill in a similar position to that in the underlying High Hazles seam (p.164b).

North and east of this belt a thin coal, the Brinsley Thin (Upper Leaf), is proved above the Brinsley Thin. It is represented by a seatearth 9 ft 3 in above the Brinsley Thin in Kennel Wood Borehole and is a 4-in coal at Annesley Park Borehole. It thickens eastwards to a 13-in seam, 12 ft 2 in above the Brinsley Thin, in Linby 52's Underground Borehole.

The Brinsley Thin Coal is variable in quality with ash content ranging from good to very inferior (3.6 to 27.0 per cent). Sulphur content varies from low to moderately high (0.5 to 2.4 per cent). The best coal was obtained from Pepper Hill Site, but wide variations occur even within one site (e.g. at Allens Green Site; ash content 5.3 to 27.3 per cent). As a result of such variability, coupled with the limited area of coal over 2 ft thick, the Brinsley Thin has never been an economic proposition for deep mining.

THE MEASURES BETWEEN THE BRINSLEY THIN (UPPER LEAF) AND LOWBRIGHT FLOOR COALS range in thickness from 50 ft at Selston Colliery in the north-west to 25 ft at Hempshill Colliery in the south-east. They comprise largely argillaceous strata containing sporadic 'mussels' including Naiadites cf. obliquus, particularly common in the roof of the Brinsley Thin, together with plant fragments and Planolites. The mudstones contain ironstone bands up to 8 in thick.

THE LOWBRIGHT (FURNACE OR ABDY) COAL is over 30 in thick in much of the district, reaching a maximum of 46 in in the south-west at Lambclose House Site (Fig. 43). It has been worked underground to a limited extent at Newstead, Watnall and Cinderhill (Babbington) collieries, but has been extensively exploited by opencast workings between Pinxton and Kimberley.

The seam thins to below 30 in around Kennel Wood Borehole (21 in) and Watnall Colliery No. 3 Shaft (12 in). These areas of thinning appear to be related to the belts of thin coal in the underlying Brinsley Thin and High Hazles seams (pp. 00 and 00). The Lowbright Coal shows a deterioration in quality, with several thick dirt partings, in the north-west of the district. The Sleet Moor Site in the Swanwick Syncline typifies this split inferior coal sequence (Fig. 43).

The Lowbright is composed largely of 'brights' with a persistent band of 'hards' about the middle of the seam, at which level the splits tend to occur. In the areas of thinner coal (see above) only that part of the seam below the 'hards' is present. In the Annesley - Linby area the bottom of the seam includes inferior pyritic coal. The ash content varies from very low (2.5 per cent) in the Hucknall - Watnall area to high (over 15 per cent) in the Portland area. Sulphur percentages range from low to moderate.

The Lowbright Coal was exposed in bridge foundation holes [5246 4302] near Broxtowe Colliery, where the following section was measured:

	ft	in
Made ground	6	0
Coal, bright, friable (Lowbright)	3	0
Seatearth, mudstone	5	9
Mudstone, dark grey, bituminous	0	3
Coal (Lowbright Floor)	1	6½
Mudstone, bituminous		

The Floor Coal usually occurs less than 10 ft beneath the Lowbright. In the north-east of the district it closely approaches the main seam and in Newstead N1 Underground Borehole a separation of only 5 in is recorded. The Floor Coal, which is commonly split by a dirt parting in the area around Hucknall and Newstead, is composed largely of 'brights' averaging some 10 in, but increases in thickness north-westwards to 16 in at Oaktree Farm Site and between 22 in and 32 in at Sleet Moor Site in the Swanwick Syncline.

THE MEASURES BETWEEN THE LOWBRIGHT AND TWO-FOOT COALS consist predominantly of interlaminated and interbedded sandstone and siltstone and are thickest, up to 44 ft, in Kennel Wood Borehole. The sandstone occurring at this locality is unusual. It contains two thin breccia horizons made up of siltstone fragments and ferruginous nodules, and shows many sedimentary structures such as current ripples, disturbed laminae, crumpled bedding and rib and furrow development. The sequence is thinnest (26 ft) in the vicinity of Selston Colliery, where the roof measures of the Lowbright Coal contain bivalves.

THE TWO-FOOT (SOUGH OR MIDDLEBRIGHT) COAL exceeds 18 in over much of the district, but varies from 29 in at Rap Site in the south-west to about 6 in in the extreme north-east (Fig. 78). In the Swanwick Syncline it averages 24 in. Sporadic sections such as Cinderhill Colliery No. 4 Shaft show a minor dirt parting within the middle of the Two-Foot Seam. The seam is of little economic value for deep mining, but its close proximity to the Mainbright Coal along most of the outcrop has led to the extraction of both seams from the same opencast workings. The seam consists largely of 'brights', but many analyses show the upper half to be rather dirty. The ash content therefore varies widely, from low (4.3 per cent, Garfit House) to very high (21.7 per cent, Whyburn House) in under two miles. Sulphur percentages range from low (0.8 per cent, Wansley Hall) to very high (4.4 per cent, Allens Green).

A thin floor coal, 2 to 9 in thick, is present some 2 ft to 10 ft below the Two-Foot between Cinderhill and Selston. In Cinderhill Colliery No. 2 Shaft a 7-in coal some 10 ft above the Two-Foot may be an upper leaf of the seam.

THE MEASURES BETWEEN THE TWO-FOOT AND MAINBRIGHT COALS range between 19 ft in Linby 52's Underground Borehole in the east and 10 ft at Willey Wood Colliery in the west. The Two-Foot Coal is invariably overlain by dark grey mudstone - 'black bind' in the old shaft records. It is slightly silty, rather fissile and contains ferruginous laminae, some with a speckled appearance, possible due to kaolinisation, whilst others are oolitic. The mudstone is richly fossiliferous, containing fish debris, 'Estheria' sp., bivalves and microfossils, together with pyritized and carbonised plant fragments (Appendix 1, p. 164d).

The Two-Foot Marine Band forms a reliable marker horizon and has been proved to be of wide lateral extent throughout the district. It comprises that part of the above-mentioned fossiliferous mudstones which contains Lingula and foraminifera, commencing immediately or close above the Two-Foot Coal and extending up to 5 ft 2 in above. The Two-Foot Marine Band is some 2 ft 6 in thick in Whyburn House Borehole with the lowest part represented by Lingula mytilloides only. The upper part contains both Lingula and Parapachites together with fish remains.

In Felley Lane Borehole the fauna contains Myalina sp. in addition to the above species an assemblage which is typical of the Myalina facies present at this horizon in the Notts-Derby Coalfield. Lingula has been recorded by Mr W. N. Edwards in the roof of the Two-Foot [4710 5140] in Bagthorpe Site and in poorly exposed mudstone [4784 5200] together with fish scales in Middle Brook near Selston. Dr J. Shirley found Lingula 2 to 3 in above the coal in Willeywood Farm Site [4707 4899]. Lingula was also recorded in Cinderhill Mainbright Drift, Cinderhill Return Drift and Bulwell Steps Drift. The marine band has been detected in shallow proving holes at Sleet Moor Site near Swanwick, and on the line of the M1 Motorway near Pinxton, just beyond the northern margin of the district. The following recent boreholes drilled by the N.C.B. passed through the Two-Foot Marine Band, the thickness of which is indicated in brackets:- Annesley Lodge (4 in), Annesley Park (13 in), Felley Lane (3 in), Garfit House (9 in), Hucknall High Main 27's (4 in), Mapplewells (38 in), Newstead N3 Underground (47 in) and Whyburn House (30 in).

The dark grey mudstones in the lower half of this cycle pass up into silty mudstones with sporadic siltstone laminae containing a few mussels including Anthraconaia cymbula and scattered plant fragments.

THE MAINBRIGHT COAL is an economically important seam, which has been worked from Hucknall, Watnall, Bulwell, Hempshill and Cinderhill collieries, but its best reserves lie beyond the eastern limits of the present district. The type section of the seam is at Hucknall No. 2 Colliery, No. 5 Shaft, where it consists of 34 in of bright coal containing a 5½-in parting of banded 'dull' and 'brights'.

In the main area of outcrop, the Mainbright Coal is thickest in the south and west with a maximum of 49 in recorded in a temporary section [5276 4307] near Broxtowe Wood (Fig. 44). It thins northwards to 36 in in the Hucknall - Watnall - Selston area and to 24 in near Annesley Park and Mexborough. The coal deteriorates in the north-east corner of the district, where it splits into thin seams with dirt partings, or is represented by cannel and inferior coal as at Newstead. A washout was encountered in Garfit House Borehole and the top of the seam is partially eroded beneath sandstone at Linby 52's Underground Borehole. In Annesley Colliery Shaft 'grey and brown rock mixed with fossils and coal' was recorded 8 ft above a 6-in coal at the Mainbright horizon and a partial washout is indicated in Mapplewells Borehole by thin coal, cannel and batt at a similar horizon.

The seam generally is of good quality with low ash and sulphur contents. It is composed largely of 'brights', with subsidiary partings of dull or banded bright and dull coal throughout (Fig. 44). In the Swanwick Syncline the Mainbright Coal is exceptionally thick (up to 50 in) but contains several dirt partings and the upper half of the

c

b

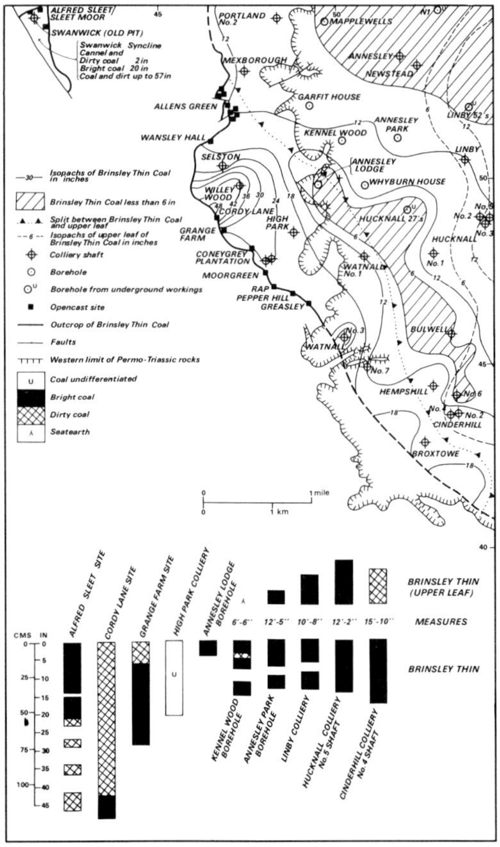

Figure 77 Plan of the Brinsley Thin Coal showing location of provings

d

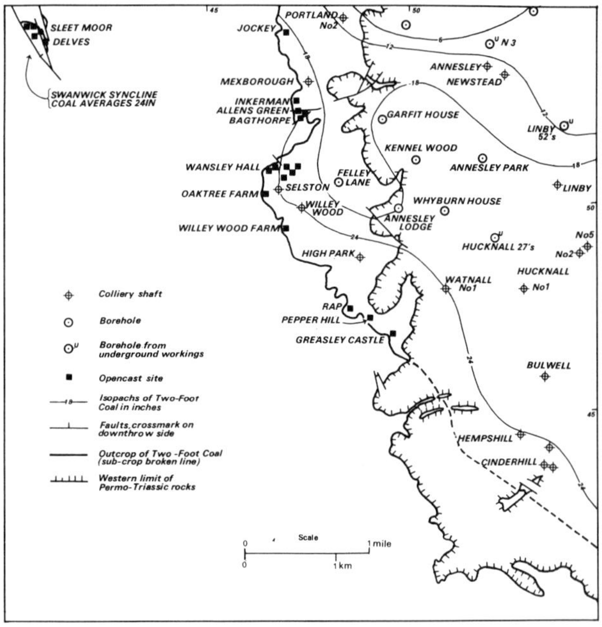

Figure 78 Plan of the Two-Foot Coal showing location of provings

a

seam is dirty coal with an ash content as high as 21.3 per cent.

THE MEASURES BETWEEN THE MAINBRIGHT COAL AND THE MANSFIELD MARINE BAND vary from 246 ft at Annesley in the north to 185 ft at Cinderhill in the south. They include several thin coals of no economic value such as the Clown and ? Swinton Pottery. Marine bands and marker horizons proved in these measures in the Chesterfield and Ollerton districts to the north tend to be less persistent and only the Haughton Marine Band has been proved with certainty in the present district. The measures are rich in ironstone nodules and ferruginous patches; sphaerosiderite is common locally, as in the seatearth of the Clown Coal. Worm burrows are much in evidence and are often lined by ferruginous residues; root structures in the seatearths are replaced by ironstone.

'Estheria' is found in the roof of the Mainbright Coal in many recent boreholes, e.g. Annesley Lodge, Annesley Park, Felley Lane, Kennel Wood and Mapplewells. The fossil is preserved mostly as irridescent films in a grey mudstone, and is associated with ostracods, small mussels and fish remains.

Distorted roof measures have been observed on the Mainbright Coal. Shirley (1955, p. 274) recorded near-vertical 'dykes' of hard sandstone 2 to 5 ft thick, in roof measures at Bagthorpe Site. He attributed this phenomenon to earthquake tremors, the effects of which he has traced in other parts of the coalfield at this horizon (Shirley in Eden and others 1957, p. 112). Shirley (1955, p. 275) noted that the roof of the Mainbright in Hucknall No. 1 Colliery, in a small area south-west of the shafts, is composed of a mixture of sandstone, mudstone and coal, which may be due to either washouts or to earthquakes.

The lowest cycle lying between the Mainbright Coal and the Manton 'Estheria' Band, contains beds of sandstone up to 50 ft thick which make a strong feature and have been mapped in the Mexborough - New Portland area [47 54].

A thin coal at the Manton 'Estheria' Band horizon has been proved in Mapplewells, Kennel Wood, Whyburn House, Annesley Park and Annesley Lodge boreholes and in Hucknall No. 2 Colliery No. 5, Linby Colliery No. 1, Newstead Colliery No. 1 and Willey Wood Colliery shafts. It is thickest (18 in) at Annesley Park Borehole, but contains many sandy inclusions and lenses. It is represented by a seatearth at Willey Wood Colliery Shaft and is probably washed out in the Cinderhill area to the south.

The Manton 'Estheria' Band horizon occurs in Annesley Park Borehole, where 'Estheria' associated with fish scales is present in 8 in of black mudstone. Nearby boreholes which showed mussels and fish debris in dark grey to black mudstones, commonly silty and carbonaceous, at this horizon are: Kennel Wood, 6 in with mussels; Mapplewells, 17 in, with fish debris and mussels 9 in above a 4-in coal; Whyburn House, 6 in with fish debris and mussels.

Some 7 ft above the ? Manton 'Estheria' Band in Grives Quarry Borehole [5015 5492], just beyond the northern edge of the district, a kaolin oolith bed was recorded. Kaolin also cements a sandstone 24 ft above the Band in Felley Lane Borehole.

The strata between the ? Manton 'Estheria' Band and the Clown Coal range from 78 ft in the west at High Park Colliery to 12 ft in the east at Annesley Park Borehole. South of Hucknall the absence of the Manton 'Estheria' Band and presence of washout sandstones at the Clown horizon make correlation uncertain.

The Clown Coal is thickest in the west, amounting to 78 in in Wansley Hall Site, but thinning rapidly eastwards. Beyond Watnall and Newstead it is either washed out, thin or represented by a seatearth (Fig. 45).

The coal commonly contains many dirt partings which render the seam economically worthless. The Coal Survey Section at Dixie Site [4765 5020], near Selston, was coal 12 in at top, dirt 165 in, coal 3 in, dirt 14 in, sandstone 2 in, coal 6 in, dirt 1 in, coal 33 in, dirt 2 in, with coal 17 in at bottom.

In the High Park area, the Clown Coal is split into two leaves by some 12 ft of measures. This split increases northwards to 35 ft at Willey Wood Colliery, where the lower leaf thins out and is represented only by seatearth. At Linby Colliery No. 1 Shaft, two coals of 10 and 9 in, split by a 15-in dirt parting, are separated by 21 ft of measures from an underlying 17-in cannel.

Leaves of the Clown Coal were exposed in the M1 Motorway cuttings [4735 5390] near Selston where the following section was recorded:

		ft	in
Boulder Clay	up to	10	0
Coal		2	1
Seatearth; mudstone, pale grey		2	0
Mudstone, grey and brown, silty		6	0
Mudstone, khaki; silty and sandy bands		4	0
Coal	9 in to	1	1
Seatearth; mudstone grey; rootlets		3	6
Mudstone, grey		1	0
Coal and dirt		1	0
Mudstone, dark grey, with coal streaks		1	0
Seatearth, mudstone; soft and grey in top 6 in, pale grey and silty below		4	6

Temporary sections in pipe trenches [4765 5237] on Selston Common proved the Clown crop, with the following partial section:

	ft	in
Mudstone, yellow sandy	3	0
Mudstone, silty, pale grey	1	0
Coal and dirt	0	6
Seatearth, brown	0	0½
Coal and dirty coal	2	0

Some 80 yd farther south the following section was recorded [4739 5381]:

	ft	in
Mudstone, yellow-brown, silty	6	0
Ironstone, oolitic, pyritic		1¾" to 11½
Coal, dirty, dull	2	0
Seatearth, pale grey	1	6
Coal	1	0 seen

b

The oolitic ironstone is very variable in thickness over a 50-yd length of cutting. The upper coal is probably the same seam as the 2-ft coal in the adjacent section.

In Selston Colliery yard [4687 5040] a temporary section showed 12 in of coal immediately overlain by a sandstone which has washed out the Clown Coal to the north in Wansley Hall Site.

Further washout phenomena were proved in Felley Lane and Annesley Lodge boreholes, Hucknall High Main 27's Underground Borehole and Hucknall No. 2 Colliery No. 5 Shaft, which lie in a clearly defined belt trending west-north-west to east-south-east (Fig. 45).

The strata overlying the Clown Coal are either dark grey mudstones or thick sandstones, the latter in places filling washouts in the seam. The sandstones produce a marked feature in the Underwood area, upon which part of the village is sited. Boreholes through the concealed measures to the east proved up to 80 ft of sandstone (Watnall Colliery No. 1 Shaft), but thicknesses vary irregularly; for instance, from 45 ft at Annesley Park Borehole, where the rock is a pale grey-green colour to only 10 ft at the nearby Linby Colliery Shaft.

The dark grey to black mudstone intermittently forming the roof of the Clown Coal contains variable amounts of fish and carbonaceous plant debris in the lowest few inches. In Mapplewells Borehole, fish scales are common, but they are rare in Annesley Park Borehole. In Whyburn House Borehole the fish scales were preserved in a 'canky' siltstone. In the Chesterfield district to the north the mudstone also contains large specimens of Lingula and forms the Clown Marine Band.

The mudstones above the horizon of the Clown Marine Band contain numerous 'mussels' typical of the Anthracosia atra fauna, together with pyritic plant fragments and ferruginous bands showing evidence of boring organisms. Some 20 ft above the Clown Coal, in Mapplewells Borehole, such mudstones and siltstones are of a distinctive pale greenish grey colour. This colouring may be penecontemporaneous, for the Permo-Triassic unconformity is some 440 ft above.

There are a number of coals or seatearths between the Clown Coal and the Haughton Marine Band. The thickest seam is 29 in at Cinderhill Colliery No. 4 Shaft. At Hucknall Colliery No. 1 Shaft a group of coals and dirts has an over-all thickness of 139 in. It is not clear whether the first coal or seatearth below the Haughton Marine Band is the Swinton Pottery of districts to the north, or whether several or all of the coal horizons between the Marine Band and the Clown represent a split Swinton Pottery.

The topmost coal lies either immediately below the Haughton Marine Band or is separated from it by up to 5 ft of variable measures including mudstones, siltstone and sandstones. The mudstones are grey to black with ironstone bands and plant fragments. At Annesley Park Borehole an oolitic ironstone, ½ in thick, immediately underlies the marine band and in Hucknall High Main 27's Borehole a 1-in kaolin oolith band was recorded in a 2-in ironstone immediately above the coal.

The Haughton Marine Band consists of dark grey mudstone with marine fossils. It is characterized by gastropods and Tomaculum sp. which occur together with Lingula and fish debris. The marine band is not seen at outcrop, but has been proved in Cinderhill Colliery No. 4 Shaft and the following boreholes: Annesley Park, Felley Lane, Hucknall High Main 27's, Kennel Wood and Mapplewells. It is only 1 in thick in Grives Quarry Borehole just north of the district boundary, and is probably washed out in Hucknall Colliery No. 5 Shaft. In Felley Lane Borehole the assemblage included Euphemites anthracinus. Bellerophontoid gastropods were also recorded in Kennel Wood Borehole, where Lingula mytilloides, Serpuloides stubblefieldi and fish debris were found near the base of the marine band. The mudstones are particularly rich in ironstone and pyrite, and listric surfaces, some of which are associated with the zone of compression around ironstone nodules, are common.

THE MEASURES BETWEEN THE HAUGHTON MARINE BAND AND THE MANSFIELD MARINE BAND comprise two main cycles totalling 57 to 78 ft in thickness. They are largely mudstones with subordinate arenaceous strata; sphaerosiderite is common. The top of the first cycle is marked by a thin, but persistent coal which has a maximum thickness of 15 in at Cinderhill Colliery No. 4 Shaft. In the north-east of the district this coal splits into several leaves, as in Annesley Park Borehole: coal 6 in at the top, dirt 64 in, coal 12 in, dirt 20 in, coal 2 in. The coal is generally of poor quality (inferior 'brights') and is interdigitated with ironstone at Kennel Wood Borehole. Just beyond the northern limit of the district, in Grives Quarry Borehole, ½ in of irregular ironstone occurs with a 7-in coal at this horizon, and in Newstead N1 Underground Borehole a 9-in ironstone with incipient cone-in-cone structure immediately overlies the coal.

This coal is overlain by the ? Sutton Marine Band in Mapplewells Borehole. The marine band may also be present in Grives Quarry Borehole and Hucknall Colliery High Main 27's Underground Borehole, but it is apparently missing over the greater part of the Derby district. At Mapplewells Borehole the possible marine band consists of 9 in of dark mudstone with pyrite and phosphatic nodules, its fauna consisting largely of fish debris.

The top of the cycle containing the Sutton Marine Band is invariably marked by a coal and/ or seatearth. The coal is thickest (7 in) in Cinderhill Colliery No. 2 Shaft. This horizon is in places separated from the overlying Mansfield Marine Band by a few inches of mudstone or siltstone with pyritised and coaly plant fragments and non-marine bivalves.

c

WESTPHALIAN C

MANSFIELD MARINE BAND TO TOP MARINE BAND

These strata, some 350 ft thick, contain only one economically important coal, the High Main. They are concealed by Permo-Triassic rocks except for a narrow strip cropping out between Greasley and Pinxton. The general characters of the strata are shown in Fig. 47.

Above the Mansfield Marine Band, the mussel fauna is less varied than in the lower beds, with Naiadites as the dominant genus. Between the Mansfield Marine Band and the Edmondia Band the species include Naiadites melvillei and N. cf. productus. Above the Edmondia Band N. cf. hindi is common. The lowest occurrence of Lioestheria vinti in this part of the sequence is in the roof measures of the first coal above the Edmondia Band, and this fossil recurs in several bands up to the Top Marine Band. Ostracods in these measures include Geisina subarcuata and Carbonita pungens.

THE MANSFIELD MARINE BAND consists of up to 15 ft of dark grey mudstone with an abundant and varied fauna. A bed of fossiliferous 'cank' (hard ankeritic siltstone) occurs in the lower part and provides a reliable and persistent lithological marker horizon which is recognisable even in the records of old shafts and sinkings. The 'cank', which is 6 in to 32 in thick, has a distinctive conchoidal fracture and is irregularly veined by calcite. Its petrography was described by Dunham (in Edwards and Stubblefield 1948, pp. 249-252). Chonetoid brachiopods and small Lingula are commonly present.

The 'cank' has been seen at outcrop in two localities near Felley. At one [4829 5217] in Middle Brook, Millington Springs, it is 6 in thick, and at the other, in a small stream [4852 4993] 100 yd S of Felley Mill, it is 2 ft thick and contains Lingula.

The Mansfield Marine Band has been recorded in the boreholes and shafts noted in Table 13.

The presence of cank has been established in the following shaft records: Annesley Colliery No. 1, 1 ft 6 in; Cinderhill Colliery No. 2, 1 ft 9 in; Cinderhill Colliery No. 6, 1 ft 9 in; Hempshill Colliery, 10 in; Hucknall Colliery No. 1, 1 ft 4 in; Hucknall Colliery No. 2 Pit No. 3, 1 ft 8 in; Hucknall No. 2 Pit No. 5, 1 ft 2 in; Linby Colliery No. 1, 2 ft 7 in; Newstead Colliery, 2 ft 4 in; Watnall Colliery No. 1, 1 ft 9 in.

The Mansfield Marine Band in Annesley Lodge Borehole is some 6 ft 9 in thick and shows three distinct phases. The basal 5 in contains a benthonic fauna of Lingula, Euphemites and Hollinella. The acme of the marine incursion is represented by the central part, which contains Coleolus and 'Anthracoceras' hindi. A regression phase in the top 4 ft has an assemblage of foraminifera and Lingula.

Annesley Park Borehole proved only a thin marine band with an impoverished fauna at this horizon. Anthracoceras is common in the lower part of the band but poorly preserved. The top part of the band contains foraminifera including Ammodiscus, together with Lingula mytilloides, Orbiculoidea cf. nitida, cf. Geisina, fish debris and fucoids.

THE MEASURES BETWEEN THE MANSFIELD MARINE BAND AND HIGH MAIN COAL vary from 158 ft in the north-west, in Mapplewells Borehole, to 115 ft in the south-east, at Hucknall Colliery No. 2 Pit, and contain up to four coals. The measures crop out just to the west of the Permo-Triassic escarpment in the Annesley Woodhouse-Selston Underwood areas, but are concealed by the younger rocks south of Watnall.

The variations of thickness in the measures are related to the presence of sandstones, particularly in the lowest and highest cycles. The sandstone beneath the High Main Coal is partially exposed in Beauvale Brook between Felley Priory and Felley Mill, where up to 10 ft of brown flaggy sandstone are visible [4857 5016].

Eden and others (1963, p. 54) record the sporadic occurrence of a kaolinite-rich band below the High Main Coal in boreholes to the south-east of the present district. The following boreholes indicate that this bed is widespread, but because of the difficulty of differentiating such tonstein-like bands from normal ironstones and ferruginous mudstones it may have been missed in many of the records. The figures in brackets indicate the depth below the High Main Coal:

	ft	in
Heatherdale Pond; 2-in tonstein	2	9
Kennel Wood; brown ? mud lenses	0	11
Washdyke Lane; 1-in pale 'ironstone'	3	0
Whyburn House; 3-in ? tonstein	2	1
Whyburn West; cream flaky hard layer	1	10

The measures between the Mansfield Marine Band and the High Main Coal are rich in ferruginous concretions, and sphaerosiderite is particularly common in the seatearths. Ironstone bands are locally oolitic and often have kaolin-filled centres. Kaolin patches and spots have been recorded in Hucknall High Main 27's and High Main 52's underground boreholes below all the minor coal seams within these measures.

The lowest of these minor coals is considered to be the probable equivalent of the Wales Coal of the East Retford area (Smith and others 1973, p. 88). It is not known whether the overlying coals are splits from the Wales or represent higher seams including the Sharlston Low of Yorkshire.

The ? Wales Coal is about 12 in thick in the north-west of the district, where it lies some 85 ft above the Mansfield Marine Band. Traced south-eastwards this interval is halved and the seam deteriorates until, in the Cinderhill area, it is represented solely by seatearth. Analysis of this coal obtained from Newstead N1 Underground Borehole just north of the district boundary, shows a medium ash content and low sulphur percentage.

The thickest single seam recorded in the measures up to the High Main Coal is 23 in at Hucknall Colliery No. 2 Pit No. 3 Shaft in the second cycle above the Mansfield Marine Band. Analysis of this seam from Babbington Pit Yard Boring showed poor quality (13 per cent ash) and high sulphur content (3.6 per cent).

d

TABLE 13

Provings in the Mansfield Marine Band

	Total thickness of marine strata		Thickness of cank	
	ft	in	ft	in
Annesley Lodge Borehole	6	9	2	8
Annesley Park Borehole	3	2	2	2½ (with 1-in silty parting)
Babbington Pit Yard Borehole	7	4+	-	
Bestwood 140's Underground Borehole*	16	11	1	9
Cinderhill Colliery No. 4 Shaft	10	0+	2	0
Garfit House Borehole	10	6	1	9
Grives Quarry Borehole*	8	0	1	4
Hucknall Colliery No. 2 Shaft	9	5+	1	2
Hucknall High Main 27's Underground Borehole	14	10	1	4
Kennel Wood Borehole	7	11	1	6
Linby 36's Slant Underground Borehole*	15	0	1	9
Linby 52's Underground Borehole	10	8	2	0
Mapplewells Borehole	8	11	1	6
Newstead N1 Underground Borehole*	16	7	1	4½
Newstead N3 Underground Borehole	4	7½	-	
Whyburn House Borehole	4	7	1	0

*Situated just beyond the limits of the district.

The highest coal in these measures occupies a fairly persistent position some 40 ft below the High Main. An analysis from Linby 36's Slant Underground Borehole showed it to consist of bright coal which becomes inferior towards the middle and base of the seam. Overall quality is poor, with an ash content of 7.7 per cent and 0.93 per cent of sulphur.

Fish scales and associated debris have been recorded above this coal at Linby 36's Slant and Linby 52's Underground boreholes. At Hucknall High Main 27's Underground Borehole a fish jaw with teeth was noted at this horizon.

Sporadic mussels occur in the measures between the Mansfield Marine Band and the High Main (Appendix 1, p. 000).

The seatearth below the High Main Coal and most of the seatearths above this horizon have a distinctive grey-green, creamy grey or pale brown coloration. Similar colours commonly occur in the stained measures below the Permo-Triassic; those found in the High Main seatearth are some 300 ft below the stained zone in this district and are considered to be a reflection of the changing conditions of deposition in the upper part of the Westphalian.

THE HIGH MAIN COAL is commonly 4 ft thick, but ranges from 43 to 62 in (Fig. 48). It is mined in Newstead and Linby collieries at depths ranging from the 150 ft cover line in the west to 500 ft in the south and east. The seam consists of two major divisions (Fig. 48) - a thin upper brights and a thick lower brights separated by a variable but generally thin dull coal or 'bluestone'

Between Annesley Woodhouse and Greasley the seam is split above the 'bluestone' over a zone some 600 to 1200 yd wide, comparable in plan with a meandering north-south channel (Fig. 48). A detailed section across the split zone is afforded by Brookshill Farm Site, where some 200 trial bores were drilled. Fig. 79 shows isopachs of the measures, largely seatearth and mudstone, separating the two leaves. A longitudinal section A - B illustrates the present configuration of the coals. The lower brights have been partially washed out in only one borehole (Whyburn South) and the upper brights are fairly uniform in thickness.

If it is assumed that the configuration of the coals in section A - B is largely of penecontemporaneous origin, then at the end of the formation of the High Main Coal, when the upper leaf should have been nearly horizontal, the lower brights seam would have the profile a - b. Such a profile suggests a NE-SW belt of rapid subsidence.

There is evidence of disturbed sandy laminae 1 ft 7 in below the seam in Kennel Wood Borehole; disturbed micaceous coaly planes 1 ft 3 in below the seam in Whyburn South Borehole; overfolded laminae 4 ft 9 in below the seam in Weavers Lane Borehole and a 12-in breccio-conglomerate in a sandstone in Allotment Gardens Borehole. These disturbances may have been caused partly by the collapse of river levees (Elliott 1969, p. 117), but the presence of presumed earthquake structures in the roof measures of the High Main seam near Calverton Colliery, south-east of the present district, strengthen the idea that the cause of the split was a short lived subsidence along a pre-existing channel and/or tectonic line.

The quality of the coal varies from good to rather poor with ash content, less dirt, ranging between 4.6 and 9.6 per cent. Sulphur content is

HIGH MAIN

BROOKBREASTING FARM

UPPER BRIGHTS

LOWER BRIGHTS

B

B

UPPER BRIGHTS

LOWER BRIGHTS

20 ft

10

0

a

b

40 ft

30

20

10

0

CROWHILL FARM

0 100 200 300 400 500 ft
0   50   100   150 m

— 4 — Isopachs of the interval (in feet) between the Upper and
Lower Brights of the High Main seam (shaded area includes
all values over 10 feet)

• Borehole

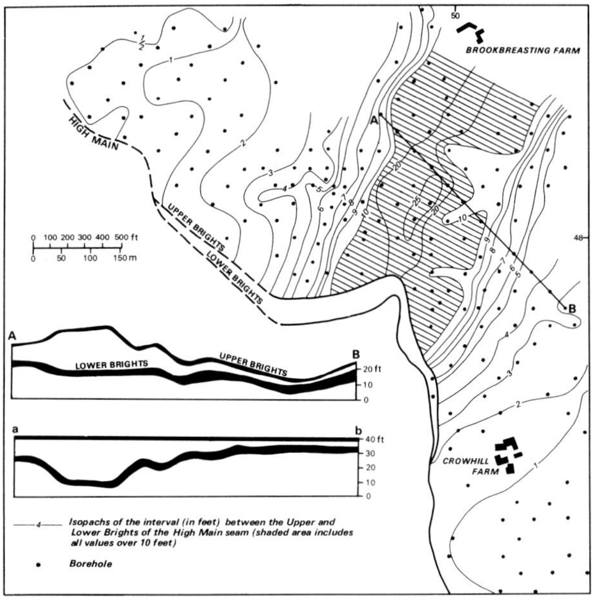

Figure 79 Isopachs of the split in the High Main Seam near Watnall

low.

The seam crops out for some 3 miles close to the foot of the Permian escarpment between Felley and Greasley. In places, westerly projecting spurs of Permian rocks, e.g. near Garfit House Farm [490 520] and Beauvale Manor Farm [490 480], conceal parts of the crop. The northernmost exposure was seen in a pipe trench [4937 5284] near Davis Bottom, where disturbed bright coal, some 4 ft thick, was recorded in the bottom of a steep-sided valley.

The High Main Seam was worked opencast only at William Wood Site where the extent of the take was limited by the close proximity of the Permian scarp. A section totalling 56 in at the north end of the site [4865 5163] was recorded as follows: brights 13½ in; seatearth 2½ in; 'bluestone' (dirty hard coal) 2½ in on hards and brights 37½ in. Dr J. Shirley recorded the following section from the middle of the same site [4881 5115]: mudstone with fish scales 5 ft 1 in; mudstone with ironstone bands and mussels including Naiadites daviesi 6 ft 7½ in; seatearth 4 ft; mudstone 24 ft on the High Main Coal.

Eden and others (1963) recorded a tonstein within the High Main Coal at several widely spaced localities in the East Midlands Coalfield. In the present district no definite occurrences are known, but the following records may be related to tonsteins or fragmental clayrocks: ironstone streaks in a dark grey mudstone 6 in above the High Main Upper Leaf in Mapplewells Borehole; an 'ironstone' lens in the roof of the High Main Upper Leaf in Washdyke Lane Borehole; several small clay-ironstone pockets 1 in above the top of the Upper Leaf in Cavendish Crescent Borehole; kaolinite vesicles (possibly former ooliths) in the roof of the Upper Leaf in Allotment Gardens Borehole and 1 in of ferruginous kaolinite rock 4 in above the High Main Upper Leaf in Annesley Lodge Borehole.

In Newstead Colliery Workings [5050 5289] an oolite bed about ¼ in thick surrounded by thinly laminated argillaceous material was collected from within the coal. Mr Siddiqui reports that: 'The ooliths (average diameter 0.35 mm) are mostly elongated and subrounded, but a few are well-rounded. They are altered to fine-grained granular well-crystalised kaolinite. The ooliths are firmly bedded in a dense matrix of micro-granular siderite. Pyritic aggregates occur in small localized patches. Carbonaceous staining is common, but is mostly restricted to the thinly bedded argillaceous horizons. The gradational contacts between oolites and shaly bands locally show ripple structures'.

THE MEASURES BETWEEN THE HIGH MAIN COAL AND EDMONDIA BAND are some 30 to 40 ft thick and consist largely of mudstone though a washout siltstone, some 23 ft thick, overlies the High Main Coal in the Whyburn South Borehole and a 10-ft sandstone overlies the same seam at crop in the Beauvale Abbey area [490 490]. They contain up to two thin coals. The lower coal, present only in the south-east of the district, where the two cycles are clearly defined in the Linby, Hucknall and Watnall areas, is possibly a split from the High Main Coal. This coal lies 10

to 20 ft below the higher coal, which is up to 11 in thick. Mussels occur locally above both coals, and they lie generally 5 ft but up to 17 ft below the Edmondia Band.

THE EDMONDIA BAND has been found in 33 boreholes and shafts in the present district. It ranges in thickness from 24 ft in Annesley Lodge Borehole to 1 ft 11 in in William Wood Borehole. It is thickest in a restricted belt of country which corresponds closely to the zone of split in the High Main Coal (Fig. 48). This belt seems to have been liable to increased subsidence intermittently throughout this short period of geological time. Edwards (1967, p. 110) noted that the band included a distinctive pale grey mudstone in the Ollerton area. Whilst pale mudstones are present within the band in this district they are variable in both thickness and position. Foraminifera are found in both pale and dark grey mudstones and were particularly well preserved in the Whyburn House Borehole.

In Annesley Lodge Borehole the Edmondia Band contained abundant foraminifera including Glomospira, Glomospirella and Tolypammina. In the middle of the band, Myalina and Hollinella occur, an assemblage typical of the Myalina facies which is the usual development shown by this band throughout the Yorkshire and East Midlands Coalfield. In Kennel Wood Borehole Curvirimula was found in association with Myalina cf. compressa. Curvirimula does not occur above the top of the C. communis Zone except in the retreat phases of some of the Westphalian B marine bands (Calver 1968a, p. 151). The basal part of the band contains cf. Planolites ophthalmoides and pyritized strap-like markings (fucoids). Thin bands and small nodules of ironstone occur throughout and pyrite is common as grains, clusters, concretions, coatings and infillings and replaces many of the plant fragments, worm burrows and fossils. The Edmondia Band contains galena in Misk Farm, Whyburn South and Kitty's Wood boreholes.

THE MEASURES BETWEEN THE EDMONDIA BAND AND SHAFTON MARINE BAND are normally about 100 ft thick, though a maximum of 156 ft was recorded at Cavendish Crescent Borehole. They include up to six variable coals. The measures are particularly sandy and are broadly equivalent to the Mexborough Rock of South Yorkshire. They are nowhere exposed, being almost wholly concealed beneath the Permo-Triassic rocks.

The lowest cycle, commonly arenaceous and averaging 50 ft in thickness, is normally topped by a thin coal. The two succeeding thin cycles contain thin coals which are overlain by mudstones with Lioestheria.

The lower of these bands which may be the "Main 'Estheria' Band" of Edwards and Stubblefield (1948, p. 231), consists of 1 in to several feet of grey mudstone. It is rich in Lioestheria sp. together with fish debris and ostracods. Pyrite is common, often replacing plant and shell fragments and infilling worm burrows. The carapaces are distinctive, appearing as abundant iridescent blotches in the mudstone. The size of the carapaces both within

the band and throughout the district is variable but averages 3.0 mm. The coal underlying this 'Estheria' band has been appropriately called the 'Bug Coal' by miners in the Ollerton area (Edwards 1951, p. 110).

The higher of the Lioestheria-bearing bands, up to 23 ft 5 in above the lower, is from 3 in to a few feet thick, and is restricted in its distribution to the areas where the High Main Coal is split and the Edmondia Band is thick. Whatever the cause of this localized increase in thickness of sediment (p. 69) it affects over 100 ft of strata.

The first cycle above the higher Lioestheria band is usually between 20 and 30 ft thick with a seatearth and/or coal up to 25 in thick, at the top. The coal, of poor quality, is composed of brights or banded brights and hards with cannel in the highest parts. It is thickest west of Linby Colliery (Fig. 80) but thins northwards, becomes cannelly and is represented by a seatearth in the northern part of the district near Annesley Woodhouse. The coal, unnamed in the Nottingham area, may be the Shafton of south Yorkshire, though Edwards (1967, p. 111) correlates the coal close below the Shafton Marine Band in the Ollerton district with the Shafton Coal. In this memoir the seam is referred to as the ?Shafton Coal.

The cycle succeeding the ?Shafton ranges from 5 ft 8 in in Annesley Hall Borehole to 25 ft 10 in in Kitty's Wood Borehole. The measures overlying the ?Shafton contain kaolin ooliths in Weavers Lane, Whyburn East and Whyburn West boreholes, and fish scales were recorded in the measures at Annesley Park Borehole. The cycle is topped by a coal which underlies the Shafton Marine Band and is apparently the equivalent of the seam in a comparable position in south Yorkshire, though evidence from the Ollerton district (see above) raises the possibility that it is the Shafton or a leaf of that seam. This coal is thickest in the Annesley Park area (Fig. 80) and thins northwards, passing into batt and cannel in Mapplewells Borehole. The over-all quality is poor with high ash and sulphur contents.

THE SHAFTON MARINE BAND recorded at fifteen localities within the district ranges in thickness from 8 in to 11 ft 7 in. The fauna includes Anthraconaia spathulata, ostracods such as Paraparchites, fish fragments, fucoids and worm burrows. Lingula is common in Washdyke Lane Borehole and has been recorded in two other bores. In Cavendish Crescent Borehole 9 ft 8 in of silty mudstones separate two bands of strata containing Lioestheria which is a common fossil at all localities.

The band usually closely overlies a coal or seatearth (see above), but may be separated by up to 14 ft 8 in of measures, as at Misk Farm Borehole.

THE MEASURES BETWEEN THE SHAFTON MARINE BAND AND TOP MARINE BAND vary from 46 ft at Washdyke South Borehole to 64 ft in Hucknall Colliery No. 2 Pit No. 5 Shaft. They contain up to three cycles with locally associated coals. The lowest cycle supports a dirty coal 6 to 13 in thick in the Annesley Park area. To the

south and east around Hucknall, Linby and Bulwell the horizon is represented by a seatearth. To the north, both the coal and the seatearth are absent.

A dark grey mudstone at the base of the second cycle contains Lioestheria in Annesley Hall and Washdyke boreholes, and fish debris in Heatherdale Pond, Annesley Park and Thurland Hall boreholes. This band is only recorded in the Annesley-Whyburn zone of thick measures (p. 69). Elsewhere the second cycle comprises some 20 ft of measures. At the top, in the southern part of the district, is a coal which has a maximum thickness of 37 in in Whyburn South Borehole. The coal splits northwards into two leaves in the Annesley Park area, the lower leaf ranging in thickness from 18 to 31 in, separated by up to 21 in of dirt and dirty coal from an upper leaf, 3 in to 5½ in thick. At Thurland Hall Borehole the seam is further split as follows: basal coal 31 in; dirt 18 in; batt 3 in; dirt 9 in; coal 3 in; dirt 6 in; overlain by batt 5 in. Analysis shows the seam to be composed throughout the district of dirty coal of poor to very poor quality with ash ranging from 8 to 17 per cent. Sulphur is variable, ranging from moderately low to moderately high (1.4 to 2.8 per cent).

The coal is overlain by mudstones containing fish scales in Annesley Park and Kennel Wood boreholes. In Whyburn East Borehole a 7-in septarian nodule with kaolinite-filled cracks was recorded 7 ft above the coal. Just beyond the eastern margin of the district Springfield Hosiery Borehole [5551 4620] at Bulwell, yielded Lioestheria sp. some 18 ft below the Top Marine Band.

The third cycle comprises some 15 ft of measures with a coal at the top, under the Top Marine Band. This coal, never more than 12 in thick, varies irregularly over the district.

Mussels occur sporadically in all of the three cycles between the Shafton Marine Band and the Top Marine Band, but are most common in the lowest (Appendix 1, p. 164c).

THE TOP MARINE BAND is a valuable and persistent marker horizon which has been proved throughout the district from Annesley Woodhouse in the north to Bulwell in the south. It usually comprises 2 to 7 ft of grey to dark grey mudstone containing a varied fauna of foraminifera, goniatites, brachiopods including Lingula, bivalves including Dunbarella, gastropods, ostracods, fish, fucoids, Planolites and Lioestheria. Pyrite is common, replacing and coating plant and organic debris.

In Annesley Park Borehole the marine band is some 6 ft 9 in thick and although poorly preserved shows the varied fauna typical of this horizon. A Planolites phase occurs at the top with fucoids and fish debris and the fully marine fauna is confined to the basal 1 ft 4 and includes Orbiculoidea sp. Aviculopecten?, Dunbarella sp., Polidevcia sp. faecal pellets including cf. Tomaculum sp.

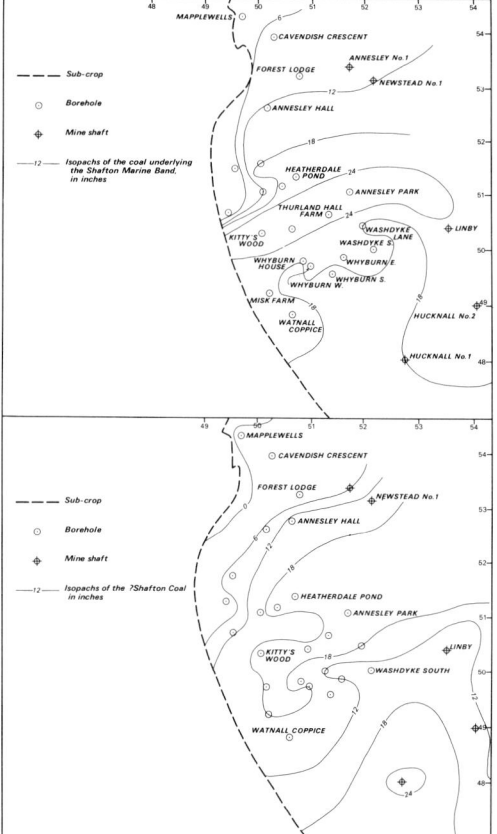

— — — Sub-crop

○ Borehole

⊕ Mine shaft

— 12 — Isopachs of the coal underlying the Shafton Marine Band, in inches

48   49   50   51   52   53   54
MAPPLEWELLS
CAVENDISH CRESCENT
FOREST LODGE
ANNESLEY No.1
NEWSTEAD No.1
ANNESLEY HALL
HEATHERDALE POND
ANNESLEY PARK
THURLAND HALL FARM
WASHDYKE LANE
KITTY'S WOOD
LINBY
WASHDYKE E
WHYBURN HOUSE
WHYBURN E
WHYBURN W
WHYBURN S
MISK FARM
WATNALL COPPICE
HUCKNALL No.2
HUCKNALL No.1

— — — Sub-crop

○ Borehole

⊕ Mine shaft

— 12 — Isopachs of the ?Shafton Coal, in inches

49   50   51   52   53   54
MAPPLEWELLS
CAVENDISH CRESCENT
FOREST LODGE
NEWSTEAD No.1
ANNESLEY HALL
HEATHERDALE POND
ANNESLEY PARK
KITTY'S WOOD
LINBY
WASHDYKE SOUTH
WATNALL COPPICE

Figure 80 Plans of the ?Shafton Coal and the coal underlying the Shafton Marine band

MEASURES ABOVE THE TOP MARINE BAND (UPPER COAL MEASURES)

Some 500 ft of measures above the Top Marine Band are calculated to have been preserved in the extreme north-eastern part of the district, but westwards they are overstepped by the Permo-Triassic rocks. The maximum thickness proved is 350 ft at Linby Colliery. In the past 10 years many boreholes have been drilled by the National Coal Board in the east of the district, but these measures are rarely cored because they lack workable coals.

The measures comprise two thick and persistent cycles near the base overlain by many thinner and variable cycles. One of the highest cycles some 220 ft above the Top Marine Band, contains a coal about 2 ft thick known as the Manor Coal (Edwards 1951, p. 75).

The measures are commonly arenaceous and mussels occur only sporadically in the argillaceous beds. A change in the non-marine bivalve fauna takes place near the base of the Upper Coal Measures with the loss of Naiadites and the incoming of Anthraconauta. The dominant form is A. phillipsii.

THE TOP MARINE BAND CYCLE reaches a maximum thickness of 78 ft in Washdyke North Borehole and is the thickest single cycle in Westphalian C. Sandstone and siltstone, containing scattered plant fragments, form a large proportion of the total, but a 15-in seatearth was recorded in Washdyke North Borehole some 19 ft below the top of the cycle. A coal, here named the Annesley Coal, is commonly present at the top of the cycle. It thickens from 9 in at Heatherdale Pond Borehole to 16 in at Annesley Park Borehole, but outside this localized east-west belt between Annesley and Watnall, variations are considerable. It is represented by a ganister at Linby and a cannel at Hucknall.

The Annesley Coal is overlain by dark grey mudstones containing 'Estheria' in Springfield Hosiery Borehole and fish debris in Hucknall No. 2 Colliery No. 5 Shaft and in Washdyke Lane, Thurland Hall, Heatherdale Pond and Annesley Park boreholes. In addition, ostracods were recorded at Heatherdale Pond Borehole and phosphatic fragments at Whyburn East Borehole. Mussels, including Anthraconauta phillipsii are also present in the lower part of this cycle. It may well be that the Annesley Coal is the correlative of the Scofton Coal of the East Retford district, which is similarly overlain by mudstones with 'Estheria', mussels and ostracods (Smith and others 1973, p. 94).

THE SECOND CYCLE ABOVE THE TOP MARINE BAND ranges in thickness from 42 ft at Springfield Hosiery Borehole to 56 ft at Annesley Park Borehole. Above it there is a complex series of seatearths, coals, batts and carbonaceous shales spread over up to an additional 40 ft of measures. One of these coals is 30 in thick in Hucknall No. 2 Colliery No. 5 Shaft.

Fish fragments were recorded at two horizons between these coals at Annesley Park Borehole and at one horizon in Washdyke North Borehole. Ostracods are also present in both boreholes but at slightly different levels. Sphaerosiderite is a common constituent of the seatearths in Springfield Hosiery, Washdyke North and Whyburn East boreholes.

Above these coals are siltstones and sandstones 30 ft thick at Newstead Colliery No. 1 Shaft to 90 ft at Hucknall Colliery No. 2 Pit. They are succeeded by a group of eight minor impersistent cycles, which in Linby Colliery No. 1 Shaft comprise 120 ft of measures.

THE MANOR COAL at the base of this group, reaches a maximum of 27 in in Hucknall No. 2 Colliery No. 5 Shaft, where it is overlain by 36 in of cannel. In the Hucknall - Linby area the coal is generally over 2 ft thick but thins northwards to 21 in at Newstead.

At Linby Colliery the Manor Coal is overlain by 110 ft of largely argillaceous strata, extending up to the unconformable Permo-Triassic cover. Sandstones are common in the Newstead and Hucknall areas. Primary red measures have not been recorded in the district.

STAINED MEASURES

The Westphalian strata underlying the Permo-Triassic rocks (Fig. 49) are normally stained to about 50 ft below the unconformity, but range from 10 ft at Annesley and Hucknall collieries to 115 ft at Whyburn East Borehole. Red, brown, pink, purple and green coloration is common in the mudstones and sandstones.

Seatearths show purple mottling, although in the measures above the Mansfield Marine Band they are commonly a pale grey or cream colour, contrasting with the darker seatearths of lower horizons (p. 24). Ironstone nodules, normally pale brown or buff, are markedly reddened. Sandstones often show staining to greater depths than do other lithologies. Staining is considered to be due to sub-aerial weathering and oxidation of the Westphalian land surface in late Carboniferous and possibly Lower Permian times (Anderson and Dunham 1953; Trotter 1953; Smith and others 1967). Because argillaceous rocks are as highly coloured as other lithologies, percolating solutions are probably less important than weathering, though the increased depths of staining in areas of porous strata would suggest that the former process took place penecontemporaneously with the denudation. Where the Permo-Triassic rocks are present, staining affects measures above the Mansfield Marine Band, but evidence from the west of the present district (p. 26) shows that the lowest sandstone in the Westphalian, i.e. the Crawshaw Sandstone, is similarly stained although the Permo-Triassic strata have long since been eroded.

The measures immediately below the unconformity are often leached to a depth of a few inches.

CHAPTER 5

Permo-Triassic

DETAILS

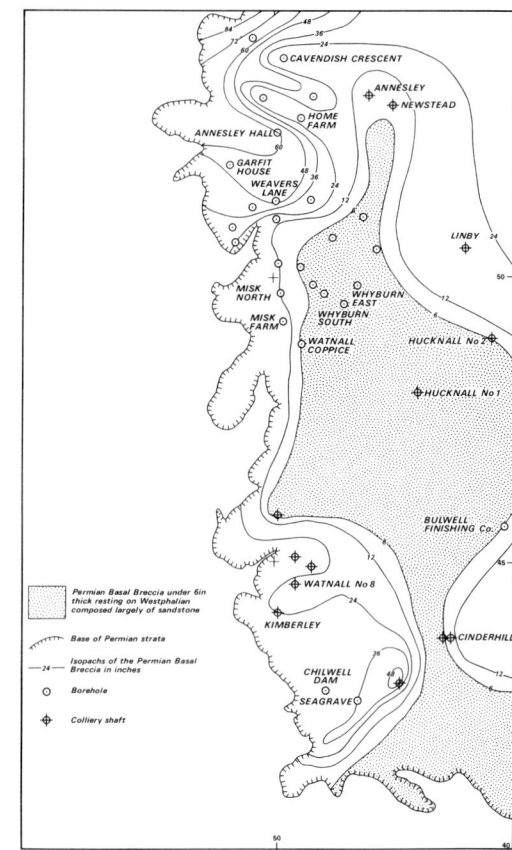

Figure 81 Thickness variations of the Permian Basal Breccia

PERMIAN BASAL BRECCIA

At the time of the resurvey the breccia was exposed in numerous temporary excavations throughout the district, but the only permanent sections were in the Kimberley railway cuttings [4510 4520], where it averaged 24 in with a maximum recorded thickness of 56 in (Sherlock 1907, p. 106). Rail closures since the resurvey have endangered the permanence of even these sections. The breccia, which rests on pink, purple and pale green stained Westphalian mudstones, consists of a grey-brown sandy calcareous matrix containing fragments of red ironstone, purplish and pale-grey quartz and Upper Carboniferous sandstones and siltstones. Some 24 in of breccia at the eastern side of Kimberley station exhibit cross-lamination with foreset beds at angles of 20° and indicate current directions from the south-south-west.

Mr K. S. Siddiqui has described a thin section (E 44773) of the 5 ft 9 in thick breccia in the Mapplewells Borehole. He found a wide range of lithoclasts averaging 0.5 mm in size including feldspathic sandstone with closely packed quartz grains, fine-grained compact sandstone with preferentially orientated sericite flakes, fine-grained dolomitic limestone showing abundant well crystallized rhombs of dolomite, hematite grains and cherty fragments. The finer clasts showed a predominance of angular to subangular fragments of quartz accompanied by orthoclase, plagioclase, a few grains of microcline and well crystallized grains of kaolinite. The cement comprised a mixture of submicroscopic carbonates (calcite-dolomite), clays (illite and kaolinite) and hematite.

Blocks of breccia, up to 16 in thick, were exposed in a temporary gas pipe trench [4890 5278] about 1 mile south-west of Annesley Woodhouse. During the war-time reconnaissance survey, Dr J. Shirley recorded a red-brown and yellow conglomerate and sandstone with pebbles and fragments up to 12 in in diameter at the northern end of the William Wood Opencast Site [4859 5175]. Large boulders of breccia are still present in the hedgerows of this area.

The basal breccia, 6 to 18 in thick, was exposed in a new road cutting at Watnall Chaworth [4973 4639].

South-east of Nuthall excavations for the M1 motorway access road provided exposures of the Permian strata faulted against Westphalian. The Permian basal breccia here is only 2 in thick; it contains a preponderance of small quartzite pebbles and ironstone nodules in a well-cemented matrix; the basal unconformity is emphasized by the truncation of a Westphalian ironstone band.

Near Broxtowe Wood [5228 4273], south-west of Cinderhill, a sewer trench encountered a hard well-cemented bed of breccio-conglomerate, variable in thickness up to a maximum of 48 in, Rounded purplish stained quartzites up to 6 inches

in diameter are common. Temporary sections on the eastern side of a new road cutting 400 to 600 yd NW of Watnall Hall [4977 4634] exposed the basal breccia which was between 2 in and 18 in thick.

The Catstone Hill Opencast site [5070 4171] provided a continuous exposure of the Permian Basal Breccia for some 400 yd. Over most of the site it comprises a massive resistant bed up to 3 ft thick, but thins south-west of Catstone Hill Farm to less than 1 ft. The 20 ft of overlying Lower Magnesian Limestone contains many conglomeratic horizons (see p.151d). Locally a distinct basal Permian breccia is absent but the lowest 3 ft of the Lower Magnesian Limestone remain relatively coarse in texture.

An examination of a thin section (E. 44835) by Mr K. S. Siddiqui from this site revealed a wide variety of lithoclasts in the breccia including slightly metamorphosed sandstone (in places stained with hematite), dolomitic limestone, hematite grains, cherty fragments, well crystal-lized kaolinite grains, quartzite, siltstone, mudstone and probably a few grains of rhyolite. The carbonate fraction of the rock has largely been recrystallized to dolomite which shows a concentration of rhombic crystals in places; numerous grains of quartz are interspersed with dolomite and feldspars. Specks of muscovite are preferentially orientated along the bedding. The silicate minerals show marked etching at the crystal boundaries and in many cases partial replacement by carbonates. The clasts are well cemented by microcrystalline carbonates slightly stained with ferric oxide.

In excavations, up to 20 ft deep, for sewers in the Beechdale Road area, near the edge of the district, a distinct basal breccia is again absent. Red-brown sandy Lower Magnesian Limestone containing interbanded purplish silty mudstones rests on red-stained Westphalian sandstones and mudstones so that the unconformity is not obvious. However, the basal 2 in of the limestone contains sporadic fragments of red and green marl together with a concentration of coarse quartz grains.

The Permian Basal Breccia has been recorded in the following boreholes:
Allotment Gardens (1 ft 10 in), Annesley Hall (5 ft 2 in), Annesley Park (1 in), Audrey Wood (2 ft 7 in), Balaklava Wood (2 ft 6 in), Cavendish Crescent (1 ft 9 in), Chalfont Drive No. 1 (3 in), Charlie's Wood Cottages (3 in) Chilwell Dam Farm (2 ft 9 in), Forest Lodge (2 ft 9 in), Garfit House (4 ft 3 in), Heatherdale Pond (2 ft 4 in), Home Farm (1 ft 4 in), Kennel Wood (1 ft 1 in), Kitty's Wood (1 ft), Mapplewells (5 ft 9 in), Misk Farm (1 ft 6 in), Misk North (1 ft), Pamela's Larches (2 ft), Seagrave (3 ft), Thurland Hall Farm (4 in), Washdyke Lane (North) (5 in), Weavers Lane

(4 ft 3 in), Whyburn East (pebbles), Whyburn House (fragments), Whyburn West (5 in), William Wood (2 ft 10 in). It has been recorded or can be recognised from sinkers' records in many of the shafts and bores of the present district (see Appendix 2).

LOWER MARL

During the resurvey of the sheet, the Lower Marl was exposed in many temporary excavations, but the only permanent and complete section is in the railway cuttings between Kimberley and Nuthall.

The most northerly exposures examined were in a gas pipe trench near Davis's Bottom [4963 5287] ½ mile S of Annesley Woodhouse, where 8 ft of grey-brown dolomitic siltstone with subordinate mudstone partings were preserved in the crest of a small anticline.

East of Felley Mill, a stream [4876 5100] and its small tributary [4982 5032] flowing through The Dumbles have incised deeply into the Permian scarp to show sporadic exposures of grey dolomitic silty mudstone with plant fragments: at one locality farther downstream [4976 5034] these are underlain by red and purple-stained silty mudstones of Westphalian C age. Excava-tions [4975 5106] along the line of the M1 motorway in this vicinity exposed 10 ft of flaggy dolomitic siltstones directly underlying the Lower Magnesian Limestone.

A spring [5030 4784] some 50 yd W of Watnall Colliery and Brickworks issues from the base of the Lower Magnesian Limestone. Beneath the spring, 3 ft of buff micaceous calcareous siltstone with carbonaceous plant fragments overlying 1 ft of grey silty mudstone are exposed.

The railway cuttings near Kimberley are now overgrown and parts are concealed by masonry. Wilson (1876) examined them during their excavation and noted a uniform thickness of about 17 ft of 'thin-bedded slate-coloured sandstones and shales' comprising no less than 75 beds from ¼ in to 8 in thick, some with annelid markings, others with abundant plant remains; sporadic casts of Schizodus were also recorded.

The junction between the Lower Marl and the Lower Magnesian Limestone is marked in the cutting by the presence of numerous springs which renders the lower part of the cutting unstable in wet seasons. After a particularly wet period in 1966 a small slip in the now disused northern cutting exposed most of the Lower Marl sequence, as follows:

	ft	in
Dolostone, grey, flaggy, fine-grained; sporadic pale grey shaly partings	10	0
Mudstone, brown	0	3
Dolostone, yellow-brown, flaggy (up to 3 in)	0	8
Mudstone, brown; dolomitic bands and laminae	0	4
Dolostone, yellow-brown; sporadic mudstone laminae	0	7
Interlaminated brown mudstone and dolostone	0	4
Mudstone, brown	0	3
Dolostone, yellow-brown; calcite-filled cavities	0	6
Mudstone, brown, yellow, silty	0	6
Clay parting, red,		
Dolostone, brown	0	3½
Mudstone, silty, vaguely laminated	0	7
Mudstone, greenish grey	0	1½
Mudstone, yellow-brown	0	1
Mudstone, greenish grey	0	0½
(Breccia 3 ft 3 in to 4 ft)		

In the cuttings [5200 4490] made for the M1 motorway and access roads near Nuthall the Lower Marl was well exposed, though affected by faults and folds associated with the Cinderhill Fault system. A few feet of brown silty dolostones at the top of the Lower Marl were exposed in the core of an anticline [5219 4391] on the roundabout excavations ½ mile SE of Nuthall. Two hundred yards to the east the top and bottom beds of the Lower Marl are brought into juxtaposition by faulting. The highest beds consist of steely blue-grey laminated siltstones in beds up to 6 in with shaly partings, irregular brown staining and carbonaceous plant fragments. The basal Lower Marl comprises 3 ft of soft dark grey-blue silty mudstone overlain by 6 ft of siltstone with individual beds normally not exceeding 2 in in thickness.

Mapping of the Lower Marl at the base of the escarpment between Felley and Kimberley revealed a variable thickness up to about 30 ft. Gibson and others (1908, p. 107) and Waring (1966, pp. 204-208) suggest that the Lower Marl is absent near Beauvale Priory.

In Annesley Lodge Borehole [4957 4992], where the Permian rocks were not cored, 30 ft of strata have been assigned to the Lower Marl on the evidence of the gamma-ray log. At the old quarry south of Shortwood Cottages [4894 4771] Waring (1966) noted cross-bedded siltstones with rolled mudstone pellets and also thin hard calcareous bands within a 10-ft sandy mudstone which passes up into Lower Magnesian Limestone. He also recorded the Permian Basal Breccia and 15 ft of marl at Bogend (Sledden Wood Spring) [499 468]. Some 4 ft of banded brown siltstone with many plant fragments were noted at the entrance to the northern rail cutting at Kimberley [4972 4499].

In the Broxtow inlier, a sewer trench [5228 4513], exposed the entire 20 ft or so of the Lower Marl succession. The highest beds, which were grey-brown, showed a marked colour contrast with the reds and buffs of the overlying limestone. The predominant lithology was a buff or pale grey dolostone/siltstone with subordinate shale partings which were stained red and purple near the top and contained many carbonaceous plant fragments, including Ullmania frumentaria. Approximately 10 ft from the base, oolites are associated with calcite-lined vugs.

The basal succession was as follows:

	ft	in
Dolostone, buff, sporadically red-stained in bands up to 6 in, usually finely crystalline, separated by shaly partings: base locally		

Figure 82 Generalised sections of the Permo-Triassic rocks in the Derby and adjacent districts

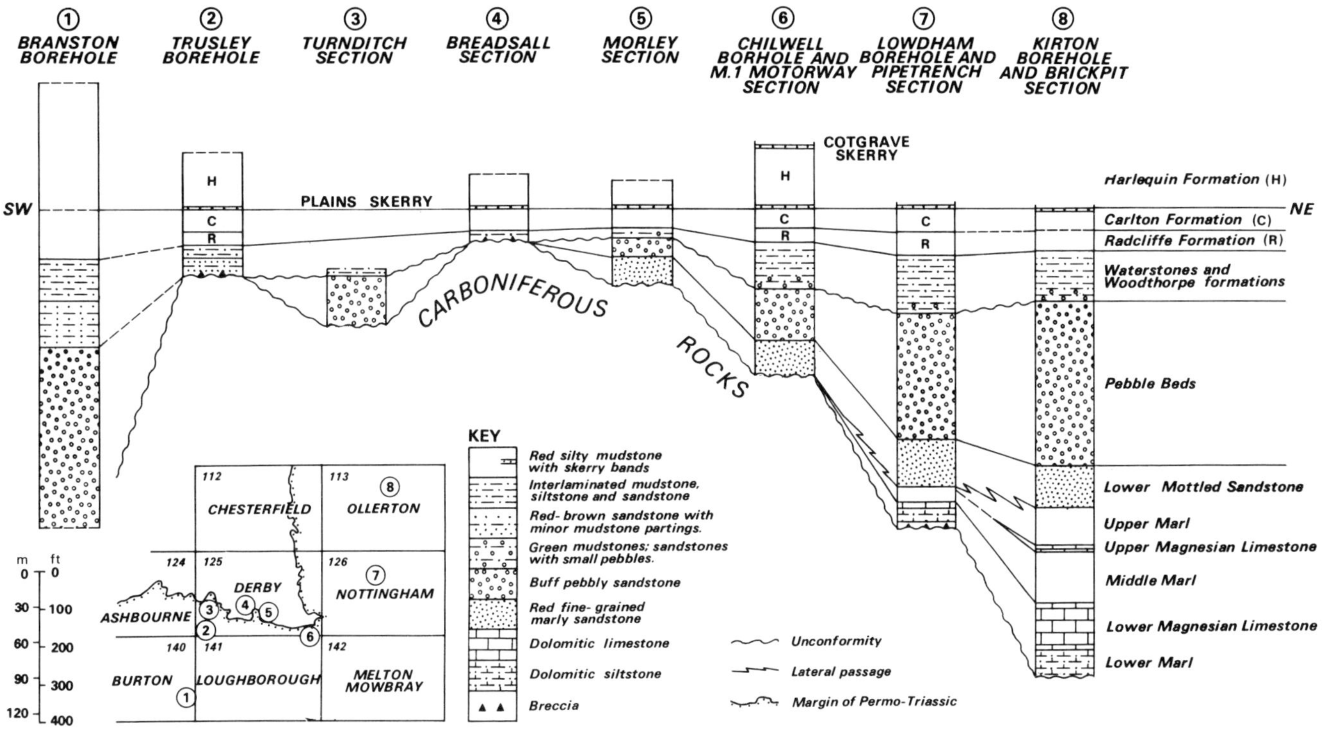

① BRANSTON BOREHOLE
② TRUSLEY BOREHOLE
③ TURNDITCH SECTION
④ BREADSALL SECTION
⑤ MORLEY SECTION
⑥ CHILWELL BORHOLE AND M.1 MOTORWAY SECTION
⑦ LOWDHAM BOREHOLE AND PIPETRENCH SECTION
⑧ KIRTON BOREHOLE AND BRICKPIT SECTION
COTGRAVE SKERRY
PLAINS SKERRY
SW
NE
H
C
R
CARBONIFEROUS ROCKS
Harlequin Formation (H)
Carlton Formation (C)
Radcliffe Formation (R)
Waterstones and Woodthorpe formations
Pebble Beds
Lower Mottled Sandstone
Upper Marl
Upper Magnesian Limestone
Middle Marl
Lower Magnesian Limestone
Lower Marl
m ft
0 0
30 100
60 200
90 300
120 400
112
113
CHESTERFIELD
OLLERTON
124
125
126
DERBY
NOTTINGHAM
ASHBOURNE
140
141
142
BURTON
LOUGHBOROUGH
MELTON MOWBRAY
KEY
Red silty mudstone with skerry bands
Interlaminated mudstone, siltstone and sandstone
Red-brown sandstone with minor mudstone partings.
Green mudstones; sandstones with small pebbles.
Buff pebbly sandstone
Red fine-grained marly sandstone
Dolomitic limestone
Dolomitic siltstone
Breccia
Unconformity
Lateral passage
Margin of Permo-Triassic

	ft	in
nodular with oolites	3	0
Mudstone, predominantly reddish brown silty with yellow-brown and grey bands: ? bivalve	4	0
Conglomerate; brown limonitic with red-purple bands, much quartzite and ironstone nodules; badly fractured;	4	0
Mudstone, pale grey-green and purplish red, silty (Westphalian)	1	0

A small exposure [4994 4836] at a spring 480 yd east of Beauvale Farm showed 3 ft of grey mudstone containing buff silty bands with plant fragments including U. frumentaria and indeterminate conifer leaves and cone scales. A 2-in buff dolomite band near the top contained shells of Bakevellia sp. and Schizodus obscurus.

The Lower Marl has been seen in the following modern boreholes (thickness to the nearest foot in brackets) (see also Fig. 70): Allotment Gardens (57), Annesley Hall (63), Annesley Park (62), Audrey Wood (56), Balaklava Wood (59), Bulwell Finishing Co. (66), Cavendish Crescent (69), Chalfont Drive No. 1 (6), Charlie's Wood Cottages (57), Forest Lodge (65), Garfit House (57), Heatherdale Pond (55), Home Farm (63), Kennel Wood (60), Kitty's Wood (52), Mapplewells (66), M1 proving hole (20), Misk Farm (52), Misk North (42), Pamela's Larches (about 53), Seagrave (9), Thurland Hall Farm (58), Washdyke Lane (North) (54), Washdyke Lane South (60), Watnall Coppice (50), Weavers Lane (69), Whyburn East (64), Whyburn House (45), Whyburn West (52), William Wood (54).

It has been recorded in, or can be recognised from, details of sinkings of many of the shafts through the Permo-Trias (see Appendix 1), although exact determination of the Lower Magnesian Limestone—Lower Marl junction is difficult. The Lower Marl was variously described as 'stone-bind', 'blue-bind', 'limestone shale with bands of stone or ironstone girdles'. Mudstones and siltstones in the Permo-Trias, whether calcitic or dolomitic, were invariably called 'marl' in older literature and in the logs of sinkings and borings.

LOWER MAGNESIAN LIMESTONE

Northern area between Annesley and Hucknall.

In many old quarries in this part of the district, yellow and buff flaggy dolomitic limestone is exposed. Half a mile north of Linby, at Quarry Banks [536 521], the Lower Magnesian Limestone is said to have been quarried by monks from Newstead Abbey in the 12th Century. The abandoned quarries were opened up again after the 2nd World War and the following section was recorded at the present working face:

	ft	in
Soil, dark red-brown loam with sporadic pebbles	2	0
Limestone, buff, with greenish grey and red staining, dolomitic, thinly bedded in posts up to 6 in thick; grey-green mudstone partings	6	0
Limestone, buff-yellow, massive, with posts up to 12 in; sporadically wedge-bedded from south and north-west; well jointed, irregular bedding planes; 4-in hard massive siliceous bed 3 ft from top. Ripple marks with crests aligned north-south	12	0
Limestone (without shales according to quarry men)	6	0

Railway cuttings to the north and south of Linby provide excellent exposures of up to 10 ft of buff flaggy dolomites dipping to the south-east and east at angles of $1\frac{1}{2}$ - 3°. The dip-slope is interrupted by gentle anticlinal 'rolls' which can be traced as features beyond the limits of the cutting (see p. 000). They appear to be a common feature of the limestone dip-slope and have been detected in gas-pipe trenches south of Newstead Abbey. An anticlinal ridge over four miles long trending north 20° west extends from Newstead through Hucknall to Bulwell and provides sporadic exposures of the Lower Magnesian Limestone (see p.161a).

Two miles west of Hucknall, sandy dolomites on the escarpment edge include breccias up to 2 ft thick about 20 ft above the base of the Limestone. They indicate an approach to the shore. Waring (1966, pp. 207-210 and pl. 11 and 12) described the breccia and figured the section in an old quarry (Abbey Wood) north-north-west of Beauvale Priory [4420 4933]. He also noted that the breccia was interbedded with the dolomite for $3\frac{1}{2}$ ft in a quarry east-south-east of the Priory [4965 4882]. Nine feet of the lowest beds of the limestone are seen in the abandoned railway cutting 200 yd N of this quarry.

The rocks exposed in quarries on the escarpment edge near Watnall Chaworth comprise flaggy sandy dolomites with harder nodular patches and cross-bedded units up to 1 ft thick. The large quarry near Bogend [4971 4655] exposes 12 ft of extensively fissured, cross-bedded flaggy dolomites. The fissures are up to 5 ft wide and are filled with mud, loam, dolomitic limestone nodules and calcite fragments.

The top of the Lower Magnesian Limestone, exposed in the M1 Motorway cutting near Misk Hill [5025 4925], comprised reddish-brown friable dolomite containing Liebea squamosa and Schizodus obscurus. Other excavations for the Motorway, 600 yd NE of Annesley Lodge, showed the lowest 16 ft of the formation to consist of buff and red-brown friable dolomite; a shell band, some 2 ft from the base, contained Bakevellia binneyi and S. obscurus.

A temporary excavation [5198 4866] on the south-west periphery of Hucknall showed the Lower Magnesian Limestone—Middle Marl junction. The top 4 ft of limestone were soft buff dolomite containing thin red 'marl' bands. Almost the entire thickness of the Lower Magnesian Limestone is exposed in the cuttings adjacent to Hucknall Station [5305 4877], where dolomitic limestone, some 30 ft thick is generally flaggy and cross-bedded from the south-west. Bedding planes up to 13 in apart show undulations up to 2 in in amplitude and 6 in wavelength. Shell bands 4 ft to 6 ft from the base contain B. binneyi and S. obscurus.

Southern area: Kimberley—Strelley—north and west Nottingham:

In the railway cuttings between Kimberley and Nuthall the Lower Magnesian Limestone is continuously exposed for over half a mile, though the northern cuttings are now disused and the branch-line to Watnall Brickworks is being infilled. Thirty feet of flaggy red-brown dolomitic limestone is recorded [5150 4500] near the road bridge to Nuthall Cemetery. The underlying Lower Marl is exposed beneath the bridge and the overlying Middle Marl occurs in an adjacent field to the north-east. The limestone is in beds up to 6 in thick and commonly displays large-scale cross-laminated units in the top 6 ft. The foreset beds dip at 12° to 20°. The limestone contains shelly lenses and one of these, about 8 ft from the base, yielded B. cf. binneyi, L. squamosa and S. obscurus.

Numerous old quarries in the Lower Magnesian Limestone scar the western side of Bulwell. One of the few still worked for 'Bulwell Stone' has a 15- to 20-ft face [5320 4560] of buff fine-grained dolomite stained dark red and pale green along the bedding planes. Fossils include cf. Bakevellia sp., and Schizodus obscurus.

Some 300 yd S of the working quarry [5336 4528] the following section was recorded using Smith's (1968) classification:

	ft	in
Soil, red clayey - approximate base of Middle Marl		
LOWER MAGNESIAN LIMESTONE (UPPER SUBDIVISION)		
Limestone, dolomitic, massive, coarse-grained, sandy; sporadic stylolitic partings	6	6
Limestone, flaggy, cross-bedded, in beds about 6 in thick	2	6
'Bulwell' Shell Band (1 ft 8 in) with B. cf. ceratophaga, L. squamosa, S. obscurus.	2	6
HAMPOLE BEDS		
Limestone, buff fine-grained dolomitic; quartz grains and many shell fragments	0	8
LOWER MAGNESIAN LIMESTONE (LOWER SUBDIVISION)		
Limestone with sporadic shells	0	8
Limestone with shells concentrated in pockets	0	4
Limestone, buff flaggy, medium-coarse, cross-bedded from south-west; ripple marks with crests aligned north-east to south-west; steepest slopes facing south-east, amplitude $1\frac{1}{2}$ in; wavelength 2 ft	8	0

The top of the Lower Magnesian Limestone is exposed some 500 yd SW of the above section where the Sankey Company have dug the overlying Middle Marl for clay. The upper surface of the Limestone is hard and well cemented with secondary silica. Scattered throughout the top few inches are irregular concentrations of galena.

The formation was well exposed in cuttings for access roads to the M1 Motorway south-east of Nuthall [517 445], where the dolomitic limestone contained Bakevellia sp., cf. Permophorus costatus and S. obscurus. In a sewer trench north of Bilborough [5249 4286] 23 ft of purplish-red and red-brown flaggy dolomitic limestone were recorded overlying the Lower Marl.

There are many small quarries and sections in the Cinderhill area which expose some part of the Limestone and show its typical buff red-brown flaggy cross-bedded dolomitic nature. At Cinderhill Junction [5379 4382] the juxtaposition of limestone and the Lower Mottled Sandstone at the Cinderhill Fault can be seen.

North-west of Strelley [5038 4252] the M1 Motorway was excavated to a depth of some 30 ft, exposing a pinkish limestone with sandy conglomeratic bands in the top part.

An old quarry [5133 4215] in Stonepit Plantation, east of Strelley Hall shows the following section:

	ft	in
Limestone, flaggy, sandy, cross-bedded (dips up to 10° to the north-west and south-east)	8	0
Limestone with many coarse quartz grains; wedges out to north and south	0	5
Limestone, massive, cross-bedded	4	0

The best sections of the Lower Magnesian Limestone were provided by the Catstone Hill Opencast site, which exposed Permian strata for some 700 yd. In the west, 16 ft of buff to red-brown dolomitic limestone underlay the Middle Marl. The top 12 in were well cemented and 7 ft from the base was a 2-ft pale brown sandy micaceous laminated mudstone. A basal breccia varied from 0 to 6 in (see p. 000) and rested unconformably on Westphalian rocks (Plate 9). At the south-eastern corner of the site [5070 4171] the limestone, 20 to 22 ft thick, contained five conglomeratic bands up to 4 in thick in the upper 8 ft. The underlying Permian Basal Breccia, up to 2 ft thick, rested on Westphalian B strata. The conglomeratic bands were mainly of quartz together with red and green mudstone fragments.

Eight hundred yards SE of the Catstone Hill site the following section was exposed in excavations for extensions to Bilborough School [5143 4139]:

	ft	in
Dolomite, flaggy buff to red-brown, with sporadic 'marl' partings	3	0
Mudstone, purplish red, laminated, silty; partings of green and red mottled clay	0	6
Dolomite, buff; with B. binneyi, cf. L. squamosa and cf. Schizodus	0	2
Mudstone, purplish red, silty, laminated	0	8
Mudstone, green	0	$0\frac{1}{2}$
Mudstone, red-brown silty, with dolomite bands	0	6
Mudstone, green silty	0	1
Mudstone, purplish red, silty; top 3 in mottled dolomitic	2	0

a

	ft	in
Clay, green (?Westphalian)	0	0½

The basal Permian beds are very variable along the southern margin of their outcrop, but they form a feature that can be followed south-eastwards for a mile or so from Catstone Hill and Bilborough to Broomhill Wood [5261 4083], where, in factory foundations, a 6-in well-cemented, massive dolomitic limestone band containing quartz grains and red and green sandstone fragments was underlain by the Lower Mottled Sandstone. Some 200 yd to the south-east [5280 4073] six feet of Lower Mottled Sandstone rest directly on Westphalian B rocks showing the locality to lie beyond the southern limit of Lower Magnesian Limestone deposition.

A sewer trench [5447 4079] 1200 yd and more to the east of the above locality, near Radford Woodhouse, showed thin representatives of the Permian basal beds. They are composed essentially of medium-grained buff to red-brown dolomitic limestone containing numerous quartz grains, mudstone and rock fragments. Fossils included Bakevellia sp., L. squamosa and Schizodus sp. The 10 ft of limestone recorded in this area are considered to represent nearly the total thickness of the formation. The Chalfont Drive Borehole [539 410], some 700 yd NW of the sewer excavations, proved 7 ft of buff and red dolomitic limestone with red and green silty mudstone bands, up to 14 in thick, overlying Westphalian mudstones. Wilson (1876) noted that at Bobbers Mill [552 415], south of Basford, the Lower Magnesian Limestone passes from a granular dolomite through a fine to a coarse brecciated rock over a distance of 200 yds. A recent water bore just beyond the eastern margin of the sheet in Nottingham [5575 4209] provided a good section in the Permian basal beds comparable with that recorded at Bilborough School (see p. 151d).

	ft	in
Conglomerate; coarse sand, green and red marl and pebbles (Base of Lower Mottled Sandstone)	0	4
Limestone, buff, well-cemented, green marl partings and calcite-lined cavities	0	8
Limestone, red-brown, with pale grey-blue and red marl laminae	1	7
Limestone, buff, red-brown mottling	1	4
Limestone, red-brown and pale grey-blue, well-cemented, with dark red micaceous silty laminae towards base	3	5
Limestone, red and grey mottled	1	0
Limestone, buff to dark-red mudstone laminae and pellets, red mottling	2	0
Mudstone, reddish purple silty micaceous; pellets; many bivalve casts; 2½-in dolomite band 6 in from top	proved 1	0

The Lower Magnesian Limestone has been proved in over fifty boreholes, and summaries and details are contained in Appendix I. Detailed sections of the limestone are provided by the following recent boreholes (thickness to the nearest foot in brackets): Annesley Hall (28), Annesley Park (36), Audrey Wood (25), Cavendish Crescent (31), Chalfont Drive (7), Charlie's Wood Cottages (30), Chilwell Dam Farm (36), Forest Lodge (34), Home Farm (33), Kitty's Wood (25), Mapplewells (34), Misk North (20), Seagrave (25+). Thurland Hall Farm (36), Whyburn West (31).

MIDDLE MARL

The few exposures of Middle Marl in the Annesley, Linby-Hucknall area show only a few feet of weathered red clay. A gas pipe trench [5446 5250], however, was excavated through the entire thickness of the Marl south of Newstead Abbey, just beyond the eastern limit of the present district. The eastern bank of the River Leen nearby is composed of 20 ft of weathered red clay with pale grey-green streaks. It becomes sandy towards the top and passes up into red sand of the Lower Mottled Sandstone at the top of the bank.

Within the district, at Newstead [5173 5258], the same gas-pipe trench exposed the lowest beds of the Middle Marl in the following section:

	ft	in
Soil, sandy clay with pebbles	1	0
Clay, red	3	0
Clay, pale grey and green	3	6
(Lower Magnesian Limestone)		

Sporadic pockets of red marl on the Lower Magnesian Limestone dip-slope between Newstead and Papplewick were cut through by this trench. Invariably the lowest few inches of 'marl' in contact with the limestone are pale grey to green in colour. Sherlock (in Gibson and others 1908, p. 110) recorded the following section, 9 ft 3 in thick, from a now disused and infilled marl pit [5260 5060] and former brickyard at the side of Wighay Road, Linby:

	ft	in
Clay, red and yellow	3	0
Marl, red	1	0
Sandstone, micaceous	2	0
Marl, red	1	0
Limestone, magnesian	1	3
Marl, red	1	0

He noted that the sequence varied considerably in the excavation.

Temporary excavations [5198 4866] on the western periphery of Hucknall showed 4 ft of dark red clay of the Middle Marl overlying the Lower Magnesian Limestone. The basal 1 in of the clay was pale green. South of Hucknall exposures of red and grey sandy marl up to 6 ft thick were seen in open graves [5376 4837] in the cemetery. The graves were floored by a buff dolomitic limestone at least 12 in thick - presumably a band within the Middle Marl.

An outlier of Middle Marl occurs some 300 yd to the east. Red clay was dug out during road work excavations [5405 4870] beneath Portland

b

Road and Hankin Street.

Excavations through Kennel Gorse Wood [4993 5054] and Watnall Coppice [5035 4900] for the M1 Motorway facilitated accurate delineation of the boundaries of the Middle Marl, but provided little in the way of detailed sections. Proving holes [5069 4759] showed that it was 19 ft thick.

Southern area

At Bulwell the complete sequence of Middle Marl exposed in the Sankey Plant Pot Company's excavations for clay [5290 4510] is as follows:

	ft	in
Soil, sandy, with pebbles at base	1	3
Sandstone, red, clayey (Lower Mottled Sandstone)	4	6
Clay, red, alternating with pale grey and brown sandstone	1	4
Clay, dark red	1	6
Sandstone, red-brown	0	2
Clay, red and grey-green, sandy	0	6
Clay, red	4	6
Clay, pale grey-green, alternating with buff dolomitic limestone bands up to 4 in	1	0
Clay, red	0	10
Sandstone, red-brown, fine-grained	5 in to 1	4
Clay, silty, and sandstone	approx 10	0
(Lower Magnesian Limestone)		

The details are variable as shown by further trial pits and excavations to the west [5243 4522]:

	ft	in
Clay, sandy, brown and yellow-brown, with pebbles at base	2	3
Sand, red, clayey		
Clay, red, with mottled grey-green silty and sandy laminae	4	0
Sandstone, fine, buff, calcareous	0	1
Clay, red	0	7
Sandstone, yellow, medium to coarse, calcareous	0	4
Clay, red, silty, micaceous, flaggy; grey-green patches	2	0

The overall clay content of the Middle Marl appears to decrease westwards thus limiting the workable area of marl.

Some 800 yd to the west in New Farm Wood [5166 4512], the M1 Motorway cut through 11 ft of beds near the base of the Middle Marl, exposing the following section:

	ft	in
Soil	0	3
Clay, yellow and pale grey	1	6
Clay, red	3	6
Sandstone, pale grey-green, soft	0	2
Clay, red	4	0
Sandstone, pale grey, soft	0	1
Clay, red	2	0

There are many old 'marl' pits in the Bulwell-Cinderhill area, most of which are now infilled.

An old record provided by Mr G. Fowler of the Brickyard at Cinderhill Colliery [approx.535 440] is as follows:

	ft	in
(Lower Mottled Sandstone)		
Marl, mottled, sandy	2	9
Marl, red and green; thin sand beds	6	3
Sand, yellow, calcareous	1	3
Marl, red	1	0
Sand, hard, calcareous	1	0
Marl and sand, hard at base	6	3
Clay, red becoming greyish green at base	13	2
(Lower Magnesian Limestone)		

This section does not agree with the following record quoted by Gibson and others (1908, p. 109) for the same Brickworks, illustrating the problems of detailed correlation within even a small area:

	ft	in
(Lower Mottled Sandstone)		
Magnesian flagstone [sic]	2	0
Marl, red	1	10
Magnesian flagstone [sic]	2	0
Marl, red	3	6
Marl, with thin magnesian flags [sic]	12	6
Marl, red, with sandy partings	5	0
(Lower Magnesian Limestone)		

The proximity of this Brick Pit to the Cinderhill Fault, usually associated with many minor dislocations, may account for the differing sections. Wilson (1876, p. 533) recorded hard red and yellow mottled and soft grey sandstones, some 40 ft thick in beds from 1 in to 6 ft thick, near Hempshill. He considered that they may be passage beds between the Permian and Bunter.

The railway viaduct [544 459] over the River Leen between Bulwell and Hucknall on the eastern margin of the district is built largely on Middle Marl. A section near the top of the formation measured in the foundations at the southern end showed:

	ft	in
Soil and made ground	3	6
Sand	2	8
Marl	1	6
Sandstone, red and white	2	6
Limestone, dolomitic	0	9
Clay and marl, white	3	3
Marl	5	0

Housing development [5425 4336] east of Basford Hall exposed red clay and sandy clay in the upper part of the Middle Marl. An outcrop of red 'marl' up to 10 ft thick and with thin dolomitic limestone bands was also discovered in a fault 'graben' which was cut through by a sewer trench at Broxtowe Wood [5273 4319] (see also p. 150d).

At Nuthall Temple, M1 Motorway excavations [5159 4400] exposed small lenses, up to 6 ft thick, of red and green clay of the Middle Marl caught up in the Cinderhill Fault Complex (see also p. 161a).

The M1 Motorway excavations provided further sections [5054 4283] of the Middle Marl north of Strelley from which the following sequence was compiled:

	ft	in
(Lower Mottled Sandstone, see p. 153d)		
Marl, red, green and silty at top	0	10
Sandstone, buff, fine-grained	0	2
Marl, red	1	5
Sandstone, massive, red-brown, fine-grained, with yellow patches	4	0
Marl, red, with green laminae	0	9
Sandstone, red-brown, with yellow band near base	0	6
Marl, red, with thin green beds 1 in and 12 in from base	2	3
Sandstone, red-brown, massive, fine-to medium-grained; yellow bands at top	1	0
Marl, red	0	4
Sandstone, red-brown, with buff spots	0	4
Marl, red	0	2
Sandstone, red-brown, ?dolomitic	0	10
Sandstone, yellow, fine-grained	0	2
Marl, green	0	1
Marl, red	1	1
Sandstone, yellow	0	1
Marl, red	2	6
Sandstone, brown and yellow, massive	2	0
(Lower Magnesian Limestone)	-	-

c

The basal yellow-brown sandstone in the Middle Marl is not easily distinguished from the Lower Magnesian Limestone. The clay beds represent some 57 per cent of the section here compared with 75 per cent at Bulwell.

One mile to the south-south-east in the Catstone Hill Opencast Site [5033 4147] the Middle Marl was only some 2 ft to 2 ft 6 in thick. The section measured in the south-west of the site was:

	ft	in
(Lower Mottled Sandstone, see p. 153d)		
Dolomite, buff	up to 0	6
Clay, red, poorly laminated, with ½-in buff dolomitic bands	0	6
Dolomite, pale brown, with sandy clay laminations	1	3
Clay, purplish red and grey-green, laminated	0	4
(Lower Magnesian Limestone, see p. 151d)		

The south-eastern corner of the site [5070 4171] showed up to 10 ft of red sandy clay which overlay and filled hollows in the Lower Magnesian Limestone (see p. 000).

In trenches and excavations at Bilborough, less than one mile to the east, the marl consists of red sandy clay and clayey sand, 5 to 10 ft thick, which passes up into the Lower Mottled Sandstone forming the outlier upon which St. Martin's Church is built.

In the Radford—Bilborough area there are patches and pockets of red clay resting upon the feather-edge of the Lower Magnesian Limestone. Patches up to 10 ft thick have been reported in places. In Chalfont Drive Borehole the driller recorded 8 ft 6 in of red clay with thin bands of yellow sandy dolomitic limestone. At Newcastle Colliery [546 422] just beyond the eastern margin of the district, Middle Marl is absent, the Lower Mottled Sandstone resting directly on Lower Magnesian Limestone. There is no evidence of Middle Marl beyond the limits of the area of Lower Magnesian Limestone deposition.

The Middle Marl has been recorded in the following boreholes and shafts (thicknesses to the nearest foot in brackets): Annesley Colliery No. 1 Shaft (27), Annesley Hall (26), Cavendish Crescent (28), Charlie's Wood Cottages (19), Cinderhill Colliery No. 6 Shaft (23), Forest Lodge (28), Heatherdale Pond (25), Hempshill Colliery Shaft (21), Home Farm (26), Mapplewells (34), Misk North (25), Motorway Proving Hole (21), Newstead Colliery No. 1 Shaft (21).

TRIASSIC

MOIRA BRECCIA

In a landslip on the east side of the Derwent valley in Croft Wood [3663 3932] near Breadsall, 2 ft of ochreous brown and red mottled breccia in a gritty matrix underlie 5 ft of Pebble Bed rocks. In the rail cutting [3798 3945] 1 ft 4 in of coarse sandy breccia with fragments, up to 6 in long, of fine sandstone, red ironstone and green-grey mudstone in a coarse sandy matrix overly and probably occupy a hollow in 'reddened' Namurian mudstones.

For 150 yd of the course of Ferriby Brook, sections [389 398] show up to 3 ft of red and ochreous mottled breccia, locally with a clayey matrix, amid small sections of friable Ashover Grit.

Two localities [3955 4089 and 3960 4074] in the road cuttings south-west and south of Morley church show up to 1 ft of Moira Breccia. At the former locality red-brown soft sandstone is exposed beneath the breccia and is assumed to be interbedded with it.

To the north of Morley church the breccia mantles part of a scarp face formed of Westphalian sandstones which underlie and overlie the Kilburn Coal; on its western side the outcrop is apparently transgressed by sandy micaceous Waterstones. The breccia is also exposed in the brook known as The Gripps [3979 4124], where 4 ft of red-brown and ochreous gritty and clayey sand overlie 2 ft + of breccia composed of ironstone, sandstone and siltstone in a red and ochreous sandy matrix. Similar rocks with quartzitic pebbles litter the soil.

Cores of the Moira Breccia have been seen from two boreholes at the Rose and Crown Opencast Site. The sequence proved below Waterstones and resting on stained Westphalian sandstone [3937 4239] was 2 ft of breccia (Plate 8) on 1 ft of red silty clay on 7 ft of purple and brown mottled sand with clayey seams near the base on 3 ft of breccia.

A thin section (E. 37601) of the breccia from this borehole was described by Mr K. S. Siddiqui. This is a reddish brown, well lithified rock with subrounded to well-rounded pebbles, up to 5 cm in diameter of fine-grained dolomitic and arenaceous limestone, mudstone, siltstone, hematitic siltstone, fine-grained sandstone quartzite, kaolinized fragments and semi-translucent rock particles heavily stained with iron oxide. The

d

fines consist, in order of abundance, of quartz, feldspars, micas (sericite) and carbonates. The feldspars include plagioclase, orthoclase and a few grains of microcline. Granules of well-crystallized kaolinite and hematite are also dispersed among the detritus. In some places the presence of well formed rhombs of dolomite indicates secondary recrystallisation of the carbonate. Microgranular calcite and some grains of chert are also scattered in the matrix. Goethite and cryptocrystalline ferric oxide constitute the bulk of the intergranular cement.

An X-ray powder photograph of the whole rock showed the presence of quartz, dolomite, kaolinite, illite, calcite, goethite and feldspar.

LOWER MOTTLED SANDSTONE

South and western areas

At the eastern end of the Morley Tunnel [3995 4009] the total thickness of Lower Mottled Sandstone is approximately 60 ft (Fig. 83), of which some 45 ft of cross-bedded pebbly sandstone are poorly exposed. Pebbles are less common than in the overlying Pebble Beds and the sandstone appears to be partly dolomitic.

Near Dunnshill, at the western end of Dale Hills [4261 3837], the Lower Mottled Sandstone is well-exposed in old quarry faces:

	ft	in
(Pebble Beds)	15	0
Sandstone, buff, fine-grained; red-stained bands; red silty and rare grey-green mudstone laminae	5	0
Sandstone, buff, fine, red-mottled; marly patches	8	0

Dale Abbey boreholes Nos.1 and 2 were drilled from the floor of the quarry and proved red marl and yellow sandrock to 48 ft 6 in. A faulted outlier north-west of Dale has been worked for sand near Hagg Farm [4310 3930]. In an old pit [4347 3889] at the foot of Arbour Hill the following section was exposed:

	ft	in
Clay, sandy, red and grey-brown	3	0
Sand, buff, fine to medium-grained; soft red clay stringers and small pebbles	4	0
Clay, pale grey-green; reddish at top and bottom	0	3
Sand, buff, medium-grained, red-stained, cross-bedded; lenticles of green clay along foreset beds	1	6
Clay, green	0	2
Sand, red-brown; green lenses	2	0
Clay, green	0	1
Clay, red, sandy; sporadic buff sandstones; rare grey-green clay lenses	4	0
Sandstone, buff and red-mottled	2	6
Clay, red and grey-green	0	9
Sandstone, pale brown, fine-grained; red staining	12	0

The junction between the Lower Mottled and Bunter Pebble Beds is exposed some 10 yd to the west, where 10 ft 6 in of buff sand and medium-grained sandstone with small pebbles are underlain by 14 ft of fine-grained red and buff mottled sandstone.

At the western extremity of the outcrop around Dale Abbey petrographical examination by Dr Berridge reveals that the clasts of the typical fine-grained sandstones are less rounded than normal for the formation; angular grains are almost as abundant as the standard subangular type. One sample (E 36090) contains a very high proportion of partly replacive interstitial baryte. It resembles the Pebble Bed lithology of the Hemlock Stone (p.154d) and it is significant that accessory interstitial baryte is present in the overlying Pebble Beds at Dale Abbey [SK 4334 3897] (E 36089). Accessory detrital minerals identified from the Lower Mottled Sandstone in this area include apatite and staurolite in addition to the more usual opaque oxide ores, zircon and tourmaline.

Farther east, near Grove Farm a small pit [4520 3858], shows 12 ft of massive, friable, yellow-brown, fine-to medium-grained, slightly micaceous sandstone; cross-bedding suggests sources to the north-east. On the opposite side of Dale Road in another pit [4522 3852], some 18 ft of similar sandstone, are overlain by 6 ft of laminated slightly silty micaceous mudstone, pale grey-green in colour, sporadically reddened, particularly at the top. Gibson (1908, p. 127, fig. 10) illustrated this region with a section showing the burying and partial re-emergence of the Pre-Triassic landscape. The resurvey shows that although the interpretation is complicated by faulting, the Crawshaw Sandstone formed a feature in the pre-Triassic landscape, for example at Thackers Wood [4570 3850].

West of Stanton are several faulted outcrops of red and buff marly sand which are considered to belong to the Lower Mottled Sandstone. In one old sand pit [4600 3820] bounded on the north by faulted Crawshaw Sandstone are 15 ft of fine-grained sandstone, predominantly red, but weathering to a yellowish colour with mottling and staining. Rare cross-bedded units indicate provenance from the north. Sporadic coarser quartz layers, 1-in bands of silty mudstone, pale grey-green in colour, and dolomitic bands also occur. There are many variations within the pit, but generally the sandstone is more marly towards the top and the southern end of the exposure.

Dr Berridge reports that tourmaline is present in all thin sections from this locality but accessory grains of zircon, andalusite, garnet and epidote are less regular in their distribution.

At Stoney Clouds, east of Stanton, the lower part of the scarp may include the topmost beds of the Lower Mottled Sandstone. The following section was measured in an old quarry [4788 3765]:

	ft	in
Sandstone, yellow-buff, massive, fine-to medium-grained, cross-bedded (?Bunter Pebble Beds)	6	0
Gap	6	0
Mudstone, red and pale green) Sandstone, massive, fine-grained) (?Lower Mottled Sandstone)	3	6

A small outlier of Lower Mottled Sandstone

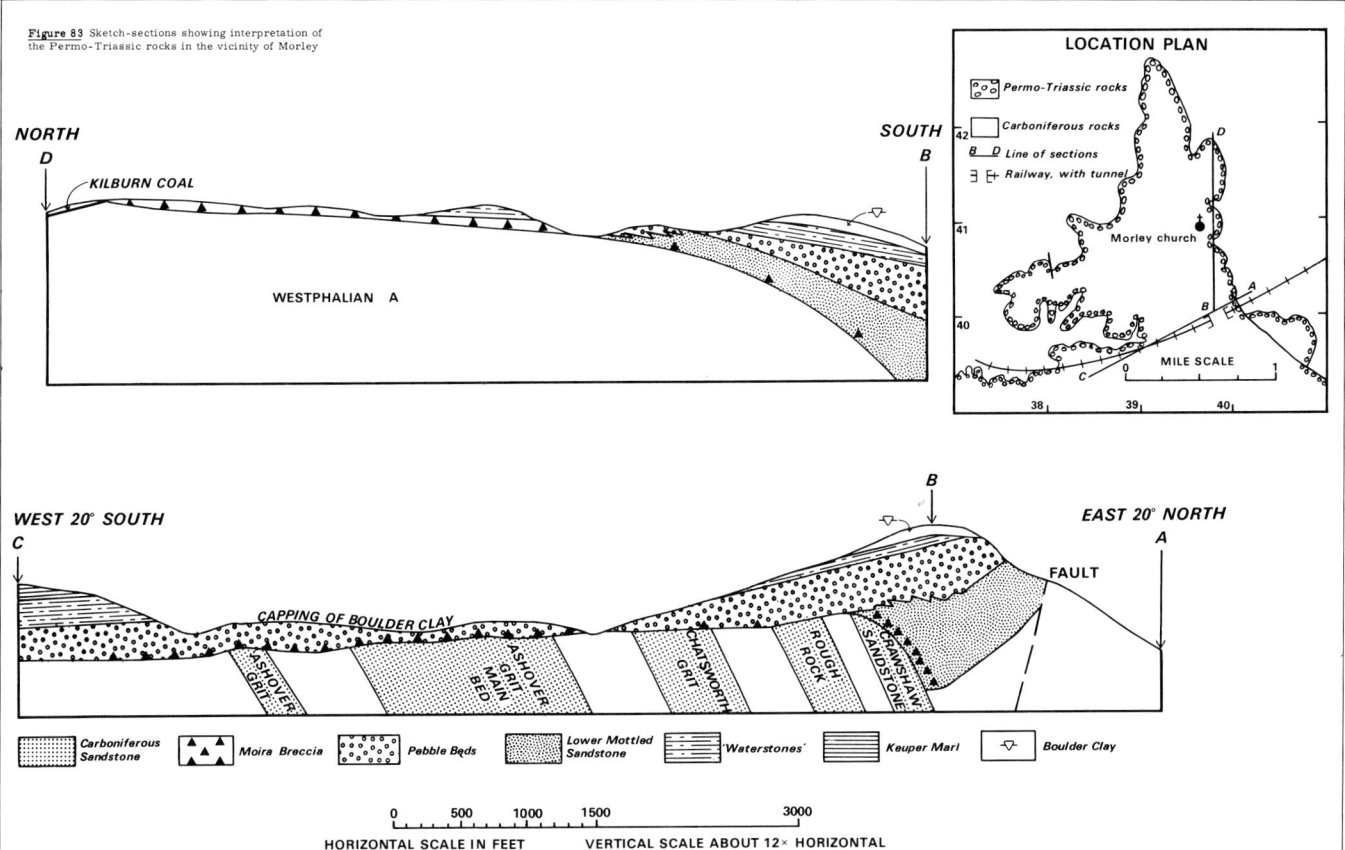

Figure 83 Sketch-sections showing interpretation of the Permo-Triassic rocks in the vicinity of Morley

occurs at New Stapleford [4950 3830]. The rock is generally less than 10 ft thick and composed of soft red fine-grained sand. Over 80 ft of massive red sandstone are seen in a disused quarry on the opposite side of Coventry Lane [4990 3885]. The lower 30 ft contain many red marly partings, and the upper 50 ft sporadic buff patches and bands together with pebbles. Cross-bedding indicates a derivation of sediment from the north-east.

In the Stapleford Hill area, Bramcote sand quarry [5030 3890] shows the following section:

	ft	in
(Pebble Beds)		
Mudstone, grey-green		1
Mudstone, red		2
Mudstone, buff with coarse quartz grains		1
Sandstone, red to dark red, fine-grained, massive, cross-bedded: sporadic discontinuous bands and patches of buff sandstone up to 12 in thick of similar lithology	50	0
Trial hole in sandstone	10	0

Taylor and Houldsworth (1973, p. 165) recorded some 40 ft of Mottled Sandstones resting unconformably on Westphalian sandstone in Swancar Quarry [491 393], which is west of the outcrop shown on the map.

Excavations for a roundabout [5050 3792] on the Sandiacre-Stapleford By-pass exposed the following section, considered to lie near the top of the Lower Mottled Sandstone:

	ft	in
Sandstone, red-brown and buff, fine- to medium-grained, cross-bedded	10	0
Marl, red	1	0
Sand, fine-grained, buff	0	1
Marl, red, sandy	0	3
Marl, red	0	1
Sand, buff with marl pebbles; red discoloration at top	0	1
Sand, red, fine-grained	0	5
Sand, buff, medium-to coarse-grained	0	2
Sand, red, fine-to medium-grained	1	0

A small outlier of Lower Mottled Sandstone occurs at Balloon Wood, where an excavation [5087 3978] showed some 10 ft of red fine-grained sandstone with sporadic small pebbles and red marly patches.

Opposite the northern entrance to Nottingham University [5462 3880] on the eastern side of the Nottingham Ring Road an old sand pit contains faces of Lower Mottled Sandstone up to 30 ft high. The sandstone is predominantly red in colour with buff patches, bands and mottling. It is commonly calcareous, partly dolomitic and was extracted for moulding sand in the 1930's.

Lower Mottled Sandstone crops out over about 1¼ square miles in the Wollaton Park area, but apart from a few excavations of red sand in Wollaton village, exposures are poor. The best exposures occur near and on the University Campus. Massive fine-grained sandstone up to 35 ft thick is seen in an old quarry face

[5440 3863] south-east of Lenton Firs. It is generally red in colour, with buff to pale brown discontinuous bands, commonest at the top of the face. Cross-bedding indicates derivation from the north-east. Swinnerton (1910) recorded lenses of dolomitic sandstone from this locality. Detailed petrological examination of the Lower Mottled Sandstone on the Nottingham University Campus indicates an absence of a coarse-grained fraction and in part it grades into siltstone. The apparent eastward trend towards an increasing degree of clast-roundness is not maintained throughout the formation. Despite the general fine-grained nature of the sediments and the presence of some comparatively well-sorted and rounded material, the rocks are subarkosic rather than protoquartzitic and, although chert or siltstone clasts probably outnumber those of metamorphic rocks, accessories include garnet, hornblende, andalusite and staurolite, none of which were identified in Catstone Hill thin sections (see p.151d). The compositional affinity is with the Stanton by Dale and Dale Abbey rocks rather than those of Catstone Hill. The Lower Mottled Sandstone at Nottingham University Campus is more consistently affected by dolomitic cementation than at the other localities discussed.

Many boreholes have proved the Lower Mottled Sandstone in the Nottingham-Derby area (Plate 10). In the Beeston area, Station Road and Stoney Street boreholes (Nos. 13 and 14, Plate 10) proved the basal breccia/conglomerate contains angular to rounded fragments of quartz, quartzite, limestone, dolomite, green igneous rocks and Westphalian mudstones; some fragments are larger than the 2 in diameter of the borehole.

The Lower Mottled Sandstone, as proved in the bores, is a little under 100 ft in the south-east of the district, just west of Nottingham, and thins westwards to about half this thickness around Dale Abbey. Exact figures cannot be established in this latter area, where the majority of the boreholes date from the 19th Century and are inadequately logged.

Eastern Area

Much of the outcrop in the north-west of this area is obscured by boulder clay, but 12 ft of brick-red, fine-grained, soft, friable sandstone were excavated [4998 5398] on a building site near Annesley Woodhouse, and 20 ft of pink cross-bedded sandstone were exposed at Nuncargate [5035 5410].

There are many small exposures of Lower Mottled Sandstone, up to 20 ft thick, in the Annesley, Newstead and Hucknall areas, as at [5105 5351], [5399 5412], [5165 5245] and [5050 5178]. Cross-bedding suggests a west and north-west origin for the sand.

The largest exposure in the Hucknall area occurs at Long Hill [5191 4913]:

	ft	in
Soil, brown, sandy, with small pebbles	1	0
Sand, yellow-brown, fine-grained with pebbles up to 3 in diameter at base; red marl bands, ironstone nodules and coaly streaks	2	0

	ft	in
Sandstone, red, fine-grained; rare small pebbles	14	0
Pebble layer, predominantly quartzit pebbles with some laminated silty mudstone fragments	2	0
Sandstone, red	2	0
Marl, dark red, with pale grey patches	2	0
Sandstone, red	1	6
Marl, dark red	1	6
Sandstone, massive, red, fine-grained, cross-bedded; beds up to 2 ft; patches of grey-green marl	10	0+

There are many small outliers of Lower Mottled Sandstone between Watnall and Hucknall, for example at Hucknall Cemetery [5288 4826] and at Butlers Hill.

At Bulwell the Lower Mottled Sandstone forms the eastern bank of the River Leen. West of the town, near the northern cemetery, the Middle Marl has been worked for clay with a cliff-like exposure of Lower Mottled Sandstone forming the back wall of the excavation. Of the 18 ft of pinkish red, fine-grained sandstone still visible, the upper 10 ft are cross-bedded from the south-west; the lower 8 ft are massive and contain red marl pellets (3 in x ½ in) in the basal 3 ft.

A large outlier of Lower Mottled Sandstone occurs between Bulwell and Nuthall. On the north side of the Cinderhill Fault there are several exposures of red sandstone containing sporadic pebbles. Up to 18 ft can be seen in a railway cutting [5295 4442] and some 39 ft of sandstone were proved in the old Hempshill Colliery Shaft.

Some 11 ft of red sandstone with a soft marly base were recorded by Mr Fowler in the old Cinderhill Brickyard [534 441] to the north of the fault. South of the dislocation he recorded:

	ft	in
Soil	7	0
Sand, mottled	8	0
Breccia	2	6
Sand, hard, yellow and red	9	7
Marl and sand, variegated	1	3
Sand, red and yellow variegated, laminated	4	8

The above section is now obscured by tipping but 6 ft of fine-grained calcareous red-buff sandstone are still visible in the nearby railway cutting at Cinderhill [5369 4379]. Red sand was dug during a new housing development east of Basford Hall [5422 4344].

A fault slice of Lower Mottled Sandstone was exposed in a sewer trench near Broxtowe Wood [5278 4314], where some 8 ft of red and buff fine-grained sandstone were seen. A similar fault trough between Kimberley and Nuthall Temple forms a ridge of Lower Mottled Sandstone, with several exposures, e.g. in Knowle Wood [5040 4443]. Nearby, a temporary excavation [5052 4412] for the Nuthall By-pass exposed the southern boundary fault, with Lower Magnesian Limestone thrown against red fine-grained sandstone.

Between Strelley Windmill and Strelley Hall,

the M1 Motorway cut through an elevated ridge of the Lower Mottled Sandstone; 15 ft of red marly sand, capped by a few feet of drift. The underlying Middle Marl (p.152b) contains many sandstone bands and the gradual passage between the two formations was well displayed.

The best exposures seen in the southern part of the eastern outcrop were at Catstone Hill Opencast Site [5033 4147] and the adjoining quarry [5076 4151]. At the top of the quarry, a lens, 10 to 12 ft thick, of buff pebbly and conglomeratic sandstone is well cemented by baryte - a lithology comparable with that of the Pebble Beds (cf. Bramcote Stone: p. 154d). The lens is aligned north-north-eastwards and probably represents a washout channel. The following section was recorded in the opencast site:

	ft	in
Sandstone, dark red, fine-grained, cross-bedded; sporadic discontinuous bands of small pebbles	12	0
Sandstone, pinkish red; variable ? dolomitic cement	1	0
Sandstone, red, fine-grained	2	11
Pebble band - pebbles up to 3 x 2"	0	2
Sandstone, red	0	5
Sand, red, clayey	0	5
Sandstone, pale grey-green and buff, medium-grained, with pebbles	0	6
Sandstone, red, fine-grained; sporadic small pebbles	1	8

At Broomhill Wood [5280 4072] temporary excavations exposed 6 ft of red, fine-grained, soft sandstone resting unconformably on stained Westphalian rocks. Nearby (see p. 152a), dolomitic limestone was recorded within the Lower Mottled Sandstone.

The lowest 50 ft or so of the Lower Mottled Sandstone have been proved in some 11 boreholes and shafts in the eastern part of the district, and old water bores [5470 4260] in Basford just beyond the margin of the sheet penetrated 100 ft of red sandstone which are assigned to the formation.

PEBBLE BEDS

Western area

The large sand and gravel quarries of the Mercaston—Muggington area provide representative sections of the Pebble Beds and are described first in the following account. Sections from other areas are described except where they show variation from this norm.

The quarries just over 1 mile SW of Turnditch [2800 4540] provide an almost complete succession through the Pebble Beds.

	ft	in
Sandstone, buff, pebbly, medium-grained,	20	0
Mudstone, red, with green margins; in lenses up to	1	0
Sandstone, buff, medium-grained	4	0
Sandstone, buff, pebbly	15	0
Sandstone in impersistent lenses	up to 5	0
Sandstone, predominantly pebbly	15	0
Mudstone, red, with green margins;		

Key to Figure 83:

Carboniferous Sandstone | Moira Breccia | Pebble Beds | Ashover Grit Main Bed | Lower Mottled Sandstone | 'Waterstones' | Keuper Marl | Boulder Clay

HORIZONTAL SCALE IN FEET   VERTICAL SCALE ABOUT 12× HORIZONTAL

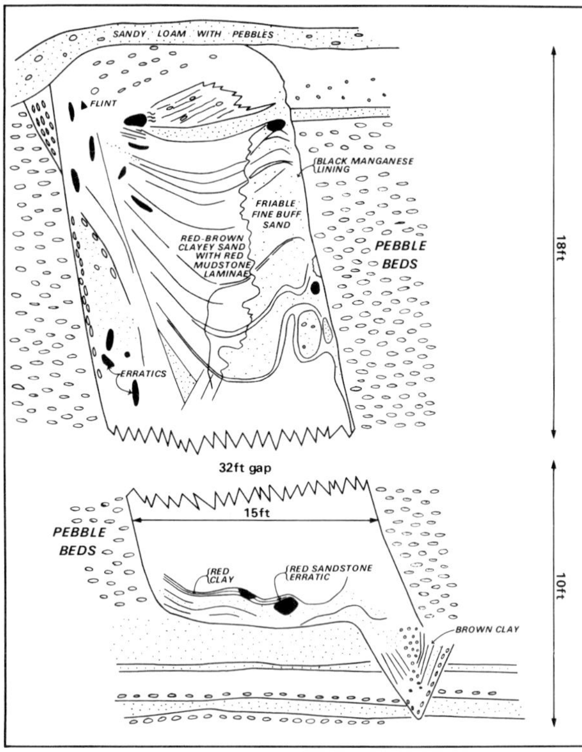

Figure 84 Sketch-section of the deposits in a large fissure in the Pebble Beds near Turnditch. This isolated structure about 1 mile S of Turnditch [2814 4549] lies at about 700 ft OD and is a vertical parallel-sided fissure some 60 ft deep. It trends ENE and has been mapped for over 200 yd length. See description on pages 93-94.

a

b

	ft	in
passes laterally into green and buff silty mudstone	1	0
Sandstone, cross-bedded, with abundant pebbles	37	0
Sandstone, buff, fine-to medium-grained massive, sporadic pebbles	30	0
Sandstone, buff, fine-to medium-grained, massive	7	0
Mudstone, green, with brown margins	0	6
Sandstone, brown, fine-to medium-grained, with rare coarse bands, massive, cross-bedded; sporadic pebbles and khaki mudstone pellets	10	0

Mr Siddiqui reports that sand samples from the top 20 ft have subangular to subrounded grains, some showing surface pitting. They are stained by pale yellow clay. The sand fraction is mainly of quartz in a matrix of chlorite, illite, fibrous kaolinite and thin flakes of muscovite. Feldspars are mainly plagioclase with traces of orthoclase and microcline; grains of chert, scattered specks of hematite and lithic particles of sandstone and quartzite are common. Heavy minerals include tourmaline, zircon, anatase and possible sphene. Manganese staining is common in bands or near joints. The gravelly portion is composed mainly of fine-grained quartzite pebbles, which are generally elongated and have conchoidal fractures, and smooth surfaces without cavities or striations. Angular fragments of milky and pink quartz and of quartzite and pebbles of sandstone, limestone, chert, mudstone, and igneous rocks are also present.

The gravelly portion is commonly cross-bedded (see Plate 11), but otherwise appears to be homogeneous and massive, except where thin layers of mudstone or siltstone pick out complex minor structures (p. 93). Sporadic joints are commonly filled with red clay.

Old disused sand pits at Bullhurst Hill show up to 20 ft of buff massive sandstone with rare pebbly layers. Red marl bands up to 12 in thick are common 20 ft or so above the base of the formation. The marl bands thicken north-eastwards and unite to form 12 ft of red, poorly laminated micaceous mudstone with thin green and yellow silty and sandy bands and laminae - the thickest 'marl' known in the Pebble Beds succession.

At Muggington Quarry [2930 4340], near Bullhurst Hill, the top 50 ft of the Pebble Beds consist of ill-sorted sand and gravel containing wedge-shaped lenses, up to 4 in thick, of yellow-brown sandstone. The lower quarry faces are formed of some 40 ft of massive buff sandstone with sporadic pebbles. The pebbles, aligned in bands, have red marl coatings.

A small old gravel pit north-west of Bullhurst Hill [2855 4390] contains about 20 ft of clayey sand and gravel, well cemented in places to form a conglomerate which was presumably responsible for the abandonment of the workings.

Exposures in a pipe trench east of Weston Underwood showed a feather edge of sand, clay and pebbles, overlying purplish red mudstones of the Widmerpool Formation. One excavation [2972 4245] gave the following sequence:

	ft	in
Clay, red	1	0
Sand, grey, with pebbles	2	0
Sand, yellow-brown; in a lens up to	3	0
Clay, red	8	0
Mudstone, purplish red, laminated, with green mottling (Widmerpool Formation)		

The sand is assumed to lie at or near the limit of deposition of the Pebble Beds.

Two boreholes [2535 4167 and 2572 4159] at Brailsford proved up to 30 ft of sand and gravel. An old gravel pit [2540 4170] contains up to 10 ft of ferruginous sand and fine gravel, worked in the past on both sides of Luke Lane.

Exposures of Pebble Beds north of the Mercaston—Mugginton area contain a greater than normal proportion of rock fragments, including chert, sandstone, siltstone and silty mudstone, e.g. at Ireton Wood [2783 4874].

Outliers of Pebble Beds occur near Bullhill [2735 4905], at The Mountain [2680 4920] and Blackwall [2560 4950]. The last was once extensively quarried for sand and gravel, but is now largely obscured by tipping and afforestation though at one place [2558 4948] 13 ft of buff, fine to medium well-sorted sand with pebbles are visibie. The lower part is soft with cross-bedding indicating a source to the east-north-east. A sample obtained from the lowest 5 ft showed a bimodal grain size distribution.

Pebble Beds form a series of irregular outliers capping the escarpment south-west of the Ecclesbourne valley between Weston Underwood and Allestree. The formation, about 90 ft thick at Gun Hills, thins rapidly south-westwards to disappear in the vicinity of Kedleston, where the Waterstones appear to rest directly on Carboniferous rocks.

In a stream section [3016 4437] near The Clouds, 2 ft of near-horizontal red mudstone with green sandy lenses are exposed beneath the main Pebble Beds feature and probably represent a local basal mudstone. Because of their relationship to the overlying Waterstones, red and green mudstones in old clay pits [3050 4158] in Brick Kiln Covert, Kedleston, are similarly assigned to the basal Pebble Beds rather than to the Keuper Marl as proposed by Wedd (in Gibson and others 1908, p. 122).

Pebbly sandstone is exposed in a few small pits near Ireton Farm [3142 4180], Cocks-hut-hill [3220 4280] and Allestree Park [3416 4070], where up to 12 ft of moderately to poorly cemented, buff, cross-bedded sandstone in beds $1\frac{1}{2}$ to 3 ft thick are visible. The sandstone contains a few rounded quartzite pebbles, up to 3 in diameter, concentrated in thin lenses or along the cross-bedding.

Southern area

The Pebble Beds form faulted outcrops which are well exposed between Nottingham and Derby. Three water boreholes in Slack Lane [336 362], Derby, proved $33\frac{1}{2}$ ft of sandstone without bottoming the sequence, and the King Howman and Co. Water Borehole [3434 3675] proved a total thickness of 36 ft (Stephens 1929, p. 90), of which the lowest $21\frac{1}{2}$ ft were without pebbles. This lithology resembles that of the Littleworth Beds of Stevenson and Mitchell (1955, p. 32) and is thus assigned to the Pebble Beds rather than to the Lower Mottled Sandstone.

A few feet of coarse, brown, cross-bedded, soft sandstone contain lenticular pebbly beds up to 1 ft thick near Allestree [5016 3950]. The total thickness of Pebble Beds hereabouts is calculated to be between 25 and 30 ft. They are a similar thickness on the eastern side of the Derwent valley, where they appear to be cambered. In Croft Wood [3663 3932], Breadsall, at the top of the stream cutting there are 6 ft of soft khaki sandstone; 15 to 20 ft below, but within a landslip, there are 5 ft of brown sandstone with 1 to 2-in angular fragments of sandstone and quartzite overlying the Moira Breccia (p. 152c). To the east of Breadsall [3761 3940], where the Pebble Beds have been dug for gravel, 6 ft of coarse brown and red-brown mottled sand and sandstone contain many pebbles and two fissures, up to 1 ft wide, filled with red mud.

A section [3836 4105] east of the roadway near Breadsall Lodge comprises 7 ft of coarse brown sand with quartzite and sandstone pebbles and cross-bedding which dips to the north-east. A pit [3837 9090] nearby shows 11 ft of cross-bedded, soft brown, sandstone with scattered quartzite, sandstone pebbles and fragments of sandstone and chert. Most of the gravel here occurs in the lowest 3 ft at the east end of the pit in beds and poorly defined masses.

Sections of a few feet of pebbly sandstone are seen in the railway cuttings west of Morley Tunnel and in a pit [3886 4006] north-east of Broomfield Hall, where they underlie the Woodthorpe Formation (p. 154d). Wedd (MS map) recorded 10 ft+ of 'coarse shingle' in a now obscured pit [3902 3986]. The total of 25 to 30 ft of Pebble Beds here is similar to the thickness in the Derwent Valley. The visible section in the cutting [3995 4009] to the east of Morley Tunnel is poor compared to that illustrated by Gibson and others (1908, p. 125). Beneath remanie Woodthorpe Formation on the north of the cutting, and according to Gibson and others (1908, p. 124) below overlapping Waterstones on the south, are some 45 ft of sands which are densely packed with pebbles of quartzite, light and dark siliceous rocks, limestone and granitic rocks. Beds of soft khaki and red-mottled micaceous sand, up to 1 ft thick on the north side of the cutting and $2\frac{1}{2}$ ft on the south, occur within the pebbly sands. These Pebble Beds rest on 60 ft or more of Lower Mottled Sandstone.

Exposures of Pebble Beds, up to 20 ft thick, occur west of Dale at Arbour Hill [4340 3855], near The Flourish [4285 3875] and Dunnshill [4210 3845]. At the last locality, bands of red-stained sandstone are common.

A faulted scarp [4540 3815] near Dale Abbey contains three exposures of up to 20 ft of massive brown medium-to coarse-grained sandstone. Dips of cross-bedded units, some pebbly, indicate derivation from the south-west, south-east and north-east.

South of Dale, a north-facing scarp of sandstones over a mile long provides excellent exposures of massive, pale brown to buff, cross-bedded, pebbly sandstone. Pebbles are up to 4 in in diameter, but average between 1 and 2 in. Cross-bedding dip directions range from south-east through south-west to north-west. The gradational contact with the underlying Lower Mottled Sandstone is exposed at the western end of the scarp near Dunnshill [4261 3836] where inclusions and lenses of pale grey-green mudstone and rare buff fine-grained sandstone fragments occur in the basal 15 ft. The uppermost beds are exposed in Ockbrook Wood [4360 3833], where the top 9 in are silicified and overlain by mudstones and sandstones of the Waterstones. The measured thickness of Pebble Beds in the Dale area is 70 ft. An old quarry [4680 3785] at Stanton by Dale shows a face of soft, massive, buff, medium-to coarse-grained sandstone. Pebbles are locally common, particularly within cross-bedded units.

At Stoney Clouds the Bunter Pebble Beds form a bold north-facing fault-scarp, some 100 ft high, of buff, pebbly, poorly cemented sandstone. At Burn Hill [5040 3730], south-west of Bramcote Church, 26 ft of red-stained, buff, cross-bedded sandstone are seen in a cliff-like feature; provenance is from the north west. Two hundred yards south of Bramcote Church, along the sides of Chilwell Lane [5070 3730], the junction between the Pebble Beds and the Woodthorpe Formation is exposed, the former comprising up to 35 ft of red-brown and buff, medium to coarse-grained, pebbly, cross-bedded sandstone. The top 2 in of the sandstone are pale grey with a concentration of small pebbles.

West of the River Erewash there are numerous old sand quarries which show the typical buff, off-white and reddish, massive sandstone, with sporadic pebbles. The red coloration is in the form of banding along the foresets of the cross-bedding.

There are many exposures of Pebble Beds in the Stapleford—Sandiacre area. A 60-ft quarry face [4933 3735] north of the cemetery exposes a mottled red and buff, massive, cross-bedded, medium-to coarse-grained sandstone. The red colour is predominant, probably due to staining from 10 ft or so of overlying Waterstones.

Within the Clifton-Highfields Fault system the highest part of the Pebble Beds is exposed at Tottle Brook [5280 3828]. A conglomerate about 4 ft thick comprises a coarse sand containing pebbles, predominantly quartzites, up to 4 in but averaging 1 in, and ironstone fragments, and is underlain by 3 ft of buff medium-grained cross-bedded sandstone. This is assumed to be the locality where Shipman (1889, p. 126) found pebbles with Orthis, Rhynchonella and Atrypa. The basal 12 ft are a reddish colour and the top 15 ft buff.

South of the Clifton-Highfields Fault system many exposures resulting from building and road construction showed brown sandy loam with pebbles overlying red-brown, buff, cross-bedded sandstone with sporadic pebbles. This rock, typical

c

d

of the Nottingham area, weathers very quickly to become friable or running sand.

An old river cliff [5410 3791] at the side of the University Lake exposes 20 ft of buff to light brown, medium-to coarse-grained sandstone. Cross-bedding is common but variable in direction with origins from the south-west and east-north-east. Layers of pebbles occur, usually quartzites with a maximum pebble diameter of 3 in, together with mudstone and other sedimentary rock fragments.

Wollaton Hall is sited on a prominent feature of Pebble Beds forming the highest ground in Wollaton Park, and two old quarries [5352 3870 and 540 390] show sections of up to 12 ft of buff, brown and reddish fine-to medium-grained sandstone.

Boreholes along the south-eastern margin of the Derby district, between Nottingham and Derby, show that the Lower Mottled Sandstone is everywhere overlain by a coarser, pebbly, red-brown, pale buff to grey sandstone of Pebble Beds lithology, which varies from 170 ft in Highfields Borehole [5439 3751] at Beeston, just beyond the margin of the district, to about 70 ft in the Dale Abbey area east of Derby (Plate 10).

Eastern area

Between Newstead and Wollaton the Pebble Beds are present largely as thin cappings or outliers on hills of Lower Mottled Sandstone; they are readily confused with Pleistocene sands and gravels, composed mainly of re-worked Triassic sandstones which are also present in this area.

The largest outcrop occurs north-east of Newstead Abbey on a spur [543 543] between Swinecotte Dale and Knightcross Dale, but exposures are few.

Another ridge [510 545] of Pebble Beds extends east of the present district to form the Shoulder of Mutton Hill, some of the highest ground in Nottinghamshire at 609 ft. All Saints' Church, New Annesley, is built on an outlier at the southern end of this ridge, where yellowish sand and pebbles have been dug out. Small outliers also occur near Castle Wood between Annesley and Newstead Abbey.

At Catstone Hill [5076 4151] a buff conglomeratic sandstone, up to 12 ft thick, occurs above a quarry in typical Lower Mottled Sandstone, (p. 153d). The presence of baryte cement in the conglomerate has been cited by Aveline (1880) as evidence of a Keuper affinity, but the resurvey suggests that the rock probably occupies a washout channel, trending north-north-east, cut into fine-grained Lower Mottled Sandstone at the onset of Pebble Beds deposition.

At Alexandrina Plantation well-cemented conglomerate is again exposed [5190 3852]. It has been argued that this conglomerate also is of basal Keuper age (Aveline 1880; Shipman 1889). If so, the deposit is little more than the reworked top of the Pebble Beds since it has few affinities with the Woodthorpe Formation (p. 154d).

A borehole [about 5015 3855] at Bramcote Hill proved 140 ft of sandstone, the lower part of which must be Lower Mottled Sandstone. At the surface 8 ft of massive cross-bedded fine-to medium-grained sandstone is predominantly red in colour with irregular buff bands and patches up to 2 ft across. Here, and to the west, distinction between the Lower Mottled Sandstone and the Pebble Beds becomes difficult. At the top of Bramcote Hill [5030 3860], 3 ft of buff, red-brown, pebbly sandstone are irregularly weathered because of uneven barytes cementation.

In Bramcote Sand Quarry [5035 3880] the formation overlies Lower Mottled Sandstone and is distinguished from it by a red-brown rather than red colour, and being composed of medium-grained sandstone with coarse and pebbly units and a low percentage of clay. Some 200 yd to the south-east an old quarry exposes 35 ft of massive sandstone, the lower part of which is predominantly red with sporadic buff bands and irregular patches. The upper part is red-brown or buff with rare pebbles. A marl band conveniently divides the face into Pebble Beds above and Lower Mottled Sandstone below. On the opposite side of the road the Pebble Beds form the impressive feature of Stapleford Hill, rising to 332 ft and providing exposures of almost the total sequence, here some 100 ft thick. The beds are fine-to medium-grained, with well-developed cross-bedding dipping to the south-west. The sandstone exposed at the top of the hill is pebbly, well-cemented and commonly red-stained. In thin section, baryte is accompanied by a significant proportion of ferric oxide cement. Feldspar and non-quartzose lithic clasts are not as common as farther west near Dale Abbey.

The Hemlock Stone [4995 3866] is a quarry-remnant of Bunter Pebble Beds left unworked because of a resistant cap of rock with a baryte cement. It is cross-bedded in large units, with current directions from the north-east and north-west. Samples from the Hemlock Stone show that the bands relatively resistant (E 36080) and susceptible to weathering (E 36081) are respectively richer and poorer in baryte. The baryte-rich rock has a texture very similar to that of heavily calcitized rocks from the top of the Pebble Beds farther west. The partially replacive poikilitic baryte crystals are individually as much as 1.2 mm in diameter and the mineral probably constitutes about 30 per cent of the rock volume. The bands in the Hemlock Stone which are more susceptible to weathering are slightly finer-grained and the baryte cement has probably replaced a lower proportion of the sand content. Other specimens from Stapleford Hill are coarser in sand grain size, but their interstitial baryte cement is more tenuous.

WOODTHORPE FORMATION

An old pit [3886 4006] on the roadside at Broomfield Hall exposes a partly disturbed section of the Woodthorpe Formation below some 4 ft of boulder clay. A basal, coarse, grey-brown, fairly hard sand with quartzite and sandstone pebbles and fragments of black chert is overlain by 8 in of greenish clayey sand with a thin red clay seam and with many pebbles which are more angular than those of the Pebble Beds. The succeeding well-bedded fine khaki sand, 10 in

a

thick, is in turn overlain by 1 ft of red clay. This is the most westerly known occurrence of the formation.

The formation is intermittently exposed in the railway cutting and associated watercourses for 600 yd west of the west portal of Morley Tunnel. One exposure [3943 3971] shows the basal 1 ft 3 in resting on a few in of Pebble Beds. In the cutting east of the tunnel an impersistent few in of greenish clay with pebbles is sandwiched between the boulder clay and the top of the Pebble Beds. Gibson and others (1908, p. 124) recorded overlap of the Pebble Beds by the Waterstones in this cutting.

Between Morley and Nottingham numerous exposures and temporary sections of the Woodthorpe Formation have been recorded.

An indication of the rapid westerly thinning of the formation was provided by an old quarry [4340 3885] at Dale Abbey:

	ft	in
(Waterstones Formation, p. 155c)		
Mudstone, dark red, silty; sporadic green laminae near base	5	6
Mudstone, grey-green, alternating with buff and red-stained sandstone; sporadic small pebbles	0	10
Pebble band	1	1
Mudstone, green, silty	0	3
Sandstone, greenish, with pebbles	0	1
(Bunter Pebble Beds, p. 154c)		

The most complete section of the formation is located in a cutting [4744 3770] on the M1 Motorway, near Stanton:

	ft	in
Sandstone, red and buff mottled, micaceous, massive, cross-bedded; small green mudstone pellets common	5	0
Mudstone, red, with green mottling	1	2
Sandstone, brown with red mottling	0	4
Mudstone, red, laminated	1	3
Sandstone, pale grey-green to buff, slightly micaceous; sporadic pebbles; mudstone inclusions at top, conglomeratic base	7	0
Mudstone, pale green, brown at top	6 in to 0	9
Siltstone/sandstone, buff	0 to 0	5
Sandstone/silty mudstone, red and green interlaminated	1	6
Mudstone, red; green top and bottom	0	2
Sandstone, buff, red mottling, fine; base irregular cutting down into underlying mudstone; small pebbles in the hollows	4	6
Mudstone, pale green and yellow, with pebbles	0 to 1	0
Sandstone and silty mudstone alternations; sandstone bands up to 5 in thick; red mudstones altered to a green colour near fault	4	0+

Exposures above Pebble Beds, along the roadside [5080 3731] at Bramcote, comprise 5 ft of red silty mudstone, interbedded with buff siltstone and fine sandstone. These beds were classified as Keuper Basement Beds by Taylor (1965, p. 183), and on the basis of one small included quartz pebble, 1/8 in in diameter, are here referred to the Woodthorpe Formation.

Two small circular outliers [5170 3850 and 5210 3840] have been mapped south of Alexandrina Plantation on the basis of a red marly soil, on which a thriving rose nursery has been established in an area surrounded by less fertile sandy Bunter soil. The presence of Woodthorpe Formation at the base of these outliers is inferred. The top of the Pebble Beds or base of the Woodthorpe Formation exposed nearby [5185 3853] consists of a hard well-cemented pebbly sandstone.

Taylor (1965, p. 189) records Keuper Basement Beds in a trench on the Nottingham University campus. As in the M1 Motorway section described above, the rocks consisted of alternating layers, up to 3½ ft thick, of red marl and fine buff sandstone. The latter was poorly consolidated and contained rare small angular pebbles. Red marl, 3 ft thick, at Beeston Lane [5365 3780] is also assigned by Taylor (1965) to the Keuper Basement Beds, but in the Tottle Brook vicinity it is difficult to differentiate the weathered solid formation from the terrace deposits of the Trent.

WATERSTONES FORMATION

To the north of Kedleston the Waterstones are poorly exposed, commonly occurring as isolated outliers capping Pebble Beds hills, as near Muggington, where a few feet of fine-grained micaceous yellow sandstone are visible at the side of the road [2850 4330]. South-east of Weston Underwood a pipe trench [2934 4180] cut through some 70 ft of red-brown sandstones with red silty mudstone laminae which rest on stained red and purple mudstones of the Widmerpool Formation. Farther south [2830 4012] the same pipeline exposed basal Waterstones comprising red-brown silty clays with conglomeratic sandstone fragments. A breccia with fragments up to 12 in long of coarse calcareous sandstone, fine sandstone and quartzites was seen nearby [2824 4002].

Waterstones crop out in the monocline lying to the west of Kirk Langley, being intermittently exposed in the streams, particularly east of Over Burrows [2667 3970]. Here 6 ft of interlaminated red micaceous mudstone and red-brown siltstone occur. South-west of Nether Burrows there are several sections showing up to 3 ft of micaceous mudstones, siltstones and sandstones.

The complete sequence, about 30 ft thick, of Waterstones was exposed in a pipe-trench north-west of Kirk Langley [2785 3930]; basal red-brown silty micaceous clays were succeeded by similarly coloured fine clayey sand, which was also present as lenses in the overlying red-brown silty and green sandy clays. The base of the Waterstones was better exposed in a road widening east of Kirk Langley [2909 3876], where brown sandstone with a few basal pebbles, including some dreikanters up to about 1½ in diameter, rested on coloured mudstones of the Widmerpool Formation. This section is about 100 yd SW of the pit illustrated by Hull (1869, p. 95). The top of the Waterstones in Trusley Borehole

b

(Appendix 1, p.174c) is drawn at the base of a 24-in green siltstone (skerry) at a depth of 239 ft. Below this bed, micaceous bedding planes are common, the ratio of sandstone to silty mudstone is greater and the sandstone is coarser. Of the samples processed and examined for palynological residues by Dr Warrington, two were productive and the palynomorph assemblages obtained were as follows:

Sample SAL. 1367 : at 278 ft

Calamospora sp.
Punctatisporites sp.
Apiculatasporites plicatus Visscher 1966
Cyclotriletes microgranifer Mädler 1964
C. oligogranifer Mädler 1964
Verrucosisporites contactus Clarke 1965
V. jenensis Reinhardt and Schmitz in Reinhardt 1964
V. krempii Mädler 1964
V. pseudomorulae Visscher 1966
V. remyanus Mädler 1964
?Spinotriletes echinoides Mädler 1964
?Perotrilites minor (Mädler) Antonescu and Taugourdeau - Lantz 1973
?Aratrisporites fimbriatus (Klaus) Mädler 1964
Accinctisporites radiatus (Leschik) Schulz 1965
Alisporites circulicorpus Clarke 1965
A. grauvogeli Klaus 1964
A. toralis (Leschik) Clarke 1965
Voltziaceaesporites heteromorpha Klaus 1964
Protodiploxypinus fastidiosus (Jansonius) Warrington 1974
P. sittleri (Klaus) Scheuring 1970
Colpectopollis ellipsoideus Visscher 1966
Illinites kosankei Klaus 1964
Triadispora crassa Klaus 1964
T. falcata Klaus 1964
T. plicata Klaus 1964
T. cf. staplini (Jansonius) Klaus 1964
Angustisulcites grandis (Freudenthal) Visscher 1966
A. klausii Freudenthal 1964

Sample SAL. 1323 : at 242 ft

(a) Miospores:
Apiculatasporites plicatus
?Cyclotriletes oligogranifer
?Perotrilites minor
Tsugaepollenites oriens Klaus 1964
Alisporites circulicorpus
A. grauvogeli
Voltziaceaesporites heteromorpha
Protodiploxypinus fastidiosus
P. sittleri
Sulcatisporites kraeuseli Mädler 1964
Colpectopollis ellipsoideus
Illinites cf. kosankei
Triadispora crassa
T. ? plicata
Angustisulcites gorpii Visscher 1966
A. klausii
Lunatisporites sp.
Striatoabieites aytugii Visscher emend Scheuring 1970
Tubantiapollenites balmei (Klaus) Visscher 1966
Cycadopites sp.

(b) Microplankton:
Dictyotidium sp.
Veryhachium reductum (Deunff) Jekhowsky 1961
tasmanitids

The above assemblages are comparable with that reported by Visscher and Commissaris (1968) from the Wellenkalk (lower Muschelkalk) of the Netherlands. In terms of the European Triassic stages this unit is conventionally regarded as Anisian in age. In terms of the British Triassic miospore reference sequence (Warrington 1974) the assemblages compare with those documented from the lower part of the Kirkham Mudstones in the west Lancashire succession. The microplankton recorded from SAL. 1323, near the top of the Waterstones, are indicative of a marine environment, a finding consonant with the occurrence of Lingula in the same facies in a neighbouring district (p.86).

The basal part of the Waterstones is poorly exposed in the stream [3013 3883] a quarter of a mile north of Bowbridgefields, and shows 2 ft of red, fine-grained sandstone. T. I. Pocock (in Gibson and others 1908, p. 118), referred sandstone with pebbles in this valley to the Bunter Pebble Beds, but they are probably at the base of the Waterstones. Pocock, on his manuscript maps, recorded 15 ft plus of massive yellow sandstone with 'seams of marl' at the bottom of an old quarry, now infilled, 250 yds W of Bowbridgefields.

The Waterstones faulted against the Widmerpool Formation (p.130d) are exposed in Mackworth [3138 3771]. The following is a composite section:

	ft	in
Soil, red silty clay, with sandstone fragments	3	0
Mudstone, red, laminated, silty, micaceous, with brown and khaki siltstone laminae and beds	9	2
Sandstone, green-brown to brown, unevenly-bedded, with mud films; soft sandstone band near top	2	2
Siltstone, soft, green, micaceous	1	8
Sandstone, brown and pale brown mottled, fine-grained, massive	2	4
Sand, soft, brown, laminated; a few red mudstone laminae	1	0
Sandstone, khaki-green		6+

Similar sandstones crop out at the north end of the road cutting in the village.

Near Quarndon and Allestree the Waterstones outcrop is bounded on the north-east by the Quarndon Fault and is characterised by fairly gentle slopes covered by a red-brown sandy clay soil. The formation is poorly exposed at and near Quarndon village, where the main street has a cutting [3325 4040] through some 3 ft of red-brown, moderately cemented fine-to medium-grained flaggy sandstone. Small exposures of laminated, red-brown and green, fine-grained sandstones and siltstone occur in stream beds [3289 4045 and 3347 4007]. Pits [3195 4148], 20 ft deep, show where the sandstone was formerly worked, but only about 1 ft of fairly soft, pinkish grey sandstone with small cavities is now exposed.

Sections in up to 4 ft of interbedded silty clay,

c

siltstone and fine-grained sandstone can be seen in a road cutting [3443 3973] and an old clay pit [3492 3939] at Allestree.

Within Derby city, at Offilers Brewery Borehole [3521 3505] 450 yd S of this district (Stephens 1929, p. 89), the strata assigned to the Waterstones lie between depths of 361 and 428 ft - a somewhat thicker sequence than might be expected from the mapping in the Derwent valley. Waterstones strata cannot be classified with accuracy in the logs of Jones and Co. Slack Lane [336 362] and King, Howman and Co. [3434 3675] water boreholes (Appendix 1 and Stephens 1929, p. 90).

On the eastern side of the Derwent valley, in Croft Wood, adjacent to the railway line [3662 3927], the basal breccia of the Waterstones rests upon about 8 ft of Pebble Beds. The breccia consists of fragments up to 1 in long tightly packed in a coarse matrix in which some grains are well-rounded and frosted.

The sequence in the degraded pit at the abandoned Racecourse Station [3639 3856] shows the basal beds of the Waterstones with less angular and less densely packed pebbly beds than those exposed in Croft Wood:

	ft	in
Mudstone, red, sandy, micaceous, interbedded with medium-grained cross-bedded sandstone	1	6
Sandstone, yellow, pebbly		1
Mudstone, red, sandy, with pebbles smaller than 1 in		3
Sandstone, brown, medium-grained, with pebbles in uppermost 3 in (Pebble Beds)		6

Gibson and others (1908, p. 111) recorded this section and although they recognized the horizon they did not separate the Waterstones hereabouts from the Keuper Marl. In Mansfield Road [3651 3855], 8 to 10 ft of red silty mudstone with rare micaceous planes and with bands, up to 4 in thick, of grey, grey-green or red fine-grained sandstone are exposed. The red sandstones appear to be dolomitic and have well-rounded quartz grains. Muddy films within the grey and grey-green sandstones show salt pseudomorphs. About 5 ft of red and green-grey sandstone, with rare silty mudstone bands up to 2 in thick overlie the red silty mudstones. At the side of the same road, 150 yd to the north-east, about 6 ft of red-brown and brownish grey-mottled cross-bedded micaceous sandstone with sporadic ½-in siliceous pebbles are overlain by an estimated 8 ft (4 ft seen) of greenish grey, finely laminated, soft sandstone with silty mudstone laminae and a thin pebble bed. These beds are succeeded by 11 in of brown, cavernous, apparently dolomitic, sandstone, which are overlain by 4 ft of red-brown and green-streaked and mottled, well-bedded silty mudstone and siltstone. The exposed sequence is considered to span the Waterstones/Keuper Marl junction, and the thin pebble bed is taken as the base of the latter.

There is an inlier of Waterstones at Chaddesden, where a few feet of red-brown and khaki micaceous siltstone and grey-green sandstone were noted in temporary sections. Some 2 ft of well-bedded sandstone overlain by 2 ft of red and green beds of clayey micaceous sand crop out in a brook [3877 3727].

A few feet of Waterstones are exposed in the road cuttings south of Morley church. Northwards from here they overlap the Pebble Beds to rest upon Moira Breccia and then directly on the Namurian and Westphalian rocks. There are no sections in the broad outcrop that extends northwards to beyond Morley Moor Farm. An auger survey proved fine sands, red micaceous clays, and sandy and stiff clays.

It is not possible to separate the Woodthorpe and Waterstones formations in the logs of the boreholes in the south-east of the district. They show the westerly thinning of the Waterstones, which are only 50 ft thick at Dale Abbey No. 3 Borehole, compared with over 150 ft in boreholes in the drift-covered country near Beeston.

Exposures along the southern margin of the district are small and scattered, the best being temporary sections seen along the route of the M1 Motorway near Sandiacre and the linking Stapleford By-pass. Typical of these are cuttings [473 375] near Stanton, where the top 40 ft or so of the formation were exposed:

	ft	in
Mudstone, green and red, with purplish tinges, silty	3	0
Mudstone, dark red, with laminated siltstone	8	0
Sandstone, green, finely laminated	2	0
Siltstone/sandstone and silty mudstone, interlaminated	6	0
Mudstone, green, silty, and green sandstone, finely interlaminated	2	0
Sandstone, green, oil-stained in beds up to 7 in thick; calcite-lined cavities	3	6
Siltstone, mudstone and sandstone, interlaminated	2	0
Sandstone, brown and green, and silty mudstone, interlaminated	3	0
Mudstone, dark red, and pale khaki sandstone with many cavities at top	7	6
Sandstone, pinkish red	0	8
Mudstone, dark red	0	1
Sandstone, pink; sporadic mudstone laminae	1	10
Mudstone, dark red	0	1
Sandstone, red-brown; rare mudstone laminae	3	0

The lowest 50 ft or so of the formation were not continuously exposed but sporadic excavations showed interbedded buff, brown and green sandstones with numerous silty mudstone laminae. Sections [4740 3763] near Stoney Clouds showed that the lowest beds rest with angular unconformity on the Woodthorpe Formation. The unconformable base is marked by a few inches of red irregularly bedded mudstone, which transgresses 5 ft of the Woodthorpe Formation over a distance of some 30 yards.

A small faulted outcrop [4960 3738] is partially exposed on the south side of the Nottingham Road at Sandicliff Garage, Stapleford. Interlaminated sandstones and mudstones dip at 15° to 20° into the

d

fault plane, where they are shown against Pebble Beds.

Some 300 yd to the west, in an old sand quarry [4933 3734], an estimated 15 ft of interbanded red mudstone and siltstone are inaccessible at the top of a 60-ft face.

The Stapleford By-pass south of the town, provided excellent exposures [492 361] of the Waterstones Formation, which, when supplemented by exposures in sewer trenches in a neighbouring estate development, gave the following section:

	ft	in
Sandstone/siltstone, pale grey-green, finely-laminated; a few silty mudstone partings	4	6
Siltstone, massive	0	4
Sandstone, siltstone and silty mudstone, interlaminated	1	2
Sandstone, dark brown, fine-grained; oil-stained cavities	0	2
Sandstone/siltstone, interbanded and interlaminated	4	6
Sandstone, red-brown to buff, fine-grained, micaceous, poorly cemented; red silty mudstone laminae	19	0
Sandstone, buff, massive	1	0
Mudstone, red, poorly laminated; silty and sandy bands	20	0

As a result of oil seepages in this cutting (Taylor 1964a) the British Petroleum Company drilled into an anticlinal structure in this neighbourhood and proved the Waterstones Formation to be at least 79 ft thick.

Some 35 ft of strata overlying the Bunter Pebble Beds crop out on the top of Bramcote Hill. A small temporary exposure [5078 3750] in the vicarage drive showed a few feet of red-brown, fine-grained, flaggy mudstone, and temporary excavations [5086 3753] at the nearby Almshouses yielded red mudstone and siltstone fragments.

In the now-disused Chilwell Brick Pit [5130 3565], the top part of the Waterstones, exposed on the south side of a fault, consists of 21 ft of buff, fine-to very fine-grained flaggy sandstone. More massive beds, up to 14 in thick, are locally pale brown with darker brown stains of hydrocarbon and contain calcite-lined cavities up to 4 in across. The flaggy beds commonly contain green silty and micaceous mudstone partings with sporadic coarse quartz grains and buff mudstone pellets. Exposures in the cuttings of the M1 north of Lady Cross [472 373] showed similar lithologies except that the sandstone was so well cemented and massive that it required blasting. The sandstone passed down into more typical Waterstones, which were exposed between Lady Cross and Stoney Clouds, where up to 20 ft of dull red, laminated, micaceous mudstones with sporadic thin sandy bands, and green micaceous partings are interbanded with buff and pale brown sandstones.

An exposure [5293 3828] at the side of the Derby Road was recorded as follows:

	ft	in
Mudstones red-brown sandy, with sporadic pebbles	3	0
Sandstone, buff, fine-grained, micaceous, massive	0	9
Sandstone with silty mudstone laminae	0	7
Sandstone, buff, massive	0	6
Sandstone, soft, flaggy; mudstone pellets	4	0

On Nottingham University campus excavations for the foundations of new halls of residence [536 383] showed a predominance of interbanded and interlaminated pale grey-green sandstones and red silty mudstones typical of the Waterstones Formation.

RADCLIFFE FORMATION

In the Weston Underwood area a pipe trench [2910 4142] provided a continuous section of 30 ft of finely laminated, red-brown siltstones and silty mudstones. Sporadic green silty beds up to 18 in thick were also present, and two purple mudstones were prominent near the base.

A detailed log of the Radcliffe Formation in Trusley Borehole is given in Appendix 1. The formation occurs between 209 ft 2 in and 239 ft. The following palynomorph assemblages (determined by Dr G. Warrington) have been obtained from the lower half of the formation in the borehole.

Sample SAL. 1364 at 234 ft 6 in.

(a) Miospores
Apiculatasporites plicatus
Cyclotriletes microgranifer
C. oligogranifer
Verrucosisporites sp.
Spinotriletes echinoides
Perotrilites minor
Aratrisporites ?fimbriatus
A. saturni (Thiergart) Mädler 1964
Alisporites circulicorpus
A. grauvogeli
Voltziaceaesporites heteromorpha
Protodiploxypinus ?doubingeri (Klaus) Warrington 1974
P. fastidiosus
P. sittleri
cf. Sulcatisporites kraeuseli
Colpectopollis ellipsoidus
Illinites chitonoides Klaus 1964
Triadispora crassa
T. falcata
Angustisulcites klausii
Lunatisporites sp.
Striatoabieites aytugii
Cycadopites sp.

(b) Microplankton
Dictyotidium sp.
Veryhachium reductum
Micrhystridium cf. stellatum Deflandre 1945
tasmanitid

Sample SAL. 1363 at 226 ft 9 in.

(a) Miospores
Calamospora sp.
Punctatisporites crassexinus Mädler 1964
Apiculatasporites plicatus
Verrucosisporites sp.
? Spinotriletes echinoides
Aratrisporites sp.
Tsugaepollenites oriens
Podocarpeaepollenites thiergartii Mädler 1964
Alisporites circulicorpus
A. grauvogeli
Voltziaceaesporites heteromorpha
Protodiploxypinus cf. doubingeri
P. fastidiosus
P. potoniei (Mädler) Scheuring 1970
P. sittleri
Granosaccus sulcatus Mädler 1964
Sulcatisporites kraeuseli
Colpectopollis ellipsoideus
Illinites ? chitonoides
I. kosankei
Triadispora crassa
T. ? plicata
Angustisulcites grandis
A. klausii
Lunatisporites cf. noviaulensis (Leschik) Scheuring 1970
Striatoabieites aytugii
Tubantiapollenites balmei
Cycadopites sp.

(b) Microplankton
Dictyotidium sp.
Micrhystridium sp.
tasmanitid

The above assemblages contain many elements in common with those from the Waterstones at lower levels in the borehole, but that from SAL. 1363 includes miospores (e.g. forms comparable with Granosaccus sulcatus) which suggest a correlation with the microfloras of the Lettenkohle (Mädler 1964). In terms of the European Triassic stages the age of that unit is conventionally regarded as Ladinian. In terms of the British Triassic miospore reference sequence (Warrington 1974), the assemblages compare with those documented from the Kirkham Mudstones around the level of the Preesall Salt.

The continued presence of microplankton in the palynomorph assemblages is indicative of a marine source during deposition of the Radcliffe Formation in the present district. Tasmanitids were particularly abundant in the assemblage from SAL. 1363.

The following section measured in the pipe trench west of Kirk Langley [2759 3878] is assigned to the upper part of the Radcliffe Formation. The top of the section lies about 30 ft below the lowest skerry.

	ft	in
Mudstone, purple and red, silty, laminated	2	0
Mudstone, red-brown, silty	1	0
Siltstone, grey-green, laminated		9
Mudstone, red-brown and purple-brown, laminated, silty, with mudcracks	5	9
Silt, green and red-brown, interlaminated	1	6
Silt, red-brown and green, poorly laminated	3	0

In the trench a little to the north of the above section, sandy Waterstones were overlain by about 4 ft of red-brown silty clay with finely divided mica. Above the clay were 20 to 30 ft of red-brown, brick red and purple silty mudstone and silt, containing a 15-in bed of green silt and a 32-in grey-green siltstone.

In the banks of a stream, west of Mackworth, a 5-ft exposure [3022 3754] of interlaminated pale red-brown silty clay and grey-green siltstone is assigned to this formation.

In Kedleston Park, a pit [3092 4022] at the base of the Keuper Marl feature, shows 2 ft of red, finely laminated mudstone with intercalations of green micaceous siltstone which are referred to the Radcliffe Formation.

Some 18 ft of the lower part of the formation exposed in the Derby Brick Pit [3297 3611] consists of interbedded and interlaminated micaceous red-brown silt, silty mudstone and grey-green siltstone with basal red siltstone and sandstone bands.

A continuous succession of the Radcliffe Formation was measured near Lady Cross [471 371], on the M1 Motorway:

	ft	in
Skerry (pale grey-green laminated siltstone)	2	6
Mudstone, red, interbanded with silty mudstone	7	0
Sandstone, buff, fine-grained	0	6
Mudstone, red	2	0
Mudstone, pale grey, interlaminated with purple mudstone; hopper outlines	1	6
Mudstone, red, laminated	10	0
Mudstone, finely laminated, dark red, red and green, with silty mudstone and sandstone beds; ripple marks	2	6
Siltstone, greenish	0	4
Mudstone, red and dark red, laminated and banded with silty mudstone and fine greenish yellow sandstone showing ferruginous staining	3	6
Mudstone, red, silty and sandy, massive; gypsiferous cavities	2	0
Sandstone, grey-green, fine-grained up to	1	0
Mudstone, red, laminated, with rare green siltstone bands up to 2 in thick; sporadic pink mudstone laminae; rare micaceous partings	10	0
Mudstone, and silty mudstone; red and purple lamination	1	0
Mudstone, silty, predominantly green	1	0
Mudstone, red and purple, with green laminae	1	0
Mudstone, and silty mudstone, micaceous; ripple-marks and mud cracks; sporadic sandstone laminae	12	0
Sandstone with calcite-lined cavities (see p. 164)	10	0+

a

b

(Waterstones Formation, p. 155c)

CARLTON FORMATION

Near Weston Underwood, in a pipe trench [2980 4100], the Carlton Formation, some 45 ft thick, consists of massively bedded mudstones with flow-type breccias some 20 ft above the base and a series of siltstones and sandstones including the Plains Skerry [2890 4110]. The section associated with the skerries was as follows:

	ft	in
Sandstone, brown, fine-grained; slump structures	0	5
Mudstone, red-brown and grey-green, silty	1	0
Siltstone, red-brown and pale grey-green; slump structures (Plains Skerry)	2	9
Mudstone, red-brown and grey-green mottled; fine-grained sandstone lenses up to 3 in thick	2	4

The Plains Skerry contained numerous cavities, presumably left after the solution of gypsum. A small syncline repeated the outcrop of the skerry farther south-west [2860 4066 and 2845 4040]. Other exposures in the Carlton Formation are rare in the district. Old pits are either water-filled or grassed over, though one [2540 4017] south of Brailsford, exposed 3 ft of red laminated mudstone with siltstone partings, overlain by 3 ft of pale green laminated sandstone skerry commonly showing small slump structures.

In the western part of the district the most complete sequence of the Carlton Formation was provided by Trusley Borehole between 150 ft and 209 ft 2 in (see Appendix 1). The lower part of the formation contained good examples of flow-breccias occurring in massive mudstones and siltstones. The Plains Skerry lies between 160 ft and 164 ft 4 in, and about 4 ft below there is a red mudstone with green patches showing purple ramifications.

A pipe trench west-south-west of Kirk Langley [2755 3872] exposed 25 to 30 ft of red-brown silty clay with a few irregular grey-green streaks and scattered sandy patches underlying a skerry. This skerry, probably the Plains Skerry, was exposed in the same trench [2751 3866], where it consisted of slumped grey-green siltstone 2 in, on red-brown and grey silty clay 12 in, on slumped red-brown and grey siltstone 10 in. This skerry is 20 in thick where exposed in an adjacent stream [2771 3856]. Ground features indicate the presence, about 30 ft higher in the sequence, of another skerry represented by grey-green coarse sand in the trench bottom below 6 ft of boulder clay [2740 3843].

A skerry, probably the Plains Skerry, crops out at the top of a steep feature in Kedleston Park about 80 ft above the base of the formation. There is a small exposure of laminated grey siltstone in red clay in a ditch [3008 4021] and a few pieces of fine pinkish grey sandstone can be seen in an old pit [3080 4066].

The topmost beds assigned to the Carlton Formation are present at the eastern entrance to the Mickleover Railway Tunnel [3154 3604]. A skerry, covered by tufa, is exposed on the north side of the cutting, and consists of 5 in of poorly bedded grey-green siltstone underlying 15 in of similar siltstone with slump structures. The overlying 4 to 5 ft of strata are not exposed, but the sequence is continued on the south side of the cutting, where the top of the formation is taken at a 1-in irregular grey-green siltstone overlying 7 ft+ of red-brown silty mudstone with a few ½-in grey-green siltstone bands and an impersistent bed (up to 4 in thick) of red-brown and grey-green mottled siltstone. Some 16 ft of overlying strata assigned to the Harlequin Formation are described on p. 000.

In the Derby Brick Pit [3305 3508] the upper face adjacent to the Ring Road exposes the following:

	ft	in
Sandstone, pale grey-brown, cross-bedded, fine-grained, with current ripples and rain pits (Plains Skerry)	1	6
Strata obscured by talus	abt 10	0
Siltstone and silty mudstone, red-brown; rare greenish patches	abt 12	0
Siltstone, green and red-brown, unevenly bedded, hardened locally	0	10
Siltstone, hard, green	1	6
Mudstone, red-brown, silty, with green spots, massive	1	4
Siltstone, grey-green, locally cavernous	0	10
Siltstone, red-brown, clayey, with green-grey mottling and streaks	7	0
Siltstone, soft, green-grey	0	4
Mudstone, red-brown, silty, with green layers and patches	2	0

At a lower level in the brickpit, strata of the Radcliffe Formation (p. 000) are exposed.

The King, Howman and Co.'s Water Bore [3434 3675] in Derby encountered 4 ft of 'hard blue stone' at a depth of 103 ft 6 in, which is interpreted as the Plains Skerry.

Little confidence can be placed in the recognition of horizons from the outline log of Oliver Williams and Co.'s Water Bore (Appendix I and Stephens 1929, p. 91). The lithologies recorded are suggestive of the Harlequin Formation, but thicknesses of 'shaly rock' are far too great. The bottom of the borehole at 200 ft probably lies close to the bottom of the Carlton Formation, in which case the Plains Skerry would be represented by the 3 ft of 'hard rock' at 150 ft depth (Plate 10).

The Plains Skerry crops out, at the top of the slope above the Derwent alluvium, from the Hospital to south of the Cemetery and can also be seen in two small exposures nearby [3685 3695 and 3762 3601]. Mr A.H. Wallace recorded trench-sections near the hospital in 1935-37 and extensive outcrops of the skerry near the hub of radiating roads [371 376] in the housing estate where the skerry is presumably brought up in an anticline.

Farther east, in cuttings of the M1 Motorway [471 365], near Lady Cross, 60 ft of the Carlton Formation were exposed. The sequence consisted of red mudstone with thin siltstone bands and included the Plains Skerry, some 2 ft 2 in thick, with 8 ft 4 in below a 15-in band of mudstone containing purple ramifying structures.

The Carlton Formation is well displayed in Chilwell Brick Pit [5130 3585], where the sequence (Plate 11) is:

	ft	in
Siltstone, pale grey	0	1
Mudstone, red-brown, silty	1	6
Mudstone, red, with interlaminated pale grey-green siltstone	1	10
Mudstone, red silty	2	3
Siltstone and mudstone, interlaminated	1	0
Mudstone, red silty	1	0
Plains Skerry		
Skerry, sandstone, fine-grained, pale grey-green; small cavities, miniature ripple marks and slump structures	1	2
Mudstone, red	0	1
Skerry, red-brown; red mudstone flakes	1	0
Skerry, pale grey; pale brown staining; massive centre, laminated margins	1	0
Mudstone, dark red	0	6
Skerry; massive centre, interlaminated mudstone margins	3	1
Mudstone, deep red, with purple patches; siltstone laminae; green patches with purple ramifying structures in basal 2 ft	6	3
Siltstone; mudstone laminae	0	8
Mudstone, red, silty; pale grey and pale green patches; sporadic pale grey-green siltstone bands and laminae up to 21 in; ripple marks; small cavities	50	0

HARLEQUIN FORMATION

The topmost 150 ft of core in Trusley Borehole are assigned to the formation. The following sparse palynomorph assemblage (determined by Dr G. Warrington) was obtained from a sample (SAL. 1315) at 136 ft:

(a) Miospores
? Cyclotriletes sp.
Alisporites sp.
Scopulisporites minor Mädler 1964
? Protodiploxypinus potoniei
Illinites chitonoides
Striatoabieites aytugii

(b) Microplankton
Dictyotidium sp.
? Micrhystridium sp.
Tasmanitid

The above assemblage is, in isolation, not diagnostic of age but, from the miospore data obtained lower in Trusley Borehole, is evidently of Ladinian or younger age. This is consonant with the situation of the assemblage slightly above the level of the Plains Skerry which elsewhere (Smith and Warrington 1971) was regarded as late Ladinian or early Carnian in age. General support for this view is available from the adjoining Burton on Trent (140) and Nottingham (126) districts where miospore assemblages of more definite Carnian aspect were obtained from the higher part of the Harlequin Formation in Bagots Park Borehole and at Wilford Hill Brickworks respectively (Warrington 1970).

Beds assigned to the base of the Harlequin Formation were measured at the eastern end of Mickeover Railway Tunnel [3154 3604].

	ft	in
Mudstone, red, silty, with silt laminae	2	6
Clay, grey-green, silty	1	2
Mudstone, red-brown, pale pink and grey-green, silty; laminated at top with a 2-in red-brown siltstone, poorly laminated below; 1-in red-brown siltstones in basal 2 ft	abt 4	0
Siltstone, grey-green, and interlaminated red-brown silty mudstone	0	3
Siltstone, grey-green	0	4
Mudstone, red-brown, silty, poorly laminated, with very silty laminae and traces of grey-green silty mudstone; 1-in grey-green laminated siltstone in middle	abt 6	5
Siltstone, grey-green, cavernous	0	8
Mudstone, red-brown, silty, mainly poorly laminated but with harder well laminated layers	2	3
Carlton Formation (see p. 000)		

Beds, probably of the Harlequin Formation, were seen in trenches for new house foundations on the north side of the Derwent valley east of Derby [385 507]; up to 20 ft of laminated, red silty mudstone with green-grey siltstone bands (up to about 12 in thick) were exposed.

In the M1 Motorway excavations [471 361] at Sandiacre, west of Nottingham, the measured thickness was 130 ft but the top part was missing due to faulting against the Cotgrave Skerry [4715 3592]. The sequence comprises some 45 cycles of siltstone/mudstone units (see p. 000). The lower, red mudstone, parts of the cycles are predominantly massive and vary in thickness from 1 in to 7 ft but average 1½ ft. The upper, pale grey-green silty mudstone and siltstone, parts of the cycles are laminated and vary in thickness from ½ in to 2 ft 2 in with an average of about 6 in.

EDWALTON FORMATION

Some 1 ft 8 in of red and pale pink, mottled, medium-grained sand overlying 2 ft of red to pale brown, medium-grained sand with small discrete hard patches, observed in a trench [2573 3510] 500 yd S of the boundary of the district, are assigned to the Cotgrave Skerry. This sand horizon lies some 60 to 80 ft above the highest strata proved in Trusley Borehole.

It is now thought probable that the skerry mapped near the Hall at Radbourne is the Cotgrave Skerry and that the mapped continuation

c

d

of its outcrop eastwards towards Mackworth Fields and Hackwood Farm is erroneous.

In the east of the district, the only exposure of the Cotgrave Skerry was seen in a faulted section [4740 3572] on the north-eastern side of the M1 - Stapleford - Sandiacre By-pass:

	ft	in
Mudstone, green	1	6
Cotgrave Skerry		
Sandstone, pale buff, fine-grained	1	6
Mudstone, grey	2	0
Sandstone, off-white, cross-bedded	0	11
Harlequin Formation (2 ft 4 in of green mudstone on 8 ft of red mudstone seen)		

CHAPTER 7

Quaternary

DETAILS

BOULDER CLAY AND GLACIAL SAND AND GRAVEL

Crich and Ambergate area

Near Crich, Wedd (in Gibson and others 1908, p. 153) noted the presence of drift between two limestone saddles at about 650 ft and saw 50 ft of drift in Hilts Quarry [3511 5438]. It consisted of brown sandy clay full of stones of all sizes and containing lenses of current-bedded sand and gravel. Most of the boulders are of local origin and the underlying limestone showed striations trending S 35° E and S 25° E. The quarry is now being filled in, but Rhys (in Smith and others 1967, p. 226) recorded up to 50 ft of sandy clay with Bunter and Carboniferous pebbles in a quarry [3558 5417] 300 yd to the south-east.

West of Nether Heage and south of Ambergate the valley between the ridges formed of Namurian grits contains at least 5 ft of glacial clay deposits with quartzite pebbles and sandstone fragments.

Wirksworth area

To the north of Hopton, there is an irregular distribution of brown sandy clay with pebbles and chert at 900 to 1000 ft OD; up to 10 ft were seen. In the Magnesium Electron quarry [2555 5475] fissures 10 ft wide in the top of dolomitic limestone were lined with brown dolomitic 'sand', containing masses, up to 5 ft wide, of pale brown clay with chert and pebbles overlain by stiff brown clay with pebbles. In Wirksworth, excavations for a school and garage [2878 5353] showed up to 6 ft of mottled grey and brown clay with pebbles and chert, limestone and sandstone fragments overlying shale of presumed Namurian age.

In the Dream Mine area [2741 5305] an old excavation near the crest of a limestone promontory contained up to 10 ft of sand, with pebbles and fragments of chert, sandstone and siltstone.

The top 24 ft in the Hopton Borehole [2677 5334] comprised a mixture of sand, pebbles and mudstone fragments overlying Namurian shales (p.168d). The deposit is a mixture of made ground, glacial sand and solifluxion material and is similar to that near the surface in the Brassington Silica Sand Pits (p. 90).

Four Lane Ends Borehole [2724 5431] encountered some 46 ft of brown clay with rare erratics of quartzite and chert. The clay becomes sandy basally and rests on fine brown calcareous sand with clay and sandstone fragments, 49 ft 3 in thick, which in turn rests on Dinantian limestone. The sand contains grains of red-stained quartz, galena, calcite, dolomite and garnet. The deposits fill a south-eastward trending valley.

Spreads of boulder clay in the bottom of the valleys of the Ecclesbourne and Scow Brook, south of Hopton, consist of brown loams and yellow clays with quartzite pebbles, limestone and sandstone fragments, calcareous nodules with gastropods and dark grey banded chert. East of Idridgehay boulder clay extends continuously from the River Ecclesbourne at abt 320 ft OD up to abt 675 ft OD near Lawn Farm. At the higher levels the boulder clay is generally of a more uniformly red-brown colour, with a higher proportion of sandstones of presumed local derivation than elsewhere.

Western area

Kirk Ireton village is situated on a plateau, 700 to 800 ft above OD, covered by at least 3 ft of ill-drained brown, grey and mottled sandy boulder clays containing pebbles and sandstone fragments. Jowett and Charlesworth (1929, p. 311) recorded occasional boulders of Eskdale granite south of Wirksworth and near Blackwall. Small pebbles of quartzite are ubiquitous in the Wirksworth-Callow area and at Blackwall.

Near Idridgehay, T. I. Pocock (MS. map) records 6 ft of brown clay containing pebbles, limestone and chert, overlying shale in the railway cutting [2890 4907]. Nearby an old gravel pit [2950 4885] over 25 ft deep still shows sand and pebbles, vaguely stratified, dipping south-eastwards. The gravel is mixed with red clay and is predominantly quartzitic, but contains some fine-grained red sandstone of Carboniferous aspect. The deposit is associated with the spread of boulder clay in the valley of the Ecclesbourne, which has been seen in temporary sections to a depth of 6 ft.

North-west of Idridgehay the boulder clay cover is rather patchy and difficult to map, but the area is particularly rich in large erratics, e.g. one of coarse-grained sandstone (possibly Ashover Grit), 3 ft x 2 ft x 1 ft; another of grit, 2 ft x 2 ft x 2 ft and a third of dark green, fine-grained igneous rock possibly from the Lake District, 3 ft x 2 ft x 2 ft.

West of Wardgate at 600 to 700 ft OD an area of boulder clay, 1 mile by up to ½ mile, is banked against and partly overlies the Pebble Bed scarp near Cross o'th'hands. It has been worked for clay in several pits [2630 4682], where up to 6 ft of red-brown sandy clay with sand lenses and containing sporadic quartzite pebbles, chert and sandstone (fine, coarse, and ganisteroid) fragments have been recorded. Sporadic large boulders of sandstone (3 ft x 2 ft x 1 ft) are present on the surface of the boulder clay.

A similar but more extensive sheet of boulder clay farther south between Muggintonlane End and Western Underwood largely overlies Pebble Beds at an altitude of 500 to 600 ft. It has been proved by boring to a depth of 35 ft and has been dug locally, presumably for brick clay. It gives rise to a brown and red-brown sandy clay soil containing sporadic pebbles, cobbles, fine and coarse-grained sandstone fragments, chert containing crinoid remains, possible chalk and ironstone nodules.

To the south-west another large patch of boulder clay, 1 mile x ½ mile, is present immediately to the north-east of Brailsford at some 500 ft OD. It is nowhere exposed, but a borehole [2572 4158] near the edge of the patch, proved 8 ft of clay. It contains many pebbles, flints and sandstone fragments.

Indistinct patches of boulder clay cap higher ground (250 to 450 ft OD) of the Permo-Triassic outcrop near Kirk Langley, Radbourne and in Derby city. Two patches near Lees [2671 3679 and 273 382], temporarily exposed in trenches, were seen to be more than 5 ft thick in places. The clay matrix is dark grey mottled with brown and the fragments include flints and Carboniferous sandstone, Pebble Beds quartzites, sideritic siltstones, coal and red clay.

A small patch of glacial sand and gravel on a bluff of Keuper Marl 1000 yd NE of Kirk Langley is exposed [2939 3950] to 3 ft. The matrix is sand and red sandy or stiff clay and the pebbles are chiefly composed of soft sandstone of Waterstones type which cannot have been transported far, but a few are of ironstone, igneous rock and quartzitic rocks from the Pebble Beds.

A temporary trench exposure north of Hall Farm, Windley [310 455] showed red-brown boulder clay with many Bunter pebbles overlain by probable Head.

A patch of boulder clay near Moseyley (306 439), exposed in a temporary trench, yielded small pieces of coal and a 3-in lump of galena as well as the more usual erratics. Much of the undulating plateau between Weston Underwood and Quarndon is probably overlain by a very thin cover of boulder clay, but it is mappable only locally.

The highest patch of boulder clay outside the Wirksworth area occurs around 800 ft OD at Sandy Ford [323 512] and contains much sand and Ashover Grit boulders as well as 'Bunter' pebbles.

The patch of glacial sand and gravel near Shottlegate was shown in a temporary exposure [314 476] to consist of irregular bands of pebbly sand with lenses of cross-laminated sand up to 1 ft thick. The deposit appears to overlie boulder clay at its southern limit and to pass into red sandy boulder clay to the north.

East and north-east of Derby

On the east of the Derwent valley near Derby patches and pockets of boulder clay cap the Keuper Marl. To the south of the Roman Road [3885 3553] some 5 ft of red-brown clay contains boulders up to 1 ft long of red-brown ironstones, skerry, fresh igneous rock, Dinantian limestone, dolomitic limestone, Carboniferous sandstone and commonly, an ochreous siltstone of Mesozoic aspect. Pebble Beds quartzitic pebbles are rare.

The flat spread of boulder clay on Chaddesden Common is overlain by a light silty loam soil up to about 2 ft thick. Some 5 ft of red silty clay with ganister, Dinantian limestone and quartzitic fragments overlie Permo-Triassic rocks exposed in the north side of Morley rail cutting [3992 4009].

The glacial sand and gravel near Chaddesden is mapped as discontinuous masses below boulder clay. Some 4 ft of pebbly sand were noted at Grove Farm [3890 3817] and Wedd (MS map) saw 12 ft of 'bedded drift shingle with thin beds of sand and lenticles and films of drift-coal passing east under red stoney boulder clay' near Stoneyflats [3841 3833]. Deeley (1886, p. 449) has described both the boulder clay and the sand and gravel of this area and recorded contortions in the surface of the latter 'by a force that seems to have come from a north-north-west direction' He also noted that the irregular base of the boulder clay contains masses of marl and that the underlying disturbed Keuper Marl contains boulders derived from the clay.

The small sand and gravel deposit north of Little Eaton appears to pass under the adjacent boulder clay; it was exposed for 2 ft at Eaton Hill [3636 4228]. The small spread south of Little Eaton was also penetrated for 4 ft 6 in below 14 ft of boulder clay in a trial bore [3678 4071].

In an old quarry [3616 4348] near Daypark the thickness of the drift formerly exposed is estimated to be about 10 ft. Wedd (in Gibson and others 1908, p. 44) recorded 17 ft of red clay with stones in Outwoods Borehole [3563 4255].

Sections proved in trial bores east of Little Eaton were noted by Wedd (in Gibson and others 1908, p. 156). An adjacent well at Glebe Farm [3717 4082] proved: soil 1 ft, red marly stoney clay 7 ft, yellow sandy marl with coal-detritus and stones 37 or 38 ft, on sand-rock with pebbles 2 or 3 ft; the 'sand-rock' may be part of the drift sequence or perhaps a concealed outcrop of Pebble Beds. Brown sandy boulder clay 2 ft thick with sandstones, quartzites, siltstone and flint pebbles up to 2 in long and a few in of ochreous gritty sand with similar pebbles were seen in a degraded pit [3691 4052] nearby.

The largest spread of boulder clay in the district occurs between Spondon, Ockbrook and Dale Abbey. It stretches from the fluvio-glacial deposits in the Trent Valley below 200 ft OD up to nearly 400 ft south-west of Dale Abbey. Temporary sections up to 10 ft deep showed red, brown, yellow and grey clay containing sporadic pebbles, sandstone and ganister fragments, ironstone nodules, and rare flint debris. Deeley (1886, pp.449-449) gave details of the boulder clay in the Spondon area, and in a brick pit [probably 4032 3633] he found the uppermost 5 or 6 ft to be a brownish or drab stiff clay with many erratic pebbles and boulders, including flints. He regarded this as the disturbed and weathered top of pale bluish or brownish boulder clay which was at least 60 ft thick in a well in the brickyard. This thickness is confirmed by Fox-Strangways (MS map) who noted 90 ft of 'blue clay' in another well [4055 3627] 129 ft deep. Isolated patches of brown, yellow and grey boulder clay, presumably erosional remnants of the same sheet, occur in Hopwell Park, where creamy coloured flints and pink granite (possibly Mount Sorrel) are common in a sandy clay loam.

Coalfield area

In the north, at Fritchley, a gas-pipe trench cut into 6 ft of yellow-brown and grey mottled boulder clay with coaly fragments and streaks. This boulder clay was seen by W. N. Edwards during the working of Chestnut Site, 600 yd SE of Fritchley, where 5 ft of sandy clay with small disc-shaped pebbles (less than 1 in diameter) were present. A hole dug by the N. C. B. Opencast Executive on this site [3645 5283] proved 5 ft 9 in of yellow sandy clay with sandstone boulders overlying 2 ft 6 in of sand and gravel.

Scattered quartzitic pebbles in a clayey soil suggest the presence of boulder clay west of Pentrich [385 525] and a similar deposit fills a hollow south of Upper Hartshay.

Opencast trial pits on Codnor Common Site [4115 4975 and 4112 4973] reputedly encountered 6 ft of yellow clay and 3 ft of sandy clay respectively.

Patches of boulder clay near Stanley, Stanley Common, West Hallam and in Heanor and Eastwood are mapped on the evidence of pebbly clay debris from shallow excavations and the soil. A small deposit at Smalley was exposed in a trench [4084 4492] and exceeded 5 ft in thickness. Farey's (1811, pp.134-142) list of gravel patches in Derbyshire includes gravel south-east of Heanor and 60 ft of more clayey gravel with flint and chalk farther south. Presumably this is the outcrop mapped in the south-eastern part of the town, but Farey's thickness appears excessive.

On Coventry Lane Site [502 407] proving holes showed up to 4 ft of gravel overlying 5 ft of brown sandy shale.

Eastern area

Boulder clay with sand and gravel occurs in the Newstead-Hucknall-Bulwell areas. The clay commonly overlies the gravels, but beds of sand and/or gravel may occur within the clay, the level of erosion determining which lithology is at surface.

The largest spreads of boulder clay occur between Selston, Annesley and Hucknall. A 50-ft section was exposed during excavations for the M1 motorway, and nearby trial bores [4931 5170] proved a maximum of 60 ft. The deposit is a stiff stony sandy clay, brown in the upper weathered part and grey below. Of the numerous pebbles, cobbles and boulders, many of local origin, the commonest are Westphalian siltstones, sandstones, ironstones and coaly debris together with fragments of Lower Marl, Lower Magnesian Limestone and Lower Mottled Sandstone. The largest boulder seen consisted of 2 cu ft of Dinantian limestone. The boulder clay rests on an irregular surface of weathered Lower Magnesian Limestone, spanning open joints and fissures considered to have resulted from mining subsidence (p.159d).

An elongated lens of yellow-buff sand trending north-east—south-west was bisected by the motorway in the vicinity of William Wood [4935 5165]. The sand is poorly cemented, fine-grained and contains only sporadic pebbles. Cross-laminated units suggest currents flowing from the south-east.

Sand deposits several of which were formerly worked, lie in a two-mile north - north-east trending belt between Underwood and Kirkby in Ashfield and lying at altitudes of between 450 ft and 500 ft. Many of the pits are overgrown or filled in but one [482 534] near Kirkbybogs Farm shows a 30-ft face of red and yellow sand containing carbonaceous laminae and sporadic small pebbles.

Such deposits, occurring on the high ground of the interfluve between the rivers Erewash and Leen, appear to have formed as lenses of sand within boulder clay.

Farther east, boulder clay and sand and gravel of similar origin overlie the Permo-Triassic rocks in the Annesley-Hucknall area. The best sections are in old disused pits [5155 5235] along Byron Walk in Annesley Park Woods. The gravel, some 30 ft thick, dips slightly to the west and is strongly cemented in parts by a ? ferruginous pan. The pebbles are largely composed of pink and brown quartzite together with ironstone, Westphalian sandstones, coal, Lower Mottled Sandstone, Dinantian limestone and chert and Permian limestone (brown sandy and pale cream and yellow dolomites of northern derivation). The maximum diameter of pebbles is about 12 in, with a norm of about 3 in. The pinkish sandy matrix is locally well cemented producing a hard conglomerate. The sand and gravel is overlain by up to 8 ft of brownish yellow sandy clay with sporadic small quartzite pebbles. The deposits of gravel form ridges elongated generally in a northerly direction. Temporary exposures in the M1 motorway cutting near Annesley Lodge showed an increase in the proportion of sand to gravel in deposits again capped by boulder clay. A proving hole [4998 4994] showed:

	Thickness ft	Thickness in	Depth ft	Depth in
Clay, brown sandy, with pebbles	4	6	4	6
Sand, clayey, brown; sand pellets	3	6	8	0
Clay, grey-brown, sandy; small 'Bunter' pebbles; Permo-Triassic and Carboniferous debris	2	6	10	6
Clay, grey; small fragments of bituminous mudstone, Lower Magnesian Limestone and Lower Marl	4	6	15	0
Clay, brown; large quartzite pebble; Westphalian sandstone and stained mudstone; Permo-Triassic fragments; sandy patches; clayey sand lens at 24 ft	11	0	26	0
Clay, dark grey; coal, mudstone and Lower Magnesian Limestone fragments at 30 ft	4	0	30	0
Clay, grey-brown, becoming red-brown at base	11	0	41	0

This section was underlain by 4 ft of glacial sand and gravel which rested on Lower Mottled Sandstone.

A cutting made in 1973 during re-alignment of the A611 road between Annesley and Hucknall provided the following generalized section:

	Thickness ft	in
Clay, brown, red-brown and dark grey, with small pebbles and erratics (up to 6 in) of Lower Magnesian Limestone, Westphalian mudstones, siltstones, sandstones, ironstones and coals	up to 20	0
Sand, fine brown marly, with coaly fragments; undulating base	0 to 2	3
Clay, red-brown and variegated, laminated, with carbonaceous rootlets	0 to 0	3
Conglomerate (up to 10 ft in places)	0 to 0	1
Sand, brown friable	0 to 0	1
Sand and gravel, variably cemented; manganese oxide along some bedding planes	0 to 12	0
(Lower Mottled Sandstone - top 8 in blackened by manganese oxide)		

The flat-topped Misk Hills [506 495] are capped by some 10 ft of red-brown and yellow clay which overlie 15 to 20 ft of sand and gravel. The gravel is exposed in the top of the quarry at the eastern end of Long Hill:

	ft	in
Soil, brown sandy, with small pebbles up to 1 in	1	0
Sand, yellow-brown, fine to medium-grained; small 'Bunter' pebbles; passes laterally into yellow sand with coaly streaks; 1-in red marl band	2	0
Pebbles up to 3 in in sandy matrix	1	0
(Lower Mottled Sandstone)		

The valleys to the north and south of Misk Hill are particularly rich in erratics, which vary in size from cobbles to boulders (4 ft x 3 ft x 2 ft) and are derived from the gravels. They are composed of granites, coarse sandstones (individual grains up to $\frac{1}{2}$ in) fine sandstones, ganisteroid sandstones, dark green igneous rocks of Lake District type and coaly shales.

In the Annesley - Hucknall area there are scattered deposits of sand and gravel. One at Hucknall [5375 4885] was worked for gravel, and up to 10 ft of sand and gravel were seen in temporary sections [5404 4620] near Bulwell.

RIVER TERRACES

Minor streams in south-west of district

Low terrace features bordering the alluvium near Trusleywood House and Lees are assigned to the First Terrace. At the former locality the terrace merges upstream with the alluvium and at the latter the stream is incised a few feet at the terrace margin.

Cutler, Markeaton and Mackworth brooks

Small patches of undifferentiated terrace in the Cutler Brook south-east of Weston Underwood and near Kedleston appear to be fan deposits at the confluence of tributary streams. Larger spreads of the same terrace at Kedleston border the alluvium, which is artificially controlled by weirs for ornamental purposes. Old pits at Kedleston show the terrace was worked for gravel but the poorly exposed deposit is very clayey and was shown partly as boulder clay on the old maps. Though classified as undifferentiated because of their remoteness from the main terraces downstream, and their relationship to knickpoints (p. 93) these deposits may equate with the First Terrace.

The First Terrace, west of Quarndon [323 403], appears to consist of red silty clay with a few pebbles but there are no good sections.

A small degraded Second Terrace feature at about 275 OD and some 50 ft above the Mackworth Brook [310 386] is covered by fine brown sand, which becomes grey at 4 ft depth.

Low First Terrace features usually rising a few, but locally up to 15 feet, above the alluvium of the Mackworth Brook from west of Kirk Langley to Markeaton appear, in the more westerly occurrences, to be composed of fine sands probably derived from the Waterstones. They reveal surface indications of gravel as Markeaton is approached. Gravels have been worked at Markeaton [330 382], and were described by Pocock (in Gibson and others 1908, p. 152) as 'strongly contorted, the pebbles in places being turned up on end'

Below Markeaton the First Terrace is more extensive on the south side of the alluvium than on the north. It was proved in trial bores for the improvement of Derby Ring Road on Queensway [3355 3702 to 3375 3724]. Here the terrace base is cut evenly into the Keuper Marl at about 175 ft OD, i.e. some 8 ft higher than the gravel base of the adjacent alluvium (p. 158d). The upper levels of the deposit comprise pebbly and sandy clays, $6\frac{1}{2}$ ft to $16\frac{1}{2}$ ft thick resting on up to 10 ft of sand and gravel.

River Ecclesbourne

Fan deposits a few feet above the alluvium at Windley [312 453] and near Farnah House [327 438] are assigned to the First Terrace. The Windley terrace has been worked for gravel, and a temporary exposure near Windley Hall [3093 4506] showed the gravel to overlie boulder clay (p. 157b). Ditch sections in the Farnah House terrace show gravelly clay with quartzitic pebbles and a few siltstone and sandstone fragments.

River Derwent

Near Lawn Farm [3455 5090] there is a thin gravelly undifferentiated terrace deposit overlying a bench feature at about 330 ft OD. There are no sections but auger holes proved brown gravelly loam to 3 ft.

At Makeney there are four remnants of a terrace-like surface. Quartzitic pebbles of 'Bunter' type are scattered over the more northerly occurrences, in which there are old diggings. Wedd (MS map) noted sandstone and grit fragments in addition. The surface lies about 250 ft above the river at about 300 to 320 ft OD, but below the level of nearby boulder clay. The northerly occurrence on Hopping Hill was noted by Farey (1811, p. 138).

Near Allestree a thin spread of pebbly sand is assigned to the Second Terrace. It covers a gently sloping bench between about 160 and 200 OD and is being eroded on its eastern side by the Derwent so that it caps a low river cliff in Carboniferous shales.

A deposit shown as Second Terrace rests upon a bench sloping gently to the east or south-east between 190 and 220 ft OD in Derby city and extending beyond the southern margin of the district. During the recent survey a gravelly loam soil was noted but no exposures were seen. Deeley (1886, p. 478) claimed the deposit as boulder clay, but Fox-Strangways (MS map) noted 6 ft of 'clay and pebbles' [3516 3566] and 'gravel' on Osmaston Road [356 354] and near Wilmot Street [3520 3558].

Several scattered terrace features above the Derwent alluvium between Ambergate and Milford are assigned to the First Terrace. Those near the confluence of the Amber, at Lawn Cottage [346 503] and near Milford [348 459], are about 5 ft, 8 ft and 12 ft respectively above the alluvium, but show no sections. The terrace at Belper is about 15 ft above the alluvium and one small section at the outer edge of the terrace exposes 4 ft of brown sand and gravel. South of the confluence with the Coppice Brook valley this terrace appears to have been covered by a head deposit, for a temporary exposure [3470 4703] showed:

	ft	in
Clay, brown sandy, with angular sandstone fragments up to 3 ft diameter (Head)	5	0
Gravel, fine clayey, with angular sandstone fragments and lenses of cross-laminated sand up to 9 in thick	2	0
Gravel, crudely bedded with rounded pebbles up to 5 in diameter, comprising mainly sandstone but with a few dolerites, limestones and cherts	3	0

Petrographic examination of two of the pebbles shows them to be olivine-dolerites closely resembling the Bonsall-Ible dolerite sills in the district to the north.

In Derby city gravel and sand of the First Terrace has been noted near the Market Place and Cathedral and appears to mantle the spur between the valleys of the Markeaton Brook and the Derwent, up to about 170 ft OD. Near Derby (Midland) Station a more or less flat gravel spread at about 150 ft OD is sharply defined by a bluff rising above the alluvium. The spread is not exposed except for a few feet of gravel near the main railway line to the south of the district.

Lees Brook, Chaddesden

Undoubted gravel within and to the south-west of Chaddesden in the Lees Brook valley lies at about 175 to 190 ft OD and up to about 25 ft above the alluvium. It is assigned to the Second Terrace; temporary sections in sandy gravel up to 2 ft deep were noted. The marked terrace form serves to distinguish the largest of these spreads from the adjacent glacial sand and gravel (p. 00).

River Erewash and tributaries

An indefinite deposit assigned to the First Terrace and showing surface indications of sand in the Erewash valley south-east of Eastwood rests upon a marked bench feature at about 190 ft OD. It appears to be a fan deposit at the mouth of a minor valley draining higher ground to the south-west. To the north of Eastwood a low terrace feature borders the alluvium of Beauvale Brook.

Two low bench features in the valley of Mapperley Brook to the south of Mapperley are shown as undifferentiated terrace. They are sharply separated from the alluvium, which is itself controlled at an artificial level by a dam in the brook. The validity of such terrace-like features lying only a few feet above the alluvium in mining areas is questionable in view of artificial base levels of the streams caused by subsidence.

Three areas of First Terrace are mapped on the west bank of the river near Sandiacre. Two small terraces some 300 x 800 yd at Sandiacre Lock [4820 3575] and another forming the site of the town [479 365] lie at 120 ft OD. A large area (1 mile x $\frac{1}{2}$ mile) of this terrace at Stanton Gate [480 382] lying at 140 ft is largely obscured by tips and works of the Stanton Iron and Steel Company.

A small terrace on the west bank of the Ock Brook [425 356] is some 5 ft above the alluvium at approximately 170 ft OD.

River Trent

A few small and isolated areas of clayey sand and gravel with flints have been mapped as Second Terrace near Spondon, just beyond the southern margin of the district [410 351]. They lie between 160 and 190 ft OD, up to 45 ft above the alluvium. A similar deposit on the west bank of the Ock Brook [420 355] lies at about 200 ft OD.

Second Terrace deposits were also mapped during the primary survey near the confluence of the Erewash and Trent at Sandiacre [480 357]. They lie at about 150 ft OD, some 30 ft above the Erewash alluvium. They are now completely built over. Above Nottingham the Hilton Terrace is in two parts (King 1966, p. 53), one some 90 ft and the other 40 - 60 ft above the alluvium.

The Beeston Terrace is 1 mile long and $\frac{1}{2}$ mile wide in its type area at Beeston. It forms a site for the town safe from the dangers of flooding, for its lowest point is some 10 to 15 ft above the alluvium. The terrace rises gradually away from the river to 145 ft OD, i.e. a rise of some 50 ft above the alluvium. As this figure is almost double that quoted by Clayton (1955) it is possible that the terrace is composite, the higher parts being of Hilton Terrace origin and affected by solifluxion.

Beeston's soil is characteristically a brown sandy loam containing many pebbles. A temporary section was seen on Midde Street [5294 3662] as follows:

	ft	in
Made Ground	3	0
Sand and gravel with pebbles up to 2 in in diameter, largely 'Bunter' quartzites with sandstone fragments and flints: lenses of ironstone up to $\frac{1}{4}$ in diameter are common and are haphazardly orientated in a poorly cemented coarse sand matrix; undulatory base	up to 2	0
Clay, off-white, sandy; sporadic pebbles	0	1
Sand and gravel, as above but pebbles larger and more common; ironstone pebbles tend to be concentrated into lenses	up to 3	0
Sand, brown, coarse, with clayey patches	up to 0	3
Clay, light brown and off-white	2	0
Sand, brown, coarse	up to 2	0
Sand and gravel	1	6

The sequence dips at approximately 3° towards the river. Boreholes [5274 3693] through the terrace have proved thicknesses of up to 14 ft.

To the east of Beeston, on the east side of Tottle Brook, the Florence Boot Hall of Residence is built upon a small area of flat ground. Red clay and pebbles were dug out of temporary excavations in this vicinity. A further area of terrace 700 x 180 yd [530 380] on the north-east bank of Tottle Brook was mapped during the primary survey, but is now completely built over.

McCullagh (1968) made an analysis of the fabric of each of the terrace deposits as classified by Clayton (1955), at nearby localities beyond the district margins, but was only able to conclude that the terraces derived their material from a single source as well as from each other. The only diagnostic factor seems therefore to be one of position relative to the river.

HEAD DEPOSITS

Head in a valley bottom west of Ravensdale Park and in a tributary valley [2845 4608] of the River Ecclesbourne near Cross o'th'Hands is largely downwash from surrounding hills formed of Pebble Beds. Sandy deposits, probably of similar origin, were mapped at The Fishpools [314 525] and near Sandy Ford [316 513].

Head deposits beneath the main scarps of the Ashover Grit are recognised by the presence of many large sandstone boulders scattered over the slopes. A temporary section [3238 5210] near Alderwasley revealed 18 ft of firm clayey sand containing many large sandstone boulders.

Site investigation boreholes [3533 5205] near Ambergate showed head deposits on the north side of the valley up to 70 ft thick. These consist of brown sandy and silty clay with shale and sandstone fragments and sandstone boulders. The proportion of rock fragments increases with depth.

Small areas of head have been mapped below the scarps of the Ashover and Chatsworth grits in Thackers Wood [3540 5045] and Dunge Wood [3514 5030].

In the southern part of Belper a spread of sandy loam with sandstone boulders extends below the outcrop of the lower leaf of the Chatsworth Grit and is poorly exposed in a bank [3522 4730] of the Coppice Brook. A temporary section [3471 4703] showed 5 ft 6 in of brown sandy clay with sandstone blocks overlying 5 ft of crudely bedded sand and gravel

Head mantles the gentler slopes of the Derwent valley near Milford and fills gullies in the steeper slopes; at Hopping Hill a temporary section [3513 4548] showed 3 ft of gravel overlying 5 ft of sandy loam. The deposit appears to be absent from the steep escarpment of the Ashover Grit south of Milford as evidenced by a temporary section [3542 4251], where 1 ft of sandy soil on 3 ft of brown sand with sandstone boulders rested upon Namurian mudstones.

The same escarpment near Breadsall, however, is mantled by head deposits that cannot be separated from the underlying grit outcrop and which are therefore not shown on the geological maps. An old quarry [3843 4139] shows 11 ft of rubbly sandstone and sand with angular sandstone blocks up to 4 x 3 ft in size, mainly at the base, resting on Ashover Grit (p. 135a).

The deposit mapped nearby at The Priory appears to consist of red and grey clayey sand. A lobe of head 200 yd S of the Priory lies at the mouth of a minor valley cut in the Ashover Grit.

The deposit filling the bottom of a minor valley south of Little Eaton is associated with springs and a little tufa. Three trial bores [3664 4086] encountered up to $11\frac{1}{2}$ ft of brown, red-brown or dark grey silty clays; one bore proved a little fine gravel within the silty clay.

Downwash consisting of brown sandy loam of unknown thickness occurs in a valley within the sandstone overlying the 'Ashgate' Coal at Little Hallam [463 406], and there are similar spreads associated with Westphalian sandstones east of Brinsley [466 490] and near Underwood [480 490]. A small tributary valley [4040 3720] of Lees Brook, north of Spondon contains 6 ft of yellow-brown clay with sporadic pebbles overlying Keuper Marl.

The irregular north-facing scarp of Dale Hills near Dale Abbey is cut by numerous valleys and gullies, all of which contain deposits of sand, pebbles and clay washed down from the Waterstones, Pebble Beds and Lower Mottled Sandstone; most are too small to be represented on the map. Sandy and pebbly soils also mask the clayey soils of the Westphalian mudstones which crop out at the base of the scarp.

The valleys of the Golden Brook and the stream near Risley are floored by head deposits composed largely of reworked boulder clays and fluvioglacial sands and gravel from the surrounding hills; a section [4617 3623] showed:

	ft	in
Brown sandy clay soil	0	9
Red-brown clay subsoil	1	3
Sand and gravel, clayey, with siltstone (skerry) fragments	5	0

Head infills gullies on the Pebble Beds outcrop north-west of Nottingham. In the Bramcote Sand Quarries [5035 3883] a cross-section through a valley fill, 20 to 30 ft wide, showed a basal lenticular bed up to 2 ft thick of grey-brown sandy clay, banded in places, with fragments of sandstone (Pebble Beds), overlain by yellow-brown slightly clayey sand up to 6 ft thick and a sandy loam soil with quartzitic pebbles near the surface.

The following section of Head was recorded at Catstone Hill [5013 4102]:

	ft	in
Sand, red-brown; clayey patches; pebbles sporadic but common in basal 12 in	4	0
Clay, red and white, in bands up to 5 in thick (on Westphalian)	1	6

Dry valleys on the dip-slope of the Lower Magnesian Limestone locally contain head deposits, e.g. at Nuthall [5210 4440], which grade downstream into alluvium.

ALLUVIUM AND FLOOD-PLAIN TERRACE

Minor streams in the south-west of the district and in Derby

A pipe trench [2555 2589] cut across Trusley Brook at the southern margin of the district encountered red-brown fine sand with traces of peat in the lowest part seen, overlain by up to 3 ft of dark grey-green silty clay with rootlets. A similar trench [2945 4198] south of Weston Underwood in the alluvium of Cutler Brook exposed $1\frac{1}{2}$ ft of brown sandy clay with rootlets, overlain by 6 in of peat with 3 ft of red-brown sandy clay at the surface.

In the bank [3249 3798] of Mackworth Brook 3 ft of red-brown silt overlies at least 2 ft of silty clay. Some 7 ft of soil and brown clay resting upon 1 ft of gravel were penetrated by a trial bore [3325 3789] at Markeaton.

Trial bores for Derby Ring-Road on Queensway [3376 3728 to 3388 3757] proved alluvium and flood-plain terrace resting on Keuper Marl at 168 to 175 ft above OD. The deeper deposits occur close to the lake in Markeaton Park, where they comprise 10 to 12 ft of soft grey-brown silt overlying 5 to 8 ft of sand and gravel.

King Howman Water Bore [3434 3675] (Stephens 1929, p. 90), sited upon the alluvium of this brook, encountered only 8 ft of deposits resting on Keuper Marl; they are recorded as made ground, 6 ft; on 'rough pebbles', 2 ft. Trial bores at the gasworks [approximately 348 364] proved up to 6 ft of soil and brown clay on up to $21\frac{1}{2}$ ft of sand and gravel; one boulder $2\frac{1}{2}$ ft across was encounterd at the base of the gravel. The total thickness of the alluvium and flood-plain terrace was $19\frac{1}{2}$ to $24\frac{1}{2}$ ft, yet downstream an old borehole at Alton's Brewery [about 3508 3619], Wardwick (Stephens 1929, p. 89), proved 5 ft of soil on 4 ft of gravel on Keuper Marl.

In Bramble Brook in the south of the district, trial bores for Derby Ring-Road [3289 3615] proved the alluvium, mainly clay and silt, to be $8\frac{1}{2}$ to 10 ft thick; one bore encountered $3\frac{1}{2}$ ft of sand at the base of the deposit. The three Slack Lane Water Bores [336 362] proved up to 10 ft of 'mixed clay' under 4 to 6 ft of made ground before entering Keuper Marl.

River Ecclesbourne

The alluvium south-east of High Cliff Lane consists generally of muds, fine silts and sands with gravel lenses. The following section [2990 4713] was exposed:

	ft	in
Loam, brown, clayey		9
Clay, brown	1	0
Clay, grey and brown mottled	2 ft to 2	6
Gravel, predominantly of quartzite pebbles with some black and brown chert and sandstone	2	0

A temporary section at Postern Lodge [3115 4612] showed dark brown sandy loam 3 ft on 1 to 2 ft of ochreous clay on dark grey clayey sand and gravel with driftwood.

River Derwent

The only sections of note recorded in the Derwent alluvium above Duffield were for the foundations of Broadholm railway bridge [348 490] where gravels predominate; the section for No. 2 pier was: river mud, $2\frac{1}{4}$ ft; fine gravel and sand, 11 ft; coarse gravel, 3 ft; fine gravel and sand 4 ft; on coarse gravel.

The eroded river bank [3491 4310] near Duffield showed $4\frac{1}{2}$ ft of brown sand overlying 2 ft 9 in of gravelly sand. Nearby at Cavendish Laundry Borehole [3494 4226] (surface level abt 180 ft OD) the full thickness of alluvial deposits proved was: soil and subsoil, 3 ft; sand and 'Bunter' pebbles, 57 ft; on Namurian shales.

Sections in trial borings through the alluvium and flood-plain terrace south of Duffield were noted by Wedd (in Gibson and others 1908, p. 162), who also (p. 157) commented on the differences in pebble content compared with adjacent sand and gravel of glacial origin. The river gravel contained a greater proportion (52 per cent) of Dinantian limestone pebbles than sandstone (24 per cent), and the glacial gravel a greater proportion of Carboniferous sandstone (48 per cent) than limestone (23 per cent). Near Burley Hill [3545 4130] Wedd recorded 4 ft of soil overlying 14 ft of gravel and sand resting upon 'grit'. Near the mouth of Bottle Brook [3601 4074] a similar section rests on blue shale. Adjacent to the Little Eaton road [3636 4077] the gravel and sand is only 10 ft thick. Trial bores for road improvements near Ford Farm [3615 3998] showed 7 ft of sandy clay on 17 ft of sand and gravel with its base at 138 ft OD. Not all the holes penetrated the full thickness of the alluvium and flood-plain terrace deposits, and the presence of a buried channel suggested by Cavendish Laundry Borehole (see above) is unconfirmed. Near the river the deposits [3589 3995] were only 14 ft thick with their base at 142 ft OD. A little to the south, near the loop in the river, a trial bore (Wedd in Gibson and others 1908) proved 4 ft of soil overlying 6 ft of gravel on an unrecorded thickness of "black 'rotten soil' with nuts and plant remains". This constitutes the only record of peaty material in the Derwent alluvium.

a

A water bore at the Cable Works [3563 3783] proved 3½ ft of soil and clay and some 23 ft of sand and gravel; the base of the deposit was at about 133 ft OD. Another water bore [3587 3700] encountered 10 ft of made ground overlying 5½ ft of blue clay on 10½ ft of gravel; 'rockhead' was at 130 ft OD.

In the river bank [3683 3520] near the southern margin of the district 6 ft of brown sand with sporadic pebbles and bands of gravel overlying 4 ft+ of gravel are exposed. A little to the east [3728 3576] the sands contain ferruginous bands, clay seams and animal bones.

River Amber

Wiremill Bridge Borehole [3789 5363] proved 10 ft of sandy clay and boulders overlying 21 ft of sandy clay with fragments of ganisteroid sandstone, ironstone and quartzitic pebbles in sandy clay and sand.

Proving holes for the Ambergate Ironworks/Gas Board Site [3525 5190] showed up to 28 ft of river deposits, which comprised an upper silty clay, up to 16 ft thick, and sand and gravel up to 19 ft thick.

Bottle Brook

Within the gorge cut through the Ashover Grit at Little Eaton a trial bore [3543 4202] a little north of the road bridge penetrated (Wedd MS. map) 2½ ft of brown soil, 7½ ft of yellow and blue clay, 6 ft of 'ratchelly gravel', 1½ ft of sand and gravel and, at the base, 3½ ft of coarse gritty sand resting upon 'rock'.

River Erewash

Opencast coal bores on Dogle Site proved at one locality [4705 4447] 1 ft of soil on 16 ft of sand and gravel, and at another [4710 4424] 9½ ft of soft sand.

Trial bores [4655 4748] through the alluvium of Beauvale Brook, north of Eastwood, proved sandy clay up to 9½ ft thick on sands with up to 4 ft of gravel, 8½ ft.

River Trent

Boreholes at Beeston [5453 3638] proved up to 26 ft of alluvial deposits. Numerous old water courses cross the alluvial flat and appear to originate from the knickpoint of Beeston Lock [5355 3535] (Clayton 1955, p. 20) on the southern margin of the Sheet. A section through the alluvium at Beeston Sewage Works [5405 3655], at 86.5 ft OD, showed:

	ft	in
Sand and gravel - largely quartzites and flints with sandstone and siltstone fragments	7	0
Sand (water bearing)	1	0
Sand and gravel (as above)	11	0
Clay, grey and brown		

Examination of the alluvium at the river's edge just beyond the southern margin of the district [5480 3567] showed:

	ft	in
Alternation of grey-brown silty sandy clay in 15-in bands with yellow-brown fine to medium sandy laminae; bedding is usually straight but has sporadic contortions: shells, commonly in a near-vertical position with pointed end downstream	8	0
Sand, coarse, fine gravel (up to 2 in long pebbles) flints, small shells; sand predominant in basal 6 in	up to 2	0
Clay, dark grey, sandy	seen 0	9

The sand and gravel within the alluvium has been worked south and east of Beeston. According to Clayton (1955, p.17) these deposits are largely Flood-plain Terrace, which occurs some 10 ft above river level, but is not exposed above and west of Beeston Lock.

PEAT

Up to 2 ft of peat occurs on sandy clay in an old drainage channel [325 522] about 1 mile SE of Alderwasley.

The Hungerhill and Black brooks, which drain southwards from Ravensdale Park, are flanked by badly drained alluvial flats on which a capping of peat up to 6 ft thick is commonly present, and is probably still being formed.

Farey (1811, p. 308) noted workings in peat near Coxbench in the Bottle Brook valley, but it cannot be traced today.

LANDSLIPS

Between Idridgehay and Wirksworth there are many landslips in the shales below and within the outcrops of the leaves of the Ashover Grit. A large slip 1000 yd by 500 yd, is present at Rough Pitty Sike [271 528] between Spink Wood and Callow. It lies below north-facing Ashover Grit separated by a depression from the steeply dipping promontory of Dinantian limestone to the north. The ground is very uneven and the backwall is cracked. Smaller slips are present at Rough Rams Carr [271 509] (300 x 200 yd), Park Hill Wood [274 516] 300 x 180 yd, Cathole Wood [274 523] 500 x 300 yd and Bennywell Brook [2685 4891] 500 x 200 yd. There are undercut landslips at Holm Brook [276 491] 200 x 100 and 300 x 100 yd, and near Blackwall [2630 4910]; at Hillcliffe Lane [288 475] there is a north-facing mudstone slip/flow caused by undercutting by the Sherbourne Brook.

At Bigginhead [269 489] there is a 300 x 80 yd slip on a steep south-facing slope of mudstone, and to the south of Biggin at Millington Green [261 476] a north-facing slope of mudstone, undercut by the Biggin Brook, has slipped over an area of 200 x 100 yd.

Mudstones of the Widmerpool Formation below the Pebble Beds outliers at Black Carr have slipped [2552 4563] over an area of 180 x 50 yd, and north-west of Mackworth [308 382] there is a small slip of 300 x 100 yd on the north-west

b

facing valley slope of a minor tributary of the Mackworth Brook.

From Washgreen [296 538] to Little Bolehill [295 545] there are slips up to 400 yd wide beneath the scarp of the main bed of the Ashover Grit. The main road to Whatstandwell locally shows signs of repeated movement in the form of patches of new road, stretches of new retaining wall and reinforcement of the old wall with metal strips. The minor road through Little Bolehill has been closed several times in recent years.

At Lane End [297 510], Ashleyhay, there is a 300 x 150 yd slip on a steep slope of mudstone between leaves of the Ashover Grit. There are two small slips a little to the south and another to the east at Bowmerlane [302 510]. Several slips are present between Hazelwood and Shottlegate where the indented scarp, capped by the lowest Ashover Grit sandstones, is at its highest elevation; the two most extensive, 530 x 160 yd [321 465] and 360 x 170 yd [323 461] are amalgamations of smaller slips and pass down into head deposits. Several smaller slips of the same type are present nearby and to the south-east of Hazelwood [332 456 and 333 450].

A small slip in which the Ashover Grit (Main Bed) has moved on the underlying shale is present on the cliff west of Milford [348 453]. Abrupt changes of dip in the old quarry here are interpreted as evidence of cambering.

East of Belperlane End the steep slopes beneath the outlier of Chatsworth Grit are uneven and probably unstable; at three localities, [338 496, 342 493 and 341 490] slips involving the base of the sandstone and the underlying shales have occurred.

Near Ambergate the flanks of an anticline which has been cut through by the River Amber have provided ideal conditions for slipping. The dip of the beds is into the valley and a water-lubricated junction is present at the base of the Ashover Grit (Main Bed) which overlies over 100 ft of shaly strata down to the river. The area of slipped strata is 220 000 sq yd. Disturbed dips up to 60° are visible in the railway cutting [3514 5166] crossing the toe of the slip. The slip provided many difficulties during construction of the railway (N. Midland Line); the cutting when part-excavated immediately filled up, and completion of this part of the line entailed the removal of 700 000 cu yd of material and took fifteen months instead of an estimated two (Smiles 1871, pp.153-154).

There are several other slips on the steep sides of the Derwent valley north of Ambergate, where Ashover Grit (Main Bed) and the underlying shales have moved down dip. The largest (1100 x 500 yd) of these slips at Crich Chase [345 526] contains huge blocks of grit that have slipped en masse and rotated so that the bedding dips into the slope. Similar slips measuring 540 x 370 yd and 400 x 200 yd are present at Shiningcliff Wood [336 527] and Beggarswell Wood [341 519] respectively. All these slips pass down into head deposits [p.].

In Belper Cemetery [352 490] the Derwent valley-side below the Chatsworth Grit displays disturbed ground features over about ¾ mile, but there is little or no sign of recent movement.

The Namurian mudstones have slipped on the steep eastern side of the Derwent valley at and to the south of Croft Wood [365 391], near Breadsall. The movements also affect the overlying Pebble Beds, which are cambered. To the north of Breadsall [375 405], there is a small slip (500 x 100 yd) in Namurian mudstones on the valley side of Boosemoor Brook.

At Dale Abbey [4430 3850] on the north-facing scarp of Dale Hills, Lower Mottled Sandstone occurs at too low an elevation to be in situ and is best explained by a landslip with a fault plane forming its backwall.

Near Stapleford [4970 3735] there is a slip 180 yd x 50 yd. Here an old quarry in Pebble Beds was worked close up to a fault bringing in Waterstones. The fault caused many rock falls near the end of the quarry's working life as the Pebble Beds were traversed by numerous joints parallel to the fault plane dipping at 75° to 80°. There has been movement recently since the building of Sandicliff Garage.

PERIGLACIAL AND SOLUTION PHENOMENA

A temporary trench section near Shottle [3151 4948] on the Namurian outcrop revealed 2 ft of boulder clay on soft weathered thinly laminated micaceous sandstone and mudstone with wedge and involution structures which are attributed to cryoturbation. Similar structures were seen in laminated mudstones of the Widmerpool Formation in a trench near Moseyley [305 437].

Two frost cracks up to 1 ft wide and infilled with red clay were noted in the Pebble Beds exposed near Breadsall [3761 3940].

A pipe-line trench [5335 5265] some 6 ft deep on the dip-slope of the Lower Magnesian Limestone south of Newstead Abbey revealed that locally the limestone surface is covered by only some 12 in of soil, but there are drift-filled hollows up to 4 ft deep (Fig. 85). The hollows are considered to have been formed by glacial meltwater running down dip into the Leen Valley. At Hucknall Aerodrome, cracking of the runway surface was discovered above such clay-filled hollows in the Lower Magnesian Limestone. The clay had absorbed water which subsequently froze and finally broke up the tarmacadam. The pattern of destruction showed that the hollows were aligned north-west - south-east, i.e. directly related to the joint pattern in the Lower Magnesian Limestone.

Frozen ground structures were exposed in Catstone Hill Opencast Site. The top of the Lower Magnesian Limestone is severely pitted; former cavities up to 4 ft deep and 3 ft wide contain dark red clay with bands of coarse sand and pebbles, which are continuous with conglomeratic horizons in the adjacent limestone (Fig. 86). The origin of these hollows is ascribed to prolonged immersion in glacial water with enhanced carbonate solubility, resulting in the carbonate matrix of the limestone being replaced by clay. Deposition of clay then sealed off the pit, preserving the structures. The cold conditions, however, persisted as evidenced by the sand pipe or ice-wedge

c

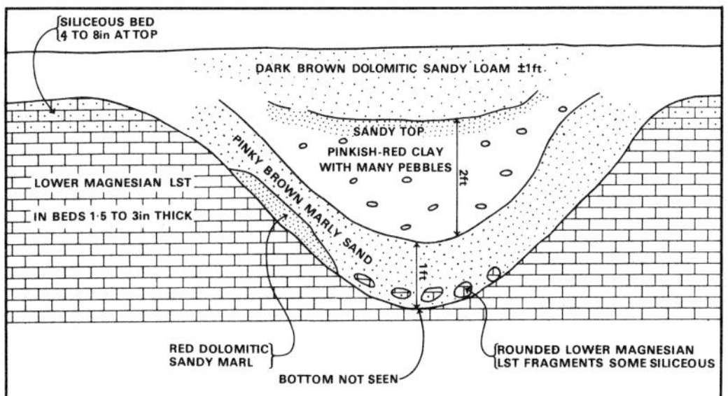

Figure 85 Clay filled hollows in the top surface of the Lower Magnesian Limestone near Newstead

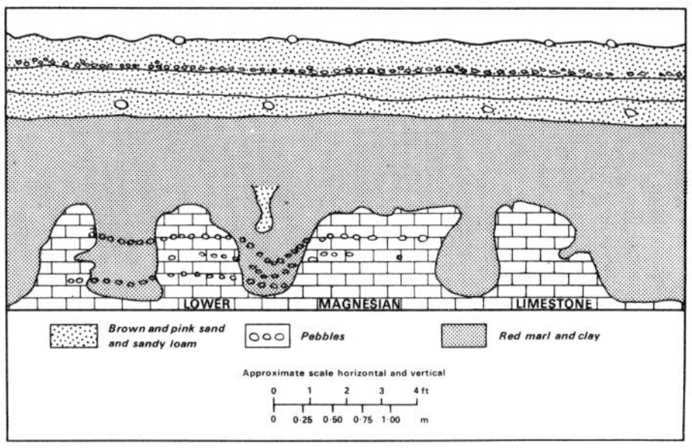

Figure 86 Sketch-section showing solution of the Lower Magnesian Limestone at Catstone Hill Opencast Site

d

pseudomorph which is preserved in the overlying glacial clay. Ameliorating conditions later led to fluviatile deposition of sandier material from the neighbouring Pebble Beds and Lower Mottled Sandstone outcrops. Eventual compaction produced downward distortion of the conglomeratic layers in the pits.

Small examples of dolomite tors at about 1000 ft OD are present at Foxcloud Plantation [2666 5392] north-east of Hopton and also at Kings Chair [2535 5380] just beyond the western margin of the district north of Carsington.

The high dolomitized areas of Dinantian limestone in Derbyshire weather into characteristic tors or crags with castellated outlines which were noted by Arnold-Bemrose (1910b). Ford (1963) noted that tors occur between 900 and 1100 ft OD and are confined to areas of dolomitic limestone. He considered that they could not have survived a glaciation and are, therefore, post-glacial. It is doubtful however whether post-glacial time has been sufficiently long for tor formation and it may be that the complex weathering process involving the removal of free calcite from the partially dolomitized limestone began in Permo-Triassic times and has continued at intervals ever since.

VALLEY BULGE AND CAMBER STRUCTURES

Examples of valley bulging are present in Ridgeway Brook [2920 4690], the upper reaches of the Holm Brook [2920 5018] and a tributary of the Biggin Brook [2624 4754] near Millington Green. At the first locality two folds trending north-north-eastwards with dips up to 52° are exposed in an area where the regional dip is 18° to 26° to the north-east.

On the eastern side of the Ecclesbourne Valley south of Wirksworth the south-south-easterly dip at some sandstone exposures [2984 5108 and 2943 5090] near Lane End is probably attributable to cambering on the steep slopes.

Surface dip readings in streams within the Westphalian rocks are frequently at variance with regional dips in underground coal workings and therefore are indicative of superficial movement. Apart from the few instances mentioned below known examples in the district are of a minor nature.

At Denby [3883 4746] the Deep Hard Coal is heaved into a tight anticline about 150 yd long and with dips on the flanks as high as 50°. Although this structure is adjacent to the outcrop of the Deep Hard Washout (p.141c) a superficial origin is likely since the sandstone overlying the coal is involved in the folding. Furthermore three monoclinal folds in the 6 ft of mudstone exposed in a trench [3794 4648] near Kilburn indicate that there is some bulging in this valley. These last folds are steeper towards the north-west, that is away from the adjacent high ground in Kilburn village.

South of Breadsall, Pebble Beds cropping out on spurs are cambered northwards towards the Derwent valley. In Croft Wood [3663 3922] there is a northerly tilted detached block of Pebble Beds with landslip below. The maximum fall traceable in the Pebble Beds towards Breadsall is about 40 ft, but the unexposed patch of 'Glacial Sand and Gravel' north of the railway line may be disintegrated remains of cambered Pebble Beds. The camber movements are related to valley-bulge structures in the Namurian mudstones in the stream at Breadsall. These continue [3815 3975] to about 1150 yd E of the church, where the dips become more orderly towards the east; this is also the easterly limit of cambering.

Temporary exposures [4973 5128] of the Lower Magnesian Limestone near Felley showed a series of cracks parallel to The Dumbles, which may be the result of cambering into this deeply incised valley.

The escarpment of the Lower Magnesian Limestone south-east of Greasley [4970 4657] appears to have camber features in the form of 'gulls', up to 5 ft wide and about 15 ft apart, infilled with red loam, calcite fragments and dolomite boulders. Similar cracks in the Lower Magnesian Limestone exposed by the M1 Motorway construction [4908 5204], near Annesley Hall, are open joints bridged by undisturbed overlying boulder clay, but which have developed since 1962 as a result of mining at depths of 600 ft and 1150 ft below OD in the Second Waterloo and Low Main seams (see also p. 157d).

CHAPTER 8

Structure

DETAILS

a

b

FOLDING IN THE CARBONIFEROUS ROCKS

The general eastward dip of the Carboniferous rocks off the eastern limb of the Pennine anticline is interrupted by four separate anticlinal structures of varying intensity and form. They are the Crich, Breadsall, Ironville and Erewash Valley anticlines (Fig. 59). Plate 12 shows contours of various horizons within the Carboniferous and an inset on Fig. 4 shows more conjectural contours of the base of Namurian rocks.

The Crich Anticline, which brings Dinantian limestone to the surface within the Namurian outcrop, is a continuation of the Ashover Anticline of the Chesterfield district to the north (Smith and others 1967, pl. 9, p. 222). The structure is asymmetrical, steeper on its eastern flank, with dips in the Ashover Grit of 30° to 50°. In the west, dips towards the Alderwasley Syncline are of the order of 7°. The axial line can be traced in the Namurian rocks on the east side of the Derwent valley as far south as Belper, but there the amplitude of the fold is much reduced.

The Breadsall Anticline, which affects the south-eastern outcrop of the Namurian rocks, is also asymmetrical, with dips of 20° towards the east compared with 10° to the west. The plunge is towards the north-north-east, also at about 20°; northwards this diminishes slightly in the Westphalian A rocks and more so in Westphalian B. The axial trace curves towards the north-north-west. At Godbers Lum Opencast Site [406 479], at the northern limit of the fold, the sequence of strata overlying the Top Soft Coal (p. 000) was seen to be fractured and discontinuous, indicating considerable tensional stress (Plate 1). These accommodation structures are present only on the west-facing limb of the anticline.

The major part of the Ironville Anticline lies within this district. Its extension to the north has been described by Smith and others (1967, p. 216). The structure is asymmetrical about a sinuous axis; dips to the east are about 8° but to the west they are 10° to 20°.

The Erewash Valley Anticline is an extension en echelon of the Ironville Anticline. The structure is only slightly asymmetrical with dips to the west of about 6° and to the east of 3° to 5°.

To the east of the Ironville and Erewash Valley structures the regional eastward dip of the Carboniferous rocks is of the order of 3 to 5°.

A number of less important structures are shown on Plate 12. The Dinantian limestone outcrop north-east of Wirksworth is part of a south-easterly tilted block truncated to the south by the east-west Yokecliffe Rake. Minor structures superimposed upon the general pattern include a broad, gently anticlinal area to the north-west of Hopton and a small anticlinal promontory of limestone showing more intense deformation in the region south of Godfreyhole. The highest dips, ranging between 30° and 65°, are on the nose of this structure. Gravity and seismic traverses over the limestone/shale margin in the Hopton area (p. 000) show that the boundary continues at a similar inclination beneath the shales for some 500 ft. As suggested by Shirley and Horsfield (1940) and Broadhurst and Simpson (1967), it seems likely that a large component of such dips is depositional. The formation of the commonly associated reef limestones implies that there were contemporaneous differential earth movements.

Two anticlines extend south-south-eastwards and east-south-eastwards respectively from the Wirksworth limestone outcrop. The first is a continuation of the limestone anticline near Godfreyhole and extends for nearly three miles along the Ecclesbourne valley. These structures are poorly defined in the surface rocks but their presence is supported by the known thickness variation in the Namurian between the Meerbrook Sough and Duffield Borehole and by geophysical evidence (p. 163d). The implied presence of limestone at relatively shallow depths in both structures could offer prospects of exploiting possible extensions of the Wirksworth mineralization.

An anticline of restricted extent with a north-south axis flanks the Horsley Fault at Farlawn, Belper [365 485]. The structure has been proved by opencast mining of the Belperlawn, Smalley and Alton coals. A weaker structure with a south-westerly axial trend at Cross Hill [420 485] is well attested by both underground mining and opencast prospecting and is interpreted as the result of local relief of pressure in the narrowest part of the syncline between the Breadsall and Ironville anticlines. A further weak, but more extensive anticline with a similar axial trend is present on the eastern flank of the Ironville Anticline [435 525] and can be traced for some 4 miles.

The structure of the Carboniferous rocks where they are concealed below Permo-Triassic strata is known in detail only in the east of the district where there is extensive information from coal mining. The 'incrops' of the individual coal seams are shown on Fig. 50.

Along the southern margin of the district the syncline to the west of the Breadsall Anticline extends at least as far as the eastern end of the Kirk Langley - Mackworth inlier of Carboniferous rocks. There is no evidence, however, of the continuation of the anticline itself in or below the Permo-Triassic rocks. Across virtually the whole of the southern margin of the district the Carboniferous rocks dip northwards and apparently occupy the northern flank of an easterly plunging monocline, the axis of which is unlocated but may, in the west, follow the line of the Kirk Langley - Mackworth inlier.

c

FOLDS IN THE PERMO-TRIASSIC ROCKS

Contours on the base of the Permo-Triassic rocks are shown on Fig. 60.

In the west the normal southerly dip of the rocks is interrupted by a gentle anticline aligned north-west to south-east through Muggintonlane End [284 449] and Quarndon with another of similar trend, which is even weaker, through Kirk Langley. The first-mentioned anticline may include elements of superficial movements (p. 93). However, the weak syncline through Kedleston which separates these two anticlines is also present in the Permo-Triassic rocks in the western parts of Derby city, and the origin of these structures thus appears to be basically tectonic.

A gentle anticline with a north-south axis begins west of Kirk Langley and extends southwards west of Trusley Borehole. The east-facing limb appears to be the steeper at the north end, but in the south the dip is apparently steeper towards the west.

East of Derby the southerly dip is maintained until the rocks become involved in the faulting near Dale and Stanton, where the dip swings towards the east. The faults are located in the anticlinal area between the southern and eastern regional dips of the Permo-Triassic rocks and they show some of the largest throws of faults affecting Permo-Triassic rocks in the district.

The Permian rocks have a regional dip towards the east. They contain a few weak anticlinal structures which appear to be the result of posthumous movements on faults in the underlying Carboniferous (see p. 98).

FAULTS

In the Dinantian limestone outcrop, faults and veins show two dominant directions approximately at right angles (Fig. 62). The predominant trend is north-westerly (320° to 330°) and comprises numerous tension gashes which are variably mineralized. The Yokecliffe Rake is an exceptionally long east-west fault like other similar structures beyond the limits of the district to the north, e.g. Gang Vein (Smith and others 1967, p. 218) and Great Rake.

The Yokecliffe Rake forms a fault-scarp nearly 100 ft high bounding the limestone for some 1000 yd west of Wirksworth. It is inclined at 70° and has a downthrow to the south, calculated to be some 100 ft in the Matlock Limestone. It has been extensively mineralized (p. 107). The Yokecliffe Rake is probably a surface reflection of one of the major basement fractures which delineate the Derbyshire 'block' from the Widmerpool Gulf.

The Gulf Fault, like the Yokecliffe Rake, forms a fault-scarp between the limestone and shale, but follows a north-westerly trend. The downthrow, to the north-east, is greater than that of the Yokecliffe Rake, but the degree of mineralization is slight, though the fault is parallel to many rich veins between Wirksworth and Bolehill (p. 108).

The Turnditch Fault brings together rocks of E_2 and R_1 age in the Biggin area [260 483] and must therefore have a northerly throw there of at least 150 ft; near Turnditch the juxtaposition of the Cravenoceras leion Band and low E Zone beds indicates the throw to have increased to 600 ft. Although the effects of valley bulging (p. 000) make the recognition of true dips difficult, the Turnditch Fault appears to be aligned along a disturbed belt with marked changes in dip directions as seen in streams near Biggin and Turnditch.

Post-Triassic movement along this fault is indicated by a drop of some 50 to 75 ft in the base of the Triassic rocks north of the Belper-Ashbourne Road [2850 4796].

Small-scale faults with alignments parallel to the major dislocations have been recorded in the Pebble Beds of the western part of the district. Fault plane dips vary from 55° to 85°. Commonly the faults are emphasized by the displacement of marl bands within the sandstone. Pebbles are sometimes aligned along the fault planes, which have crush zones up to 12 in wide in clays and mudstones, even where vertical displacements are as little as 6 ft.

The Quarndon and Kirk Langley - Mackworth faults lie on the north and south limbs respectively of the Kedleston Syncline. Their throws, which can only be estimated in the Permo-Triassic rocks are a maximum of 50 ft. On the evidence of the road cutting at Mackworth [3138 3769], the Mackworth Fault consists of two sub-parallel fractures about 30 ft apart with minor cross-fractures. The fault contact between the Carboniferous and Waterstones is not exposed but the second fracture plane, within the Waterstones, dips to the north at 83°.

The Fishponds, Sandyford and Crowtrees faults are strike faults associated with the axial line of one of the postulated anticlines in the Dinantian rocks (p. 160b). At the surface the faults repeat and extend the outcrop of Ashover Grit (Main Bed) on the south limb of the Alderwasley Syncline.

The Sandyford Fault divides south of Streets Rough [337 504] to produce a minor graben structure, the topographic effect of which is clearly visible as a broad valley crossing the dip-slope of the Ashover Grit (Main Bed).

The Markeaton Fault cannot be mapped from surface features but is inferred from water borehole information within Derby city and by trial bores on the Derby Ring Road. The throw increases towards the south-east to over 200 ft to make it one of the major fractures in the Permo-Triassic rocks.

The Southern Crich Fault brings shales below the Chatsworth Grit against Dinantian limestone, suggesting a downthrow in excess of 1000 ft towards the south-west. This is the largest known throw of a fault within the district. A proportion of the apparent throw is possibly accounted for by unconformity and/or thinning of the lowest Namurian shales around the Crich Anticline.

The Horsley Fault in the north is a north-south strike-fault of variable throw affecting the Namurian rocks on the eastern flank of the Crich Anticline. North of Belper it curves south-eastwards and forms the boundary fault of the Westphalian A rocks with a downthrow to the east

d

of about 450 ft increasing to about 650 ft near Pinchom's Hill, where its position becomes less well defined. Between Pinchom's Hill and Horsley it apparently consists of two parallel faults; the throw of the south-westerly fault diminishes towards Horsley, and that of the north-easterly fault increases to about 450 ft south of that village. Although the amount of the throw of the latter was not proved, evidence of its presence was noted in drivages for the Kilburn coal [3760 4516]. The fault-plane near Belper and Openwoodgate is nearly vertical.

The Smalley Fault-complex consists of a series of normal strike-faults, perhaps with a wrenching element. Much of the information on this complex is based upon opencast prospecting for coal, but faulting has also been proved underground in the Kilburn Coal. The overall throw of the complex is slight, yet within the complex, repetition of the Low Main and Blackshale coal outcrops favours opencast working, although exploitation has been local. Throws of individual faults are generally less than 100 ft, but reach a maximum of 315 ft in the Kilburn Coal in the north-eastern fault near Stanley [4200 4085]. This last fault is vertical or perhaps reversed. The complex displaces the axial trace of the Breadsall Anticline.

The Ridgeway Fault trends north-westwards, affecting Westphalian rocks below the Blackshale Coal in the east and increasing in throw westwards to 200 ft in the Ambergate area where it bends northwards to affect Namurian rocks in the steeply-dipping eastern limb of the Crich Anticline. Between Heage and Ripley it is one of a series of en echelon faults of which the Porter Barn Fault (see below) is dominant.

The Pentrich - Porters Barn and Parkgate Fault system is similar to the Smalley Fault-complex, but on a broader scale. The Pentrich Fault has a trend just north of west. Its throw increases eastwards and in the vicinity of Pentrich Colliery is of the order of 170 ft down northwards. There it is joined by a northerly branch-fault and the combined structure swings round to a north-westerly trend on the east side of Ripley, where a throw of 162 ft has been proved in Low Main workings [4068 5024]. The dip of the fault-plane is 75°. The Porters Barn fault-plane is inclined about 60° towards the south-west. It trends west-north-west through Ripley and has proved downthrows to the south of 96 ft in the Deep Soft Coal [3933 5050] and 105 ft in the Low Main [4115 4849]. The dip of the Parkgate Fault is 55° to the north-east with a downthrow in the same direction of 168 ft in the Deep Soft [3967 4848].

The Swanwick Fault has a throw of 185 ft in the Low Main Coal between Swanwick and Alfreton; it increases to 250 ft [4172 5267] south of Swanwick. Its surface position was proved in Tramway Site [4160 5278], where workings in the Top Hard and Comb seams, on the upthrow side, were adjacent to those in the Lowbright seam on the downthrow side. It has a dip of 73° between the Blackshale and Low Main seams. The fault continues south-wards beyond the Swanwick area and appears to be deflected by the Cromford Fault [4178 5193] which it crosses at almost 90°, passing into an area of complex faulting east of Codnor.

The Riddings, or Ridge Fault trends north-north-westwards on the western flank of the Ironville Anticline. It has a downthrow to the west of 294 ft proved in Low Main workings [4175 5460] near Swanwick Colliery. The throw decreases to 180 ft southwards over a distance of 400 yds. The fault-plane has a dip of 73° between the Blackshale and Low Main seams. The fault passes southwards into the axial area of the Ironville Anticline and breaks up into a series of small faults, one of which was seen in Agnes Site [4315 4995] displacing the Low Main seam by some 6 ft. The fault-plane, inclined at an average of 70°, was slightly steeper in the overlying sandstone than in the underlying mudstones and coals.

The Aldercar and Godkin faults continue the graben structure between the Riddings and Swanwick faults. They displace the axial trace of the Ironville Anticline to the west and the eastward continuation of the Godkin Fault displaces that of the Erewash Valley Anticline slightly to the east. The dip of the Godkin fault-plane appears to be about 70° at its union with the Ilkeston Fault, where the throw is 450 ft to the north-east. About ½ mile to the north-west, where the surface position is marked by subsidence features, the fault is nearly vertical. The Aldercar Fault is less well known underground, but appears also to have a steeply dipping fault-plane.

The Mapperley Fault is well documented both at the surface and underground with a maximum throw in the Kilburn Coal of 285 ft [4376 4080]. The fault-plane dips eastwards at about 55°.

The Lowcote Fault is similarly well proved and broadly follows the axis of the Shipley Syncline. Near Lowcote [4467 4168], where it is proved at surface and in the Low Main (throw 81 ft) and Kilburn coals, it is steepest near the surface with a dip of about 72° but flattens at depth to about 63°. In Shipley Park [4378 4356], however, it dips at 58° from the surface to the Deep Hard Coal and steepens to 65° between the Deep Hard and Kilburn coals.

The Ilkeston Fault has a throw of 147 ft at Granby Colliery [4653 4328], where the dip of the fault-plane between the Low Main and Kilburn coals is 50° towards the east.

The Cromford Fault has a downthrow to the north of some 162 ft in the west [3974 5253] near Swanwick with a fault-plane dip of only 27°. The surface position was proved in Tramway Site [4100 5229]. Where the fault crosses the Ironville Anticline, a fault-plane dip of 39°, giving a throw of some 200 ft at the surface, has been calculated. At Westwood Bents a 30° dip is present between the Deep Hard and Blackshale coals. This steepens to 75° at the southern end of the fault.

The Annesley Park Fault increases in throw eastwards from 30 ft near Selston to 70 ft near Garfit House [4865 5275]. One branch trends north-eastwards to intersect the Mapplewells Fault (see below). The main fracture is offset to the south and continues past Annesley Hall, Newstead and Annesley collieries to Newstead

Abbey. A throw of 75 ft proved in the Deep Soft Coal [5096 5277] decreases upwards to 57 ft in the Top Hard. The average dip of the fault-plane between the Deep Soft and the surface is calculated to be 63°.

The Mapplewells Fault trends north-westwards from its intersection with the Annesley Park Fault near Annesley Hall, through Annesley Woodhouse towards Kirkby. A downthrow to the east of 70 ft was recorded in the Deep Soft seam [5035 5335]. The calculated dip of the fault-plane varies from 68° (Deep Soft to Top Hard) to 60° (Top Hard to surface). The fault was seen in the cores of Mapplewells Borehole (p. 000) and the details may well be typical of the many other faults in the district. The main fault-plane was intersected at a depth of about 650 ft, but the first indications of dislocation appear 550 ft above, in the form of irregular and oblique calcite-and ankerite-filled joints. Minor faulting with slickensided and listric surfaces occurs 225 to 325 ft above the main fault-plane. From some 80 ft above to 140 ft below the main fault-plane, a complex zone of disturbance includes fractures inclined between 10° and 27° together with slickensides.

The Hucknall Fault is unusual in that a displacement of some 70 ft in the Top Hard and Dunsil seams is represented at the surface in the Lower Magnesian Limestone by a small faulted anticlinal or monoclinal fold (Gibson and others 1908, p. 108). This fold produces a marked topographic ridge about 10 ft high and 70 yd wide which can be traced from Newstead in the north for over 4 miles through Hucknall to Bulwell. The dip of the fault between the Top Hard and High Main seams is 70° but decreases to 60° between the High Main and the surface. The main fault is intersected by a cross-fault of small displacement near Linby (12 ft in the High Main workings), which displaces the anticlinal structure to the east. Throughout the outcrop of the Lower Magnesian Limestone similar anticlinal rolls are recorded with a trend (E 38° N) parallel to the Linby Fault, but their downward extensions as faults have not always been proved.

The Cinderhill Fault is in fact a north-westerly belt of arcuate faults which are characterized by low-dipping fault-planes. Near Eastwood the exact surface position of the main fault is known only in City Site, but there is abundant evidence underground, where the dip of the fault-plane is 39° to 50°. Near Giltbrook the fault appears to throw about 50 ft in the Permian rocks compared to 192 ft in the Deep Soft seam, yet the same fault-plane near Kimberley does not displace the Permian rocks. There is thus good evidence that the post-Permian period of movement only affected part of this fault. On the promontory of Permian rocks near Giltbrook there is an arcuate counter fault on the downthrow side of the main fault. East of Kimberley the main fault has its maximum proved throw (213 ft down to the south in the Top Hard seam). The overall dip of the fault is calculated at 45°. Information from Cinderhill Colliery workings shows that the throw decreases upwards to between 160 and 170 ft in the Main-bright seam. In the Permo-Triassic rocks the vertical displacement is less than 40 ft.

Exposures made during excavations for the motorway link-road at Nuthall showed a complex fan of faults of small displacement at the southern end of one of the north-west echelons of the Cinderhill Fault. Two folds of small amplitude in the Permian strata trending just west of north were also exposed. The main fault was exposed [5158 4402] near Nuthall Temple along the line of the motorway where Middle Marl and Lower Magnesian Limestone are juxtaposed. Farther west, at Knowle Wood, the southernmost fracture of the fault system was exposed [5050 4413], showing a downthrow to the north bringing Lower Mottled Sandstone against Lower Magnesian Limestone. There is a vertical displacement of 36 ft in both the Permo-Triassic rocks and the Deep Soft workings at this locality.

The Strelley Fault was extensively proved in opencast workings; the throw is of the order of 100 ft to the north-east and the dip of the fault-plane between the surface and the Deep Soft Coal is about 63°.

The Broxtowe Fault system is closely related to the Cinderhill Fault and forms a complex area of trough faults with displacements of 51 ft in the north and 30 ft in the south, as proved in Top Hard and Deep Soft workings respectively. Evidence at surface was provided by temporary exposures in a sewer trench in Broxtowe Wood, which showed a 150-yd belt of faulted Permo-Triassic and Westphalian rocks. There are three normal faults with dips of about 62°, and a reversed fault dipping at 55° which separates the Westphalian from Lower Magnesian Limestone.

The Dale Abbey - Stanton Fault-system is visible at surface, but there is no information at depth. The fault-plane exposed in a quarry [4603 3832] near Stanton dips at 80°, and strongly developed slickensides show that the last phase of movement was in a vertical direction. The Crawshaw Sandstone along the fault-plane is brecciated. A minor parallel fault-plane within the adjacent Lower Mottled Sandstone dips at 55°. This fault system together with the Clifton - Highfields graben farther north (see below) effectively forms the boundary between the Westphalian to the north and the Permo-Triassic rocks to the south.

The Stanton Fault, showing a dip of 60° to 70°, was exposed in motorway excavations [4747 3775] near Stoney Clouds. Between it and a parallel fault some 100 yd to the south are many short north-westerly trending faults and joints which divide up the Pebble Beds scarp at Stoney Clouds into a series of fault blocks. One such short fault on the motorway was seen to dip at 45°.

Several dislocations similar in trend to the Stanton Fault affect the Triassic rocks to the south. In motorway excavations [4710 3675] west of Sandiacre, a gentle anticline, with dips of up to 10° and an axial trend parallel to the main boundary faults was exposed. The crest of the anticline was dislocated by a series of small faults parallel to the axis and with displacements of less than 6 ft. On the northern limb, the faults throw down to the north (60° dip of fault-planes) and on the southern limb to the south. The core of the fold is marked by a small graben.

The Clifton (Lenton) and Highfields faults form a downfaulted block of country 300 yd 700 yd wide extending from Wollaton to Beeston. Near Trowell, dips of 40° indicate the close proximity of the plane of the Clifton Fault. Near Bramcote a working quarry [5030 3896] in the Lower Mottled Sandstone is said to have proved the fault-plane with Westphalian on its north side. East of the quarry the Pebble Beds are faulted against the Westphalian with a marked scarp feature some 50 ft high. A downthrow to the south of 185 ft has been proved in the Deep Soft Coal [5217 3855]. Extensions to the buildings of the Nottingham University campus enabled Taylor (1965, p. 189) to trace the Clifton Fault in detail. The plane of dislocation was exposed in Cutthrough Lane [5432 3837].

The Clifton Fault has been proved by drilling and underground coal workings near West Bridgeford to have a dip of 68° to 83° in the Trias and 75° in the Westphalian. Elliott (1961) and Taylor (1965) noted that the displacements in the Westphalian rocks are the same as those in the Permo-Triassic rocks, and the fault is thus of post-Permian age.

The Highfields Fault has a topographic expression similar to that of the Clifton Fault. It has been proved just beyond the eastern margin of the district in workings of the Top Hard seam [5542 3861] from Clifton Colliery, where there is a downthrow of 180 ft to the north.

a

b

c

d

CHAPTER 10

Geophysical investigations

DETAILS

a

b

The association in the Derby district of low density Westphalian and Permo-Triassic sediments with older, higher density limestones of Dinantian age would be expected to produce pronounced Bouguer anomalies due to near-surface density contrasts. While this is generally true the Bouguer anomaly map also reflects more deeply buried structures such as the thick Namurian sequence in the Widmerpool Gulf and a largely unproven ridge of Lower Carboniferous, or older, rocks which extends southwards from the Derbyshire Dome.

Magnetic surveys reveal the existence of extensive zones of magnetic material at depth beneath a cover of non-magnetic sediments. The magnetic zones bear little overall relationship to the structures revealed by the gravity data, although there is a certain correspondence in minor features; they probably represent magnetic bodies at a deeper level in the crust.

The gravity and magnetic anomalies discussed below extend beyond the sheet boundaries, since in the Derby district nearly all the major features seem to be related to large scale structural features. Complete interpretation will therefore depend to some extent upon future examination of the extensions in critical areas outside those described below.

PHYSICAL PROPERTIES OF THE MAIN ROCK TYPES

Although no systematic sampling has been made in the Derby district, sufficient information is available from published work and from surveys by the Applied Geophysics Unit of the Institute of Geological Sciences to compile values for the main rock types (Table 13). The values for the densities are based mainly on sample measurements and those for the velocities on both sample measurements and the results of seismic refraction investigations. Resistivity values were obtained entirely from geophysical borehole logs.

Samples from the Dinantian limestone of Derbyshire give consistent values for both density and sonic velocity (Table 13), but these are possibly not typical of the entire sequence due to the effect of weathering at the surface, jointing and thin horizons of different lithology. Although the limestone at depth in the Burton upon Trent area was correlated with a 5.95 km s^{-1} layer, seismic studies on limestone outcrops near Wirksworth produced velocities as low as 2.5 km s^{-1} within 100 m of the present land surface.

Density values for the Namurian shales are important both in the reduction of gravity data and in its interpretation since they are likely to represent a major density contrast with older sediments. Typical sample densities are difficult to obtain but an estimate of 2.42 g cm^{-3} was obtained from a gravity meter profile across a valley in the Belper area. The representative velocity for the Namurian obtained from refraction studies in the Burton upon Trent area (Bullerwell in Stevenson and Mitchell 1955) is considerably greater than other values (Table 13) and must be regarded as a maximum value, perhaps due to thin, high velocity horizons.

Susceptibility determinations on samples indicate values of about 4 x 10^{-4} emu for the dolerite intrusion at Ible, 5 km north-west of Wirksworth, and 1 x 10^{-4} emu for the Hopton agglomerate, 3 km west of Wirksworth. Sedimentary rocks in the area would be expected to be practically non-magnetic, with susceptibilities less than 1 x 10^{-5} emu.

Sonic velocity data for the main rock types in the Nottingham area have been obtained from well velocity surveys, mostly in oil boreholes, including those at Eakring and Widmerpool. Wyrobeck (1959) reported that the increase with depth of interval velocities for the 'Bunter', Westphalian and Namurian rocks were all approximately proportional to the depth of burial to the power of one fifth, but that the 'Keuper' and Permian velocities increased according to the power of one tenth. Typical interval velocities (at 300 and 600 m depths) derived from the above relationship and average vertical velocities to the top of the corresponding formation were reported by Wyrobeck (1959) as follows:

	Interval velocities		Mean velocity to top of division km s^{-1}
	300 m	600 m	
'Keuper'	3.2	3.4	
'Bunter' (East Midlands)	2.8	3.2	
Permian	3.1	3.4	2.9
Westphalian	2.8	3.1	3.2
Namurian	-	3.2	3.4
Dinantian	-	-	3.6

GRAVITY SURVEYS

The gravity surveys for this area were made by the Geological Survey with a station density of about 1 per 1.3 km^2.

The combined free air and Bouguer corrections were made using a variable density determined by the rock type underlying the stations. The densities used were: Dinantian limestone 2.65 g cm^{-3}, Namurian and Westphalian 2.45 g cm^{-3} and Trias 2.40 g cm^{-3}. With elevation ranging from +50 to +350 m above OD, the choice of the correct density is of considerable importance for the correct computation of the Bouguer anomalies, particularly in the area of dissected topography formed by the Namurian outcrops. The Bouger anomaly values calculated require a correction of about 2.63 mGal. to make them compatible with National gravity reference net (1973) values.

The Bouguer anomaly map of the Derby district (Fig. 64) is dominated by a'high', reaching a maximum of +23 mGal north-west of Wirksworth and showing a decrease towards the east and south-east. Two elongated Bouguer anomaly 'highs' run south-eastwards and due south from the Wirksworth high. Various residual Bouguer anomaly maps were made to separate anomalies such as these ridges from broader regional variations, and one of these is shown in Fig. 88. The 6 km by 6 km grid interval used to estimate the background values for the residual anomalies results in an emphasis on anomalies with widths less than 6 km.

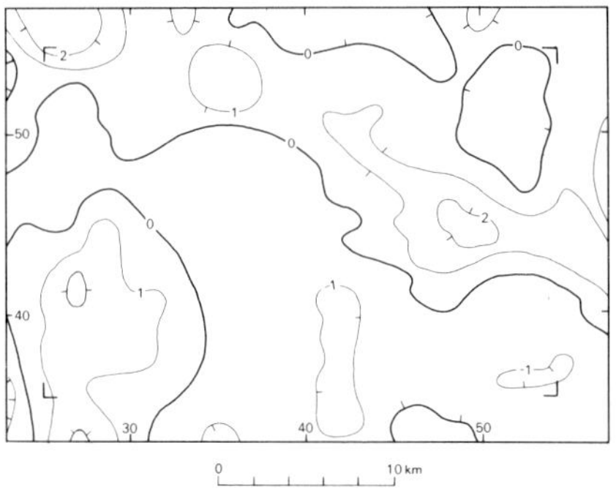

Figure 87 Residual Bouguer anomaly map based on a 6-km² grid for part of the area shown in Figure 64

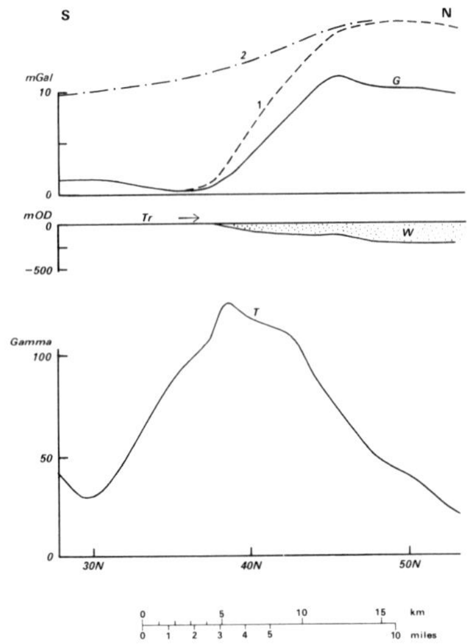

Figure 88 Bouguer anomaly (G) and aeromagnetic (T) profiles along National Grid Line 47E

c

d

The south-western part of the Bouguer anomaly map (Fig. 64) is dominated by a north to south trending'high' which passes through Trusley. As the anomaly level of this 'high' is only some 3 mGal less than that over the main Derbyshire limestone outcrop, the feature can most conveniently be explained by the existence of buried limestone. However the northern part of the high north of Trusley coincides with an area of shaly limestones of the Widmerpool Formation, and a borehole at Mickleover Hospital, 7 km south-west of Derby encountered similar strata below a depth of -69 m OD (Stephens 1929). This shaly limestone facies should be less dense than the main massive limestone sequence and it seems likely that it is underlain by denser limestone or older rocks responsible for the Bouguer anomaly 'high'. South of the Derby district the Bouguer anomaly ridge changes direction to south-south-east and coincides with the Ashby Anticline between the Leicestershire and South Derbyshire coalfields. The change in the trend of the Bouguer anomaly from north to south to the south-south-east to north-north-west occurs along the line of the fault near Hopton. A north to south trend can also be traced southwards as a Bouguer anomaly feature south of Burton upon Trent and possibly through to the western margin of the Warwickshire Coalfield.

In the northern part of the Derby district the regional decrease in Bouguer anomaly values to the east (Fig. 64) is largely due to the increasing thickness of low density Westphalian rocks and, farther east, to the appearance of Permo-Triassic sediments. Closer examination of the relationship between the Bouguer anomaly map (Fig. 64 and Plate 12 suggests, however, that the known thickness of Westphalian would need to have an improbably high density contrast of about - 0.33 g cm^{-3} to explain the observed gradient, and therefore a deeper mass deficiency must also exist in this eastern area.

South of a line from Wirksworth to Nottingham an explanation of the Bouguer anomaly decrease to the east in terms of low density Westphalian sediments fails completely. If the gravitational effect of known Westphalian rocks is removed the southward decrease of the Bouguer anomaly steepens across an east-west line coinciding more or less with the southern boundary of the coalfield (Fig. 88). The observed maximum gradient of 1.5 mGal km^{-1} suggests an anomalous step-like structure which cannot be deeper than 2.3 km (7600 ft). It is highly probable therefore that this anomaly represents the effect of the southward thickening of the Namurian sediments across the northern margin of the Widmerpool Gulf (Falcon and Kent 1960, Kent 1966). The full extent of this anomaly is difficult to judge but it would seem that the amplitude must be at least 11 mGal (Fig. 89), assuming that the regional anomaly continues southwards to a level of about +10 mGal near Charnwood Forest. With a density contrast against the Dinantian limestone of - 0.25 g cm^{-3} an additional thickness of at least 1.05 km (3450 ft) of Namurian sediments is necessary to explain this anomaly. As the Namurian sequence is about 0.3 km (1,000 ft) thick at Ironville (Kent 1966) away from the gulf development, the total depth to the base of the Namurian would have to be at least 1.4 km (4,600 ft), allowing for the comparatively thin Trias. The estimate of 1.4 km (4600 ft) exceeds the Namurian thickness of 0.8 km (2600 ft) at Widmerpool and 0.8 km (2600 ft) in the Duffield area. The southward decrease in Bouguer anomaly values in this area is therefore not due solely to the thickening of lower density sediments in the Namurian but probably also to a thickening of the shaly facies in the upper part of the Dinantian. In both Widmerpool (Kent 1966) and Duffield boreholes the Widmerpool Formation is considerably thicker than equivalent strata in the Ironville area, containing mudstone and sandstones which are less dense than the massive limestone and therefore tend to cause Bouguer anomaly 'lows'. Any gravitational effect of the igneous intrusion which is presumably responsible for the magnetic anomaly west of Nottingham (Fig. 65) is obscured by the steep Bouguer anomaly gradient.

The Wirksworth-Nottingham Bouguer anomaly 'high' can be traced from near Wirksworth through the limestone inlier of Crich (Fig. 64) and the Ironville Anticline. The south-eastward trend of the 'high' is seen clearly on the residual anomaly map (Fig. 87). In the extreme south-east the appearance of a local anomaly is due mainly to the abrupt change in the direction of the Bouguer anomaly contours from north-south (reflecting the increasing thickness of Westphalian) to east-west through Ilkeston (largely reflecting the increased thickness of Namurian in the Widmerpool Gulf). There is a small Bouguer anomaly peak along the crest of the high of about 1 or 2 mGal (Fig. 88), which can be explained in places by structures such as the Ironville Anticline. At about 45.5 N on the profile along grid line 47E (Fig. 88) a north-west to south-east ridge in the Lower Carboniferous may be due to faulting which is not carried through into the Westphalian. The belt of faults in the Westphalian (Plate 12) coinciding with the Bouguer anomaly 'high' may reflect the presence of the buried limestone ridge.

The highest Bouguer anomaly value (+23 mGal) in the area is located over the main outcrop of limestone north-west of Wirksworth at the intersection point of the trends of the two elongated highs described above. The amplitude of the closure in this area may be exaggerated by local features such as a thicker development of dolomite in the Dinantian sequence or high density basic intrusions similar to that at Ible (SK 252 575).

Superimposed on the broad major anomalies in

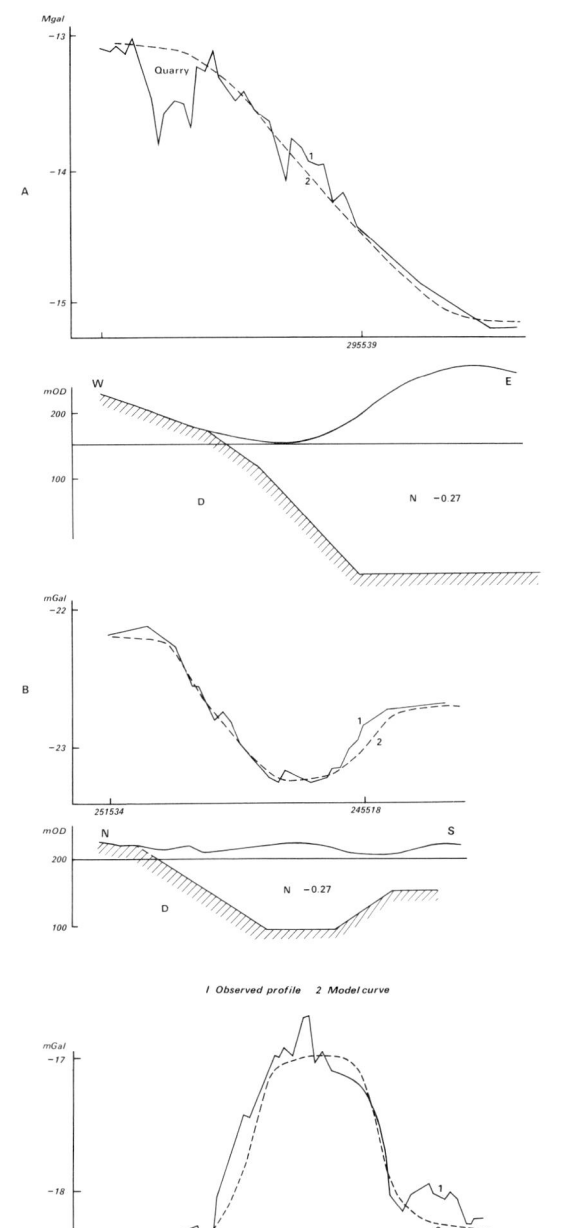

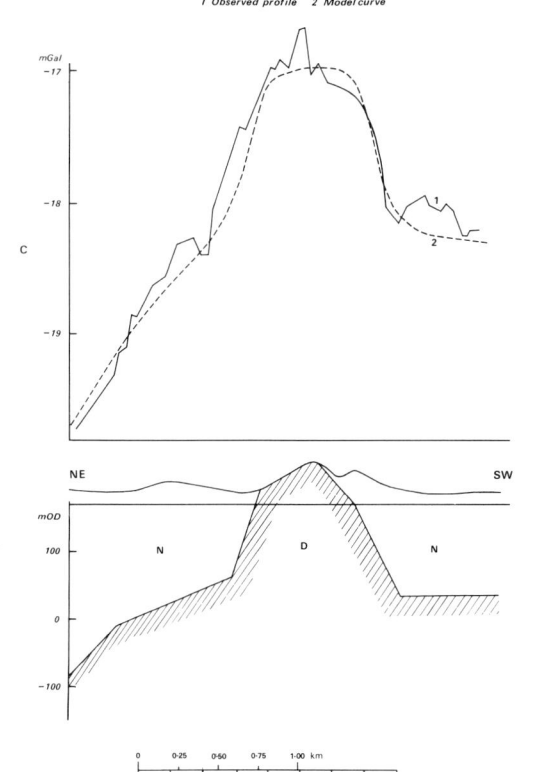

1 Observed profile   2 Model curve

Figure 89 Bouguer anomaly profiles and interpretations across the boundary between the Dinantian limestone and Namurian at A-Wirksworth, B-Carsington [SK 250 533] and C-the Crich inlier (combined profile). Bouguer anomaly values have been reduced to local datum levels. (NA = Namurian, D = Dinantian)

Fig. 64 are local small-amplitude Bouguer anomalies which can usually be related directly to mapped features at the surface. These are not easily recognizable in Fig. 64 and need more detailed measurements such as those (Mrs J Allsop, private communication) across the boundary of the Namurian shales and the main limestone outcrop in the Wirksworth, Brassington and Crich areas (Fig. 89). In both cases the steepness of the 2-3 mGal step-like anomalies indicates that the limestone dips at angles of up to 20° near the contact but then levels off to a more gentle slope. Around Wirksworth (Fig. 65a), the zone of steep gradients occurs near the outcrop of the limestone shale contact. This zone swings to the south-west, indicating that the limestone has a thick shale cover south of of the inlier at Hognaston [225 510]. The shale between the inlier and Carsington to the north-east is apparently only about 130 m (425 ft) thick and the Bouguer anomaly profile in this area (Fig. 89 b) indicates an expected southerly rise in the limestone top as the inlier is approached. The south-westerly belt of steeply dipping rocks could be explained by a fault line, perhaps that responsible for the 1904 Derby earthquake (Davison 1905).

A Bouguer anomaly profile across the Crich inlier (Fig. 89c) generally indicates shale thicknesses comparable with those estimated on geological evidence although there is some discrepancy in the area immediately west of the inlier where the steep Bouguer anomaly gradient is taken to indicate a slope of the limestone surface steeper than expected.

MAGNETIC SURVEYS

The Derby district was included in the aeromagnetic survey flown in 1955 with a mean terrain clearance of 1000 ft (308 m). The map shown in Fig. 65 is taken from the 1:625 000 scale aeromagnetic map (Sheet 2) of Great Britain (Geological Survey 1965). The total field measurements were made with a fluxgate magnetometer along east to west flight lines 1.61 km (1 mile) apart and north to south tie lines 9.65 km (6 miles) apart.

The magnetic field in the Derby district is dominated by a broad magnetic 'high', about 30 km wide and at least 80 km long, trending south-eastwards from Wirksworth through Nottingham. The smooth, widely spaced magnetic contours in the south-west suggest that the structure causing the anomaly probably lies at a considerable depth. Superimposed on the main magnetic high are several sharply defined anomalies suggesting smaller magnetic bodies at comparatively shallow depths of 1 or 2 km (3300 - 6600 ft). The pronounced circular anomaly at Hucknall is the most obvious of these, but other examples are the circular feature 10 km east of Belper and the ridge trending north-west from near the Hucknall anomaly. Broader magnetic 'highs' west of Nottingham and north-west of Wirksworth suggests magnetic bodies at intermediate depths of 3 to 4 km (10 000 - 13 000 ft).

Several types of igneous rock are known in the area but their relationship to the magnetic bodies responsible for the anomalies in Fig. 65 is not clear. In the Dinantian limestones, cropping out in the north-west corner of the map, there are dolerite intrusions and basalts in addition to bedded tuffs and agglomerates. Igneous rocks are not uncommon in the Namurian in the area and include the dolerite and tuffs at Widmerpool (Falcon and Kent 1960) and at Bottesford (Edwards 1951). Lavas and tuffs are also common in the Westphalian east of Nottingham within an area of at least 700 km[2] (Falcon and Kent 1960). There is no obvious connection between the distribution of the igneous rocks and the magnetic anomalies but the effect of thin horizontal magnetic horizons, such as sills, would not be noticeable except at their margins.

The main north-west trending magnetic zone in Fig. 65 can be interpreted as a broad horizontal slab extending from a depth of - 3 km (10 000 ft) down to - 6 km (20 000 ft) with a susceptibility of 2×10^{-4} emu. The depth extent however can be varied without invalidating the interpretation and the susceptibility changed accordingly, but it seems that the south-western margin must be inclined in all cases to account for the low gradient west of Derby. The steeper north-eastern margin crosses the corner of the area shown in Fig. 65 north of Hucknall but it is necessary to postulate the existence of some magnetic material at depth north-east of this line to reproduce the higher background value in this area. The depth and shape of this model, its elongate form and the lack of corresponding structures in the Carboniferous suggest that it could be due to a belt of volcanic or metamorphic rocks without an unusually high density in Lower Palaeozoic or Precambrian basement rocks.

Superimposed on the broad magnetic belt are more localized anomalies of probable origin which may be due to intrusions some of which may rise through the Lower Carboniferous strata. The shallowest of these is the anomaly 9 km east of Belper (Fig. 65) which is probably due to a plug-like intrusion rising to less than -0.7 km (-2300 ft) OD. The anomaly lies on an extended belt of faulting on the northern flank of the south-east trending Bouguer anomaly 'high', but the slightly larger and deeper source of the anomaly at Hucknall lies beneath Permian sediments and has no surface indication. The east-west anomaly west of Nottingham coincides in position and direction with the Bouguer anomaly gradients associated with the northern margin of the Widmerpool Gulf (Fig. 88). The positive magnetic anomaly near the north-west corner of the area shown in Fig. 65 coincides with the outcrop of Dinantian limestones on the south-east corner of the Derbyshire Dome and represents a broad rise in the magnetic material.

A comparison of the structures suggested by the magnetic map with those deduced from the main Bouguer anomalies (Fig. 64) reveals little in common. The broad magnetic zone appears to lie below the Carboniferous rocks responsible for almost all the Bouguer anomaly features, but its north-west to south-east trend is repeated by the Bouguer anomaly 'high' interpreted as a ridge in the

Lower Carboniferous.

The broad magnetic zone has a Charnian trend and a Pre-Cambrian or Lower Palaeozoic age is suggested. Farther to the south, outside the Derby area, a distinct group of magnetic anomalies, probably due to intrusions related to the ?Caledonian Mountsorrel granodiorite, occur within the broad magnetic zone. From this evidence it would appear that the zone has been the scene of repeated igneous activity extending from Lower Palaeozoic or even Pre-Cambrian times through to the late Carboniferous.

North-south trending structures of Hercynian age are prominent on the Bouguer anomaly map, which mainly reflects density contrasts within the Carboniferous. There are only two places where this trend can be seen on the magnetic map. One is the slight bulge in the contours west of Wirksworth, which is the only magnetic feature correlating with the pronounced north-south Bouguer anomaly 'high'. The other is the north-south trending western margin of the shallower magnetic anomalies west and north-west of Nottingham which coincides at the surface with the eastern flank of the anticline beneath Denby (398 462). On the Bouguer anomaly map this feature is shown only as a weak local anomaly of -1 mGal in Fig. 87.

SEISMIC SURVEYS

A seismic reflection and refraction survey was carried out by the Applied Geophysics Unit of IGS to trace the continuation of the Dinantian limestones beneath the Namurian shales in the Ecclesbourne Valley area south of Wirksworth. The contact of the shale with the limestone was expected to be a good reflecting horizon, but it was found that several, poorly defined reflectors existed, making correlation difficult in an area without any borehole control. The survey confirmed the existence of the small anticlinal structure trending SSE from near Wirksworth (Plate 12), but it appeared that the mean velocity of the Namurian shales could not be greater than 2.5 km s^{-1} if depths consistent with the gravity and geological evidence were to be produced.

From refraction data obtained in the same survey, drift thicknesses were found to be mostly less than 12 m in the area and the bedrock velocities varied between 1.5 and 3.5 km s^{-1} with an average of 2.46 km s^{-1} for the Namurian shales. Sandstone (Ashover Grit) horizons in the sequence gave a slightly higher mean velocity of 2.71 km s^{-1}. It has been mentioned in the section on physical properties that the seismic observations located on the limestone outcrop in this area revealed velocities much lower than those determined from samples (6.1 km s^{-1}). This decrease is most likely due to the increased permeability and enlarged jointing resulting from weathering at the present land surface. Weathering could also be responsible for the abnormal Namurian shale velocities noted above, which are lower than the interval velocities reported by Wyrobeck (1959).

APPENDIX 1

RECORDS OF BOREHOLES, SHAFTS AND DRIFTS

The most important boreholes, shafts and drifts are summarized in alphabetical order below and their locations are plotted on Plate 13. All the records have been abridged to some degree, but a few (those underlined on Plate 13) are given in some detail. Most of the coalfield records show only the depths to the bases of the major coals together with simplified seam sections and skeletal stratigraphical data. Only a few of the numerous opencast coal boreholes are included; there are, for example, over 6000 such bores on sheet SK 44 SW alone.

In each record the name is followed by the National Grid reference, the date of construction, the OD elevation in feet of the top, or in the case of oil bores of the Rotary Table, the names of the recording geologists and lastly cross-references to other parts of this memoir or to other publications. All depths are in feet and inches and, unless otherwise stated, to the base of the formation or bed. Coal seam details are given in inches in downward sequence using the following abbreviations: b = batt (carbonaceous mudstone with or without interlaminated coal); c = coal; ca = cannel; d = dirt; dc = dirty coal; m = measures. Other abbreviations are: abt = about; BH = Borehole; LL = Lower Leaf; MB = Marine Band; UBH = Underground Borehole; TD = Total Depth where definitely recorded; TS = Tuffaceous Siltstone; UL = Upper Leaf.

ALLOTMENT GARDENS BH, ANNESLEY WOODHOUSE [4993 5325] 1968, 557 ft.
Cores from 10 ft 2, examined by R. Draper and C. Beal. See Fig. 82 for graphic section of Permo-Triassic rocks. Boulder clay to 30 ft 3; Lower Mottled Sandstone to 40 ft 3; Middle Marl to 72 ft 7; Lower Magnesian Limestone to 103 ft 2; Lower Marl to 159 ft 9; Permian Basal Breccia to 161 ft 7; Main 'Estheria' Band (UL) at 194 ft 3; Main 'Estheria' Band (LL) at 214 ft 10; Coal, c 3, d 17 on c 12 at 218 ft 9; Edmondia Band at 275 ft 4; High Main (UL), d c 1 on c 10 at 300 ft; High Main (LL), d c 5 on c 34 at 308 ft 6; TD 338 ft 10.

ALTON'S BREWERY WATER BH, DERBY [abt 3508 3619] 1877, abt 160 ft. See Stephens 1929, pp. 89 and 134.
Alluvium and gravel to 9 ft; Keuper Marl to TD at 57 ft 11 in.

AMBER DYE WORKS BH No. 1, BULLBRIDGE [3563 5206] 1936, 250 ft.
Coring commenced at 58 ft. Recorded by Professor H. H. Swinnerton. See Swinnerton 1946 and Plate 4 for graphic section.

	Thickness ft	in	Depth ft	in
ALLUVIUM				
Top Soil	1	0	1	0
Clay and boulders	7	0	8	0
Boulders and gravel	10	0	18	0
NAMURIAN				
CHATSWORTH GRIT				
Sandstone	37	6	55	6
Ironstone, clay and boulders	2	6	58	0
Sandstone	1	6	59	6
Shale	0	2	59	8
Sandstone	0	7	60	3
Shale	5	0	65	3
Sandstone with shale bands	4	10	70	1
Shale	40	11	111	0
Reticuloceras superbilingue MARINE BAND				
Shale with Lingula sp., Catastroboceras sp., Gastrioceras spp. nov., Reticuloceras superbilingue	7	0	118	0
Shale	29	0	147	0
Shale with non-marine bivalves	3	0	150	0
Shale, sandy	20	0	170	0
Shale	30	6	200	6
COAL	1	6	202	0
ASHOVER GRIT (Main Bed)				
Sandstone coarse-grained to TD	65	0	267	0

AMBER DYE WORKS BH No. 2, 65 yd NW of No. 1 [3559 5210] 1950.
Proved a similar sequence to No. 1. The R. superbilingue MB at 156 ft contained Gastrioceras sp. nov., Homoceratoides aff. divaricatus and R. superbilingue. The Ashover Grit (Main Bed) was proved to be 96 ft thick and underlain by alternating shale and sandstone bands for a further 34 ft.

ANNESLEY, BRITISH RAILWAYS WATER BH [5259 5313] 1948 abt 390 ft.
Middle Marl to 5 ft; Lower Magnesian Limestone to 31 ft; Lower Marl to TD at 86 ft.

ANNESLEY COLLIERY No. 1 SHAFT [5179 5329] 1867.
Deepened from Top Hard 1914. See also Edwards 1951 pp. 119-121, shown graphically on Fig. 49. Base of Permo-Trias at 200 ft; ?Manor, b 13 at 229 ft 11; Coal 10 at 323 ft 5; Annesley, c 13 at 392 ft 2; inferred Top MB at 457 ft 2; Coal 5 at 469 ft 7; inferred Shafton MB, at 532 ft 1; Coal 9 at 532 ft 10; Shafton, c 6 at 543 ft 1; inferred Main 'Estheria' Band at 584 ft 10; inferred Edmondia Band, at 653 ft 1; High Main, c 48 at 680 ft 6; Coal, c 12, d 3 on c 14 at 775 ft 1; Mansfield MB cank at 820 ft 2; inferred Haughton MB at ?887 ft; ?Clown, c 12 at 963 ft 5; ?Mainbright (partly washout), c 6 at 1067 ft; inferred Two-Foot MB at 1082 ft 6; Two-Foot, c 15 at 1083 ft 9; Lowbright, c 31 at 1122 ft 3; Lowbright Floor, c 6, b 3 on c 7 at 1125 ft 4;

a

b

High Hazles, c 26, d 19 on c 8 at 1220 ft 8; horizon of Cinderhill at 1251 ft; Mainsmut (UL), c 9 at 1356 ft; Mainsmut (LL), c 2 at 1365 ft 9; Comb, c 25 at 1413 ft 3; Top Hard, c 48 at 1418 ft 1; Coal 21 at 1486 ft 2; Coal, batt 1 on c 12 at 1495 ft 1; First Waterloo, c & d 11, d 7 on c 16 at 1524 ft 9; Waterloo Marker, ca 6, b 18 on ca 6 at 1551 ft 10; ? passed through fault abt 1590 ft; Fourth Waterloo, c 10 at 1598 ft 11; First Ell, b 12 on c 10 at 1618 ft 8; Second Ell, c 14 at 1657 ft 6; inferred Clay Cross MB at 1700 ft; Brown Rake, c 3 on b 4 at 1735 ft 4; Deep Soft, c 36 at 1858 ft; Deep Hard, c 38 at 1911 ft 3; TD 1913 ft 2.

ANNESLEY COLLIERY No. 3 (B 3's) UBH [4957 5117] 1958.
Down from Deep Soft at -985 ft. Recorded by S. Brunskill. Deep Hard, c 32 at 35 ft 5; First Piper, ca 2 on c 11 at 75 ft; Second Piper horizon at 159 ft 3; Hospital horizon at 172 ft 2; Cockleshell, c 3 at 209 ft 1; Low Main, ca 6 on c 23 at 231 ft 5; Threequarters, c 14, d 1 on d c 13 at 247 ft 2; Yard and Blackshale, c 24, d 2, c 23, d 6, c 19, d 3, c 9, d 2 on c 16 at 345 ft; TD 348 ft 10.

ANNESLEY COLLIERY B4 UBH [5066 5242] 1959.
Down from Deep Soft at -1151 ft to Yard and Blackshale coals. Recorded by S. Brunskill. TD 345 ft 10. See Fig. 23 for section of Threequarters Coal and Fig. 28 for section of Cockleshell and Low Main coals.

ANNESLEY COLLIERY 7's UBH [5283 5384] 1972.
Down from Deep Soft at -1584 ft. Recorded by C. Beal. Deep Hard, c 25 at 47 ft 9; First Piper, c & d 25 at 85 ft 7; Cockleshell, d c 8 at 214 ft 7; Low Main, c 37 at 218 ft 3; Threequarters, c & d 18 at 246 ft 4; Yard and Blackshale, c & d 8 ft 8 at 324 ft 1; TD 340 ft 7.

ANNESLEY COLLIERY 10's UBH [5109 5071] 1955.
Down from Deep Soft at -1207 ft. See Fig. 28 for details of Cockleshell and Low Main coals. TD 326 ft 10.

ANNESLEY COLLIERY 204's UBH [5277 5290]. 1966.
Down from Deep Soft at -1470 ft. Recorded by S. Brunskill. See Figs. 28 & 31 for sections of Cockleshell and Low Main coals, and Deep Hard Coal respectively. TD 323 ft to Yard/Blackshale.

ANNESLEY COLLIERY 214's UBH [5375 5319] 196[illegible].
Down from Deep Soft at -1585 ft. Mainly open hole; coals and adjacent strata cored. Recorded by S. Brunskill. See Fig. 28 for Tupton Coal section. Deep Hard, ca 6 on c 32 at 33 ft 11; First Piper, c & d 26 at 97 ft 10; Tupton, c 8, d 7 on c 37 at 215 ft 10; Threequarters c 11, m 5 ft 11 on c 6 at 235 ft 1; Yard, c 8 at 312 ft 11; Blackshale, c & d 89 at 340 ft 5; TD 342 ft.

ANNESLEY COLLIERY (DEEP HARD) UBH [5167 5289] 1948.
Down from Deep Hard at -1366 ft. See Figs. 18 and 20 for graphic section Kilburn to Threequarters and Fig. 28 for Low Main Coal. First Piper, ca 1, d 6 on c 21 at 46 ft 9; Second Piper horizon at 97 ft 5; Hospital horizon at 121 ft 6; Cockleshell, ca 1, d 4 on ca 2 at 165 ft 5; Low Main, c 32 at 172 ft 6; Threequarters, c 7, m 26 on d c 3 at 190 ft 9; Blackshale, c 12, d 2, c 12, d 4 on c 21 at 294 ft 4; 'Ashgate' (UL), c 2 at 362 ft; 'Ashgate' (LL), c 2 at 389 ft 7; Mickley Thin horizon at 440 ft 4; Lower Brampton horizon at 483 ft 7; Morley Muck, c 4 at 563 ft 8; Kilburn, c 7 at 677 ft 1; Coal 3 at 736 ft 8; TD 818 ft.

ANNESLEY COLLIERY WATER BH [5173 5325] 1934, 464 ft.
Lower Mottled Sandstone to TD at 51 ft.

ANNESLEY HALL BH [5018 5267] 1968, 499 ft.
Recorded by T. Draper & D. V. Frost. See Fig. 82 for graphic section of Permo-Triassic rocks and Fig. 48 for High Main Coal section. Permian Basal Breccia at 133 ft 10; Coal 13 at 195 ft 6; Shafton, c 4 at 202 ft 9; ?Main 'Estheria' Band (UL) at 234 ft 3; Edmondia Band at 315 ft 1; High Main (UL), c 10 at 330 ft 4; High Main (LL), c 30 at 350 ft; TD 368 ft 9.

ANNESLEY, HOOPER WATER BH [Approx. 5174 5268] 1949, abt 450 ft.
Lower Mottled Sandstone to 31 ft; Middle Marl to 52 ft; Lower Magnesian Limestone to about 84 ft; inferred Lower Marl to TD at 100 ft.

ANNESLEY LODGE BH [4954 4991] 1965, 492 ft.
See Figs. 42 and 77 for High Hazles and Brinsley Thin coal sections respectively. Cores begin at 91 ft. Base of boulder clay from gamma log at abt 10 ft; base of Permian from gamma log at abt 60 ft; Main 'Estheria' Band (?LL) at 127 ft; Edmondia Band at 199 ft; High Main, c 16, d 5 on c 33 at 231 ft 9; Mansfield MB at 372 ft 1; ?Sutton MB at 412 ft 4; Coal 10 at 413 ft 2; inferred Haughton MB at 437 ft 1; Clowne washout at abt 470 ft; inferred Manton 'Estheria' Band at 543 ft 8; Mainbright, c 30 at 592 ft 6; Two-Foot MB at 605 ft; Two-Foot, c 18 at 606 ft 6; Lowbright, c 29 at 647 ft 6; Lowbright Floor, c 4, m 18 on c 1 at 658 ft 1; Brinsley Thin, c 4 at 698 ft 9; High Hazles, c 14 at 721 ft 2; TD 753 ft.

ANNESLEY PARK BOREHOLE [5172 5111] 1963, 407 ft.
Cores, from 18 ft, examined by D. V. Frost. See also Figs. 43, 44, 46 and 49.

	Thickness ft	in	Depth ft	in
PERMIAN				
MIDDLE MARL				
Soil and clay	3	0	3	0
Sandstone, buff, fine-grained	1	0	4	0
Clay, red, calcareous	8	0	12	0
LOWER MAGNESIAN LIMESTONE				
Limestone, red-brown	6	0	18	0
Limestone, red-brown dolomitic, sandy	18	0	36	0
Mudstone, silty, calcareous; plants	1	0	37	0
Siltstone, red-brown at top becoming grey, calcareous, nodular; plants	11	0	48	0
LOWER MARL				
Mudstone, grey, silty, calcareous bands; bands of siltstone with dolomitic cement below 65 ft; galena; plants; foraminifera, brachiopods, gastropods, bivalves at 103	62	5	110	5
PERMIAN BASAL BRECCIA	0	1	110	6
WESTPHALIAN C				
Mudstone, silty, stained; bivalves, ostracods	21	0	131	6
COAL, d c 1	0	1	131	7
Seatearth	4	3	135	10
Mudstone, carbonaceous	0	1	135	11
Seatearth	2	1	138	0
Mudstone	1	3	139	3
Siltstone and sandstone	12	3	151	6
Mudstone, silty with plants	9	1	160	7
Mudstone, carbonaceous at base; ironstone; Anthraconauta phillipsii, Carbonita sp., fish debris	4	1	164	8
COAL, ca 2	0	2	164	10
Seatearth	3	11	168	9
Mudstone	1	10	170	7
Mudstone, carbonaceous	0	7	171	2
Mudstone	2	1	173	3
COAL, d c 13	1	1	174	4
Seatearth	2	8	177	0
Siltstone	6	6	183	6
Mudstone; ironstone; fish debris	3	5	186	11
COAL, c 3	0	3	187	2
Seatearth	3	5	190	7
Mudstone, sporadic siltstone and sandstone bands; Anthraconauta phillipsii, Cochlichnus kochi	52	6	243	1
ANNESLEY c 16	1	4	244	5
Seatearth	4	1	248	6
Mudstone with siltstone; Anthraconaia cf. spathulata	9	0	257	6
Sandstone	20	6	278	0
Siltstone	2	0	280	0
Mudstone; plants	14	0	294	0
TOP MARINE BAND				
Mudstone; Orbiculoidea sp., Aviculopecten?, Dunbarella sp., Polidevcia sp., Planolites ophthalmoides, Tomalculum sp., 'fucoids', fish debris	6	9	300	9
Mudstone; Lioestheria vinti	0	7	301	4
COAL, c 1	0	1	301	5
Seatearth	4	7	306	0
Mudstone; sphaerosiderite; plants, fish at base	10	7	316	7
Mudstone, with coal streaks	0	5	317	0
Mudstone; plants	3	6	320	6
Seatearth	1	1	321	7
COAL, c 5	0	5	322	0
Seatearth	1	7	323	7
COAL, c 18	1	6	325	1
Seatearth	0	11	326	0
Mudstone	0	6	326	6
Siltstone and sandstone; plants and macrospores	9	9	336	3
Mudstone; Naiadites hindi, fish debris	6	3	342	6
Seatearth	1	6	344	0
Sandstone	0	9	344	9
Seatearth	0	3	345	0
COAL, c 6	0	6	345	6
Seatearth, sandstone	0	6	346	0
Siltstone	4	2	350	2
Mudstone; ironstone; plants;	3	4	353	6
Mudstone; Naiadites daviesi, Geisina subarcuata, Lioestheria vinti, fish tooth	5	5	358	11
SHAFTON MARINE BAND				
Mudstone; ?Anthraconaia spathulata, Paraparchites sp.	8	1	367	0
Mudstone with siltstone band	4	1	371	1
COAL, c 3, d 5 on c 20	2	4	373	5
Seatearth	2	7	376	0
Mudstone; sphaerosiderite; plants, macrospores and fish debris at base	12	7	388	7
SHAFTON, ca 24	2	0	390	7
Mudstone; ironstone	9	5	400	0
?MAIN 'Estheria' BAND				
Mudstone; Naiadites cf. hindi, Lioestheria vinti	5	2	405	2
COAL, c 7	0	7	405	9
Seatearth	2	3	408	0
Mudstone; ironstone; plants, Naiadites sp., Geisina subarcuata, Lioestheria vinti	7	10	415	10
Mudstone, canneloid; Lioestheria vinti	0	4	416	2
COAL, c 8, d 3 on c 4	1	3	417	5
Seatearth	3	7	421	0
Mudstone; ironstone; plants	3	7	424	7
Siltstone with sandstone and mudstone	13	2	437	9
Mudstone; Naiadites cf. hindi,	29	1	466	10
Edmondia BAND				
Mudstone; Ammodiscus sp., Glomospirella sp., Lituotuba sp., Myalina cf. compressa, pyritised plant fragments	4	7	471	5
Mudstone; Spirorbis sp., Naiadites melvillei, N. sp. cf. triangularis	8	7	480	0
Seatearth	1	6	481	6
Sandstone	8	6	490	0
Mudstone; bivalves, ostracods	5	2	495	2
Seatearth	0	8	495	10
Mudstone, ironstone	12	4	508	2
HIGH MAIN (UPPER LEAF), c 12	1	0	509	2
Mudstone	1	6	510	8
HIGH MAIN, c 30	2	6	513	2
Seatearth	2	10	516	0
Mudstone	1	4	517	4
Sandstone	2	7	519	11
Mudstone; ironstone; plants	31	0	550	11
COAL, c 12	1	0	551	11
Seatearth, sandstone	1	1	553	6
Mudstone with siltstone and sandstone bands	8	1	561	7
Mudstone, carbonaceous; plants, Naiadites melvillei, fish including Rhabdoderma sp.	0	10	562	5
COAL, c 1	0	1	562	6
Seatearth	4	4	566	10
Siltstone, plants	35	2	602	0
Mudstone, carbonaceous	4	4	606	4
Seatearth	0	3	606	7
COAL, c 2	0	2	606	9
Seatearth, siltstone	4	3	611	0
Mudstone, carbonaceous; Naiadites sp., cf. Carbonita pungens	5	0	616	0
Sandstone	15	0	631	0
Mudstone; Naiadites cf. productus, worm tracks	24	11	655	11
MANSFIELD MARINE BAND				
Mudstone; Ammodiscus sp., Lingula mytilloides, cf. Geisina sp.	1	8	657	7
Siltstone, calcareous; Lingula mytilloides, Orbiculoidea cf. nitida, fish debris including a spine of Listracanthus wardi	2	2	659	9
Mudstone; Lingula mytilloides, gastropods, Anthracoceras cf. hindi, Hindeodella	1	6	661	3
WESTPHALIAN B				
Mudstone; plants	0	3	661	6
Seatearth, brown	5	9	667	3
Mudstone; sphaerosiderite; Naiadites aff. angustus, fish at base (? SUTTON MARINE BAND HORIZON)	26	2	693	5
COAL, c 6	0	6	693	11
Mudstone, carbonaceous	0	1	694	0
Seatearth	5	3	699	3
COAL, c 12	1	0	700	3
Seatearth	1	8	701	11
COAL, c 2	0	2	702	1
Seatearth	6	11	709	0
Mudstone; gastropods, bivalves, fish	9	7	718	7
HAUGHTON MARINE BAND				
Mudstone; oolitic ironstone; Lingula cf. elongata, Euphemites ?	1	0	719	7
Sandstone and siltstone	5	0	724	7
?SWINTON POTTERY c 2	0	2	724	9
Mudstone, carbonaceous	0	2	724	11
Seatearth, brown	4	4	729	3
COAL, c & d 13	1	1	730	4
Sandstone and siltstone	1	5	731	9
Mudstone	2	6	734	3
COAL, c & d 5	0	5	734	8
Seatearth	1	9	736	5
COAL, c 12	1	0	737	5
Seatearth, brown	2	3	739	8
Sandstone; oolitic ironstones	45	4	785	0
Mudstone; Anthracosia atra	2	10	787	10
? CLOWN MARINE BAND				
Mudstone, carbonaceous; fish debris; ironstone	2	0	789	10
CLOWN, c 17, d 17 on c 7	3	5	793	3
Seatearth; sphaerosiderite	4	4	797	7
Mudstone; Anthracosia sp., plants	7	9	805	4
? MANTON 'Estheria' BAND				
Mudstone and siltstone, carbonaceous; ?'Estheria' sp., fish debris	1	9	807	1
COAL, c 18	1	6	808	7
Seatearth	1	11	810	6
Siltstone	26	4	836	10
Mudstone with ironstone; Lepidodendron sp., Naiadites sp., Lioestheria vinti, fish debris	1	8	838	6
Mudstone; plants, ironstone	3	8	842	2
Sandstone and siltstone	3	5	845	7
Mudstone; Naiadites sp., Lioestheria vinti	9	3	854	10
MAINBRIGHT, c 29	2	5	857	3
Seatearth	2	9	860	0
Mudstone; plants, Anthraconaia cymbula, fish debris	5	6	865	6
TWO-FOOT MARINE BAND				
Mudstone; Lingula sp., bivalves	0	5	865	11
Mudstone; Anthracosia atra, A. simulans, Naiadites cf. obliquus, Lioestheria vinti	2	3	868	2
TWO-FOOT, c 21	1	9	869	11
Seatearth	1	9	871	8
Siltstone	2	0	873	8
Mudstone	9	6	883	2
Sandstone and siltstone	27	7	910	9
LOWBRIGHT, c 32	2	8	913	5
Seatearth	3	1	916	6
Sandstone	3	0	919	6
Mudstone; plants	3	10	923	4
LOWBRIGHT FLOOR, c 5, d 1, c 4, d 8 on c 1	1	7	924	11
Seatearth	2	7	927	6
Sandstone	6	6	934	0
Mudstone, carbonaceous near base; ironstone; bivalves	26	3	960	3
BRINSLEY THIN (UPPER LEAF), c 4	0	4	960	7

c

d

	Thickness ft	in	Depth ft	in
Seatearth	2	9	963	4
Sandstone and siltstone	5	5	968	9
Mudstone; ironstone; plants, Naiadites cf. obliquus	4	2	972	11
BRINSLEY THIN, c 7, d 3 on c 3	1	1	974	0
Seatearth	0	6	974	6
Siltstone and sandstone	4	11	979	5
Mudstone, siltstone at top; ironstone; bivalves at base	22	8	1002	1
HIGH HAZLES, c 30	2	6	1004	7
Seatearth	5	6	1010	1
Mudstone and siltstone	7	5	1017	6
Seatearth	2	10	1020	4
Sandstone with siltstone	19	2	1039	6
Mudstone; ironstone; Anthracosia cf. caledonica, Anthracosia sp. cf. fulva, ostracods	13	0	1052	6
CINDERHILL (UPPER LEAF), c 6	0	6	1053	0
Seatearth	1	5	1054	5
Mudstone; Anthracosia sp. cf. fulva, fish debris at base	14	11	1069	4
CINDERHILL, c 5	0	5	1069	9
Seatearth	1	3	1071	0
Sandstone with siltstone	18	10	1089	10
Mudstone, rare siltstones; shelly ironstones; Anthracosia cf. planitumida, A. sp. aff. phrygiana, A. pulchella, A. simulans	24	1	1113	11
? SECOND ST. JOHNS, c 3	0	3	1114	2
Seatearth	2	3	1116	5
Siltstone and sandstone	18	7	1135	0
Mudstone	10	8	1145	8
MAINSMUT (UPPER LEAF), c 3	0	3	1145	11
Seatearth	4	7	1150	6
Mudstone; ironstone; plants, bivalves	12	8	1163	2
MAINSMUT (LOWER LEAF), c 1	0	1	1163	3
Seatearth	0	3	1163	6
Sandstone and siltstone, mudstone bands	38	0	1201	6
Mudstone	6	6	1208	0
Open hole (TOP HARD workings at abt 1217 ft)	12	0	1220	0
Seatearth	4	1	1224	1
Siltstone	8	1	1232	2
Mudstone	13	7	1245	9
TOP HARD FLOOR, ca 3 on c 1	0	4	1246	1
Mudstone, carbonaceous	0	5	1246	6
Seatearth	1	6	1248	0
Mudstone; ironstone; plants, Anthracosia disjuncta, A. aff. phrygiana, Carbonita humulis, C. cf. scalpellus	7	0	1255	0
Sandstone with siltstone	9	0	1264	0
Mudstone; Anthracosia phrygiana, Anthracosphaerium sp.	14	9	1278	9
Mudstone, carbonaceous	0	8	1279	5
DUNSIL, c 29, d 2 on c 2	2	9	1282	2
Seatearth	2	10	1285	0
Mudstone; plants	8	3	1293	3
Sandstone	1	8	1294	11
Mudstone; ironstone; Anthracosia disjuncta, Naiadites sp.	11	7	1306	6
Mudstone, carbonaceous	0	3	1306	9
FIRST WATERLOO, c 16, d 2 on c 19	3	1	1309	10
Seatearth	1	5	1311	3
Mudstone	0	8	1311	11
Sandstone and siltstone	4	1	1316	0
Mudstone; Naiadites sp., fish debris including teeth of Cochliodont sp.	12	8	1328	8
WATERLOO MARKER, c 4, d 11 on c 1	1	4	1330	0
Seatearth	1	9	1331	9
Sandstone and siltstone	12	3	1344	0
Mudstone; bivalves	9	10	1353	10
TOP SECOND WATERLOO, c 23, d 6 on c 11	3	4	1357	2
Seatearth	2	4	1359	6
Mudstone	4	9	1364	3
BOTTOM SECOND WATERLOO, c 1, d 4 on c 1	0	6	1364	9
Seatearth	0	6	1365	3
Mudstone	4	3	1369	6
Siltstone and sandstone	1	6	1371	0
Mudstone; Planolites montanus	12	5	1383	5
Mudstone, carbonaceous	0	1	1383	6
THIRD WATERLOO, c 16	1	4	1384	10
Seatearth	4	2	1389	0
Mudstone; ironstone; plants, Anthracosia disjuncta, to TD	6	6	1395	6

ANNESLEY SIDINGS BH [5158 5435] 1946, abt 450 ft.
Lower Mottled Sandstone to Lower Marl. TD 84 ft.

ANNESLEY WELL (SELSTON PUMPING STATION) [5140 5443] 1895, 484 ft.
See also Lamplugh and Smith 1914, pp. 47 and 151. Lower Mottled Sandstone to 38 ft; Middle Marl to 58 ft; Lower Magnesian Limestone to 78 ft; Lower Marl to TD at 87 ft.

AUDREY WOOD BH [4957 5164] 1972, 540 ft.
Cores, from 10 ft, examined by C. Beal. Boulder clay to 11 ft 2; Middle Marl to 22 ft; Lower Magnesian Limestone to abt 47 ft; Lower Marl to 102 ft 10; Permian Basal Breccia to 105 ft 5; Shafton, c 9 at 139 ft 2; Main 'Estheria' Band (UL) at 180 ft 6; Main 'Estheria' Band (LL) at 189 ft 4; Edmondia Band to 244 ft 11; High Main (UL), c 15 at 281 ft 10; High Main (LL), c 36 at 285 ft 10; TD 286 ft 9 (NB. High Main Tonstein recorded at 286 ft 9).

AWSWORTH COLLIERY SHAFT [Approx. 479 444], abt 215 ft.
Deep Soft, c 58 at 16 ft; Deep Hard workings at 57 ft 9; Deep Hard Floor, c 6 at 68 ft 5; First Piper, c 44 at 108 ft 6; Hospital, c 11 at 152 ft 11; Cockleshell, c 20 at 179 ft 8; Low Main, c 38 at 206 ft 11; Threequarters, c 17 at 215 ft 10; Blackshale, c 15, d 9, d c 9, d 6 on c 15 at

a

b

381 ft 6; Coal 1 at 413 ft 7; 'Ashgate', c 81 at 486 ft 4; recorded to 509 ft 4.
A second shaft record with, for instance, Low Main at 225 ft and Blackshale at 402 ft 10 is incomplete below 402 ft 10 but with Kilburn, c 30 at 802 ft 1 and a TD of 856 ft 5. There was a boring below to 983 ft 11.

BABBINGTON (CINDERHILL) PIT YARD BH [5321 4384] 1957, 200 ft.
Base of Permian at 80 ft; Wales, c 21 at 96 ft 10; Mansfield MB at 162 ft 1; TD 162 ft 1.

BAILEY BROOK COLLIERY WEST SHAFT AND HEADING [4352 4743], 233 ft.
See Figs 13, 14, 16, 68 and 20. Fourth Waterloo, c 18 at 39 ft 6; Coal and d 12 at 55 ft; First Ell, c 36 at 92 ft; Second Ell, b 9, 69, ca 28, m 14 ft, b 12 on c 24 at 152 ft 5; base of Westphalian B estimated at 243 ft; Brown Rake, c 19 at 256 ft 6; False Ell, c 9, d 8 on d c 36 at 291 ft 5; Deep Soft, c 46 at 344 ft 8; Deep Hard, ca 5, d c 5, c 34, d c 14, m 24 on c 1 at 408 ft 7; First Piper, c 34, d c 7 on ca 12 at 447 ft 3; Hospital, c 17 at 502 ft 2; Cockleshell, d c 5 at 526 ft 2; Low Main, c 46 at 577 ft 1; Threequarters, at 619 ft 11; Yard, c 5, d 5 on c & d c 23 at 670 ft 10; Blackshale, d c 10, b 20, c 27, d 5, c 28, d 6, c 13, d 4 on c 17 at 706 ft; Coal, c & b 10 at 745 ft 8; Coal 13 at 779 ft 6; 'Ashgate' (UL), c 21 at 802 ft 4; 'Ashgate' (LL), c 26 at 807 ft; Mickley Thin, c 11 at 859 ft 4; Lower Brampton, c 4, m 6 ft 4 on b 1 at 916 ft 6; Morley Muck, c & d c 14, m 26 on b 3 at 992 ft 3; Coal, c & b 3 at 1074 ft 2; Kilburn, d c 3 on c 33 at 1101 ft 6. Sequence continued in 'Stone Heading'. Upper Band horizon at 1479 ft 2; Norton, c 18 at 1507 ft 11; Forty-Yards (UL), c 5 at 1533 ft; Forty-Yards (LL), c 11 at 1543 ft 11; Alton MB at 1611 ft 2; Alton, c 10 at 1612 ft 1; First Smalley, c trace, d 2 on c trace at 1626 ft 4; Second Smalley, c 2 at 1636 ft 5; Belperlawn, d c 4 on c 13 at 1671 ft 2; driven to 1678 ft 4.

BALAKLAVA WOOD BH [5043 5117] 1973, 562 ft.
Recorded by C. Beal. Boulder clay with sand and gravel horizons to 113 ft (? Channel fill); Lower Magnesian Limestone to 124 ft; Lower Marl to 183 ft; Permian Basal Breccia to 185 ft 6; Top MB at 276 ft 4; Coal 19 at 337 ft 3; Shafton, c 19 at 352 ft 1; Edmondia Band at 443 ft 5; High Main (UL), c 13 at 475 ft 10; High Main (LL), c 40 at 486 ft 5; TD 491 ft 11.

BARLOW & SON, WATER BH, BEESTON [5274 3694] 1938, abt 110 ft.
First Terrace gravel to 14 ft; Triassic sandstone to TD at 84 ft.

BEECHDALE ROAD (ROBINS WOOD) BH, NOTTINGHAM [5361 4113] 1948, 157 ft.
Cores, from 89 ft, examined by R. E. Elliott and R.A. Eden. See Figs.14, 16, 20, 25, 37, 68, 75, Plate 5 and Eden 1954.

	Thickness ft	in	Depth ft	in
PERMIAN				
Lower Magnesian Limestone	10	0	10	0
WESTPHALIAN B				
Red measures	5	0	15	0
Grey measures	33	0	48	0
HIGH HAZLES?, c traces 12, d 48 on c 5	5	5	53	5
Mudstone	35	7	89	0
Seatearth	1	6	90	6
Mudstone; ironstone nodules and bands; bivalves	12	10	103	4
CINDERHILL, c 30	2	6	105	10
Seatearth	3	8	109	6
Siltstone and sandstone; plants	35	6	145	0
Mudstone; ironstone bands	6	3	151	3
Mudstone, shaly, black, pyritic; plants, bivalves	1	9	153	0
Siltstone; ironstone and coaly lenses	5	0	158	0
Sandstone; plants	0	6	158	6
Mudstone, black near base; ironstone lenses; plants, bivalves	12	2	170	8
Siltstone, black; ironstone bands; bivalves, fish debris	2	4	173	0
Mudstone; plants	0	10	173	10
Sandstone; plants	1	6	175	4
Siltstone; plants	13	8	189	0
Mudstone and siltstone; plants	10	0	199	0
Mudstone; plants, bivalves at 202 ft 5	3	8	202	8
Siltstone, black pyritic; plants, bivalves	0	6	203	2
MAINSMUT (UPPER LEAF), c 10	0	10	204	0
Seatearth, silty carbonaceous	11	1	215	1
Siltstone, black; plants	0	8	215	9
MAINSMUT (LOWER LEAF), c 3	0	3	216	0
Seatearth, silty	5	6	221	6
Core lost	5	0	226	6
Siltstone; plants	8	6	235	0
Mudstone, silty; plants	12	6	247	6
? TOP HARD (cores fragmentary), c 8	0	8	248	2
Mudstone; ironstone; plants	1	1	249	3
COAL, c 4	0	4	249	7
Mudstone; plants	0	9	250	4
Sandstone; coal streaks	4	0	254	4
Seatearth	3	8	258	0
Mudstone, silty	19	0	277	0
Mudstone; ironstone; plants, bivalves	7	6	284	6
Mudstone, black, shaly, pyritic; bivalves, fish debris	0	1	284	7
Seatearth	3	5	288	0
Siltstone and sandstone	5	11	293	11
Mudstone	2	1	296	0
Sandstone	11	0	307	0

Beechdale Road (Robins Wood) BH cont.

	Thickness ft	in	Depth ft	in
Mudstone	13	0	320	0
Mudstone, black, shaly; ironstone; bivalves	0	11	320	11
DUNSIL, c 29	2	5	323	4
Mudstone, carbonaceous	0	11	324	3
Seatearth, silty in lower part	4	9	329	0
Mudstone; plants	7	6	336	6
Siltstone; plants	5	9	342	3
Mudstone; galena in ironstone; plants, bivalves	11	6	353	9
FIRST WATERLOO, c 29	2	5	356	2
Seatearth, silty in lower part	6	4	362	6
Siltstone and sandstone; plants	13	3	375	9
Mudstone; Spirorbis sp., bivalves, fish debris	3	2	378	11
TOP SECOND WATERLOO, c 18	1	6	380	5
Seatearth, mudstone	8	7	389	0
Sandstone; plants	0	4	389	4
Mudstone, shaly at base; bivalves, fish debris	7	2	396	6
BOTTOM SECOND WATERLOO, c 4	0	4	396	10
Seatearth, mudstone	8	2	405	0
Siltstone; plants	10	6	415	6
Mudstone; ironstone bands; plants, bivalves	15	10	431	4
Mudstone, shaly, carbonaceous; bivalves, pyritic	0	2	431	6
THIRD WATERLOO, c 7	0	7	432	1
Seatearth	3	11	436	0
Siltstone with sandy laminae; plants	21	0	457	0
Mudstone, laminated; ironstone; plants, bivalves, ostracods	3	3	460	3
Sandstone and siltstone; plants	8	6	468	9
Mudstone; plants, bivalves	11	8	480	5
FOURTH WATERLOO, c 8, d 1 on d c 22	2	7	483	0
Seatearth, silty	2	6	485	6
Siltstone, with sandstone; plants	11	0	496	6
Mudstone; bivalves	5	0	501	6
Siltstone; plants	3	1	504	7
FIRST ELL, ca 7 on c 1	0	8	505	3
Seatearth, silty	1	3	506	6
Sandstone and siltstone; ironstone	17	0	523	6
Mudstone; ironstone; plants, bivalves, ostracods	23	9	547	3
SECOND ELL, c 4, d 9 on c 6	1	8	548	11
Seatearth, carbonaceous	5	1	554	0
Siltstone; plants	11	0	565	0
Mudstone; ironstone bands; plants, pyritic bivalves	44	1	609	1
CLAY CROSS MARINE BAND				
Mudstone, pyritous; foraminifera, Lingula sp., bivalves, goniatites, ostracods, fish	8	7	617	8
WESTPHALIAN A				
Mudstone, dark; plants, bivalves, ostracods	7	11	625	7
BROWN RAKE, IRONSTONE; shelly	0	4	625	11
Mudstone; ironstone bands; plants, bivalves, ostracods	4	10	630	9
BROWN RAKE, c 2 d 3 on c 1	0	6	631	3
Mudstone, carbonaceous	0	2	631	5
Seatearth	7	1	638	6
Mudstone; plants, bivalves	6	5	644	11
BLACK RAKE T.S.	0	7	645	6
Mudstone; Spirorbis sp., bivalves, ostracods, fish debris	21	7	667	1
ROOF SOFT (UPPER LEAF), c 2	0	2	667	3
Seatearth	2	6	669	9
Sandstone; plants	14	9	684	6
Mudstone, silty	4	1	668	7
ROOF SOFT HORIZON	-	-	668	7
Seatearth	7	5	696	0
Siltstone and sandstone; plants	33	3	729	3
Mudstone; plants	2	9	732	0
Mudstone, carbonaceous	3	0	735	0
DEEP SOFT (old workings)	4	0	739	0
Seatearth	0	9	739	9
Ironstone with rootlets; galena; kaolin	1	3	741	0
Core lost	8	6	749	6
Mudstone; bivalves	18	0	767	6
DEEP HARD, c 45	3	9	771	3
Seatearth	6	0	777	3
Siltstone and sandstones	16	7	793	10
Mudstone; bivalves, ostracods, fish debris	13	7	807	5
FIRST PIPER, c 22	1	10	809	3
Seatearth	4	9	814	0
Sandstone and siltstone	8	6	822	6
Mudstone; Spirorbis sp., bivalves, ostracods, fish debris	8	6	831	0
Mudstone, dark; bivalves common	3	4	834	4
SECOND PIPER HORIZON	-	-	834	4
Seatearth	2	2	836	6
Sandstone; plants	3	9	840	3
Mudstone; ironstone	5	9	846	0
HOSPITAL HORIZON	-	-	846	0
Seatearth	4	0	850	0
Mudstone; bivalves	2	0	852	0
HOSPITAL HORIZON	-	-	852	0
Seatearth	3	0	855	0
Mudstone; bivalves	1	0	856	0
Seatearth	6	0	862	0
Siltstone	31	0	893	0
Sandstone with siltstone	36	0	929	0
Siltstone; plants	5	0	934	0
Sandstone with siltstone	8	2	942	2
LOW MAIN, c 6	0	6	942	8
Core lost	2	2	944	10
Seatearth	7	10	952	8
Sandstone with ironstone nodules	0	4	953	0
THREEQUARTERS, c 11	0	11	953	11
Seatearth, mudstone and siltstone	7	10	961	9
Sandstone and siltstone	15	6	977	3
Sand	12	9	990	0
Mudstone, silty	15	3	1005	3
Siltstone	2	5	1007	8

c

d

Beechdale Road (Robins Wood) BH cont.

	Thickness ft	in	Depth ft	in
Mudstone; plants, fish debris	6	1	1013	9
Siltstone, black; plants, fish debris	0	3	1014	0
YARD HORIZON	-	-	1014	0
Seatearth, cream; sphaerosiderite	5	6	1019	6
Siltstone and sandstone	16	3	1035	9
Mudstone; plants, bivalves, fish debris	1	3	1037	0
BLACKSHALE HORIZON	-	-	1037	0
Seatearth	7	3	1044	3
Siltstone with sandstone; plants	6	9	1051	0
Mudstone; ironstone bands; bivalves	13	10	1064	10
COAL, c 1	0	1	1064	11
Mudstone, carbonaceous; plants	0	7	1065	6
Mudstone; bivalves	6	3	1071	9
Mudstone, carbonaceous	0	3	1072	0
Seatearth, cream	3	3	1075	3
Mudstone; ironstone	32	7	1107	10
Mudstone, carbonaceous; fish debris	0	2	1108	0
Seatearth, with sandstone and carbonaceous bands	19	0	1127	0
COAL, c 1	0	1	1127	1
Seatearth	0	11	1128	0
Mudstone; plants, fish debris	12	3	1140	3
'ASHGATE' (UPPER LEAF) HORIZON	-	-	1140	3
Seatearth	6	3	1146	6
Siltstone and sandstone; plants	14	0	1160	6
Mudstone; plants, bivalves	4	0	1164	6
'ASHGATE' (LOWER LEAF) HORIZON	-	-	1164	6
Seatearth	3	9	1168	3
Siltstone and sandstone, interbedded	5	3	1173	6
Siltstone and sandstone; plants	41	6	1215	0
Mudstone; ironstone; ostracods, fish debris	10	0	1225	0
MICKLEY THIN, c 3	0	3	1225	3
Seatearth, cream	15	9	1241	0
Mudstone; ironstone; bivalves, ostracods, fish debris	32	1	1273	1
LOWER BRAMPTON, c 10	0	10	1273	11
Seatearth	8	1	1282	0
Siltstone, passing to mudstone near base; plants, bivalves, fish	16	10	1298	10
Seatearth	2	2	1301	0
Mudstone	2	6	1303	6
Sandstone and siltstone, interlaminated	6	0	1309	6
Mudstone; shelly ironstone; plants, bivalves, fish debris	23	0	1332	6
MORLEY MUCK HORIZON	-	-	1332	6
Seatearth, carbonaceous	6	0	1338	6
Sandstone with mudstone partings; plants	10	9	1349	3
Mudstone; ironstone bands; plants, bivalves	6	3	1355	6
Ironstone, shelly	0	3	1355	9
Sandstone with siltstone bands near top; plants	4	9	1360	6
Siltstone and sandstone	24	6	1385	0
Mudstone, silty; ironstone; plants, fish debris	22	8	1407	8
Siltstone, oolitic; fish debris	0	3	1407	11
COAL, c 7	0	7	1408	6
Seatearth	1	6	1410	0
Mudstone with siltstone; plants (core lost 1433 ft 10 - 1437 ft 9)	32	3	1442	3
KILBURN HORIZON (WASHOUT): sandstone and conglomerate; coal fragments	1	0	1443	3
Seatearth, silty	4	9	1448	0
WINGFIELD FLAGS				
Core lost	10	5	1458	5
Sandstone with siltstone; seatearth 1 ft 6 at 1568 ft; conglomeratic horizons; plants	139	7	1598	0
Mudstone; bivalves, ostracods and fish debris near base	175	2	1773	2
UPPER BAND HORIZON	-	-	1773	2
Seatearth	2	7	1775	9
Mudstone	2	6	1778	3
Seatearth	5	3	1783	6
Mudstone, dark; fish debris	23	3	1806	9
NORTON, c & d 3	0	3	1807	0
Mudstone; fish debris	19	6	1826	6
FORTY-YARDS MARINE BAND				
Mudstone; Lingula sp.	5	4	1831	10
Mudstone	1	2	1833	0
FORTY-YARDS, ca 2, d 4 on c 2	0	8	1833	8
Seatearth	9	10	1843	6
LOXLEY EDGE ROCK				
Sandstone, grit at base; plants	18	11	1862	5
Mudstone	10	7	1873	0
PARKHOUSE MARINE BAND				
Mudstone; foraminifera, Lingula sp., fish debris	3	8	1876	8
Mudstone	0	10	1877	6
Mudstone, (core lost)	7	6	1885	0
Mudstone, silty	11	2	1896	2
ALTON MARINE BAND				
Mudstone, Lingula sp.	0	2	1896	4
Mudstone	3	8	1900	0
Mudstone; Lingula sp., goniatites, bivalves, ostracods, fish debris towards base	4	0	1904	0
ALTON, c trace	0	1	1904	1
Seatearth	2	0	1906	1
Mudstone; plants	6	1	1912	2
Core lost	5	10	1918	0
Mudstone	1	9	1919	9
Mudstone, coaly	0	2	1919	11
FIRST SMALLEY HORIZON	-	-	1919	11
Seatearth (ganister)	2	4	1922	3
Siltstone	3	3	1925	6
Mudstone; bivalves, fish debris	3	6	1929	0
SECOND SMALLEY, c 2	0	2	1929	2
Seatearth	7	1	1936	3
HOLBROOK Lingula BAND				
Mudstone; pyrite at top; Lingula sp. near base, fish debris	2	6	1938	9

a

Beechdale Road (Robins Wood) BH cont.

	Thickness ft in	Depth ft in
Core lost	1 7	1940 4
HOLBROOK HORIZON	- -	1940 4
Seatearth (ganister)	4 5	1944 9
Siltstone	13 3	1958 0
Mudstone with fish	6 0	1964 0
BELPERLAWN HORIZON		
(no core)	3 0	1967 0
Seatearth	4 0	1971 0
CRAWSHAW SANDSTONE		
Siltstone, with sandstone	9 3	1980 3
Sandstone, coarse; small pebbles	52 9	2033 0
Sandstone, fine	24 0	2057 0
Mudstone; plants	1 6	2058 6
Mudstone; fish debris	7 10	2066 4
POT CLAY MARINE BAND		
Mudstone, shaly; ironstone bands; plants, Lingula sp., bivalves, goniatites, fish debris	2 4	2068 8
NAMURIAN		
Mudstone; fish debris	0 4	2069 0
POT CLAY, c 2	0 2	2069 2
Seatearth (ganister)	0 4	2069 6
ROUGH ROCK		
Sandstone	20 6	2090 0
Mudstone, silty, dark grey	19 6	2109 6
Core lost to TD	0 6	2110 0

BELPER (MEADOWS) WATER BH [3399 4776] 1895, 212 ft.
Bore replaced by a well. See Stephens 1929, pp. 80-82 and Plate 4 for graphic section. Soil, gravel and broken sandstone to 19 ft; Ashover Grit (Main Bed) to 171 ft 3; shale to 246 ft 9; leaf of Ashover Grit to 359 ft; shale to TD at 369 ft 8.

BENNERLEY COLLIERY SHAFT [4709 4402] abt 175 ft.
Brown Rake, d c 14 at 55 ft 9; Top Soft, c 13 at 90 ft 7; Roof Soft (UL), d c 14 at 105 ft 11; Roof Soft (LL), c 27 at 141 ft 5; Deep Soft, c 44 at 158 ft 1; Foot, b 12 at 176 ft 3; Deep Hard, c & ca 16, c 35, d 6 on c 18 at 212 ft 9; First Piper, c 60 at 270 ft 3; Low Main, c 42 at 374 ft 9; Threequarters, c 24 at 391 ft 9; Blackshale, c 54 at 540 ft 9; TD 540 ft 9.

BENTINCK COLLIERY 27's UBH [5019 5161] 1965.
Up from Second Waterloo at - 750 ft. Examined by S. Brunskill. See Fig. 36. Dunsil c 24 at 78 ft 9; First Waterloo, c 22, d 6 on c 22 at 46 ft 7; length drilled 81 ft 3.

BENTINCK COLLIERY 31's UBH [4973 5185] 1965.
Up from Second Waterloo at - 654 ft. Dunsil, c 33 at 78 ft 5; First Waterloo, c & d c 16, d 15 on c 22 at 47 ft 8; length drilled 83 ft.

BENTINCK COLLIERY W1 UBH [4975 5252] 1961.
Up from Second Waterloo at -652 ft. Dunsil horizon at 117 ft; First Waterloo (See Fig. 36), c 19, d 30 on c 21 at 54 ft 1; length drilled 134 ft.

BIRCHWOOD COLLIERY UPCAST SHAFT [4373 5470] 1915.
Section from 35 ft below Blackshale. Coal 10 at 22 ft 4; 'Ashgate' (UL), c 14 at 49 ft 6; 'Ashgate' (LL), ca 6 at 78 ft 10; Mickley Thin, c 15 at 132 ft 10; Lower Brampton, d c 3 at 188 ft 7; Morley Muck, ca 4 on d c 6 at 277 ft 1; Kilburn, c 22, m 6 on d c 9 at 415 ft 11; recorded to 427 ft 4.

BIRCHWOOD COLLIERY SHADY SHAFT [4329 5439], 369 ft.
Base of Westphalian B estimated at 47 ft 2; Brown Rake, c 12, m 4 ft 2 on c 9 at 74 ft 0; Black Rake, c 6 at 92 ft 9; Top Soft, c 18 at 101 ft 3; Roof Soft (UL), c 6 at 106 ft 9; Roof Soft (LL), c 36 at 131 ft 3; Deep Soft workings at 176 ft 3; Deep Hard workings at 235 ft 3; First Piper (Upper Coal), c 24 at 317 ft 7; First Piper (Lower Coal), c 6, m 6 ft 1 on c 8 at 330 ft 8; Second Piper, c 1 at 352 ft 8; ? Hospital, c 1 at 394 ft 2; ? Tupton, c 48 at 429 ft 5; Threequarters, c 25 at 442 ft 10; Yard, c 25 at 534 ft 3; Blackshale, c 17, d 3, c 9, d 2, c 16, d 7 on d c 6 at 617 ft 7; recorded to 617 ft 7.

BLACKBROOK WATER BH, BELPER [Approx. 3308 4775] 1895, 265 ft.
See Stephens 1929, pp. 82-83 for section. Namurian shales with some interbedded sandstones to TD at 231 ft 2.

BONDLAND COLLIERY No. 4 SHAFT [3733 5095] abt 420 ft.
No details to 295 ft 10; See Figs. 14, 16, 68 and Eden 1954. Kilburn, c 36, d 6 on c 10 at 300 ft 2; Upper Band horizon at 630 ft 6; Norton, c 19 at 672 ft 11; Forty-Yards, c 11 at 690 ft 11; Alton MB at 769 ft 2; Alton, c 22 at 771 ft 10; First Smalley, c 9, d 1, c & d 6, m 17 on c 2 at 787 ft 7; Second Smalley, c 2 at 798 ft 9; Belperlawn, c 33 at 832 ft 3; TD 832 ft 3.

BRAILSFORD BH [2572 4158] abt 490 ft.
Boulder clay (gravelly clay) to 8 ft; Pebble Beds (gravelly sand) to 35 ft; gravelly clay to TD at 36 ft.
Another record of a well in Brailsford (47 ft deep containing gravel and sand to the bottom) cannot be accurately located. It is considered to be the village pump [2539 4162] (See also Express Dairies BH).

BRANDS COLLIERY [4138 5196] 1851, abt 360 ft.
See also Figs. 25 and 33, Plate 5. The depths below are by addition from the complete record and differ slightly from those on the record. Bottom First Waterloo, c 15 at 37 ft 6; Waterloo Marker, c 8 at 61 ft 6; Top Second Waterloo, c 30 at 95 ft 2; Bottom Second Waterloo, ca 16 at 114 ft 10; Third Waterloo, c 17 at 141 ft 11; Fourth Waterloo, c 24 at 195 ft 2; First Ell, c 12 at 234 ft 8; Second Ell, ca 9 on c 8 at 295 ft; base of Westphalian B estimated at 383 ft; Brown Rake, c 7, d 31 on c 14 at 424 ft 2; Black Rake TS at 443 ft 11; False Ell, c 25, m 3 ft, c 8, m 10,

b

c & b 14, m 8 on c 10 at 459 ft; Deep Soft, c 42 at 494 ft 2; Deep Hard, c 47 at 571 ft 4; First Piper, c 27, d 27 on c 10 at 639 ft 8; Hospital (UL), c 5 at 698 ft 1; ? Hospital (LL), c 8 at 731 ft 1; Low Main, c 48 at 775 ft 5; Threequarters, c 19, m 3 ft 6 on c 6 at 794 ft 5; Yard, c 36 at 884 ft 7; Blackshale, c 12, d 6, c & b 3, d & b 15 on c 64 at 914 ft 10; recorded to 932 ft 10.

BRINSLEY COLLIERY No. 1 SHAFT [4643 4878], 289 ft.
Cinderhill (UL), c 18 at 37 ft 11; Cinderhill, c 15 at 57 ft 8; Mainsmut, d c 18 at 154 ft 8; Comb, c 25 at 181 ft 9; Top Hard, c 52 at 213 ft 1; Top Hard Floor, c 15 at 235 ft 10; Dunsil, c 18 at 273 ft 9; First Waterloo, ca 23, m 14 on c 24 at 327 ft 3; Waterloo Marker, ca 15 at 350 ft 9; Top Second Waterloo, c 26, d 4 on c 4 at 373 ft 10; Bottom Second Waterloo, c 18 at 399 ft 4; Third Waterloo, c 21 at 423 ft 7; Fourth Waterloo, c 27 at 469 ft; First Ell, d c 15, c 20, d 5 on c 1 at 504 ft 11; Second Ell, ca 26 at 552 ft 10; base of Westphalian B estimated at 600 ft; Brown Rake, c 12 at 619 ft; Black Rake horizon at 643 ft 9; Top Soft, c 9 at 656 ft 7; Roof Soft (UL), c 11 at 659 ft 6; Roof Soft (LL), c 20 at 679 ft 3; Deep Soft, c 46 at 731 ft 1; Deep Hard, ca 11 on c 39 at 779 ft 10; First Piper, c 27 at 824 ft 7; TD 826 ft 7. Section continued in Selston Colliery Brinsley Drift.

BROXTOWE COLLIERY No. 1 DOWNCAST OR WESTERLY SHAFT [5236 4293] 1863, 265 ft.
See Fig. 74 and Plate 5. Record below Deep Soft possibly from nearby staple pit.
Lowbright, c 36 at 27 ft 7; Lowbright Floor, c 18, d 42 on c 1 at 38 ft 2; Brinsley Thin, c 17 at 74 ft; High Hazles, c 18 at 101 ft 1; Cinderhill, c 36 at 171 ft; Mainsmut, c 20 at 248 ft 9; Comb, c 14, d 11 on c 16 at 293 ft 9; Top Hard, c 60 at 299 ft 2; Top Hard Floor, c 17 at 321 ft; Dunsil, c 24 at 348 ft; First Waterloo, c 36 at 379 ft 2; Waterloo Marker, c 4 at 397 ft 2; Top Second Waterloo (UL), c 15 at 413 ft 5; Top Second Waterloo (LL), c 11 at 424 ft 4; Bottom Second Waterloo, c 9 at 429 ft 5; Coal 3 at 457 ft 6; Third Waterloo, c 21 at 474 ft; Fourth Waterloo, c 3 (roof coal), m 47 on c 30 at 525 ft 1; First Ell, c 14 at 538 ft 2; Second Ell, c 12 at 575 ft 6; base of Westphalian B estimated at 630 ft; Brown Rake, c 2 at 646 ft; Top Soft horizon at 666 ft 4; Roof Soft (UL), c 5 at 681 ft 7; Roof Soft, c 12 at 706 ft 3; Deep Soft, c 30 at 735 ft 10; Deep Hard, c 26 at 765 ft 11; recorded to 766 ft 7.

BULWELL COLLIERY [5306 4586] 1868, 222 ft.
Plumbed depths only; from NCB records. Note details obtained by W. N. Edwards during Wartime Reconnaissance Survey. See also Bulwell Steps. Mainbright c 38 at 546 ft; Two-Foot, c 22 at 556 ft; Lowbright, c 26 on d c 17 at 603 ft; Lowbright Floor, c 18 at 605 ft; High Hazles, c 32 at 672 ft; Top Hard, c 60 at 921 ft; TD 921 ft.

BULWELL FINISHING CO. WATER BH [5427 4564] 1956, abt 150 ft.
Cores examined by I. P. Stevenson below 138 ft depth. Alluvium to 16 ft; Lower Magnesian Limestone to ? 47 ft; Lower Marl to ? 113 ft; Shafton MB at 163 ft; Shafton, ca 3 on c 9 at 168 ft; Shafton, c and d 9 at 179 ft 6; Coal 12 at 219 ft; Edmondia Band at 260 ft; High Main, c 66 at 305 ft 6; TD 363 ft.

BULWELL STEPS (1 in 2 STONEHEAD) [Approx. 5321 4583].
Measured by Dr J. Shirley in 1944. Mainbright, c 36 at 3 ft; Two-Foot MB at 16 ft; Two-Foot, c 20, d 23 on c 2 at 19 ft 9; Lowbright, c 26 at 44 ft 10; Brinsley Thin (UL), c 1 at 100 ft 7; Brinsley Thin, c 6 at 116 ft 8; High Hazles, c 32 at 152 ft 5; Cinderhill, c 13 at 248 ft 8; Mainsmut (UL), c 19 at 330 ft 9; Mainsmut (LL), c 2 at 348 ft; Comb, c 12, b 1 on c 18 at 383 ft 10; Top Hard, c 60 at 393 ft 5; recorded to 393 ft 5.

BURROWS WATER BH, BRAILSFORD [2588 3904] 1944, abt 420 ft.
Keuper Marl to 94 ft; Pebble Beds to TD at 130 ft.

BUTTERLEY PARK ENGINE SHAFT [4183 5131] 1837, abt 345 ft.
See Plate 5. Top Second Waterloo, c 26 at 29 ft 2; Bottom Second Waterloo, ca 10 at 45 ft 1; Third Waterloo, c 18 at 84 ft; Fourth Waterloo, c 18 at 112 ft 6; First Ell, c 18 at 168 ft; estimated base of Westphalian B at 309 ft; Brown Rake, c 15 at 338 ft; False Ell, c 27, d 9, c 6 on d c 27 at 363 ft 9; Deep Soft, c 41 at 393 ft 9; Deep Hard, c 44 at 446 ft 10; recorded to 446 ft 10.

CALLOW BH, WIRKSWORTH [2665 5282] 1967, abt 800 ft.
Examined by D. V. Frost.

	Thickness ft in	Depth ft in
NAMURIAN		
Mudstone & clay	19 0	19 0
Reticuloceras gracile MARINE BAND		
Mudstone, dark grey, finely laminated, ferruginous	1 0	20 0
Mudstone, grey; ostracods	3 6	23 6
Mudstone, pale grey-purple, laminated, pyritic	5 6	29 0
Mudstone, dark grey, phosphatic spots, ferruginous laminae; rare plants, fish debris at base	30 9	59 9
Mudstone, grey, laminated with 5-in cank at 74 ft; Dunbarella sp., fragmentary Reticuloceras co-reticulatum ostracods	17 3	77 0
Mudstone, grey, silty, pyritic; bivalves	4 0	81 0
Mudstone, grey, pyritic, with 2-in cank at 83 ft; spat ostracods to TD	9 0	90 0

CATSTONE PIPER 3's (B) UBH [5053 4149] 1958.
Cores, below 90 ft, examined by G. D. Gaunt. Down from First Piper at -169 ft.

Low Main, c 39 at 150 ft 3; Threequarters (UL), c 2 d 3 on c 1 at 164 ft 10; Threequarters (LL), c 1, d 3 on c 1 at 165 ft 2; Yard horizon at 202 ft 7; Denby leaf of Blackshale, c 1 at 237 ft 2; TD 253 ft 6.

CAVENDISH CRESCENT BH [5029 5400] 1971, 574 ft.
Cores, from 20 ft 5, examined by C. Beal. See Fig. 82 for section of Permo-Triassic rocks. Lower Mottled Sandstone to abt 85 ft 5; Middle Marl to 113 ft 7; Lower Magnesian Limestone to 145 ft; Lower Marl to 214 ft; Permian Basal Breccia to 215 ft 9; Top MB at 269 ft 6; Shafton MB at 333 ft 8; Coal 11 at 360 ft 5; Shafton, c 3, ca 23 on b 2 at 381 ft 1; ? Main 'Estheria' Band at 419 ft 5; Coal 11 at 429 ft 2; Edmondia Band at 490 ft; High Main (UL), c 10 at 519 ft 8; High Main (LL), d c 6 on c 34 at 524 ft 3; TD 524 ft 11.

CAVENDISH LAUNDRY WATER BH, DUFFIELD [3494 4226] 1961, abt 180 ft.
Soil to 3 ft; 'sand and pebbles' (Flood Plain Terrace) to 60 ft; Namurian shales to TD at 63 ft.

CHALFONT DRIVE No. 1 BH, NOTTINGHAM [539 410] 1967, abt 150 ft.
Recorded by D. V. Frost. Middle Marl to 8 ft 6; Lower Magnesian Limestone to 15 ft 5; Lower Marl to 21 ft 10; Permian Basal Breccia to 22 ft 1; ? Brinsley Thin horizon at 25 ft 6; High Hazles, c 36 at 51 ft 6; TD 100 ft.

CHARLIE'S WOOD COTTAGES BH [5062 5036] 1970, 504 ft.
Recorded by C. Beal. See Fig. 82 for graphic section of Permo-Triassic rocks. Permian on Westphalian to High Main. TD 427 ft 10.

CHILWELL BOREHOLE [5285 3592] 1872, 95 ft
See Plate 10 for graphic section of Permo-Triassic rocks, Lamplugh and Smith 1944, pp. 61-62 and R. Br. Assoc. for 1890, p. 366.

	Thickness ft in	Depth ft in
ALLUVIUM		
Earth	1 6	1 6
Gravel, clay and sand	12 2	13 8
Marl, white	1 0	14 8
KEUPER MARL (HARLEQUIN FORMATION)		
Clay, red	22 6	37 2
? PLAINS SKERRY		
Sandrock	6 0	43 2
CARLTON & RADCLIFFE FORMATIONS		
Clay and hard bands; gypsum	70 10	114 0
WATERSTONES		
Stone, rock bands, red and white; thin clays	120 9	234 9
PEBBLE BEDS		
Sandrock, red and white; thin clays and gypsum	109 7	344 4
Conglomerate and red bands with pebbles	15 8	360 0
Conglomerate	15 0	375 0
LOWER MOTTLED SANDSTONE		
Sandrock, red; partings of marl	59 9	434 9
Sandstone, soft	28 11	463 8
WESTPHALIAN recorded to	876 9	1340 5

CHILWELL DAM FARM BH [5105 4277] 1957, 393 ft.
Base of Permian at 41 ft 4. Westphalian from Cinderhill to Second Waterloo, TD 375 ft 8.

CINDERHILL COLLIERY Nos. 1 AND 2 SHAFTS* [5332 4367] and [5331 4366] 1841-3, 189 ft.
For details of these shafts see Edwards 1951, p. 153. Minor differences in depths occur in the two records below Lowbright Coal. Permian and base of Westphalian C to Top Hard at 663 ft 1. Section quoted below Top Hard, presumably a shaft deepening: Top Hard Floor, c 10 and 30 ft 10; Dunsil, c 28 at 71 ft 4; TD No. 1 Shaft 744 ft. TD No. 2 Shaft and BH 1330 ft.

CINDERHILL COLLIERY No. 4 SHAFT [5320 4371] 1861, 193 ft.
See Figs. 36, 42 and 46. Deepening from Top Hard (1945-6) recorded by Dr J. Shirley. Base of Permian at 36 ft 3; Mansfield MB at 106 ft 9; Coal 2 at 106 ft 11; Coal, 15 at 157 ft; Haughton MB at 176 ft 2; ? Swinton Pottery group, c 29 at 178 ft 7; c 11, d 8 on c 18 at 194 ft 6; Coal, c & b 11 at 204 ft; ? Clown at 247 ft 6; Mainbright, c 43 at 295 ft 3; Two-Foot MB at 310 ft 3; Two-Foot, c 13, d 2 on c 18 at 313 ft; Coal, c & b 6 at 319 ft 4; Lowbright and Lowbright Floor, c 44, d 3, c 15, d 3 on c 2 at 361 ft 4; Brinsley Thin (UL), c 9 at 389 ft 1; Brinsley Thin, c 17 at 406 ft 5; High Hazles (see Fig. 36), c 35 at 434 ft 8; Cinderhill, c 41, d 1 on c 3 at 515 ft 1; Mainsmut, c 26 at 601 ft 6; Comb, c 11, d 4 on c 14 at 647 ft 11; Top Hard, c 56 at 652 ft 10; Top Hard Floor, c 12 at 682 ft 9; Dunsil, ca 2 on c 27 at 729 ft 11; First Waterloo, c 37 at 765 ft 1; Waterloo Marker, c 3 at 789 ft 5; Top Second Waterloo (UL), c 22 at 812 ft 11; Top Second Waterloo (LL), c 11 at 828 ft 6; Bottom Second Waterloo, c 7 at 833 ft 5; Cinderhill Fault (dip of plane 37½° SW) at 856 ft 4; Top Soft, c 1 at 871 ft 11; Roof Soft (UL), c 4 at 884 ft 2; Roof Soft, ca 2 on c & b 12 at 916 ft 4; Deep Soft, c 31 at 938 ft 1; Deep Hard, c 14 at 961 ft; First Piper, c 14 at 1003 ft 4; Hospital, c & b 9 at 1043 ft 6; Low Main washout at 1122 ft 9; Threequarters, c 11 at 1135 ft 5. Boring in the bottom of the shaft proved the following sequence from -1076 ft: ? Blackshale washout at 1215 ft; ? 'Ashgate' (UL), c 2 at 1318 ft; ? 'Ashgate' (LL), c 4 at 1362 ft 8; TD 1376 ft.

*Shafts at Cinderhill Colliery are numbered up to 6. There is no record or site known for No. 5 Shaft. It is thought that No. 3 Shaft is now described as Hempshill No. 3.

c

d

CINDERHILL COLLIERY No. 6 DOWNCAST (OR NORTHERLY) SHAFT [5331 4422] 1941, 200 ft.
Lower Mottled Sandstone to 21 ft 5; Middle Marl to 43 ft 11; Lower Magnesian Limestone to 69 ft 11; Lower Marl to 101 ft; base of Westphalian C to Lowbright; TD 418 ft 3. Details of measures (Lowbright-High Hazles) were recorded in 1947 from the Main Drift near No. 6 Shaft and (First Piper (Fig. 30) to Low Main) from Tupton Intake Drift [5317 4399 - 5325 4443].

CINDERHILL 1's MAIN GATE UBH [5321 4593] 1954.
Down from High Hazles at - 503 ft. Cinderhill (UL), c 14 on b 4 at 82 ft 5; Cinderhill, c 2 at 108 ft 7; Mainsmut (UL), c 24 at 160 ft 2; Comb, c 27 on b 1 at 217 ft 7; Top Hard, c 69 at 225 ft 5; Top Hard Floor, c 15 at 242 ft 5; Dunsil, ca 15 on c 13 at 283 ft 4; First Waterloo, c 21 at 312 ft 1; TD 313 ft 10.

CINDERHILL 1's UBH [5305 4439] 1955.
Down from Deep Soft at - 955 ft. Cores examined by R. E. Elliott. Deep Hard, c 28 at 38 ft 10; First Piper, c 29 at 84 ft 2; Second Piper horizon at 105 ft 3; Hospital, c 2 at 125 ft 10; Cockleshell, c 8, d 9, c 5 on b 3 at 171 ft 10; Low Main, c 36 at 198 ft 8; Threequarters, c 11 at 204 ft 7; Yard horizon at 255 ft 9; Denby leaf of Blackshale, c 9 at 304 ft 11; Blackshale, c 12, d 16, c 15, d 6 on c 18 at 313 ft 9; TD 326 ft 10.

CINDERHILL 2's HEADING UBH [5220 4755] 1953-55.
Down from Deep Soft at - 1228 ft to Blackshale. TD 376 ft 5. See Fig. 30 for section of First Piper Coal.

CINDERHILL (BABBINGTON) 2's MAIN INBYE BELT ROAD UBH [5187 4896] 1960.
Down from Deep Soft at - 1308 ft to Blackshale. TD 391 ft 11. See Fig. 23 for Threequarters Coal section.

CINDERHILL OLD 2's LEFT GATE INBYE UBH [5305 4416] 1955.
Down from Deep Soft at - 909 ft to Blackshale. TD 339 ft 6. See Fig. 23 for Threequarters Coal section.

CINDERHILL 6's MAIN GATE UBH [5284 4505] 1955.
Down from Deep Soft at - 972 ft to Blackshale. TD 358 ft 4.

CINDERHILL 9's LEFT GATE UBH [5217 4535] 1953.
Down from Deep Soft at - 892 ft to Blackshale. TD 380 ft 1.

CINDERHILL 10's FINISHING FACE UBH [5396 4605] 1954.
Down from Deep Soft at - 1278 ft to Blackshale. TD 387 ft 10.

CINDERHILL 10's INBYE UBH [5319 4586] 1954.
Down from Deep Soft at - 1180 ft to Blackshale. TD 381 ft 6.

CINDERHILL 10's MAIN GATE UBH [5276 4575] 1953.
Down from Deep Soft at - 1106 ft to Blackshale. TD 380 ft 11.

CINDERHILL 14's FINISHING FACE UBH [5381 4651] 1954.
Down from Deep Soft at - 1318 ft to Blackshale. TD 388 ft 7.

CINDERHILL 14's (MAIN GATE) UBH [5259 4619] 1953.
Down from Deep Soft at - 1170 ft.

	Thickness ft in	Depth ft in
WESTPHALIAN A		
DEEP SOFT workings (no cores)	9 4	9 4
Mudstone; plants	3 11	13 3
Siltstone	3 0	16 3
Mudstone, dark grey at base; bivalve	13 10	30 1
DEEP HARD, ca 1 on c 48	4 1	34 2
Seatearth	3 4	37 6
Sandstone and siltstone	7 9	45 3
Mudstone, carbonaceous	1 11	47 2
Seatearth	0 5	47 7
Mudstone	1 7	49 2
Siltstone and sandstone	27 4	76 6
Mudstone; bivalves, ostracods, fish debris	8 7	85 1
FIRST PIPER, c 22, d 1 on c 21	3 8	88 9
Seatearth, siltstone	4 3	93 0
Siltstone	5 4	98 4
Mudstone; bivalves, ostracods, fish debris	4 6	102 10
SECOND PIPER HORIZON	- -	102 10
Seatearth	3 2	106 0
Siltstone	10 9	116 9
Mudstone; bivalves, ostracods	10 3	127 0
HOSPITAL, c 13	1 1	128 1
Seatearth	1 11	130 0
Mudstone; bivalves, plants	1 3	131 3
Core lost	0 7	131 10
HOSPITAL (LOWER LEAF) HORIZON	- -	131 10
Seatearth	1 2	133 0
Mudstone; ironstone; bivalves	22 5	155 5
COCKLESHELL, c 21	1 9	157 2
Seatearth, siltstone	0 11	158 1
Sandstone with siltstone	5 9	163 10
Mudstone; plants, bivalves common	16 1	179 11
LOW MAIN, c 37	3 1	183 0
Seatearth	3 0	186 0
Mudstone	3 0	189 0
THREEQUARTERS, c 16	1 4	190 4
Seatearth siltstone	3 5	193 9
Siltstone with sandstone	30 0	223 9
Mudstone; bivalves, fish debris	14 4	238 1
YARD HORIZON	- -	238 1
Seatearth	6 8	244 9
Mudstone	1 9	246 6
? DENBY LEAF OF BLACKSHALE HORIZON	- -	246 6

a

	Thickness ft	in	Depth ft	in
Seatearth	1	9	248	3
Sandstone and siltstone; breccia at 300 ft	78	6	326	9
BLACKSHALE, c 6, d 1, c 35, d 2 on c 3	3	11	330	8
Seatearth	3	10	334	6
Mudstone; ostracod	9	2	343	8
Siltstone	5	1	348	9
Strata not seen	14	8	363	5
Mudstone; fish debris	0	4	363	9
COAL, c 17	1	5	365	2
Seatearth to TD	1	6	366	8

CINDERHILL 19's LEFT GATE UBH [5190 4617] 1953.
Down from Deep Soft at -1031 ft to Blackshale; TD 374 ft.

CINDERHILL 25's LEFT UBH [5218 4735] 1959.
Down from Deep Soft at -1216 ft to Deep Hard, ca 1, c 45 on b 16 at 42 ft 7; TD 61 ft 6.

CINDERHILL 25's RIGHT GATE SLANT UBH [5213 4757] 1959.
Down from Deep Soft at -1218 ft to Deep Hard, ca 2, c 42 on b 14 at 47 ft 11; TD 61 ft 6.

CINDERHILL 27's MAIN INBYE UBH [5148 4765] 1960.
Down from Deep Soft -1148 ft. Deep Hard, ca 2 on c 41 at 44 ft 3; First Piper (See Fig. 30), c & d 53 at 91 ft; Hospital, c 12 at 134 ft 6; Cockleshell, c 15 at 161 ft 10; Low Main, c 37 at 190 ft 1; Threequarters, c 11 at 200 ft; Yard, b 1 at 258 ft 9; Blackshale (partial washout), c 11 at 344 ft 4; 'Ashgate', c & d 19 at 377 ft 11; TD 387 ft 10.

CINDERHILL 27's MAIN OUTBYE UBH [5196 4778] 1959.
Down from Deep Soft at -1218 ft 6 to Deep Hard, c 50 on b 15 at 48 ft 2; TD 59 ft 10.

CINDERHILL S 32's UBH [5184 4963] 1971.
Down from Deep Soft at -1288 ft. Recorded by M. Allen and C. Beal. Drilled to Blackshale at 357 ft; TD 367 ft.

CINDERHILL H 68's UBH [5283 4977] 1971.
Down from Deep Soft at -1388 ft. Cores, below 149 ft, examined by J. Whalley and M. Allen. Low Main, c 37 at 198 ft 11; Threequarters (UL) ca 1 on c 14 at 215 ft 8; Threequarters (LL) c & d c 20 at 222 ft 10; Yard/Blackshale, c & d 111 at 310 ft 3; TD 334 ft.

CINDERHILL T 10's UBH [5278 4640] 1971.
Down from Low Main at -1412 ft to Blackshale at 143 ft 8. Recorded by D. E. Raisbeck. TD 146 ft 6.

CINDERHILL T 20's UBH [5159 4443] 1971.
Down from Low Main at -884 ft. Recorded by J. Allen. Blackshale, c & d c 15, d 4, c & d c 9, d 3 on c 16 at 138 ft; TD 147 ft 1.

COLLIERS' REST BH [3690 4683] 1960, 350 ft.
Recorded by D. V. Frost. See Figs. 14, 16 and 68. Upper Band horizon at 266 ft 3; Norton, c 26 at 314 ft 5; Forty-Yards, c 15 at 345 ft; Parkhouse MB (?Lower) at 440 ft 6; Alton MB at 457 ft 6; Alton, c 30 at 460 ft; First Smalley Lingula Band at 484 ft 6; First Smalley, c 9 at 485 ft 3; Belperlawn, c 36 at 539 ft 4; TD 586 ft 4.

COPPICE COLLIERY No. 1 SHAFT [4318 4510] 1875, 318 ft.
Surface to First Piper Coal; TD 661 ft 2. See Figs. 33 and 34, for Deep Soft group of coals, and Plate 5.

COPPICE COLLIERY No. 2 (UPCAST) SHAFT [4315 4510] 1876, 298 ft.
Deepening 1924 below Deep Hard Coal (only outline record available). All the depths quoted omit 11 ft 4 of made ground. See also Fig. 22 for Blackshale Coal sequence. Dunsil, c 25 at 37 ft 5; Top First Waterloo, c 6 at 75 ft 6; Bottom First Waterloo, c 16 at 97 ft 10; Waterloo Marker, c 7 at 120 ft 11; Top Second Waterloo, c 43 at 148 ft 5; Bottom Second Waterloo, c 12 at 164 ft 3; Third Waterloo, c 20 at 199 ft 11; Fourth Waterloo, c 6 at 239 ft 11; First Ell, c 23 at 298 ft; Second Ell, ca 31 at 353 ft 8; base of Westphalian B estimated at 441 ft; Brown Rake, c 24 at 458 ft 5; Black Rake TS at 475 ft; Black Rake, c 6 at 481 ft 5; Top Soft, c 32 at 506 ft 6; Coal 1 at 528 ft 2; Roof Soft & Deep Soft, c 50, d 26, c 53, b & m 20 on c 2 at 544 ft 1; Deep Hard, d c 8 on c 44 at 594 ft 9; First Piper, at 637 ft; Low Main, at 759 ft; Threequarters, c 24 at 774 ft; Yard, c 2 at 838 ft 2; Blackshale, c 19, d 9, c 5, d 1, c 6, d 5 on c 17 at 947 ft 9; Coal 2 at 967 ft 9; Coal 16 at 1023 ft 1; 'Ashgate', c 27, b 1, m 29 on c 27 at 1057 ft 1; Mickley Thin, c 10 at 1109 ft 7; Lower Brampton, c 6 at 1170 ft 1; Morley Muck, c 16, m 18 on c 6 at 1253 ft 11; Kilburn, c 43 at 1360 ft 6; recorded to 1360 ft 6.

COPPICE COLLIERY WATERLOO SHAFT [4322 4513], abt 320 ft.
Top Hard workings at 36 ft 3; Dunsil, c 27 at 78 ft 6; Top First Waterloo, c 6 at 112 ft 1; Bottom First Waterloo, c 17 at 133 ft 6; Waterloo Marker, c 9 at 157 ft; Top Second Waterloo, c 43 at TD 181 ft 1.

COPPICE COLLIERY UBH [4303 4512] 1951.
Down from Kilburn at -1049 ft. Cores, from 43 ft, examined by I. P. Stevenson. See Figs. 14, 15, 16 and 68. Upper Band horizon at 368 ft 6; Norton MB at 400 ft 2; Norton, c 18 at 402 ft; Forty-Yards MB at 431 ft 7; Forty-Yards, c 6 at 432 ft 1; Parkhouse MB at 505 ft 1; Alton MB at 533 ft; Alton, c 30 at 535 ft 6; First Smalley horizon at 542 ft; Belperlawn, c 16 at 587 ft 2; TD 650 ft.

COSSALL COLLIERY No. 1 SHAFT [4783 4268], abt 190 ft.
First Piper at 32 ft 6; Low Main at 138 ft 6; Blackshale at 293 ft 10; fault 528-98 ft; Kilburn at 667 ft 6; TD 681 ft.

b

COSSALL COLLIERY No. 2 SHAFT [4777 4266], 187 ft.
See also Fig. 20. First Piper, c 42 at 29 ft 8; Hospital, c 4 at 68 ft 9; Cockleshell, c 21 at 108 ft 9; Low Main, c 47 at 136 ft 11; Threequarters, c 17 at 146 ft 3; Yard horizon at 203 ft 3; Blackshale (?partial washout), c 14 at 300 ft 7; 'Ashgate' washed out; Mickley Thin horizon at 488 ft 4; Lower Brampton, c 6 at 541 ft 3; Coal 4 at 581 ft 10; Morley Muck, c 6, d 9, c 6, d 10 on c 1 at 631 ft 8; Kilburn, c 46 at 762 ft 10; TD 798 ft.

COSSALL 3's UBH [5094 4315] 1965.
Down from Low Main at -491 ft to Blackshale. Recorded by J. Chilton and D. V. Frost. TD 165 ft.

COSSALL 40's UBH [5111 4183] 1964.
Down from Deep Hard to Blackshale. TD 290 ft. See Fig. 70 for sequence of Blackshale Coal.

COSSALL 40's UBH (rebore) [5118 4186] 1964-65.
Down from Deep Hard at -311 ft. First Piper old workings at 38 ft; Low Main, c 38 at 171 ft 9; Threequarters, c 24 at 183 ft; Denby Leaf of Blackshale, c 9, d 6 on c 6 at 258 ft 9; Blackshale, c 15, d c 2, d 11, d c 9, d 10 on c & d c 16 at 285 ft 5; 'Ashgate' (UL) horizon at 360 ft 9; 'Ashgate' (LL), ca 2 on c 2 at 382 ft 4; Mickley Thin, c 3 at 435 ft 2; TD 442 ft 6.

COTESPARK* COLLIERY No. 2 UBH [4203 5458] 1955.
Down from Yard at abt 290 ft (datum abt 1 ft below the Yard Coal). Recorded by G. H. Rhys. Blackshale workings at 58 ft 10; 'Ashgate' (UL) horizon at 150 ft 10; 'Ashgate' (LL), ca 9, d 3 on ca 2 at 184 ft 5; Mickley Thin, c 8 at 226 ft 6; Lower Brampton, c 4 at 285 ft 1; Coal 3 at 327 ft 9; Morley Muck horizon at 367 ft 9; Kilburn, c 26 at 489 ft 6; TD 492 ft 8.

COTESPARK COLLIERY No. 3 UBH [4371 5413] 1955.
Down from Yard at -176 ft; TD 530 ft.

COTMANHAY COLLIERY [4646 4503], abt 190 ft.
See Fig. 31 for section of Deep Hard Coal. First Ell, c 30 at 8 ft 6; Second Ell, c & ca 20 at 65 ft 5; base of Westphalian B estimated at 152 ft; Brown Rake, c 14 at 158 ft 11; Black Rake TS at 175 ft 5; Black Rake horizon at 180 ft 6; Top Soft, c 14 at 188 ft 2; Roof Soft (UL), c 16 at 202 ft 6; Roof Soft (LL), c 30 at 238 ft; Deep Soft, c 38 at 258 ft 10; Deep Hard, c 45 at 304 ft 11; TD 319 ft 11.

CROW WOOD BH, WOLLATON PARK [Approx. 539 391] 1786, abt 160 ft.
Trias to 18 ft on Westphalian to Top Hard Coal. TD 324 ft 6.

CYCLE WORKS, BEESTON WATER BH [5365 3709] 1913, abt 90 ft.
See Lamplugh and Smith 1914, p. 51. Alluvium to 21 ft; Waterstones to 23 ft; Pebble Beds to 167 ft; Lower Mottled Sandstone to TD at 173 ft.

DALE ABBEY No. 2* BH [4254 3842] abt 1877, abt 350 ft.
See Fig. 16.

	Thickness ft	in	Depth ft	in
Soil	0	6	0	6
Sand, loamy	1	0	1	6
PERMO-TRIASSIC				
LOWER MOTTLED SANDSTONE				
Sandrock, yellow and red marl	42	0	43	6
Marl, red	5	0	48	6
WESTPHALIAN A				
Clod	29	6	78	0
UPPER BAND HORIZON	-	-	78	0
Fireclay, with ironstone	2	0	80	0
Clod	0	6	80	6
Fireclay, with ironstone	2	0	82	6
Clunch, blue	4	9	87	3
Bind, with ironstone	27	0	114	3
Rock, dark	0	6	114	9
NORTON, c 24	2	0	116	9
Bind, with ironstone	30	3	147	0
FORTY-YARDS (UPPER LEAF), c 12	1	0	148	0
Bind	0	6	148	6
Rock	4	0	152	6
Clod	6	0	158	6
FORTY-YARDS (LOWER LEAF), c 27	2	3	160	9
Black batt and bind	62	9	223	6
ALTON, c 39	3	3	226	9
Fireclay, rocky	1	0	227	9
Fireclay, recorded to	7	3	235	0

Clod is soft, dark mudstone; bind is mudstone; rock is sandstone.

DALE ABBEY No. 3 BOREHOLE [4288 3817] abt 1877, abt 375 ft.
For graphic section of Permo-Triassic rocks see Plate 10. Keuper Marl (Radcliffe Formation) to 5 ft 6; Waterstones to 74 ft 6; Pebble Beds to 141 ft 10; Lower Mottled Sandstone to 194 ft 10; Forty-Yards (UL), c 4, m 6 on c 6, at 210 ft 2; Forty-Yards (LL), c 4, m 6 on c 6 at 220 ft 2; recorded to 263 ft 11.

DEEPFIELDS SHAFT. See Shipley Colliery, Deepfields Shaft.

DENBY (DRURY-LOWE) COLLIERY, ALTON No. 1 UBH [3849 4622] 1956.
Down from Alton at -712 ft. From skeleton log. First Smalley, c & d 24 at 26 ft; Second Smalley, c & d 24 at 34 ft; Belperlawn, c 30 at 80 ft 6; TD 150 ft.

DENBY (DRURY-LOWE) COLLIERY, KILBURN SHAFT [3841 4622] 1905, 257 ft.
See Figs. 14, 16, 19 and 68. 'Ashgate' (LL), c 25 at 55 ft; Mickley Thin, c 18 at 121 ft 6; Lower Brampton, c 2 at 180 ft; Morley Muck, c & d 20 at 270 ft 1; Kilburn, c 36 at 387 ft 5; Upper Band horizon at 782 ft 10; Norton, c 23 at 831 ft 8; Forty-Yards (UL), c 4, d 2 on b 14 at 862 ft 8; Forty-Yards (LL), c 12 at 877 ft 1; Alton MB at 959 ft 2; Alton, c 26 at 961 ft 4; First Smalley Lingula Band at 982 ft 1; First Smalley, c 13, d 1, c 6, m 21 on c & b 8 at 986 ft 2; Second Smalley, c & d 12 at 994 ft 9; Holbrook, c & d 4 at 1015 ft 6; Belperlawn, b 5 on c 27 at 1037 ft 4; recorded to 1042 ft 11.

*For Cotespark Colliery No. 1 UBH see Smith and others 1967, p. 316.

*No. 1 BH [4255 3840] was drilled to only 12 ft depth in 'wet sand and gravel'

c

DENBY (DRURY-LOWE) COLLIERY SHAFT [3825 4709] 1839, 284 ft.
See Figs 20 and 22. Coal 9 at 56 ft 11; 'Ashgate' (UL), c 18 at 100 ft 10; 'Ashgate' (LL), c 24 at 140 ft 10; Mickley Thin, c 18 at 202 ft; Lower Brampton, ca 12 on c 9 at 271 ft; Morley Muck, c 3, d 3 on c 9 at 370 ft 8; Kilburn, c 45, d 4 on c 13 at 480 ft 3; TD 484 ft 6.

DENBY (DRURY-LOWE) COLLIERY UBH [4032 4649] 1950.
Down from Kilburn at -215 ft. Cores, from 46 ft 2, examined by R. A. Eden. See Figs. 14, 16 and 68 for graphic sections. Upper Band horizon at 369 ft; Norton MB at 418 ft 3; Norton, c 16 at 420 ft 4; Forty-Yards MB at 453 ft 10; Forty-Yards (UL), c 8 at 455 ft 2; Upper Parkhouse MB at 481 ft 9; Lower Parkhouse MB at 514 ft 8; Alton MB at 531 ft 1; Alton, c 31 at 533 ft 8; First Smalley Lingula Band at 545 ft 5; First Smalley, c 21 at 547 ft 11; Second Smalley, c 16 at 562 ft 1; Holbrook, d c 2 at 583 ft 5; Belperlawn, c 29 at 604 ft; TD 650 ft 9.

DENBY COLLIERY, NEW WINNINGS SHAFT [3951 4700] 1867, abt 360 ft.
See also Gibson and others 1908, p. 90. Deep Soft, c 43 at 44 ft 5; Deep Hard, c 60 at 145 ft 7; First Piper, c 35, d 8, c 9, d 5 on c 19 at 196 ft 2; Hospital, c 26 at 237 ft 4; Low Main, c 50 at 317 ft 7; Threequarters, c 12, m 6 ft 8 on c 12 at 341 ft 8; Yard, c 7 at 411 ft 11; Blackshale (?partial washout) c 17, d 3 on c 11 at 453 ft 7; Coal 7 at 527 ft 1; 'Ashgate' (UL), c 28 at 542 ft 6; 'Ashgate' (LL), c 31 at 550 ft 10; Mickley Thin, c 16 at 607 ft 7; Lower Brampton, c 17 at 671 ft 7; Coal 3 at 700 ft 5; Morley Muck, c 3, d 3, c 9 on b 19 at 760 ft 6; Kilburn, c 45, d 4 on c 12 at 872 ft 6; TD 914 ft 1.

DENBY HALL COLLIERY [3968 4821] 1881, abt 357 ft.
See also Fig. 22 for sequence of Yard and Blackshale coals. Second Ell, ca 24 at 22 ft 11; Brown Rake, c 24 at 155 ft 6; base of Westphalian B estimated at 112 ft; False Ell, c 22, m 26, c 1, m 20, c 14, m 16 on c 30 at 203 ft 3; Deep Soft, c 52 at 247 ft 9; Deep Hard workings at 351 ft; First Piper, c 33, d 17 on c 31 at 399 ft; Hospital, c 27 at 439 ft 8; Low Main, c 48 at 505 ft; Threequarters, c 12, m 6 ft 8 on b 12 at 529 ft 6; Yard, c 7 at 612 ft 11; Blackshale, b 24, d 4, c 16, d 5, c 30, d 15, c 10, d 15 on c 19 at 662 ft 9; Coal 7 at 737 ft 10; 'Ashgate' (UL), c 28 at 753 ft 3; 'Ashgate' (LL), c 31 at 761 ft 10; Mickley Thin, c 14 at 816 ft 3; Lower Brampton, c 17 at 876 ft 6; Coal, c & d 6 at 906 ft 3; Morley Muck, b 59 at 967 ft 3; Kilburn, b 5, c 36, d 3 on c 9 at 1071 ft 9; recorded to 1071 ft 9.

DIGBY COLLIERY NEW SHAFT (SPEEDWELL PIT) [4916 4478] 1870, 254 ft.
?Bottom Second Waterloo, c 18 at 37 ft 6; Third Waterloo, c 18 at 51 ft (?truncated by fault in shaft); Coal, d c 3 at 72 ft 3; Fourth Waterloo, c 12 at 91 ft; First Ell, c & d 15 at 110 ft 11; Second Ell, c & d 22 at 163 ft 1; base of Westphalian B estimated at 253 ft; Brown Rake, d c 6 at 275 ft 5; Top Soft, c 6 at 285 ft 10; Roof Soft (UL), c 10 at 298 ft 8; Roof Soft (LL), c 22 at 310 ft 8; Deep Soft, c 34 at 363 ft; Deep Hard, c 36 at 403 ft; TD 421 ft.

DOVECOTE ROAD BH, see Greasley (Dovecote Road) Borehole.

DUFFIELD BH [3428 4217] 1967, 202 ft.
Examined by N. Aitkenhead, A. Crosby, J. G. O. Smart and D. V. Frost. See also Aitkenhead (in press); Fig. 4 and Plate 4.

	Thickness ft	in	Depth ft	in
SUPERFICIAL				
Soil, clayey gravel and silt	2	10	2	10
NAMURIAN				
CHOKIERIAN STAGE (H_1)				
HOMOCERAS SUBGLOBOSUM (H_{1a}) ZONE				
Mudstone; interbedded mudstone, siltstone and sandstone from 2 ft 10 to 35 ft; H. subglobosum bands at 86 ft 11, 92 ft and 109 ft 9	106	11	109	9
ARNSBERGIAN STAGE (E_2)				
NUCULOCERAS NUCULUM (E_{2c}) ZONE				
Mudstone; interbedded mudstone, siltstone and sandstone from 130 ft 4 to 146 ft 1, 158 ft 4 to 165 ft 5, 234 ft 3 to 239 ft and 250 ft 10 to 281 ft 1; N. nuculum bands at 123 ft 2, 127 ft 10, 155 ft 7 and 197 ft 3; N. cf. stellarum band at 285 ft 10	176	3	286	0
CRAVENOCERATOIDES NITIDUS (E_{2b}) ZONE				
Mudstone, calcareous in places, with limestone bullions; K-bentonitic laminae at 294 ft 8, 321 ft 7 and 407 ft 10; Ct. nititoides band at 296 ft 4; Eumorphoceras rostratum band associated with cherty dolomitic limestone at 299 ft 10 and at 309 ft 5 (repeated by a fault); Ct. nitidus and E. leitrimense band at 421 ft 4; Ct. edalensis band at 450 ft 8, Cravenoceras holmesi band at 462 ft 10	176	10	462	10

d

Duffield BH cont.

	Thickness ft	in	Depth ft	in
EUMORPHOCERAS BISULCATUM (E_{2a}) ZONE				
Mudstone, calcareous in places; graded sharp-based mudstone-siltstone and sandstone units from 495 ft 9 to 546 ft 8, 558 ft to 624 ft, 636 ft 2 to 671 ft 4; K-bentonitic laminae at 496 ft 6, 470 ft 6, 494 ft 6, 561 ft 4, 708 ft 1, 734 ft 9 and 760 ft; E. yatesi band at 487 ft 3, E. bisulcatum band at 689 ft 4, C. cowlingense band at 756 ft 8	302	4	765	2
PENDLEIAN STAGE (E_1)				
CRAVENOCERAS MALHAMENSE (E_{1c}) ZONE				
Mudstone, calcareous in places; graded sharp-based mudstone/siltstone and sandstone units from 765 ft 2 to 944 ft 10; K-bentonitic laminae at 949 ft 5; C. malhamense band at 988 ft 5	235	11	1001	1
EUMORPHOCERAS PSEUDOBILINGUE (E_{1b}) ZONE				
Mudstone, calcareous in places; graded sharp-based mudstone/siltstone and sandstone units from 1001 ft 1 to 1021 ft 11 and 1084 ft 6 to 1099 ft 10; E. pseudobilingue bands at 1080 ft 7 and 1143 ft 11	142	10	1143	11
CRAVENOCERAS LEION (E_{1a}) ZONE				
Mudstone, calcareous with a few thin beds of argillaceous limestone in basal 23 ft 5; C. leion band at 1331 ft 3	187	4	1331	3
DINANTIAN (WIDMERPOOL FORMATION)				
UPPER POSIDONIA (P_2) ZONE				
Sandstone in sharp-based graded units with siltstone and mudstone	39	11	1371	2
Mudstone, mainly calcareous, with graded sharp-based mudstone/silstone/sandstone or mudstone/calcareous siltstone/bioclastic limestone units; graded calcareous mudstone/bioclastic limestone/limestone conglomerate unit 7 ft 8 at 1717 ft 3; tuffs, 6 ft 4 at 1535 ft 5, 1 ft 7 at 1629 ft 6, 10 ft 3 at 1690 ft 2 and 2 ft 10 at 1735 ft; Sudeticeras sp., Neoglyphioceras sp.	380	8	1751	10
Dolerite, fine-to medium-grained	27	9	1779	7
Mudstone with graded sharp-based mudstone/siltstone/sandstone or limestone units; tuffs, 1 in at 1843 ft 6, 3 in at 1864 ft 10 and 24 ft 5 at 1891 ft 2; Sudeticeras sp.	146	0	1925	7
LOWER POSIDONIA (P_1) ZONE				
Mudstone, mainly calcareous, with graded sharp-based mudstone/siltstone/sandstone or limestone units; tuff 4 ft 6 at 1830 ft 1; Posidonia becheri, Goniatites sp., G. cf. koboldi (at 1960 ft 10), G. sphaericostriatus (at 2379 ft 2), G. falcatus (at 2428 ft 10), G. striatus (below 2525 ft 7), G. crenistria (below 2624 ft 9)	1104	11	3030	6
UPPER BEYRICHOCERAS (B_2) ZONE				
Mudstone, mainly calcareous, with graded sharp-based units of mudstone/calcareous siltstone/sandstone or limestone, baked below 3077 ft 11; G. sp. [maximus group] (at 3040 ft 6)	82	6	3113	0
Hornfels, laminated dark to pale grey, mottled	8	4	3121	4
Dolerite, passing down to gabbro between 3128 ft 8 and 3263 ft 3 and then to picrite; chilled margins	215	11	3337	3
Hornfels, dark to pale grey, spotted	3	5	3340	8
Dolerite, fine-grained; chilled margins	1	6	3342	2
Hornfels, dark grey, spotted	0	6	3342	8
Mudstone, mainly dark grey, baked and indurated	9	4	3352	0
Mudstone, calcareous, with graded sharp-based calcareous mudstone/siltstone/limestone units	16	5	3368	5
Sandstone in sharp-based graded units with siltstone				

a

Duffield BH cont.

	Thickness ft	in	Depth ft	in
and mudstone; G. maximus (at 3991 ft 7)	68	5	3436	10
Core not recovered to TD	16	2	3453	0

DUNSTEAD BH, LANGLEY MILL [4453 4747] 1913, 296 ft.
Mainsmut, c 18, d 5 on c 13 at 31 ft 6; ?Comb, c 49 at 66 ft 6; Top Hard workings at ? 78 ft; ?Third Waterloo, c 6 at 200 ft 4; ?Fourth Waterloo, c 18 at 252 ft; First Ell, c 14 at 301 ft 1; Second Ell, c 12 at 342 ft 10; base of Westphalian B estimated at 460 ft; False Ell, c 30, d 14 on c 14 at 543 ft 10; Deep Soft, c 48 at 576 ft 11; Deep Hard, c 36 at 626 ft 6; recorded to 631 ft 4.

EASTWOOD COLLIERY No. 3 SHAFT [Approx. 4621 4577], abt 190 ft.
First Ell, c & m 68 at 8 ft 5; Second Ell, ca 15 at 54 ft 9; base of Westphalian B estimated at 126 ft; Brown Rake, c 10 at 140 ft 11; Top Soft, c & b 10 at 171 ft 11; Roof Soft (UL), c 8 at 180 ft 1; Roof Soft (LL), c 22 at 201 ft 11; Deep Soft, c 43 at 241 ft 10; Deep Hard, c 45 at 293 ft 3; recorded to 308 ft 3. A second record labelled 'Eastwood Colliery', perhaps the same or a closely adjacent shaft, continues the section: Deep Soft at 249 ft; Deep Hard, c 45 at 301 ft 3, First Piper, c 39 at 340 ft 7; Hospital, c 17 at 401 ft 3; Cockleshell, c 22 at 418 ft; Low Main, c 42 at 452 ft 8; Threequarters, c 19 at 461 ft 11; Coal 2 at 838 ft; Morley Muck, c 13, m 6 ft 8 on c & d 2 at 918 ft 3; Kilburn, ca 17 at 1025 ft; recorded to 1102 ft 2, drilled below to 1134 ft 2.

EASTWOOD HALL BH [4652 4753] 1963-64,219 ft.
Cores, below 27 ft, examined by J. G. O. Smart and J. Chilton. See Figs.19, 25, 28, 34, 37 and Plate 5. Top Hard, c 36 at 22 ft 8 (from drillers log); Dunsil, c 20, d 3 on d c 5 at 63 ft 8; Top First Waterloo, c 2, m 1 ft 9 on d c 3 at 89 ft 10; Bottom First Waterloo, ca 6 on c 13 at 103 ft; Waterloo Marker, c 4, d 2 on c 2 at 124 ft 4; Top Second Waterloo, c 25, d 2 on c 2 at 153 ft 6; Bottom Second Waterloo, c 12 at 167 ft; Third Waterloo, c 15 at 191 ft 3; ?Third Waterloo (LL) horizon at 198 ft 10; Coal 2 at 233 ft 2; Fourth Waterloo, ca 1, c 14 on c & d 6 at 238 ft 9; First Ell, c 1, d & b 4 on c 17 at 271 ft 8; Second Ell, ca 11 at 322 ft 5; Clay Cross MB at 387 ft 3; Brown Rake, c 5 at 404 ft; Black Rake TS at 435 ft 10; Black Rake horizon at 435 ft 10; Top Soft, c 12 at 449 ft 8; Roof Soft (UL), c 12 at 462 ft; Roof Soft (LL), c 18 at 475 ft 6; Deep Soft workings at 520 ft 6; Deep Hard workings at 570 ft 8; First Piper, c 24 on d c 18 at 603 ft 6; Hospital, c 18 at 666 ft 2; Cockleshell, c & d 27 at 694 ft 7; Low Main, c 40 at 722 ft 10; Threequarters, c 12, d 6, c 1, m 2 ft 7 on c 5 at 738 ft; Yard, c 4 at 796 ft 6; Blackshale, c 8, d 17, d c 1, d 3, c 16, d 4, c 9, d 4 on c 23 at 845 ft; 'Ashgate' (LL), c 24 at 960 ft 6; Mickley Thin, d c 3 on c 2 at 1011 ft 11; Morley Muck horizon at 1139 ft 1; TD 1193 ft 6. Borehole deflected and proved Kilburn, c 18 at 1272 ft 6; TD 1284 ft.

ERICSSON LTD WATER BHs, BEESTON [5328 3590] 1902, 1906, 1911, abt 1945 and 1947, closely adjacent.
See Plate 10,and Lamplugh and Smith 1914, p. 51. Record of the 1911 BH: Made ground to 2 ft; alluvium and Flood Plain Terrace to 19 ft; Keuper Marl to 31 ft; Waterstones to 184 ft; Pebble Beds to 333 ft; Lower Mottled Sandstone to TD at 400 ft (?base of Triassic rocks).

EXHIBITION PIT, BUTTERLEY PARK [4264 5068] 1851, abt 460 ft.
No detailed record to 204 ft 6. Low Main at 60 ft; Blackshale at 204 ft 6; Coal, c & b 12 at 264 ft 10; 'Ashgate' (UL) horizon at 280 ft 5; 'Ashgate' (LL), c 8 at 309 ft 8; Mickley Thin, c 15 at 375 ft; Lower Brampton, c 4 at 428 ft 6; Coal 2 at 467 ft 10; Morley Muck, c 6 at 514 ft 1; Kilburn, c 26 at 635 ft 2; recorded to 919 ft 2; bored below for 111 ft to Alton, no known details.

EXPRESS DAIRIES BH, BRAILSFORD [2535 4167], abt 480 ft.
Pebble Beds ('gravel') to 30 ft, ?Widmerpool Formation (rocky marl, hard rock, sandstone and shale) to TD at 255 ft.

FELLEY LANE BH, UNDERWOOD [4822 5053] 1964, 375 ft.
Cores, from 50 ft, examined by D. V. Frost. See Figs.43 and 44.

	Thickness ft	in	Depth ft	in
BOULDER CLAY				
Soil and clay with pebbles	25	0	25	0
WESTPHALIAN B				
Mudstone; ironstone	38	0	63	0
Seatearth, pale brown	1	0	64	0
COAL, c 6, d 2 on c 2	0	10	64	10
Seatearth	3	2	68	0
Mudstone; ironstone; plants, Anthracosphaerium aff. propinquum	10	1	78	1
HORIZON OF SUTTON MB	-	-	78	1
Seatearth	0	4	78	5
Mudstone, carbonaceous	0	1	78	6
COAL, c 6	0	6	79	0
Seatearth, brown	2	0	81	0
Sandstone, with mudstone	1	10	82	10
Mudstone; ironstone	9	10	92	8
HAUGHTON MB				
Mudstone; Euphemites anthracinus, Lingula sp.	0	1	92	9
Mudstone, wormy; pyritised plants	4	9	97	6
Seatearth	3	3	100	9
Mudstone, carbonaceous	0	3	101	0
Seatearth	0	5	101	5
Mudstone, carbonaceous	0	1	101	6
?SWINTON POTTERY, c & d 12	1	0	102	6
Seatearth	0	6	103	0
Mudstone; ironstone	5	0	108	0
Siltstone and sandstone	26	0	134	0

b

	Thickness ft	in	Depth ft	in
Seatearth	4	0	138	0
Mudstone	9	0	147	0
Sandstone	8	8	155	8
Mudstone; plants	6	0	161	8
Sandstone	22	6	184	2
Mudstone; plants	7	10	192	0
Siltstone	3	0	195	0
Mudstone	9	6	204	6
?HORIZON OF MANTON 'Estheria' BAND	-	-	204	6
Seatearth	0	6	205	0
Mudstone	11	8	216	8
Sandstone; siltstone laminae	26	10	243	6
Mudstone; bivalves, 'Estheria' sp., fish debris	13	0	256	6
MAINBRIGHT, c 30	2	6	259	0
Seatearth	3	0	262	0
Mudstone; ironstone; plants	7	9	269	9
TWO-FOOT MB				
Mudstone; Lingula mytilloides, Myalina sp., ostracods including Paraparchites sp., fish debris	0	9	270	6
TWO-FOOT, c 15	1	3	271	3
Seatearth	3	6	274	9
COAL, c 3	0	3	275	0
Seatearth	0	9	275	9
Siltstone and sandstone	11	3	287	0
Mudstone	9	11	296	11
LOWBRIGHT, c 30	2	6	299	5
Seatearth	3	7	303	0
Mudstone; ironstone	2	8	305	8
Siltstone and sandstone	4	11	310	7
Mudstone; ironstone; plants	5	5	316	0
LOWBRIGHT FLOOR, c 6	0	6	316	6
Seatearth	1	6	318	0
Mudstone; plants	15	5	333	5
Sandstone	1	2	334	7
Mudstone; siltstone laminae	13	8	348	3
Mudstone; ironstone; bivalves	11	0	359	3
BRINSLEY THIN, c 12	1	0	360	3
Seatearth	2	9	363	0
Mudstone; ironstone	19	4	382	4
HIGH HAZLES, c 20	1	8	384	0
Seatearth	1	9	385	9
Mudstone; plants	1	6	387	3
Siltstone to TD	3	0	390	3

FORD'S BH, MAREHAY [3923 4932] 1963, 495 ft.
Recorded by J. G.O. Smart and J. Chilton. See Fig. 34. Fourth Waterloo, c & d c 22 at 66 ft; First Ell, c 27 at 115 ft 9; Second Ell, cannel fragments at 162 ft; Clay Cross MB at 247 ft 2; Brown Rake, c & d 24 at 284 ft 6; False Ell, c 30, m 7 ft 6, c & d c 6, m 39 on c & d 24 at 325 ft 6; Deep Soft workings at 400 ft; Deep Hard workings at 473 ft, First Piper, c 62 at 530 ft 5; Hospital, c 24 at 572 ft 9; TD 574 ft 2.

FORD'S PIT, MAREHAY [3925 4929] (No. 5 SHAFT MAREHAY MAIN COLLIERY) before 1905, 498 ft.
Fourth Waterloo, c 16 on b 9 at 62 ft 3; First Ell, c 24 at 111 ft 1; Second Ell, b 24 at 141 ft 9; base of Wesphalian B estimated at 238 ft; Brown Rake, c 13 at 274 ft 7; False Ell, c 22, m 36 on c 33 at 309 ft 7; Deep Soft workings at 383 ft; Deep Hard and Floor Coal, c 60 at 460 ft 6; Piper, c 23, d 1, c 12, d 36 on c 16 at 524 ft 11; Hospital, c 27 at 567 ft 2; Cockleshell (UL) horizon at 589 ft 7; Cockleshell (LL) horizon at 602 ft 7; Low Main, c 52 at 631 ft 9; recorded to 631 ft 9.

FOREST LODGE BH, ANNESLEY WOODHOUSE [5080 5317] 1969, 517 ft.
Recorded by T. Draper, C. Beal and L. Fusez. See Fig. 82 for graphic section of Permo-Triassic rocks. Permian on Westphalian to High Main. TD 563 ft 2.

FRITCHLEY No. 1 BH [3639 5272] 1952, 280 ft.
See Fig. 14 for sequence Belperlawn to Alton coals.

	Thickness ft	in	Depth ft	in
WESTPHALIAN A				
Core lost	10	0	10	0
Mudstone	1	6	11	6
ALTON MB				
Mudstone; Lingula	6	2	17	8
ALTON, c 10	0	10	18	6
Seatearth	4	6	23	0
Ganister	1	0	24	0
Clay, sandy at base	8	6	32	6
Sandstone	5	0	37	6
Mudstone; plants, bivalves	3	0	40	6
SECOND SMALLEY Lingula BAND				
Mudstone; Lingula	0	7	41	1
SECOND SMALLEY, c 3	0	3	41	4
Seatearth	0	8	42	0
Sandstone	7	4	49	4
Mudstone; bivalves	3	11	53	3
HOLBROOK Lingula BAND				
Mudstone; Lingula	0	3	53	6
HOLBROOK HORIZON	-	-	53	6
Seatearth	0	3	53	9
Sandstone and siltstone	6	3	60	0
Seatearth	4	0	64	0
Siltstone and sandstone	22	3	86	3
BELPERLAWN, c 12, d 1 on c 5	1	6	87	9
Seatearth	0	2	87	11
CRAWSHAW SANDSTONE				
Sandstone; siltstone and mudstone partings; plants	21	5	109	4
Mudstone	2	8	112	0
Siltstone; mudstone partings	3	10	115	10
Mudstone; plants	15	10	131	8
Sandstone and mudstone interbanded to TD	56	4	188	0

GARFIT HOUSE BH [4934 5204] 1966, 589 ft.
Cores, from 50 ft, examined by D.E. Raisbeck. See Figs.42 and 48. Boulder clay to 70 ft; Lower Magnesian Limestone to 80 ft; Lower Marl to 136 ft 8; Permian Basal Breccia at 140 ft 11; ?Main 'Estheria' Band at 150 ft 7; Edmondia Band at 214 ft 5; High Main, d c 2, c 14, d 5 on c 27 at 254 ft 5; Coal 9 at 301 ft 5; Coal 10 at 316 ft 5; Mansfield MB at 418 ft 9; Coal 4 at 419 ft 1; ?Sutton MB at 468 ft; Steep dip and evidence of faulting between 510 ft 10 and 541 ft 8; ?Clown, c & d 51 at 519 ft 6; Mainbright horizon (washout) at 615 ft 9; Two-Foot MB at 625 ft 11; Two-Foot, c 22 at 628 ft 4; Lowbright, c 22, d 1 on c 12 at 673 ft 9; Lowbright Floor, c & d 11 at 679 ft 2; High Hazles, ca 2 on c 16 at 758 ft 8; TD 784 ft 6.

GRAMMER STREET BH [4164 4769] 1963, 377 ft.
Recorded by J. G.O. Smart and J. Chilton. See Fig. 34. ?Second Ell horizon at 21 ft; Clay Cross MB at 118 ft 5; Brown Rake, c 16 at 143 ft 2; False Ell, d c 7, c 29, m 3 ft 7, d c 3, c 9, d 3, c 29, d 4 on d c 4 at 194 ft; Deep Soft workings at 240 ft 5; TD 244 ft.

GREASLEY CASTLE BH [4933 4678] 1961-62, 273 ft.
Cores, 20-30 ft, 95-102 ft, 264-369 ft and 380 ft to TD, examined by J. Chilton. See Fig. 44. Coal 10 at 8 ft 10; High Hazles, c 24 at 100 ft 6; Cinderhill, c 18 at 203 ft 6; Mainsmut, c 22 at 308 ft 10; Comb, c 36 at 360 ft; Top Hard workings at 367 ft 6; Dunsil, c 31 at 397 ft 3; Top First Waterloo, c 15, d 6 on c 1 at 422 ft; Bottom First Waterloo, d c 8 at 453 ft 8; TD 470 ft.

GREASLEY (DOVECOTE ROAD) BH [4780 4675] 1967, 355 ft.
Recorded by J. G.O. Smart and D. E. Raisbeck. See also Plate 5. Cinderhill Fault at 74 ft 3; Top Hard workings at 92 ft 2; Dunsil, c 28, d 1 on d c & c 6 at 135 ft 11; First Waterloo, c 10, m 5 ft 2, c 2, d 11 on c 15 at 168 ft 10; Waterloo Marker, c 2 at 190 ft 1; Second Waterloo, c 29, d 4, c 5, m 2 ft 9 on c 19 at 221 ft 2; Coal 2 on b 1 at 223 ft 5; Third Waterloo, c 16 at 243 ft 4; TD 254 ft 5.

GREENHILL LANE No. 1 BH* [4228 5217] 1913, abt 430.
From Second Waterloo to Deep Soft, TD 516 ft 4.

HANSON FARM WATER BH, ALDERWASLEY [3233 5459], abt 405 ft.
See Plate 4 for graphic section.

	Thickness ft	in	Depth ft	in
'Soil and stone'	5	6	5	6
NAMURIAN				
ASHOVER GRIT (MAIN BED)				
Stone, red, pink and brown	170	6	176	0
Clay, soft reddish brown	15	6	191	6
Stone and clay seams	2	6	194	0
Clay, reddish brown and bands of stone (tough)	64	0	258	0
Clay, hard grey	36	6	294	6
Stone, grey	4	0	298	6
Clay, hard grey with small stones	287	6	586	0
Clay, dark with traces of limestone to TD	14	0	600	0

*Another record known as 'Sinking at Green Hill Lane' is considered to be unreliable and is possibly a built-up section from Dunsil to Kilburn. TD 1335 ft. Site unknown.

HEANOR GATE BH [4266 4601] 1962-63,364 ft.
Cores, from 19 ft, examined by R.S. Arthurton, J.G.O. Smart and J. Chilton. See Figs. 33 and 36 and Plate 5. Dunsil, c 28 at 50 ft 4; Top First Waterloo, ca 7 at 90 ft 6; Bottom First Waterloo, c 18 at 112 ft 3; Waterloo Marker, c 1, m 7 on c 3 at 132 ft 9; Top Second Waterloo workings at 163 ft 6; Bottom Second Waterloo, c 12 at 184 ft; Third Waterloo, d c 3, d 2 on c 25 at 218 ft 4; Fourth Waterloo, c 16 at 272 ft 10; First Ell, d c 6, d 3, c 27, m 9 ft 11 on c 1 at 327 ft; Second Ell, ca 12, m 19 on c 6 at 369 ft; Clay Cross MB at 464 ft 4; Brown Rake, c 28 at 488 ft 2; Black Rake TS at 513 ft 7; False Ell, c 12, m 18 on c 36 at 536 ft 3; Deep Soft, c 58, d 2 on c 6 at 545 ft 9; TD, in Deep Hard workings, 590 ft.

HEATHERDALE POND BH, ANNESLEY [5068 5141], 514 ft.
Recorded by T. Draper, C. Beal and L. Fusez. See Fig. 82 for graphic section of Permo-Triassic rocks. Permian to High Main, TD 514 ft 4.

HEMPSHILL COLLIERY No. 3 SHAFT [5265 4449] 1850-51, abt 250 ft.
See Figs.40 and 46 and Lamplugh and Smith 1914, p. 77. Depths calculated from top of pit which is 8 ft 6 above ground level. Lower Mottled Sandstone to 52 ft 1; Middle Marl to 73 ft 1; Lower Magnesian Limestone to 97 ft 1; Lower Marl to 133 ft 2; inferred Mansfield MB cank at 178 ft 9; Coal, c 21, b 19 on c 18 at 227 ft 4; Coal, c & b 19, d 68 on c 15 at 251 ft 4; Clown, c & d 31 at 289 ft 3; Mainbright, c 40 at 352 ft 11; Two-Foot, c 24 at 367 ft 2; Coal 3 at 370 ft 11; Lowbright, c 27 at 403 ft 3; Lowbright Floor, c 66 at 412 ft 5; Brinsley Thin (UL) horizon at 439 ft 10; Brinsley Thin, c 14 at 456 ft 7; High Hazles, c 26 at 479 ft 2; Cinderhill, c 25 at 586 ft 2; ?St. Johns, c 5 at 615 ft 8; Mainsmut, c 20 at 660 ft 5; Comb, c 13, b 4 on c 17 at 699 ft 3; Top Hard, c 58 at 705 ft 3; Top Hard Floor, c 11 at 727 ft 1; Dunsil, c 25 on b 6 at 770 ft 5; First Waterloo, c 39 at 807 ft 6; TD 829 ft 1.

HEMPSHILL HEADING (Main or Northerly Drift) 1909-11, close to No. 3 Shaft (above). From Top Hard at -458 ft.
Second Waterloo, c 18, m 66 on c 28 at 187 ft; Third Waterloo, c 4, d 4, c 1, m 15 on c 17 at 234 ft 11; Fourth Waterloo, c 1 (roof coal), m 31 on c 10 at 270 ft 8; First Ell, d c 8, d 18 on d c 16 at 312 ft 6; Second Ell, c 21 at 345 ft 9; inferred base of Westphalian B at 406 ft; Brown Rake, d c 2 at 419 ft 2; Top Soft, c 3 at 444 ft 7; Roof Soft (UL), c 7 at 455 ft 6; Roof Soft (LL), ca 1 on c 9 at 480 ft 3; Deep Soft, c 33 at 508 ft 6; Deep Hard, c 20 at 541 ft 8; TD 541 ft 8.

HIGH LEYS BH, HUCKNALL [5299 4880] 1970, 304 ft.
Cores, from 1200 ft 5, examined by C. Beal. See Fig. 37 for Dunsil Coal section. Top Hard Floor, c 10 at 1213 ft 10; Dunsil, ca 6, c 4, d 3 on c 19 at 1251 ft 11; First Waterloo, c 20, d 15 on c & d 24 at 1276 ft 11; Waterloo Marker, c 7 on b 2 at 1294 ft 2; Top Second Waterloo, b 1, ca 2, c 25, d 4 on d c 10 at 1314 ft 7; Bottom Second Waterloo, c 4, b 3 on b 5 at 1322 ft; Third Waterloo, c 13 at 1343 ft 8; Coal, b 1 at 1358 ft 1; Fourth Waterloo, c 13 at 1407 ft; ?First Ell (UL), c 1 at 1417 ft 7; ?First Ell (LL), ca 2 on c 2 at 1420 ft 4; Second Ell, ca 1, c 16, d 9 on c 3 at 1486 ft 2; TD 1495 ft 0. c

d

HIGH PARK COLLIERY No. 1 SHAFT [4864 4866] 1854-60, 338 ft.
See also Figs.46 and 77. ?Swinton Pottery, c 13 at 46 ft 9; Clown, c 24, m 12 ft on c & d c 28 at 91 ft 5; Coal 10 at 170 ft 3; Mainbright, c 35 at 210 ft; Two-Foot, c 27, m 4 ft 3 on c & b 5 at 231 ft 3; Lowbright, c 36 at 267 ft 3; Lowbright Floor, c & d 14 at 277 ft 3; Brinsley Thin, c 20 at 328 ft 4; High Hazles, c 28 on b 14 at 362 ft 1; Cinderhill (UL), c & b 19 at 412 ft 9; Cinderhill, c 10 at 426 ft 11; Mainsmut, c 21 at 532 ft 2; Comb, c 30 at 570 ft; Top Hard, c 61 at 582 ft 1; Top Hard Floor, c 13 at 599 ft 10; Dunsil, c 25 at TD, 633 ft 5.

HILLTOP COLLIERY [4764 4677], abt 368 ft.
High Hazles (See Fig. 42), c 36 at 15 ft 6; Coal, d c 8 at 34 ft 6; Cinderhill (UL), c 14 at 86 ft 6; Cinderhill, c 18 at 106 ft; ?St. John's, d 15 at 133 ft 2; Mainsmut, c 30 at 205 ft 5; Comb, c 37 at 254 ft 7; Top Hard, c 92 at 294 ft 9; TD 297 ft 9.

HOME FARM (ANNESLEY) BH [5068 5282] 1968-69, 528 ft.
Recorded by T. Draper, C. Beal, L. Fusez. See Fig. 82 for graphic section of Permo-Triassic rocks and Fig. 48 for section of High Main Coal. Lower Mottled Sandstone to 92 ft 11; Middle Marl to 119 ft 4; Lower Magnesian Limestone to 152 ft 5; Lower Marl to 215 ft 3; Permian Basal Breccia to 216 ft 7; Top MB at 312 ft 8; Edmondia Band at 500 ft 11; High Main, c 13, d 7 on c 35 at 539 ft 3; TD 577 ft 8.

HOPTON BH, WIRKSWORTH [2677 5334] 1967, abt 700 ft.
Cores, from 23 ft, examined by D. V. Frost.

	Thickness ft	in	Depth ft	in
SUPERFICIAL				
Clay and sand, grey and brown; mudstone fragments	23	0	23	0
Clay, with quartzite pebbles, sandstone and ironstone fragments	1	0	24	0
NAMURIAN				
ARNSBERGIAN STAGE				
Sandstone and cank	3	0	27	0
Mudstone, dark grey, silty, pyritic; bivalves; dip 40° at 35 ft, 55° at 40 ft; fish debris and ostracods at 40 ft 6, Cravenoceratoides sp. at 43 to 49 ft (E_{2b}), bivalves, fish debris at 49 ft and plants at 50 ft. Dip 60° at 59 ft. Vertical fault plane at 60 ft. Dip 80° at 64 ft and 30° at 65 ft. Bivalves and shell fragments at base	41	0	68	0
Siltstone, calcareous	1	0	69	0
Siltstone, carbonaceous, plant and shell debris	1	0	70	0
Mudstone, dark grey, silty; 6-in cank at 72 ft 4; plants	15	0	85	0
Mudstone, dark grey with silty laminae at 88 ft; Lingula sp. at 89 ft, bivalves, including Leioptera sp. (common)	5	0	90	0
Mudstone, dark grey silty; pyrite; plants	3	6	93	6
Mudstone; Lingula sp. bivalves, including Leioptera sp.	4	6	98	0
Mudstone; plants, ostracods and fish at base (dip slight)	12	0	110	0
Mudstone, dark grey, silty; pyritic plants, brachiopods, bivalves, including Leioptera sp. (common), goniatite fragments, ostracods and fish	4	2	114	2
Mudstone; pyritic plants, spat at base	6	10	121	0
Mudstone; Lingula sp., Leioptera sp., fish	16	8	137	8
Mudstone, pyritic, dip 55°; cank nodules at 138 ft, 1-in cank at base, to TD	11	0	148	8

HOPWELL HALL No. 2 BH [4387 3610] abt 1891, abt 325 ft.
For graphic section see Plate 10. Keuper Marl to 162 ft; Waterstones to 265 ft 6; Pebble Beds recorded to 322 ft 6.

HUCKNALL No. 1 COLLIERY (TOP PIT) No. 2 SHAFT* [5270 4802] 1861, deepened below Top Hard 1954-55, 310 ft.
See Plate 5, Lamplugh and Smith 1914, p. 79 and also Edwards 1951, p. 187 for record to Top Hard Coal. Lower Magnesian Limestone to 23 ft 7; Lower Marl to 64 ft 6; Permian Basal Breccia to 64 ft 10; Westphalian: Top Hard Coal at 1161 ft 4; First Waterloo, c 19, d 7 on c 26 at 1257 ft; Waterloo Marker, c 9 at 1276 ft 10; Second Waterloo, c 33, d 2, c 8, d 23 on c & d 19 at 1307 ft 2; Third Waterloo, c 24 at 133 ft; Fourth Waterloo, c 1 (roof coal), m 7 ft 3, c 18 (main leaf), d 14, on c 1, b 4 (floor coal) at 1380 ft 9; First Ell, c 2, d 42, ca 10 on c 11 at 1407 ft 6; Second Ell, ca 9 on c 15 at 1454 ft 5; Clay Cross MB at 1515 ft 3; Brown Rake, b 3 on c 10 at 1531 ft 8; Top Soft (UL), c 6 at 1560 ft 7; Top Soft (LL), b 1, d 5 on c 5 at 1570 ft 3; Roof Soft, ca 11 at 1613 ft 8; Deep Soft, c 37 at 1635 ft 7;

* No.1 Shaft site unknown.

a

Deep Hard, ca 2 on c 41 at 1680 ft 10; TD ?1699 ft 6.

HUCKNALL No. 2 COLLIERY No. 3 SHAFT* [5408 4896] 1865-66, 238 ft.
Deepened 1963 below Top Hard Coal (Samples examined by D.E. Raisbeck). Lower Magnesian Limestone to 32 ft 6; Lower Marl to 85 ft 6; Westphalian from Manor Coal to Top Hard Coal at 1235 ft; First Waterloo, c 18, d 6 on c 15 at 1313 ft 5; Waterloo Marker, c 8 at 1335 ft 4; Top Second Waterloo, c 22 at 1351 ft; Bottom Second Waterloo, c 8 at 1361 ft 3; Third Waterloo, c 14 at 1387 ft 2; Third Waterloo (LL) horizon at 1412 ft; Fourth Waterloo, c 8, c & d 4 on c 5 (floor coal) at 1455 ft 2; First Ell, c 1 at 1475 ft 9; Second Ell, c & ca 22 at 1514 ft 1; Clay Cross MB at 1566 ft; Brown Rake, c 9 at 1585 ft 6; Black Rake TS at 1595 ft 5; Top Soft horizon at 1611 ft 3; Roof Soft, c 16 at 1670 ft 7; Deep Soft workings at 1702 ft 6; Deep Hard, c 67 (abnormal swilley section) at 1761 ft 10; TD 1795 ft.

HUCKNALL No. 2 COLLIERY No. 5 SHAFT [5403 4899] 1959/60, 237 ft.
See Figs. 31, 40, 44, 46 and 49, and Plate 5. Lower Magnesian Limestone to 33 ft 3; Lower Marl to abt 88 ft; Manor ca 36 on c 27 at 136 ft 3; Coal 30 at 229 ft 10; Coal, c 8, d 2 on c 8 at 249 ft 7; Annesley, ca 22 at 293 ft 10; Top MB at 352 ft 6; Coal 16 at 369 ft 5; inferred Shafton MB at 416 ft 3; Coal 22 at 418 ft 1; Shafton, c 11 at 436 ft 5; ?Main 'Estheria' Band at 455 ft 6; Edmondia Band at 522 ft; Coal 3 at 557 ft 8; High Main, c 60 at 568 ft 9; Coal, c 15, d 8 on c 4 at 603 ft 9; Coal 11 at 634 ft 7; Mansfield MB at 683 ft 7; Clown (part washout), c 5 at 830 ft; inferred Manton 'Estheria' Band at 847 ft 11; Mainbright, c 36 at 880 ft 3; inferred Two-Foot MB at 890 ft; Two-Foot, c 20 at 894 ft 4; Lowbright, c 30 at 930 ft; Lowbright Floor, c 12 at 938 ft; Brinsley Thin (UL), c 12 at 976 ft 8; Brinsley Thin, c 15 at 990 ft 1; High Hazles, c 30 at 1015 ft 1; Cinderhill (UL), c 8 at 1094 ft 6; Cinderhill, b 30 at 1118 ft 9; Mainsmut, c 13 at 1161 ft 2; Comb, c 22 at 1225 ft 9; Top Hard, c 47 on ca 27 at 1232 ft; Top Hard Floor, c 2 on ca 6 at 1246 ft 8; Dunsil, ca 10 on c 15 at 1278 ft 3; First Waterloo, c 18 on d c 22 at 1306 ft 8; Waterloo Marker, c 5 at 1328 ft 6; Top Second Waterloo, c 22 at 1344 ft 9; Bottom Second Waterloo, c 6 at 1354 ft 10; Third Waterloo, c 12 at 1382 ft 7; low leaf of Third Waterloo horizon at 1405 ft 3; Fourth Waterloo, c 10 at 1442 ft 5; First Ell, b 6, d 42 on c 7 at 1465 ft 1; Second Ell, ca 8 on c 13 at 1514 ft 9; Clay Cross MB at abt 1570 ft; Brown Rake, c 9 at 1585 ft 7; Roof Soft, ca 4 at 1676 ft 9; Deep Soft, c 33 at 1702 ft 1; Deep Hard†, c 22 on c & d 17 at 1738 ft 2; First Piper washed out at abt 1804 ft; Second Piper, c 9, d 9 on c 3 at 1862 ft; TD 1896 ft.
NB. Records also exist of drifts No. 1 [546 478] and No. 2 [537 484].

*Nos. 1 and 2 Shafts sites unknown. No 4 upcast or southerly shaft deepened 1889-90 from Top Hard-Deep Hard; TD 1781 ft.
†Varies from 25 to 76 in 'Swilley'.

HUCKNALL No. 2 COLLIERY UBH [5363 4861] 1956.
Down from No. 1 Top Hard - Deep Soft Drift at -1170 ft. See Fig. 25. First Ell, ca 9 on c 11 at 58 ft 4; Clay Cross MB at 114 ft 3; Brown Rake, c 8 at 129 ft 8; Black Rake TS at 139 ft 2; Deep Soft, c 37 on b 1 at 245 ft 4; Deep Hard, c 23 at 278 ft; First Piper, c 24 at 342 ft 10; Hospital, c 5, d 29 on d c 6 at 405 ft 3; Cockleshell, c 17 at 436 ft 1; ?detrital coal, c 1 at 454 ft 9; Low Main, c 37 at 466 ft 9; Threequarters, ca 5 on c 6 at 481 ft; Yard, c 10 at 553 ft 7; Blackshale, c & d 44 at 600 ft 1; TD 611 ft 8.

HUCKNALL C's HEADING UBH [5412 5018] 1966.
Down from Deep Soft at -1508 ft. Recorded by D.E. Raisbeck & T. Draper. Deep Hard, c 29 at 34 ft 11; First Piper, ca 2, c 10 (Upper Coal), m 5 ft 8, b 4, m 2 ft 11 on b 4 at 83 ft 9; Tupton, c 9, d 47, c & d 20, d 15 on c 42 at 226 ft 9; Threequarters, c 14, m 99 on c 21 at 254 ft 1; Yard/Blackshale, c 12, d 14, d c 4, c 29, d 2, c 8, d 4, c 10, d 1, d c 4, c 7, d 2 on c 17 at 341 ft 10; Coal 8 at 369 ft 6; Coal 4, m 24 on b 2 at 383 ft 4; 'Ashgate' (UL), c 5 at 399 ft 1; 'Ashgate' (?LL) horizon at 405 ft 3; Coal 6 at 417 ft 10; TD 442 ft.

HUCKNALL C's MAIN OUTBYE UBH [5430 4965] 1967.
Down from Deep Soft at -1494 ft. Recorded by D.E. Raisbeck & T. Draper. Deep Hard, c 20 at 29 ft 8; First Piper, ca 2 on c 5 at 77 ft; Second Piper horizon at 84 ft 8; ?Cockleshell, c 6, d 11 on b 3 at 198 ft 8; Low Main, c 36 at 252 ft; Threequarters (UL), c 1, d c 9 on c 1 at 269 ft; Threequarters (LL), c 14, d 8 on c 5 at 282 ft 9; Yard & Blackshale, d c 8, d 3, d c 1, c 32, d 8, c 22, d 1 on c & d c 32 at 365 ft 6; Coal, c 9, d 6 on c 2 at 396 ft 2; Coal, c 2, d 5, c 3, d 17 on b 6 at 410 ft 9; 'Ashgate' (UL), c 11 at 427 ft 8; 'Ashgate' (LL), c 5 at 439 ft 1; TD 442 ft 6.

HUCKNALL 27's MAIN UBH [5190 4929] 1962/63.
Down from High Main at 102 ft. Recorded by D.E. Raisbeck. Coal 15 at 38 ft 4; Coal 9 at 72 ft 1; Mansfield MB at 129 ft 10; Coal, d c 2 at 130 ft 4; ?Sutton MB at 156 ft 9; Coal 15 at 184 ft 8; Haughton MB at 190 ft 5; ?Swinton Pottery, c 1, d 2 on c 2 at 200 ft 9; ?Clown (part washout) c 5 at 297 ft 2; Mainbright, c 30 at 342 ft 7; Two-Foot MB at 354 ft 3; Two-Foot, c 22 at 356 ft 1; Lowbright, c 35 at 407 ft 5; Lowbright Floor, c 7, d 2, c 6, d 1 on d c 1 at 412 ft 3; Brinsley Thin horizon at 451 ft 4; High Hazles, c 22 at 474 ft 2; TD 489 ft 6.

HUCKNALL SOUTH UBH [5438 4797] 1955.
Down from Mainbright at -624 ft to High Hazles; TD 128 ft 6.

b

IRONVILLE No. 1 OIL BORE [4307 5130] 1919, 335 ft.
Chipped samples examined by the late C.N. Bromehead and K.S. Siddiqui. See Figs. 4 and 6 and Plate 4 for graphic section of the Namurian sequence. The samples are labelled to a total depth of 3655 ft, the bore record finishes at 3630 ft.

	Thickness ft	in	Depth ft	in
WESTPHALIAN				
No samples	50	0	50	0
Shale, grey, carbonaceous and ferruginous; traces of coal; bed of sand at base	200	0	250	0
Clay, sandy and shaly sandstone	55	0	305	0
Shale, grey	63	0	368	0
?KILBURN, c 24	2	0	370	0
Shale, grey	30	0	400	0
WINGFIELD FLAGS				
Shale, grey, sandy and sandstone beds	45	0	445	0
Shale, grey, micaceous	15	0	460	0
Shale and shaly sandstone; trace oil at 510 ft	55	0	515	0
Shale, grey, dark grey and black, sandy in parts; ironstone	245	0	760	0
?NORTON HORIZON (coal traces)	-	-	760	0
Shale, dark grey	25	0	785	0
Shale, pale grey	5	0	790	0
Shale and sandstone	15	0	805	0
Shale, dark grey	74	0	879	0
ALTON, c 36	3	0	882	0
Ganister, fireclay and sandstone, partly coarse	53	0	935	0
Shale and sandstone	40	0	975	0
Shale, coal and ganister with inferred POT CLAY MB	10	0	985	0
NAMURIAN				
Shale and sandy shale	21	0	1006	0
ROUGH ROCK				
Sandstone, buff, micaceous and shale; traces oil at 1015 - 1035	44	0	1050	0
Shale, sandy, micaceous	50	0	1100	0
Shale and shaly sandstone	10	0	1110	0
Shale	65	0	1175	0
CHATSWORTH GRIT				
Sand, white, with feldspar, mica; oil trace; shaly partings	60	0	1235	0
Shale	155	0	1390	0
ASHOVER GRIT (Main Bed)				
Sandstone, coarse, feldspathic	65	0	1455	0
Shale and sandy shale with 5 ft of dark grey micaceous sand at 1640 ft; brown calcareous grit chippings 1715-1735 ft (inferred base of Marsdenian at 1715 ft)	350	0	1805	0
Shale, becoming 'calcareous'; sporadic calcareous grits at top. Thin grey pyritic limestone at 1925 ft. Oil-stained at base	230		2035	0
DINANTIAN				
Limestone, grey-brown; sporadic chert and dolomite; oil stained at top	20	0	2055	0
Limestone, dark grey, oil-stained	15	0	2070	0
Shale	5	0	2075	0
Limestone, buff-grey; tuff & shale fragments	30	0	2105	0
Tuff with limestone	10	0	2115	0
Limestone, grey; sporadic green pumiceous chloritic tuff and glauconite; black shale fragments 2195 - 2200 ft	100	0	2215	0
Limestone, with tuff	5	0	2220	0
Limestone, grey; sporadic shale; pyrite and quartz. Oil-stained 2280 - 2290 ft	75	0	2295	0
Limestone with tuff	5	0	2300	0
Olivine dolerite	10	0	2310	0
Limestone, grey, slightly dolomitic	10	0	2320	0
Olivine dolerite	5	0	2325	0
Limestone	5	0	2330	0
Olivine dolerite	25	0	2355	0
Limestone, with associated calcite and glauconite, and tuff. Tuff particularly common at base	35	0	2390	0
Limestone, whitish, slightly dolomitic; abundance of tuff fragments 2600 - 2630 ft. Oil-stained at 2915 - 2920 ft, 3100 - 90 ft	800	0	3190	0
Limestone, pale grey and pinkish yellow; pyrite, quartz; pale yellow-creamy partings at base; dark grey oil-stained limestones 3440 to 3610 ft	465	0	3655	0

IRONVILLE No. 3* OIL BORE [4325 5232] 1956, 407 ft.
See Fig. 6 and Plate 4. Low Main workings at abt 150 ft; Threequarters workings at abt 168 ft; Yard horizon at 210 ft; Blackshale (coal traces) at 275 ft; 'Ashgate' (coal traces) at 428 ft; Morley Muck horizon at 545 ft; Kilburn workings at 697 ft; Wingfield Flags, 120 ft at 875 ft; Forty-Yards, c 12 at 1075 ft; inferred Parkhouse MB (gamma 'high') at 1147 ft; inferred Alton MB (gamma 'high') at 1180 ft; Alton, c 27 at 1197 ft; Crawshaw Sandstone, 80 ft at 1305 ft; inferred Pot Clay MB (gamma 'high') 1304 - 1308 ft. Rough Rock abt 1325 ft to 1378 ft (from lithological log); inferred G. cumbriense MB (gamma 'high') 1422 ft; inferred G. cancellatum MB (gamma 'high') at 1436 ft; Chatsworth Grit, 35 ft at 1510 ft; inferrred R. superbilingue MB (gamma 'high') 1578 ft; Ashover Grit (Main Bed), 78 ft at 1738 ft; ?base of Marsdenian at abt 1894 ft; ?base of Kinderscoutian at abt 2057 ft; ?base of Chokierian at abt 2108 ft; ?base of Arnsbergian at abt 2184 ft; base of Namurian at abt 2220 ft. Shale, black, calcareous with thin limestones to 2233 ft; limestone black, argillaceous with corals to 2259 ft; limestone, black-grey, brown to 2299 ft; 'white trap' to 2300 ft; limestone, pale brown, sandy and pyritic at base, some fluorspar veining to 2400 ft; igneous rock to 2412 ft; ash, grey, brown, black, red and green mottled, calcite veined to 2580 ft; agglomerate to 2620 ft; tuff, green and red to 2698 ft; limestone, pale brown to off-white to TD at 2742 ft.

*For Ironville No. 2 Oil Bore, See Smith and others 1967, p. 337.

c

IRONVILLE No. 4 OIL BORE [4318 5193] 1958, 336 ft.
Namurian cores 1145-1448 ft and 1499-1514 ft examined by Dr T.R.W. Hawkins. See also Plate 4. Low Main and Threequarters workings at 75 ft; Yard, c? 6 ft at 155 ft; Blackshale, c & d 12 ft at 250 ft; 'Ashgate' (UL), c 36 at 322 ft; 'Ashgate' (LL), c 36 at 352 ft; Kilburn workings at 665 ft; Wingfield Flags, 115 ft at 845 ft; faulted junction with Namurian at abt 1030 ft. Rough Rock, 44 ft at 1086 ft.

	Thickness ft	in	Depth ft	in
Start of coring	1145	0	1145	0
G. CUMBRIENSE MARINE BAND				
Mudstone, dark grey, slightly silty, more silty basally with numerous small irregular pyrite concretions. Lingula mytilloides and poorly preserved Gastrioceras spp. above 1155 ft, G. cumbriense at 1159 ft, fish scales	15	0	1160	0
Sandstone, pale grey with abundant thin bands and laminae of darker silty mudstone and numerous pyrite concretions. Passage	6	0	1166	0
Mudstone, dark grey, silty, micaceous, with numerous thin bands and laminae of paler sandstone (less frequent below 1167 ft 5) and small pyrite concretions; 1-in ironstone band at 1179 ft 6; scattered plants below 1167 ft, rare below 1170 ft. Small Carbonicola [juv.] at 1167 ft 2, numerous Carbonicola deansi at 1177 ft 8 and 1177 ft 11 and below, small fish scale at 1173 ft 6.	14	0	1180	0
Mudstone, dark grey, silty, with sporadic ferruginous patches, becoming less silty downwards; ironstone band, 1 in at 1182 ft 4; sporadic pyritous plants below 1183 ft, numerous small fish scales. Passage	5	6	1185	6
Mudstone, dark grey, slightly shaly to 1186 ft, more shaly below; numerous fish scales	3	1	1188	7
G. CANCELLATUM MARINE BAND				
Mudstone, dark grey, shaly, silty from 1191 ft to 1191 ft 4 with Lingula mytilloides (large), and numerous from 1192 ft to base	5	11	1194	6
Mudstone, dark grey, very silty, with a few small L. mytilloides and some fish scales	0	6	1195	0
Mudstone, dark grey, slightly silty, with small pyritised conretions; abundant L. mytilloides from 1196 ft 4 to base, fish scales	2	2	1197	2
MARSDENIAN				
Mudstone, grey-green, cemented, seatearth-like	0	10	1198	0
Gritstone, pale grey, coarse-grained, angular quartz	4	10	1202	10
Mudstone, dark grey, silty, micaceous more carbonaceous towards the base with pyritised markings; 3-in ironstone band at 1205 ft 8; rare plant impressions, increasing below 1213 ft, fish scales at 1210 ft 6 and rarely below (Lingula sp., recorded by the oil geologists)	12	0	1214	10
Mudstone, dark grey, shaly, silty and micaceous; fish scales	0	8	1215	6
Mudstone, rather dark grey, silty, micaceous, with small irregular pyrite concretions; large plants from 1216 ft to the base, fish scales	1	3	1216	9
CHATSWORTH GRIT				
Sandstone, pale grey with wispy laminae of micaceous carbonaceous silty mudstone; many mudstone laminae in the top 3 in showing turbulent bedding	9	3	1226	0
Mudstone, grey, silty, micaceous; 1-in sandstone band at 1227 ft 1	1	5	1227	5
Sandstone, pale grey, fine-grained, with up to 1-in darker silty mudstone bands in the top 12 in; numerous laminae of micaceous and carbonaceous silty mudstone; sporadic ferruginous patches	6	7	1234	0
Mudstone, grey, silty and micaceous	1	1	1235	1
Sandstone, pale grey, fine-grained, with numerous thin bands and wisps of darker silty, micaceous, mudstone showing some disturbed bedding	2	5	1237	6
Mudstone, dark grey, silty, micaceous; thin ironstone bands and lenses; paler sandy laminae and thin bands 1241 ft 9 to 1242 ft 1; 1-in ferruginous band at 1242 ft and 2-in ironstone band at 1252 ft 1; scattered plants at top; fish debris at 1243 ft 3	14	7	1252	1
Mudstone, dark grey; shaly pyritized ?Carbonicola sp. at 1253 ft	1	1	1253	2
Mudstone, dark grey, slightly silty; 1-in ironstone band at 1254 ft	1	10	1255	0
Mudstone, dark grey, silty, with siltstone and sandstone bands up to 1-in thick; 3-in sandstone band at 1260 ft 3; scattered plants	5	10	1260	10
Mudstone, dark grey, silty; ironstone nodules above 1263 ft 10; 1-in ironstone band at 1266 ft 1; plants, scattered fish-scales at top	10	8	1271	6
Mudstone, dark grey, slightly silty, with more silty patches; laminae of paler fine-grained sandstone at base; scattered plants	3	6	1275	0
Mudstone, dark grey, silty; sporadic thin ferruginous ribs; 1-in ironstone band at 1280 ft 1; small ironstone nodules at 1280 ft 6; fragmentary plants	7	4	1282	4
R. SUPERBILINGUE MARINE BAND				
Mudstone, dark grey, shaly, with fairly abundant well preserved Lingula sp.	3	8	1286	0
Mudstone, dark grey shaly, slightly silty, with numerous small pyrite crystals and some concretions; poorly preserved goniatites at 1289 ft 6	4	0	1290	0
Mudstone, dark grey, silty, micaceous, with thin ironstone ribs	2	5	1292	5
Ironstone	0	1	1292	6
Mudstone, dark grey, slightly silty, shaly; Lingula sp. 1295 ft 9 to 1297 ft, fragmentary Gastrioceras sp. at 1296 ft, fish scale at 1295 ft 8	5	6	1298	0
Mudstone, dark grey, rather blocky, with slickensided surfaces	1	8	1299	8
Mudstone, dark grey, slightly silty, pyritous towards the base; fish debris at 1299 ft 9	0	11	1300	7
Mudstone, dark grey, silty, micaceous, with numerous pyritous patches in top 3 in; small tube-like pyrite concretions 1301 ft 10 to 1302 ft	1	8	1302	3
Mudstone, dark grey, shaly; L. mytilloides 1302 ft 10 to base	1	3	1303	6
Mudstone, dark grey, silty, shaly and micaceous; L. mytilloides 1305 ft 6 to base, cf. R. superbilingue and Gastrioceras sp. from 1304 ft 1 to 1305 ft 3, fish remains at 1303 ft 11	17	6	1321	0
Mudstone, dark grey silty; plants, ironstone bands, 1 in at 1312 ft 9, 2 in at at 1314 ft 8; ?Anthraconaia at 1310 ft 10, ?Carbonicola at 1316 ft 3, sporadic below 1316 ft 5, worm markings below 1320 ft 5	10	3	1331	3
Mudstone, dark grey, slightly silty; ironstone nodules and bands; ?Lingula fragments and fish debris	8	9	1340	0
Mudstone, dark grey, silty, micaceous; sporadic poorly preserved plants; large bivalve at 1342 ft 5	4	0	1344	0
Sandstone, pale buff-grey, ferruginous; veins and drusy cavities filled with fluorspar; plant debris	1	4	1345	4
Mudstone, grey, silty, with micaceous carbonaceous and sandy laminae; plant debris	11	8	1357	0
Mudstone, grey, silty, with darker carbonaceous partings; a few paler sandy laminae; abundant plant debris	3	2	1360	2
Mudstone, dark grey, silty, micaceous, with sporadic ribs of ferruginous sandstone below 1363 ft; 1-in ironstone band at 1366 ft; plants; passage.	19	10	1380	0
Mudstone, dark grey, silty, micaceous; numerous ferruginous patches; trace of coal at base	6	2	1386	2
ASHOVER GRIT				
Ganister, pale buff-grey	1	10	1388	0
Sandstone, pale grey, medium- to coarse-grained ferruginous, with large mica flakes	5	3	1393	3
Gritstone, pale buff-grey	24	9	1418	0
Sandstone, pale buff-grey, medium-grained; sporadic bands up to 18 in of coarser				

d

	Thickness ft	in	Depth ft	in
gritstone	20	0	1438	0
Mudstone, dark grey, very silty, micaceous; thin bands and laminae of paler sandstone	2	1	1440	1
Sandstone, pale grey, fine-grained, micaceous, with numerous laminae of darker silty mudstone	1	11	1442	0
Mudstone, dark grey, silty, micaceous, with some pyritous and paler sandy laminae	6	0	1448	0
Sandstone, medium- to coarse-grained (from chipping samples)	20	0	1468	0
Sandstone, fine, buff-brown (from chipping samples)	31	0	1499	0
San stone, pale buff-grey, medium-grained	3	2	1502	2
Sandstone, pale grey, fine-grained, with wavy laminae of darker silty micaceous mudstone	0	2	1502	4
Mudstone, dark grey, silty, micaceous, with wavy laminae of paler sandstone	4	0	1506	4
Sandstone, pale grey, fine-grained, with darker silty mudstone laminae showing disturbed bedding	0	10	1507	2
Mudstone, dark grey, silty, micaceous. Numerous paler sandstone laminae from 1508 ft 6 to 1509 ft 6 and 2-in predominantly sandy band at 1510 ft 9; bivalves at 1509 ft 10 in; seen to	6	10	1514	0
Mudstone, dark grey, silty, micaceous, with paler sandstone laminae and bands (from chipping samples) to TD	34	0	1548	0

KENNEL WOOD BH, ANNESLEY [5008 5114] 1966, 515 ft.
Recorded by D.V. Frost and D.E. Raisbeck. See Figs.43, 46 and 49. Middle Marl to 23 ft 9; Lower Magnesian Limestone to 50 ft; Lower Marl to 110 ft 5; Permian Basal Breccia to 111 ft 6; inferred Top MB at 150 ft 2; Shafton MB at 214 ft 2; Coal, ca 2 on c 10 at 215 ft 2; Shafton, c 16 at 238 ft 10; ? Main 'Estheria' Band (UL) at 259 ft 5; ? Main 'Estheria' Band (LL) at 268 ft; Edmondia Band at 319 ft 6; High Main (UL), c 18 at 346 ft 6; High Main (LL), c 40 at 366 ft 5; Mansfield MB at 508 ft 8; Haughton MB at 578 ft; Clown, c & d 36 at 624 ft 6; Mainbright, c 22 at 721 ft 4; Two-Foot MB at 730 ft 10; Two-Foot, c & d 18 at 732 ft 4; Lowbright, c 21 at 779 ft; Low Bright Floor, c 8 on d c 1 at 785 ft 7; Brinsley Thin (UL) horizon at 833 ft 3; Brinsley Thin c 8, d 3 on c 4 at 843 ft 10; High Hazles, c 18 at 866 ft 10; TD 890 ft 9.

KIMBERLEY COLLIERY No. 1 SHAFT [5014 4416] abt 1840, abt 420 ft.
Permian, High Hazles to Deep Hard. See Edwards 1951, p. 191, Figs. 23 and 40, and Plate 5. TD 855 ft 8. Another shaft and borehole reported in same colliery yard as No. 1 Shaft give coals below Deep Hard, including: ? 'Ashgate' (LL), c 19 at 420 ft 6; ? Mickley Thin, c 5 at 468 ft 2; ? Lower Brampton, c 2 at 521 ft 5. Overall TD 1377 ft 1.

KIMBERLEY WATER BH [4983 4499] abt 1890, 320 ft.
See Lamplugh and Smith 1914, p. 83. TD 224 ft.

KING, HOWMAN WATER BH, DERBY [3434 3675] 1914, abt 164 ft.
See also Plate 10 and Stephens 1929, p. 90. Made Ground to 6 ft; base of drift at 8ft; Plains Skerry at 103 ft 6; Keuper Marl to ? 193 ft 6; Waterstones to 273 ft; Pebble Beds to 308 ft 10; Carboniferous to TD at 502 ft.

KITTY'S WOOD BH [5005 5040] 1970, 498 ft.
Recorded by T. Draper. See Fig. 82. Middle Marl to 20 ft 2; Lower Magnesian Limestone to 45 ft, Lower Marl to 96 ft 7; Permian Basal Breccia to 97 ft 7; Top MB at 124 ft 6; Coal, 27 at 181 ft 8; Edmondia Band at 296 ft 2; High Main (UL), c 13 at 33 ft 1; High Main (LL), c 41 at 335 ft 5; TD 362 ft 10.

LADY FLATTS MINE SHAFT (THE BLOBBER) [2811 5322] abt 480 ft.
176 ft deep in Dinantian limestone. Top few feet in Namurian mudstones of the ? E_2 Stage.

LANE END BH, ASHLEYHAY [2971 5105] 1967, abt 700 ft.

	Thickness ft	in	Depth ft	in
NAMURIAN				
Shales, brown, silty, micaceous (partly landslip)	24	0	24	0
Shales, grey; ferruginous patches	23	0	47	0
Shale, grey, sandy; hard thin sandstone bands to TD	13	0	60	0

LINBY COLLIERY No. 1 SHAFT [5356 5044] 1873, 301 ft.
See Fig. 77 for section of Brinsley Thin Coal, Fig. 49 for graphic section of measures above Mansfield MB and Edwards 1951, p. 199. Lower Magnesian Limestone to 25 ft; Lower Marl to 84 ft, Permian Basal Breccia to 85 ft 8; Westphalian proved from Manor Coal to First Waterloo Coal; TD 1374 ft.

LINBY COLLIERY (ENGINE HOUSE) UBH [5365 5019] 1954.
Down from Top Hard at -1008 ft. Cores, below 47 ft 3, examined by R.E. Elliott (See also Figs. 25, 33, 37 and 74 and Plate 5. Dunsil, ca 3 on c 13 at 53 ft; Top First Waterloo, c 14 at 70 ft 6; Bottom First Waterloo, c 11 at 104 ft 4; Waterloo Marker, c 10 at 122 ft 5; Second

a

b

Waterloo, c 21, d 5 on b 6 at 139 ft 7; Third Waterloo, c 20 at 181 ft; Coal 1 at 195 ft 9; Fourth Waterloo, b 2 (roof coal), m 18 ft 1, c 9 at 234 ft; First Ell, c 2, m 20 on c 19 at 261 ft 5; Second Ell, c 22 on b 9 at 295 ft 8; Clay Cross MB at 334 ft 5; Brown Rake, c 14 at 368 ft 11; Black Rake TS at 393 ft 2; Top Soft horizon at 396 ft; Roof Soft, ca 6 at 442 ft 8; Deep Soft, c 36 at 479 ft 3; Deep Hard, c 36 at 517 ft 3; First Piper, ca 2 on c 15 at 554 ft 1; Hospital horizon at 641 ft 10; Cockleshell, b 1 at 678 ft 9; Low Main, c 36 at 690 ft 9; Threequarters, c 14 at 706 ft 10; TD 709 ft 6.

LINBY COLLIERY (PIT BOTTOM) UBH [5356 5043] 1954.
Down from Top Hard at -1012 ft, 11 yards south of No. 1 Shaft. See Fig. 34. Top Hard Floor, c 10 at 19 ft 11; Dunsil, ca 2, c 16, d 5 on c 2 at 50 ft 8; Top First Waterloo, c 16 at 70 ft 3; Waterloo Marker, c 2, d 4 on c 1 at 121 ft 8; Second Waterloo, c 20, d 8 on c 3 at 144 ft 2; Third Waterloo, c 20, at 185 ft 9; Fourth Waterloo, c 1 (roof coal), m 11 ft 9, c 9, d 7 on c 1 (floor coal) at 234 ft 5; First Ell, c 1, d 24, ca 2 on c 17 at 262 ft 10; Second Ell, c & d 23 at 297 ft 11; Clay Cross MB at 353 ft 5; Brown Rake, c 9 at 370 ft 9; Black Rake TS at 380 ft 8; Black Rake horizon at 383 ft 5; Top Soft (UL), b 2 at 393 ft 6; Top Soft (LL) horizon at 396 ft 8; Roof Soft, ca 8 on d c 2 at 446 ft 11; Deep Soft, c 30 at 480 ft 10; Deep Hard, ca 1 on c 33 at 520 ft 2; First Piper, c 14 at 556 ft 2; Hospital, c & d 34 at 643 ft 11; Cockleshell horizon at 652 ft 11; Low Main, c 3 at 681 ft 5; TD 681 ft 5.

LINBY COLLIERY 52's UBH [5372 5192] 1957.
Down from High Main at -391 ft 2. See Figs. 42 and 43 for sections of High Hazles and Low Bright respectively. Mansfield MB at 123 ft 1; Haughton MB at 182 ft 4; Mainbright (part washout), c 20 at 312 ft 10; Two-Foot MB at 320 ft 2; Two-Foot, c 11 at 333 ft; Lowbright, c 36 at 382 ft 11; High Hazles, c 32, d 5 on b 4 at 459 ft 3; TD 469 ft 9.

LITTLE EATON WATER BH's [3684 4249] 1934, 1936 and shaft 1938.
Record of 1934 BH (See Plate 4 for graphic section), 197 ft:

	Thickness ft	in	Depth ft	in
SUPERFICIAL				
Clay, yellow and sand (? Head)	14	0	14	0
NAMURIAN				
Shale	7	0	21	0
Sandstone, grey and yellow	7	0	28	0
Shales and sandy shales	45	0	73	0
ASHOVER GRIT (MAIN BED)				
Sandstone	215	0	288	0
Shale	10	0	298	0
Sandstone	129	0	427	0
Shale to TD	18	0	445	0

LITTLE HALLAM WATER BH [4621 4048] 1898-99, 160 ft.
See Figs.16 and 68, and Plate 4; Lamplugh and Smith 1914, pp. 79-80, 130; Stephens 1929, pp. 95-98 for detailed section; Eden 1954 and Falcon and Kent 1960, p. 19. Lower Brampton, m & c at 70 ft; Morley Muck, c 9 at 164 ft 10; Kilburn ? workings at 275 ft 4; Upper Band horizon at 604 ft 5; Norton, c 27 at 638 ft 1; Forty-Yards (UL), c 6 at 668 ft 8; Forty-Yards (LL), c 5 at 678 ft 1; lton, c 22 at 761 ft; Coal 12 at 781 ft 9; Belperlawn, c & d 21 on c 10 at 806 ft 4; Crawshaw Sandstone to 920 ft 7; base of Westphalian at 939 ft 7; base of Rough Rock, 1051 ft 7; Simmondley, c 4 at 1097 ft 11; Ringinglow, c 8 at 1153 ft 7; base of Chatsworth Grit at 1262 ft; Ashover Grit 1406 ft to 1792 ft; TD 1801 ft.

LODGE COLLIERY A1 UBH [4708 4644] 1955.
Down from First Piper at -288 ft.
Blackshale at about 254 ft; TD 289 ft 5.

LODGE COLLIERY SHAFT [4742 4531], 236 ft.
Record is of doubtful reliability. Deep Soft, c 39 at 14 ft 3; Deep Hard, c 36 at 37 ft 3; First Piper, c 39 at 91 ft 4; Hospital, c 24 at 148 ft 3; Cockleshell, c 20 at 165 ft 3; Low Main, c 42 at 193 ft 8; Threequarters, c 17 at 205 ft 10; Yard horizon at 263 ft 9; Blackshale horizon at 347 ft 3; Morley Muck, c & b 9 at 679 ft 7; Kilburn, c 27 at 790 ft 6. The record continues to 1373 ft 6, but appears to be another section of the measures between the Blackshale and Kilburn coals. TD is therefore not known.

LOWER BEAUVALE (GREASLEY) BH [4735 4757] 1967, 229 ft.
See Plate 5 for graphic section from Top Hard to Third Waterloo coals, Fig. 37 for section of Dunsil Coal and Fig. 75 for Mainsmut Coal section. Mainsmut to Third Waterloo; TD 345 ft 3.

MALT SHOVEL INN BH, WIRKSWORTH MOOR [2995 5413] 1965, abt 820 ft.
Ashover Grit (Main Bed) to 32 ft; mudstones to TD at 80 ft.

MANCHESTER WOOD BH, SMALLEY [4160 4417] 1943-44, 401 ft.
Cores, below 23 ft 4, examined by W.N. Edwards. Clay Cross MB at 114 ft; Brown Rake, c 37 at 142 ft 6; Top Soft, c 48 at 185 ft 2; Roof Soft and Deep Soft, c 51, d 11, c 57, m 11 on c 5 at 201 ft 8; Deep Hard workings at 250 ft 10; First Piper, ca 4 on c 44 at 294 ft 8; Hospital, c 8, d 13 on c 11 at 358 ft 5; Cockleshell horizon at 374 ft 9; Low Main, c 41 at 419 ft 2; Threequarters, c 22 at 432 ft 3; Yard horizon at 498 ft 7; Blackshale washed out; 'Ashgate' (UL), c 33 at 739 ft 1; 'Ashgate' (LL), c 28 at 751 ft 5; Mickley Thin, c 10 at 804 ft 4; Lower Brampton, c 1, d 9 on c 1 at 868 ft 2; TD 882 ft 2.

MANNERS COLLIERY BH [abt 4581 4244] 1958-59, 264 ft.
Cores, below 641 ft, examined by G.D. Gaunt. See Figs.14, 16 and 68. Denby leaves of Blackshale, c 3, d 9, c 1, m 50, c 10, d 5 on d c 7 at 657 ft 9; Blackshale, c 25, d 21, d c 4, d 17, d c 3 on c 9 at 687 ft 7; Coal (dirty), 16 at 741 ft 11; 'Ashgate' (UL) horizon at 773 ft 6; 'Ashgate' (LL), c 3 at 817 ft 1; Mickley Thin horizon at 874 ft; Lower Brampton, c 11 at 931 ft 11; Coal 3 at 976 ft 6; Morley Muck, c 3, m 10 ft 10 on c 24 at 1041 ft 8; Kilburn, c 45 at 1134 ft 9; Upper Band horizon at 1476 ft 3; Norton, c 16 at 1514 ft 4; Forty-Yards MB at 1544 ft 5; Forty-Yards, c 4 at 1544 ft 10; Upper Parkhouse MB at 1616 ft 6; Lower Parkhouse MB at 1620 ft; Alton MB at 1646 ft 10; Alton, c 28 at 1649 ft 2; Second Smalley, c 8 at 1667 ft 8; Holbrook, c 14 at 1686 ft 6; Belperlawn, c 27 at 1694 ft 5; TD 1750 ft.

MANNERS COLLIERY SHAFT [4588 4245] 264 ft.
See Fig. 25. Fourth Waterloo, c 17 at 59 ft 6; First Ell, ca 10, m 30 on c 17 at 100 ft 4; base of Westphalian B estimated at 214 ft; Brown Rake, c 14 at 229 ft 9; Top Soft, c 24 at 258 ft 3; Roof Soft (UL), c & d 43 at 271 ft 4; Roof Soft (LL), c & d 9 at 290 ft 8; Deep Soft, c 66 at 333 ft 7; Foot, b 7 at 348 ft 11 Deep Hard, c 36 at 386 ft 5; First Piper, c 56 at 432 ft 7; Hospital, c 7, d 13 on c 8 at 472 ft 1; Cockleshell, c 14 at 502 ft 4; Low Main, c 41 at 546 ft 3; Threequarters, c 29 at 560 ft 2; Yard horizon at 635 ft 4; Denby leaves of Blackshale, c 3, m 42 on c & d 25 at 670 ft 9; Blackshale, c 21, m 26, c 10, m 16 on c 9 at 700 ft 8; Coal 12 at 751 ft 8; 'Ashgate' (UL), b 13 at 786 ft 5; 'Ashgate' (LL), c 19 at 827 ft 2; Mickley Thin, ca 9 at 884 ft 11; Lower Brampton, c 9 at 925 ft 1; Coal 5 at 981 ft 9; Morley Muck, ca 38, at 1037 ft 11; Kilburn, b 8, c 38, d 2, c 3 at 1148 ft 4. Recorded to 1148 ft 4. There is also an outline record labelled No. 2 Shaft with slightly different depths and thicknesses and TD 1152 ft 4.

MAPPERLEY COLLIERY No. 2 SHAFT [4230 4334], abt 400 ft.
See also Figs.18 and 31, and Plate 5. For glossary of old miners' rock terms see next page.

	Thickness ft	in	Depth ft	in
WESTPHALIAN B				
Pit top raised	4	0	4	0
Soil and clay	8	2	12	2
Shale and blue bind	6	10	19	0
BOTTOM SECOND WATERLOO, c 14	1	2	20	2
Clunch	5	9	25	11
Stone bind and rock	15	3	41	2
Bind, blue and dark	10	0	51	2
THIRD WATERLOO, c 2, b 4 on c 24	2	6	53	8
Clunch	3	5	57	1
Rock	9	8	66	9
Bind, blue and dark; ironstone	25	4	92	1
FOURTH WATERLOO, c 10	0	10	92	11
Clunch	0	2	93	1
Bind, blue	17	0	110	1
Stone clunch	12	3	122	4
Bind, blue	14	6	136	10
FIRST ELL, c 42	3	6	140	4
Clunch, stone clunch and bind	4	6	144	10
Rock and bind	3	0	147	10
Bind, blue and dark	24	10	172	8
Shale, strong dark	3	2	175	10
Bind, dark	4	8	180	6
SECOND ELL, c 18	1	6	182	0
Clunch	4	3	186	3
Stone bind with cank	22	1	208	4
Bind, blue and dark	67	2	275	6
WESTPHALIAN A (inferred top)				
Bind, dark; ironstone	27	1	302	7
BROWN RAKE, c 35	2	11	305	6
Batt	0	6	306	0
Clunch and stone bind	9	0	315	0
Bind, blue and dark; ironstone	9	10	324	10
Cank (BLACK RAKE TS)	0	9	325	7
Bind, dark	18	7	344	2
TOP SOFT, c 51	4	3	348	5
Clunch, soft	5	8	354	1
ROOF SOFT and DEEP SOFT, c 54, b 8 on c 59	10	1	364	2
Batt, clunch and stone clunch	7	9	371	11
Bind, blue	40	4	412	3
DEEP HARD, ca 14 on c 56	5	10	418	1
Clunch	3	0	421	1
Stone bind and rock	20	9	441	10
Bind and shale	13	3	455	1
FIRST PIPER, c 28, d 2 on c 13	3	7	458	8
Spavin and stone spavin	5	4	464	0
Rock, bind and stone bind	26	3	490	3
Shale, carbonaceous		8	490	11
SECOND PIPER HORIZON	-	-	490	11
Spavin	0	6	491	5
Rock and rock bind	2	2	493	7
Bind, blue and stone bind	29	4	522	11
HOSPITAL, c 7, d 7, c 6, d 1 on c 3	2	0	524	11
Spavin	5	2	530	1
Bind with ironstone; mussels	20	8	550	9
Stone bind and rock	34	9	585	6
Bind, blue	2	0	587	6
LOW MAIN, c 41	3	5	590	11
Spavin	4	1	595	0
Clod, dark	6	7	601	7
THREEQUARTERS, c 22	1	10	603	5
Spavin	2	11	606	4
Stone bind	6	11	613	3
Bind, blue; ironstone	3	0	616	3
Stone bind, rock and cank	108	0	724	3
Bat	1	6	725	9
YARD, c 9	0	9	726	6
Spavin and bat	6	11	733	5
DENBY LEAF OF BLACKSHALE, c 5	0	5	733	10
Spavin	0	11	734	9
Stone bind	18	8	753	5
BLACKSHALE, c 26, d 3,				

c

d

	Thickness ft	in	Depth ft	in
c, 5 d 20 on c 15	5	9	759	2
Spavin and bind	18	10	778	0
Bind, dark and blue; ironstone	15	2	812	8
COAL, c 14	1	2	813	10
Spavin	6	0	819	10
COAL, c 3	0	3	820	1
Stone bind	14	9	834	10
Bind, blue	3	10	838	8
'ASHGATE' (UPPER LEAF), c 11	0	11	839	7
Stone spavin	9	6	849	1
Stone bind, rock and cank	50	2	899	3
Bind blue; ironstone		6	899	9
Rock with ironstone	2	0	901	9
'ASHGATE' (LOWER LEAF), c 16	1	4	903	1
Spavin and stone spavin	2	9	905	10
Stone bind and rock	35	5	941	3
Bind, with ironstone	16	8	957	11
MICKLEY THIN, c 8	0	8	958	7
Spavin	4	6	963	1
Stone bind and galliard	10	0	973	1
Bind blue and dark; ironstone	48	6	1021	7
Batt		3	1021	10
LOWER BRAMPTON, ca 7, m 3 on c 4	1	2	1023	0
Spavin	3	11	1026	11
Rock and stone bind	5	6	1032	5
Bind, blue and black	36	1	1068	6
COAL, c 5	0	5	1068	11
Spavin	5	0	1073	11
Rock on blue bind	21	7	1095	6
Bind, dark and blue; mussels	15	0	1110	6
MORLEY MUCK, c 42, m 12, c 9, m 9 on c 4	6	4	1116	10
Stone spavin	1	6	1118	4
Stone bind and rock	80	11	1199	3
Bind blue and dark	20	2	1219	5
KILBURN, c 6, d 1 on c 62	5	9	1225	2
Recorded to			1225	2

Records also exist for Mapperley Colliery No. 1 (Recorded to 1219 ft, see also Fig. 25) and No. 3 shafts (Recorded to 1214 ft 4)

Batt	=	carbonaceous shale, interlaminated carbonaceous shale and coal
Bind	=	mudstone
Cank	=	cemented sandstone/siltstone
Clod	=	soft mudstone
Clunch	=	seatearth underclay or fireclay
Galliard	=	ganister
Rock	=	sandstone, or any hard massive rock
Rock bind	=	stone bind
Spavin	=	clunch
Stone bind	=	banded silty mudstone or silty sandstone
Stone clunch	=	hard seatearth

MAPPERLEY COLLIERY UBH [4229 4339] 1950.
Down from Kilburn at -818 ft.
Cores examined by R.A. Eden. See Figs. 14, 16 and 68, and Eden 1954.

	Thickness ft	in	Depth ft	in
WESTPHALIAN A				
Open hole	44	0	44	0
Mudstone, grey and dark grey, silty in part; plant debris, Spirorbis sp., Carbonicola sp., Curvirimula trapeziforma, Geisina arcuata and fish debris	34	5	78	5
Seatearth, coarse silty clay	2	1	80	6
WINGFIELD FLAGS				
Sandstone and siltstone interbedded	22	9	103	3
Mudstone, dark grey; ironstone; rare mussels	11	0	114	3
Siltstone with mudstone beds	105	6	219	9
Sandstone	2	3	222	0
Siltstone, coarse sandy	73	0	295	0
Mudstone and shale, partly pyritic; rare thin siltstones; fish debris	70	8	365	8
UPPER BAND HORIZON	-	-	365	8
Fireclay, brownish grey	4	8	370	4
Seatearth, mudstone, dark grey	3	8	374	0
Mudstone, dark grey, silty in part; ironstone; Carbonicola crispa, C. proxima and Naiadites sp.	31	7	405	7
NORTON MARINE BAND				
Shale, black, pyritic; Ammodiscus sp., Lingula mytilloides, Elonichthys	0	10	406	5
Batt	0	1	406	6
NORTON, c 6	0	6	407	0
Seatearth, mudstone, dark grey; ironstone	4	7	411	7
Mudstone and shale; fish debris; ironstone	22	3	433	10
Shale, black; with Ammodiscus sp., Lingula mytilloides, Elonichthys sp. and Rhadinichthys sp. (FORTY-YARDS MARINE BAND) at 434 ft	1	10	435	8
FORTY-YARDS (UPPER LEAF), c 8	0	8	436	4
Fireclay	2	3	438	7
Siltstone, coarse with rootlets	5	8	444	3
Seatearth, mudstone; ironstone	3	10	448	1
FORTY-YARDS (LOWER LEAF), b 14	1	2	449	3
Seatearth and siltstone, light grey	1	2	450	5
Sandstone, micaceous, interbedded with siltstone	22	1	472	6
Siltstone, coarse, micaceous	19	4	491	10
Mudstone and shale, dark grey to grey; Lingula mytilloides at 512 ft 4 and 513 ft 11 (UPPER AND LOWER PARKHOUSE MB's)	43	4	535	2

	Thickness ft	Thickness in	Depth ft	Depth in
ALTON MARINE BAND				
Mudstones, dark grey; Dunbarella papyracea, Cancyella multirugata Posidonia gibsoni Anthracoceratites sp., Gastrioceras listeri, Hindeodella sp.	4	3	539	5
ALTON, c 28	2	4	541	9
Fireclay, brown and green-grey	3	5	545	2
Mudstone, silty and shale, dark grey	9	6	554	8
FIRST SMALLEY Lingula BAND				
Shale, black, pyritic; Lingula mytilloides and fish debris	1	4	556	0
Shale, carbonaceous; coal streaks	0	1	556	1
FIRST SMALLEY, c 11, d 6 on c 12	2	5	558	6
Fireclay, siltstone and seatearth	9	9	568	3
SECOND SMALLEY Lingula BAND				
Shale, black; Lingula mytilloides and fish debris	0	3	568	6
SECOND SMALLEY, c & d 1, d 17 on c 3	1	9	570	3
Sandstone and siltstone	6	6	576	9
Shale, black, micaceous		4	577	1
HOLBROOK Lingula BAND				
Shale, black; Lingula mytilloides	1	5	578	6
Shale, black	0	6	579	0
HOLBROOK HORIZON	-	-	579	0
Seatearth, sandy	3	8	582	8
Sandstone and siltstone, micaceous; plants	21	0	603	8
BELPERLAWN, c 28	2	4	606	0
CRAWSHAW SANDSTONE				
Siltstone, coarse, grey, ganisteroid	3	2	609	2
Sandstone, micaceous, medium-grained, red and grey bands and green streaks to TD	16	3	625	5

MAPPLEWELLS BH, ANNESLEY WOODHOUSE [4973 5432] 1967, 550 ft.
Recorded by D.V. Frost and R. Draper. See Figs.42, 49 and 82. Lower Mottled Sandstone to 24 ft 1; Middle Marl to 58 ft 1; Lower Magnesian Limestone to 92 ft; Lower Marl to 158 ft 2; Permain Basal Breccia to 163 ft 11; Shafton, c 1 at 203 ft 2; ? Main 'Estheria' Band (UL) at 208 ft 8; ? Main 'Estheria' Band (LL), at 223 ft 9; Coal 10 at 229 ft 11; Edmondia Band at 286 ft 10; High Main (UL), c & d c 11 at 318 ft 3; High Main (LL), d c 7 on c 35 at 325 ft 1; Coal 15 at 378 ft 11; Coal 10 at 392 ft 6; Mansfield MB at 492 ft 3; ? Sutton MB at 528 ft 3; Haughton MB at 569 ft 10; Clown, b 5 on c 14 at 621 ft 4; Manton 'Estheria' Band at 641 ft 1; Mainbright (partly washout), ca 6, m 50, c 4 on b 2 at 723 ft; Two-Foot MB at 735 ft 5; Two-Foot horizon at 737 ft 4; Fault at 750 ft; Brinsley Thin horizon (UL) at 756 ft 4; Brinsley Thin horizon at 780 ft 1; High Hazles, ca 2, c 24, d 55, d c 3, d 1, d c 3, d 11 on c on c & d 1 at 823 ft 10; possible small-scale faulting c. 850 ft - 884 ft; TD 942 ft 9.

MEXBOROUGH COLLIERY SHAFT [4742 5304], 500 ft.
Mainbright, c 24 at 43 ft 5; Two-Foot, c 19 at 61 ft 4; Lowbright, c 30 at 107 ft 10; Lowbright Floor, c 7 at 112 ft 9; Brinsley Thin (UL), c 2 at 157 ft 3; Brinsley Thin, c 9 at 168 ft 1; High Hazles (Simpsons), c 17 at 197 ft 6; Cinderhill (UL), c 10 at 253 ft 6; Cinderhill, c 2 at 261 ft 10; Mainsmut, c 3 at 371 ft 6; Comb, c 31 at 442 ft 8; Top Hard, c 55 at 452 ft 4; Top Hard Floor, c 8 at 471 ft; Dunsil, c 23 at 512 ft 11; Top First Waterloo, c 12 at 535 ft 4; TD 539 ft 2.

MISK FARM BH [5027 4927] 1956, 445 ft.
Base of Permian at 88 ft 2; High Main (UL) c 13 at 295 ft 4; High Main (LL) c 37 at 300 ft 10; TD 312 ft 3.

MISK NORTH BH [5018 4975] 501 ft.
Cores, below 20 ft, examined by C. Beal. Lower Mottled Sandstone to 29 ft 6; Middle Marl to 54 ft 8; Lower Magnesian Limestone to 74 ft 6, Lower Marl to 116 ft 9; Permian Basal Breccia to 117 ft 9; Westphalian from close below Top MB to High Main. TD 368 ft 3.

MODEL FARM No. 1 BH [5143 3923] 1953, 178 ft.
Cores, below 49 ft 11, examined by R.E. Elliott. First Ell at 21 ft; Second Ell, c 21 at 74 ft 6; Clay Cross MB at 41 ft 10; Brown Rake horizon at 153 ft; Black Rake TS at 184 ft 6; Top Soft, d c 3 at 192 ft 9; Roof Soft (UL) horizon at 209 ft 4; Roof Soft (LL), ca 2 on c 7 at 229 ft 11; Deep Soft, c 41 at 280 ft 8; Deep Hard, ca 1, c 33 on ca 6 at 319 ft; First Piper, c 21 at 355 ft; Second Piper horizon at 392 ft 8; Hospital, c 4 on d c 5 at 410 ft 5; Low Main washed out; TD 500 ft 6.

MODEL FARM No. 2 BH [5153 3893] 1954, 157 ft.
Cores, below 26 ft 6 recorded by R.E. Elliott. Second Ell at 16 ft; Deep Hard, c 60 at 253 ft 2; First Piper, c 22 at 293 ft 5; Second Piper horizon at 324 ft 6; Hospital, c 1 at 339 ft 6; Low Main washed out; Threequarters, c 21 at 454 ft 4; TD 490 ft 8.

MODEL FARM No. 3 BH [5165 3885] 1954, 150 ft.
Recorded by R.E. Elliott. Clay Cross MB at 86 ft 9; sequence (See Fig. 25) similar to Nos. 1 and 2 BHs, but: Cockleshell horizon at 411 ft 8; Low Main, ca 1 on c 38 at 437 ft 7; Threequarters, 27 in at 450 ft 2; TD 453 ft 6.

MOOR FARM BH [5062 4004] 1958, 241 ft.
Cores, 19 ft 6 to 249 ft, and below 300 ft, examined by G.D. Gaunt. See Figs.14, 16, 19, 22 and 68. Second Ell (UL), c 15 at 33 ft; Second Ell (LL), c 12 at 40 ft; Clay Cross MB at 106 ft 8; Brown Rake, c 8 at 119 ft 8; Top Soft, c 8 at 151 ft 10; Roof Soft (UL), c 3 at 169 ft 1; Roof Soft (LL), c 12 at 181 ft 7; Deep Soft workings at 233 ft 6; First Piper, c 18 at 304 ft 6; Hospital

a

b

horizon at 348 ft 2; Cockleshell, c 1 at 471 ft; Low Main, c 33, d 9 on c 16 at 488 ft 10; Threequarters, c 16 at 499 ft 10; Yard, c 8, d 19, c 4, d 3 on c 4 at 567 ft 6; Denby leaves of Blackshale, c 2, d 8 on c 31 at 576 ft 10; Blackshale, c 12, d 43, c 1, d 4, c 7, d 32 on c 15 at 599 ft 6; Coal 10 at 638 ft 7; 'Ashgate' washout; Mickley Thin, ca 1 on c 1 at 779 ft 6; Lower Brampton, c 9 at 831 ft 11; Morley Muck, c 10, d 7 on c 2 at 915 ft 9; Coal 1 at 988 ft 7; Kilburn, c & d c 25, m 5 ft 7, d c 4, c 3, d 4, c 3, d 2 on d c 3 at 1049 ft 6; Upper Band horizon at 1358 ft 2; Norton, c 26 at 1391 ft 5; Forty-Yards MB at 1418 ft 1; Forty-Yards, c 2 at 1420 ft; Upper Parkhouse MB, upper bed 2 in at 1480 ft 11; lower bed 6 in at 1484 ft 8; Lower Parkhouse MB 16 in at 1496 ft 8; Alton MB at 1524 ft 1; Alton, c 36 at 1527 ft 1; Holbrook horizon at 1553 ft 2; Belperlawn, c 24 at 1563 ft 8; TD 1617 ft 6.

MOORGREEN COLLIERY (LOCO ROAD) UBH [4835 4739] 1959.
Down from Low Main at -709 ft. Cores, below 13 ft, examined by G.D. Gaunt. Yard, c 1, d 4 on c 1 at 59 ft 6; Blackshale, c 33 at 127 ft 9; 'Ashgate' (LL) horizon at 302 ft 3; Mickley Thin horizon at 365 ft; Kilburn, c 9 at 563 ft 9; TD 600 ft.

MOORGREEN COLLIERY LOW MAIN 2's UBH [4823 4743] 1961.
Down from Low Main at -699 ft to Blackshale at 129 ft 2; TD 129 ft 11.

MOORGREEN COLLIERY LOW MAIN 16's UBH [4923 4816] 1961.
Down from Low Main at -994 ft to Blackshale at 138 ft 6; TD 139 ft 6.

MOORGREEN COLLIERY LOW MAIN 17's UBH [4951 4788] 1962.
Down from Low Main at -1020 ft to Blackshale (see Fig. 27) at 138 ft 10; TD 140 ft.

MOORGREEN COLLIERY LOW MAIN 25's UBH [4986 4734] 1962.
Down from Low Main at -1000 ft to Blackshale at 148 ft; TD 150 ft.

MOORGREEN COLLIERY MAIN DIPS UBH [4889 4797] 1959.
Down from Low Main at -908 ft to Blackshale at 132 ft 11; TD 133 ft 11.
See Figs. 25 and 34, and Plate 5. Brinsley Thin,
MOORGREEN COLLIERY No. 1 SHAFT [4795 4786], 1865 to Deep Hard, deepened to Threequarters. 267 ft.

c 24 at 48 ft 11; High Hazles, c 36 at 89 ft 11; Mainsmut, d c 22, m 6 ft on d 18 at 274 ft 2; Comb, c 30 at 297 ft 8; Top Hard, c 67 at 339 ft 11; Top Hard Floor, c 20 at 350 ft 4; Dunsil, c 30, b 2, c 2, b 4 on c 1 at 392 ft 5; First Waterloo, c 16, m 16 on c 24 at 423 ft 4; Waterloo Marker, ca 46 at 460 ft 7; Second Waterloo, c 27, d 3, c 4, m 12 on c 22 at 473 ft 4; Third Waterloo, c 22 at 498 ft 6; Coal 2 at 513 ft 8; Fourth Waterloo, ca 8 at 544 ft 8; First Ell, c & b 28 at 576 ft 7; Second Ell, c 4, d 12 on c 4 at 633 ft 4; base of Westphalian B estimated at 704 ft; Brown Rake, c 10 at 712 ft 8; Black Rake TS at 732 ft 1; Top Soft, c 10 at 743 ft 8; Roof Soft (UL), c 13 at 757 ft 4; Roof Soft (LL), c 14 at 768 ft 11; Deep Soft, c 41, b 8 on c 3 at 813 ft 8; Deep Hard, c 44 at 858 ft 1; Floor Coal, c 15 at 860 ft 11; First Piper, c 35 at 896 ft 10; Hospital, c 20 at 962 ft 11; Cockleshell, c 21 at 986 ft 7; Low Main, c 41 at 1015 ft 10; Threequarters, c 13, d 3, c & d 5 on c 3 at 1018 ft; recorded to 1018 ft. See Moorgreen Colliery UBH for details of borehole in shaft bottom.

MOORGREEN COLLIERY No. 2 SHAFT [4791 4784] 1866, deepened 1940-45. 267 ft.
To Threequarters; TD 1086 ft 9.

MOORGREEN COLLIERY UBH [4795 4786] 1951.
In centre of No. 1 Shaft, depths below base of Low Main at -740 ft. Cores, below 20 ft 2, examined by R.A. Eden. Yard horizon at 71 ft; horizons of Denby leaves of Blackshale at 91 ft 6 and 108 ft 10; Blackshale, c with thin d 46 at 125 ft 3; TD 185 ft 8.

MOORGREEN COLLIERY 2's DRIFT [5008 4714] 1960.
Up from 2nd Waterloo at -468 ft; measurements from a level 2 ft below roof of seam. See Fig. 36 for section of First Waterloo. Top First Waterloo, c 18, d 2 on c 2 at 70 ft; Bottom First Waterloo (LL), ca 6 at 38 ft 4; Waterloo Marker, c 8 at 21 ft 9; Second Waterloo, c & d 49; Driven to 73 ft 10.

MOORGREEN COLLIERY 2's/10's UBH [4987 4758] 1960.
Up from Second Waterloo at -476 ft. Length Drilled 64 ft. See Fig. 36 for section of First Waterloo Coal.

MOORGREEN COLLIERY 3's UBH [5070 4626] 1959.
Up from Second Waterloo at -397 ft. Recorded by J. Chilton. See Fig. 36 for section of First Waterloo Coal. First Waterloo, c & d 41 at 44 ft; length drilled 48 ft.

MOORGREEN COLLIERY 3's UBH [5074 4619] 1960-61.
Down from Second Waterloo at -392 ft. Cores, below 200 ft, examined by D.V. Frost, E.G. Smith and J. Chilton. Clay Cross MB at 233 ft; Deep Soft, c 42 at 342 ft; Low Main, c 36 at 536 ft 6; TD 638 ft 6.

MOORGREEN COLLIERY W66's/21's CROSS CUT No. 4 UBH [5136 4817] 1972.
Up from Second Waterloo at 826 ft. Top First Waterloo, c 23 at 40 ft 8; Bottom First Waterloo, c & d 24 at 34 ft; length drilled 45 ft.

MOORGREEN COLLIERY W66's/28's CROSS CUT UBH [5156 4782] 1972.
Up from Second Waterloo at -821 ft. Recorded by M.J. Allen. Top First Waterloo, c & d 22 at 40 ft 10; Bottom First Waterloo, c & d 25 at 37 ft 7; length drilled 43 ft 6.

MOORGREEN COLLIERY W66's/33's CROSS CUT MAIN GATE No. 1 UBH [5173 4752] 1972.
Up from Second Waterloo at -822 ft. Recorded by M.J. Allen. First Waterloo, c & d 64 at 44 ft 3; length drilled 54 ft 9.

MOORGREEN COLLIERY WATERLOO 10's R.H. UBH [5040 4555] 1964.
Up from Second Waterloo at -168 ft. Top Hard, c 45 at 122 ft 11; Top Hard Floor, c 12 at 115 ft 11; Dunsil, c 28 at 80 ft 11; First Waterloo, c 20, d 8 on c 31 at 46 ft 3; Waterloo Marker, c 7 at 26 ft; length drilled 126 ft 8.

MOORGREEN COLLIERY WATERLOO 71's UBH [4979 4920] 1963.
Up from First Waterloo at -605 ft. Drilled 105 ft to Top Hard workings.

MOTORWAY PROVING HOLE 513B [5071 4758] 1962, abt 425 ft.
Cores examined by D.V. Frost.

	Thickness ft	Thickness in	Depth ft	Depth in
LOWER MOTTLED SANDSTONE				
Sand, red and sandstone	12	0	12	0
MIDDLE MARL				
Clay, red-brown; sandy lenses	21	0	33	0
LOWER MAGNESIAN LIMESTONE				
Dolomite, red-brown, sandy; red mudstone partings; quartz grains common at 57 ft 7	27	0	60	0
LOWER MARL				
Dolostone, grey; mudstone laminae; plants	17	0	77	0
Mudstone, grey, micaceous; plant fragments to TD	3	0	80	0

NETHER HEAGE BH [3599 5119] 1974, abt 318 ft.
Cores examined by D.V. Frost. See Plate 4 for graphic section.

	Thickness ft	Thickness in	Depth ft	Depth in
MADE GROUND				
Ridgeway Site restoration	46	1	46	1
WESTPHALIAN A				
Seatearth; mudstone with ironstone	2	4	48	5
Sandstone, fine-grained	0	11	49	4
Mudstone; sandstone laminae at base	10	1	59	5
CRAWSHAW SANDSTONE				
Sandstone, fine- to medium-grained, grey to pale grey, iron-stained and locally pyritised; coarse-grained below 88 ft. Numerous oblique joints and fractures	64	9	124	2
Mudstone, dark grey, silty	22	8	146	10
POT CLAY MARINE BAND				
Mudstone, dark grey, with Lingula sp. and goniatites	1	2	148	0
NAMURIAN				
Seatearth; ganister	13	7	161	7
ROUGH ROCK				
Sandstone, pale grey, massive; mudstone partings at base	28	10	190	5
Mudstone, siltstone and sandstone laminae at top, becoming less silty and dark grey at base	99	5	289	10
G. CUMBRIENSE MARINE BAND				
Mudstone, shaly	1	10	291	8
Sandstone, ganisteroid	11	7	303	3
Siltstone, sandstone laminae mudstone partings	9	1	312	4
Mudstone, dark grey, silty, pyritic	14	9	327	1
G. CANCELLATUM MARINE BAND				
Mudstone, dark grey, silty, pyritic, with Lingula sp.	3	6	330	7
Sandstone, ganisteroid	7	6	338	1
Siltstone	3	2	341	3
Mudstone, silty	13	4	354	7
CHATSWORTH GRIT				
Sandstone, ganisteroid; mudstone bands at top	3	4	357	11
Sandstone and mudstone alternation	9	0	366	11
Mudstone, micaceous, silty	0	3	367	2
Sandstone, massive, brown and pink-stained, cross-laminated; oblique fractures common to TD	9	0	376	2

NEW LANGLEY WATER BH [4443 4632], abt 300 ft.
TD 220 ft 7. See Stephens 1929, p. 95 for record.

NEW LANGLEY COLLIERY BH [4455 4640] 1959, 245 ft.
Cores, below 424 ft, examined by G.D. Gaunt and C.G. Godwin. See Figs.14, 15, 16, 19 and 68. Hospital, c 6 at 451 ft; Cockleshell (UL), c 1 at 482 ft 1; Low Main, c 36 at 516 ft 6; Threequarters*, c 6 at 527 ft 9; Yard horizon at 590 ft 8; Blackshale, c 5, d c 3, d 4, d c 5, c 4, d 5 on c 17 at 660 ft 2; 'Ashgate', c 4, d 7 on c 24 at 755 ft 2; Mickley Thin, c 9 at 807 ft 3; Lower Brampton, c 2 at 853 ft 9; Morley Muck, c 8, d 32 on c 2 at 934 ft 5; Kilburn, c 35 at 1045 ft 7; Upper Band MB at 1361 ft 6; Upper Band horizon at 1366 ft; Norton MB at 1400 ft 2; Norton, c 19 at 1402 ft 2; Forty-Yards MB at 1434 ft 9; Forty-Yards (UL), c 11 at 1436 ft 6; Parkhouse MB at 1509 ft 3; Alton MB at 1533 ft 8; Alton, c 16 at

*Section on Colliery records of Threequarters Coal is c 23, d 2 on c 1. The exact locality is not known

c

d

1535 ft 9; First Smalley horizon at 1544 ft 9; Belperlawn, c 18 at 1571 ft 9; TD 1641 ft 11.

NEW LANGLEY COLLIERY SHAFT [abt 4462 4641], 246 ft.
The record contains an error in addition, 9 ft has been added to the recorded depths below the Deep Hard Coal.
First Ell, c 30 at 36 ft 6; Second Ell, ? ca 25 at 77 ft 10; base of Westphalian B estimated at 161 ft 7; Brown Rake, c 10 at 179 ft 3; Black Rake TS at 190 ft 3; False Ell, c 16, d 5 1, c 8, d 4 on c 24 at 229 ft 11; Deep Soft, c 38 at 285 ft 3; Deep Hard, c 37 at 334 ft 7; First Piper, c 24, m 7 ft on c 36 at 380 ft 7; Second Piper horizon at 407 ft 1; Hospital, c 12 at 444 ft 1; Cockleshell, c, d & m 6 ft 3 at 472 ft 9; Low Main, c 39, d 15 on c 16 at 509 ft 8; recorded to 511 ft 11. For sequence below see New Langley Colliery Borehole.

NEW SELSTON COLLIERY No. 1 DOWNCAST SHAFT [4577 5277] 1892, 363 ft.
See Figs.31 and 34, and Plate 5. Top First Waterloo, ca 12 at 21 ft 2; Bottom First Waterloo, c 20 at 39 ft 5; Waterloo Marker, c 16 at 75 ft 5; Second Waterloo, c 32, d 3, c 7, d 25 on c 18 at 109 ft 2; Third Waterloo, c 18 at 135 ft 6; Coal 1 at 150 ft 2; Fourth Waterloo, c 26 (main leaf), d 15 on c 5 (floor coal) at 183 ft 7; First Ell, c 8 at 213 ft 5; Second Ell, c 7 at 257 ft 8; base of Westphalian B estimated at 335 ft; Brown Rake, c 4, m 5 ft 2 on c 10 at 384 ft 3; Black Rake TS at 404 ft 1; Black Rake, c 10 at 409 ft 6; False Ell, c 15, m 27, c 10, m 8, c 4, m 48 on c 27 at 430 ft 2; Deep Soft, c 46, b 5, c 4, d 7 on c 3 at 470 ft; Deep Hard, c 5, d 2 on c 49 at 517 ft 7; First Piper (Upper Coal), c 10, b & c 14 on c 16 at 531 ft 11; First Piper (Lower Coal), c 24, d 8 on c 23 at 541 ft 1; TD 541 ft 8.

NEWSTEAD COLLIERY No. 1 (EASTERLY) SHAFT [5217 5314] 1875, 422 ft.
See Figs.43 and 44. Lower Mottled Sandstone to 20 ft; Middle Marl to 41 ft; Lower Magnesian Limestone to 78 ft; Lower Marl to ? 160 ft, Permian Basal Breccia to 161 ft; Manor, c 21 at 239 ft 8; Coal, c 24, b 3 on c 7 at 355 ft 1; Annesley, c 10 at 421 ft 11; inferred Top MB at 485 ft; inferred Shafton MB at 545 ft 2; Coal 12 at 562 ft 2; Shafton, c 15 at 575 ft 11; High Main, c 48 at 716 ft 4; Coal, c 12, d 24 on c & b 15 at 781 ft 10; inferred Mansfield MB at 829 ft 1; ? Clown, c & b 21 at 956 ft 6; Mainbright, ca 8 on c 11 at 1036 ft 1; Two-Foot, c 14 at 1051 ft 1; Lowbright, c 31 at 1080 ft 1; High Hazles, c 30 at 1162 ft 7; ? Cinderhill, b 2 at 1210 ft 7; ? St Johns, b 1 at 1265 ft 11; Mainsmut, c 10 at 1304 ft 4; Comb, c 23 at 1367 ft 11; Top Hard, c 51 at 1373 ft 2; Top Hard Floor, ca 4 at 1421 ft 5; Dunsil, c 29, d 3 on c 3 at 1456 ft 2; Top First Waterloo, c 20 at 1468 ft 7; Bottom First Waterloo, c 24 at 1482 ft 2; TD 1497 ft 3.

NEWSTEAD COLLIERY N1 UBH [5291 5464] 1962.
Down from High Main at -402 ft. Recorded by E.G. Smith and S. Brunskill. See Figs.42 and 44 for sections in High Hazles and Mainbright coals; Smith and others 1967, pp. 357-359 for record. To High Hazles. TD 538 ft 6.

NEWSTEAD COLLIERY N3* UBH [5184 5383] 1965-66.
Down from the High Main at -235 ft. Recorded by S. Brunskill. Mansfield MB, 55 at 132 ft 11; Clown, c 15 on c & d 50 at 297 ft 9; Mainbright, ca 9, m 15 on c 4 at 397 ft 10; Two-Foot MB at 409 ft 8; Two-Foot, c 8, d 38, c 6, d 5 on c 2 at 420 ft 2; Lowbright, c 30 at 455 ft 7; Lowbright Floor, c & d 16 at 458 ft 9; Brinsley Thin, c & d ?7 at 519 ft; High Hazles (core lost) at 553 ft 8; TD 555 ft.

NEWSTEAD COLLIERY N4 UBH [5403 5328] 1966.
Down from High Main at -472 ft. Cores, 351 ft to 381 ft and below 461 ft, examined by S. Brunskill. For section of High Hazles Coal see Fig. 42. Lowbright, c 31 at 380 ft 7; High Hazles, c 35 at 478 ft 6; TD 479 ft 4.

NORTH FARM BH, BRAILSFORD [2561 4332] 1962, abt 460 ft.
Pebble Beds (sand and pebbles with a 30-in clay band) to 41 ft; ? Widmerpool Formation (red and green marls) to bottom at 88 ft.

NUTBROOK AIR PIT [4506 4308] 1867, abt 230 ft.
See also Plate 5. Top First Waterloo, c 12 at 22 ft; Waterloo Marker, c 48 at 70 ft; Top Second Waterloo, c 39 at 101 ft 3; Bottom Second Waterloo, c 18 at 117 ft 9; Third Waterloo, c 28 at 142 ft 1; Coal, b 9 at 157 ft 2; Fourth Waterloo, c 27 at 198 ft 11; First Ell, c 12, b 3 on c & b 19 at 221 ft 5; Second Ell, d c 21 at 266 ft 2; base of Westphalian B estimated at 350 ft; Brown Rake, c 16 at 362 ft 7; Top Soft, c 8 at 407 ft; Roof Soft and Deep Soft, c 51, d 9 on c 51 at 488 ft 6; Deep Hard, d c 5 on c 48 at 543 ft 11; First Piper, c 29, d 14 on c 9 at 583 ft 8; Hospital, d c 24 at 644 ft; Cockleshell, b 9 at 668 ft 10; Low Main, c 43 at 707 ft 4; Threequarters, c 22 at 720 ft 2; Yard horizon at 791 ft 9; Blackshale, c 13, b 16, c 8, b 13, c 10, b 4 on c 2 at 887 ft; recorded to 954 ft 7.

NUTHALL WOOD ENGINE PIT [5168 4343] 1850, 326 ft.
Lowbright Floor, c 22 at 99 ft 6; High Hazles, c 34 at 179 ft 7; Cinderhill, c 33 at 261 ft 11; Mainsmut, c 23 at 358 ft 1; Comb, c 14, d 7 on c 15 at 403 ft; Top Hard, c 61 at 408 ft 7; TD 408 ft 7.

OAKWELL COLLIERY [abt 462 413], abt 275 ft.
Not completely recorded to Threequarters Coal at 134 ft 8; Yard, c 13 at 193 ft 4; Denby leaves of Blackshale, c 13, m 14 ft on c 5 at 213 ft 1; ? Blackshale (part washout), c 8 at 255 ft 5; 'Ashgate' (UL) horizon at 364 ft 2; 'Ashgate' (LL), c 19 at 388 ft; Mickley Thin, c 4, m 36 on b 18 at 446 ft 2; Lower Brampton, c 10 at 505 ft 11; Coal 2 at 551 ft 8; Morley Muck, c 9, ca 18,

*Newstead Colliery N2 UBH lies to the east of the present district.

a

m 31 on b 5 at 614 ft 4; Kilburn, c 50 at 703 ft 8; recorded to 703 ft 8.

OAKWELL COLLIERY STAPLE PIT AND BH [4377 4081].
Down from Kilburn through fault at abt -530 Details to floor of Alton from old written log; below Alton from Abandonment Plan 5969. See also Vernon 1909. Upper Band horizon at 84 ft 8; Norton, c 21 at 126 ft 5; Forty-Yards (UL), c 13 at 160 ft 3; Forty-Yards (LL), c 9 at 168 ft 6; Alton, c 28 at 249 ft; Belperlawn, c9, m 18 on c 9 at 303 ft; staple pit bottom at 326 ft, drilled to 338 ft.

OCKBROOK No. 1 BOREHOLE [4320 3678] 1891, abt 220 ft.
For graphic section see Plate 10. Keuper Marl to 29 ft 3; Waterstones to 123 ft 3; Pebble Beds to TD at 151 ft 6.

OLD RESERVOIR BH [5103 4869] 1956, 381 ft.
Open hole to 117 ft; Edmondia Band at 282 ft 8; High Main (UL), c 13 at 308 ft 11; High Main (LL), c 39 at 313 ft 8; TD 351 ft 3.

OLIVER WILLIAMS WATER BH, DERBY [3582 3599] 1911, abt 155 ft.
See Plate 10 and Stephens 1929, p. 91. Not recorded to 32 ft; Keuper Marl to TD at 200 ft.

ORMONDE COLLIERY SOUTH SHAFT [4274 4761], 255 ft.
See Figs. 25 and 30. Made ground to 12 ft; Third Waterloo horizon at 17 ft 11; Fourth Waterloo, c 19 at 95 ft 6; First Ell, c 28 at 145 ft 1; Second Ell, ca 31, m 4 ft 9 on ca 24 at 189 ft 10; base of Westphalian B estimated at 282 ft; Brown Rake, c 13 at 295 ft 4; False Ell, c 6, d 9, c 30, m 28, on c 36 at 347 ft 1; Deep Soft workings at 397 ft; Deep Hard workings at 458 ft 2; First Piper, c 28, d 15 on c 10 at 499 ft 5; Hospital, c 15 at 546 ft 11; Cockleshell horizon at 560 ft 5; Low Main, c 47 at 640 ft 2; Threequarters, c 13, m 7 ft 8 on c 12 at 664 ft 3; Yard, c & d 33 at 739 ft 10; Blackshale, d c 21, d 11, d c 21, d 8, c 30, m 24 on c 21 at 789 ft 9; 'Batty Coal', 28 at 831 ft 8; Coal 18 at 838 ft 9; 'Ashgate', c 24, m 36 on c 28 at 868 ft 1; Mickley Thin, c 12 at 918 ft 11; Lower Brampton, c 3 at 969 ft 3; Kilburn, c 39 at 1164 ft 9; recorded to 1201 ft 11.

ORMONDE COLLIERY BLACKSHALE 70's UBH [4414 4838] 1966.
Down from Blackshale at -148 ft to Kilburn. TD 379 ft.

ORMONDE COLLIERY WOODLINKIN No. 1 DRIFT [4392 4791] 1964-65, 326 ft.
Bottom First Waterloo to First Piper. See Plate 5 for graphic section of Westphalian B measures and Fig. 74 for section in Second Waterloo Coal.

ORMONDE COLLIERY 6's RHG UPBORE [4517 4881] 1968.
Up at 45° on bearing 158° from Low Main at -427 ft to Deep Hard. Length drilled 250 ft.

OUTWOODS BH, DUFFIELD [3583 4256] abt 400 ft.
See also Gibson and others 1908, p. 44. Red clay with stones (Boulder clay) to 17 ft; Ashover Grit (Main Bed) to TD at 320 ft.

PAMELA'S LARCHES BH, [4945 5085] 1972, 513 ft.
Cores examined by C. Beal. Open Hole (including Lower Magnesian Limestone) to 10 ft; Lower Marl to 84 ft 6; Permian Basal Breccia to 86 ft 6; Shafton, b 12 at 117 ft; Edmondia Band at 209 ft 3; High Main, c 13, d 10 on c 33 at 251 ft 8; TD 277 ft.

PARK BROOK BH, HORSLEY [3830 4385] 1959, 240 ft.
Cores examined by G.D. Gaunt. See Fig. 14.

	Thickness ft	in	Depth ft	in
WESTPHALIAN A				
Not cored: Coal at 45 ft 3; Norton at 131 ft 3; Forty-Yards at 160 ft 6	200	0	200	0
Sandstone, grey and brown; interbedded siltstone	10	7	210	7
Mudstone; thin ironstone bands	60	1	270	8
ALTON MARINE BAND				
Mudstone, dark grey; ironstone bands; marine fossils	6	10	277	6
ALTON, c 36	3	0	280	6
Seatearth	10	6	291	0
Mudstone	10	5	301	5
FIRST SMALLEY, c 2, m 35 on c 2	3	3	304	8
Seatearth mudstone, light grey	7	4	312	0
Mudstone grey	4	10	316	10
SECOND SMALLEY Lingula BAND				
Mudstone, dark grey; Lingula sp. and fish debris	0	8	317	6
Mudstone, black	0	4	317	10
SECOND SMALLEY, c 8	0	8	318	6
Seatearth siltstone, grey	2	6	321	0
Sandstone, light grey	2	2	323	2
Mudstone, silty, grey; dark grey and less silty basally	3	5	326	7
HOLBROOK Lingula BAND				
Mudstone, dark grey; Lingula sp. and fish debris	1	3	327	10
Seatearth mudstone, grey, silty	4	2	332	0
Sandstone, light grey and siltstone	12	0	344	0
Mudstone, grey, silty	4	8	348	8
Sandstone, light grey, massive	7	11	356	7
Mudstone, grey	0	11	357	6
BELPERLAWN, c 38	3	2	360	8
Seatearth mudstone, grey	1	4	362	0
Ganister, white	2	6	364	6
CRAWSHAW SANDSTONE				
Sandstone, grey, red, pink and purple-stained; partly cross-bedded	96	6	461	0

b

	Thickness ft	in	Depth ft	in
Mudstone, grey, silty at top; 6-in siltstone at base; occasional fish scales	40	0	501	0
POT CLAY MARINE BAND				
Mudstone, dark grey; marine fossils and fish debris	4	8	505	8
NAMURIAN				
Mudstone, dark grey; mussels and fish debris	1	6	507	2
Seatearth siltstone, grey; pink-stained patches	2	0	509	2
ROUGH ROCK				
Sandstone, light grey, massive, locally pink-stained, to TD	10	10	520	0

PENTRICH COLLIERY, HARTINGTON (DOWNCAST) SHAFT [3938 5184] 1880.
? Black Rake, c 24 at 165 ft; Deep Soft old workings at 236 ft; Deep Hard, c 36 at 288 ft; First Piper (upper coal), c 27 at 329 ft 3; First Piper (lower coal, upper part), c 12 at 337 ft 6; First Piper, lower coal, lower part, c 9 at 349 ft 3; Hospital, c 15, d 75 on c 12 at 388 ft; Low Main (Furnace), c 48 at 454 ft; Threequarters, c 27 at 479 ft; TD 479 ft.
Another record named 'Downcast Shaft' (1841) proved measures comparable with Speedwell Shaft (see below) but with differences in coal thicknesses.

PENTRICH COLLIERY, SPEEDWELL SHAFT [3948 5187] 1841, abt 325 ft.
Deepened from Deep Soft to Low Main in 1880. See also Fig. 30. Second Ell, c 30 at 59 ft 9; inferred Clay Cross MB at abt 130 ft; Brown Rake, c 15 at 158 ft 6; ? Black Rake, c 15 at 191 ft; Deep Soft, c 46 at 286 ft; Deep Hard, c 48 at 343 ft; First Piper (upper coal), c 27 at 384 ft 9; First Piper (Lower Coal, upper part), c 4 at 392 ft 4; First Piper (Lower Coal, lower part), c 6 at 404 ft 6; Hospital, c 18, d 84 on c 9 at 443 ft 9; ? Cockleshell horizon at 455 ft 6; Low Main, c 48 at 509 ft; TD at 509 ft.
A third shaft named Pentrich Colliery is probably sited nearby [3944 5182] and proved similar measures.

PINXTON COLLIERY No. 1 SHAFT [4499 5424], 293 ft.
Deep Soft at 350 ft 11; Deep Hard at 400 ft 6; Blackshale, c 45 at 776 ft 6; TD not known.

PINXTON COLLIERY No. 2 SHAFT [4487 5450] before 1904, 296 ft.
Record incomplete to 759 ft 3; Deep Soft at 339 ft 4; ? Tupton, c 45 at 588 ft; Threequarters, c 25 at 601 ft 10; Yard, c 27 at 683 ft 1; Blackshale, c 15, b 2, c 8, d 3, c 16 on ca 24 at 765 ft; Coal 8 at 796 ft 8; 'Ashgate' (UL), c 10 at 826 ft; 'Ashgate' (LL), c 6 at 844 ft 6; Mickley Thin, c 12 at 913 ft 3; Lower Brampton, b 41 at 977 ft 6; Morley Muck, c 5 at 1051 ft 9; Coal 12 at 1172 ft 7; Kilburn in faulty ground at 1263 ft 2; recorded to 1263 ft 2.

PINXTON COLLIERY No. 3 SHAFT [4462 5465] 1836, 302 ft.
Coal 13 at 7 ft 7; ? First Ell horizon at abt 40 ft 7; Second Ell, c 18 at 79 ft 9; base of Westphalian B estimated at 166 ft; Brown Rake, c 12, m 7 ft on c & b 17 at 214 ft 8; Black Rake TS at abt 231 ft 11; Black Rake, c 8 at 236 ft 1; Top Soft, c 16 at 242 ft 2; Roof Soft (UL), c 6 at 246 ft; Roof Soft (LL), d c 36 at 258 ft 3; Deep Soft, c 48 at 296 ft 6; Deep Hard at 356 ft 6; First Piper at 443 ft 3; ? Tupton, c 45 at 561 ft 4; Threequarters, c 25 at 575 ft 2; Yard, c 17 at 648 ft 5; Denby leaf of Blackshale horizon at 733 ft 10; Blackshale, c 15, b 3, c 8, d 3, c 15, m 10 on ca 10 at 754 ft 8; recorded to 754 ft 8.

PLASTIC No. 1 BH [4369 4814] 1956, 363 ft.
Rotary chipped from Third Waterloo to Hospital. TD 367 ft.

PLASTIC No. 2 BH [4378 4818] 1956, 337 ft.
Rotary chipped from First Ell to Deep Hard. TD 350 ft.

PLASTIC No. 3 BH [4387 4823] 1956, 312 ft.
Rotary chipped from Bottom Second Waterloo to Deep Soft. TD 331 ft.

PLUMTREE (PLUMPTRE) COLLIERY [461 480]
Three records exist of measures between the Mainsmut and First Piper coals. They differ somewhat in detail and cannot be precisely located. See Plate 5 for one section Top Hard Coal to base of Westphalian B.

POLLINGTON COLLIERY No. 1 SHAFT [4555 5013] 1876, abt 395 ft.
See Plate 5. Mainsmut (UL), c & b 17 at 45 ft 1; Mainsmut (LL), c 8 at 55 ft 8; Comb, c 30 at 82 ft 1; Top Hard workings at 108 ft 4; Top Hard Floor, c 17 at 131 ft 7; Dunsil, c 24, b 6 on c 7 at 170 ft 11; First Waterloo, c 12, d 42 on c 24 at 206 ft 5; Waterloo Marker, c & d 14 at 232 ft 3; Second Waterloo, c 38, m 10 ft on c 15 at 268 ft 10; Third Waterloo, c 22, d 68 on b 10 at 345 ft 6; Fourth Waterloo, c 22 at 339 ft; First Ell, c 15 at 372 ft 4; Second Ell, c & d 20 at 416 ft 10; base of Westphalian B estimated at 490 ft; Brown Rake, c 8 at 508 ft; Black Rake, c 4 at 533 ft 6; Top Soft, c 9 at 542 ft 4; Roof Soft (UL), c 10 at 545 ft; Roof Soft (LL), c 18 at 564 ft; Deep Soft, c 44 at 606 ft 3; Deep Hard, c 41 at 656 ft 9; TD ? 672 ft 6. Some records give 667 ft 1.

PORTLAND COLLIERY No. 1 SHAFT [4789 5479].
See Smith and others 1967, p. 367. Proved measures below the Mainbright Coal to the Kilburn Coal. Recorded to 1914 ft.

PORTLAND COLLIERY No. 2 SHAFT [4830 5447] 1887, 355 ft.
Clown, c 12 at 32 ft; Mainbright not recorded; Two-Foot, c 1 at 154 ft 9; Lowbright, c 11 at 196 ft 4; Lowbright Floor, c 3 at 201 ft; Brinsley Thin (UL), b 8 at 249 ft 10; Brinsley Thin, c 8 at

c

265 ft; High Hazles, b 4 on c 22 at 312 ft; Cinderhill (UL), c 10 at 361 ft 6; Cinderhill, c 3 at 378 ft 5; ? St Johns, c 2 at 428 ft 11; Mainsmut, c 1, m 65, c 6, m 59 on c 4 at 481 ft 9; Comb, c 6, b 4 on c 17 at 530 ft 6; Top Hard, c 54 at 536 ft 6; Top Hard Floor, c 2 at 561 ft 10; TD 581 ft 6.

PORTLAND COLLIERY No. 3 SHAFT, site unknown; No. 4 SHAFT [4853 5378], abt 375 ft.
Top Hard at 540 ft; recorded to 600 ft.

PORTLAND COLLIERY No. 5 PIT [4857 5357] 1851, abt 420 ft.
Coal 14 at 49 ft 6; Coal 18 at 59 ft; ? Clown, c 34 at 75 ft 10; TD 186 ft.

PORTLAND COLLIERY No. 6 SHAFT [4775 5387] and No. 7 SHAFT [4779 5374], no known details.

PYE HILL COLLIERY DOWNCAST SHAFT [4474 5228] 1874, 292 ft.
See also Figs. 22, 25 and 30. First Ell, c 18, d 19 on c 2 at 49 ft 2; Second Ell, c 12 at 96 ft 10; base of Westphalian B estimated at 183 ft; Brown Rake, c 6, m 4 ft 10 on c 14 at 210 ft 11; Black Rake TS at 232 ft 8; Black Rake, c 10 at 238 ft 8; False Ell, c 18, m 3 ft 2, c 10, m 18 on c 27 at 253 ft 1; Deep Soft, c 50 at 287 ft 8; Deep Hard, c 36 at 333 ft 2; First Piper (Upper Coal and Lower Coal, upper part), c 27, c & d 21* at 394 ft 11; First Piper (Lower Coal, lower part), c 4 at 417 ft 8; Second Piper horizon at 442 ft; Low Main, c 48 at 552 ft 4; Threequarters, c 27 at 567 ft 10; Yard, c 36 on c & b 23 at 655 ft 6; Blackshale, c 12, m 23, c 24, d 3, c 10, d 7 on c 16 at 682 ft 5; TD 697 ft 5.

RADFORD 5's SLANT OUTBYE UBH [5426 4099] 1956.
Down from Deep Hard at -640 ft. First Piper, c 29 at 38 ft 5; Second Piper horizon at 60 ft 9; Hospital, ca 4 on c 7 at 78 ft 2; Cockleshell, c 10 at 121 ft 6; Low Main, c 37 at 151 ft 5; Threequarters, c 24, m 32 on c 3 at 163 ft 5; TD at 163 ft 6.

RADFORD 5's SLANT UBH [5414 4077] 1953.
Down from Deep Hard at -637 ft to Threequarters. TD 173 ft.

RADFORD 7's LEFT GATE UBH [5392 4104] 1952.
Down from Deep Hard at -633 ft 10 Threequarters. TD 173 ft 7.

RIPLEY COLLIERY No. 1 SHAFT [4042 4998] before 1905, 472 ft.
See Plate 5. Bottom First Waterloo, c 15 at 53 ft 3; Waterloo Marker, c 6 at 68 ft 7; Top Second Waterloo, c 29 at 87 ft; Bottom Second Waterloo, ca 11 at 119 ft 4; Third Waterloo, c 12 at 144 ft 4; Fourth Waterloo, c 15 at 179 ft 11; First Ell, c 18 at 231 ft 9; Second Ell, c 18 at 280 ft 3; base of Westphalian B estimated at 349 ft; Brown Rake, c 12 at 391 ft 9; False Ell, c 27, d 48, d c 8, d 10 on d c 24 at 434 ft 9; Deep Soft, c 55 at 491 ft 8; Deep Hard, c 51 at 574 ft 4; First Piper, c 72 at 627 ft 2; TD 637 ft 8.

*A more detailed section in adjacent workings is shown on Fig. 30

SANDIACRE AND STAPLEFORD STATION WATER BHs [4824 3669] 1948-49 and 1952, abt 120 ft.
Record of 1948-49 bore. Clay, sand and gravel to 15 ft; Pebble Beds to abt 120 ft; Lower Mottled Sandstone to 188 ft; Westphalian to TD at 203 ft.

SANDY LANE BH, BRAMCOTE [5164 3811] 1922, abt 150 ft.
Pebble Beds to 37 ft; Lower Mottled Sandstone to 121 ft; Deep Soft, c 46, d 8 on c 2 at 181 ft 8; Deep Hard, c 49, ca 7 on d c 3 at 225 ft 11; First Piper, c 38 at 266 ft 7; TD 269 ft.

SEAGRAVE BOREHOLE [5159 4255] 1958, 340 ft.
Cores examined by G. D. Gaunt. See also Figs. 25, 72 and 74, and Plate 5.

	Thickness ft	in	Depth ft	in
PERMIAN				
Soil and clay (no cores)	5	3	5	3
LOWER MAGNESIAN LIMESTONE				
Limestone (no core)	4	9	10	0
Limestone, red-brown, dolomitic, sandy	9	2	19	2
Marl, red-brown	1	2	20	4
Limestone, yellow-brown, dolomitic, sandy; coalified plants	1	1	21	5
LOWER MARL				
Marl; bivalves	1	1	22	6
Limestone	3	8	26	2
Marl	1	10	28	0
Limestone	2	0	30	0
PERMIAN BASAL BRECCIA				
Breccia	3	0	33	0
WESTPHALIAN B				
Mudstone, stained	2	0	35	0
Siltstone, stained; plants	1	0	36	0
Core lost	2	0	38	0
Mudstone, stained	2	6	40	6
Siltstone, stained	2	6	43	0
Core lost	2	0	45	0
Mudstone, stained	1	0	46	0
Siltstone, brown; plants	3	0	49	0
Core lost	5	0	54	0
HIGH HAZLES FLOOR HORIZON	-	-	54	0
Seatearth with mudstone bands; carbonaceous in part	6	3	60	3
Sandstone with mudstone and siltstone	13	9	74	0
Mudstone; plants	9	3	83	3
Mudstone, carbonaceous	0	7	83	10
Seatearth sandstone	2	5	86	3
Sandstone, mudstone and siltstone bands; plants	7	11	94	2
Mudstone; plants	6	4	100	6
CINDERHILL, c 24	2	0	102	6
Seatearth, silty at base	8	6	111	0
Mudstone; plants	3	1	114	1
Sandstone; thin mudstone laminae	33	8	147	9

d

Seagrave Borehole cont.

	Thickness ft	in	Depth ft	in
Mudstone; fish debris at base	2	6	150	3
COAL, c 1	0	1	150	4
Seatearth	1	7	151	11
Mudstone	0	4	152	3
Siltstone and sandstone	6	9	159	0
Mudstone; ironstone; bivalves	46	3	205	3
MAINSMUT, c 28	2	4	207	7
Seatearth, silty, sandy at top	9	9	217	4
Mudstone, with siltstone bands; plants	16	8	234	0
No cores; driller recorded Comb, c 16 at 245 ft 11 and Top Hard, c 45 at 251 ft	32	6	266	6
Mudstone; bivalves	3	0	269	6
TOP HARD FLOOR, c 10	0	10	270	4
Seatearth, sandy at base	3	5	273	9
Siltstone and sandstone	5	3	279	0
Mudstone	6	0	285	0
Siltstone	2	2	287	2
Mudstone; ironstone; bivalves	15	7	302	9
DUNSIL, c 9, d 3 on c 5	1	5	304	2
Seatearth	5	10	310	0
Siltstone with mudstone	31	2	341	2
Sandstone	1	4	342	6
Mudstone	5	9	348	3
Sandstone	0	5	348	8
FIRST WATERLOO, c 34	2	10	351	6
Seatearth	2	2	353	8
Sandstone and siltstone	8	1	361	9
Mudstone; bivalves at base	11	1	372	10
WATERLOO MARKER, c 5	0	5	373	3
Seatearth	7	1	380	4
Mudstone	0	5	380	9
Mudstone, carbonaceous; plants	0	2	380	11
TOP SECOND WATERLOO (UPPER LEAF), c 15	1	3	382	2
Seatearth	4	1	386	3
Sandstone	3	1	389	4
Mudstone	5	7	394	11
TOP SECOND WATERLOO (LOWER LEAF), c 8, d 2 on c 3	1	1	396	0
Seatearth, sandy at base	4	8	400	8
BOTTOM SECOND WATERLOO, c 7	0	7	401	3
Seatearth	1	9	403	0
Sandstone	10	2	413	2
Mudstone; plants	0	8	413	10
Siltstone	9	2	423	0
Mudstone, carbonaceous at base	9	4	432	4
THIRD WATERLOO				
COAL, c 2	0	2	432	6
Seatearth	3	4	435	10
COAL, c 1, d 3 on c 22	2	2	438	0
Seatearth	2	0	440	0
Sandstone	2	7	442	7
Sandstone, siltstone bands	11	11	454	6
Mudstone, carbonaceous	0	1	454	7
THIRD WATERLOO COAL (LOWER LEAF) HORIZON	-	-	454	7
Seatearth	0	8	455	3
Mudstone	19	3	474	6
FOURTH WATERLOO, c 18	1	6	476	0
Seatearth	7	3	483	3
Mudstone, siltstone and sandstone laminae at top; bivalves near base	33	0	516	3
FIRST ELL, c 21	1	9	518	0
Seatearth	6	3	524	3
Siltstone with mudstone laminae	4	1	528	4
Mudstone; bivalves, ostracods, fish	34	1	562	5
SECOND ELL, ca 1, c 13, d c 7, d 4 on d c 6	2	7	565	0
Seatearth	6	1	571	1
Sandstone and siltstone	3	8	574	9
Mudstone	43	5	618	2
CLAY CROSS MARINE BAND				
Mudstone; Lingula sp., goniatites, fish debris, fucoids; interbanded with non-marine bivalves (see Fig. 72)	8	2	626	4
WESTPHALIAN A				
Seatearth	0	8	627	0
Mudstone; ironstone; bivalves, ostracods, fish debris	11	2	638	2
BROWN RAKE, c 2	0	2	638	4
Seatearth	9	2	647	6
Mudstone, with rare siltstone; bivalves, ostracods	13	0	660	6
TOP SOFT (UPPER LEAF) HORIZON	-	-	660	6
Seatearth	5	6	666	0
Siltstone	3	0	669	0
Mudstone	4	6	673	6
Seatearth	2	4	675	10
ROOF SOFT (UPPER LEAF), c 4	0	4	676	2
Mudstone; plants	2	11	679	1
Siltstone	5	11	685	0
Mudstone; ironstone	1	6	686	6
Seatearth	0	6	687	0
Mudstone	0	6	687	6
ROOF SOFT, c 12	1	0	688	6
Seatearth	7	6	696	0
Sandstone and siltstone	28	6	724	6
Mudstone	10	0	734	6
DEEP SOFT OLD WORKINGS	3	2	737	8
Seatearth	2	10	740	6
Open hole, driller's log:				
Mudstone	31	0	771	6
DEEP HARD OLD WORKINGS	3	4	774	10
Mudstone	25	2	800	0
Mudstone; plants, bivalves, ostracods, fish debris	3	0	803	0
FIRST PIPER, c 21	1	9	804	9
Seatearth, sandy and silty	4	9	809	6
Sandstone and siltstone with mudstone	28	0	837	6
Mudstone; ironstone; bivalves, ostracods	29	6	867	0
HOSPITAL, c 43	3	7	870	7
Mudstone, carbonaceous	0	2	870	9
Seatearth	0	9	871	6
Sandstone and siltstone	49	5	920	11
Mudstone	0	7	921	6
LOW MAIN, ca 7 on c 28	2	11	924	5

Seagrave Borehole cont.

	Thickness ft	in	Depth ft	in
Seatearth	5	9	930	2
THREEQUARTERS, c 16	1	4	931	6
Seatearth siltstone	3	9	935	3
Sandstone with siltstone	41	9	977	0
Mudstone; ostracods, fish debris	10	0	987	0
YARD HORIZON	-	-	987	0
Seatearth	7	2	994	2
Mudstone	2	2	996	4
?DENBY LEAF OF BLACKSHALE HORIZON	-	-	996	4
Seatearth siltstone	0	10	997	2
Mudstone; ironstone; plants	10	10	1008	0
Siltstone with sandstone	18	6	1026	6
Mudstone; ironstone	26	1	1052	7
BLACKSHALE, c 11, d 1, d c 1, d 9, d c 11, d 4 on c 9	3	10	1056	5
Seatearth	2	1	1058	6
Mudstone, sandstone and ironstone; plants at base, fish debris	28	6	1087	0
Seatearth	2	2	1089	2
COAL, c 1	0	1	1089	3
Seatearth	0	8	1089	11
Mudstone, coaly partings	0	10	1090	9
Seatearth mudstone and siltstone	6	0	1096	9
Siltstone and mudstone	1	11	1098	8
Core not recovered to TD	2	4	1101	0

SELSTON COLLIERY BLACKSHALE 80's RH UBH [4871 4957].
Up from Blackshale at -1125 ft to Yard at 37 ft 6; length drilled 50 ft.

SELSTON COLLIERY BRINSLEY DRIFT [4654 4906] 1963.
Up from Blackshale workings to Deep Hard Coal in Brinsley Colliery Shaft. Deep Hard, c 40 at drift datum; First Piper, c 34 on c & d 8 at 48 ft 6; Hospital, c 10, d 12 on c 14 at 103 ft 7; Cockleshell, c 15 at 121 ft 4; Low Main, c 40 at 149 ft 2; Threequarters, c 13, d 3 on c 1 at 164 ft 6; Yard, c 27 at 230 ft 9; Blackshale, c 8 (Denby leaf), d 4, c 25, d 4, c 9, d 4 on c 18 at 312 ft 4; driven to 312 ft 4.

SELSTON COLLIERY No. 1 DRIFT [4765 4999 (top) to 4750 4988].
Down from Low Main at -875 ft to Blackshale; TD 132 ft 5.

SELSTON COLLIERY No. 2 DRIFT [4697 5034], from Deep Soft Workings at -570 ft to Blackshale [at 4731 4994].
See Fig. 28 for section of Low Main Coal.
Deep Hard, ca 5 on c 29 at 38 ft 2; First Piper, Upper Coal, c 33 at 87 ft 3; ?Second Piper, d c 12 at 147 ft 7; ?Hospital, d c 3 at 171 ft 9; Cockleshell (UL), c 16 at 198 ft; Cockleshell (LL), d c 1 at 207 ft 10; Low Main, c 44 at 228 ft 2; Threequarters, c 18, d 1, c 10, m 4 ft 4 on c 15 at 249 ft 4; Yard, c 35, ca 8 on d c 7 at 333 ft 6; Blackshale, c 11 (Denby Leaf), d 5, c 25, d 3, c 10, d 3 on c 16 at 353 ft 7; bottom of Drift at 353 ft 7.
Nearby are PYE HILL-SELSTON DRIFT [4680 5051 to 4661 5080], Deep Soft to Blackshale and PYE HILL DRIFT [4606 5022 - 4590 5010] First Piper to Low Main. They show similar sections.

SELSTON COLLIERY STAPLE PIT AND UBH [4702 4982] 1926-27.
Down from Low Main at abt -774 ft. See Fig. 68.

	Thickness ft	in	Depth ft	in
WESTPHALIAN A				
Mudstone; ironstone nodules	13	6	13	6
THREEQUARTERS, c 25	2	1	15	7
Shale, grey and dark grey	6	5	22	0
Sandstone	26	7	48	7
Shale, grey and blue; ironstone nodules	44	2	92	9
YARD, c 31	2	7	95	4
Shale and sandy shale, grey and blue; ironstone	26	2	121	6
BLACKSHALE, c 66	5	6	127	0
Fireclay	2	4	129	4
Shale, sandy, and sandstone (base of Staple Pit at 135 ft)	18	6	147	10
Shale, dark grey; ironstone nodules	18	8	166	6
Shale, dark grey; bands of dirty coal	3	2	169	8
Shale, grey and dark grey	12	4	182	0
Fireclay	4	3	186	3
Shale, dark grey; thin shaly sandstones	37	2	223	5
'ASHGATE' (?LOWER LEAF), c 17	1	5	224	10
Shale, dark grey, and grey sandy shale with a thin sandstone	51	5	276	3
MICKLEY THIN, c 1	0	1	276	4
Shales, light and dark grey; ironstone	48	8	325	0
LOWER BRAMPTON HORIZON	-	-	325	0
Fireclay	6	0	331	0
Shale, sandy, and shaly sandstone	25	11	356	11
Shale, grey, and sandy shale; ironstone	35	0	391	11
?MORLEY MUCK HORIZON	-	-	391	11
Fireclay	2	1	394	0
Sandstone and sandy shale	11	2	405	2
Shaly sandstone	40	2	445	4
Shale, grey and dark grey	12	0	457	4
Fireclay	0	6	457	10
Shale, sandy, and shaly sandstone	29	11	487	9
Shale, blue-grey; ironstone	24	11	512	8
KILBURN, c 7	0	7	513	3
Fireclay	0	9	514	0
WINGFIELD FLAGS				
Shale, sandy, and sandstone	114	0	628	0
Shale and sandy shale, recorded to	49	0	677	0

a

b

SELSTON COLLIERY (UNDERWOOD) SHAFT [4680 5042] 1876, 444 ft.
See Plate 5. Mainbright, c 32 at 102 ft; Two-Foot, c 24, m 40 on c 9 at 121 ft 2; Lowbright, c 32 at 150 ft; Lowbright Floor, c 9 at 159 ft 5; Brinsley Thin, c 21 at 211 ft 7; High Hazles, c 22 at 239 ft 3; Cinderhill (UL), c 12 at 318 ft 8; Cinderhill, c 12 at 336 ft 7; ?St Johns, b 6 at 366 ft 9; Mainsmut, c 17 at 435 ft 1; Comb, c 30 at 469 ft 1; Top Hard, c 48 at 479 ft 7; Top Hard Floor, c 13 at 505 ft 8; Dunsil, c 20 at 531 ft 1; First Waterloo, c 16, d 14 on c 21 at 587 ft 1; Waterloo Marker, c 1, m 98 on c 1 at 614 ft 1; Second Waterloo, c 27, d 4, c 5, d 12 on c 20 at 655 ft 9; Third Waterloo, c 20 at 680 ft; Coal 1 at 695 ft 1; Fourth Waterloo, c 2, b 3 on c 20 at 724 ft 5; First Ell, c 17 at 756 ft 5; Second Ell, c & ca 14 at 800 ft 2; base of Westphalian B estimated at 877 ft; Brown Rake, c 8 at 897 ft 9; Top Soft, c 8 at 933 ft 10; Roof Soft (UL), c 9 at 936 ft 9; Roof Soft (LL), c 18 at 962 ft 3; Deep Soft, c 44, d 1 on c 1 at 996 ft 3; Deep Hard, ca 4 on c 31 at 1035 ft 5; TD 1048 ft 11.

SHIPLEY COLLIERY, DEEPFIELDS SHAFT [4440 4480] ?1817, abt 280 ft.
See Plate 5 and Stephens 1929, pp. 112-113 Section below Deep Hard, from old section book of A. Lupton, not available during preparation of memoir text. Coal 18 at 29 ft 6; Mainsmut, c 34 at 108 ft 10; Upper Comb, Lower Comb and Top Hard, c 24, d 6, c 22, d 4 on c 88 at 153 ft 1; Dunsil, c 36 at 210 ft 1; ?Top First Waterloo, 18 of 'Blackshale'* at 231 ft 7; Bottom First Waterloo, c 18 at 253 ft 7; Waterloo Marker, c 9 at 279 ft 4; Top Second Waterloo, c 44 at 311 ft 3; Bottom Second Waterloo, c 18 at 330 ft 9; Third Waterloo, c 24 at 357 ft; Fourth Waterloo, c 16 at 399 ft 4; First Ell, c & d 36 at 444 ft 4; Second Ell, ca 24 at 489 ft 10; base of Westphalian B estimated at 574 ft; Brown Rake, c 16 at 588 ft 3; Top Soft, c & d 36 at 636 ft 3; Roof Soft and Deep Soft, c 51, d 57 on c 51 at 684 ft 9; Foot, c 9 at 704 ft 9; Deep Hard, c 66 at 734 ft 6; Piper, c & d 48 at 790 ft 3; Hospital, c 24 at 822 ft 3; Low Main, c 48 at 925 ft 3; Threequarters, c 24 at 945 ft 3; Blackshale, d & c 36, d 3, c 36, d 12 on c 27 at 1059 ft 9; Coal 24 at 1269 ft 9; Kilburn, c 46 at 1499 ft 7; TD 1499 ft 7.

SHIPLEY PARK BH [4293 4416] 1961, 281 ft.
Cores, below 58 ft, examined by J.G.O. Smart and J. Chilton. Top Second Waterloo, c ?24 at 41 ft; Bottom Second Waterloo, c 21 at 63 ft 3; Third Waterloo, d c 5 on c 17 at 98 ft 8; Third Waterloo (LL) horizon at 134 ft 3; Fourth Waterloo, d c 6 on c 15 at 155 ft 9; First Ell, d c 14 on c 22 at 193 ft; Second Ell, ca 9, m 3, c 5, m 5 on d c 1 at 244 ft; Clay Cross MB at 326 ft 4; Brown Rake, c 26 at 343 ft 2; Black Rake, d c 6 at 364 ft 6; Top Soft, c 6, d c 7 on c 21 at 394 ft 3; TD 396 ft.

SHORTWOOD BH [4964 4053] 1958, 248 ft.
Cores, 10 ft to 27 ft 6 and below 37 ft, examined by G.D. Gaunt. Roof Soft at 1 ft; Deep Soft workings at 37 ft; Deep Hard workings at 76 ft 6; First Piper, c 33 at 117 ft 3; ?Hospital, d c 2 at 149 ft 3; Low Main, c 42 at 271 ft; Threequarters, c 33, m 8 ft 2 on c 8 at 294 ft 4; Yard and Denby leaf of Blackshale, c 11, m 4 ft 9, c 1, m 6, c 2, (Yard), m 8 on c 25 at 368 ft 9; Blackshale, c 18, m 11 ft 2 on c 8 at 402 ft 8; TD 420 ft 6.

SKEAVINGTONS LANE BH, COTMANHAY [4655 4399] 1954, 191 ft.
Outline information only, cored only from 422 ft to 485 ft and 486 ft to 526 ft 6. Second Waterloo at 76 ft 4; base of Westphalian B estimated at 345 ft; Brown Rake, c 18 at abt 360 ft 6; Top Soft, c 24 at 403 ft 6; Roof Soft (UL), c 12 at 424 ft; Deep Soft workings at 469 ft 2; Deep Hard workings at 517 ft 5; TD 550 ft 5.

SLACK LANE WATER BH 'A' DERBY [3360 3619] 1886, abt 195 ft.
Made ground to 4 ft; base of drift at 14 ft; Keuper Marl to ?25 ft; Waterstones to ?92 ft; ?Pebble Beds to TD at 95 ft.

SLACK LANE WATER BH 'B' [3353 3619] 1902, abt 195 ft.
Made ground to 6 ft; base of drift at 8 ft; Keuper Marl and Waterstones to 92 ft; Pebble Beds to TD at 125 ft 6.

SLACK LANE WATER BH 'C' [3371 3619] 1906, abt 195 ft.
Made ground to 6 ft; 'marl', 'rock' and 'stone' to TD at 125 ft.

SLACK LANE WATER BH RIPLEY [3948 5029], 180 ft.
In Westphalian B, no details recorded - see Stephens 1929, pp. 111 and 140.

SMALLEY GREEN BH [4071 4314] 1961, 394 ft.
Cores, below 70 ft, examined by J.G.O. Smart and J. Chilton. Low Main workings at 81 ft; Threequarters, c 18 at 90 ft 6; Yard horizon at 167 ft 11; Denby leaf horizon at 183 ft 2; Blackshale, c 33, m 6 ft 9 on c 15 at 271 ft 9; Coal 12 at 289 ft; Coal 8 at 333 ft 1; 'Ashgate' (UL), c 24 at 380 ft 1; 'Ashgate' (LL), c 36 at 384 ft 9; TD 385 ft 6.

SMALLEY MILL BH & WELL [3949 4309], 299 ft.
Section in well, 1897 (for discrepant record of BH see Stephens 1929, pp. 113-115):
Upper Band horizon at 48 ft; Norton, c 22 at 80 ft 9; Forty-Yards (?UL), c trace at 107 ft 4; Alton, c 29 at 215 ft 5; First Smalley, c 'with 1 ft rock parting' 3 ft 11 at 239 ft 5; Second Smalley, d c & c 18 at 250 ft 5; Holbrook, c 4 at abt 266 ft; Belperlawn, c & d c 24 at 283 ft 4; Crawshaw Sandstone to 390 ft; TD 435 ft.

SPRAY & BURGASS LTD WATER BH's, BULLWELL [5418 4406] & [5413 4418] 1969, abt 140 ft.
River Deposits to 7 ft; Middle Marl to 10 ft; Lower Magnesian Limestone to 39 ft 6; Lower Marl to ?72 ft; Westphalian C to TD at 90 ft.

*Recorded as coal in one record. There are two records of this section which differ somewhat in the depths.

c

STANLEY COLLIERY SHAFT No. 1 [4253 4093] 1892, 263 ft.
See Fig. 20. For a different, apparently incomplete, record see Stephens 1929, pp. 88-89. Threequarters, c 28 at 24 ft 2; Yard horizon at 85 ft 5; Blackshale, d c 6, c 20, d 21, c 6, d 24 on c 24 at 111 ft 1; Coal 18 at 119 ft 5; Coal 3 at 139 ft 10; 'Ashgate' (UL) horizon at 256 ft 8; 'Ashgate' (LL), ca 18 at 307 ft; Mickley Thin, ca 10 on c 6 at 365 ft 4; Lower Brampton, ca 9, m 18 on c 6 at 434 ft 3; Coal 10 at 480 ft 3; Morley Muck, c 22, d 9, c 10, d 2 on c 9 at 528 ft 6; Kilburn, c 4, d 3 on c 66 at 633 ft 1; recorded to 672 ft 8.

STANTON (IRONWORKS) SINKING [4730 3919], abt 140 ft.
See Figs. 16 and 68. Kilburn, c 57 at surface; Upper Band horizon at 316 ft 3; Norton, c 27 at 352 ft 2; Forty-Yards, c 12, m 8 ft 9 on c 6 at 388 ft 11; recorded to Alton, c 33 at 479 ft 2. A second record exists with slightly different depths.

STAPLEFORD No. 1 OIL BORE [4907 3595] 1966, 180 ft.
Mainly chipped bore, recorded by D.I. Iosson.
See Plate 10 for graphic section of Permo-Triassic.

	Thickness ft	in	Depth ft	in
WATERSTONES				
Sandstone, off-white to pale yellow, fine-grained	8	0	8	0
Mudstone, red, silty	3	0	11	0
Sandstone, fine, calcareous	9	0	20	0
Mudstone, red, silty	6	0	26	0
Sandstone, fine-grained, calcareous	11	0	37	0
Mudstone and siltstone interbedded, sandstone bands	42	0	79	0
PEBBLE BEDS				
Sandstone, medium- to coarse-grained; pebble horizons; sporadic bands of red mudstone	110	0	189	0
LOWER MOTTLED SANDSTONE				
Siltstone, grey-green	4	0	193	0
Mudstone, red	18	0	211	0
Sandstone, off-white, pale grey and red-brown, micaceous. Conglomeratic band 232 ft - 238 ft	80	0	291	0
WESTPHALIAN A				

Alton MB at 431 ft; top of Crawshaw Sandstone at 470 ft; drilled in Crawshaw Sandstone to TD at 539 ft.

STATION ROAD WATER BH, BEESTON [5327 3643] 1925, abt 90 ft.
See Plate 10 for graphic section in Permo-Triassic rocks. Alluvium to 20 ft; Keuper Marl to fault at 42 ft 6; Waterstones to 62 ft 6; Pebble Beds to 190 ft; Lower Mottled Sandstone to 278 ft; Westphalian to TD at 302 ft.

STONEYFORD BH [4460 4916] 1962, 233 ft.
Cores, below 27 ft, examined by J.G.O. Smart and J. Chilton. See Figs. 25 and 34. Clay Cross MB at 108 ft 7; Brown Rake, c 8 at 116 ft 10; Black Rake TS at 137 ft; Black Rake, c 1 at 138 ft 9; False Ell, c 24, m 20, c & d c 12, d 5 on c 25 at 173 ft 2; Deep Soft workings at 235 ft 2; Deep Hard workings at 281 ft 6; First Piper, c 20 on d c 18 at 319 ft 2; Second Piper horizon at 351 ft 3; Hospital, c 3, m 16, c 1, m 7 on c 7 at 380 ft 1; Cockleshell, c & d 1 at 402 ft 9; Low Main workings at 443 ft; Threequarters, c & d c 28 at 452 ft 10; Seatearth at 514 ft 8; Yard, c 5 at 525 ft 2; Blackshale, c 4, d 4, c 19, d c 4, d 2, d c 7, d 3 on c 23 at 578 ft 6; Coal 5 at 621 ft 2; 'Ashgate' (UL), c 5 at 660 ft 1; 'Ashgate' (LL), c 29 at 686 ft 5; Mickley Thin, c 18 at 742 ft 5; Kilburn, c 36 at 967 ft 2; TD 980 ft 2.

STONEY STREET QUARRY BH, BEESTON [5281 3730] 1959, 95 ft.
Figs. 31 and 33, and Plate 10 for graphic section of Permo-Triassic rocks. Pebble Beds to 106 ft 6; Lower Mottled Sandstone to 199 ft 9; Roof Soft (UL), d c 2 at 227 ft 8; Roof Soft (LL), ca 34 at 232 ft 6; Deep Soft, c 36, d 5 on d c 7 at 252 ft 7; Coal 1 at 261 ft 3; Deep Hard, c 54 at 292 ft 1; Deep Hard Floor, ca 8 at 293 ft 5; First Piper, c 34 at 341 ft 9; Second Piper horizon at 366 ft; Hospital, c & d 44 at 394 ft 8; Cockleshell, c 10, c & d c 21 on c 9 at 410 ft 2; Low Main, c 42 at 440 ft 8; Threequarters, c 20 at 449 ft 2; TD 450 ft 6.

STRELLEY BH [probably 4933 4321], abt 292 ft.
Fourth Waterloo, c 36 at 3 ft; First Ell horizon at 79 ft 2; Second Ell, d c 48 on c 19 at 126 ft 3; base of Westphalian B estimated at 189 ft; Brown Rake horizon at 204 ft 7; Top Soft horizon at 251 ft 3; Roof Soft (UL), c 13 at 263 ft 3; Roof Soft (LL), c 20 at 273 ft 11; Deep Soft, c 40 at 310 ft 8; Deep Hard, c 39 at 353 ft 9; First Piper, c 42 at 393 ft 5; ?Cockleshell, c 24 at 471 ft 7; Low Main, c 38 at 520 ft 2; Threequarters, c 15 at 526 ft 8; Blackshale, c 55 at 704 ft 8; TD 704 ft 8.

SWANWICK COLLIERY COMMON PIT [4047 5441], 383 ft.
Mainsmut, c 14 at 11 ft 4; Comb, c 33 at 48 ft 3; Top Hard, c 84 at 109 ft 7; fault in shaft 144-152 ft truncates Dunsil, c 36; Top First Waterloo, ca 4 at 170 ft 2; Bottom First Waterloo, ca 30 at 212 ft 2; Waterloo Marker, c 9 at 221 ft 9; Top Second Waterloo, c 20 on ca 42 at 256 ft 9; Bottom Second Waterloo, c 11 at 266 ft; recorded to 268 ft.

SWANWICK COLLIERY DEEP PIT [4131 5457] 1867, 413 ft.
Not recorded to 246 ft; Top Hard, c 88 at 253 ft 4; Dunsil, c 40 at 295 ft 11; Bottom First Waterloo, c 12, d 3 on c and d 30 at 341 ft 11; Waterloo Marker, c 12 at 375 ft; Top Second Waterloo, c 28, ca 36 on b 1 at 410 ft 10; Bottom Second Waterloo, c 16 at 417 ft 7; Third Waterloo, c 17

d

at 457 ft 9; Third Waterloo (LL), c 1 at 468 ft 5; Fourth Waterloo, c 18 at 504 ft 10; First Ell, c 18 at 555 ft 11; Second Ell, c 18 at 618 ft 8; base of Westphalian B estimated at 703 ft; Brown Rake, c 11, m 6 ft 3 on c 8 at 755 ft 11; Black Rake TS at 772 ft 3; ?Black Rake, c 19 at 778 ft 6; ?Top Soft, c 2 at 785 ft 5; ?Roof Soft (LL), c 8 at 846 ft 10; Deep Soft, c 48, b 2, c 3, d 6 on c 3 at 887 ft 5; Deep Hard, c 55 at 967 ft 3; recorded to 993 ft 6.

SWANWICK COLLIERY ENGINE SHAFT [about 418 526], abt 440 ft.
To Top Hard. TD 363 ft 6.

SWANWICK COLLIERY NEW PIT [4150 5460] 1915, 435 ft.
See Figs. 18, 25, 33, 34 and 74, and Plate 5.
Mainsmut, c 16 at 16 ft 10; Comb, c 34 at 57 ft 2; Top Hard workings at 128 ft 6; Dunsil, c 40 at 169 ft; Top First Waterloo, ca 11 at 193 ft 6; Bottom First Waterloo, c 15 at 208 ft 9; Waterloo Marker, c and m 48 at 243 ft 9; Top Second Waterloo, c 19 on ca 42 at 278 ft 8; Bottom Second Waterloo, c 11 at 285 ft 9; Coal 1 at 294 ft 11; Third Waterloo, c 21 at 333 ft 9; Coal 2 at 349 ft; Fourth Waterloo, c 29 at 378 ft 7; First Ell, c 19 at 424 ft 2; Second Ell, c 14 at 483 ft; base of Westphalian B estimated at 568 ft 6; Brown Rake, c 1, m 6 ft 2, c 4, m 6 ft 2 on c 8 at 616 ft 7; ?Black Rake, c 19 at 636 ft 4; ?Roof Soft (UL), c 3 at 644 ft 6; ?Roof Soft (LL), c 9 at 696 ft 2; Deep Soft, c 48, b 6 on c 2 at 729 ft 2; Deep Hard, c 56 at 809 ft; Deep Hard Floor, c and b 16 at 818 ft 4; First Piper (Upper Coal), c 15 at 847 ft 9; First Piper (Lower Coal), c 9, d 8 ft 2 on c 9 at 889 ft 3; Second Piper horizon at 910 ft 5; Hospital, ca 17 at 962 ft 6; ?Tupton, c 50 at 1008 ft 6; Threequarters, c 25, m 6 ft 2 on c 3 at 1031 ft 10; Yard, c 38 at 1123 ft 6; Denby leaf of Blackshale, c and d 14 at 1131 ft 3; Blackshale, c 8, d 48* at 1179 ft 1; recorded to 1288 ft 5.

SWANWICK COLLIERY No. 1 UBH [4098 5452] 1955.
Down from Blackshale at -1063 ft. Recorded by G.H. Rhys. See Figs. 18 and 20. Coal 12 at 53 ft 4; Coal 10 at 65 ft 5; 'Ashgate' (UL), c 17 at 91 ft 8; 'Ashgate' (LL), c 9 at 125 ft 8; Mickley Thin, c 17 at 182 ft 5; Lower Brampton, c 3 at 240 ft 11; Morley Muck, c 4 at 333 ft 1; Coal, d c 2, d 1 on d c 8 at 415 ft 11; Kilburn, c 29 at 473 ft 1; TD 477 ft 11.

SWANWICK COLLIERY OLD PIT [4087 5453], 385 ft.
See also Figs. 42 and 46.

	Thickness ft	in	Depth ft	in
WESTPHALIAN B				
Pit top raised	6	0	6	0
Soil and clay	9	3	15	3
Sandstone	1	2	16	5
Mudstone	2	0	18	5
LOWBRIGHT FLOOR, c 6	0	6	18	11
Seatearth	1	0	19	11
Mudstone	3	1	23	0
Sandstone	1	4	24	4
Mudstone; ironstone	36	2	60	6
BRINSLEY THIN, c 21	1	9	62	3
Seatearth and coal	4	3	66	6
Seatearth	2	3	68	9
Mudstone	13	9	82	6
Siltstone	7	4	89	10
Mudstone	10	5	100	3
HIGH HAZLES, c 26, m 4 ft 9, c 27	9	2	109	5
Seatearth	1	11	111	4
Mudstone, sandstone and siltstone near base	13	3	124	7
COAL, c 12	1	0	125	7
Seatearth	6	4	131	11
Siltstone and sandstone	21	10	153	9
Mudstone; ironstone	10	3	164	0
CINDERHILL, c 17	1	5	165	5
Seatearth	1	10	167	3
Seatearth, sandy and sandstone	3	9	171	0
Siltstone	4	0	175	0
Mudstone; ironstone	13	2	188	2
Mudstone, carbonaceous	0	7	188	9
Seatearth sandstone	4	10	193	7
Mudstone	11	11	205	6
Siltstone and sandstone	22	11	228	5
Mudstone and siltstone	19	2	247	7
Sandstone, carbonaceous	0	4	247	11
MAINSMUT (UPPER LEAF), c 11	0	11	248	10
Seatearth and sandstone	12	2	261	0
Mudstone	2	6	263	6
MAINSMUT (LOWER LEAF), c 10	0	10	264	4
Seatearth	1	2	265	6
Sandstone	24	6	289	10
Mudstone	12	10	302	8
COMB, c 34	2	10	305	6
Seatearth, part sandy	7	0	312	6
Mudstone	6	5	318	11
Siltstone and sandstone	21	2	340	1
Mudstone	16	1	356	2
TOP HARD, c 87	7	3	363	5
Seatearth	5	0	368	5
Sandstone	3	4	371	9
Mudstone	30	11	402	8
DUNSIL, c 40	3	4	406	0
Seatearth and sandstone	6	1	412	1
Mudstone	15	7	427	8
Mudstone, carbonaceous	1	6	429	2
TOP FIRST WATERLOO HORIZON	-	-	429	2
Seatearth	0	8	429	10
Mudstone	18	0	447	10
Batt	0	5	448	3
BOTTOM FIRST WATERLOO, c 12	1	0	449	3
Seatearth	2	9	452	0
Siltstone	29	8	481	8
Mudstone	2	6	484	2
WATERLOO MARKER, c 12	1	0	485	2
Seatearth; ironstone	6	0	491	2
Mudstone; ironstone	24	4	515	6

*Another section in this seam is c 20, d 1, c 9, d 1 on c 17.

	Thickness ft	in	Depth ft	in
SECOND WATERLOO, c 64	5	4	520	10
Recorded to			520	10

SYCAMORE FARM WATER BH [3274 4947] 1967, abt 615 ft.
Soil to 1 ft 6; sandy clay and shales with thin hard bands to 74 ft; sandstones (leaf of Ashover Grit) to TD (where 'grey clay' encountered) at 91 ft.

THURLAND HALL FARM BH [5125 5064] 1970, 449 ft.
Recorded by C. Beal. See Fig. 82 for graphic section of Permo-Triassic rocks. Permian on Westphalian to High Main. TD 503 ft 3.

TROWELL MOOR COLLIERY DOWNCAST (EASTERLY) SHAFT [4932 3908] abt 1880, abt 178 ft.
See Figs. 68 and 18, see below. Low Main, c 26 at 26 ft 6; Threequarters, c 3 at 34 ft; Yard, c 66 at 129 ft 8; Blackshale, c 52 at 142 ft; Coal 15 at 210 ft; 'Ashgate', c 30, m 2 ft 11 on c 24 at 261 ft 2; Mickley Thin, c 4 at 315 ft 6; Lower Brampton, c 12 at 370 ft 3; ? Morley Muck, c 3, m 37 ft 9, c 9, m 12 ft 3 on c 7 at 464 ft 10; Coal 27 at 544 ft 3; Kilburn, c 57 at 585 ft 3; Upper Band horizon at 891 ft 3; Norton, c 24 at 918 ft 6; Forty-Yards, c 17 at 959 ft 5; Alton, c 32 at 1059 ft 4; recorded to 1064 ft.
There is a second record with slightly different depths and thicknesses of the strata above the Kilburn Coal which is shown graphically on Fig. 18.

TROWELL MOOR COLLIERY UBH [4942 3962] 1923.
Down from Kilburn at -410 ft. See Figs. 14 and 16, and Plate 4. Upper Band horizon at 300 ft; Norton, c 9 at 326 ft 3; Forty-Yards, c 10 at 364 ft; Alton, c 41 at 461 ft 11; Belperlawn, c 14 at 504 ft 11; base of Westphalian at 643 ft 3; Pot Clay, c 2 at 643 ft 5; Base of Rough Rock at 723 ft 6; ? Simmondley, c 6 at 776 ft; Chatsworth Grit from 831 ft 6 to TD at 897 ft 6.

TRUSLEY BH [2548 3588] 1969, 260 ft.
Cores examined by D. V. Frost, J. G. O. Smart and N. Aitkenhead. See also Fig. 54 and Plate 10.

	Thickness ft	in	Depth ft	in
TRIASSIC				
HARLEQUIN FORMATION				
No core	10	0	10	0
Mudstone, red-brown, silty with thin siltstone bands and laminae	11	0	21	0
Core lost	5	0	26	0
Mudstone, red-brown, silty; sporadic green mottling; sporadic red-brown and pale grey-green siltstone bands; rare mica flakes	12	6	38	6
Core lost	1	9	40	3
Siltstone, pale red-brown, massive at top; undulating current laminae near base with small cavities	0	2	40	5
Mudstone, red-brown; black ferruginous globules	1	4	41	9
Siltstone, pale grey-green; ripple-marked	2	6	44	3
Mudstone, red-brown, silty; sporadic silt laminae	4	11	49	2
Siltstone, pale grey-green; small cavities; ripple marks, wave length $2\frac{1}{2}$ in, amplitude $\frac{1}{2}$ in	0	4	49	6
Mudstone, red-brown, silty, interlaminated with grey-green and red-brown siltstone, purple-stained at base	7	9	57	3
Core lost	1	3	58	6
Mudstone, red-brown, silty	3	6	62	0
Siltstone, grey-green and red-brown, laminated, with 1/8-in horizontal gypsum veins	0	8	62	8
Mudstone, red-brown silty, laminated, with siltstone at top; numerous gypsum veins	4	2	66	10
Core lost	1	10	68	8
Siltstone, grey-green	0	9	69	5
Gypsum band	0	2	69	7
Mudstone, red-brown, silty, interlaminated with siltstone	2	1	71	8
Mudstone, red-brown, silty; gypsum bands up to $1\frac{1}{2}$ in	13	6	85	2
Siltstone/sandstone, pale grey-green, irregularly laminated with dark green mudstone in top 3 in; massive central part; basal 3 in with current structures, ripple-marks	0	9	85	11
Mudstone, red, silty, massive; small irregular gypsum nodules; green mottling towards base	1	3	87	2
Siltstone, pale green; sporadic dark green mudstone laminae and red mudstone lenses; gypsum bands up to $1\frac{1}{2}$ in near base	4	4	91	6
Mudstone, red-brown silty massive; pale grey siltstone pouches; numerous gypsum veinlets and spots	0	11	92	5
Siltstone, pale grey-green, massive	0	11	93	4
Mudstone, dark red and pale red interlaminated; sporadic green silty patches; gypsum bands	1	3	94	7
Siltstone, pale green	0	4	94	11
Mudstone, dark red, silty; small gypsum nodules	2	2	97	1
Siltstone, pale grey-green; dark green mudstone laminae	0	2	97	3

a

b

	Thickness ft	in	Depth ft	in
Gypsum	0	2	97	5
Mudstone, dark red and pale red interlaminated; disrupted laminae at base	1	7	99	0
Siltstone, pale grey; red to dark red mudstone laminae, disrupted and arched over gypsum nodules	0	11	99	11
Siltstone, pale grey-green; dark green mudstone laminae; sandstone pouches 1 in from base	0	8	100	7
Gypsum	0	1	100	8
Siltstone, slump structures	0	4	101	0
Mudstone, dark red, silty, massive; gypsum veinlets; sporadic quartz grains	3	4	104	4
Mudstone, dark red and siltstone, pale red, pale grey and dark green interlaminated	0	8	105	0
Siltstone, pale grey-green; delicate current structures; salt pseudomorphs at base	0	8	105	8
Gypsum	0	1	105	9
Mudstone, dark red; irregular lamination	2	3	108	0
Mudstone, dark red, silty, massive; thin gypsum veinlets	1	0	109	0
Gypsum	0	2	109	2
Mudstone, dark red, interlaminated with pale grey-red siltstone	0	5	109	7
Mudstone, dark red, massive	0	3	109	10
Mudstone, dark red; disturbed laminae of pale red mudstone; thin gypsum veinlets	0	5	110	3
Siltstone, pale grey-green	1	1	111	4
Mudstone, dark red; disturbed laminae	0	9	112	1
Gypsum	0	1	112	2
Mudstone; disrupted laminae	0	4	112	6
Mudstone, red, silty, massive; irregular nodules of gypsum	4	11	117	5
Siltstone, red-brown	0	2	117	7
Mudstone, pale red, silty; siltstone laminae; 3-in gypsum band near top	1	10	119	5
Mudstone, dark red, massive	1	3	120	8
Siltstone, grey-green; 3-in red-brown centre; current structures at base	0	6	121	2
Gypsum	0	2	121	4
Mudstone, red, laminated at top and bottom, with massive central portion; silt-filled desiccation cracks	1	8	123	0
Siltstone, brown	0	1	123	1
Mudstone, pale grey-red and dark red interlaminated	0	5	123	6
Gypsum	0	1	123	7
Mudstone, dark red, silty, massive	0	2	123	9
Mudstone, dark red and siltstone, pale red, interlaminated; gypsum veins up to 3-in	8	8	132	5
Siltstone, pale grey-green and grey-brown interlaminated	2	0	134	5
Mudstone, dark red, massive; 1-in gypsum at base	2	5	136	10
Siltstone, pale grey-green; dark green mudstone interlaminated; slump structures at base	2	2	139	0
Mudstone, dark red, silty, massive; green spots	1	0	140	0
Siltstone, pale grey-green; ripple-marks	1	7	141	7
Mudstone, massive	1	1	142	8
Siltstone, massive; slump structures	0	2	142	10
Mudstone, dark red, silty, massive; sporadic gypsum veins; rare siltstone laminae and lenses	7	2	150	0
CARLTON FORMATION				
Mudstone, red, silty, massive, with sporadic siltstone bands	8	2	158	2
Siltstone, red-brown; pale green areas; flow-type breccias with dark red mudstone flakes	1	10	160	0
PLAINS SKERRY				
Siltstone, pale grey-green; sporadic red staining; dark green mudstone laminae; major slump pouches	4	4	164	4
Silstone, red-brown; pale green bands and stains; 2-in mudstone band at 164 ft 6; slight slumping at 164 ft 8	1	8	166	0
Siltstone, green	0	8	166	8
Siltstone, red-brown; flow-breccia; gypsum	1	10	168	6
Mudstone, red, with green patches; purple ramifications	2	6	171	0
Core lost	8	3	179	3
Mudstone, red-brown, silty, micaceous; gypsum nodules	2	11	182	2
Sandstone, pale grey-green, red-stained; cavities	0	1	182	3
Siltstone, grey-green, massive	1	0	183	3
Mudstone, red-brown, silty, massive; sporadic green mottling; traces of mica	6	9	190	0
Siltstone, grey-green, red mottled; gypsum nodules	1	0	191	0
Mudstone, red-brown, silty; sandy patches	1	0	192	0
Sandstone, pale grey, fine-grained; red-brown staining	0	3	192	3
Siltstone, red-brown, massive; red mudstone flakes scattered throughout	3	9	196	0
Siltstone, red-brown; flow-type breccia	0	7	196	7
Mudstone, red-brown, silty, slightly brecciated	1	2	197	9

c

	Thickness ft	in	Depth ft	in
Siltstone, red-brown and grey-brown; slumped and disturbed laminae	1	0	198	9
Mudstone, grey-green and red-brown mottled	0	3	199	0
Siltstone, red-brown at top, grey-green and laminated at base	2	2	201	2
Mudstone, red-brown, silty; disrupted laminae	2	0	203	2
Siltstone, red-brown; silty mudstone bands and laminae	6	0	209	2
RADCLIFFE FORMATION				
Mudstone, pinkish-red, silty and dark red interlaminated; slight disruption of laminae; current structures; salt pseudomorphs; sun cracks	9	2	218	4
Siltstone, red-brown; hopper outlines	0	4	218	8
Mudstone/siltstone interlaminated; pink siltstone common; rare hopper structures; vague purplish beds; gypsum; rare fine even lamination at 223 ft; disturbed, irregularly slumped and brecciated at 226 ft; 6-in pale green sandstone at 330 ft 6;	20	4	239	0
WATERSTONES				
Siltstone, red-brown and mudstone, dark red-brown, interlaminated and interbanded; sun-cracks causing disruption of laminae; micaceous partings common; sporadic green laminated siltstone bands; worm burrows; gypsum veins and nodules common	24	0	263	0
Sandstone, pale-grey, fine-grained, micaceous, interbanded and interlaminated with dark-red silty mudstone and red-brown siltstone	10	8	273	8
Sandstone, pale-grey and red-brown stained; finely micaceous, massive	0	10	274	6
Siltstone, red-brown	0	4	274	10
Mudstone, green to dark green, interbanded and interlaminated with siltstone, pale-green; slumped pouches at 275 ft 6 - 275 ft 10; sand-filled sun-cracks	1	10	276	8
'SANDY FACIES of WATERSTONES'				
Sandstone, red-brown, fine-grained micaceous, largely massive; thin dark red mudstone partings and siltstone bands; gypsum veins, bands and nodules; some bioturbation structures; 13 in soft, silty, micaceous, mottled red-brown mudstone at 321 ft 1.	46	6	323	2
Sandstone, red-brown and pale grey-green, banded and mottled, fine-grained, micaceous; quartz grains and irregular fragments up to $\frac{1}{2}$ inch below 326 ft 5; dark-green mudstone laminae at 328 ft 11	6	8	329	10
Breccia; sandstone, pale grey, coarse; pale green mudstone fragments; gypsum nodules; basal $2\frac{1}{2}$ in red-brown siltstone with sand filled burrows	0	10	330	8
CARBONIFEROUS				
DINANTIAN				
Mudstone, purplish-red, silty; oblique listric planes with gypsum	2	3	332	11
Sandstone, pinkish-red, massive, canky, graded; gypsum veinlets; sole structures at base	0	7	333	6

The underlying sequence is composed essentially of these two lithologies together with siltstone forming a series of turbidite units varying from a few inches to several feet in thickness.

Fault at 336 ft; fault plane at 336 ft 5 to 337 ft 3; Sandstone, slightly calcareous at 344 ft, with fault plane with listric surfaces 347 ft to 347 ft 10 and subsidiary fractures, common to 361 ft 7; small fault at 368 ft; dip at 375 ft, 12°; at 379 ft, 30°; at 428 ft, 68°; and at 450 ft, 50°. Calcareous below 450 ft; gypsum veinlets and bands throughout. No reduction or change in the reddening of the beds at the base of the hole. TD 507 ft 2.

TURKEY FIELD COLLIERY [4990 4240] 1848, 330 ft.
Bottom Second Waterloo horizon at 19 ft 8; Third Waterloo horizon at 42 ft 3; Fourth Waterloo, c 14 at 79 ft 5; First Ell, c & d 45 at 124 ft 2; Second Ell, c & ca 36 at 192 ft 8; base of Westphalian B estimated at 258 ft; Brown Rake, c 18 at 273 ft 10; Top Soft, c 18 at 300 ft 8; Roof Soft (UL), c 18 at 316 ft 8; Roof Soft (LL), c 18 at 326 ft 5; Deep Soft, c 42, d 7 on c 3 at 370 ft 6; Deep Hard, ca 2 on c 46 at 415 ft 10; recorded to 416 ft 1.

UPPER HARTSHAY COLLIERY (SOUTH PIT) [3873 5031] 1903, 485 ft.
See also Figs. 25 and 34.

	Thickness ft	in	Depth ft	in
Clay	5	0	5	0
WESTPHALIAN B				
FIRST ELL, c 22	1	10	6	10
Seatearth, sandy at base	6	0	12	10
Sandstone	6	6	19	4
Mudstone	44	8	64	0
SECOND ELL, c 28, m 26, c 1	4	7	68	7

d

	Thickness ft	in	Depth ft	in
Seatearth and sandstone	33	11	102	6
Mudstone; ironstone	37	6	140	0
WESTPHALIAN A (estimated top)				
Mudstone; ironstone; bivalves at base	31	11	171	11
Siltstone	1	0	172	11
Mudstone; ironstone	1	10	174	9
BROWN RAKE IRONSTONE	0	8	175	5
Sandstone	2	0	177	5
Mudstone	3	8	181	1
BROWN RAKE, c 13	1	1	182	2
Sandstone	2	4	184	6
Seatearth, silty	4	2	188	8
Sandstone and siltstone; ironstone	13	3	201	11
Mudstone; bivalves	3	6	205	5
? BLACK RAKE AND TOP SOFT, c 26	2	2	207	7
Seatearth	11	8	219	3
Ironstone	0	6	219	9
? ROOF SOFT (UPPER LEAF), b 4, m 3 on c 1	0	8	220	5
Seatearth, coal partings	13	2	233	7
Siltstone and mudstone	24	3	257	10
ROOF SOFT (? LOWER LEAF) HORIZON	-	-	257	10
Seatearth	5	8	263	6
Siltstone	11	9	275	3
Mudstone	17	5	292	8
DEEP SOFT (OLD WORKINGS)	2	10	295	6
Seatearth with rooty siltstone	18	11	314	5
Siltstone	16	4	330	9
Mudstone	8	6	339	3
Batt	0	4	339	7
Mudstone; ironstone	32	4	371	11
DEEP HARD, c 10 on old workings 42	4	4	376	3
Seatearth siltstone	0	6	376	9
Sandstone and siltstone	25	9	402	6
Mudstone	22	8	425	2
FIRST PIPER, upper coal, c 24	2	0	427	2
Shale, carbonaceous	1	0	428	2
FIRST PIPER, lower coal (upper part), d c 4	0	4	428	6
Seatearth	4	1	432	7
FIRST PIPER, lower coal (lower part), c 11	0	11	433	6
Seatearth mudstone and rooty siltstone	5	8	439	2
Siltstone	5	2	444	4
Mudstone; ironstone	1	8	446	0
Sandstone	11	5	457	5
Mudstone; ironstone	21	4	478	9
HOSPITAL, c 22	1	10	480	7
Seatearth	1	10	482	5
Sandstone	0	8	483	1
Siltstone	10	1	493	2
Mudstone; bivalves	7	2	500	4
Siltstone	17	4	517	8
Mudstone	1	0	518	8
COCKLESHELL, c 9	0	9	519	5
Seatearth and carbonaceous shale	1	9	521	2
Siltstone	5	3	526	5
Sandstone	15	7	542	0
Mudstone; ironstone	19	2	561	2
LOW MAIN, c 50	4	2	565	4
Seatearth	6	9	572	1
Mudstone; ironstone	8	9	580	10
THREEQUARTERS, c 30, m 8 ft 8 on c 4	11	6	592	4
Seatearth	0	3	592	7
Sandstone and siltstone	29	3	621	10
Mudstone	21	1	642	11
Siltstone and sandstone	5	1	648	0
Mudstone	14	6	662	6
Sandstone and siltstone	6	3	668	9
Mudstone	4	4	673	1
YARD, c 23, d 4 on c 9	3	0	676	1
Seatearth	4	5	680	6
Mudstone	30	1	710	7
Siltstone	12	4	722	11
Mudstone	8	9	731	8
Seatearth	1	9	733	5
BLACKSHALE, c 20, d 6, c 40, d 9, c 12, d 6, c 24, d 3 on c 10	10	10	744	3
Seatearth	8	10	753	1
Mudstone	18	11	772	0
COAL, c 16	1	4	773	4
Seatearth, silty and sandy	6	1	779	5
Sandstone and siltstone	20	4	799	9
Mudstone	27	5	827	2
'ASHGATE', c 24, d 38, c 31	7	9	834	11
Seatearth	4	3	839	2
Siltstone	11	2	850	4
Mudstone	38	3	888	7
MICKLEY THIN, c 17	1	5	890	0
Seatearth	3	3	893	3
Siltstone and sandstone	7	0	900	3
Mudstone	52	10	953	1
LOWER BRAMPTON, c 8	0	8	953	9
Seatearth	6	9	960	6
Siltstone	13	5	973	11
Mudstone	24	6	998	5
Seatearth siltstone	1	2	999	7
Sandstone with siltstone	18	7	1018	2
Mudstone; ironstone; bivalves	29	0	1047	2
MORLEY MUCK HORIZON	-	-	1047	2
Seatearth	9	10	1057	0
Sandstone and siltstone	37	7	1094	7
Mudstone	3	0	1097	7
Siltstone	6	9	1104	4
Mudstone; ironstone	5	6	1109	10
Siltstone; ironstone	16	3	1126	1
Mudstone; ironstone	7	2	1133	3
Siltstone; ironstone	18	9	1152	0
Mudstone; ironstone	18	7	1170	7
Batt	0	5	1171	0
KILBURN, c 35, d 6 on c 11	4	4	1175	4
Seatearth	6	5	1181	9
Mudstone to TD	14	11	1196	8

UPPER HOUSE FARM BH, KIRK IRETON [2689 5207] 1968-69, abt 800 ft.
Mudstone to 10 ft; Sandstone with mudstone to 21 ft; Mudstone to 32 ft; Sandstone with mudstone partings (Ashover Grit) to TD at 90 ft.

WAINGROVES COLLIERY [4104 4893] 1859, 442 ft.
? Bottom Second Waterloo horizon at 9 ft 4; ? Third Waterloo at 34 ft 1; Porter Barn Fault from 77 ft 11 to 118 ft 7; Brown Rake, c 17 at 133 ft 4; False Ell, c 34, m and b 39 on c 47 at 182 ft 10; Deep Soft, c 42 at 217 ft 1; Deep Hard, c 54 at 281 ft 10; TD 281 ft 10.

WASHDYKE LANE (NORTH) BH [5197 5053] 1955, 364 ft.
Cores, below 16 ft, examined by R. E. Elliott. See Fig. 49. Lower Magnesian Limestone to 25 ft; Lower Marl to 79 ft; Permian Basal Breccia at 79 ft 5; Coal 30 at 165 ft 10; Annesley, c 16 at 217 ft 4; Top MB at 294 ft 10; Shafton MB at 350 ft; Coal 18 at 353 ft; Shafton, c 19 at 370 ft 9; Main 'Estheria' Band (UL) at 388 ft 11; Main 'Estheria' Band (LL) at 401 ft 6; Edmondia Band at 455 ft 5; High Main (UL), c 13 at 493 ft 2; High Main (LL), c 42 at 498 ft 5; TD 507 ft 6.

WASHDYKE LANE (SOUTH) BH [5219 5006] 1955, 353 ft.
Permian on Westphalian to High Main. TD 505 ft.

WATNALL COLLIERY No. 1 (SOUTHERLY) SHAFT [5073 4801] 1873, 379 ft.
See Figs. 46 and 49. See also Lamplugh and Smith 1914, p. 124 and Edwards 1951, p. 256. Middle Marl to 23 ft; Lower Magnesian Limestone to 45 ft 3; Lower Marl to 108 ft 11; Permian Basal Breccia to 109 ft 5; Westphalian from horizon of Main 'Estheria' Band to Dunsil; recorded to 1026 ft 5.

WATNALL COLLIERY No. 3 PIT [501 458], abt 410 ft.
Middle Marl to 9 ft; Lower Magnesian Limestone to 45 ft; Lower Marl to 63 ft; Permian Basal Breccia to 65 ft; Lowbright, c 12 at 84 ft; Lowbright Floor, c 10 at 92 ft 10; Brinsley Thin, c 36 at 146 ft 10; High Hazles, c 24 at 172 ft 10; Cinderhill (UL), c & d 16 at 251 ft 4; Cinderhill, c & d 24 at 267 ft 8; Mainsmut, c 33 at 350 ft 4; Comb, c 40 at 403 ft 2; Top Hard, c 72 at 425 ft 10; TD 425 ft 10.

WATNALL COLLIERY No. 6 PIT [? 5044 4509] abt 350 ft.
Lower Magnesian Limestone to 31 ft; Lower Marl to 43 ft 6; Permian Basal Breccia to 45 ft 9; High Hazles, c 16 at 101 ft 8; Cinderhill, c 18 at 208 ft; Mainsmut, c 27 at 292 ft 2; Comb, c 36 at 323 ft 2; Top Hard, c 54 at 349 ft 8; TD 349 ft 8.

WATNALL COLLIERY No. 7 PIT [5078 4486] abt 350 ft.
Lower Magnesian Limestone to 33 ft; Lower Marl to 43 ft; Permian Basal Breccia to 45 ft 3; Brinsley Thin, c 12 at 112 ft 4; High Hazles, c 20 at 137 ft 8; Cinderhill (UL), c 6 at 237 ft 1; Cinderhill, c 17 at 239 ft 3; Mainsmut, c 19 at 326 ft 6; Comb, c 31 at 360 ft 1; Top Hard, c 48 at 374 ft 6; TD 374 ft 6.

WATNALL COLLIERY No. 8 PIT [5050 4459] abt 350 ft.
Middle Marl to 12 ft; Lower Magnesian Limestone to 54 ft; Lower Marl to 65 ft 2; Permian Basal Breccia to 66 ft 11; ? High Hazles Floor, d c 5 at 100 ft 6; Mainsmut (UL), c 18 at 289 ft 10; Comb, c 22 at 318 ft 5; Top Hard, c 49 at 333 ft; TD 333 ft.

WATNALL COLLIERY INSET UBH [5067 4798] 1971.
Down from Top Hard in Waterloo Drift at -513 ft. Cores, 649 ft 10 to 670 ft 2 and below 686 ft, examined by D. E. Raisbeck. Low Main, c 30 on d c 4 at 711 ft 8; Blackshale, c & d 46 at 859 ft 1; TD 861 ft 6.

WATNALL COPPICE BH [5063 4888], 1956, 403 ft.
See Fig. 49.

	Thickness		Depth	
	ft	in	ft	in
PERMIAN				
Open hole	32	5	32	5
LOWER MARL				
Siltstone, grey; pyritic coaly plant fragments; mudstone partings	20	8	53	1
Mudstone; siltstone with 3-in limestone near base	28	11	82	0
WESTPHALIAN C				
Seatearth mudstone, silty	1	9	83	9
Siltstone and mudstone	8	1	91	10
Seatearth mudstone	2	6	94	4
Siltstone	3	8	98	0
COAL, c 1	0	1	98	1
Seatearth mudstone, sandy at base	12	8	110	9
Sandstone, siltstone and silty mudstone	6	3	117	0
Mudstone	0	6	117	6
SHAFTON MARINE BAND				
Mudstone, blue-grey, with bivalves	3	6	121	0
Mudstone	13	2	134	2
COAL, c 21	1	9	135	11
Seatearth mudstone, silty	3	8	139	7
Sandstone and siltstone	20	4	159	11
SHAFTON, c 14	1	2	161	1
Seatearth mudstone	2	6	163	7
Mudstone, silty	6	11	170	6
Sandstone	2	3	172	9
Mudstone, silty	10	3	183	0
? MAIN 'Estheria' BAND (UPPER LEAF)				
Mudstone; 'Estheria' and fish scales	3	5	186	5
Shale, dark	0	6	186	11
Seatearth mudstone	0	10	187	9
COAL, c 4	0	4	188	1
Seatearth mudstone	2	3	190	4
Siltstone and mudstone	2	11	193	3
? MAIN 'Estheria' BAND (LOWER LEAF)				
Mudstone, 'Estheria'	1	4	194	7
Mudstone	3	7	198	2
COAL, c 8	0	8	198	10

a

b

	Thickness		Depth	
	ft	in	ft	in
Seatearth mudstone, silty	6	10	205	8
Sandstone and siltstone	27	6	233	2
Mudstone	15	5	248	7
Edmondia BAND				
Mudstone; foraminifera	7	5	256	0
Mudstone	9	6	265	6
COAL, c 3	0	3	265	9
Seatearth mudstone	6	5	272	2
Mudstone	0	11	273	1
HIGH MAIN (UPPER LEAF), c 9	0	9	273	10
Seatearth mudstone	5	9	279	7
Mudstone	17	10	297	5
HIGH MAIN (LOWER LEAF), c 26	2	2	299	7
Seatearth mudstone, silty in lower part	5	11	305	6
Sandstone and siltstone to TD	5	6	311	0

WEAVERS LANE BH [5006 5188] 1968, 541 ft.
Recorded by C. Beal and T. Draper. Permian on Westphalian to High Main. TD 407 ft 5.

WESTERN PIT SHAFT [4154 5178] 1843-45 deepened 1917, 392 ft.
See Fig. 30. Not recorded to Deep Soft at 489 ft; Deep Hard, c 44 at 563 ft 8; First Piper, c 30, m 39 on b & c 10 at 616 ft 11, Coal 1 at 624 ft 3; Hospital, c 9 at 691 ft 11; Low Main, c 46 at 737 ft 4; Threequarters, c 32, m 3 ft 11 on c 9 at 757 ft 10; Yard, b and c 51 at 842 ft 2; Denby leaf of Blackshale, c 10 at 849 ft 2; Blackshale, c 3, b 18 on c 61 at 865 ft 2; 'Ashgate' (UL), c 9 at 933 ft 4; 'Ashgate' (LL), c 14 at 984 ft 11; Mickley Thin, c 15 at 1033 ft 11; Lower Brampton c 4 at 1088 ft 7; Kilburn, c 31 at 1302 ft 10; recorded to 1327 ft.

WEST HALLAM COLLIERY BH [4406 4246] 1959-60, 240 ft.
Cores, 600 ft to 1102 ft 9 and below 1316 ft, examined by G. D. Gaunt. See Figs. 13 and 14. Yard, c 5 at 605 ft 8; Blackshale, c 25, d 12, d c 12, m 18 on d c 9 at 654 ft 5; 'Ashgate' (UL) washed out; 'Ashgate' (LL), ca 14 at 801 ft 2; Mickley Thin, ca 14 at 850 ft 6; Lower Brampton, c 2, d 4 on c 4 at 911 ft 6; Coal 6 at 959 ft; Morley Muck, c 6, d 2, c 10, d 6 on c 8 at 1006 ft 3; Kilburn workings at 1102 ft 9; Upper Band horizon at 1435 ft; Norton, c 8 at 1472 ft 8; Forty-Yards MB at 1500 ft 1; Forty-Yards (UL), c 6 at 1501 ft 6; Forty-Yards (LL), c 6 at 1512 ft 6; Alton MB at 1600 ft 5; Alton, c 24 at 1602 ft 5; First Smalley Lingula Band at 1610 ft 9; First Smalley, c 29 at 1613 ft 2; Second Smalley Lingula Band at 1619 ft 9; Holbrook Lingula Band at 1630 ft 6; Holbrook, c 6 at 1631 ft 4; Belper-lawn, d c 16 on c 7 at 1655 ft 3; TD 1705 ft 5.

WEST HALLAM COLLIERY No. 1 (KILBURN) SHAFT [4414 4247] 240 ft.
See Fig. 34. Fourth Waterloo, c 21 at 14 ft 3; First Ell, c 33 at 39 ft 3; Second Ell, ca 4 on c 10 at 109 ft 8; base of Westphalian B estimated at 190 ft; Brown Rake, c 15 at 203 ft 1; Black Rake TS at 220 ft 9; Black Rake, b 24 at 223 ft 11; Top Soft b 18, ca 9, d 9 on c 3 at 243 ft 2; Roof Soft and Deep Soft, c 48, d 5 on c 55 at 311 ft 2; Foot, c 3 at 329 ft 3; Deep Hard, c 60 at 367 ft 7; First Piper, c 50 (with 18b) at 402 ft; ? Second Piper, b 9 at 427 ft 4; Hospital, c 7 at 454 ft 5; Cockleshell, c 8 at 479 ft 7; Low Main, c 43 (with 9b) at 518 ft 2; Threequarters, c 24 at 530 ft 2; Yard, c 12 at 614 ft 10; Blackshale, c 18, b 18, c 12, b 8 on c 12 at 649 ft 4; 'Ashgate' (UL) washed out; 'Ashgate' (LL), ca 17 at 803 ft 9; Mickley Thin, c 7 at 898 ft 7; Lower Brampton, c & b 9 at 950 ft 3; Morley Muck, c 16, d 15, c 7, d 7 on c 4 at 1011 ft 11; Kilburn, c 11, d and c 12 on c 57 at 1128 ft 4; TD 1158 ft 4.
Depths to principal seams in unnamed shaft at this colliery in Gibson and others (1908, p. 89) differ slightly from the above.

WHYBURN EAST BH [5161 4993] 1956-57, 381 ft 7.
Base of Permian at 99 ft; Westphalian from Top MB to High Main; TD 458 ft 8.

WHYBURN HOUSE BH [5080 4992] 1965, 442 ft.
Cores, below 330 ft, cutting samples and gamma log examined by J.A. Chilton, N. Aitkenhead and D.V. Frost. See Fig. 49. Middle Marl to 15 ft; Lower Magnesian Limestone to 55 ft; Lower Marl to 100 ft; Permian Basal Breccia to 105 ft; Top MB at 164 ft 4; inferred Shafton MB at 228 ft 6; Coal 18 at 230 ft; Shafton, ca 6 at 245 ft; Edmondia Band at 353 ft 4; High Main (UL), c 14 at 381 ft 8; High Main (LL), c 46 at 388 ft 2; Coal, c & d 30 at 434 ft 7, Coal, c 12, d 4 on c 1 at 445 ft 5; Mansfield MB at 529 ft 3; Coal, c & d 14 at 521 ft 5; ? Sutton MB at 558 ft 3; inferred Haughton MB at 593 ft 11; ? Swinton Pottery, c 21 at 600 ft 9; Clown, c 14 at 661 ft 6; inferred Manton 'Estheria' Band at 684 ft 7; Mainbright, c 29 at 730 ft 3; Two-Foot MB at 743 ft 2; Two-Foot, c 22 at 745 ft; Lowbright, c 32 at 782 ft 8; Lowbright Floor, c 7 at 788 ft 7; Brinsley Thin, c 14 at 839 ft; High Hazles, c 21 at 862 ft 6; TD 882 ft 6.

WHYBURN NORTH BH [5129 5004] 1955, 416 ft.
Permian on Westphalian to High Main. Open hole to 334 ft; TD 462 ft 4.

WHYBURN SOUTH BH [5136 4960] 1955, 376 ft.
Open hole to 297 ft 8; High Main partly washed out at 404 ft 8; TD 453 ft.

WHYBURN WEST BH [5099 4978] 1957, 411 ft.
Cores, below 17 ft, examined by R. E. Elliott and A. W. Woodland. Lower Magnesian Limestone to 30 ft 9; Lower Marl to 82 ft 10; Permian Basal Breccia to 83 ft 3; Top MB at 167 ft 10; Coal 8 at 168 ft 6; Coal 5, d 6 on c 27 at 192 ft 4; Shafton MB at 227 ft 9; Coal 21 at 231 ft 7; Shafton (LL), c 18 at 251 ft 1; ? Main 'Estheria' Band (UL) at 276 ft 7; Coal 10 at 277 ft 5; ? Main 'Estheria' Band (LL) at 286 ft 6; Edmondia Band at 351 ft 9; High Main (UL), c 13 at 379 ft 8; High Main (LL), c 35 at 387 ft; TD 406 ft 8.

WIGWELL WATER BH [3170 5465] abt 500 ft.
Soil to 6 ft; Ashover Grit (Main Bed) to 169 ft; shale to 175 ft; leaf of Ashover Grit to 198 ft; shale to TD at 201 ft.

WILLEY WOOD COLLIERY [4724 5000] 420 ft.
See also Fig. 46. Clown, c 48 at 59 ft 4; Clown (LL) horizon at 93 ft 10; Manton horizon at 139 ft 2; Mainbright, c 36 at 175 ft 2; Two-Foot, c 24, m 6 ft 6 on c 4 at 194 ft; Lowbright, c 36 at 222 ft; Lowbright Floor, c 12 at 229 ft 8; Brinsley Thin, c 43 at 283 ft 4; High Hazles, c 25 at 316 ft 4; Cinderhill (UL), c and d 17 at 379 ft 5; Cinderhill, c 9 at 393 ft 5; St. John's, b 7 at 420 ft 1; Mainsmut (UL), c 18 at 493 ft 2; Mainsmut (LL), c and d 36 at 506 ft 9; Comb, c 36 at 535 ft 9; Top Hard, c 55 at 547 ft 1; TD 571 ft 1.

WILLIAM WOOD BH [4938 5127] 1972, 537 ft.
Open hole to 10 ft 6; ? Middle Marl to 11 ft 4; Lower Magnesian Limestone to abt 42 ft; Lower Marl to 96 ft 2; Permian Basal Breccia to 99 ft; Shafton horizon at 109 ft; ? Main 'Estheria' Band (UL) at 149 ft 2; ? Edmondia Band at 208 ft; High Main, c 9, c and d 10 on c 30 at 249 ft 5; TD 270 ft 6.

WIREMILL BRIDGE BH [3789 5363] 1954, 255 ft.
Recorded by P. McL. Duff. See Figs. 16 and 68.

	Thickness		Depth	
	ft	in	ft	in
ALLUVIUM				
Boulders and sandy clay coring commenced at 10 ft	10	0	10	0
Clay, sandy, and sand with pebbles of quartzite, ironstone and sandstone	21	2	31	2
WESTPHALIAN A				
Mudstone; coal and seatearth fragments at base; fish	22	2	53	4
KILBURN, c 49, d 33 on c 7	7	5	60	9
Seatearth	3	5	64	2
Mudstone; ironstone; bivalves	40	10	105	0
Seatearth	4	0	109	0
WINGFIELD FLAGS				
Sandstone	1	5	110	5
Mudstone	3	11	114	4
Siltstone and sandstone	8	8	123	0
Mudstone with siltstone and sandstone	63	1	186	1
Sandstone with siltstone	72	7	258	8
Mudstone, sandstone bands	78	4	337	0
Mudstone; ironstone; ostracods, fish debris and fucoids near base	90	0	427	0
UPPER BAND HORIZON	-	-	427	0
Seatearth	3	3	430	3
Siltstone	13	0	443	3
Mudstone; bivalves common	36	1	479	4
Cannel; ironstone; plants bivalves	0	8	480	0
NORTON, c 22	1	10	481	10
Seatearth	9	8	491	6
Mudstone; fish	16	5	507	11
FORTY-YARDS (UPPER LEAF), c 7	0	7	508	6
Ganister	2	3	510	9
Sandstone	4	3	515	0
Mudstone	1	3	516	3
FORTY-YARDS (LOWER LEAF), c 3	0	3	516	6
Seatearth sandstone	3	4	519	10
Sandstone, with mudstone bands	20	8	540	6
Mudstone; plants	16	8	557	2
UPPER AND LOWER PARKHOUSE MARINE BANDS				
Mudstone; Lingula and fish debris	1	4	558	6
Mudstone silty; ironstone; bivalves, fish debris	3	11	562	5
Seatearth, sandy	0	2	562	7
Sandstone and siltstone	22	2	584	9
Mudstone; fish debris at base	16	7	601	4
ALTON MARINE BAND				
Mudstone; bivalves and fish debris	0	5	601	9
ALTON, c 22	1	10	603	7
Ganister	1	1	604	8
Seatearth	0	8	605	4
Ganister	1	0	606	4
Mudstone	2	3	608	7
Siltstone	1	3	609	10
Mudstone	4	4	614	2
Siltstone to TD	2	4	616	6

WOLLATON CANAL BH [5220 4036] 1952, 193 ft.
Comb, c 24 at 33 ft 9; Top Hard, c 40 at 45 ft 4; Top Hard Floor, c 15 at 72 ft 10; Dunsil, c & ca 25 at 106 ft 8; First Waterloo, c 34 at 149 ft 1; TD 164 ft 6.

WOLLATON COLLIERY No. 1 SHAFT [5214 4038] 1873, abt 195 ft.
Comb, c 24 at 27 ft 2; Top Hard, c 24 and old workings at 38 ft 4; Top Hard Floor, c 15 at 67 ft 5; Dunsil, c 23 at 87 ft 7; First Waterloo, c 34 at 140 ft 1; Top Second Waterloo (UL), c 32 at 192 ft 3; Top Second Waterloo (LL), c 7 at 211 ft 4; Bottom Second Waterloo, c 7 at 236 ft; Third Waterloo, c 1, d 4 on c 24 at 254 ft 5; Coal, d c 6 at 269 ft 1; Fourth Waterloo, c 30 at 289 ft 4; First Ell, c and b 12, b 12, m 14, ca 6, m 4 on c 10 at 3 10 ft; Second Ell, c 20 at 354 ft 8; base of Westphalian B estimated at 420 ft; Brown Rake horizon at 435 ft 2; Black Rake TS at 460 ft 3; Top Soft, c & b 6 at 471 ft 2; Roof Soft (UL) horizon at 486 ft 9; Roof Soft (LL), c 16 at 501 ft 9; Deep Soft, c 36, b 21 on ca 3 at 556 ft 9; Deep Hard, c 58 at 596 ft 1; First Piper, c 24 at 625 ft 9; TD at 637 ft 3.

WOLLATON COLLIERY No. 2 SHAFT [5218 4040] 1873, abt 195 ft.
TD 619 ft. See Plate 5 for graphic section Top Hard Coal to Clay Cross Marine Band.

WOLLATON COLLIERY No. 3 SHAFT and BH [5207 4034] 1876, 199 ft.
BH (1877) commences 1416 ft. Outline log only from Comb to First Piper at 611 ft 6; Second Piper horizon at 638 ft; Hospital, b 6 at 655 ft; Low Main, c 26 at 792 ft 1; Threequarters, c 12 at 800 ft 1; Yard, c 21 at 853 ft 8; Denby leaves

c

d

of Blackshale, c & b 12, d 15 on c & b 20 at 865 ft 11; Blackshale, c 14, b 15 on c 13 at 890 ft 11; Coal 4 at 924 ft 7; 'Ashgate' washed out; Mickley Thin, c 4 at 1068 ft 2; Lower Brampton, c 12 at 1107 ft 8; Coal 1 at 1145 ft 5; Morley Muck, c 6 at 1176 ft 9; Coal, b 2 at 1247 ft 5; ? Norton horizon at 1650 ft 4; ? Alton, c 10 at 1721 ft 1; TD 1725 ft 1.

WOLLATON COLLIERY DEEP HARD 21's LEFT AIRWAY UBH [5377 3991] 1952.
Down from Deep Hard at -604 ft. Recorded by R. E. Elliott. First Piper, c 41 at 52 ft 1; Hospital, c 9 at 91 ft 7; Cockleshell (interbedded coal, batt and measures 5 ft) at 121 ft 6; Low Main, c 35 at 156 ft; Threequarters, c 27, m 6 ft 1 on c 4 at 172 ft 10; TD 175 ft 6.

WOLLATON COLLIERY JUNCTION OF S MAIN & S DIPS UBH [5217 4019] 1952.
Down from Deep Hard at -375 ft. Deep Hard, c 45 on ca 10 at top of BH; First Piper, c 18 at 27 ft 6; Hospital, b 3 at 76 ft 1; Low Main horizon at 214 ft 4; Threequarters, c 5, d 1, c 12, c 2 on ca 11 at 230 ft 6; Yard, c 16 on b 2 at 283 ft 2; Denby leaves of Blackshale, c 2, d 16 on c 7 at 290 ft 8; TD at 299 ft 7.

WOLLATON COLLIERY SOUTH UBH [5253 4013] 1954.
Down from Deep Hard at -443 ft to Threequarters. TD 203 ft 2.

WOLLATON COLLIERY SOUTH DIPS UBH [5239 4016] 1955.
Down from Deep Hard at -410 ft to Threequarters. TD 207 ft 5.

WOLLATON COLLIERY 15's MAIN UBH [5324 4041] 1952.
Down from Deep Hard at -568 ft to Threequarters. TD 201 ft 3.

WOLLATON COLLIERY 30's N UBH [5195 4063] 1939.
Down from Deep Hard at -375 ft to Threequarters. TD 212 ft 8.

WOLLATON COLLIERY 114's MAIN INBYE UBH [5265 3916] 1952.
Down from Deep Hard at -374 ft. First Piper, c 36 at 38 ft 7; Hospital, c 7, ca 2, m 15 on c and d 2 at 83 ft 9; Cockleshell horizon at 108 ft 4; Low Main, ca 2 on c 36 at 147 ft; Threequarters, c 22, m 4 ft 6, c 1 on ca 1 at 161 ft 3; TD 165 ft 8.

WOLLATON COLLIERY 114's MAIN OUTBYE UBH [5229 3959] 1952.
Down from Deep Hard at -353 ft. Recorded by R. E. Elliott. First Piper, c 26 at 37 ft 8; ? Hospital, c 2 at 80 ft 10; ? Cockleshell, d c 9 at 157 ft 1; Low Main, ca 7 on c 29 at 186 ft 5; Threequarters, c 25, m 9 ft 9 on ca 9 at 204 ft 5; Yard, c 14 at 260 ft 4; ? Denby leaf, c 9 at 269 ft 5; Blackshale, c 14, m 21, c 1, d c & c 16, c 9 on d c 8 at 304 ft 6; TD 310 ft.

WOODLINKIN DRIFT See Ormonde Colliery, Woodlinkin No. 1 Drift.

WOODSIDE COLLIERY SHAFT [448 444], 289 ft.
See Fig. 30 and Plate 5. Gibson and others 1908, p. 87. Mainsmut, c 24 at 68 ft 8; Upper Comb, c 24 at 102 ft 2; Lower Comb and Top Hard workings, c 18, d 3 on c 30+ at 111 ft 5; Dunsil, c 31 at 161 ft; Top First Waterloo, c 12 at 181 ft 7; Bottom First Waterloo, ca 24 at 209 ft 7; Waterloo Marker, c 12 at 235 ft 7; Top Second Waterloo, c 35 at 262 ft 6; Bottom Second Waterloo, c 12 at 278 ft 6; Third Waterloo, c 24 at 305 ft; Coal 24 at 343 ft 6; Fourth Waterloo, c 12 at 362 ft 6; First Ell, c 30 at 404 ft 10; Second Ell, ca 18 at 454 ft 7; base of Westphalian B estimated at 536 ft; Brown Rake, c 12 at 550 ft 7; Top Soft, c & b 18 at 594 ft 7; Roof Soft and Deep Soft, c 60, d 24 on c 48 at 652 ft 11; Deep Hard workings at 705 ft 11; First Piper, c 48 at 745 ft 11; Hospital, b 24 at 812 ft 3; Cockleshell, b 12 at 835 ft 3; Low Main, c 42 at 870 ft 9; Threequarters, c 24 at 882 ft 9; Yard, b 6 at 939 ft 3; Blackshale, c & d 69 at 1039 ft; 'Ashgate' (UL), c 24 at 1142 ft 4; 'Ashgate' (LL), c 24 at 1153 ft 4; Mickley Thin, b 12 at 1207 ft 4; Lower Brampton, c 6 at 1263 ft 10; Coal 6 at 1316 ft 10; Morley Muck, c 9, b 3, c 12, d 12 on c 6 at 1364 ft 10; Kilburn, c 54 at 1463 ft 4; recorded to 1538 ft 6.

WOODSIDE No. 1 UBH [4356 4338] 1958.
Down from First Piper at -177. Cores, 10 ft -110 ft and below 125 ft, examined by G.D. Gaunt. Hospital, c 10, d 1 on c 2 at 55 ft 2; Cockleshell horizon at 71 ft; Low Main workings at 116 ft; ? Yard horizon at 204 ft 6; Blackshale (part washout), c 1, d 6 on c 1 at 299 ft 10; 'Ashgate' (UL) horizon at 373 ft; 'Ashgate' (LL), c 12 at 391 ft; TD 413 ft.

WOODSIDE COLLIERY STAPLE PIT [4491 4460] 1951.
From First Piper (c 48) at -428 ft. Hospital, c & d 24 at 61 ft 4; Cockleshell, c & d 12 at 82 ft 10; Low Main, c 42 at 115 ft 10; TD 115 ft 10.

a

APPENDIX 2

DETAILED SECTIONS OF THE WESTPHALIAN

The following selected list of sections is inserted to illustrate both the variable nature and cyclical sequences of the Westphalian rocks of the district. The sections are arranged alphabetically using the initial letter of the name of their adjacent town or village. Many of the sections were temporary.

AMBERGATE - SAWMILL RAIL CUTTING AND QUARRY [3600 5214]
Measured by D. V. Frost

	ft	in
WESTPHALIAN		
CRAWSHAW SANDSTONE		
Sandstone, brown, massive, cross-bedded, ferruginous	22	10
Mudstone and sandstone, interlaminated	0	2
Sandstone	0	9
Mudstone silty; sandstone laminae	0	2
Sandstone	0	10
Mudstone silty; sandstone laminae	0	4
Sandstone	1	0
Mudstone, silty; sandstone laminae	0	10
Sandstone; ferruginous staining	1	0
Mudstone silty; sandstone laminae	1	4
Sandstone, grey-brown, micaceous	1	6
Mudstone	3	0
Sandstone	1	0
Mudstone, silty; sandy laminae	10	0
Sandstone	1	0
Mudstone, silty	2	0
Sandstone, grey-brown, micaceous, fine-grained	1	0
Mudstone, laminated; fish scales towards base	11	0
Mudstone, listric	0	2 to 4
Mudstone, grey, mottled purple; pyritic fish scales (Horizon of Pot Clay Marine Band)	0	6
NAMURIAN		
Ganister, white, fine	2	6
Seatearth, siltstone	1	0
Siltstone, pale grey	1	0
ROUGH ROCK		
Sandstone, brown, fine-grained		

AMBERGATE, RIDGEWAY QUARRY [3583 5146]
Measured by D. V. Frost

	ft	in
Soil and overburden	3	0
WESTPHALIAN		
CRAWSHAW SANDSTONE		
Sandstone, yellowish, medium-grained with coarse bands showing cross-bedding dipping mainly towards the north-east; ripple-marks, Stigmaria	50	0
Mudstone, micaceous	6	0
Mudstone, grey, micaceous, laminated	14	0
Mudstone, black, fissile (Horizon of Pot Clay Marine Band)	0	1
NAMURIAN		
Ganister, grey, very hard	2	6
Mudstone	0	6
ROUGH ROCK		
Sandstone	1	3+

AMBERGATE, BRICKWORKS QUARRY [3605 5150]
Measured by D. V. Frost

	ft	in
WESTPHALIAN		
BELPERLAWN		
Coal	3	0
Seatearth, mudstone; siltstone bands	1	6
Ganister, variable	3	6
Sandstone and mudstone interlaminations	1	6
Sandstone, fine-grained; planty partings	5	0
Mudstone, grey; siltstone laminae	25	0
Sandstone/siltstone and mudstone interlaminations	25	0
Mudstone, grey, silty; basal 6 in listric with poorly preserved fish scales (Horizon of Pot Clay Marine Band)	12	0
NAMURIAN		
Ganister, top inch dark grey-blue with purple staining	3	0
Seatearth, pale grey to off white, purple staining		

AMBERGATE, RIDGEWAY SITE
Measured by R. A. Eden
Section (1) north of stream [3620 5165]

	ft	in
ALTON COAL		
Mudstone, (inaccessible)	abt 6	0
Siltstone	6 to 15	0
FIRST SMALLEY Lingula BAND		
Shale, black; ironstones, fish and Lingula	0	7
Mudstone, black, micaceous; mussels	0	7
Seatearth, mudstone, black, micaceous	0	3
Ganister	1	6
Seatearth, siltstone, coarse	1	0
Siltstone, coarse	5	0
Shale, micaceous, dark; fish debris	1	2

b

	ft	in
Siltstone, black; coaly with coal bands and pyrite	0	7
BELPERLAWN COAL		

Section (2) south of stream [3603 5133]

	ft	in
ALTON COAL		
Gap in section	abt 6	0
Siltstone, coarse	23	0
Mudstone, black	0	6
Sandstone, coarse		0 to 2
? HOLBROOK		
Coal	0	2
Seatearth, mudstone, dark; pale bands, coal streaks	0	9
Seatearth, silty	1	0
Siltstone; rootlets	3	0
Mudstone; fish debris and rootlets	3	0
BELPERLAWN COAL		

See also Eden 1954, p. 87

BRACKLEY GATE, RYKNELD COVERT SITE [386 430]
Measured by R. A. Eden

	ft	in
Siltstone, buff	10	0
Mudstone, grey, silty	8	0
Ironstone band; 'Estheria'	0	2
Mudstone, darkish grey	1	6
PARKHOUSE MARINE BAND		
Mudstone, dark grey; pyritised foraminifera and rare Lingula sp.	1	6
Mudstone, dark grey; sporadic ironstones; fish	11	10
Ironstone	0	2
ALTON MARINE BAND		
Mudstone, dark grey, Lingula sp. and fish	1	6
Ironstone; Lingula sp. and fish		0 to 2
Mudstone, dark grey; Lingula sp. and fish	1	3
Mudstone, dark grey; small pyritised goniatites and mussels	0	4
Shale, black, fissile, carbonaceous; goniatites	0	1
Mudstone, dark grey; small pyritised goniatites and mussels	0	6
Shale silty, black fissile pyritous, goniatites and Dunbarella sp.	0	2
ALTON		
Coal		21 to 27
Fireclay	2	0+

BRINSLEY, BRINSLEY SIDINGS SITE [450 498]
Measured by R. A. Eden

	ft	in
Shale, dark; ironstone	2	11
TOP FIRST WATERLOO		
Cannel	1	0
Shale, dark; plants	1	2
Mudstone, grey; ironstone lenses; plants, rootlets basally	5	0
BOTTOM FIRST WATERLOO		
Coal	1	8
Seatearth, silty, yellowish-grey	2	6
Siltstone, coarse and sandstone interbedded	6	0
Siltstone, grey; yellow stains	3	0
Mudstone, grey and dark grey; few mussels	6	6
WATERLOO MARKER		
Coal 6 in		
Seatearth, mudstone, dark grey 4 in		
Coal 2 in	1	0
Seatearth, mudstone, dark grey	1	0
Seatearth, siltstone, grey	0	5
Sandstone, laminated; rootlets	2	9
Siltstone, grey; small ironstone nodules	5	0
Siltstone and sandstone, laminated	1	6
Mudstone, grey, silty and siltstone interbedded	9	0
Mudstone, soapy, grey	3	0
Mudstone, dark grey, laminated	0	1
TOP SECOND WATERLOO COAL		

CODNOR, AGNES SITE
Measured by J. G. O. Smart
Section (1) [4300 4949]

	ft	in
Mudstone, inaccesible	abt 20	0
Mudstone, dark grey	2	0
Mudstone, thinly bedded and flaky; impersistent ironstone beds up to 2½ in thick	3	0
Mudstone, flaky	1	5
CLAY CROSS MARINE BAND		
Mudstone, darkish grey, thinly bedded; goniatites and Lingula sp. Lowest 3½ in with ? Lingula only	abt 1	3
Mudstone, dark grey, thinly bedded; thin (1 in) ironstone bands (shelly)	2	5
Mudstone, grey, silty; ironstone bands to abt 1½ in	7	9
Mudstone, darkish-grey; mussels	2	0
Mudstone, grey, silty; ironstone bands up to 3 in thick	8	9
Mudstone, broken	2	1
Mudstone, laminated, silty; plants, mussels and shelly nodular ironstone up to 4 in thick (2 ft up)	2	6
Mudstone, brown, slightly carbonaceous; fusainous plants	0	10
BROWN RAKE		
Coal	1	3
Mudstone, very listric	0	1½
Seatearth; silty mudstone	1	3
Sandstone, grey, fine; rootlets	1	10

c

	ft	in
Seatearth; siltstone; ironstone nodules and cank	0	5
Sandstone, massive, greyish brown	2	10
Mudstone, evenly laminated, blue-grey, silty; sporadic thin (1 in) planty siltstone and fine sandstone beds	11	0

Section (2), composite from faces due W of Codnor Castle [4350] Measured by D. V. Frost

	ft	in
DEEP HARD		
Coal	3	6
Small gap in section		
Sandstone, massive, pale grey-yellow, brown, fine-grained, carbonaceous/micaceous partings	6	0
Mudstone, grey silty; ironstone nodules and bands	12	0
Sandstone, massive	2	6
Mudstone, grey, silty	abt 5	0
FIRST PIPER		
Coal and dirt	4	6
Seatearth; sandstone	3	0
Sandstone, flaggy	5	0
Sandstone, pale brown, massive	10	0
Mudstone, grey; ironstone bands and nodules; ? local unconformity	6	6
Sandstone, pale grey, fine-grained; silty mudstone bands	20	0
HOSPITAL		
Coal 6 in)		
Ganister 17 in)		
Coal 4 in)	2	3
Seatearth, mudstone, grey brown; plant fragments	3	0
Mudstone, grey to dark grey; mussels in basal 4 in	17	0
Mudstone, grey; small ironstone nodules, coal streaks	2	0
COCKLESHELL		
Coal	0	8
Seatearth, mudstone, grey, listric; plants	1	0
Sandstone; silty mudstone partings (up to 18 in thick), carbonaceous/micaceous partings	15	0
Sandstone, massive	4	0
Sandstone, thinly laminated; carbonaceous/micaceous partings	1	0
Sandstone; basal part massive, pale grey, micaceous	20	0
Mudstone, grey; ironstone nodules and bands	4	0
LOW MAIN		
cannel 4 in		
Coal 41 in	3	9
Mudstone, dark grey to black, listric; seatearth at top	11	0
THREEQUARTERS		
Coal 16 in		
dirt 1 in		
Coal 13 in	2	6
Seatearth; mudstone grey, very silty; plants	2	0+

Section (3) [425 504], Measured by D. V. Frost

	ft	in
Shale, tough carbonaceous	abt 5	0
Sandstone, flaggy beds up to 3 ft thick	abt 40	0
Mudstone, grey, silty; ironstone nodules	10	0
Mudstone, dark grey; ironstone bands; small mussels at base Clay Cross Marine Band Horizon	15	0
Mudstone, grey	2	6
Shale, black; mussels	1	6
Mudstone, grey; ironstone bands	9	0
Mudstone, black; mussels, ostracods and Spirorbis sp.	0	2
Mudstone, grey; ironstone bands	3	0
Mudstone, dark grey	0	6
BROWN RAKE		
Coal	1	3
Seatearth, mudstone, soft	0	7
Seatearth, mudstone, grey, silty; plants; ironstone nodules	2	0
Sandstone, fine-grained, yellowish brown	1	0
Mudstone, grey, silty, siltstone bands; many ironstone bands of up to 3 in	12	6
Mudstone, dark grey to black; many ostracods and mussels	2	6
BLACK RAKE TUFFACEOUS SILTSTONE		
Ironstone	0	8
Mudstone, dark, grey to black, disturbed bedding (? old ironstone workings) mussels	3	0
Mudstone, dark-grey; rare mussels and ostracods	0	9
FALSE ELL		
Coal 3 in		
Mudstone, dark grey; coaly partings 15 in		
Coal, variable thickness, 18 in		
Seatearth, mudstone, grey, variable thickness, 48 in		
Coal 36 in	10	0
Seatearth, mudstone; small ironstone nodules	1	6
Seatearth, siltstone	2	3
Siltstone, pale grey mudstone partings; ironstone nodules, plant fragments	abt 30	0
DEEP SOFT		
Coal	abt 5	0

DENBY, HIGH BANK SITE [4055 4678]
Measured by R. A. Eden

	ft	in
Flagstones, medium-grained	2	0
Sandstone, fine, brown, laminated, micaceous	1	2
Flagstones, medium-grained	1	10
Sandstone, fine, brown, laminated, micaceous	4	2

d

DENBY, HIGH BANK SITE [4055 4678]
Measured by R. A. Eden

	ft	in
Flagstones, medium-grained	2	0
Sandstone, fine, brown, laminated, micaceous	1	2
Flagstones, medium-grained	1	10
Sandstone, fine, brown, laminated, micaceous	4	2

	ft	in
Flagstones, medium-grained	1	10
Sandstone, light grey; abundant small coal fragments	0	4
Sandstone, hard, light grey; coal fragments and ironstone lenses	0	2½
Mudstone, buff	0	1½
Shale, light coloured at top, ironstone nodules	18	6
Shale, black with fish debris	0	1
FIRST PIPER		
Coal 28½ in		
Dirt 6 to 8 in		
Coal 9½ in		
Dirt 2 to 4 in		
Coal 18½ in	5	4½ - 8½
Not seen (section continued at 4059 4662)	abt 9	0
Mudstone, light grey	0	6
Shale; mussels and fish debris	1	0
Clay, grey	0	3½
Shale, dark; mussels and fish debris	0	2½
Mudstone	4	0
Not seen	4	0
Flagstone, light-grey and buff	8	9
Mudstone; ironstone nodules	3	9
Shale, grey; rare mussels	1	3
Shale, dark grey	0	1
Shale, grey; rare mussels	4	9
Mudstone, dark; mussels	0	3
Siltstone, dark; mussels and ostracods	0	7
Mudstone, dark; ironstone; mussels and ostracods	0	11
Shale, dark; mussels and ostracods	1	5
HOSPITAL		
Coal	2	0

EASTWOOD, MANNERS BRICK PIT [469 459]
Composite section measured by J. G. O. Smart

	ft	in
Mudstone, brown and grey, very silty; siltstone bands	15	0
Mudstone, brown and grey laminated	2	0
BOTTOM SECOND WATERLOO		
Coal 16 in		
Clay, silty, grey and brown; plants 9 in		
Coal 1 to 4 in	2	2 to 5
Clay, grey and black	0	1 to 4
Seatearth, mudstone, grey; ironstone nodules	4	0
Sandstone and siltstone interlaminated	0 to 2	0
Mudstone, grey, silty; ironstone nodules	8	0
Not seen	5	0
Mudstone, black, carbonaceous	0	2
THIRD WATERLOO		
Coal	1	11
Seatearth, siltstone, pale grey and ochreous mottled	6	0
Sandstone, fine and mudstone interlaminated	3	6
Sandstone, fine, khaki-brown, interbanded with silty mudstone	5	0
Section continues in adjacent face		
Siltstone, grey-brown	abt 6	0
Mudstone, silty, grey-brown, laminated	0	8
? LOW LEAF OF THIRD WATERLOO		
Mudstone, dark grey; carbonaceous streaks	0	2
Mudstone, silty, grey-brown; rootlets at top	4	0
Siltstone, sandy, brown; sandstone beds	1	10
Section continues in adjacent face		
Clay, grey and ochreous	5	0
Sand, silty; ferruginous layers	1	0
Sandstone, nodular	1	0
Sandstone, brown; hard bands	4	0
(Fourth Waterloo, 8 in of cannel, exposed 10 to 15 ft below in adjacent face)		

HEAGE, MANOR FARM SITE [3770 5026]
Measured by R. A. Eden

	ft	in
Shale, dark grey	1	0
Siltstone, dark grey	0	4
Siltstone, coarse, light grey; sandstone bands and laminae	6	6
Siltstone, light grey	7	0
Sandstone, laminated	1	6
Siltstone, grey; ironstone lenses and sandstone laminae at top	4	6
Shale, dark grey; plants and fish debris and ironstone band (4 in)	3	0
? DENBY LEAF OF BLACKSHALE		
Coal and black carbonaceous shale	1	10
Seatearth, banded grey and buff	2	10
Siltstone, fine, brown; plants	4	0
Sandstone and siltstone, interbedded; wedges out locally - average thickness	10	0
Siltstone, light grey; plants	4	0
Shale, grey; plants	1	6
Shale, dark, carbonaceous	0	7
Shale, grey; many plants; small ironstone nodules	4	9
Siltstone, light grey; small ironstone nodules	5 to 12	0
Mudstone, light grey; plants and ironstone nodules	3	2
Shale, grey, many plants	1	3
Mudstone, dark grey; rootlets and slickensides	0	1
BLACKSHALE COAL		

a

HEANOR, WHITELEYS PLANTATION SITE [4218 4500]
Measured by Dr J. Shirley

	ft	in
Mudstone; ironstone nodules	12	0
BOTTOM FIRST WATERLOO		
Cannel and coal	up to 5	0
Mudstone, grey; mussels	9	6
Bind, dark grey; mussels and ironstone	0	5
WATERLOO MARKER		
Coal	0	8
Clunch, carbonaceous	1	0
Clunch, grey, sandy in basal 3 ft	5	0
Mudstone, sandy, laminated	abt 20	0
Shale, black	0	1
TOP SECOND WATERLOO		
Coal	4	0

HEANOR, LOSCOE BRICK WORKS, CLAY PIT [426 471]
Measured by J. G. O. Smart

	ft	in
Mudstone, dark grey, fissile; mussels	6	0
? TOP FIRST WATERLOO HORIZON		
Mudstone, black, fissile, locally canneloid	0	9
Mudstone, grey, fissile	1	11
Mudstone, silty, grey; more silty beds and ironstone nodules	16	8
Not seen	abt 6	0
BOTTOM FIRST WATERLOO		
Coal	0	10
Seatearth, silty, ochreous and grey	1	10
Sandstone, fine, khaki and brown	5	0
Not seen	abt 15	0
Mudstone, grey with ironstone nodules; darker and with mussels at base	2	9
WATERLOO MARKER		
Coal, with dirt lenses 6 in	0	6
Seatearth, mudstone, listric; coaly streaks	2	6
Mudstone, dark grey, carbonaceous	0	3
Mudstone, grey	abt 4 to 5	0
Siltstone, grey; fine sandstone ripples and laminae	12	0
Sandstone, fine, khaki and brown, cross-bedded, developed as lenses	0 to 13	0
Mudstone, silty; laminae of fine siltstone	5 to 20	0
Mudstone, grey; bands of ironstone nodules, prominent 4 in nodules at top	5	10
TOP SECOND WATERLOO		
Coal	3	11
Seatearth; mudstone, grey	2	6
Mudstone, silty, grey; coal streaks and small ironstone nodules	2	0
Sandstone, hard, brown and grey; plants; passing into siltstone locally	2	7
Mudstone, grey, silty; ironstone nodules	5	0
BOTTOM SECOND WATERLOO		
Cannel 4 in		
Coal 14 in	1	6
Seatearth, mudstone	0	4

HEANOR, CINDERHILL COPPICE SITE [4338 4465]
Measured by R. A. Eden

	ft	in
Shale, dark grey, many ironstone nodules and mussels	1	0
? ST. JOHN'S		
Coal 8 in		
Mudstone, carbonaceous 1 in		
Coal 3 in	1	0
Seatearth, mudstone, light grey	abt 3	6
Sandstone, massive, light grey	12	0
Shale, grey and dark grey; ironstone bands	7	7
Shale, black; fish debris	1	6
Siltstone, light grey; rootlets at top	5	0
Sandstone	1	0
Shale, grey and dark grey; ironstone bands and mussels	21	4
Siltstone, black	1	7
MAINSMUT		
Coal	2	9
Seatearth, silty, dark grey	2	3
Seatearth, light grey; ironstone nodules	2	6
Siltstone, dark grey; rootlets at top becoming fewer basally	16	0
Mudstone, grey	1	9
UPPER COMB		
Coal	2	4
Shale, dark, coaly	0	1
Seatearth, dark, carbonaceous	0	7
LOWER COMB		
Coal	1	6
Seatearth, dark, carbonaceous	0	4
TOP HARD		
Coal	7	2
Not seen	abt 20	0
Sandstone, hard, weathering to flags	18	0
Siltstone, light grey; ironstone nodules and bands	7	0
Mudstone, light grey; thin ironstone beds	5	0
Shale, dark grey; many mussels	0	11
Mudstone, light grey	0	3
Shale, silty, dark grey	0	1
DUNSIL		
Coal	2	4

b

HEANOR, HEANOR GATE SITE [4305 4556]
Measured by Dr J. Shirley and Nottingham Coal Survey Laboratory

	ft	in
Shale, black	15	0
Ironstone band	0	3½
Shale, grey; thin ironstone band	9	7
Shale, black	0	1
UPPER COMB, LOWER COMB AND TOP HARD		
Coal 28 in		
Dirt 3 in		
Coal 19 in		
Dirt 10 in		
Coal 87 in	12	3

HEANOR, THE FALL SITE [4366 4703]
Combined section measured by Dr J. Shirley and Nottingham Coal Survey Laboratory

	ft	in
Soil and subsoil	abt 7	0
WATERLOO MARKER		
Coal 9 in		
Clunch 31 in		
Coal and shaley coal 10 in	4	2
Clunch	1	9
Mudstone, sandy; ironstone nodules	2	0
Shale, coaly and sandy, base erosional	0	6
Sandstone, muddy	0	10
Coal, weathered	0	2
Stone bind	6	0
Mudstone, grey	5	6
TOP SECOND WATERLOO		
Coal 35 in		
Dirt 6 in		
Dirt and Coal 3 in		
Coal 4 in	4	0

HEANOR, TOP DUMBLES SITE
Section (1) [413 463] measured by R. A. Eden

	ft	in
Subsoil	3	0
Shale, grey and dark grey; bands of ironstone nodules; mussels and ostracods	15	0
Shale, grey; many mussels and ostracods	0	6
Ironstone band; mussels and ostracods	0	4
Shale, dark; mussels and ostracods	0	9
Shale, black, carbonaceous; ostracods	0	8
Ironstone	0	6
Shale, dark; mussels and ostracods	4	5
Mudstone, light grey; ironstone nodules	2	7
Shale, dark; mussels and ostracods	0	6
BROWN RAKE		
Coal	3	2
Seatearth, silty, light grey	2	4
Sandstone; rootlets	1	1
Siltstone, coarse; rootlets	0	9
Sandstone	4	1
Shale; abundant plants	0	8
Shale, silty, light grey	2	6
Sandstone and siltstone interlaminated	0	10
Sandstone and shale, silty, interbanded	6	6
Shale, grey and dark grey	11	8
BLACK RAKE TUFFACEOUS SILTSTONE		
Ironstone band	0	11
Shale, dark; mussels, ostracods, fish debris, Spirorbis; ironstones	14	0
TOP SOFT		
Coal 45 in		
Seatearth 21 in		
Coal 1 in	5	7
Seatearth	1	11
ROOF SOFT		
Coal	4	6½
Mudstone, dark grey	1	3
DEEP SOFT		
Coal	3	0+

Section (2) [410 461] Overburden of First Piper Coal measured by R. A. Eden

	ft	in
Subsoil	4	0
Flagstone, soft, crumbling	4	0
Sandstone, soft, bright orange weathering	1	0
Sandstone, soft, fine, massive	6	0
Sandstone	8	3
Siltstone, light grey; ironstone bands	7	0
Sandstone; siltstone laminae	0	10
Siltstone, light grey; ironstone bands	6	0
Mudstone; ironstone lenses	7	0
Shale, dark, carbonaceous, coal streaks and fish debris	0	2
FIRST PIPER		
Coal	4	0+

Section (3) [407 461] Overburden of Threequarters Coal measured by R. A. Eden

	ft	in
Subsoil	2	0
HOSPITAL		
Coal, shaley 11½ in		
Seatearth 14 in		
Coal and clay 6½ in	2	8
Seatearth	3	10
Siltstone, light grey	3	9
Mudstone, light and dark grey; ironstones and abundant mussels	17	5
Siltstone; beds of mudstone	16	0
Mudstone, grey; ironstones, rare mussels	6	0
LOW MAIN		
Coal	3	4
Seatearth, silty	5	0
Mudstone, dark grey; rare mussels	5	0
Mudstone, black; plants	1	0
THREEQUARTERS COAL		

c

HOLBROOK, BROWNS ROAD SITE [3675 4542]
Measured by R. A. Eden

	ft	in
Shale, dark grey; fish debris	0	6
SECOND SMALLEY		
Coal	0	9½
Seatearth, sandy, soft, yellow	0	9 to 1 11
Sandstone, white, hard, flaggy; rootlets	0	9
Sandstone, soft, flaggy	2	0
Shale, dark-grey; mussels, Spirorbis, ostracods, fish debris	0	8
Sandstone, thin, flaggy	0	0 to 4
Siltstone, shaly; thin ironstone lenses	6	3
Shale, dark grey; fish debris	0	7
Shale, dark grey, fish debris and Lingula sp.	0	5
Shale, dark grey, fish debris and mussels	1	0
Siltstone, grey, coarse, coal streaks	0	9
HOLBROOK		
Coal	0	4
Seatearth, siltstone, micaceous, coal streaks	0	9
Sandstone, coarse, thin siltstone partings; rootlets at top	6	0
Siltstone, grey; ironstone bands and lenses, micaceous plants	18	0
Siltstone, dark grey, micaceous; fish debris and plants	1	0
BELPERLAWN		
Coal	2	9
Seatearth; sandy	1	6
Sandstone, purple and yellow, soft	6	0+

ILKESTON, NUTBROOK SITE [457 430]
Composite section measured by Dr J. Shirley

	ft	in
Subsoil	abt 5	0
Shale, dark; ironstone bands, rare mussels	6	4
TOP FIRST WATERLOO		
Coal	1	4
Shale grey with ironstone nodules	6	0
BOTTOM FIRST WATERLOO		
Coal 22 in		
Dirt ½ in		
Coal 4½ in	2	3
Clunch		
Not exposed	abt 18	0
Subsoil	3	0
Mudstone	5	0
WATERLOO MARKER		
Coal 6 in		
Clunch 9 in		
Mudstone 24 in		
Coal, shaley 1½ in	3	4½
Mudstone, grey; ironstone nodules	7	0
Mudstone, grey; sandy laminae	1	6
Sandstone	1	4
Mudstone, grey; ironstone nodules	3	4
Shale, dark grey	0	1
TOP SECOND WATERLOO		
Coal	3	6
Clunch		

IRETON HOUSES (east of) CLAYPIT [3783 4725]
Measured by J. G. O. Smart

	ft	in
Sandstone, fine, pale brown; ferruginous layers	10	0+
Mudstone, silty, grey-brown; ironstone nodules	6	0
Gap in section	abt 4	0
Mudstone, grey; ironstone nodules	11	0
Mudstone, dark grey; fish debris	0	2
Mudstone, grey; prominent 1-in ironstone band	3	1
Mudstone, dark grey; carbonaceous	1	7
Mudstone, grey; ironstone nodules	7	0
MICKLEY THIN		
Coal	1	10
Mudstone, dark grey; carbonaceous	0	9
Seatearth, just exposed		

LITTLE MATLOCK, IRONVILLE OLD QUARRY [4345 5137]
Measured by D. V. Frost

	ft	in
Sandstone, massive	15	0
Shale	8	0
LOW MAIN		
Coal	4	6
Seatearth; mudstone, grey, silty; rootlets	2	0
Mudstone, grey; ironstone nodules	12	0
THREEQUARTERS		
Coal	2	2
Seatearth; mudstone, grey, silty; rootlets	1	0
Shale	10	0+

MAREHAY, GODBERS LUM SITE [4067 4807]
Section above Second Ell Coal measured by J. G. O. Smart

	ft	in
Clay	8	0+
Mudstone, dark grey and grey; mussels	0	6
Mudstone, grey; ironstone nodules	1	10
Mudstone, dark grey, carbonaceous, mussels and fish debris	2	10
Mudstone, silty, grey; sporadic ironstone nodules	6	7
Mudstone, dark grey, carbonaceous	0	5
Mudstone, dark grey, canneloid	0	10
Siltstone, ferruginous, brown and dark grey interlaminated	0	1
Mudstone, silty, dark grey, carbonaceous mussels	0	2
SECOND ELL		
Coal	0	5
Clay, yellow and grey mottled	0	3+
Mudstone, grey; rootlets	4	0
Shale, carbonaceous, dark grey; coal streaks	0	2
Mudstone, grey; ironstone nodules, plants	2	3
Shale, grey; coal streaks	0	6
Shale, grey; 2 in ironstone bed	1	10
Sandstone, fine, silty, slumped	2	6
Siltstone, fine; ferruginous beds	9	0
Ironstone	0	3
Mudstone; 1-in ironstone beds	3	0
Shale, black; Spirorbis sp. and fish debris	0	1
TOP SECOND WATERLOO COAL		

d

ROWSON GREEN [3738 4686]
Temporary section measured by J. G. O. Smart.

	ft	in
Sandstone, fine, khaki, rubbly	abt 15	0
Siltstone, interlaminated fine and coarse, khaki	19	6
Sandstone, fine, hard, khaki; ripples	2	0
Siltstone, ochreous	8	6
Mudstone, silty, grey; ironstone nodules	15	0
Mudstone, grey	2	0
Ironstone, nodules	0	1
Mudstone, pale grey, fissile; ironstone nodules	4	7
Mudstone, dark grey, fissile, ferruginous stained	abt 8	0
KILBURN COAL (just exposed)		

SMALLEY, CARRINGTON COPPICE SITE [413 456]
Measured by R. A. Eden

	ft	in
Shale, grey; thin ironstone lenses and mussels	4	0
Ironstone	0	2
Mudstone, dark grey; thin ironstone lenses and mussels	2	10
Mudstone, dark grey	0	5
CLAY CROSS MARINE BAND		
Mudstone, dark grey; Lingula	1	2
Shale, dark; Lingula	0	9
Shale, dark; Lingula and goniatites	1	1
Shale, dark; goniatites, Dunbarella sp. and orthocones	0	6
Shale, dark; Lingula	0	2
Mudstone, dark grey; mussels	1	3
Mudstone; ostracods and mussels	10	4
Mudstone, grey to black; ironstones and mussels	18	3
BROWN RAKE		
Coal (just exposed)		

SMALLEY, ABBOTTS ROUGH SITE [4258 4414]
Measured by Dr J. Shirley

	ft	in
Mudstone, grey; ironstone nodules	10	0
Ironstone band	0	2
Shale, dark with mussels	0	5
WATERLOO MARKER		
Coal 10 in		
Clunch; ironstone nodules 27 in		
Coal 0½ in		
Clunch; ironstone nodules 8 in		
Coal 0½ in	3	10
Clunch	2	0
Sandstone	6	0
Mudstone, grey	13	0
TOP SECOND WATERLOO		
Coal	4	5

SMALLEY COMMON, SWINEHILL WOOD SITE [4065 4287]
Measured by R. A. Eden

	ft	in
Ironstone, lenticular, cone-in-cone, many mussels	up to 0	11½
Siltstone, light grey; ironstone lenses	7	0
Shale, dark grey; ironstone nodules and lenses	abt 6	7
Shale, dark grey; ironstones, abundant mussels	1	5
Siltstone, light grey	1	10
Shale, silty, grey	0	7
Siltstone, pale orange-brown	0	4
COCKLESHELL		
Coal streaks in carbonaceous mudstone 1 to 12 in	0	6
Seatearth; light grey, coarse, silty	1	9
Sandstone, soft, buff, laminated	5	9
Cank, friable matrix; plants	0	8
Siltstone, light grey; plants	17	6
Mudstone, grey	3	0
LOW MAIN		
Coal	3	2
Clay, black; streaks of coal	0	3
Seatearth; silty, black	2	4
Seatearth; siltstone, fine, grey; ironstone lenses at top	2	6
Seatearth; clayey, dark grey	8	6
THREEQUARTERS		
Coal	2	3½
Seatearth; dark buff	1	0+

SOMERCOTES, OLD CLAY PIT [4340 5332]
Measured by D. V. Frost

	ft	in
Soil	1	0
? BLACK RAKE TUFFACEOUS SILTSTONE		
Ironstone	0	6
Shale, finely laminated; Spirorbis and ostracods	0	7
BLACK RAKE		
Coal	0	9
Mudstone, khaki staining; ironstone nodules	4	0

a

	ft	in
carbonaceous mussels	0	2

SECOND ELL

Coal	0	5
Clay, yellow and grey mottled	0	3

MAREHAY, GODBERS LUM SITE [4058 4848]
Measured by J.G.O. Smart

	ft	in
Coal traces, weathered	0	6
Sandstone, fine, brown; impersistent	0 - 3	0
Mudstone, silty, brown-grey; ironstone nodules	15 -18	0
Mudstone, grey; ironstone nodules	7	6
Mudstone, dark grey; pyritic mussels and fish debris	0	8
Mudstone, dark grey; ironstone nodules	0	4

FOURTH WATERLOO

Coal	0	11
Seatearth; silty mudstone, grey and brown	2	0
Sandstone, fine, brown; rootlets	4	0
Siltstone and sandstone, fine; interlaminated	8	6
Sandstone, brown, hard; impersistent	0 to 5	0
Mudstone, silty and sandstone, fine, brown interbedded	16	0
Mudstone, silty, grey; ironstone band near top	8	0
*Ironstone, brown, pyritic, kaolinitic	0	1½
Mudstone, dark grey, carbonaceous	0	11

FIRST ELL

Coal	2	4

*The kaolinitic ironstone band in a second section [402 482] varied from 0 to 8 in and extended through the underlying carbonaceous mudstone (there up to 2 in thick) into the underlying coal. It was also replaced locally by ironstone nodules up to 2½ in thick.

RIPLEY, BUTTERLEY PARK SITE [4138 5125]
Measured by R.A. Eden

	ft	in
Mudstone, grey; ironstone lenses, Spirorbis sp. and mussels	8	0
Ironstone band; abundant mussels	0	1
Mudstone; abundant mussels, some ostracods	0	2

WATERLOO MARKER

Coal	0	6
Seatearth; grey	2	0

	ft	in
Ironstone band	0	3
Mudstone, yellowish grey; ironstone nodules	7	6

TOP SOFT

Coal 22 in
Seatearth 4 in
Mudstone, grey; plant fragments 24 in

Coal ½	4	2½
Mudstone	2	1

? ROOF SOFT UPPER LEAF

Coal 12 in
Dirt 6 in

Coal shaly 4 in	1	10
Seatearth; grey	3	0
Mudstone, grey; ironstone nodules	7	6

ROOF SOFT ? LOWER LEAF

Coal	3	8
Shale; carbonaceous	0	10
Seatearth; rootlets	2	0
Shale, grey, silty; sandstone and siltstone bands up to 2ft 6in, rich in plant debris	31	0

DEEP SOFT

Coal	abt 5	0
Seatearth; grey, rootlets		

STANLEY, HOME FARM SITE [4167 3945]

	ft	in
Soil and subsoil	abt 5	0
Mudstone, brown weathered	6	0
Shale, grey	7	6
Shale, black	3	0
Ironstone	0	1½
Shale, black; fish teeth	2	0
Dirt with coal	0	5

KILBURN COAL

STANLEY, MOAT WOOD SITE [4211 3942]
Measured by R.A. Eden

	ft	in
Sandstone, rusty-brown; flaggy at top	20	0
Siltstone, grey	18	0
Mudstone, grey	4	6
Shale, dark grey; ironstone lenses	4	0
Ironstone	0	2
Shale, dark grey, silty at base, fish debris	1	6
Coaly shale	0	7

KILBURN

Coal	5	6

STRELLEY, CATSTONEHILL SITE [5018 4110]
Section (1) measured by D.V. Frost

	ft	in
HEAD		
Sand, red-brown; clayey patches, sporadic pebbles	3	0
Pebble layer	1	0
Clay, red and white bands up to 5 in; red ironstone nodules	1	6
Mudstone, grey, silty and sandy laminae; ironstone nodules	20	0
Mudstone, grey	7	0

c

	ft	in
Coal 12 in		
Dirty coal 1 in	2	10
Seatearth; mudstone, grey	2	0
Mudstone, grey, silty	6	0
Sandstone, pale grey	1	0
Mudstone, grey	2	0
Sandstone, grey, many coaly plants	1	0
Mudstone, grey	11	0+

STRELLEY, CATSTONEHILL SITE [4990 4140]
Section (4) measured by J.G.O. Smart

	ft	in
FOURTH WATERLOO		

Coal 25 in
Dirt 2 in
Seatearth, pale grey 16 in

Coal 6 in	4	1
Siltstone, brown	12	0
Mudstone, grey-brown, silty at top; ferruginous layers	6	0

FIRST ELL

Coal 13 in
Dirt 10 in

Coal 17 in	3	4
Seatearth; mudstone, silty	3	9
Sandstone, fine, brown, micaceous	4	0
Siltstone, brown ferruginous; lenses to 6ft thick of fine sandstone	abt 28	0
Mudstone, silty, grey; ironstone nodules and mussels	22	0
Mudstone, dark grey; mussels and shelly ironstones	0	4
Mudstone grey and dark grey, ironstone, mussels and fish debris	2	1
Seatearth; mudstone, incipient	0	4
Mudstone, grey	1	4

SECOND ELL

Coal	2	7

STRELLEY, CATSTONEHILL SITE [4974 4151]
Section (5) measured by J.G.O. Smart

	ft	in
Mudstone, grey; ironstone nodules and bands	8	0
Mudstone, dark grey; fucoidal markings	1	0

CLAY CROSS MARINE BAND

Mudstone, dark-grey, pyritic; Lingula sp., Dunbarella sp.	2	0
Mudstone, pale grey; ironstone nodules and bands contorted; mussels	5	10
Mudstone, dark grey; ironstones and mussels	3	6
Mudstone, pale grey; ironstones, seatearthy	1	5
Ironstone, brown; shelly	0	3 to 4
Mudstone, dark grey; ironstones and mussels	4	0
Mudstone, dark grey; many mussels	0	3
Mudstone, pale grey	0	2

BROWN RAKE

Coal 13 in	1	1

TROWELL, M1 MOTORWAY CUTTING [4857 3912]
Measured by D.V. Frost

	ft	in
Sandstone, fine- to medium-grained, yellow-brown, massive, discontinuous silty mudstone bands up to 2ft thick. Base of sandstone irregular and cuts out locally the top 8ft of underlying beds	12	0
Mudstone, grey, silty, laminated	3	0
Siltstone and sandstone bands and laminae interbedded with grey mudstone containing ironstone nodules	5	0
Mudstone, grey, silty; laminated, khaki staining	5	0

'ASHGATE' UPPER LEAF

Coal 44 in	3	8
Coal and dirt	2	8
Seatearth; mudstone, pale grey	0	8

'ASHGATE' LOWER LEAF

Coal, cannelly 35 in	2	11

TROWELL M1 MOTORWAY [4865 3942]
Measured by D.V. Frost

	ft	in
Soil and clay; buff sandstone fragments	3	0
Mudstone, grey-brown, silty, laminated	1	6
Sandstone, buff, ferruginous; well-jointed	1	5
Mudstone, yellow-brown, silty, laminated	1	0
Sandstone as above	1	0
Mudstone, grey-brown, silty; with numerous ironstone bands and nodules	22	5
Mudstone, black, carbonaceous, laminated	2	0

BLACKSHALE

Coal pyritic at base 34 in
Dirt 4 in
Coal 1½ in
Dirt 6 in
Coal 1 in
Dirt 5 in
Coal 11 in
Coal 2 in
Coal 8 in

Dirt 1½ in	6	2
Seatearth; mudstone pale grey		

WEST HALLAM COMMON, WHITE HART SITE [4225 4172]
Measured by Dr J. Shirley

	ft	in
Soil and subsoil	3	6
Shale, black	0	6

DEEP HARD

Coal 66 in	5	6
Clunch	2	0
Mudstone, grey	2	9
Sandstone	1	2
Mudstone, sandy, laminated	4	0

b

	ft	in
Mudstone, dark grey	0	3
Mudstone, black, carbonaceous	0	1½

THIRD WATERLOO

Coal 7 in
Mudstone 4 in
Coal 2½ in
Mudstone 3½ in

Coal 24 in	3	5

small gap

Mudstone, grey; sporadic ferruginous bands and nodules	15	0
Mudstone, grey	1	0
Mudstone, dark grey	0	4
Mudstone, carbonaceous	0	2
Mudstone, grey, silty; plant fragments	1	0
Mudstone, grey, brown weathering; interlamination of sandy and silty laminae	15	0
Mudstone, grey; ironstone bands common	9	0

FOURTH WATERLOO

Coal abt 24 in
Seatearth, mudstone, pale grey 12 in

Coal 6 in	3	6
Seatearth; mudstone, grey, silty; rootlets and coal streaks	3	5
Mudstone, grey, silty laminae; plant fragments	8	6
Mudstone, grey, with mussels at base	6	0

FIRST ELL

Coal and cannelloid coal, 24 in	2	0
Seatearth; pale grey		

STRELLEY, CATSTONEHILL SITE [5032 4165]
Section (2) measured by D.V. Frost

	ft	in
PERMIAN		
Lower Magnesian Limestone	16	0
Permian Basal Breccia 10 in to	2	0
WESTPHALIAN		
Sandstone, pale green, fine-grained; purple-stained at base	2	5
Mudstone, purple, silty; ferruginous bands	4	0
Sandstone, green, red and brown staining	4	0
Mudstone, grey, silty, faint purple staining at top; small mussels in basal 2 ft	10	0
Mudstone, black, carbonaceous, cannelloid; pyritised mussels	0	8

DUNSIL

Coal 11 in
Cannel 2 in

Coal 11 in	2	0
Mudstone, carbonaceous	0	1
Seatearth; mudstone, grey	4	1
Mudstone, grey, silty; ironstone nodules, plants	6	0
Sandstone, fine-grained, ferruginous; passage by intercalation to	2	0
Mudstone, grey, silty; ironstone bands at top	9	0

	ft	in
Sandstone, ferruginous	1	0
Mudstone, grey, silty; ironstone nodules	3	0
Siltstone, canky; carbonaceous plant fragments up to	0	7
Mudstone, grey, silty; many ironstone nodules and bands	12	0
Gap (largely mudstone with ?1-in coal a few ft above base)	7	0

FIRST WATERLOO COAL

Coal (1 in only exposed)		
Seatearth; mudstone, grey, silty	1	8
Siltstone, pale grey; carbonaceous partings, ironstone nodules	2	0
Sandstone/siltstone alternation	0	9
Siltstone, carbonaceous, micaceous laminae; slump structures at base	3	6
Sandstone/siltstone alternation	2	0
Siltstone	2	0
Mudstone, grey, silty; many ironstone bands	4	0
Mudstone, grey; ironstone bands and mussels	6	0

WATERLOO MARKER

Coal 7 in
Seatearth, pale grey, up to 12 in
Coal and cannel 8 in

Coal, cannely bands 24 in	4	0+

STRELLEY, CATSTONEHILL [5070 4175]
Section (3) measured by D.V. Frost

	ft	in
PERMO-TRIASSIC		
Lower Mottled Sandstone?		
Sandstone with clayey patches	5	0
Lower Magnesian Limestone		
Red-brown sandy dolomite	20	0
?Lower Marl		
Mudstone, deep purple and red, silty	0	8
Mudstone pale green	0	1
Permian Basal Breccia 8 in to	2	0
WESTPHALIAN		
Mudstone, grey, stained purple and red, silty	17	0

TOP HARD FLOOR COAL

Coal 23 in
Dirt 1 in

Coal 1 in	2	1
Seatearth; mudstone, pale grey; rooty	1	0
Mudstone, grey, silty	0	6
Sandstone and silty mudstone interlaminations	2	10
Sandstone, massive, fine-grained	15	0
Mudstone, grey; many ironstone bands	10	0
Mudstone, black, carbonaceous, silty; pyritic mussels	0	6

DUNSIL

Coal with pyrites 5 in
Coal 15 in
Dirt 1 in

d

	ft	in
Mudstone, silty, brown	5	6
Ironstone nodules	to 0	4
Mudstone, grey	5	0
Sandstone, ferruginous, blue-hearted	2	0
Mudstone, sandy, blocky	6	0
Mudstone, grey; ironstone nodules	abt 17	6
Shale, black	0	8

FIRST PIPER

Coal 59 in	4	11
Clunch		

APPENDIX 3
LOCATIONS OF OPENCAST SITES

In this list the name of each site is followed by the National Grid Reference and an approximate bearing and distance from an adjacent town or village. Because of the diversity in size and in shape of the sites the Grid Reference is approximate and refers to the centre of the worked areas or in the case of unworked sites the centre of the prospected area. There are measured sections in Appendix 2 of those sites marked with an asterisk.

*ABBOTTS ROUGH [426 441]
1 mile ESE of Smalley.

*AGNES [429 501]
½ mile NE of Codnor.

ALFRED-SLEET See Sleetmoor Lane

ALLEN'S GREEN [468 520]
1/3 mile NW of Bagthorpe.

ALMA [474 518]
immediately NW of Bagthorpe.

ARGYLL HOUSE [423 464]
immediately W of Heanor.

AVENUE [401 440]
½ mile SW of Smalley.

BABBINGTON [493 438]
1 mile NE of Cossall.

BACK LANE [394 534]
½ mile W of Swanwick.

BACON LANE [384 535]
¾ mile NNW of Pentrich.

BAGTHORPE [471 512]
¼ mile SW of Bagthorpe.

BEAL [456 386]
2/3 mile NW of Stanton-by-Dale.

BEAUVALE [492 489]
1 mile N of Greasley.

BEECH HILL FARM [370 532]
1 mile NE of Bullbridge.

BIRDSWOOD [416 536]
2/3 mile east of Swanwick.

BOTTLE [390 453]
2/3 mile ESE of Kilburn.

BRAKE HOUSE [450 522]
1/3 mile NE of Pye Hill.

BRINSLEY MOOR [459 494]
1/3 mile N of Brinsley.

*BRINSLEY SIDINGS [450 498]
1 1/3 miles NW of Brinsley.

BRINSLEY WHARF [450 483]
1/3 mile NE of Aldercar, part of Cromford Canal Site.

BROAD LANE [452 504]
1/3 mile W of New Brinsley.

BROADOAK FARM [398 522]
2/3 mile SW of Swanwick.

BROOKSHILL FARM [496 480]
½ mile NNE of Greasley.

*BROWN'S ROAD [367 453]
2/3 mile WSW of Kilburn.

BUCKLAND HOLLOW [370 514]
½ mile N of Heage.

BUNKERHILL [442 410]
½ mile ESE of West Hallam.

*BUTTERLEY PARK [414 512]
¾ mile NE of Ripley.

*CARRINGTONS COPPICE [413 456]
1 mile SW of Heanor Gate.

CAT AND FIDDLE [425 405]
½ mile SW of West Hallam.

*CATSTONE HILL [501 417]
½ mile WSW of Strelley.

CHALFONT [445 395]
2/3 mile NE of Dale.

CHESTNUT [363 527]
1/3 mile NE of Bullbridge.

*CINDERHILL COPPICE [432 446]
1 mile NNW of Mapperley.

CITY [489 461]
1/3 mile ESE of Newthorpe.

CLUB ROOM [413 433]
2/3 mile SSE of Smalley.

CODNOR COMMON [415 494]
immediately E of Codnor.

CONEYGREY PLANTATION [471 481]
2/3 mile SE of Brinsley, continuous with Grange Farm.

CORDY LANE [463 492]
immediately NE of Brinsley.

CORFIELD [432 495]
2/3 mile E of Codnor, worked with Agnes Site.

COVENTRY LANE [503 405]
1 mile NW of Wollaton.

CROMFORD CANAL [455 483]
1 mile NW of Eastwood, includes Waggon and Brinsley Wharfe sites.

CROSSLEY-WILLIAM (CROSSLEY BANKS) [489 507]
continuous with William Wood, 1 mile SE of Bagthorpe.

DALE MOOR [447 388]
½ mile E of Dale.

DAMSTEAD LILY [403 540]
1/3 mile N of Swanwick.

DELVES, THE [424 457]
immediately S of Heanor Gate, includes Station Site.

DEVONSHIRE [378 518]
1 mile SW of Pentrich.

DIXIE [475 514]
at Bagthorpe.

DOBBS [426 543]
immediately NW of Somercotes.

ELLERSLIE [461 464]
½ mile SW of Eastwood.

EXHIBITION [426 510]
1 mile south of Riddings.

*FALL, THE [437 470]
immediately W of Langley Mill.

FAR LAWN [366 489]
1 mile N of Openwoodgate.

FELLEY MILL [489 500]
1 mile ESE of Underwood.

FEZ; see Macfez.

FORECLOSE FARM [363 494]
1 mile SSE of Nether Heage.

FOXHOLE FARM [445 402]
1 mile SE of West Hallam.

GILT HILL [487 455]
¾ mile SSE of Newthorpe.

GIPSY [399 445]
1/3 mile SE of Horsley Woodhouse.

*GODBERS LUM [406 482]
½ mile E of Marehay.

GRANGE FARM [467 486]
1/3 mile SE of Brinsley, continuous with Coneygrey Plantation.

GREASLEY CASTLE [491 467]
immediately SE of Greasley.

GUTTER SLANG [513 404]
½ mile NW of Wollaton.

HEAD HOUSE FARM [440 430]
1/3 mile E of Mapperley.

*HEANOR GATE [430 456]
immediately SW of Heanor.

*HIGH BANK [406 466]
½ mile NE of Denby.

HIGH PARK WOOD [488 494]
1 mile N of Moorgreen.

HOBSIC [453 538]
1/3 mile NW of Selston.

HOLLIES FARM [428 390]
½ mile WNW of Dale.

HOLLYDENE [487 482]
immediately N of Moorgreen.

HOLMES [450 516]
½ mile E of Ironville.

*HOME FARM [417 394]
½ mile S of Stanley.

INKERMAN [472 524]
1/3 mile NNW of Bagthorpe.

JOCKEY [469 538]
2/3 mile NW of Selston.

JOHNSON FARM [444 451]
1 mile SE of Heanor.

JOPLE [431 429]
immediately W of Mapperley.

KATIE [478 467]
immediate W of Newthorpe.

KIDSLEY PARK FARM [416 455]
2/3 mile NE of Smalley.

KIM [492 443]
immediately SW of Kimberley.

KIRKBY PARK [474 545]
1 mile NE of Selston.

LAMBCLOSE HOUSE [473 485]
½ mile NW of Moorgreen, continuous with Willeywood Farm.

MACFEZ [496 430]
1 mile ENE of Cossall.

*MANOR FARM [377 503]
immediately S of Heage.

MARY [405 404]
¾ mile ESE of Morley.

MILL FARM [386 522]
1/3 mile S of Pentrich.

MILTON [483 518]
½ mile NE of Bagthorpe.

MORLEY HAYES [402 418]
¾ mile NE of Morley.

MORLEY PARK FARM [382 497]
1 mile WSW of Ripley.

MORRELLS WOOD [378 481]
½ mile NE of Openwoodgate.

MOSES LANE [407 410]
¾ mile E of Morley.

NEW COVERT [518 397]
1 mile WNW of Wollaton Hall.

NEWTHORPE FARM [478 448]
½ mile NW of Awsworth, now part of Shilo Site.

NIX [395 542]
½ mile NW of Swanwick.

NODINHILL [358 508]
immediately W of Nether Heage.

*NUTBROOK [456 429]
1¼ miles E of Mapperley.

OAKTREE FARM [464 502]
immediately E of New Brinsley.

OFFICE [437 447]
¾ mile SSW of Marlpool.

OPENWOOD [384 475]
2/3 mile ENE of Openwoodgate.

PARKFIELD FARM [451 436]
1 mile ENE of Mapperley.

PARK MEADOW FARM [400 474]
2/3 mile W of Denby Common.

PENNYTOWN [424 545]
immediately N of Summercotes.

PEPPER HILL [486 470]
1/3 mile NW of Greasley.

PIPPINHILL [401 462]
1/3 mile E of Denby.

PLASTIC [436 488]
1 mile ESE of Codnor.

POPLAR FARM [465 445]
immediately NE of Cotmanhay.

PORTWAY [368 438]
½ mile SSE of Holbrook.

PURDY HOUSE [458 451]
1 mile ESE of Marlpool.

RAP [483 476]
immediately W of Moorgreen.

*RIDGEWAY [361 512]
½ mile SE of Bullbridge.

ROSE AND CROWN [392 423]
½ mile N of Morley.

ROSEDALE [488 538]
2/3 mile W of Annesley Woodhouse.

ROWSON-IRETON [373 470]
1/3 mile SE of Openwoodgate.

*RYKNELD STREET (COVERT) [388 430]
1 mile SE of Horsley.

ST HELEN'S [461 531]
immediately SE of Selston.

SALTERWOOD [380 485]
1 mile NE of Openwoodgate.

SHILO [473 444]
in Erewash valley, 1 mile W to 2 miles S of Eastwood.

SHIPLEY HALL [441 439]
2/3 mile NNE of Mapperley.

SHIPLEY LAKE [447 438]
1 mile NE of Mapperley.

SHORTWOOD [490 413]
¾ mile SE of Cossall.

SHREW (WINGSHREW) [378 544]
1½ miles NW of Swanwick.

SLACKFIELDS FARM [385 445]
1/3 mile N of Horsley.

SLEET MOOR worked with Alfred-Sleet [409 543]
2/3 mile NNE of Swanwick.

SPANKER [363 502]
1/3 mile WSW of Heage.

STANLEY FOOTRIL [409 400]
½ mile WSW of Stanley.

STARVEHIMVALLEY [378 510]
1 mile NW of Ripley.

STATION; see The Delves Site.

SWANCAR FARM [491 397]
¾ mile E of Trowell.

*SWINEHILL WOOD [406 429]
1 mile S of Smalley.

TAVERN HOUSES [410 468]
1/3 mile S of Denby Common.

TINKLERS [442 413]
½ mile E of West Hallam.

*TOP DUMBLES [413 463]
1 mile E of Denby.

TRAMWAY [412 525]
½ mile SW of Alfreton.

TROWELL [481 408]
1 mile S of Cossall.

WAGGON [455 500]
2/3 mile NNW of Brinsley, part of Cromford Canal Site.

WANSLEY HALL [460 514]
1 1/3 miles S of Selston.

WATER TOWER [435 444]
2/3 mile N of Mapperley.

WESTWOOD BENTS [462 522]
2/3 mile S of Selston.

WHITE HART [422 417]
immediately E of West Hallam Common.

*WHITELEYS PLANTATION [422 450]
1 mile E of Smalley.

WHITEMOOR [361 484]
2/3 mile NE of Belper.

WILLEYWOOD FARM [470 490]
2/3 mile ENE of Brinsley, continuous with Lamclose House Site.

WILLIAM IV [397 513]
½ mile NNW of Ripley.

WILLIAM WOOD [488 515]
continuous with Crossley;William, ¾ mile E of Bagthorpe.

a

b

c

d

APPENDIX 4
LIST OF GEOLOGICAL SURVEY PHOTOGRAPHS

Copies of these photographs are available for reference in the library of the Institute of Geological Sciences, Exhibition Road, South Kensington, London SW7 2DE, and in the library of the Institute's Northern England Office, Ring Road Halton, Leeds LS15 8TQ. Prints and lantern slides may be supplied at a fixed tariff.

All these photographs belong to Series L, unless stated otherwise. Those marked with an asterisk are available in colour; those with a cross are available as 2 x 2 in colour slides.

The National Grid References, all in 100-km square SK, are those of the viewpoints.

Dinantian

878[+] – 879[*+]	Quarries at Wirksworth and Middleton viewed from the east side of the Ecclesbourne valley [2995 5390 and 2968 5352].
901[*+]	Southern limit of Dinantian limestone at Wirksworth viewed from the escarpment of a low leaf of Ashover Grit [267 526].
907	Dinantian limestone sequence exposed in Dale Quarry, Wirksworth [285 543].
929[*+] – 930[*+]	Middlepeak Quarry, Wirksworth [2805 5485].
931[*+]	Baileycroft Quarry, Wirksworth. Cawdor Limestone resting unconformably on Hoptonwood Limestone [2869 5415].
932[*+]	Dale Quarry, Wirksworth, showing penecontemporaneous thrust and old rake workings in Matlock and Cawdor limestones [2820 5425].
933[*+]	Dale Quarry, Wirksworth, showing Cawdor, Matlock and Hoptonwood limestones [2820 5425].
934[*+]	Dale Quarry, Wirksworth, showing Matlock and Hoptonwood limestones in a pillar left unworked [2820 5425].
A1186	Hilts Quarry, Crich [353 543].
A1187	Eastern Quarry, Crich [357 542].
A9120 – 9121	Dale Quarry, Wirksworth [283 541].

Namurian

880	Pillar of Ashover Grit left in Alport Quarry [3038 5157].
881[+]	Baryte crystals on joint surface of Ashover Grit at Alport Quarry [3041 5157].
882 – 884	Ashover Grit in the rail cutting south of Ambergate [347 507].
885[+]	Ashover Grit in Chevin Quarry, Hazelwood [3367 4584].
887[*+] – 888[*+]	Steeply dipping Ashover Grit and underlying mudstones at the East Midlands Gas Board site at Ambergate [354 520].
892[+]	Quarried sandstone columns (probably Rough Rock) in the front portico of Kedleston Hall.
893	Steeply dipping Namurian mudstones (probably high E Zone) in Mill Plantation, Breadsall [3758 3954].
894[*+]	Steeply dipping Namurian (R Zone) mudstones in Ferriby Brook, Breadsall [3800 3973].
896[+]	Namurian (E Zone) shales in the bank of the River Ecclesbourne [317 452].
897[*]	Anticlinal closure possibly due to valley bulging in Shipley Brook, Belper [3186 4791].
898	Cross-bedded Ashover Grit in Milford Quarry [352 452].
903[+] – 904[*+]	Cross-bedded Ashover Grit in Manor Quarry, Duffield Bank [354 433].
909[*+]	Slumping in the Rough Rock in Coxbench Wood Quarry [370 432].
910[*]	Ferruginous nodule near top of Rough Rock, Coxbench Quarry [374 433].
928[*+]	Undulating topography of Namurian shales and sandstone below Alport Hill; Dinantian limestone in distance [2975 5145].
A9122	Namurian near Wirksworth; Dinantian rocks in distance [28 54].

Westphalian

865[+]	Subsidence caused by mining in churchyard at Codnor [419 486].
867[+]	Catstonehill Opencast Site, Strelley; folding and faulting in Westphalian B [506 418].
869	Laminated sandstone overlying Top Second Waterloo Coal in Loscoe Brick Pit, near Heanor [426 471].
870	Impersistent lens of coal with well-developed cleat in Loscoe Brick Pit, near Heanor [426 471].
871[+]	Face-shovel excavating Top Hard Coal at Cromford Canal Opencast Site [453 485].
872 – 874	Detrital coal within 'washout' sandstone in the Aglite (Midlands) Ltd quarry near Denby [388 475].
875	Deep Hard Coal with overlying sandstone in the Aglite (Midlands) Ltd quarry near Denby [388 475].
876	Blackshale Coal in the M1 Motorway cutting at Trowell [487 399].
889[*+]	Trowell Moor motorway service station at a late stage in construction. It is built upon reinstated opencast coal workings. [496 406].
895[*+]	Denby Hall Colliery, near Ripley [398 478].
898 – 900[+]	Crawshaw Sandstone with thrust, in the Butterley and Blaby Ltd quarry at Ambergate [359 518].
911, 912[*], 913[*+], 914[*+], 915[*+]	Disturbed strata above the False Ell Coal at Godbers Lum Opencast Site [406 482].
916	Detail of the Black Rake Ironstone and Tuffaceous Siltstone at Godbers Lum Opencast Site [406 482].
917[*+]	Disturbed strata above the False Ell Coal at Godbers Lum Opencast Site [406 482].
918[*+]	False Ell Coal and overlying strata at Godbers Lum Opencast Site [406 482].
919[*+] – 920[*+]	Distant views of disturbances above the False Ell Coal at Godbers Lum Opencast Site [406 482].
921[*]	False Ell Coal and overlying strata at Godbers Lum Opencast Site [406 482].
922	Detail of fossil tree, Cromford Canal Opencast Site [453 493].
923[*], 924[*], 925[*]	Fossil tree, Cromford Canal Opencast Site [453 498].
926[*+], 927[*+]	General view of Annesley and Newstead Collieries [512 536].
A7985, 7986, 7987, 7988, 7989	Thacker Barn Opencast Site [448 410].
A7991	Tavern Houses Opencast Site [410 468].
A7993 – 5	Carrington Farm Opencast Site [413 456].
A7996	Drilling at Thacker Barn Opencast Site [448 410].
A8001	Thacker Barn Opencast Site [448 410].

Permo-Triassic

851, 852, 853, 855[*+]	Anticline in the topmost beds of Lower Marl. Temporary exposure at roundabout linking A610 with M1 at Nuthall [521 439].
854[*], 856[+]	Temporary exposure on west side of M1 near Sandiacre showing laminated siltstones and mudstones of the Harlequin Formation [471 351].
857[+], 858[+]	Temporary exposure on the M1 near Sandiacre showing minor faulting and attenuation in Keuper Marl below the Plains Skerry horizon [471 366].
859[+], 860[+]	Plains Skerry, forming the topmost beds exposed in Chilwell Brick Pit [512 358].
861, 862	Ripple-marked bedding-plane in Lower Magnesian Limestone, Quarry Banks, Linby [5370 5230].
863	Flaggy dolomite of the Lower Magnesian Limestone overlying Lower Marl in rail cutting near Watnall [4509 3451].
864	Lower Marl and Basal Breccia in rail cutting near Watnall [4509 3451].
866[+]	Catstonehill Opencast Site, Strelley; Permian rocks overlying Westphalian B [506 417].
868[+]	Sand quarry, Catstone Hill, Strelley; cemented coarse pebbly sandstone of Pebble Beds [507 415].
877[+]	Cross-laminated pebbly sandstone of the Pebble Beds in Hilton Quarry, near Mercaston [280 457].
886[+]	Pebble Beds in quarry near Cocks Hat Hill Farm, Windley [3219 4290].
890[+]	Keuper Marl in Derby Brick Pit [331 359].
891[*]	Skerry band (Plains Skerry group) at Derby Brick Pit [331 359].
905[+]	Pebble Beds showing variation of grade and cross-lamination at Derbyhills Quarry, near Hulland Ward [281 455].
908	Sequence of Waterstones exposed in road cutting at Mackworth [314 377].
A5033-4	Linby Quarry; ornamented work in 'Linby Stone' [535 522].
A5035-6	Lower Magnesian Limestone in Annesley Woodhouse Quarry [490 534].
A5023-4	Lower Magnesian Limestone in Bulwell Quarries [? 533 459].
A5025	Method of building lime-kilns, Jackson's Quarry, Bulwell [? 533 459].
A5026-7	Weathering of Lower Magnesian Limestone in mineral railway bridge near Robin Hood's Well [? 5094 5472].
A5028-9	Lower Magnesian Limestone in Linby Quarry [535 522].
A5030	Method of working Lower Magnesian Limestone, Linby Quarry [535 522].
A5031-2	Ornamental work in "Linby Stone", Linby Quarry [535 522].

Quaternary

902[*+]	Glacially filled pipe in Pebble Beds, Derbyhills Quarry, near Hulland Ward [281 455].
906[+]	Detail of left hand margin of the pipe shown in L.902.
A1188	Boulder Clay resting on fine sand ¾ mile south of Annesley Woodhouse church [? 496 524].

a

b

c

d

INDEX OF FOSSILS

No distinction is made between a positively determined species and a variant of the species or examples doubtfully referred to it (e.g. with the qualifications aff., cf. or ?)

Fossils identifiable at generic level only (e.g. *Anthracoceras sp.*) are listed after the named species.

GENERAL INDEX

HER MAJESTY'S STATIONERY OFFICE

Government Bookshops
49 High Holborn, London WC1V 6HB
13a Castle Street, Edinburgh EH2 3AR
41 The Hayes, Cardiff CF1 1JW
Brazennose Street, Manchester M60 8AS
Southey House, Wine Street, Bristol BS1 2BQ
258 Broad Street, Birmingham B1 2HE
80 Chichester Street, Belfast BT1 4JY
Government publications are also available through
booksellers

INSTITUTE OF GEOLOGICAL SCIENCES
Exhibition Road, London SW7 2DE

Murchison House, West Mains Road,
Edinburgh EH9 3LA

The full range of Institute publications is
displayed and sold at Murchison House and at
the Institute's Bookshop at the Geological
Museum, Exhibition Road, London SW7 2DE

The Institute was formed by the incorporation of
the Geological Survey of Great Britain and the Geological
Museum with Overseas Geological Surveys and is a
constituent body of the Natural Environment
Research Council.